Principles of Anatomy and Physiology

Fourth Edition

Gerard J. Tortora
Bergen Community College

Nicholas P. Anagnostakos
late, Bergen Community College

1817

HARPER & ROW, PUBLISHERS, New York
Cambridge, Philadelphia, San Francisco,
London, Mexico City, São Paulo, Sydney

Sponsoring Editor: Claudia M. Wilson
Development Editor: Robert Ginsberg
Project Editor: Robert Greiner
Design Supervision: Helen Iranyi
Production Assistant: Debi Forrest-Bochner
Manager, New Book Production: Kewal Sharma
Photo Researcher: Mira Schachne
Compositor: Kingsport Press
Printer and Binder: Kingsport Press
Art Studio: Vantage Art, Inc.
Medical Illustrators: Leonard Dank, Charles Bridgman, Paul Melloni,
            Marsha J. Dohrmann, Helen Gee Jeung
Cover Design and Page Layout: Caliber Design Planning, Inc.

PRINCIPLES OF ANATOMY AND PHYSIOLOGY, Fourth Edition

Library of Congress Cataloging in Publication Data

Tortora, Gerard J.
  Principles of anatomy and physiology.

  Bibliography: p.
  Includes index.
  1. Human physiology.   2. Anatomy, Human.
I. Anagnostakos, Nicholas Peter, 1924–1981.   II. Title.
[DNLM: 1. Anatomy.   2. Homeostasis.   3. Physiology.
QS 4 T712p]
QP34.5.T67   1984   612   83–18426
ISBN 0–06–046656–1

Harper International Edition
ISBN 0–06–350734X
Australian Edition
ISBN 0–06–3507315

COVER ILLUSTRATIONS

Front: The heart by Leonard Dank
Back: Diagram of a lobule of the lung by Leonard Dank

The fourth edition of *Principles of Anatomy and Physiology* is dedicated to
NICHOLAS P. ANAGNOSTAKOS
Friend, Coauthor, and Colleague

Contents in Brief

Contents in Detail

Preface

AUDIENCE

Principles of Anatomy and Physiology, Fourth Edition, is designed for the introductory course in anatomy and physiology. The text is geared to students in health-oriented, medical, and biological programs. Among the students specifically served by this text are those aiming for careers as nurses, medical assistants, medical laboratory technologists, radiologic technologists, inhalation therapists, dental hygienists, physical therapists, morticians, and medical record keepers. Because of its scope, the text is also useful for students in the biological sciences, science technology, liberal arts, physical education, and in premedical, predental, and prechiropractic programs.

OBJECTIVES

The objectives of the fourth edition are the same as those of previous editions: (1) to provide a basic understanding and working knowledge of the human body and (2) to present this essential material at a level that average students can handle.

Throughout, our goal has been to eliminate barriers to a ready comprehension of the structure and function of the human body. We recognize, however, that some technical vocabulary and difficult concepts are vital to the course. Such material is developed in step-by-step, easy-to-understand explanations that avoid needlessly difficult nontechnical vocabulary and syntax.

THEMES

Two major themes still dominate the book—*homeostasis* and *pathology.* Throughout, the book shows students how dynamic counterbalancing forces maintain normal anatomy and physiology. Pathology is viewed as a disruption in homeostasis. Accordingly, we present a number of clinical topics and contrast them with specific normal processes.

ORGANIZATION

The book follows the same unit and topic sequence as its three earlier editions. It is divided into five principal areas of concentration: organization, support and move-

ment, control systems, maintenance, and continuity. Unit I, Organization of the Human Body, provides an understanding of the structural and functional levels of the body, from molecules to organ systems. Unit II, Principles of Support and Movement, analyzes the anatomy and physiology of the skeletal system, articulations, and the muscular system. Unit III, Control Systems of the Human Body, emphasizes the importance of the nerve impulse in the immediate maintenance of homeostasis, the role of receptors in providing information about the internal and external environment, and the significance of hormones in maintaining long-range homeostasis. Unit IV, Maintenance of the Human Body, illustrates how the body maintains itself on a day-to-day basis through the mechanisms of circulation, respiration, digestion, cellular metabolism, urine production, and buffer systems. Unit V, Continuity, covers the anatomy and physiology of the reproductive systems, sexual development, and the basic concepts of genetics.

In the fourth edition, the strengths of the previous editions have been maintained. Revisions made for the new edition have focused on updating certain topics and strengthening the coverage of physiology. Among the specific changes made in topic coverage in the fourth edition are the following.

UNIT I. ORGANIZATION OF THE HUMAN BODY

Chapter 1 has been updated to include basic principles and clinical applications of the dynamic spatial reconstructor (DSR). In Chapter 2, new sections have been added on nuclear magnetic resonance (NMR); positron emission tomography (PET); hydrogen bonding; saturated, unsaturated, and polyunsaturated fats; structural levels of protein organization; and left-handed DNA. Chapter 3 includes revised discussions of the structure, chemistry, and functions of plasma membranes; facilitated diffusion; active transport; ribosomes; lysosomes; cell division; and cells and cancer. New to Chapter 3 are discussions on nucleosomes, peroxisomes, the cellular cytoskeleton, protein synthesis, DNA repair, recombinant DNA, and laetrile and cancer. In Chapter 4, the discussion of the basement membrane has been revised, and the section on the inflammatory response has been updated. Also in Chapter 4, the exhibits on epithelial and connective tissue have again been improved by the addition of new color photomicrographs. Changes in Chapter 5 consist of new material on lines of cleavage; epidermal ridges and grooves; hair color, growth, and replacement; apocrine and eccrine sweat glands; ceruminous glands; collagen implants; pruritis; and over-the-counter corticosteroids. The discussions on skin color, systemic lupus erythematosus (SLE), and sun-screening agents have been revised.

UNIT II. PRINCIPLES OF SUPPORT AND MOVEMENT

New to Chapter 6 is a section on pulsating electromagnetic fields (PEMFs) and fracture repair. In Chapter 7 there is a new section on sternal puncture. Chapter 9 has been expanded to include discussions of arthroscopy and DMSO and arthritis. In Chapter 10 there are new sections on rigor mortis and recruitment. The section on the histology of cardiac muscle tissue has been revised. Chapter 11 includes new exhibits that describe and illustrate the muscles of the pharynx and larynx and sports injuries related to muscles.

UNIT III. CONTROL SYSTEMS OF THE HUMAN BODY

Chapter 12 has been updated to include a new section on neuronal intracellular transport and revised sections on factors that alter synaptic transmission and membrane potentials. In Chapter 14, discussions of drugs and the blood–brain barrier, brain injuries, brain electrical activity mapping (BEAM), rabies, Reye's syndrome (RS), and senility have been added. A new exhibit in this chapter summarizes cranial nerves, and material on the thalamus, brain lateralization, the cerebellum, and multiple sclerosis (MS) has been revised. Chapter 15 now contains a revised discussion of cutaneous reception and new sections on the somatosensory cortex and motor cortex. In Chapter 17 coverage has been expanded to include corneal transplants, color blindness, light and dark adaptation, and deafness. The treatment of static and dynamic equilibrium, olfactory and gustatory sensations, and sound waves has been revised. Chapter 18 contains revised material on hormone chemistry, prostaglandins, the mechanism of hormone action, the control of hormone secretion through feedback systems, and the histology of the adenohypophysis. Endocrine disorders are now integrated throughout the chapter as clinical applications and several new summary exhibits are included

UNIT IV. MAINTENANCE OF THE HUMAN BODY

Chapter 19 contains a revised discussion of hemostasis and new sections on blood substitutes and hemophilia. The section on the cardiac cycle and stroke volume in Chapter 20 has been extensively expanded; new material on the artificial heart, the Holter monitor, and defibrillation have been added. Updated and expanded topics in Chapter 21 are the histology of capillaries, atherosclerosis, hypertension, and hemodynamics. New topics in the chapter are coronary artery disease (CAD), coronary artery spasm, and the cardiovascular effects of exercise.

The section on capillary exchange has been moved from Chapter 27 to Chapter 21. Chapter 22 has been revised by the addition of discussions of immunology and cancer, monoclonal antibodies, and acquired immune deficiency syndrome (AIDS). The discussion of nonspecific resistance, hypersensitivity, and autoimmune disease have been revised. Chapter 23 contains additional coverage on pulmonary ventilation, the nervous control of respiration, and factors that modify respiration. There are now new sections on pulmonary embolism, compliance, and airway resistance, nitrogen narcosis, hyperbaric oxygenation, bronchitis, the common cold, and influenza. In Chapter 24, the sections on deglutition, the control of digestive secretions and motility, and the absorption of nutrients have been expanded. New material has been included on the absorption of water, electrolytes, and vitamins; hepatitis; bulimia; mumps; heartburn; vomiting; diarrhea and constipation; and liver biopsy. Several new summary exhibits have been added and the discussion of the regulation of food intake has been moved from Chapter 25 to Chapter 24. Chapter 25 contains expanded and reorganized discussions of metabolism and updated mineral and vitamin exhibits. The chapter also includes new material on absorptive and postabsorptive states. Chapter 26 provides new material on tests for renal function, bilirubin, urobilinogen, microbes in urine, and continuous ambulatory peritoneal dialysis (CAPD).

UNIT V. CONTINUITY

New or expanded topics in Chapter 28 include actions of inhibin and relaxin, diagnosis of pregnancy, premenstrual syndrome (PMS), and toxic shock syndrome (TSS). In Chapter 29, the sections on embryonic development, hormones of pregnancy, and lactation have been expanded. There are also new sections on morning sickness and twinning.

SPECIAL FEATURES

As in previous editions, the book contains numerous learning aids. Users of the book have cited the pedagogical aids as one of the book's many strengths. All of the tested and successful learning aids of previous editions have been kept in the fourth edition, and several new ones have been added. These special features are:

1. **Student Objectives.** Each chapter opens with a comprehensive list of Student Objectives. Each objective describes a knowledge or skill students should acquire while studying the chapter. (See **Note to the Student** for an explanation of how the objectives can be used.)

2. **Study Outline.** A Study Outline at the end of each chapter provides a brief summary of major topics. This section consolidates the essential points covered in the chapter so that students can recall and relate the points to one another. Page numbers have been added to the outlines so that topics within chapters can be located more easily.

3. **Review Questions.** Review Questions at the end of each chapter provide a check to see if the objectives stated at the beginning of the chapter have been met. After answering the questions, students should reread the objectives to determine whether they have met the goals.

4. Exhibits. Health-science students are generally expected to learn a great deal about the anatomy of certain organ systems, specifically, skeletal muscles, articulations, blood vessels, and nerves. In order to avoid interrupting the discussion of concepts and to organize the data, anatomical details have been presented in tabular form in exhibits, most of which are accompanied by illustrations. New summary exhibits have been added to the fourth edition, especially in relation to physiological principles.

5. Disorders: Homeostatic Imbalances. Abnormalities of structure or function are grouped at the end of appropriate chapters in sections entitled "Disorders: Homeostatic Imbalances." These sections provide a review of normal body processes as well as demonstrate the importance of the study of anatomy and physiology to a career in any of the health fields. All disorders have been updated and many new ones have been added.

6. Phonetic Pronunciations. Throughout the text, phonetic pronunciations are provided in parentheses for selected anatomical and physiological terms. These pronunciations are given at the point where the terms are introduced and are repeated in the Glossary of Terms. The **Note to the Student** explains the pronunciation key.

7. Medical Terminology. Glossaries of selected medical terms appear at the end of appropriate chapters; these listings are entitled "Medical Terminology." All of the glossaries have been revised for the fourth edition.

8. Drugs. Drugs associated with various body systems are discussed at the end of selected chapters. They have been revised for the fourth edition.

9. Clinical Applications. Throughout the text, Clinical Applications are boxed off for greater emphasis. Many new ones have been added.

10. Line Art. The line drawings in the book are large so that details are easily seen. In the fourth edition, many new illustrations have been added and full color is used throughout to differentiate structures and regions.

11. Photographs. The photographs amplify the narrative and the line drawings. Numerous photomicrographs (in full color), newly added scanning electron micrographs, and transmission electron micrographs enhance the histological discussions. Color photographs of specimens clarify gross anatomy discussions and have been added throughout, especially in Chapter 13 (spinal cord), Chapter 14 (brain), and Chapter 17 (special senses). Photographs of regional dissections, some in full color, have been added to the fourth edition in chapters dealing with the spinal cord, brain, heart, respiratory, digestive, urinary, and reproductive systems. A unique feature of the photographic component of the fourth edition is the addition of colored cross sections of various regions of the body in Chapter 11 (muscular system).

12. Appendixes. Appendix A, Measurements, summarizes U.S., metric, and apothecary units of length, mass, volume, and time. Appendix B, Abbreviations, is new to the fourth edition. It is an alphabetical list of commonly encountered medical abbreviations.

13. Glossary. Two glossaries appear at the end of the book. The first deals with prefixes, suffixes, and combining forms. The second is a comprehensive glossary of terms and has been greatly expanded for the fourth edition.

14. Bibliography. The extensive bibliography lists current suggested readings that correspond to the unit organization of the text.

SUPPLEMENTS

The following supplementary items are available to accompany the fourth edition of *Principles of Anatomy and Physiology:*

1. Learning Guide. By Kathleen S. Prezbindowski and Gerard J. Tortora, the *Learning Guide* is designed to help students *learn* anatomy and physiology. Rather than asking for mere repetition of the text material, the *Learning Guide* requires students to perform activities such as labeling a drawing or coloring parts of a diagram, as well as to answer multiple-choice, matching, and fill-in questions. The book's emphasis is on *using* new information in order to learn it. To enhance the effectiveness of the exercises, answers are provided for key-concept exercises so that the student has immediate feedback. A Mastery Test at the end of each chapter provides the student with a means of evaluating his or her learning of the chapter material and also gives practice for classroom testing situations.

2. Test Bank. The complimentary *Test Bank* contains 55 test items—multiple-choice, true-false, and matching questions—for each of the 29 chapters in the book. The test bank is available in standard printed format as well as on ACCESS (a computer-based test generator) and MICROTEST (for use with a microcomputer).

3. Instructor's Manual. Each chapter in the complimentary *Instructor's Manual,* prepared by Professor Tortora, consists of a chapter overview, a list of instructional concepts relating to the chapter, suggested problem-solving essay questions, and lists of audiovisual materials relating to the chapter topic.

4. Transparencies. Seventy-five full-color transparencies will be available. Illustrations from the text that are often shown and discussed in class are the subjects for the transparencies, and special care has been taken to make them clear and usable as overhead projections.

ACKNOWLEDGMENTS

I wish to thank the following people for their contribution to the fourth edition of *Principles of Anatomy and Physiology:*

Michael J. Autuori, University of Bridgeport
Frank Baker, Golden West College
Anita Been, University of Wisconsin
Mary Jane Burge, Cuyahoga Community College
Kenneth Carpenter, Shelby State Community College
Daniel Fertig, East Los Angeles College
Elvis Holt, Indiana University–Purdue University
Emron A. Jensen, Weber State College
R. Bruce Judd, Edison Community College
Michael C. Kennedy, Hahnemann Medical College
Donald Kisiel, Suffolk County Community College
Katherine Klein, Community College of Allegheny County
Gloria Jeanne Rogillio, Hinds Junior College
Paul Spannbauer, Hudson Valley Community College
Kent Van De Graaff, Brigham Young University
Edward P. Wallen, University of Wisconsin
Donald Wheeler, Cuyahoga Community College

Continuing thanks go to those people who worked on the three previous editions. Gratitude is also extended for the contributions of those people whose names appear with the photographs in the text. Special thanks to Dr. Michael H. Ross of the University of Florida for providing the excellent color photomicrographs. The exceedingly complex task of coordinating the art program was carried out smoothly and efficiently by Phil Alkana and Ellen Meek Tweedy, both of whom have my deepest thanks. Finally, for typing drafts of the manuscript and numerous other clerical duties associated with the task of putting together a textbook, thanks to Geraldine C. Tortora.

As the acknowledgments indicate, the participation of many individuals of diverse talent and expertise is required in the production of a textbook of this scope and complexity. For this reason, readers and users of the fourth edition are invited to send their reactions and suggestions so that plans can be formulated for subsequent editions.

Gerard J. Tortora

Biology Department
Bergen Community College
400 Paramus Road
Paramus, NJ 07652

Note to the Student

At the beginning of each chapter is a listing of **Student Objectives.** Before you read the chapter, please read the objectives carefully. Each objective is a statement of a skill or knowledge that you should acquire. To meet these objectives, you will have to perform several activities. Obviously, you must read the chapter carefully. If there are sections of the chapter that you do not understand after one reading, you should reread those sections before continuing. In conjunction with your reading, pay particular attention to the figures and exhibits; they have been carefully coordinated with the textual narrative.

At the end of each chapter are two and sometimes three other learning guides that you may find useful. The first, **Study Outline,** is a concise summary of important topics discussed in the chapter. This section is designed to consolidate the essential points covered in the chapter, so that you may recall and relate them to one another. The second guide, **Review Questions,** is a series of questions designed specifically to help you master the objectives. A third aid, **Medical Terminology,** appears in some chapters. This is a listing of terms designed to build your medical vocabulary. After you have answered the review questions, you should return to the beginning of the chapter and reread the objectives to determine whether you have achieved the goals.

As a further aid, we have included pronunciations for many terms that may be new to you. These appear in parentheses immediately following the new words, and they are repeated in the glossary of terms at the back of the book. (Of course, since there will always be some conflict among medical personnel and dictionaries about pronunciation, you will come across variations in different sources.) Look at the words carefully and say them out loud several times. Learning to pronounce a new word will help you remember it and make it a useful part of your medical vocabulary. Take a few minutes now to read the following pronunciation key, so it will be familiar as you encounter new words. The key is repeated at the beginning of the glossary of terms.

PRONUNCIATION KEY

1. The strongest accented syllable appears in capital letters, for example, bilateral (bī-LAT-er-al) and diagnosis (dī-ag-NŌ-sis).
2. If there is a secondary accent, it is noted by a single quote mark('), for example, constitution (kon'-sti-TOO-shun) and physiology (fiz'-ē-OL-ō-jē). Any additional secondary accents are also noted by a single quote mark, for example, decarboxylation (dē'-kar-bok'-si-LĀ-shun).
3. Vowels marked with a line above the letter are pronounced with the long sound as in the following common words.

   ā as in *māke*
   ē as in *bē*
   ī as in *īvy*
   ō as in *pōle*

4. Vowels not so marked are pronounced with the short sound as in the following words.

   e as in *bet*
   i as in *sip*
   o as in *not*
   u as in *bud*

5. Other phonetic symbols are used to indicate the following sounds:

   a as in *above*
   oo as in *sue*
   yoo as in *cute*
   oy as in *oil*

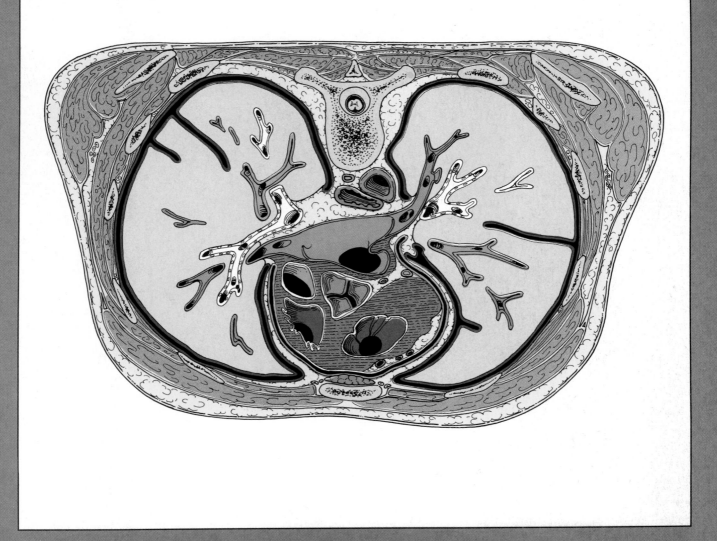

UNIT I

Organization of the Human Body

This unit is designed to show you how your body is organized at different levels. After you study the various regions and parts of your body, you will discover the importance of the chemicals that make it up. You will then find out how your cells, tissues, and organs form the systems that keep you alive and healthy.

- Define anatomy, with its subdivisions, and physiology.

- Explain the relationship between structure and function.

- Define each of the following levels of structural organization that make up the human body: chemical, cellular, tissue, organ, system, and organismic.

- Identify the principal systems of the human body, list the representative organs of each system, and describe the function of each system.

- Describe the general anatomical characteristics of the structural plan of the human body.

- Define the anatomical position.

- Compare common and anatomical terms used to describe various regions of the human body.

- Define several directional terms used in association with the human body.

- Define the common anatomical planes that may be passed through the human body.

- Distinguish a cross section, frontal section, and midsagittal section.

- Define the relationship of radiographic anatomy to the diagnosis of disease.

- List by name and location the principal body cavities and their major organs.

- Explain how the abdominopelvic cavity is divided into nine regions and four quadrants.

- Contrast the principles employed in conventional radiography and in computed tomography (CT) scanning.

- Point out the differences between a roentgenogram and a computed tomography (CT) scan.

- Explain how the dynamic spatial reconstructor (DSR) operates and its diagnostic importance in radiographic anatomy.

- Define homeostasis and explain why homeostasis is a state that results in normal body activities and why the inability to achieve homeostasis leads to disorders.

- Define a stress and identify the effects of stress on homeostasis.

- Describe the interrelationships of body systems in maintaining homeostasis.

- Compare the role of the endocrine and nervous systems in maintaining homeostasis.

- Contrast the homeostasis of blood pressure (BP) through nervous control and of blood sugar (BS) level through hormonal control.

- Define a feedback system and explain its role in homeostasis.

- Compare negative and positive feedback systems.

1 An Introduction to the Human Body

You are about to begin a study of the human body in order to learn how your body is organized and how it functions. The study of the human body involves many branches of science. Each contributes to a comprehensive understanding of how your body normally works and what happens when it is injured, diseased, or placed under stress.

ANATOMY AND PHYSIOLOGY DEFINED

Two branches of science that will help you understand your body parts and functions are anatomy and physiology. **Anatomy** (*anatome* = to dissect) refers to the study of *structure* and the relationships among structures. Anatomy is a broad science, and the study of structure becomes more meaningful when specific aspects of the science are considered. **Surface anatomy** is the study of the form (morphology) and markings of the surface of the body. **Gross** or **macroscopic anatomy** deals with structures that can be studied without a microscope. Another kind of anatomy, **systemic** or **systematic anatomy,** covers specific systems of the body, such as the system of nerves, spinal cord, and brain or the system of heart, blood vessels, and blood. **Regional anatomy** deals with a specific region of the body, such as the head, neck, chest, or abdomen. **Developmental anatomy** is the study of development from fertilized egg to adult form. **Embryology** is generally restricted to the study of development from the fertilized egg through the eighth week in utero. Other branches of anatomy are **pathological** (*patho* = disease) **anatomy,** the study of structural changes caused by disease; **histology** (*histio* = tissue), the microscopic study of the structure of tissues; and **cytology** (*cyt, cyto, cyte* = cell), the study of cells. **Radiographic anatomy,** the study of the structure of the body using radiographic (x-ray) techniques, is discussed in detail later in the chapter.

Whereas anatomy and its branches deal with structures of the body, **physiology** (fiz'-ē-OL-ō-jē) deals with *functions* of the body parts, that is, how the body parts work. Since physiology cannot be completely separated from anatomy, you will learn about the human body by studying its structures and functions together, and you will see how each structure of the body is custom-modeled to carry out a particular function. The structure of a part often determines the functions it will perform. In turn, body functions often influence the size, shape, and health of the structures.

LEVELS OF STRUCTURAL ORGANIZATION

The human body consists of several levels of structural organization that are associated with one another in several ways. Here we will consider the principal levels that will help you to understand how your body is organized (Figure 1-1). The lowest level of organization, the **chemical level,** includes all chemical substances essential for maintaining life. All these chemicals are made up of atoms joined together in various ways.

The chemicals, in turn, are put together to form the next higher level of organization: the **cellular level. Cells** are the basic structural and functional units of an organism. Among the many kinds of cells in your body are muscle cells, nerve cells, and blood cells. Figure 1-1 shows several isolated cells from the lining of the stomach. Each has a different structure, and each performs a different function.

The next higher level of structural organization is the **tissue level. Tissues** are made up of groups of similar cells and their intercellular material (substance between cells) that perform certain special functions. When the isolated cells shown in Figure 1-1 are joined together, they form a tissue called epithelium, which lines the stomach. Each cell in the tissue has a specific function. Mucous cells produce mucus, a secretion that lubricates food as it passes through the stomach. Parietal cells produce acid in the stomach. Chief cells produce enzymes needed to digest proteins. Other examples of tissues in your body are muscle tissue, connective tissue, and nervous tissue.

In many places in the body, different kinds of tissues are joined together to form an even higher level of organization: the **organ level. Organs** are structures of definite form and function composed of two or more different tissues. Organs usually have a recognizable shape. Examples of organs are the heart, liver, lungs, brain, and stomach. Figure 1-1 shows three of the tissues that make up the stomach. The serosa is a layer of connective tissue and epithelium around the outside that protects the stomach and reduces friction when the stomach moves and rubs against other organs. The muscle tissue layers of the stomach contract to mix food and pass it on to the next digestive organ. The epithelial tissue layer lining the stomach produces mucus, acid, and enzymes.

The next higher level of structural organization in the body is the **system level. A system** consists of an association of organs that have a common function. The digestive system, which functions in the breakdown of food, is composed of the mouth, saliva-producing glands called salivary glands, pharynx (throat), esophagus (gullet), stomach, small intestine, large intestine, rectum, liver, gallbladder, and pancreas.

The highest level is the **organismic level.** All the parts of the body functioning with one another constitute the total **organism**—one living individual.

In the chapters that follow, you will examine the anatomy and physiology of the major body systems. Exhibit 1-1 illustrates these systems, their representative organs, and their general functions. Unless otherwise stated, male and female bodies are similar. The systems are presented in the exhibit in the order in which they are discussed in later chapters.

STRUCTURAL PLAN

The human body has certain general **anatomical characteristics** that will help you to understand its overall structural plan. For example, humans have a **backbone,** or **vertebral column,** a characteristic that places them in a large group of organisms called *vertebrates.* Another characteristic is the body's **tube within a tube** construction. The outer tube is formed by the body wall; the inner tube is the digestive tract. Moreover, humans are for the most part **bilaterally symmetrical,** that is, externally the left and right sides of the body are mirror images.

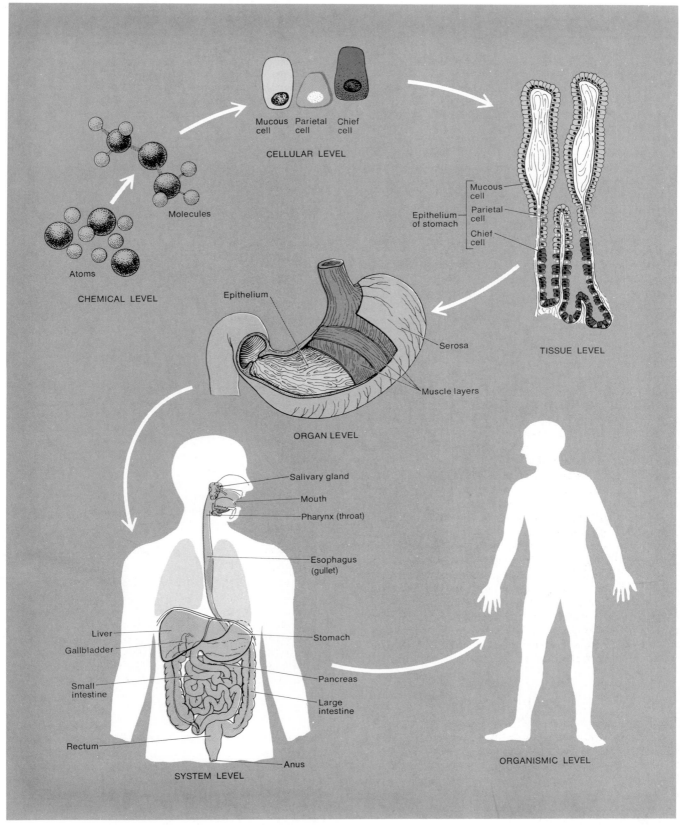

FIGURE 1-1 Levels of structural organization that compose the human body.

EXHIBIT 1-1

PRINCIPAL SYSTEMS OF HUMAN BODY, REPRESENTATIVE ORGANS, AND FUNCTIONS

1. Integumentary

Definition: The skin and structures derived from it, such as hair, nails, and sweat and oil glands.

Function: Helps regulate body temperature, protects the body, eliminates wastes, synthesizes vitamin D, and receives certain stimuli such as temperature, pressure, and pain.

2. Skeletal

Definition: All the bones of the body, their associated cartilages, and the joints of the body.

Function: Supports and protects the body, gives leverage, produces blood cells, and stores minerals.

11. Reproductive

Definition: Organs (testes and ovaries) that produce reproductive cells (sperm and ova) and organs that transport and store reproductive cells.

Function: Reproduces the organism.

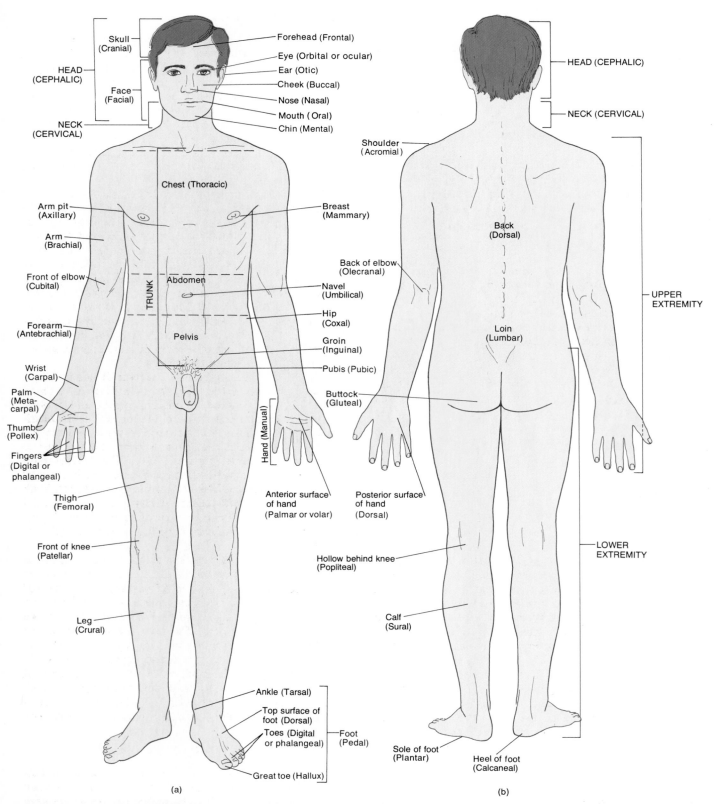

FIGURE 1-2 Anatomical position. The common name and anatomical terms, in parentheses, are indicated for many of the regions of the body. **(a)** Anterior view. **(b)** Posterior view.

ANATOMICAL POSITION AND ANATOMICAL NAMES

In all anatomical texts and charts, descriptions of any region or part of the human body assume that the body is in a specific position, called the **anatomical position,** so that directional terms are clear and any part can be related to any other part. In the anatomical position, the subject is standing erect facing the observer, the upper extremities are placed at the sides, and the palms of the hands are turned forward (Figure 1-2). Once the body is in the anatomical position, it is easier to visualize and understand how it is organized into various regions. The common and anatomical terms, in parentheses, of the principal body regions are also presented in Figure 1-2.

DIRECTIONAL TERMS

In order to explain exactly where various body structures are located in relationship to each other, anatomists use certain **directional terms.** If you want to point out the sternum (breastbone) to someone who knows where the clavicle (collarbone) is, you can say that the sternum is inferior (farther away from the head) and medial (toward the middle of the body) to the clavicle. As you can see, using the terms *inferior* and *medial* avoids a great deal of complicated description. Many directional terms are defined in Exhibit 1-2 and the parts of the body referred to in the examples are labeled in Figure 1-3. Studying the exhibit and the figure together should make clear to you the directional relationships among various body parts.

PLANES

The structural plan of the human body may also be discussed with respect to **planes** (imaginary flat surfaces) that pass through it. Several of the commonly used planes are illustrated in Figure 1-4. A **midsagittal (median) plane** is a vertical plane that passes through the midline of the body and divides the body or organs into equal right and left sides. A **sagittal** (SAJ-i-tal) **plane,** also called a **parasagittal** (*para* = near) **plane,** is a plane parallel to a midsagittal plane that divides the body or organs into unequal left and right portions. A **frontal (coronal;** kō-RŌ-nal) **plane** is a plane at a right angle to a midsagittal (or sagittal) plane that divides the body or organs into anterior and posterior portions. Finally, a **horizontal (transverse) plane** is a plane that is parallel to the ground, that is, at a right angle to the midsagittal, sagittal, and frontal planes. It divides the body or organs into superior and inferior portions.

When you study a body structure, you will often view it in section, that is, look at the flat surface resulting from a cut made through the three-dimensional structure. It is important to know the plane of the section so that you can understand the anatomical relationship of one part to another. Figure 1-5 indicates how three different sections—a *cross section,* a *frontal section,* and a *midsagittal section*—are made through different parts of the brain.

BODY CAVITIES

Spaces within the body that contain internal organs are called **body cavities.** Specific cavities may be distinguished if the body is divided into right and left halves. Figure 1-6 shows the two principal body cavities. The **dorsal body cavity** is located near the dorsal (posterior) surface of the body. It is further subdivided into a **cranial cavity,** which is a bony cavity formed by the cranial (skull) bones and contains the brain, and a **vertebral (spinal) canal,** which is a bony cavity formed by the vertebrae of the backbone and contains the spinal cord and the beginnings of spinal nerves.

The other principal body cavity is the **ventral body cavity** or **coelom** (SĒ-lōm). This cavity is located on the ventral (anterior) aspect of the body. The organs inside the ventral body cavity are called the **viscera** (VIS-er-a). Its walls are composed of skin, connective tissue, bone, muscles, and a membrane called the peritoneum. Like the dorsal body cavity, the ventral body cavity has two principal subdivisions—an upper portion, called the **thoracic** (thō-RAS-ik) **cavity** (or chest cavity), and a lower portion, called the **abdominopelvic** (ab-dom'-i-nō-PEL-vik) **cavity.** The anatomical landmark that divides the ventral body cavity into the thoracic and abdominopelvic cavities is the muscular diaphragm.

The thoracic cavity contains several divisions. There are two **pleural cavities** (Figure 1-7). Each is a small potential space between the visceral pleura and parietal pleura, the membranes covering the lungs. The **mediastinum** (mē'-dē-as-TĪ-num), is a mass of tissue between the pleurae of the lungs that extends from the sternum to the vertebral column (Figure 1-7a, b). The mediastinum includes all of the contents of the thoracic cavity, except the lungs themselves. The **pericardial** (per'-ē-KAR-dē-al; *peri* = around, *cardi* = heart) **cavity** is a small potential space between the visceral pericardium and parietal pericardium, the membranes covering the heart (Figure 1-7b).

The abdominopelvic cavity, as the name suggests, is divided into two portions, although no wall separates them (see Figure 1-6). The upper portion, the **abdominal cavity,** contains the stomach, spleen, liver, gallbladder, pancreas, small intestine, most of the large intestine, the kidneys, and the ureters. The lower portion, the **pelvic cavity,** contains the urinary bladder, sigmoid colon, rectum, and the internal male or female reproductive organs. One way to mark the division between the abdominal and pelvic cavities is to draw an imaginary line from the symphysis pubis (anterior joint between hipbones) to the superior border of the sacrum (sacral promontory).

ABDOMINOPELVIC REGIONS

To describe the location of organs easily, the abdominopelvic cavity may be divided into the **nine regions** shown in Figure 1-8. Although some unfamiliar terms are used in describing the nine regions and their contents, follow the descriptions as well as you can. When the organs are studied in detail in later chapters, they will have more meaning. The **epigastric** (*epi* = above; *gaster* = stomach) **region** contains the left lobe and medial part of the right

EXHIBIT 1-2
DIRECTIONAL TERMS

TERM	DEFINITION	EXAMPLE
Superior (cephalad or **craniad)**	Toward the head or the upper part of a structure; generally refers to structures in the trunk.	The heart is superior to the liver.
Inferior (caudad)	Away from the head or toward the lower part of a structure; generally refers to structures in the trunk.	The stomach is inferior to the lungs.
Anterior (ventral)	Nearer to or at the front of the body.	The sternum is anterior to the heart.
Posterior (dorsal)	Nearer to or at the back of the body.	The esophagus is posterior to the trachea.
Medial	Nearer to the midline of the body or a structure.	The ulna is on the medial side of the forearm.
Lateral	Farther from the midline of the body or a structure.	The ascending colon of the large intestine is lateral to the urinary bladder.
Intermediate	Between two structures, one of which is medial and one of which is lateral.	The ring finger is intermediate between the little (medial) and middle (lateral) fingers.
Ipsilateral	On the same side of the body.	The gallbladder and ascending colon of the large intestine are ipsilateral.
Contralateral	On the opposite side of the body.	The ascending and descending colons of the large intestine are contralateral.

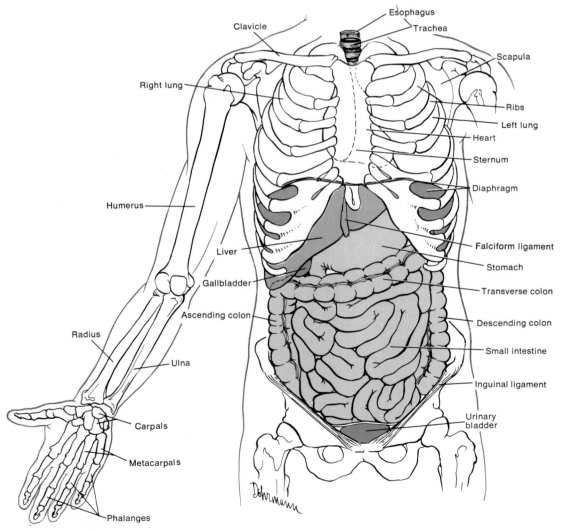

FIGURE 1-3 Anatomical and directional terms. By studying Exhibit 1-2 with this figure, you should gain an understanding of the meanings of the terms *superior, inferior, anterior, posterior, medial, lateral, intermediate, ipsilateral, contralateral, proximal,* and *distal.*

Proximal	Nearer to the attachment of an extremity to the trunk or a structure; nearer to the point of origin.	The humerus is proximal to the radius.
Distal	Farther from the attachment of an extremity to the trunk or a structure; farther from the point of origin.	The phalanges are distal to the carpals (wrist bones).
Superficial	Toward or on the surface of the body.	The muscles of the thoracic wall are superficial to the viscera in the thoracic cavity. (See Figure 1-7b.)
Deep	Away from the surface of the body.	The muscles of the arm are deep to the skin of the arm.
Parietal	Pertaining to or forming the outer wall of a body cavity.	The parietal pleura forms the outer layer of the pleural sacs that surround the lungs. (See Figure 1-7b.)
Visceral	Pertaining to the covering of an organ (viscus).	The visceral pleura forms the inner layer of the pleural sacs and covers the external surface of the lungs. (See Figure 1-7b.)

FIGURE 1-4 Planes of the human body.

(a)

FIGURE 1-5 Sections through different parts of the brain. (a) Cross section. (Courtesy of Stephen A. Kieffer and E. Robert Heitzman, *An Atlas of Cross-Sectional Anatomy,* Harper & Row, Publishers, Inc., New York, 1979.) (b) Frontal section. (Courtesy of C. Yokochi and J. W. Rohen, *Photographic Anatomy of the Human Body,* 2nd ed., 1979, IGAKU-SHOIN, Ltd., Tokyo, New York.) (c) Midsagittal section. (Courtesy of C. Yokochi and J. W. Rohen, *Photographic Anatomy of the Human Body,* 2nd ed., 1979, IGAKU-SHOIN, Ltd., Tokyo, New York.)

(b)

(c)

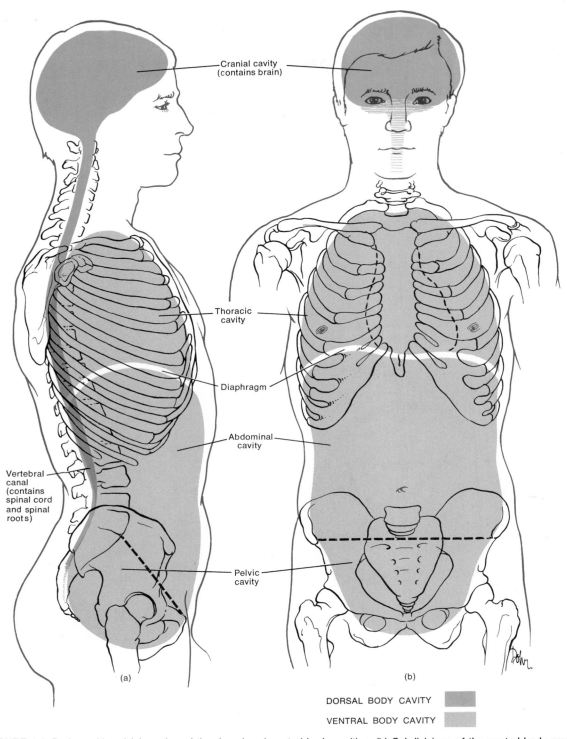

FIGURE 1-6 Body cavities. (a) Location of the dorsal and ventral body cavities. (b) Subdivisions of the ventral body cavity.

lobe of the liver, the pyloric part and lesser curvature of the stomach, the superior and descending portions of the duodenum, the body and upper part of the head of the pancreas, and the two adrenal (suprarenal) glands. The **right hypochondriac** (*hypo* = under; *chondro* = cartilage, of ribs) **region** contains the right lobe of the liver, the gallbladder, and the upper third of the right kidney. The **left hypochondriac region** contains the body and fun-

dus of the stomach, the spleen, the left colic (splenic) flexure, the upper two-thirds of the left kidney, and the tail of the pancreas. The **umbilical region** contains the middle of the transverse colon, the inferior part of the duodenum, the jejunum, the ileum, the hilar regions of the kidneys, and the bifurcations (branching) of the abdominal aorta and inferior vena cava. The **right lumbar** (*lumbus* = loin) **region** contains the superior part of the

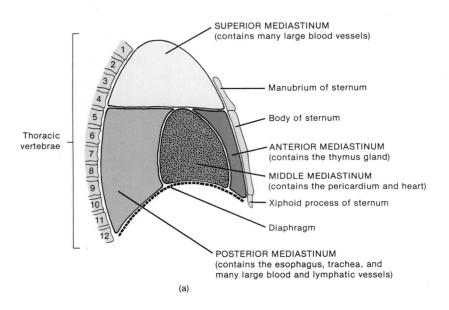

SUPERIOR MEDIASTINUM
(contains many large blood vessels)

Manubrium of sternum

Body of sternum

ANTERIOR MEDIASTINUM
(contains the thymus gland)

MIDDLE MEDIASTINUM
(contains the pericardium and heart)

Xiphoid process of sternum

Diaphragm

Thoracic vertebrae

POSTERIOR MEDIASTINUM
(contains the esophagus, trachea, and
many large blood and lymphatic vessels)

(a)

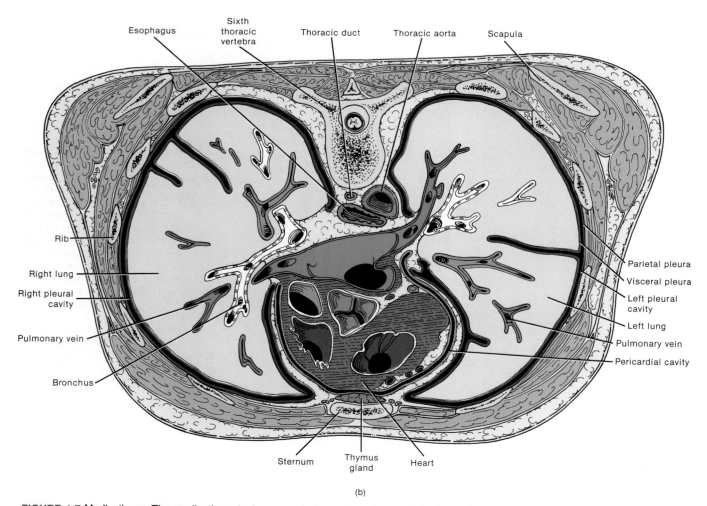

Esophagus

Sixth thoracic vertebra

Thoracic duct

Thoracic aorta

Scapula

Rib

Right lung

Right pleural cavity

Pulmonary vein

Bronchus

Parietal pleura

Visceral pleura

Left pleural cavity

Left lung

Pulmonary vein

Pericardial cavity

Sternum

Thymus gland

Heart

(b)

FIGURE 1-7 Mediastinum. The mediastinum is the space between the pleurae of the lungs that extends from the sternum to the vertebral column. (a) Subdivisions of the mediastinum seen in right lateral view. (b) Mediastinum seen in a cross section of the thorax.

- Identify by name and symbol the principal chemical elements of the human body.

- Explain, by diagraming, the structure of an atom.

- Describe the principle of nuclear magnetic resonance (NMR) and its diagnostic importance.

- Define a chemical reaction as a function of electrons in incomplete outer energy levels.

- Describe ionic bond formation in a molecule of sodium chloride (NaCl).

- Discuss covalent bond formation as the sharing of outer energy level electrons.

- Explain the nature and importance of hydrogen bonding.

- Define a radioisotope and explain the principle of positron emission tomography (PET) scanning and its diagnostic importance.

- Explain the basic differences among synthesis, decomposition, exchange, and reversible chemical reactions.

- Discuss how chemical reactions are related to metabolism.

- Define the type of energy involved when chemical bonds are formed and broken.

- Define and distinguish between inorganic and organic compounds.

- Discuss the functions of water as a solvent, suspending medium, chemical reactant, heat absorber, and lubricant.

- List and compare the properties of acids, bases, and salts.

- Define pH as the degree of acidity or alkalinity of a solution.

- Explain the role of a buffer system as a homeostatic mechanism that maintains the pH of a body fluid.

- Compare the structure and functions of carbohydrates, lipids, and proteins.

- Differentiate between dehydration synthesis and hydrolysis of organic molecules.

- Define the homeostatic role of enzymes as catalysts.

- Contrast the structure of deoxyribonucleic acid (DNA) and ribonucleic acid (RNA).

- Define the roles of DNA and RNA in heredity and protein synthesis.

- Identify the function and importance of adenosine triphosphate (ATP) and cyclic adenosine-3′,5′-monophosphate (cyclic AMP).

2 The Chemical Level of Organization

M any of the common substances you eat and drink—water, sugar, table salt, cooking oil—play vital roles in keeping you alive. In this chapter you will learn something about how the molecules of these substances function in your body. Fundamental to this study is a knowledge of basic chemistry and chemical processes. To understand the nature of the matter you are made from and the changes this matter goes through in your body, you will need to know which chemical elements are present in the human organism and how they interact.

INTRODUCTION TO BASIC CHEMISTRY

CHEMICAL ELEMENTS

All living and nonliving things consist of **matter,** which is anything that occupies space and has mass. Matter may exist in a solid, liquid, or gaseous state. All forms of matter are made up of a limited number of building units called **chemical elements,** substances that cannot be decomposed into simpler substances by ordinary chemical reactions. At present scientists recognize 106 different elements, of which 92 are naturally occurring. Elements are designated by letter abbreviations, usually derived from the first or first and second letters of the Latin or English name for the element. Such letter abbreviations are called **chemical symbols.** Examples of chemical symbols are H (hydrogen), C (carbon), O (oxygen), N (nitrogen), Na (sodium), K (potassium), Fe (iron), and Ca (calcium).

Approximately 24 elements are found in the human organism. Carbon, hydrogen, oxygen, and nitrogen make up about 96 percent of the body's weight. These 4 elements together with phosphorus and calcium constitute approximately 99 percent of the total body weight. Eighteen other chemical elements, called trace elements, are found in low concentrations and compose the remaining 1 percent.

STRUCTURE OF ATOMS

Each element is made up of units of matter called **atoms.** An element is simply a quantity of matter composed of atoms all of the same type. A handful of the element carbon, such as pure coal, contains only carbon atoms. A tank of oxygen contains only oxygen atoms. Measurements indicate that the smallest atoms are less than 0.00000001 cm (1/250,000,000 inch) in diameter, and the largest atoms are 0.00000005 cm (1/50,000,000 inch) in diameter. In other words, if 50 million of the largest atoms were placed end to end, they would measure approximately 1 inch in length.

An atom consists of two basic parts: the nucleus and the electrons (Figure 2-1). The centrally located **nucleus** comprises most of the atomic mass and contains positively charged particles called **protons** (p^+) and uncharged (neutral) particles called **neutrons** (n^0). Because each proton has one positive charge, the nucleus itself is positively charged. The second basic part of an atom contains the **electrons** (e^-). These negatively charged particles spin around the nucleus. The number of electrons in an atom of an element always equals the number of protons. Since each electron carries one negative charge, the negatively

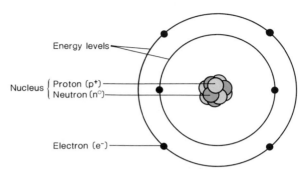

FIGURE 2-1 Structure of an atom. In this highly simplified version of a carbon atom, note the centrally located nucleus. The nucleus contains six neutrons and six protons, although all are not visible in this view since some are behind others. The six electrons orbit the nucleus at varying distances from its center.

charged electrons and the positively charged protons balance each other, and the atom is electrically neutral.

What makes the atoms of one element different from those of another? The answer lies in the number of protons. Figure 2-2 shows that the hydrogen atom contains one proton. The helium atom contains two. The carbon atom has six, and so on. Each different kind of atom has a different number of protons in its nucleus. The number of protons in an atom is called the atom's **atomic number.** Therefore we can say that each kind of atom, or element, has a different atomic number.

being noninvasive (not involving puncture or incision of the skin or insertion of an instrument or foreign material into the body), not utilizing radiation, and gathering biochemical information without time-consuming chemical analyses.

In clinical and preclinical trials, NMR studies have confirmed the findings of other diagnostic techniques such as CT scans and, in some cases, have added more detail. Thus, NMR has proved useful in identifying existing pathologies. Scientists hope that NMR can also be used to perform a "biopsy" on tumors without an operation, assess mental disorders, measure blood flow, study the evolution of hematomas and infarctions in the brain following a stroke, identify the potential for developing a stroke, assess treatment for conditions such as heart disease and stroke, monitor the progress and treatment of a disease, study the effects of toxic drugs on tissues, measure intracellular pH, study metabolism, and determine how donor organs from cadavers may function after transplantation.

ATOMS AND MOLECULES

When atoms combine with or break apart from other atoms, a **chemical reaction** occurs. In the process, new products with different properties are formed. Chemical reactions are the foundation of all life processes.

The electrons of the atom actively participate in chemical reactions. The electrons spin around the nucleus in orbits, shown in Figure 2-2 as concentric circles lying at varying distances from the nucleus. We call these orbits **energy levels.** Each orbit has a maximum number of electrons it can hold. For instance, the orbit nearest the nucleus never holds more than two electrons, no matter what the element. This orbit can be referred to as the first energy level. The second energy level holds a maximum of eight electrons. The third level of atoms whose atomic number is less than 20 also can hold a maximum of eight electrons. The third level of more complex atoms can hold a maximum of 18 electrons.

An atom always attempts to have its outermost orbit with the maximum number of electrons it can hold. To do this, the atom may give up an electron, take on an electron, or share an electron with another atom—whichever is easiest. Take a look at the chlorine atom. Its outermost energy level, which happens to be the third level, has seven electrons. Since the third level of an atom can hold a maximum of eight electrons, chlorine can be described as having a shortage of one electron. In fact, chlorine usually does try to pick up an extra electron. Sodium, by contrast, has only one electron in its outer level. This again happens to be the third energy level. It is much easier for sodium to get rid of the one electron than to fill the third level by taking on seven more electrons. Atoms of a few elements, like helium, have completely filled outer energy levels and do not need to gain or lose electrons. These are called **inert** elements.

Atoms with incompletely filled outer energy levels, like sodium and chlorine, tend to combine with other atoms

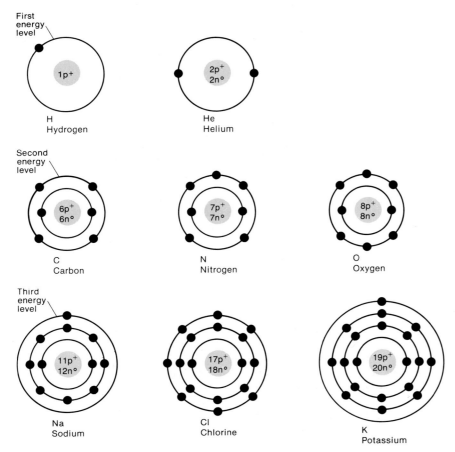

FIGURE 2-2 Atomic structures of some representative atoms.

FIGURE 2-3 Nuclear magnetic resonance (NMR) image of the female pelvis. (Courtesy of John Cassese, Director of Marketing, Fonar Corporation, Melville, New York.)

in a chemical reaction. During the reaction, the atoms can trade off or share electrons and, thereby, fill their outer energy levels. Atoms that already have filled outer levels generally do not participate in chemical reactions for the simple reason that they do not need to gain or lose electrons. When two or more atoms combine in a chemical reaction, the resulting combination is called a **molecule** (MOL-e-kyool). A molecule may contain two atoms of the same kind, as in the hydrogen molecule: H_2. The subscript 2 indicates that there are two hydrogen atoms in the molecule. Molecules may also be formed by the reaction of two or more different kinds of atoms, as in the hydrochloric acid molecule: HCl. Here an atom of hydrogen is attached to an atom of chlorine. A molecule that contains at least two different kinds of atoms is called a **compound.** Hydrochloric acid, which is present in the digestive juices of the stomach, is a compound. A molecule of hydrogen is not.

The atoms in a molecule are held together by forces of attraction called **chemical bonds.** Here we will consider ionic bonds, covalent bonds, and hydrogen bonds.

Ionic Bonds

Atoms are electrically neutral because the number of positively charged protons equals the number of negatively charged electrons. But when an atom gains or loses electrons, this balance is destroyed. If the atom gains electrons, it acquires an overall negative charge. If the atom loses electrons, it acquires an overall positive charge. Such a negatively or positively charged atom or group of atoms is called an **ion** (Ī-on).

Consider the sodium ion (Figure 2-4a). The sodium atom (Na) has 11 protons and 11 electrons, with 1 electron in its outer energy level. When sodium gives up the single electron in its outer level, it is left with 11 protons and

only 10 electrons. The atom now has an overall positive charge of one (+1). This positively charged sodium atom is called a sodium ion (written Na^+).

Another example is the formation of the chloride ion (Figure 2-4b). Chlorine has a total of 17 electrons, 7 of them in the outer energy level. Since this energy level can hold 8 electrons, chlorine tends to pick up an electron that has been lost by another atom. By accepting an electron, chlorine acquires a total of 18 electrons. However, it still has only 17 protons in its nucleus. The chloride ion therefore has a negative charge of one (−1) and is written as Cl^-.

The positively charged sodium ion (Na^+) and the negatively charged chloride ion (Cl^-) attract each other—unlike charges attract each other. The attraction, called an **ionic bond,** holds the two atoms together, and a molecule is formed (Figure 2-4c). The formation of this molecule, sodium chloride (NaCl) or table salt, is one of the most common examples of ionic bonding. Thus an ionic bond is an attraction between atoms in which one atom loses electrons and another atom gains electrons. Generally, atoms whose outer energy level is less than half-filled lose electrons and form positively charged ions called **cations** (KAT-ī-ons). Examples of cations are potassium ion (K^+), calcium ion (Ca^{2+}), iron ion (Fe^{2+}), and sodium ion (Na^+). By contrast, atoms whose outer energy level is more than half-filled tend to gain electrons and form negatively charged ions called **anions** (AN-ī-ons). Examples of anions include iodine ion (I^-), chloride ion (Cl^-), and sulfur ion (S^{2-}).

Notice that an ion is always symbolized by writing the chemical abbreviation followed by the number of positive (+) or negative (−) charges the ion acquires.

Hydrogen is an example of an atom whose outer level is exactly half-filled. The first energy level can hold two electrons, but in hydrogen atoms it contains only one.

GENERAL NATURE   SPECIFIC EXAMPLE

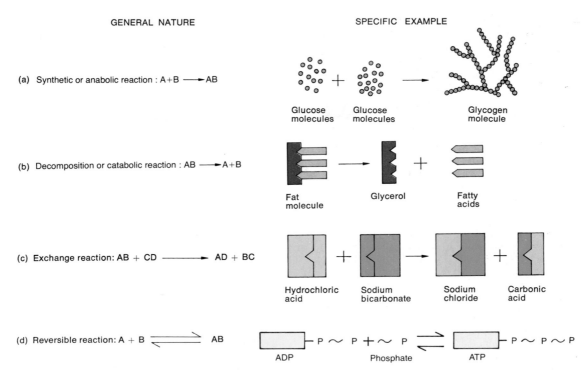

(a) Synthetic or anabolic reaction : A+B ——→ AB

Glucose molecules   Glucose molecules   Glycogen molecule

(b) Decomposition or catabolic reaction : AB ——→ A+B

Fat molecule   Glycerol   Fatty acids

(c) Exchange reaction: AB + CD ————→ AD + BC

Hydrochloric acid   Sodium bicarbonate   Sodium chloride   Carbonic acid

(d) Reversible reaction: A + B ⇌ AB

ADP   Phosphate   ATP

FIGURE 2-7 Kinds of chemical reactions. (a) Synthetic or anabolic reaction. When linked together as shown, molecules of glucose form a molecule of glycogen. Glucose is a sugar that is the primary source of energy. Glycogen is a storage form of that sugar found in the liver and skeletal muscles. (b) Decomposition or catabolic reaction. The example shown is a molecule of fat breaking down into glycerol and fatty acids. This reaction occurs whenever a food that contains fat is digested. (c) Exchange reaction. In this reaction, atoms of different molecules are exchanged with each other. Shown is a buffer reaction in which the body eliminates strong acids to help maintain homeostasis. (d) Reversible reaction. ATP (adenosine triphosphate) is an important source of stored energy. When the energy is needed, the ATP breaks down into ADP (adenosine diphosphate) and PO_4 (phosphate group), and energy is released in the reaction. The phosphate group is symbolized as P. The cells of the body reconstruct ATP by using the energy of foods to attach ADP to PO_4.

Chemical reactions in the bodies of people with low metabolism that is due to a disruption of homeostasis proceed more slowly than normal. Food is broken down slowly. Much of it is only partially broken down and then stored. Such people tend to gain weight easily, have little energy, and are often cold. Because their synthesis reactions are also slowed down, their bodies build up new structures very slowly. Wounds, for instance, often take a long time to heal.

Energy and Chemical Reactions

Energy is involved whenever bonds between atoms in molecules are formed or broken during the chemical reactions taking place in the body. When a chemical bond is formed, energy is required. When a bond is broken, energy is released. This means that synthesis reactions need energy in order to occur, whereas decomposition reactions give off energy. The building processes of the body—the construction of bones, the growth of hair and nails, the replacement of injured cells—occur basically through synthesis reactions. The breakdown of foods, on the other hand, occurs through decomposition reactions. When foods are decomposed, they release energy that can be used by the body for its building processes. Re-

leased energy can also be used to warm the body by taking the form of **heat energy.** Foods can be partially broken down into compounds that can be stored in the body. Later, when additional energy is required, the body finishes breaking down these reserve compounds. This stored energy is called **potential energy.**

Another form of energy that is important to life processes, **kinetic energy,** is discussed in Chapter 3.

CHEMICAL COMPOUNDS AND LIFE PROCESSES

Most of the chemicals in the body exist in the form of compounds. Biologists and chemists divide these compounds into two principal classes: inorganic compounds, which usually lack carbon, and organic compounds, which always contain carbon. **Inorganic compounds** are usually small, ionically bonded molecules that are vital to body functions. They include water, many salts, acids, and bases. **Organic compounds** are held together mostly or entirely by covalent bonds. They tend to be very large molecules and are therefore good building blocks for body structures. Organic compounds present in the body include carbohydrates, lipids, proteins, nucleic acids, and ATP.

INORGANIC COMPOUNDS

Water

One of the most important, as well as the most abundant, inorganic substances in the human organism is **water.** In fact, with a few exceptions, such as tooth enamel and bone tissue, water is by far the most abundant material in all tissues. About 60 percent of red blood cells, 75 percent of muscle tissue, and 92 percent of blood plasma is water. The following functions of water explain why it is such a vital compound in living systems:

1. Water is an excellent solvent and suspending medium. A *solvent* is a liquid or gas in which some other material (solid, liquid, or gas), called a *solute,* has been dissolved. The combination of solvent plus solute is called a **solution,** and one common example of a solution is salt water. A solute, such as salt in water, typically does not settle out of its solution. The solute can be retrieved through a chemical reaction or, in some cases, by boiling off the solvent. In a **suspension,** by contrast, the suspended material mixes with the liquid or suspending medium, but it will eventually settle out of the mixture. An example of a suspension is cornstarch and water. If the two materials are shaken together, a milky mixture forms. After the mixture sits for a while, however, the water clears at the top and the cornstarch settles to the bottom.

The solvating property of water is essential to health and survival. For example, water in the blood forms a solution with a small portion of the oxygen you inhale, allowing the oxygen to be carried to your body cells. Water in the blood also dissolves a small portion of the carbon dioxide that is carried from the cells to the lungs to be exhaled. Furthermore, if the surfaces of the air sacs in your lungs are not moist, oxygen cannot dissolve and, therefore, cannot move into your blood to be distributed throughout your body. Water, moreover, is the solvent that carries nutrients into and wastes out of your body cells.

As a suspending medium, water is also vital to your survival. Many large organic molecules are suspended in the water of your body cells. These molecules are consequently able to come in contact with other chemicals, allowing various essential chemical reactions to occur.

2. Water can participate in chemical reactions. During digestion, for example, water can be added to large nutrient molecules in order to break them down into smaller molecules. This kind of breakdown is necessary if the body is to utilize the energy in nutrients. Water molecules are also used in synthesis reactions. Such reactions occur in the production of hormones and enzymes.

EXHIBIT 2-1
DISSOCIATION OF REPRESENTATIVE SALTS INTO IONS THAT PROVIDE ESSENTIAL CHEMICAL ELEMENTS FOR THE BODY

SALT	DISSOCIATES INTO	CATION		ANION
NaCl Sodium chloride	$\longrightarrow$	Na^+ Sodium ion	+	Cl^- Chloride ion
KCl Potassium chloride	$\longrightarrow$	K^+ Potassium ion	+	Cl^- Chloride ion
$CaCl_2$ Calcium chloride	$\longrightarrow$	Ca^{2+} Calcium ion	+	$2Cl^-$ Chloride ions
$MgCl_2$ Magnesium chloride	$\longrightarrow$	Mg^{2+} Magnesium ion	+	$2Cl^-$ Chloride ions
$CaCO_3$ Calcium carbonate	$\longrightarrow$	Ca^{2+} Calcium ion	+	CO_3^{2-} Carbonate ion
$Ca_3(PO_4)_2$ Calcium phosphate	$\longrightarrow$	$3Ca^{2+}$ Calcium ions	+	$2PO_4^{3-}$ Phosphate ions
Na_2SO_4 Sodium sulfate	$\longrightarrow$	$2Na^+$ Sodium ions	+	SO_4^{2-} Sulfate ion

3. Water absorbs and releases heat very slowly. In comparison to other substances, water requires a large amount of heat to increase its temperature and a great loss of heat to decrease its temperature. Thus the presence of a large amount of water moderates the effects of fluctuations in environmental temperature and thereby helps to maintain a homeostatic body temperature.

4. Water requires a large amount of heat to change from a liquid to a gas. When water (perspiration) evaporates from the skin, it takes with it large quantities of heat and provides an excellent cooling mechanism.

5. Water serves as a lubricant in various regions of the body. It is a major part of mucus and other lubricating fluids. Lubrication is especially necessary in the chest and abdomen, where internal organs touch and slide over each other. It is also needed at joints, where bones, ligaments, and tendons rub against each other. In the digestive tract, water moistens foods to ensure their smooth passage.

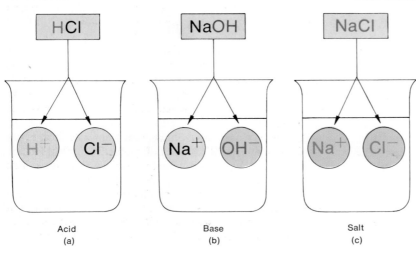

FIGURE 2-8 Ionization of acids, bases, and salts. (a) When placed in water, hydrochloric acid (HCl) dissociates into H^+ ions and Cl^- ions. Acids are proton donors. (b) When the base sodium hydroxide (NaOH) is placed in water, it dissociates into OH^- ions and Na^+ ions. Bases are proton acceptors. (c) When table salt (NaCl) is placed in water, it dissociates into positive and negative ions (Na^+ and Cl^-), neither of which is H^+ or OH^-.

Acids, Bases, and Salts

When molecules of inorganic acids, bases, or salts are dissolved in water in the body cells, they undergo **ionization** (ī'-on-i-ZĀ-shun) or **dissociation** (dis'-sō-sē-Ā-shun), that is, they break apart into ions. Such molecules are also called **electrolytes** (ē-LEK-trō-līts) because the solution will conduct an electric current (the chemistry and importance of electrolytes are discussed in detail in Chapter 27). An **acid** may be defined as a substance that dissociates into one or more *hydrogen ions* (H^+) and one or more negative ions (anions). An acid may also be defined as a proton (H^+) donor. A **base,** by contrast, dissociates into one or more *hydroxyl ions* (OH^-) and one or more positive ions (cations). A base may also be viewed as a proton acceptor. Hydroxyl ions, as well as some other negative ions, have a strong attraction for protons. A **salt,** when dissolved in water, dissociates into cations and anions, neither of which is H^+ or OH^- (Figure 2-8). Acids and bases react with one another to form salts. For example, the combination of hydrochloric acid (HCl), an acid, and sodium hydroxide (NaOH), a base, produces sodium chloride (NaCl), a salt, and water (H_2O).

Many salts are found in the body. Some are in cells, whereas others are in the body fluids, such as lymph, blood, and the extracellular fluid of tissues. The ions of salts are the source of many essential chemical elements. Exhibit 2-1 shows how salts dissociate into ions that provide these elements. Chemical analyses reveal that sodium and chloride ions are present in higher concentrations than other ions in extracellular body fluids. Inside the cells, phosphate and potassium ions are more abundant than other ions. Chemical elements such as sodium, phosphorus, potassium, or iodine are present in the body only in chemical combination with other elements or as ions. Their presence as free, un-ionized atoms could be instantly fatal. Exhibit 2-2 lists representative elements found in the body.

Acid–Base Balance: The Concept of pH

The fluids of your body must maintain a fairly constant balance of acids and bases. In solutions such as those found in body cells or in extracellular fluids, acids dissociate into hydrogen ions (H^+) and anions. Bases, on the other hand, dissociate into hydroxyl ions (OH^-) and cations. The more hydrogen ions that exist in a solution, the more acid the solution; conversely, the more hydroxyl ions, the more basic (alkaline) the solution. The term **pH** is used to describe the degree of *acidity* or *alkalinity* (*basicity*) of a solution. Biochemical reactions—reactions that occur in living systems—are extremely sensitive to even small changes in the acidity or alkalinity of the environment in which they occur. In fact, H^+ and OH^- ions are involved in practically all biochemical processes, and the functions of cells are modified greatly by any departure from narrow limits of normal H^+ and OH^- concentrations. For this reason, the acids and bases that are constantly formed in the body must be kept in balance.

A solution's acidity or alkalinity is expressed on a **pH scale** that runs from 0 to 14 (Figure 2-9). The pH scale is based on the number of H^+ ions in a solution expressed

EXHIBIT 2-2
REPRESENTATIVE CHEMICAL ELEMENTS FOUND IN THE BODY

CHEMICAL ELEMENT	COMMENT
Oxygen (O)	Constituent of water and organic molecules; functions in cellular respiration.
Carbon (C)	Found in every organic molecule.
Hydrogen (H)	Constituent of water, all foods, and most organic molecules.
Nitrogen (N)	Component of all protein molecules and nucleic acid molecules.
Calcium (Ca)	Constituent of bone and teeth; required for blood clotting, intake (endocytosis) and output (exocytosis) of substances through plasma membranes, motility of cells, movement of chromosomes prior to cell division, glycogen metabolism, synthesis and release of transmitter substances, and contraction of muscle.
Phosphorus (P)	Component of many proteins, nucleic acids, ATP, and cyclic AMP; required for normal bone and tooth structure; found in nerve tissue.
Chlorine (Cl)	Cl^- is an anion of NaCl, a salt important in water movement between cells.
Sulfur (S)	Component of many proteins, especially the contractile protein of muscle.
Potassium (K)	Required for growth and important in conduction of nerve impulses and muscle contraction.
Sodium (Na)	Na^+ is a cation of NaCl; structural component of bone; essential in blood to maintain water balance; needed for conduction of nerve impulses.
Magnesium (Mg)	Component of many enzymes.
Iodine (I)	Vital to functioning of thyroid gland.
Iron (Fe)	Essential component of hemoglobin and respiratory enzymes.

in chemical units called moles per liter. A pH of 7 means that a solution contains one ten-millionth (0.0000001) of a mole of H^+ ions per liter. The number 0.0000001 is written 10^{-7} in exponential form. To convert this value to pH, the negative exponent (-7) is converted into the positive number 7. A solution with a concentration of 0.0001 (10^{-4}) of H^+ ions per liter has a pH of 4, a solution with a concentration of 0.000000001 (10^{-9}) has a pH of 9, and so on.

A solution that is zero on the pH scale has many H^+ ions and few OH^- ions. A solution that rates 14, by contrast, has many OH^- ions and few H^+ ions. The midpoint in the scale is 7, where the concentration of H^+ and OH^- ions is equal (see Exhibit 2-3). A substance with a pH of 7, such as pure water, is neutral. A solution that has more H^+ ions than OH^- ions is an *acid solution* and has a pH below 7. A solution that has more OH^-

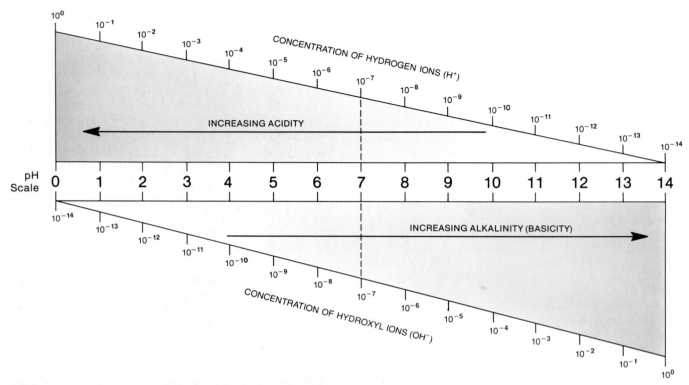

FIGURE 2-9 pH scale. At pH 7 (neutrality), the concentration of H^+ and OH^- ions is equal. A pH value below 7 indicates an acid solution; that is, there are more H^+ ions than OH^- ions. The lower the numerical value of the pH, the more acid the solution is because the H^+ ion concentration becomes progressively greater. A pH value above 7 indicates an alkaline (basic) solution, that is, there are more OH^- ions than H^+ ions. The higher the numerical value of the pH, the more alkaline the solution is because the OH^- ion concentration becomes progressively greater. A change in one whole number on the pH scale represents a tenfold change from the previous concentration.

ions than H^+ ions is a *basic* or *alkaline solution* and has a pH above 7. A change of one whole number on the pH scale represents a tenfold change from the previous concentration. That is, a pH of 2 indicates 10 times fewer H^+ ions than a pH of 1. A pH of 3 indicates 10 times fewer H^+ ions than a pH of 2 and 100 times fewer H^+ ions than a pH of 1.

Maintaining pH: Buffer Systems

Although the pH of body fluids may differ, the normal limits for the various fluids are generally quite specific and narrow. Exhibit 2-4 shows the pH values for certain body fluids compared with common substances. Even though strong acids and bases are continually taken into

EXHIBIT 2-3

RELATIONSHIP OF pH SCALE TO RELATIVE CONCENTRATIONS OF HYDROGEN (H^+) AND HYDROXYL (OH^-) IONS

CONCENTRATION OF H^+ IONS		pH	CONCENTRATION OF OH^- IONS	
1.0	10^0	0	10^{-14}	0.00000000000001
0.1	10^{-1}	1	10^{-13}	0.0000000000001
0.01	10^{-2}	2	10^{-12}	0.000000000001
0.001	10^{-3}	3	10^{-11}	0.00000000001
0.0001	10^{-4}	4	10^{-10}	0.0000000001
0.00001	10^{-5}	5	10^{-9}	0.000000001
0.000001	10^{-6}	6	10^{-8}	0.00000001
0.0000001	10^{-7}	Neutrality 7 Neutrality	10^{-7}	0.0000001
0.00000001	10^{-8}	8	10^{-6}	0.000001
0.000000001	10^{-9}	9	10^{-5}	0.00001
0.0000000001	10^{-10}	10	10^{-4}	0.0001
0.00000000001	10^{-11}	11	10^{-3}	0.001
0.000000000001	10^{-12}	12	10^{-2}	0.01
0.0000000000001	10^{-13}	13	10^{-1}	0.1
0.00000000000001	10^{-14}	14	10^0	1.0

EXHIBIT 2-4
NORMAL pH VALUES OF REPRESENTATIVE SUBSTANCES

SUBSTANCE	pH VALUE
Gastric juice (digestive juice of the stomach)	1.2–3.0
Lemon juice	2.2–2.4
Grapefruit juice	3.0
Cider	2.8–3.3
Pineapple juice	3.5
Tomato juice	4.2
Clam chowder	5.7
Urine	5.0–7.8
Saliva	6.35–6.85
Milk	6.6–6.9
Pure (distilled) water	7.0
Blood	7.35–7.45
Semen (fluid containing sperm)	7.35–7.50
Cerebrospinal fluid (fluid associated with nervous system)	7.4
Pancreatic juice (digestive juice of the pancreas)	7.1–8.2
Eggs	7.6–8.0
Bile (liver secretion that aids in fat digestion)	7.6–8.6
Milk of magnesia	10.0–11.0
Limewater	12.3

the body, the pH levels of these body fluids remain relatively constant. The mechanisms that maintain these homeostatic pH values in the body are called **buffer systems.**

The essential function of a buffer system is to react with strong acids or bases in the body and replace them with weak acids or bases that can change the normal pH values only slightly. Strong acids (or bases) ionize easily and contribute many H^+ (or OH^-) ions to a solution. They therefore change the pH drastically. Weak acids (or bases) do not ionize so easily. They contribute fewer H^+ (or OH^-) ions and have little effect on the pH. The chemicals that change strong acids or bases into weak ones are called **buffers** and are found in the body's fluids. Of the several buffer systems, we will examine only the *carbonic acid–bicarbonate buffer system*—the most important one found in extracellular fluid.

The carbonic acid–bicarbonate buffer system consists of a *pair* of compounds. One of them is a *weak acid,* and the other is a *weak base.* The weak acid of the buffer pair is *carbonic acid* (H_2CO_3); the weak base is *sodium bicarbonate* ($NaHCO_3$). The carbonic acid is a proton (H^+) donor and the bicarbonate ion of the sodium bicarbonate is a proton acceptor. In solution, the members of this buffer pair dissociate as follows:

Acidic component: $H_2CO_3 \rightleftharpoons H^+ + HCO_3^-$
Carbonic acid   Hydrogen ion   Bicarbonate ion

Basic component: $NaHCO_3 \rightleftharpoons Na^+ + HCO_3^-$
Sodium bicarbonate   Sodium ion   Bicarbonate ion

Each member of the buffer pair has a specific role in helping the body maintain a constant pH. If the body's pH is threatened by the presence of a strong acid, the weak base of the buffer pair goes into operation. If the body's pH is threatened by a strong base, the weak acid goes into play.

Consider the following situation. If a strong acid, such as HCl, is added to extracellular fluid, the weak base of the buffer system goes to work and the following acid-buffering reaction occurs:

$HCl + NaHCO_3 \rightleftharpoons NaCl + H_2CO_3$
Hydrochloric acid (strong acid)   Sodium bicarbonate (weak base of buffer system)   Sodium chloride (salt)   Carbonic acid (weak acid)

The chloride ion of HCl and the sodium ion of sodium bicarbonate combine to form NaCl, a substance that has no effect on pH. The hydrogen ion of the HCl could greatly lower pH by making the solution more acid, but this H^+ ion combines with the bicarbonate ion (HCO_3^-) of sodium bicarbonate to form carbonic acid, a weak acid that lowers pH only slightly. In other words, because of the action of the weak base of the buffer system, the strong acid (HCl) has been replaced by a weak acid and a salt, and the pH remains relatively constant.

Now suppose a strong base, such as sodium hydroxide (NaOH), is added to the extracellular fluid. In this instance, the weak acid of the buffer system goes to work and the following base-buffering reaction takes place:

$NaOH + H_2CO_3 \rightleftharpoons H_2O + NaHCO_3$
Sodium hydroxide (strong base)   Carbonic acid (weak acid of buffer system)   Water   Sodium bicarbonate (weak base)

In this reaction, the OH^- ion of sodium hydroxide could greatly raise the pH of the solution by making it more alkaline. However, the OH^- ion combines with an H^+ ion of carbonic acid and forms water, a substance that has no effect on pH. In addition, the Na^+ ion of sodium hydroxide combines with the bicarbonate ion (HCO_3^-) to form sodium bicarbonate, a weak base that has little effect on pH. Thus, because of the action of the buffer system, the strong base is replaced by water and a weak base, and the pH remains relatively constant.

Whenever a buffering reaction occurs, the concentration of one member of the buffer pair is increased while the concentration of the other decreases. When a strong acid is buffered, for example, the concentration of carbonic acid is increased, but the concentration of sodium bicarbonate is decreased. This happens because carbonic acid is produced and sodium bicarbonate is used up in the acid-buffering reaction. When a strong base is buffered, the concentration of sodium bicarbonate is increased, but the concentration of carbonic acid is decreased because sodium bicarbonate is produced and carbonic acid is used up in the base-buffering reaction. When

FIGURE 2-10 Dehydration synthesis and hydrolysis of a molecule of sucrose. In the dehydration synthesis reaction (read from left to right), the two smaller molecules, glucose and fructose, are joined to form a larger molecule of sucrose. Note the loss of a water molecule. In hydrolysis (read from right to left), the larger sucrose molecule is broken down into the two smaller molecules, glucose and fructose. Here, a molecule of water is added to sucrose for the reaction to occur.

the buffered substances, HCl and NaOH in this case, are removed from the body via the lungs or kidneys, the carbonic acid and sodium bicarbonate formed as products of the reactions function again as components of the buffer pair. Do you now understand why buffers are sometimes called "chemical sponges"?

ORGANIC COMPOUNDS

Organic compounds contain carbon and usually hydrogen and oxygen as well. Carbon has several properties that make it particularly useful to living organisms. For one thing, it can react with one to several hundred other carbon atoms to form large molecules of many different shapes. This means that the body can build many compounds out of carbon, hydrogen, and oxygen. Each compound can be especially suited for a particular structure or function. The relatively large size of most carbon-containing molecules and the fact that they do not dissolve easily in water make them useful materials for building body structures. Carbon compounds are mostly or entirely held together by covalent bonds and tend to decompose easily. This means that organic compounds are also a good source of energy. Ionic compounds are not good energy sources because they form new ionic bonds as soon as the old ones are broken.

Carbohydrates

A large and diverse group of organic compounds found in the body is the **carbohydrates,** also known as sugars and starches. The carbohydrates perform a number of major functions in living systems. A few even form structural units. For instance, one type of sugar (deoxyribose) is a building block of genes, the molecules that carry hereditary information. Some carbohydrates are converted to proteins and to fats or fatlike substances, which are used to build structures and provide an emergency source of energy. Other carbohydrates function as food reserves. One example is glycogen, which is stored in the liver and skeletal muscles. The principal function of carbohydrates, however, is to provide the most readily available source of energy to sustain life.

Carbon, hydrogen, and oxygen are the elements found in carbohydrates. The ratio of hydrogen to oxygen atoms is always 2 to 1, the same as in water. This ratio can be seen in the formulas for carbohydrates such as ribose

$(C_5H_{10}O_5)$, glucose $(C_6H_{12}O_6)$, and sucrose $(C_{12}H_{22}O_{11})$. Although there are exceptions, the general formula for carbohydrates is $(CH_2O)_n$, where n symbolizes three or more CH_2O units. Carbohydrates can be divided into three major groups: monosaccharides, disaccharides, and polysaccharides.

1. Monosaccharides. Monosaccharides (mon-ō-SAK-a-rīds), or simple sugars, are compounds containing from three to seven carbon atoms. Simple sugars with three carbons in the molecule are called trioses. The number of carbon atoms in the molecule is indicated by the prefix *tri*. There are also tetroses (four-carbon sugars), pentoses (five-carbon sugars), hexoses (six-carbon sugars), and heptoses (seven-carbon sugars). Pentoses and hexoses are exceedingly important to the human organism. The pentose called deoxyribose is a component of genes. The hexose called glucose is the main energy-supplying molecule of the body.

2. Disaccharides. A second group of carbohydrates, the **disaccharides** (dī-SAK-a-rīds), consists of two monosaccharides joined chemically. In the process of disaccharide formation, two monosaccharides combine to form a disaccharide molecule and a molecule of water is lost. This reaction is known as **dehydration synthesis** (*dehydration* = loss of water). The following reaction shows disaccharide formation. Molecules of the monosaccharides glucose and fructose combine to form a molecule of the disaccharide sucrose (table sugar):

$$C_6H_{12}O_6 \ + \ C_6H_{12}O_6 \ \rightarrow \ C_{12}H_{22}O_{11} + H_2O$$

Glucose	Fructose	Sucrose	Water
(monosaccharide)	(monosaccharide)	(disaccharide)	

You may be puzzled to see that glucose and fructose have the same chemical formulas. Actually, they are different monosaccharides, since the relative positions of the oxygens and carbons vary in the two different molecules (see Figure 2-10). The formula for sucrose is $C_{12}H_{22}O_{11}$ and not $C_{12}H_{24}O_{12}$, since a molecule of H_2O is lost in the process of disaccharide formation. In every dehydration synthesis a molecule of water is lost. Along with this water loss, there is the synthesis of two small molecules, such as glucose and fructose, into one large, more complex molecule, such as sucrose (Figure 2-10). Similarly, the dehydration synthesis of the two monosaccharides glucose and galactose forms the disaccharide lactose (milk sugar).

Disaccharides can also be broken down into smaller, simpler molecules by adding water. This reverse chemical reaction is called **digestion** or **hydrolysis,** which means to split by using water. A molecule of sucrose, for example, may be digested into its components of glucose and fructose by the addition of water. The mechanism of this reaction also is represented in Figure 2-10.

EXHIBIT 2-5

RELATIONSHIPS OF REPRESENTATIVE LIPIDS TO HUMAN ORGANISM

LIPIDS	RELATIONSHIP
FATS	Protection, insulation, source of energy.
PHOSPHOLIPIDS	
Lecithin	Major lipid component of cell membranes; constituent of plasma.
Cephalin and sphingomyelin	Found in high concentrations in nerves and brain tissue.
STEROIDS	
Cholesterol	Constituent of all animal cells, blood, and nervous tissue; suspected relationship to heart disease and atherosclerosis; precursor of bile salts, vitamin D, and steroid hormones.
Bile salts	Substances that emulsify or suspend fats before their digestion and absorption; needed for absorption of fat-soluble vitamins (A, D, E, K).
Vitamin D	Produced in skin on exposure to ultraviolet radiation; necessary for bone growth, development, and repair.
Estrogens	Sex hormones produced in large quantities by females.
Androgens	Sex hormones produced in large quantities by males.
PORPHYRINS (lipid portions of organic molecules)	
Hemoglobin	Oxygen-transporting pigment in red blood cells.
Bile pigments	Bilirubin, a reddish pigment, and biliverdin, a greenish pigment, both formed from the heme group of hemoglobin; bilirubin responsible for the brown color of feces and both pigments contribute to the color changes associated with a bruise.
Cytochromes	Coenzymes involved in the respiration of all cells.
OTHER LIPOID SUBSTANCES	
Carotenes	Pigment in egg yolk, carrots, and tomatoes; vitamin A is formed from carotenes; retinene, formed from vitamin A, is a photoreceptor in the retina of the eye.
Vitamin E	May promote wound healing, prevent scarring, and contribute to the normal structure and functioning of the nervous system; deficiency causes sterility in rats and muscular dystrophy in monkeys; deficiency in humans believed to cause the oxidation of certain fats (unsaturated) resulting in abnormal structure and function of certain parts of cells (mitochondria, lysosomes, and plasma membranes); may reduce the severity of visual loss associated with retrolental fibroplasia (eye disease in premature infants caused by too much oxygen in incubators) by functioning as an antioxidant.
Vitamin K	Vitamin that promotes blood clotting and prevents excessive bleeding.
Prostaglandins	Membrane-associated lipids that stimulate uterine contractions, induce labor and abortions, regulate blood pressure, transmit nerve impulses, regulate metabolism, regulate stomach secretions, inhibit lipid breakdown, and regulate muscular contractions of the gastrointestinal tract.

3. Polysaccharides. The third major group of carbohydrates, the **polysaccharides** (pol'-ē-SAK-a-rīds), consists of three or more monosaccharides joined together through dehydration synthesis. Polysaccharides have the formula $(C_6H_{10}O_5)_n$. Like disaccharides, polysaccharides can be broken into their constituent sugars through hydrolysis reactions. Unlike monosaccharides or disaccharides, however, they usually lack the characteristic sweetness of sugars like fructose or sucrose and are usually not soluble in water. One of the chief polysaccharides is glycogen.

Lipids

A second group of organic compounds that is vital to the human organism is the **lipids.** Like carbohydrates, lipids are composed of carbon, hydrogen, and oxygen, but they do not have a 2:1 ratio of hydrogen to oxygen.

Most lipids are insoluble in water, but they readily dissolve in solvents such as alcohol, chloroform, and ether. Since lipids are a large and diverse group of compounds, we will only discuss two types in detail at this point: fats and prostaglandins. Pertinent information regarding other lipids is provided in Exhibit 2-5.

A molecule of **fat** (triglyceride) consists of two basic components: **glycerol** and **fatty acids** (Figure 2-11). A single molecule of fat is formed when a molecule of glycerol combines with three molecules of fatty acids. This reaction, like the one described for disaccharide formation, is a dehydration synthesis reaction. During hydrolysis, a single molecule of fat is broken down into fatty acids and glycerol.

In subsequent chapters, we will be talking about saturated, unsaturated, and polyunsaturated fats. These terms

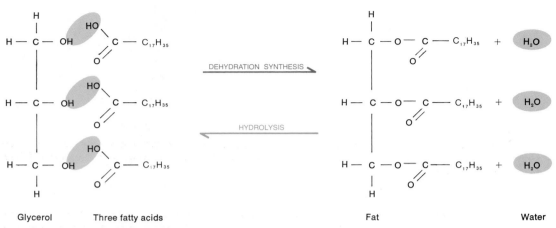

Glycerol   Three fatty acids   Fat   Water

FIGURE 2-11 Dehydration synthesis and hydrolysis of a fat. In the dehydration synthesis reaction (read from left to right), one molecule of glycerol combines with three fatty acid molecules, and there is a loss of three molecules of water. In hydrolysis (read from right to left), a molecule of fat is broken down into a single molecule of glycerol and three fatty acid molecules upon the addition of three molecules of water. The fatty acid shown is stearic acid, a component of corn oil, coconut oil, beef fat, and pork fat.

have the following meanings. A *saturated fat* contains only single covalent bonds between its carbon atoms and all the carbon atoms are bonded to the maximum number of hydrogen atoms; it is saturated with hydrogen atoms. Saturated fats occur mostly in animal foods already high in cholesterol such as beef, pork, butter, whole milk, eggs, and cheese. They also occur in some plant products such as cocoa butter, palm oil, and coconut oil. An *unsaturated fat* contains at least one double covalent bond between its carbon atoms; it is not completely saturated with hydrogen atoms. Examples are olive oil and peanut oil and

they have no significant effect on cholesterol level. A *polyunsaturated fat* is one in which each carbon atom can still bond to two or more hydrogen atoms. Corn oil, safflower oil, sesame oil, and soybean oil are examples of polyunsaturated fats. Many researchers believe that they help to reduce cholesterol in the blood.

Fats represent the body's most highly concentrated source of energy. They provide more than twice as many calories (a measure of energy) per weight as either carbohydrates or proteins. In general, however, fats are about 10 to 12 percent less efficient as body fuels than are carbohydrates. A great amount of the fat calorie is wasted and thus not available for the body to use.

Prostaglandins (pros'-ta-GLAN-dins), also called **PG**, are a large group of membrane-associated lipids composed of 20-carbon fatty acids containing 5 carbon atoms joined to form a ring (cyclopentane ring). Prostaglandins were first discovered in prostate gland secretions, but they are now known to be produced in all nucleated cells in the body and to be able to influence the functioning of any type of cell.

Prostaglandins are produced in cell membranes and are rapidly decomposed by catabolic enzymes. Although synthesized in minute quantities, they are potent substances and exhibit a wide variety of effects on the body. Basically, prostaglandins mimic hormones. They are involved in the modulation of many hormonal responses (Chapter 18), inducing menstruation, and inducing second-trimester abortions. They are also involved in the inflammatory response (Chapter 4), preventing peptic ulcers, opening bronchial and nasal passages, and platelet aggregation and inhibition of aggregation.

Proteins

A third principal group of organic compounds is **proteins.** These compounds are much more complex in structure than the carbohydrates or lipids. They are also responsible for much of the structure of body cells and are related to many physiological activities. For example, proteins in the form of enzymes speed up many essential biochemi-

EXHIBIT 2-6
CLASSIFICATION OF PROTEINS BY FUNCTION

TYPE OF PROTEIN	DESCRIPTION
Structural	Proteins that form the structural framework of various parts of the body. Examples: keratin in the skin, hair, and fingernails and collagen in connective tissue.
Regulatory	Proteins that function as hormones and regulate various physiological processes. Examples: insulin, which regulates blood sugar, and adrenaline, which regulates the diameter of blood vessels.
Contractile	Proteins that serve as contractile elements in muscle tissue. Examples: myosin and actin.
Immunological	Proteins that serve as antibodies to protect the body against invading microbes. Example: gamma globulin.
Transport	Proteins that transport vital substances throughout the body. Example: hemoglobin, which transports oxygen and carbon dioxide in the blood.
Catalytic	Proteins that act as enzymes and function in controlling biochemical reactions. Examples: salivary amylase, pepsin, and lactase.

FIGURE 2-12 Protein formation. When two or more amino acids are chemically united, the resulting bond between them is called a peptide bond. In the example shown, glycine and alanine, the two amino acids, are joined to form the dipeptide glycylalanine. At the point where water is lost, the peptide bond is formed.

cal reactions. Other proteins assume a necessary role in muscular contraction. Antibodies are proteins that provide the human organism with defenses against invading microbes. And some hormones that regulate body functions are also proteins. A classification of proteins on the basis of function is shown in Exhibit 2-6.

Chemically, proteins always contain carbon, hydrogen, oxygen, and nitrogen. Many proteins also contain sulfur and phosphorus. Just as monosaccharides are the building units of sugars, and fatty acids and glycerol are the building units of fats, **amino acids** are the building blocks of proteins. In protein formation, amino acids combine to form more complex molecules, while water molecules are lost. The process is a dehydration synthesis reaction, and the bonds formed between amino acids are called *peptide bonds* (Figure 2-12).

When two amino acids combine, a **dipeptide** results. Adding another amino acid to a dipeptide produces a **tripeptide.** Further additions of amino acids result in the formation of **polypeptides,** which are large protein molecules. At least 20 different amino acids are found in proteins. A great variety of proteins is possible because each variation in the number or sequence of amino acids can produce a different protein. The situation is similar to using an alphabet of 20 letters to form words. Each letter could be compared to a different amino acid, and each word would be a different protein.

Proteins have four levels of structural organization. The *primary structure* is the sequence of amino acids making up the protein. An alteration in primary structure can have serious consequences. For example, a single substitution of an amino acid in a blood protein can result in a deformed hemoglobin molecule that produces sickle-cell anemia. The *secondary structure* of a protein is its coiling or zigzag arrangement along one dimension, such as clockwise spirals and pleated sheets. The *tertiary structure* refers to the bending or folding of a protein into a three-dimensional shape. The *quaternary structure* of a protein refers to two or more tertiary patterns bonded to each other (see Figure 19-3).

Enzymes and Homeostasis

Enzymes (EN-zīms) are specialized proteins that are produced by living cells to catalyze reactions in the body. A *catalyst* is a chemical substance that alters the speed of a chemical reaction without becoming part of the prod-

ucts of the reaction or being used up. Some enzymes consist entirely of proteins. Many also contain a nonprotein portion called a **cofactor,** without which the enzyme will not function. Some cofactors are ions; others are complex organic molecules called **coenzymes.** Vitamins function as coenzymes.

For a chemical or biochemical reaction to occur, a certain amount of energy is required: the **activation energy.** As stated earlier, energy can be transformed from one state to another. When heat energy is added to molecules, some of the heat is transformed to kinetic energy and the energy of the molecules is increased. One way to speed up molecules to activation-level energy is to heat them. Unfortunately, moderate heat denatures (coagulates and destroys the activity of) many body proteins. The role of an enzyme, then, is to decrease the amount of energy needed to start the reaction.

Exactly how enzymes lower activation energies is not fully understood. However, it is known that an enzyme attaches itself to one of the reacting molecules, called a **substrate,** and the two form a temporary **enzyme-substrate complex.** The surface of the substrate makes contact with a specific region on the surface of the enzyme called the **active site.** Thousands of enzymes exist, but each kind can attach to only one kind of substrate. Apparently, the enzyme molecule must fit perfectly with the substrate molecule like pieces of a jigsaw puzzle (Figure 2-13). If the enzyme and substrate do not fit properly, no reaction occurs.

After the reaction is completed, the products of the reaction move away from the enzyme and the enzyme is free to attach to another substrate molecule. The whole process of attachment, reaction, and detachment takes place very quickly. Most enzymes are capable of interacting with up to 5 million substrate molecules per minute at 0°C and, within limits, about double that for every 10°C increase in temperature above 0°C.

Many enzymes are named by adding the suffix *ase* to the name of their substrates. For example, the enzyme involved in breaking down sucrose is called *sucrase.* Enzymes that hydrolyze proteins are classified as *proteases.* Other enzymes are named by their action. *Dehydrogenases,* for example, are enzymes that remove hydrogen atoms from a substrate.

Enzymes are clearly essential to the body's overall homeostasis. They quickly catalyze chemical reactions, and they also govern the reactions that occur.

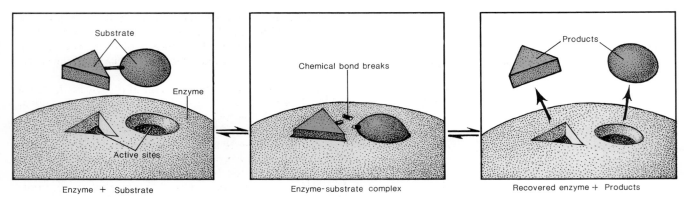

FIGURE 2-13 Enzyme action in a decomposition reaction. The enzyme and substrate molecules combine to form an enzyme-substrate complex. During combination, the substrate is changed into products. Once the products are formed, the enzyme is recovered and may be used again to catalyze a similar reaction.

Key:

G = Guanine

C = Cytosine

A = Adenine

T = Thymine

S = Deoxyribose sugar

P = Phosphate group

Strand 1   Strand 2

(b)

Nucleic Acids: Deoxyribonucleic Acid (DNA) and Ribonucleic Acid (RNA)

Nucleic (nu-KLĒ-ic) **acids,** compounds first discovered in the nuclei of cells, are exceedingly large organic molecules containing carbon, hydrogen, oxygen, nitrogen,

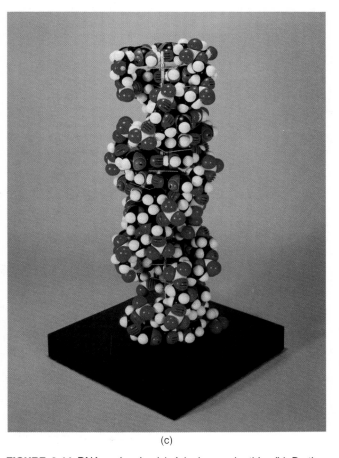

(c)

FIGURE 2-14 DNA molecule. (a) Adenine nucleotide. (b) Portion of an assembled DNA molecule. (c) Space-filling (three-dimensional) model to show relative size and location of atoms. (Courtesy of Ealing Corp.)

- Define and list a cell's generalized parts.
- Explain the chemistry, structure, and functions of the plasma membrane.
- Describe how materials move across plasma membranes by diffusion, facilitated diffusion, osmosis, filtration, dialysis, active transport, phagocytosis, and pinocytosis.
- Define the structure and function of several modified plasma membranes.
- Identify the chemical composition and functions of cytoplasm.
- Describe the structure and two general functions of a cell nucleus.
- Define the structure and function of ribosomes.
- Distinguish between agranular and granular endoplasmic reticulum with regard to structure and function.
- Describe the structure and functions of the Golgi complex.
- Discuss the structure and function of mitochondria as "powerhouses of the cell."
- Explain why a lysosome in a cell is called a "suicide packet."

- Discuss the role of peroxisomes.
- Distinguish between the structure and function of microfilaments and microtubules as components of the cytoskeleton.
- Discuss the structure and function of centrioles in cellular reproduction.
- Differentiate between cilia and flagella in terms of function.
- Define a cell inclusion and give several examples.
- Define extracellular material and give several examples.
- Discuss the stages and events involved in cell division.
- Explain the significance of cell division.
- Define a gene and explain the sequence of events involved in protein synthesis.
- Describe the principle and importance of genetic engineering.
- Explain the relationship of aging to cells and the effects of aging on various body systems.
- Describe cancer as a homeostatic imbalance of cells.
- Define medical terminology associated with cells.

3 The Cellular Level of Organization

The study of the body at the cellular level of organization is important because many activities essential to life occur in cells and many disease processes originate there. A **cell** may be defined as the basic, living, structural and functional unit of the body and, in fact, of all organisms. **Cytology** is the branch of science concerned with the study of cells. This chapter concentrates on the structure, functions, and reproduction of cells.

A series of illustrations accompanies each cell structure that you study. A diagram of a generalized animal cell shows the location of the structure within the cell. An electron micrograph shows the actual appearance of the structure.* A diagram of the electron micrograph clarifies some of the small details by exaggerating their outlines. Finally, an enlarged diagram of the structure shows its details.

GENERALIZED ANIMAL CELL

A **generalized animal cell** is a composite of many different cells in the body. Examine the generalized cell illustrated in Figure 3-1, but keep in mind that no such single cell actually exists.

For convenience, we can divide the generalized cell into four principal parts:

1. Plasma (cell) membrane. The outer, limiting membrane separating the cell's internal parts from the extracellular fluid and external environment.
2. Cytoplasm. The substance between the nucleus and the plasma membrane.
3. Organelles. The cellular components that are highly specialized for specific cellular activities.
4. Inclusions. The secretions and storage areas of cells.

Extracellular materials, which are substances external to the cell surface, will also be examined in connection with cells.

PLASMA (CELL) MEMBRANE

The exceedingly thin structure that separates one cell from other cells and from the external environment is called the **plasma membrane** or **cell membrane.** The membrane ranges from 65 to 100 angstroms (Å) in thickness.

CHEMISTRY AND STRUCTURE

Proteins are the major component of nearly all plasma membranes, comprising from 50 to 70 percent by weight. Next in abundance are phospholipids. Other constituents in lesser amounts include cholesterol, water, carbohydrates, and ions. Recent studies of membrane structure suggest a new concept regarding the arrangement of the various molecules. This new concept is referred to as the **fluid mosaic model** (Figure 3-2b).

The phospholipid molecules are arranged in two parallel rows, forming a **phospholipid bilayer.** A phospholipid

* An **electron micrograph (EM)** is a photograph taken with an electron microscope. The electron microscope can magnify an object more than 200,000 times. In comparison, the light microscope that you probably use in your laboratory magnifies objects up to 1,000 times their size.

molecule consists of a polar, phosphate-containing "head" which mixes with water and nonpolar fatty acid "tails" that do not mix with water. The molecules are oriented in the bilayer so that "heads" face outward on either side and the "tails" face each other in the membrane's interior.

The membrane proteins are classified into two categories: integral and peripheral. **Integral proteins** are embedded in the phospholipid bilayer among the fatty acid "tails." Some of the integral proteins lie at or near the inner and outer membrane surfaces; others penetrate the membrane completely. Since the phospholipid bilayer is somewhat fluid and flexible and the integral proteins have been observed moving from one location to another in the membrane, the relationship has been compared to icebergs (proteins) floating in the sea (phospholipid bilayer). The subunits of some integral proteins form minute channels through which substances could be transported into and out of the cell (described shortly). Other integral proteins carry branching chains of carbohydrates. Such combinations of carbohydrate and protein, called *glycoproteins,* provide receptor sites that enable a cell to recognize other cells of its own kind so that they can associate to form a tissue, to recognize and respond to foreign cells that might be potentially dangerous, and to recognize and attach to hormones, nutrients, and other chemicals. Red blood cells also have glycoprotein receptors that prevent them from clumping and producing unwanted blood clots. Diabetes mellitus is a disease believed to be caused by faulty cell receptor sites.

Peripheral proteins are loosely bound to the membrane surface and easily separated from it. Far less is known about them than integral proteins and their functions are not yet completely understood. For example, some peripheral proteins are believed to serve as enzymes that catalyze cellular reactions. An example is cytochrome c, involved in cellular respiration. Other peripheral proteins, such as spectrin of red blood cells, are believed to have a mechanical function by serving as a scaffolding to support the plasma membrane. It is also believed that peripheral proteins may assume a role in changes in membrane shape during such processes as cell division, locomotion, and ingestion.

FUNCTIONS

Based upon the discussion of the chemistry and structure of the plasma membrane, we can now describe its several important functions. First, the plasma membrane provides a flexible boundary that encloses the cellular contents and separates them from the external environment. Second, the membrane facilitates contact with other body cells or with foreign cells or substances. Third, the membrane provides receptors for chemicals such as hormones, enzymes, nutrients, and antibodies. Fourth, the plasma membrane mediates the entrance and exit of materials. The ability of a plasma membrane to permit certain substances to enter and exit, but to restrict the passage of others is called **selective permeability.** Let us look at this mechanism in a bit more detail.

A membrane is said to be *permeable* to a substance if it allows the free passage of that substance. Although

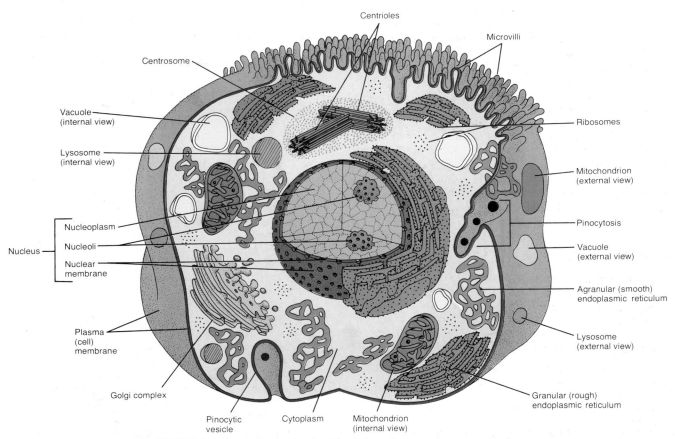

FIGURE 3-1 Generalized animal cell based on electron microscope studies.

plasma membranes are not actually freely permeable to any substance, they do permit some substances to pass more readily than others. For example, water passes more readily than most other substances. The permeability of a plasma membrane appears to be a function of several factors.

1. Size of molecules. Large molecules cannot pass through the plasma membrane. Water and amino acids are small molecules and can enter and exit the cell easily. However, most proteins, which consist of many amino acids linked together, seem to be too large to pass through the membrane. Many scientists believe that the giant-sized molecules do not enter the cell because they are larger than the minute channels in the integral proteins.

2. Solubility in lipids. Substances that dissolve easily in lipids pass through the membrane more readily than other substances, since a major part of the plasma membrane consists of lipid molecules. Examples of lipid-soluble substances are oxygen, carbon dioxide, and steroid hormones.

3. Charge on ions. The charge of an ion attempting to cross the plasma membrane can determine how easily the ion enters or leaves the cell. The protein portion of the membrane is capable of ionization. If an ion has a charge opposite that of the membrane, it is attracted to the membrane and passes through more readily. If the ion attempting to cross the membrane has the same charge as the membrane, it is repelled by the membrane and its passage is restricted. This phenomenon conforms to the rule of physics that opposite charges attract, whereas like charges repel each other.

4. Presence of carrier molecules. Some integral proteins called carriers are capable of attracting and transporting substances across the membrane regardless of size, ability to dissolve in lipids, or membrane charge. Their purpose is to modify membrane permeability. The mechanism by which carriers do this will be described shortly.

MOVEMENT OF MATERIALS ACROSS PLASMA MEMBRANES

The mechanisms whereby substances move across the plasma membrane are essential to the life of the cell. Certain substances, for example, must move into the cell to support life, whereas waste materials or harmful substances must be moved out. Plasma membranes mediate the movements of such materials. The processes involved in these movements may be classed as either passive or active. In **passive processes,** substances move across plasma membranes without assistance from the cell. Their movement involves the kinetic energy of individual molecules. The substances move, on their own, down a concentration gradient from an area where their concentration is greater to an area where their concentration is less. The substances may also be forced across the plasma membrane by pressure from an area where the pressure is greater to an area where it is less. In **active processes,** the cell contributes energy in moving the substance across the membrane since the substance moves against a concentration gradient.

FIGURE 3-2 Plasma membrane. (a) Electron micrograph of portions of two plasma membranes separated by an intercellular space at a magnification of 200,000X. (Copyright © Dr. John Fawcett, Photo Researchers.) (b) Enlargement of the plasma membrane showing the latest concept of the relationship of the phospholipid bilayer and protein molecules. The separation of the bilayer is for illustrative purposes only.

Passive Processes

● *Diffusion* A passive process called **diffusion** occurs when there is a *net* or greater movement of molecules or ions from a region of high concentration to a region of low concentration. The movement from high to low concentration continues until the molecules are evenly distributed. At this point, they move in both directions at an equal rate. This point of even distribution is called *equilibrium.* The difference between high and low concentrations is called the *concentration gradient.* Molecules moving from the high-concentration area to the low-concentration area are said to move *down* or *with* the concentration gradient. If a dye pellet is placed in a beaker filled with water, the color of the dye is seen immediately around the pellet. At increasing distances from the pellet, the color becomes lighter (Figure 3-3). Later, however, the water solution will be a uniform color. The dye mole-

cules possess kinetic energy (energy of motion) and move about at random. The dye molecules move down the concentration gradient from an area of high dye concentration to an area of low dye concentration. The water molecules also move from their high-concentration to their low-concentration area. When dye molecules and water molecules are evenly distributed among themselves, equilibrium is reached and diffusion ceases, even though molecular movements continue. As another example of diffusion, consider what would happen if you opened a bottle of perfume in a room. The perfume molecules would diffuse until an equilibrium is reached between the perfume molecules and the air molecules in the room.

In the examples cited, no membranes were involved. Diffusion may occur, however, through selectively permeable membranes in the body. Large and small lipid-soluble molecules pass through the phospholipid bilayer of the membrane. A good example of this is the movement of

FIGURE 3-3 Principle of diffusion. Molecules of dye (solute) in a beaker of water (solvent) move down the concentration gradient from a region of high concentration to a region of low concentration.

FIGURE 3-4 Principle of osmosis. (a) Apparatus at the start of the experiment. (b) Apparatus at equilibrium. In (a), the cellophane tube contains a 20-percent sugar solution and is immersed in a beaker of distilled water. The 20 percent sugar solution contains 20 parts sugar and 80 parts water, while the distilled water contains 0 parts sugar and 100 parts water. The arrows indicate that water molecules can pass freely into the tube, but that sugar (sucrose) molecules are held back by the selectively permeable membrane. As water moves into the tube by osmosis, the sugar solution is diluted and the volume of the solution in the cellophane tube increases. This increased volume is shown in (b), with the sugar solution moving up the glass tubing. The final height reached (FH) occurs at equilibrium and represents the osmotic pressure. At this point, the number of water molecules leaving the cellophane tube is equal to the number of water molecules entering the tube.

oxygen from the blood into the cells and the movement of carbon dioxide from the cells back into the blood. This is essential in order for body cells to maintain homeostasis. It ensures that cells receive adequate amounts of oxygen and eliminate carbon dioxide as part of their normal metabolism. Small molecules that are not lipid-soluble, such as certain ions (sodium, potassium, chloride), are able to diffuse through channels formed by integral proteins in the membrane.

● *Facilitated Diffusion* Another type of diffusion through a selectively permeable membrane occurs by a process called **facilitated diffusion.** This process is accomplished with the assistance of integral proteins in the membrane that serve as carriers. Although some chemical substances are large molecules and insoluble in lipids, they can still pass through the plasma membrane. Among these are different sugars, especially glucose. In the process of facilitated diffusion, it is believed that glucose is picked up by a carrier. The combined glucose-carrier is soluble in the phospholipid bilayer of the membrane, and the carrier moves the glucose to the inside of the membrane and then inside the cell. The carrier makes the glucose soluble in the phospholipid bilayer of the membrane so it can pass through the membrane. By itself, glucose is insoluble and cannot penetrate the membrane. In the process of facilitated diffusion, the cell does not expend energy, and the movement of the substance is from a region of its higher concentration to a region of its lower concentration.

The rate of facilitated diffusion is considerably faster than that of simple diffusion and depends on (1) the difference in concentration of the substance on either side of the membrane, (2) the amount of carrier available to transport the substance, and (3) how quickly the carrier and substance combine. The process is greatly accelerated by insulin, a hormone produced by the pancreas. One

of insulin's functions is to lower the blood glucose level by accelerating the transportation of glucose from the blood into body cells. This transportation, as we have just seen, is by facilitated diffusion.

● *Osmosis* Another passive process by which materials move across membranes is **osmosis.** It is the net movement of water molecules through a selectively permeable membrane from an area of higher water concentration to an area of lower water concentration. The water molecules pass through channels in integrated proteins in the membrane. Once again, a simple apparatus may be used to demonstrate the process. The apparatus shown in Figure 3-4 consists of a tube constructed from cellophane, a selectively permeable membrane. The cellophane tube is filled with a colored, 20 percent sugar (sucrose) solution. Such a solution consists of 20 parts sugar and 80 parts water. The upper portion of the cellophane tube is plugged with a rubber stopper through which a glass tube is fitted. The cellophane tube is placed into a beaker containing distilled (pure) water. Pure water consists of 0 parts sugar and 100 parts water. Initially, the concentrations of water on either side of the selectively permeable membrane are different. There is a lower concentration of water inside the cellophane tube than outside. Because of this difference, water moves from the beaker into the cellophane tube. The force with which the water moves is called osmotic pressure. Very simply, **osmotic pressure** is the force under which a solvent, usually water, moves from a solution of higher water (lower solute) concentration to a solution of lower water (higher solute) concentration when the solutions are separated by a selectively permeable membrane. There is no movement of sugar from the cellophane tube inside the beaker, since the cellophane is impermeable to molecules of sugar—sugar molecules are too large to go through the pores of the membrane. As water moves into the cellophane tube, the sugar solution becomes increasingly diluted and the increased volume forces the mixture up the glass tubing. In time, the water that has accumulated in the cellophane tube and the glass tube exerts a downward pressure that forces water molecules back out of the cellophane tube and into the beaker. When water molecules leave and enter the cellophane tube at the same rate, equilibrium is reached. Osmotic pressure is an important force in the movement of water between various compartments of the body.

● *Isotonic, Hypotonic, and Hypertonic Solutions* Osmosis may also be understood by considering the effects of different water concentrations on red blood cells. If the normal shape of a red blood cell is to be maintained, the cell must be placed in an **isotonic solution.** This is a solution in which the total concentrations of water molecules and solute molecules are the same on both sides of the semipermeable cell membrane. The concentrations of water and solute in the extracellular fluid outside the red blood cell must be the same as the concentration of the intracellular fluid. Under ordinary circumstances, a 0.85 percent NaCl solution is isotonic for red blood cells. In this condition, water molecules enter and exit the cell at the same rate, allowing the cell to maintain its normal shape.

A different situation results if red blood cells are placed in a solution that has a lower concentration of solutes and, therefore, a higher concentration of water. This is called a **hypotonic solution.** In this condition, water molecules enter the cells faster than they can leave, causing the red blood cells to swell and eventually burst. The rupture of red blood cells in this manner is called **hemolysis** (hē-MOL-i-sis) or **laking.** Distilled water is a strongly hypotonic solution.

A **hypertonic solution** has a higher concentration of solutes and a lower concentration of water than the red blood cells. One example of a hypertonic solution is a 10 percent NaCl solution. In such a solution, water molecules move out of the cells faster than they can enter. This situation causes the cells to shrink. The shrinkage of red blood cells in this manner is called **crenation** (krē-NĀ-shun). Red blood cells may be greatly impaired or destroyed if placed in solutions that deviate significantly from the isotonic state.

● *Filtration* A third passive process involved in moving materials in and out of cells is **filtration.** This process involves the movement of solvents such as water and dissolved substances such as sugar across a selectively permeable membrane by gravity or mechanical pressure, usually hydrostatic (water) pressure. Such a movement is always from an area of higher pressure to an area of lower pressure and continues as long as a pressure difference exists. Most small to medium-sized molecules can be forced through a cell membrane.

An example of filtration occurs in the kidneys, where the blood pressure supplied by the heart forces water and small molecules like urea through thin cell membranes of tiny blood vessels and into the kidney tubules. In this basic process, protein molecules are retained by the body since they are too large to be forced through the cell membranes of the blood vessels. The molecules of harmful substances such as urea are small enough to be forced through and eliminated, however.

● *Dialysis* The final passive process to be considered is **dialysis.** Dialysis is the diffusion of solute particles across a selectively permeable membrane and involves the separation of small molecules from large molecules. Suppose a solution containing molecules of various sizes is placed in a tube that is permeable only to the smaller molecules. The tube is then placed in a beaker of distilled water. Eventually, the smaller molecules will move from the tube into the water in the beaker and the larger molecules will be left behind.

The principle of dialysis is employed in artificial kidney machines. It does not occur within the human body. The patient's blood is passed across a dialysis membrane outside the body. The dialysis membrane takes the place of the kidneys. As the blood moves across the membrane, small-particle waste products pass from the blood into a solution surrounding the dialysis membrane. At the same time, certain nutrients can be passed from the solution into the blood. The blood is then returned to the body.

Active Processes

When cells actively participate in moving substances across membranes, they must expend energy. Cells can even move substances against a concentration gradient. The active processes considered here are active transport, phagocytosis, and pinocytosis.

● *Active Transport* The process by which substances, usually ions, are transported across plasma membranes typically from an area of lower concentration to an area of higher concentration is called **active transport.** In order to transport a substance against a concentration gradient, the membrane uses energy supplied by ATP. In fact, a typical body cell probably expends up to 40 percent of its ATP for active transport. Although the molecular events involved in active transport are not completely understood, integral proteins in the plasma membrane do assume a role.

As just noted, glucose can be transported across cell membranes via facilitated diffusion from areas of higher to lower concentration. Glucose can also be moved by the cells lining the digestive tract from the cavity of the tract into the blood even though blood concentration of glucose is higher. This involves active transport. One proposed mechanism is that glucose (or other substance) enters a channel in an integral membrane protein (Figure

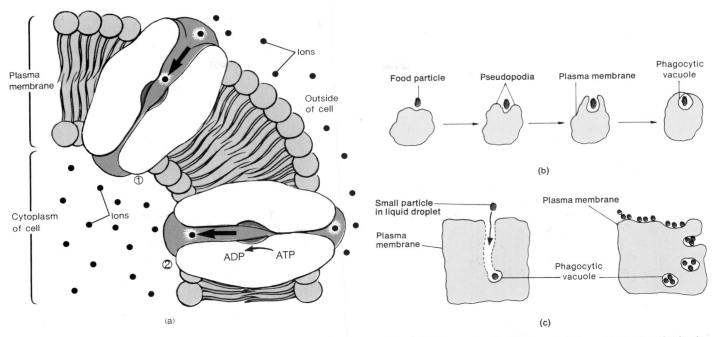

FIGURE 3-5 Active processes. (a) Proposed mechanism of active transport. (b) Phagocytosis. (c) Two variations of pinocytosis. In the variation on the left, the ingested substance enters a channel formed by the plasma membrane and becomes enclosed in a vacuole at the base of the channel. In the variation on the right, the ingested substance becomes enclosed in a vacuole that forms and detaches at the surface of the cell.

3-5a). When the glucose molecule makes contact with an active site in the channel, the energy from ATP induces a change in the membrane protein that expels the glucose on the opposite side of the membrane. As you will see later, kidney cells also have the ability to actively transport glucose back into the blood so that it is not lost in the urine.

Active transport is also an important process in maintaining the concentrations of some ions inside body cells and other ions outside body cells. For example, before a nerve cell can conduct an impulse, the concentration of potassium (K^+) ions must be considerably higher inside the cell than outside, even when the intracellular K^+ concentration is higher than the extracellular concentration. This is accomplished by a "potassium pump" which consists of an integral protein serving as the "pump," powered by ATP (see Figure 12-5a). Nerve impulse conduction also requires that nerve cells have a higher concentration of sodium (Na^+) ions outside than inside, even when the extracellular Na^+ concentration is higher than the intracellular concentration. This is accomplished by a "sodium pump," whose operation is coupled with that of the "potassium pump." The "sodium pump" is also an integral protein powered by ATP (see Figure 12-5a).

● *Phagocytosis*   Another active process by which cells take in substances across the plasma membrane is called **phagocytosis** (fag'-ō-sī-TŌ-sis), or "cell eating" (Figure 3-5b). In this process, projections of cytoplasm, called *pseudopodia* (soo'-dō-PŌ-dē-a), engulf solid particles external to the cell. Once the particle is surrounded, the membrane folds inwardly, forming a membrane sac around the particle. This newly formed sac, called a *phagocytic vacuole,* breaks off from the outer cell mem-

brane, and the solid material inside the vacuole is digested. Indigestible particles and cell products are removed from the cell by a reverse phagocytosis. This process is important because molecules and particles of material that would normally be restricted from crossing the plasma membrane can be brought into or removed from the cell. The phagocytic white blood cells of the body constitute a vital defense mechanism. Through phagocytosis, the white blood cells engulf and destroy bacteria and other foreign substances (see Figure 4-4).

● *Pinocytosis*   In **pinocytosis** (pi'-nō-sī-TŌ-sis), or "cell drinking," the engulfed material consists of a liquid rather than a solid (Figure 3-5c). Moreover, no cytoplasmic projections are formed. Instead, the liquid is attracted to the surface of the membrane. The membrane folds inwardly, surrounds the liquid, and detaches from the rest of the intact membrane. Few cells are capable of phagocytosis, but many cells carry on pinocytosis. Examples include cells in the kidneys and urinary bladder.

The movement of materials into the cell by either phagocytosis or pinocytosis is referred to as **endocytosis.** The export of material from the cell by the reverse processes is called **exocytosis.**

MODIFIED PLASMA MEMBRANES

Electron-microscope studies have revealed that plasma membranes of certain cells are modified in various ways for specific purposes. For example, the membranes of some cells lining the small intestine have small, cylindrical projections called **microvilli** (see Figure 3-1). These fingerlike projections enormously increase the absorbing area of the cell surface. A single cell may have as many as 3,000 microvilli, and a 1 sq mm (0.00155 sq inch) area of small intestine may contain as many as 200 million microvilli, which would increase the surface area for absorption by 20 times.

Another membrane modification is found in the rods and cones of the eye. They serve as photoreceptors, or light-receiving cells. The upper portion of each rod contains two-layered, disc-shaped membranes called **sacs** that contain the pigments involved in vision.

Another example of a membrane modification is the **stereocilia.** They are found only in cells lining a duct (ductus epididymis) of the male reproductive system. They appear by light microscopy as long, slender, branching processes at the free surfaces of the lining cells (see Figure 28-6). Electron micrographs show stereocilia to be microvilli.

A final example of a membrane modification is the **myelin sheath** that surrounds portions of certain nerve cells (see Figure 12-3). It is thought that the myelin sheath increases the velocity of impulse conduction, protects the portion of the nerve cell it surrounds, is related to the nutrition of the nerve cell, and is necessary for its repair and regeneration.

CYTOPLASM

The substance inside the cell's plasma membrane and external to the nucleus is called **cytoplasm** (Figure 3-6a, b). It is the matrix, or ground substance, in which various cellular components are found. Physically, cytoplasm may be described as a thick, semitransparent, elastic fluid containing suspended particles and a series of minute tubules and filaments that form a cytoskeleton. Chemically, cytoplasm is 75 to 90 percent water plus solid components. Proteins, carbohydrates, lipids, and inorganic substances compose the bulk of the solid components. The inorganic substances and most carbohydrates are soluble in water and are present as a true solution. The majority of organic compounds, however, are found as colloids—particles that remain suspended in the surrounding ground substance. Since the particles of a colloid bear electrical charges that repel each other, they remain suspended and separated from each other.

Functionally, cytoplasm is the substance in which chemical reactions occur. The cytoplasm receives raw materials from the external environment and converts them into usable energy by decomposition reactions. Cytoplasm is also the site where new substances are synthesized for

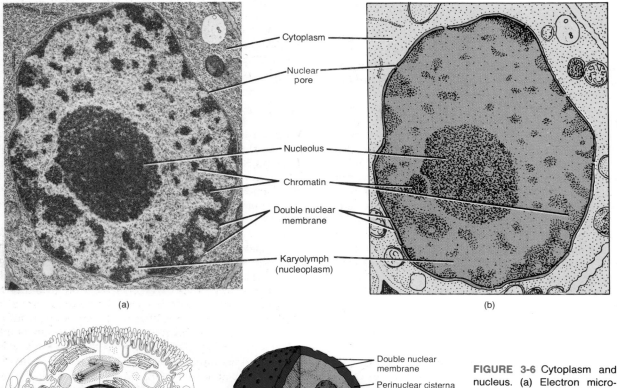

(a)

(b)

Cytoplasm

Nuclear pore

Nucleolus

Chromatin

Double nuclear membrane

Karyolymph (nucleoplasm)

(c)

Double nuclear membrane

Perinuclear cisterna

Karyolymph (nucleoplasm)

Nucleolus

Nuclear pore

Chromatin

FIGURE 3-6 Cytoplasm and nucleus. (a) Electron micrograph of cytoplasm and the nucleus at a magnification of 31,600×. (Courtesy of Biophoto Associates/Dr. Myron C. Ledbetter, Brookhaven National Laboratory.) (b) Diagram of the electron micrograph. (c) Diagram of a nucleus with two nucleoli.

cellular use. It packages chemicals for transport to other parts of the cell or other cells of the body and facilitates the excretion of waste materials.

ORGANELLES

Despite the myriad chemical activities occurring simultaneously in the cell, there is little interference of one reaction with another. The cell has a system of compartmentalization provided by structures called **organelles.** These structures are specialized portions of the cell that assume specific roles in growth, maintenance, repair, and control.

NUCLEUS

The **nucleus** is generally a spherical or oval organelle and is the largest structure in the cell (Figure 3-6a–c). It contains hereditary factors of the cell, called genes, which control cellular structure and direct many cellular activities. Mature red blood cells do not have nuclei. These cells carry on only limited types of chemical activity and are not capable of growth or reproduction.

The nucleus is separated from the cytoplasm by a double membrane called the *nuclear membrane* or *envelope* (Figure 3-6c). Between the two layers of the nuclear membrane is a space called the *perinuclear cisterna*. Each of the nuclear membranes resembles the structure of the plasma membrane. Minute pores in the nuclear membrane allow the nucleus to communicate with a membranous network in the cytoplasm called the endoplasmic reticulum. Substances entering and exiting the nucleus are believed to pass through the tiny pores.

Three prominent structures are visible internal to the nuclear membrane. The first of these is a gel-like fluid that fills the nucleus called *karyolymph* (*nucleoplasm*). One or more spherical bodies called the *nucleoli* are also present. These structures are composed of protein, DNA, and RNA. DNA synthesizes the RNA which is stored in the nucleoli. The RNA, as you will see shortly, assumes a function in protein synthesis. Finally, there is the *genetic material* consisting principally of DNA. When the cell is not reproducing, the genetic material appears as a threadlike mass called *chromatin*. Prior to cellular reproduction the chromatin shortens and coils into rod-shaped bodies called *chromosomes.*

Studies of chromosomes have provided a great deal of information about their chemistry and detailed structure. Although it has been known for some time that chromosomes consist of DNA and proteins called *histones,* their organization and arrangement in chromosomes have only recently been determined. In specially prepared electron micrographs, it can be seen that chromosomes consist of elementary subunits of structure called **nucleosomes.** Each nucleosome has a diameter of about 100 Å. A nucleosome consists of two each of four different histone proteins (collectively called an *octamer*) associated with a relatively fixed length of DNA (about 200 nitrogenous base pairs). Present evidence indicates that the DNA is wrapped around the histones in some way. A fifth type of histone maintains adjacent nucleo-

somes in a helical coil (*solenoid*). Nucleosomes are separated by stretches of DNA only so that the arrangement resembles beads on a string. The functional significance of nucleosome structure is still speculative. It is believed that histones may facilitate changes in chromosomal structure that expose activated genes (DNA) to perform a specific task in the cell.

RIBOSOMES

Ribosomes are tiny granules, 250 Å at their largest dimension, that are composed of a type of RNA called ribosomal RNA (rRNA) and a number of specific ribosomal proteins. The rRNA is manufactured by DNA in the nucleolus. Ribosomes were so named because of their high content of rRNA. Structurally, a ribosome consists of two subunits, one about half the size of the other. Recently scientists have provided three-dimensional models of the structure of a ribosome (Figure 3-7d). Functionally, ribosomes are the sites of protein synthesis: they receive genetic instructions and translate them into protein. The mechanism of ribosomal function is discussed later in the chapter.

Some ribosomes, called *free ribosomes,* are scattered in the cytoplasm; they have no attachments to other parts of the cell. The free ribosomes occur singly or in clusters, and they are primarily concerned with synthesizing proteins for use inside the cell. Other ribosomes are attached to a cellular structure called the endoplasmic reticulum. These ribosomes are concerned with the synthesis of proteins for export from the cell.

ENDOPLASMIC RETICULUM (ER)

Within the cytoplasm, there is a system of pairs of parallel membranes enclosing narrow cavities of varying shapes. This system is known as the **endoplasmic reticulum,** or **ER** (Figure 3-7). The ER, in other words, is a network of channels (*cisternae*) running through the cytoplasm. These channels are continuous with the nuclear membrane.

On the basis of its association with ribosomes, ER is distinguished into two types. **Granular (rough) ER** is studded with ribosomes; **agranular (smooth) ER** is free of ribosomes. Recent studies using radioactive tracers suggest that agranular ER is synthesized from granular ER.

Numerous functions related to homeostasis are attributed to the ER. It contributes to the mechanical support and distribution of the cytoplasm. It also conducts intracellular nerve impulses, such as in muscle cells, where it is called sarcoplasmic reticulum. The ER is involved in the intracellular exchange of materials with the cytoplasm and provides a surface area for chemical reactions. Various products are transported from one portion of the cell to another via the ER, so the ER is considered an intracellular circulatory system. The ER also serves as a storage area for synthesized molecules. And, together with a cellular structure called the Golgi complex, the ER assumes a role in the synthesis and packaging of molecules.

Ribosomes

Cisternae of
granular (rough)
endoplasmic
reticulum

(a)

(b)

FIGURE 3-7 Endoplasmic reticulum and ribosomes. (a) Electron micrograph of the endoplasmic reticulum and ribosomes at a magnification of 76,000×. (Courtesy of Biophoto Associates/Myron C. Ledbetter, Brookhaven National Laboratory.) (b) Diagram of the electron micrograph. (c) Diagram of the endoplasmic reticulum and ribosomes. See if you find the agranular (smooth) endoplasmic reticulum in Figure 3-8. (d) Diagram of the three-dimensional structure of ribosomes.

Ribosomes

Cisternae of
granular (rough)
endoplasmic
reticulum

(c)

Large subunit

Small subunit

Complete functional ribosome

(d)

GOLGI COMPLEX

Another structure found in the cytoplasm is the **Golgi (GOL-jē) complex.** This structure consists of four to eight flattened baglike channels, stacked upon each other with expanded areas at their ends. Like those of the ER, the stacked elements are called *cisternae,* and the expanded, terminal areas are *vesicles* (Figure 3-8). Generally, the Golgi complex is located near the nucleus.

One function of the Golgi complex is the packaging of secreted proteins. *Secretion* is the production and re-

(a)

(b)

Vesicles

Cisternae

Vesicles

Cisternae

(c)

FIGURE 3-8 Golgi complex. (a) Electron micrograph of the two Golgi complexes at a magnification of 78,000×. (Courtesy of Biophoto Associates/ Myron C. Ledbetter, Brookhaven National Laboratory.) (b) Diagram of the electron micrograph. (c) Diagram of a Golgi complex.

lease from a gland cell of a fluid that usually contains a variety of substances. Proteins synthesized at ribosomes associated with granular ER are transported into the ER cisternae (Figure 3-9). The proteins then pass in membrane sacs pinched off from the ER into the Golgi complex. As proteins accumulate and concentrate in the cisternae of the Golgi complex, the cisternae expand to form vesicles. After a certain size is reached, the vesicles pinch off from the cisternae. The protein and its associated vesicle is referred to as a *secretory granule*. The secretory granule then moves toward the surface of the cell where the protein is secreted. The contents of the granule are discharged into the extracellular space, and the granule membrane is incorporated into the plasma membrane. Cells of the digestive tract that secrete protein enzymes utilize this mechanism. The secretory granule prevents "digestion" of the cytoplasm of the cells by the enzymes as it moves toward the cell surface. Other vesicles that pinch off from the Golgi complex are loaded with special digestive enzymes and remain within the cell. They become cellular structures called lysosomes.

Another function of the Golgi complex is associated with lipid secretion. It occurs in essentially the same way as protein secretion, except the lipids are synthesized by the agranular ER. The lipids pass through the ER into the Golgi complex. As in the mechanism just described,

the lipids migrate into the cisternae and vesicles and are discharged at the surface of the cell. In the course of moving through the cytoplasm, the vesicle may release lipids into the cytoplasm before being discharged from the cell. These appear in the cytoplasm as lipid droplets. Among the lipids secreted in this manner are the steroids (see Exhibit 2-5).

The Golgi complex also functions in the synthesis of carbohydrates. Recent evidence indicates that carbohydrates synthesized by the Golgi complex are combined with proteins synthesized by ribosomes associated with granular ER to form glycoproteins. As these carbohydrate-protein complexes are assembled, they accumulate in the flattened channels of the Golgi complex. The channels expand and form vesicles. After a critical size is reached, the vesicles pinch off from the channel, migrate through the cytoplasm, and discharge their contents through the plasma membrane. The Golgi complex is well developed and highly active in secretory cells such as those found in the pancreas and the salivary glands.

MITOCHONDRIA

Small, spherical, rod-shaped, or filamentous structures called **mitochondria** (mī'-tō-KON-drē-a) appear throughout the cytoplasm. When sectioned and viewed under

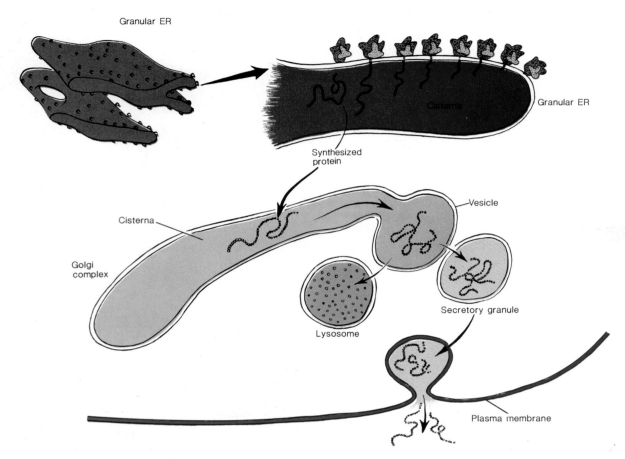

Granular ER

Granular ER

Cisterna

Synthesized protein

Cisterna

Golgi complex

Vesicle

Lysosome

Secretory granule

Plasma membrane

FIGURE 3-9 Packaging of synthesized protein for export from the cell.

an electron microscope, each reveals an elaborate internal organization (Figure 3-10). A mitochondrion consists of two membranes, each of which is similar in structure to the plasma membrane. The outer mitochondrial membrane is smooth, but the inner membrane is arranged in a series of folds called *cristae*. The center of the mitochondrion is called the *matrix*.

Because of the nature and arrangement of the cristae, the inner membrane provides an enormous surface area for chemical reactions. Enzymes involved in energy-releasing reactions that form ATP are located on the cristae. Mitochondria are frequently called the "powerhouses of the cell" because they are the sites for the production of ATP. Active cells, such as muscle and liver cells, have a large number of mitochondria because of their high energy expenditure.

Mitochondria are self-replicative, that is, they can divide to form new ones. The replication process is controlled by DNA which is incorporated into the mitochondrial structure. Self-replication usually occurs in response to increased cellular need for ATP.

LYSOSOMES

When viewed under the electron microscope, **lysosomes** (*lysis* = dissolution; *soma* = body) appear as membrane-enclosed spheres (Figure 3-11). They are formed from Golgi complexes, have a double membrane, and lack detailed structure. They contain powerful digestive enzymes capable of breaking down many kinds of molecules. You will see in Chapter 14 that Tay-Sachs disease results from

a deficiency of a lysosomal enzyme. These enzymes are also capable of digesting bacteria that enter the cell (see Figure 22-8). White blood cells, which ingest bacteria by phagocytosis, contain large numbers of lysosomes.

Scientists have wondered why these powerful enzymes do not also destroy their own cells. Perhaps the lysosome membrane in a healthy cell is impermeable to enzymes so they cannot move out into the cytoplasm. However, when a cell is injured, the lysosomes release their enzymes. The enzymes then promote reactions that break the cell down into its chemical constituents. This process of self-destruction by cells is known as **autolysis.** The chemical remains are either reused by the body or excreted. Because of this function, lysosomes have been called "suicide packets."

Lysosomes are crucial in the removal of cell parts, whole cells, and even extracellular material, which may be the process underlying bone removal. In bone reshaping, especially during the growth process, special bone-destroying cells called osteoclasts, secrete extracellular enzymes that dissolve bone. Bone tissue cultures given excess amounts of vitamin A remove bone apparently through an activation process involving lysosomes.

CLINICAL APPLICATION

Animals overfed on vitamin A interestingly develop spontaneous fractures, suggesting greatly increased lysosomal activity. On the other hand, cortisone and hydrocortisone, steroid hormones produced by the adrenal gland, have a stabilizing effect on lysosomal mem-

Double
mitochondrial
membrane

Matrix

Cristae

(a)

(b)

Inner mitochondrial
membrane

Outer mitochondrial
membrane

Cristae

Matrix

(c)

FIGURE 3-10 Mitochondria. (a) Electron micrograph of an entire mitochondrion and a portion of another. (Courtesy of Lester V. Bergman & Associates.) (b) Diagram of the electron micrograph. (c) Diagram of a mitochondrion.

branes. The steroid hormones are well known for their antiinflammatory properties, which suggests that they reduce destructive cellular activity by lysosomes.

PEROXISOMES

Organelles similar in structure to lysosomes, but smaller, are called **peroxisomes** (pe-ROKS-i-sōms). They are abundant in liver cells and contain several enzymes related to the metabolism of hydrogen peroxide (H_2O_2), a substance that is toxic to body cells. One of the enzymes

in peroxisomes, called *catalase,* immediately breaks down H_2O_2 into water and oxygen:

$$\underset{\substack{\text{Hydrogen} \\ \text{peroxide}}}{H_2O_2} \xrightarrow{\text{Catalyase}} \underset{\text{Water}}{H_2O} + \underset{\text{Oxygen}}{O_2}$$

MICROFILAMENTS AND MICROTUBULES: THE CYTOSKELETON

In recent years, electron microscopy has revealed that the cytoplasm of cells is more than just a structureless

FIGURE 3-11 Lysosome. (a) Electron micrograph of a lysosome at a magnification of 55,000×. (Courtesy of F. Van Hoof, Université Catholique de Louvain.) (b) Diagram of the electron micrograph.

Lysosome

(a)

(b)

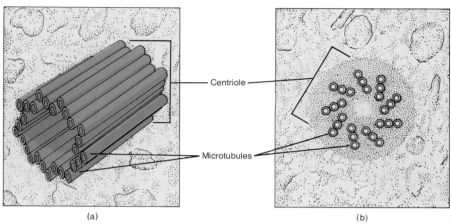

(a)

Centriole

Microtubules

(b)

FIGURE 3-12 Centrosome and centrioles. (a) Diagram of a centriole in longitudinal section. (b) Diagram of a centriole in cross section. (c) Electron micrograph of a centriole in cross section at a magnification of 67,000×. (Courtesy of Biophoto Associates.)

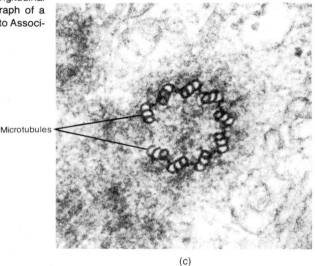

Microtubules

(c)

medium in which the various cellular components are suspended. It has been shown that cytoplasm actually has a complex internal structure, consisting of a series of exceedingly small microfilaments and microtubules, together referred to as the **cytoskeleton.**

Microfilaments are rodlike structures ranging from 30 to 120 Å in diameter. They are of variable length and may occur in bundles, randomly scattered throughout the cytoplasm, or arranged in a meshwork, depending on the type of cell in which they are found. Some microfilaments consist of a protein called *actin;* others consist of a protein called *myosin.* In muscle tissue, actin microfilaments (thin myofilaments) and myosin microfilaments (thick myofilaments) are involved in the contraction of muscle cells. This mechanism is described in Chapter 10. In nonmuscle cells, microfilaments help to provide support and shape and assist in the movement of entire cells (phagocytes and cells of developing embryos) and movements within cells (secretion, phagocytosis, pinocytosis).

Microtubles are relatively straight, slender, cylindrical structures that range in diameter from 180 to 300 Å, usually averaging about 240 Å. They consist of a protein called *tubulin.* Microtubules dispersed in the cytoplasm, together with microfilaments, help to provide support and shape for cells. Some evidence suggests that microtubules may form conducting channels through which various substances can move throughout the cytoplasm. This mechanism has been studied extensively in nerve cells. Microtubules also assist in the movement of pseudopodia that are characteristic of phagocytes. As you will see shortly, microtubules form the structure of flagella and cilia (cellular appendages involved in motility), centrioles (organelles that may direct the assembly of microtubules), and the mitotic spindle (structures that are involved in cell division).

CENTROSOME AND CENTRIOLES

A dense area of cytoplasm, generally spherical and located near the nucleus, is called the **centrosome (centrosphere).** Within the centrosome is a pair of cylindrical structures:

the **centrioles** (Figure 3-12). Each centriole is composed of a ring of nine evenly spaced bundles. Each bundle, in turn, consists of three microtubules. The two centrioles are situated so that the long axis of one is at right angles to the long axis of the other. Centrioles assume a role in cell reproduction by serving as centers about which microtubules involved in chromosome movement are organized. This will be described shortly as part of cell division. Certain cells, such as most mature nerve cells, do not have a centrosome and so do not reproduce. This is why they cannot be replaced if destroyed. Like mitochondria, centrioles contain DNA that controls their self-replication.

FLAGELLA AND CILIA

Some body cells possess projections for moving the entire cell or for moving substances along the surface of the cell. These projections contain cytoplasm and are bounded by the plasma membrane. If the projections are few and long in proportion to the size of the cell, they are called **flagella.** The only example of a flagellum in the human body is the tail of a sperm cell, used for locomotion (see Figure 28-4). If the projections are numerous and short, resembling many hairs, they are called **cilia.**

Microtubules

Flagellar or ciliary membrane

Cell membrane

FIGURE 3-13 Structure of a flagellum or cilium.

In humans, ciliated cells of the respiratory tract move mucus that has trapped foreign particles over the surface of the tissue (see Figure 23-5b). Electron microscopy has revealed no fundamental structural difference between cilia and flagella (Figure 3-13). Both consist of nine pairs of microtubules that form a ring around two microtubules in the center.

CELL INCLUSIONS

Cell inclusions are a large and diverse group of chemical substances, some of which have recognizable shapes. These products are principally organic and may appear or disappear at various times in the life of the cell. *Melanin* is a pigment stored in certain cells of the skin, hair, and eyes. It protects the body by screening out harmful ultraviolet rays from the sun. *Glycogen* is a polysaccharide that is stored in the liver, skeletal muscle cells, and the vaginal mucosa. When the body requires quick energy, liver cells can break down the glycogen into glucose and release it. *Lipids,* which are stored in fat cells, may be decomposed for producing energy. A final example of an inclusion is *mucus,* which is produced by cells that line organs. Its function is to provide lubrication and protection.

The major parts of the cell and their functions are summarized in Exhibit 3-1.

EXTRACELLULAR MATERIALS

The substances that lie outside cells are called **extracellular materials.** They include the body fluids, which provide a medium for dissolving, mixing, and transporting substances. Among the body fluids are interstitial fluid, the fluid that fills the microscopic spaces (interstitial spaces), and plasma, the liquid portion of blood. Extracellular materials also include secreted inclusions like mucus and special substances that form the matrix in which some cells are embedded.

The matrix materials are produced by certain cells and deposited outside their plasma membranes. The matrix supports the cells, binds them together, and gives strength and elasticity to the tissue. Some matrix materials

EXHIBIT 3-1
CELL PARTS AND THEIR FUNCTIONS

PART	FUNCTIONS
Plasma membrane	Protects cellular contents; makes contact with other cells; provides receptors for hormones, enzymes, and antibodies; mediates the entrance and exit of materials.
Cytoplasm	Serves as the ground substance in which chemical reactions occur.
Organelles **Nucleus**	Contains genes and controls cellular activities.
Ribosomes	Sites of protein synthesis.
Endoplasmic reticulum (ER)	Contributes to mechanical support; conducts intracellular nerve impulses in muscle cells; facilitates intracellular exchange of materials with cytoplasm; provides a surface area for chemical reactions; provides a pathway for transporting chemicals; serves as a storage area; together with Golgi complex synthesizes and packages molecules for export.
Golgi complex	Packages synthesized proteins for secretion in conjunction with endoplasmic reticulum; forms lysosomes; secretes lipids; synthesizes carbohydrates; combines carbohydrates with proteins to form glycoproteins for secretion.
Mitochondria	Sites for production of ATP.
Lysosomes	Digest substances and foreign microbes; may be involved in bone removal.
Peroxisomes	Contains several enzymes, such as catalase, related to hydrogen peroxide metabolism.
Microfilaments	Form cytoskeleton with microtubules; involved in muscle cell contraction; provide support and shape; assist in cellular and intracellular movement.
Microtubules	Form cytoskeleton with microfilaments; provide support and shape; form intracellular conducting channels; assist in cellular movement; form the structure of flagella, cilia, centrioles, and spindle fibers.
Centrioles	Help organize mitotic spindle during cell division.
Flagella and cilia	Allow movement of entire cell (flagella) or movement of particles along surface of cell (cilia).
Inclusions	Melanin (pigment in skin, hair, eyes) screens out ultraviolet rays; glycogen (stored glucose) can be decomposed to provide energy; lipids (stored in fat cells) can be decomposed to produce energy; mucus provides lubrication and protection.

are *amorphous* (*a* = without, *morpho* = shape); they have no specific shape. These include hyaluronic acid and chondroitin sulfate. **Hyaluronic** (hī'-a-loo-RON-ik) **acid** is a viscous, fluidlike substance that binds cells together, lubricates joints, and maintains the shape of the eyeballs. **Chondroitin** (kon-DROY-tin) **sulfate** is a jellylike substance that provides support and adhesiveness in cartilage, bone, heart valves, the cornea of the eye, and the umbilical cord.

Other matrix materials are *fibrous,* or threadlike. Fibrous materials provide strength and support for tissues. Among these are **collagenous** (*kolla* = glue) **fibers** consisting of the protein *collagen.* These fibers are found in all types of connective tissue, especially in bones, cartilage, tendons, and ligaments. **Reticular** (*rete* = net) **fibers** consisting of the protein collagen and a coating of glycoprotein, form a network around fat cells, nerve fibers, muscle cells, and blood vessels. They also form the framework or stroma for many soft organs of the body such as the spleen. **Elastic fibers,** consisting of the protein *elastin,* give elasticity to skin and to tissues forming the walls of blood vessels.

CELL DIVISION

Most of the cell activities mentioned thus far maintain the life of the cell on a day-to-day basis. However, cells become damaged, diseased, or worn out and then die. New cells must be produced as replacements and for growth.

Cell division is the process by which cells reproduce themselves. It consists of a nuclear division and a cytoplasmic division. Because nuclear division can be of two types, two kinds of cell division are recognized. The first is the mechanism by which sperm and egg cells are produced, preliminary to the formation of a new organism. The process consists of a nuclear division called **meiosis** plus a cytoplasmic division called **cytokinesis.** It is often called reproductive cell division and is discussed in detail in Chapter 29.

In the second kind of division, often called somatic cell division, a single parent cell duplicates itself. This process consists of a nuclear division called **mitosis** and cytokinesis. The process ensures that each new daughter cell has the same *number* and *kind* of chromosomes as the original parent cell. After the process is complete, the two daughter cells have the same hereditary material and genetic potential as the parent cell. This kind of cell division results in an increase in the number of body cells. In a 24-hour period, the average adult loses trillions of cells from different parts of the body. Obviously, these cells must be replaced. Cells that have a short life span— the cells of the outer layer of skin, the cornea of the eye, the digestive tract—are continually being replaced. Mitosis and cytokinesis are the means by which dead or injured cells are replaced and new cells are added for body growth.

When a cell reproduces, it must replicate its chromosomes so its hereditary traits may be passed on to succeeding generations of cells. A **chromosome** is a highly coiled DNA molecule that is partly covered by protein. The protein causes changes in the length and thickness of the chromosome. Hereditary information is contained in the DNA portion of the chromosome in units called **genes.** Each human chromosome consists of about 20,000 genes.

When a cell is carrying on every life process except division, it is said to be in **interphase** or the **metabolic phase.** It is during this stage that the replication of chromosomes occurs.

When DNA replicates, its helical structure partially uncoils (Figure 3-14). Those portions of DNA that remain coiled stain darker than the uncoiled portions. This unequal distribution of stain causes the DNA to appear as a granular mass called **chromatin** (see Figure 3-15a). During uncoiling, DNA separates at the points where the nitrogen bases are connected. Each exposed nitrogen base then picks up a complementary nitrogen base (with associated sugar and phosphate group) from the cytoplasm of the cell. This uncoiling and complementary base pairing continues until each of the two original DNA strands is matched and joined with two newly formed DNA strands. The original DNA molecule has become two DNA molecules.

During interphase the cell is also synthesizing most of its RNA and proteins. It is producing chemicals so that all cellular components can be doubled during division. A microscopic view of a cell during interphase shows a clearly defined nuclear membrane, nucleoli, karyolymph, chromatin, and a pair of centrioles. Once a cell completes its replication of DNA and its synthesis of RNA and proteins during interphase, mitosis begins.

MITOSIS

The succession of events that takes place during mitosis and cytokinesis is plainly visible under a microscope after the cells have been stained in the laboratory.

The process called **mitosis** is the distribution of the two sets of chromosomes into two separate and equal nuclei following the replication of the chromosomes of the parent nucleus. For convenience, biologists divide the process into four stages: prophase, metaphase, anaphase, and telophase. These are arbitrary classifications. Mitosis is actually a continuous process, one stage merging imperceptibly into the next.

Prophase

During **prophase** (Figure 3-15b), the first stage of mitosis, chromatin shortens and coils into chromosomes. The nucleoli become less distinct and the nuclear membrane disappears. Each prophase "chromosome" is actually composed of a pair of structures called **chromatids.** Each chromatid is a complete chromosome consisting of a double-stranded DNA molecule and each is attached to its chromatid pair by a small spherical body called a **centromere.** During prophase, the chromatid pairs assemble near the equatorial plane region or equator of the cell.

Also during prophase, the paired centrioles separate and each pair moves to an opposite pole (end) of the cell. Between the centrioles, a series of microtubules is organized into two groups of fibers. The *continuous* (*inter-*

polar) *microtubules* originate from the vicinity of each pair of centrioles and grow toward each other. Thus, they extend from one pole of the cell to another. As they grow toward each other, the second group of microtubules develops. These are called *chromosomal microtubules* and grow out of the centromeres; they extend from a centromere to a pole of the cell. Together, the continuous and chromosomal microtubules constitute the **mitotic spindle** and, with the centrioles, are referred to as the **mitotic apparatus.**

Metaphase

During **metaphase** (Figure 3-15c), the second stage of mitosis, the centromeres of the chromatid pairs line up on the equatorial plane of the cell. The centromeres of each chromatid pair form a chromosomal microtubule that attaches the centromere to a pole of the cell.

Anaphase

The third stage of mitosis, **anaphase** (Figure 3-15d), is characterized by the division of the centromeres and the movement of complete identical sets of chromatids, now called chromosomes, to opposite poles of the cell. During this movement, the centromeres attached to the chromosomal microtubules seem to drag the trailing parts of the chromosomes toward opposite poles.

The mechanism by which the chromosomes move to opposite poles is not completely understood. Several theories have been presented and the key points of two of them may be summarized as follows. One suggests that the movement of chromosomes is caused by the assembly

FIGURE 3-14 Replication of DNA. The two strands of the double helix separate by breaking the bonds between nucleotides. New nucleotides attach at the proper sites, and a new strand of DNA is paired off with each of the original strands. After replication, the two DNA molecules, each consisting of a new and an old strand, return to their helical structure.

Key:

- G = Guanine
- C = Cytosine
- A = Adenine
- T = Thymine

Old   New   New   Old

FIGURE 3-15 Cell division: mitosis and cytokinesis. Photomicrographs and diagrammatic representations of the various stages of cell division in whitefish eggs. Read the sequence starting at (a), and move clockwise until you complete the cycle. (Courtesy of Carolina Biological Supply Company.)

and disassembly of microtubules. As the chromosomal microtubules are shortened by the removal of tubulin subunits, the chromosomes are pulled toward the poles. Simultaneously, the continuous microtubules are lengthened by the addition of tubulin subunits, further contributing to chromosomal movement.

A second theory is that chromosomal movement and pole separation are caused by mechanochemical interactions between continuous and chromosomal microtubules. This is accomplished by temporary cross bridges between microtubules. As a result of reactions that use energy, the cross bridges cause the continuous and chromosomal microtubules to slide past each other in a manner similar to the closing of an extension ladder.

Telophase

Telophase (Figure 3-15e), the final stage of mitosis, consists of a series of events nearly the reverse of prophase. By this time, two identical sets of chromosomes have reached opposite poles. As telophase progresses, new nuclear membranes begin to enclose the chromosomes, the chromosomes start to assume their chromatin form, nucleoli reappear, and the mitotic spindles disappear. The centrioles also replicate so that each cell has two centriole pairs. The formation of two nuclei identical to those of cells in interphase terminates telophase. A mitotic cycle has thus been completed (Figure 3-15f).

Time Required

The time required for mitosis varies with the kind of cell, its location, and the influence of factors such as temperature. Furthermore, the different stages of mitosis are not equal in duration. Prophase is usually the longest stage, lasting from one to several hours. Metaphase is considerably shorter, ranging from 5 to 15 minutes. Anaphase is the shortest stage, lasting from 2 to 10 minutes. Telophase lasts from 10 to 30 minutes. These lengths of time are only approximate, however.

CYTOKINESIS

Division of the cytoplasm, called **cytokinesis** (sī'-tō-ki-NĒ-sis), often begins during late anaphase and terminates at the same time as telophase. Cytokinesis begins with the formation of a *cleavage furrow* that extends around the cell's equator. The furrow progresses inward, resembling a constricting ring, and cuts completely through the cell to form two separate portions of cytoplasm (Figure 3-15d–f).

GENE ACTION

PROTEIN SYNTHESIS

We have already noted that genes consist of DNA and that genes determine the traits that a cell inherits and control the activities that take place in a cell. Mitosis and cytokinesis are vital processes because they ensure that daughter cells have identical sets of chromosomes and, therefore, identical genes. In this way, hereditary information passes from one generation of cells to another.

Although cells synthesize numerous chemicals in order to maintain homeostasis, much of the cellular machinery is concerned with producing proteins. Some of the proteins are structural, helping to form plasma membranes, microfilaments, microtubules, centrioles, flagella, cilia, the mitotic spindle, and other parts of cells. Other proteins serve as hormones, antibodies, and contractile elements in muscle tissue. Still other proteins serve as enzymes that regulate the myriad chemical reactions that occur in cells. Basically, cells are protein factories that constantly synthesize large numbers of diverse proteins that determine the physical and chemical characteristics of cells and, therefore, of organisms.

The genetic instructions for making proteins are found in DNA. Cells make proteins by translating the genetic information encoded in DNA into specific proteins. In the process, the genetic information in a region of DNA is copied to produce a specific molecule of RNA. Through a complex series of reactions, the information contained in RNA is translated into a corresponding specific sequence of amino acids in a newly produced protein molecule. Let us take a look at how DNA directs protein synthesis by considering the two principal steps in protein synthesis: transcription and translation.

Transcription

Transcription is the process by which genetic information encoded in DNA is copied by a strand of RNA called **messenger RNA (mRNA)**. It is called transcription because it resembles the transcription of a sequence of words from one tape to another. By using a specific portion of the cell's DNA as a template, the genetic information stored in the sequence of nitrogen bases of DNA is rewritten so that the same information appears in the nitrogen bases of mRNA. As in DNA replication (see Figure 3-14), a cytosine (C) in the DNA template dictates a guanine (G) in the mRNA strand being made; a G in the DNA template dictates a C in the mRNA strand; and a thymine (T) in the DNA template dictates an adenine (A) in the mRNA. Since RNA contains uracil (U) instead of T, an A in the DNA template dictates a U in the mRNA. As an example, if the template portion of DNA has the base sequence ATGCAT, the transcribed mRNA strand would have the complementary base sequence UACGUA (Figure 3-16). Note that only one of the two DNA strands serves as a template for RNA synthesis. This strand is referred to as the *sense strand*. The other strand, the one not transcribed, is the complement of the sense strand and is called the *antisense strand*.

In addition to serving as the template for the synthesis of mRNA, DNA also synthesizes two other kinds of RNA. One is called **ribosomal RNA** or **rRNA** which, together with ribosomal proteins, makes up ribosomes. The other is called **transfer RNA (tRNA)**. Once synthesized, mRNA, rRNA, and tRNA leave the nucleus of the cell. In the cytoplasm, they participate in the next principal step in protein synthesis—translation.

Translation

The process by which information in the nitrogen base sequence of mRNA is used to specify the amino acid sequence of a protein is called **translation.** The key events involved in translation are as follows (Figure 3-17).

1. In the cytoplasm, the small ribosomal subunit binds one end of the mRNA molecule (Figure 3-17a). In this respect, ribosomes are the sites of protein synthesis.

2. There are 20 different amino acids in the cytoplasm that may participate in protein synthesis. Whichever amino acids participate to form a particular protein are picked up by tRNAs (Figure 3-17b). For each different amino acid there is a different type of tRNA. One end of a tRNA molecule couples with a specific amino acid. This amino acid activation requires energy

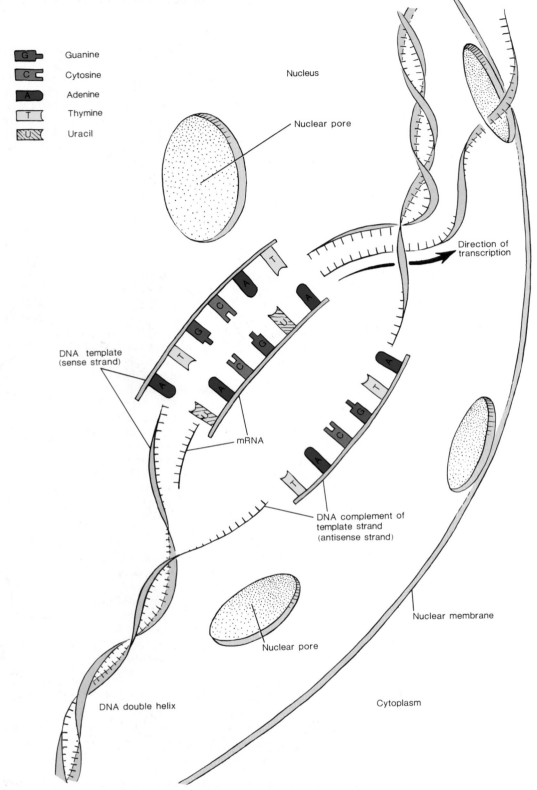

Legend:

- G — Guanine
- C — Cytosine
- A — Adenine
- T — Thymine
- U — Uracil

Nucleus

Nuclear pore

Direction of transcription

DNA template (sense strand)

mRNA

DNA complement of template strand (antisense strand)

Nuclear membrane

Nuclear pore

DNA double helix

Cytoplasm

FIGURE 3-16 Transcription. Transcription is the process whereby an mRNA molecule is produced that is complementary in base sequence to a section of one of the DNA strands (the sense strand) in a DNA double helix. The process occurs in the nucleus. The diagram shows a DNA double helix, partly unwound, with transcription occurring from the DNA template in the unwound region. The upper enlargement shows in more detail a six-base segment of the template and transcript; note the nitrogen-base pairing (A with T and G with C), although in RNA U replaces T. The lower enlargement shows the base sequence of the nontemplate DNA strand (antisense strand) for the same six-base segment as detailed in the upper enlargement. In addition to mRNA, DNA templates can also synthesize rRNA and tRNA. Note here, first, that the base sequence is complementary to the base sequence of the DNA template strand and, second, that the base sequence is the same as that found in the RNA transcript from the template strand (except, of course, that in RNA, U replaces T). The process of transcription is carried out by a special enzyme—RNA polymerase.

Guanine

Cytosine

Adenine

Thymine

Uracil

(a) mRNA becomes associated with small ribosomal subunit

(b) Specific tRNA picks up a specific amino acid

(c) tRNA anticodon attaches to complementary mRNA codon

(d) The next tRNA with its amino acid moves into position on mRNA

(e) Amino acids are joined by a peptide bond and the first tRNA detaches

(f) As more amino acids are detached by their tRNA's, the protein gets progressively longer

(g) Termination codon stops synthesis and protein is released

(h) Following protein synthesis, ribosomal subunits separate

(i) Summary of movement of ribosome along mRNA

FIGURE 3-17 Translation. Note that during protein synthesis the ribosomal subunits join together. When not involved in protein synthesis, the two subunits are separate entities.

EXHIBIT 3-2
EFFECTS OF AGING ON VARIOUS BODY SYSTEMS

SYSTEM	AGING EFFECTS
Integumentary	General changes include wrinkling; loss of subcutaneous (under the skin) fat; atrophy (wasting away) of sweat glands; reduction of blood flow to skin, decrease in number of functioning pigment-producing cells, resulting in gray hair and atypical skin color; increase in size of some pigment cells, producing pigmented blotching; atrophy of sebaceous (oil) glands, producing dry and broken skin that is susceptible to infection. Aged skin becomes susceptible to pathological conditions such as senile pruritus (itching), decubitus ulcers (bedsores), and herpes zoster (shingles).
Skeletomuscular	There is loss of muscle mass and diminishing of muscle reflexes. Common disorders include osteoarthritis, rheumatoid arthritis, gout, osteoporosis (decrease in bone mass), and Paget's disease.
Nervous	Nerve cells (neurons) are lost. Associated with this decline is a decreased capacity for sending nerve impulses to and from the brain. Conduction velocity decreases, voluntary motor movements slow down, and the reflex time for skeletal muscles is decreased. Degenerative changes and disease states involving the sense organs can alter vision, hearing, taste, smell, and touch. The disorders that represent the most common visual problems and may be responsible for serious loss of vision are presbyopia (inability to focus on nearby objects), cataracts (cloudiness of the lens), and glaucoma (excessive fluid pressure in the eyeball). Impaired hearing associated with aging, known as presbycusis, is usually the result of changes in important structures of the inner ear. Parkinson's disease is the most common movement disorder involving the central nervous system.
Endocrine	The endocrine system exhibits a variety of changes, and many researchers look to this system with the hope of finding the key to the aging process. Disorders of the endocrine system are not frequent, and when they do occur, most often they are related to pathological changes rather than age. Diabetes and thyroid disorders are the most important endocrine problems that have a significant effect on health and function.
Cardiovascular	General changes include loss of extensibility of the aorta, reduction in cardiac muscle cell size, progressive loss of cardiac muscular strength, and a reduced output of blood by the heart. There is an increase in blood pressure. Coronary artery disease increases and represents the major cause of heart disease and death in older Americans. Congestive heart failure occurs and is viewed as a set of symptoms associated with the impaired pumping performance of the heart. Changes in blood vessels such as hardening of the arteries and cholesterol deposits in arteries that serve brain tissue can reduce its nourishment and result in the malfunction or death of brain cells, which is called cerebrovascular disease.
Respiratory	The cardiovascular and respiratory systems operate as a unit, and damage or disease in one of these organ systems is often secondarily reflected in the other. In pulmonary heart disease the right side of the

from ATP breakdown. The other end has a specific sequence of three nitrogen bases (triplet) known as an *anticodon.* The complementary triplet on an mRNA strand is called a *codon.*

3. By base pairing, the anticodon of a specific tRNA recognizes the corresponding codon of mRNA and attaches to it. If the tRNA anticodon is UAC, the mRNA codon would be AUG (Figure 3-17c). In the process, the tRNA also brings along the specific amino acid. The pairing of anticodon and codon occurs only where mRNA is attached to a ribosome.

4. Once the first tRNA attaches to mRNA, the ribosome moves along the mRNA, and the next tRNA with its amino acid moves into position (Figure 3-17d).

5. The two amino acids are joined by a peptide bond and the first tRNA detaches itself from the mRNA strand. The larger ribosomal subunit contains the enzymes that join the amino acids together (Figure 3-17e). The released tRNA can now pick up another similar amino acid if necessary.

6. As the proper amino acids are brought into line, one by one, peptide bonds are formed between them and the protein gets progressively longer (Figure 3-17f).

7. When the specified protein is completed, further synthesis is stopped by a special *termination codon.* The assembled protein is released from the ribosome and the ribosome comes apart into its component subunits (Figure 3-17g,h).

As each ribosome moves along the mRNA strand, it "reads" the information coded in mRNA and synthesizes

a protein according to that information; the ribosome synthesizes the protein by translating the codon sequences into an amino acid sequence.

Protein synthesis progresses at the rate of about 15 amino acids per second. As the ribosome moves along the mRNA and before it completes translation of that gene, another ribosome may attach and begin translation of the same mRNA strand, so that several ribosomes may be attached to the same mRNA strand. Such an mRNA strand with its several ribosomes attached is called a **polyribosome.** Several ribosomes moving simultaneously in tandem along the same mRNA molecule permit the translation of a single mRNA strand into several identical proteins simultaneously.

Based upon the description of protein synthesis just presented, we can appropriately define a **gene** as a group of nucleotides on a DNA molecule that serves as the master mold for manufacturing a specific protein. Genes average about 1,000 pairs of nucleotides, which appear in a specific sequence on the DNA molecule. No two genes have exactly the same sequence of nucleotides and this is the key to heredity.

Remember that the base sequence of the gene determines the sequence of the bases in the mRNA. The sequence of the bases in the mRNA then determines the

● Define a tissue.

● Classify the tissues of the body into four major types and define each type.

● Discuss the distinguishing characteristics of epithelial tissues.

● Contrast the structural and functional differences between covering and lining epithelium and glandular epithelium.

● Compare the layering arrangements and cell shapes of covering and lining epithelium.

● List the structure, location, and function for the following types of epithelium: simple squamous, simple cuboidal, simple columnar (nonciliated and ciliated), stratified squamous, stratified cuboidal, stratified columnar, transitional, and pseudostratified.

● Define a gland and distinguish between exocrine and endocrine glands.

● Classify exocrine glands according to structural complexity and function and give an example of each.

● Identify the distinguishing characteristics of connective tissue.

● Contrast the structural and functional differences between embryonic and adult connective tissues.

● Discuss the ground substance, fibers, and cells that constitute connective tissue.

● List the structure, function, and location of loose (areolar) connective tissue, adipose tissue, and dense, elastic, and reticular connective tissue.

● List the structure, function, and location of the three types of cartilage.

● Distinguish between interstitial and appositional growth of cartilage.

● Define an epithelial membrane.

● List the location and function of mucous, serous, cutaneous, and synovial membranes.

● List and describe the symptoms of tissue inflammation.

● Outline the stages involved in the inflammatory response.

● Describe the conditions necessary for tissue repair.

● Explain the importance of nutrition, adequate circulation, and age to tissue repair.

4 The Tissue Level of Organization

Cells are highly organized units, but they do not function in isolation. They work together in a group of similar cells called a tissue.

TYPES OF TISSUES

A **tissue** is a group of similar cells and their intercellular substance functioning together to perform a specialized activity. Certain tissues function to move body parts. Others move food through body organs. Some tissues protect and support the body. Others produce chemicals such as enzymes and hormones. The various tissues of the body are classified into four principal types according to their function and structure:

1. **Epithelial** (ep'-i-THĒ-lē-al) **tissue,** which covers body surfaces or tissues, lines body cavities, and forms glands.
2. **Connective tissue,** which protects and supports the body and its organs and binds organs together.
3. **Muscular tissue,** which is responsible for movement.
4. **Nervous tissue,** which initiates and transmits nerve impulses that coordinate body activities.

EPITHELIAL TISSUE

The tissues in this principal type perform many activities in the body, ranging from protection to secretion. **Epithelial tissue,** or more simply **epithelium,** may be divided into two subtypes: (1) *covering and lining epithelium* and (2) *glandular epithelium.* Covering and lining epithelium forms the outer covering of external body surfaces and the outer covering of some internal organs. It lines the body cavities and the interiors of the respiratory and digestive tracts, blood vessels, and ducts. It makes up, along with nervous tissue, the parts of the sense organs that are sensitive to the stimuli that produce smell and hearing sensations. And, it is the tissue from which gametes (sperm and eggs) develop. Glandular epithelium constitutes the secreting portion of glands.

Both types of epithelium consist largely or entirely of closely packed cells with little or no intercellular material between adjacent cells. (Such intercellular material is also called the matrix.) Epithelial cells are arranged in continuous sheets that may be either single or multilayered. Nerves may extend through these sheets, but blood vessels do not. They are *avascular* (*a* = without, *vascular* = blood vessels). The vessels that supply nutrients and remove wastes are located in underlying connective tissue.

Both types of epithelium overlie and adhere firmly to the connective tissue, which holds the epithelium in position and prevents it from being torn. The surface of attachment between the epithelium and the connective tissue is a thin extracellular layer called the **basement membrane.** With few exceptions, epithelial cells secrete along their basal surfaces a material consisting of a special type of collagen and glycoproteins. This structure averages 500 to 800 Å in thickness and is referred to as the **basal lamina.** Frequently, the basal lamina is reinforced by an underlying *reticular lamina* consisting of reticular fibers and glycoproteins. This lamina is produced by cells in the underlying connective tissue. The combination of the basal lamina and reticular lamina constitutes the basement membrane.

Since all epithelium is subjected to a certain degree of wear, tear, and injury, its cells must divide and produce new cells to replace those that are destroyed.

COVERING AND LINING EPITHELIUM

Arrangement of Layers

Covering and lining epithelium is arranged in several different ways related to location and function. If the epithelium is specialized for absorption or filtration and is in an area that has minimal wear and tear, the cells of the tissue are arranged in a single layer. Such an arrangement is called **simple epithelium.** If the epithelium is not specialized for absorption or filtration and is found in an area with a high degree of wear and tear, then the cells are stacked in several layers. This tissue is referred to as **stratified epithelium.** A third, less common, arrangement of epithelium is called **pseudostratified.** Like simple epithelium, pseudostratified epithelium has only one layer of cells. However, some of the cells do not reach the surface—an arrangement that gives the tissue a multilayered, or stratified, appearance. The pseudostratified cells that do reach the surface either secrete mucus or contain cilia that move mucus and foreign particles for eventual elimination from the body.

Cell Shapes

In addition to classifying covering and lining epithelium according to the number of its layers, we may also categorize it by cell shape. The cells may be flat, cubelike, or columnar or may resemble a cross between shapes. **Squamous** (SKWĀ-mus) cells are flattened and scalelike. They are attached to each other like a mosaic. **Cuboidal** cells are usually cube-shaped in cross section. They sometimes appear as hexagons. **Columnar** cells are tall and cylindrical, appearing as rectangles set on end. **Transitional** cells often have a combination of shapes and are found where there is a great degree of distention or expansion in the body. Transitional cells in the bottom layer of an epithelial tissue may range in shape from cuboidal to columnar. In the intermediate layer, they may be cuboidal or polyhedral. In the superficial layer they may range from cuboidal to squamous, depending on how much they are pulled out of shape during certain body functions.

Classification

Considering layers and cell type in combination, we may classify covering and lining epithelium as follows:

Simple
1. Squamous
2. Cuboidal
3. Columnar

Stratified
1. Squamous
2. Cuboidal
3. Columnar
4. Transitional

Pseudostratified

Each of the epithelial tissues described in the following sections is illustrated in Exhibit 4-1.

EXHIBIT 4-1

EPITHELIAL TISSUES

COVERING AND LINING EPITHELIUM

Simple Squamous (250×)

Description: Single layer of flat, scalelike cells; large, centrally located nuclei.

Location: Lines air sacs of lungs, glomerular capsule of kidneys, and inner surfaces of the membranous labyrinth and tympanic membrane of the ear. Called endothelium when it lines heart, blood and lymphatic vessels, and forms capillaries. Called mesothelium when it lines the ventral body cavity and covers viscera as part of a serous membrane.

Function: Filtration, absorption, and secretion in serous membranes.

Cell boundary

Cytoplasm

Nuclei of simple
squamous cells

Simple Cuboidal (450×)

Description: Single layer of cube-shaped cells; centrally located nuclei.

Location: Covers surface of ovary, lines anterior surface of capsule of the lens of eyes, forms pigmented epithelium of retina of eye, and lines kidney tubules and smaller ducts of many glands.

Function: Secretion and absorption.

Nuclei of simple
cuboidal cells

Simple cuboidal
epithelium

Lumen of tubule

Simple Epithelium

● *Simple Squamous Epithelium*   This type of simple epithelium consists of a single layer of flat, scalelike cells. Its surface resembles a tiled floor. The nucleus of each cell is centrally located and oval or spherical. Since simple squamous epithelium has only one layer of cells, it is highly adapted to diffusion, osmosis, and filtration. Thus it lines the air sacs of the lungs, where oxygen is exchanged with carbon dioxide. It is present in the part of the kidney that filters the blood. It also lines the inner surfaces of the membranous labyrinth and tympanic membrane of the ear. Simple squamous epithelium is found in body parts that have little wear and tear.

EXHIBIT 4-1
(*Continued*)

Simple Columnar (Nonciliated) (600×)
Description: Single layer of nonciliated rectangular cells; contains goblet cells; nuclei at bases of cells.
Location: Lines the digestive tract from the cardia of the stomach to the anus, excretory ducts of many glands, and gallbladder.
Function: Secretion and absorption.

Nuclei of
absorptive cells

Mucus–
containing
part of
goblet cells

Connective
tissue

Simple columnar
(nonciliated)
epithelium

Simple Columnar (Ciliated) (400×)
Description: Single layer of ciliated columnar cells; contains goblet cells; nuclei at bases of cells.
Location: Lines a few portions of upper respiratory tract, uterine (fallopian) tubes, uterus, some paranasal sinuses, and central canal of spinal cord.
Function: Moves mucus by ciliary action.

Connective
tissue

Arteriole

Nuclei of simple
columnar (ciliated)
cells

Nuclei of simple
columnar (ciliated)
cells

One form of stratified squamous epithelium is called **nonkeratinized stratified squamous epithelium.** This tissue is found on wet surfaces that are subjected to considerable wear and tear and does not perform the function of absorption—the lining of the mouth, the tongue, the esophagus, and the vagina. Another form of stratified squamous epithelium is called **keratinized stratified squamous epithelium.** The surface cells of this type are modified into a tough layer of material containing keratin. **Keratin** is a protein that is waterproof and resistant to friction and helps to resist bacterial invasion. The outer layer of skin consists of keratinized tissue.

● *Stratified Cuboidal Epithelium* This relatively rare type of epithelium is found in the ducts of the sweat glands of adults, fornix of the conjunctiva of the eye, cavernous urethra of the male urogenital system, pharynx, and epiglottis. It sometimes consists of more than two layers of cells. Its function is mainly protective.

● *Stratified Columnar Epithelium* Like stratified cuboidal epithelium, this type of tissue is also uncommon in the body. Usually the basal layer or layers consist of shortened, irregularly polyhedral cells. Only the superficial cells are columnar in form. This kind of epithelium lines part of the male urethra, some larger excretory ducts such as lactiferous (milk) ducts in the mammary glands, and small areas in the anal mucous membrane. It functions in protection and secretion.

● *Transitional Epithelium* This kind of epithelium is very much like nonkeratinized stratified squamous epithelium. The distinction is that cells of the outer layer in transitional epithelium tend to be large and rounded rather than flat. This feature allows the tissue to be stretched (distended) without the outer cells breaking apart from one another. When stretched, they are drawn out into squamouslike cells. Because of this arrangement, transitional epithelium lines hollow structures that are subjected to expansion from within, such as the urinary bladder. Its function is to help prevent a rupture of the organ.

Pseudostratified Epithelium

The third category of covering and lining epithelium consists of columnar cells, and is called pseudostratified epithelium. The nuclei of the cells are at varying depths. Even though all the cells are attached to the basement membrane in a single layer, some do not reach the surface. This feature gives the impression of a multilayered tissue, the reason for the designation *pseudo*stratified epithelium. It lines the larger excretory ducts of many glands, parts of the male urethra, and the auditory tubes. Pseudostratified epithelium may be ciliated and may contain goblet cells. In this form, it lines most of the upper respiratory tract and certain ducts of the male reproductive system.

GLANDULAR EPITHELIUM

The function of glandular epithelium is secretion, accomplished by glandular cells that lie in clusters deep to the covering and lining epithelium. A **gland** may consist of one cell or a group of highly specialized epithelial cells that secrete substances into ducts, onto a surface, or into the blood. The production of such substances always requires active work by the glandular cells and results in an expenditure of energy.

All glands of the body are classified as exocrine or endocrine according to whether they secrete substances into ducts (or directly onto a free surface) or into the blood. **Exocrine glands** secrete their products into ducts (tubes) that empty at the surface of covering and lining epithelium or directly onto a free surface. The product of an exocrine gland may be released at the skin surface or into the lumen of a hollow organ. The secretions of exocrine glands include mucus, perspiration, oil, wax, and digestive enzymes. Examples of exocrine glands are sweat glands, which eliminate perspiration to cool the skin, salivary glands, which secrete a digestive enzyme, and goblet cells, which produce mucus. **Endocrine glands** are ductless and secrete their products into the blood. The secretions of endocrine glands are always hormones, chemicals that regulate various physiological activities. The pituitary, thyroid, and adrenal glands are endocrine glands.

Structural Classification of Exocrine Glands

Exocrine glands are classified into two structural types: unicellular and multicellular. **Unicellular glands** are single-celled. A good example of a unicellular gland is a goblet cell (Exhibit 4-1). Goblet cells are found in the epithelial lining of the digestive, respiratory, urinary, and reproductive systems. They produce mucus to lubricate the free surfaces of these membranes.

Multicellular glands occur in several different forms (Figure 4-1). If the secretory portions of a gland are tubular, it is referred to as a **tubular gland.** If they are flasklike, it is called an **acinar** (AS-i-nar) **gland.** If the gland contains both tubular and flasklike secretory portions, it is called a **tubuloacinar gland.** Further, if the duct of the gland does not branch, it is referred to as a **simple gland;** if the duct does branch, it is called a **compound gland.** By combining the shape of the secretory portion with the degree of branching of the duct, we arrive at the following structural classification for exocrine glands:

I. **Unicellular.** Single-celled gland that secretes mucus. Example: goblet cell of the digestive and respiratory systems.

II. **Multicellular.** Many-celled glands.
  A. *Simple.* Single, nonbranched duct.
   1. **Tubular.** The secretory portion is straight and tubular. Example: intestinal glands.
   2. **Branched tubular.** The secretory portion is branched and tubular. Examples: gastric and uterine glands.
   3. **Coiled tubular.** The secretory portion is coiled. Example: sudoriferous (sweat) glands.
   4. **Acinar.** The secretory portion is flasklike. Example: seminal vesicle glands.
   5. **Branched acinar.** The secretory portion is branched and flasklike. Example: sebaceous (oil) glands.
  B. *Compound.* Branched duct.
   1. **Tubular.** The secretory portion is tubular. Examples: bulbourethral glands, testes, and liver.

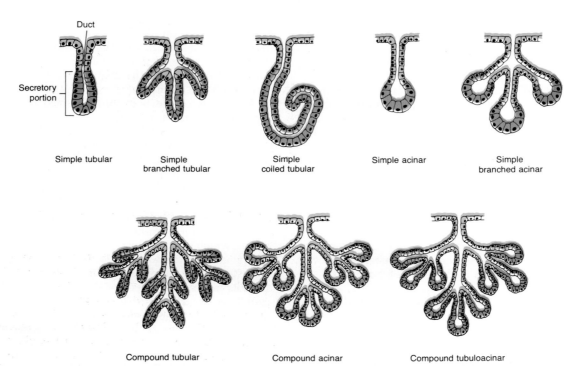

FIGURE 4-1 Structural types of multicellular exocrine glands. The secretory portions of the glands are indicated in purple. The blue areas represent the ducts of the glands.

2. Acinar. The secretory portion is flasklike. Examples: salivary glands (sublingual and submandibular).

3. Tubuloacinar. The secretory portion is both tubular and flasklike. Examples: salivary glands (parotid) and pancreas.

Functional Classification of Exocrine Glands

The functional classification of exocrine glands is based on how the gland releases its secretion. The three recognized categories are holocrine, merocrine, and apocrine glands. **Holocrine glands** accumulate a secretory product in their cytoplasm. The cell then dies and is discharged with its contents as the glandular secretion (Figure 4-2a). The discharged cell is replaced by a new cell. One example of a holocrine gland is a sebaceous gland of the skin. **Merocrine glands** simply form the secretory product and discharge it from the cell (Figure 4-2b). Examples of merocrine glands are the salivary glands and pancreas. **Apocrine glands** accumulate their secretory product at the apical (outer) margin of the secreting cell. That portion of the cell pinches off from the rest of the cell to form the secretion (Figure 4-2c). The remaining part of the cell repairs itself and repeats the process. An example of an apocrine gland is the mammary gland.

CONNECTIVE TISSUE

The most abundant tissue in the body is **connective tissue.** This binding and supporting tissue usually is highly vascular and thus has a rich blood supply. An exception is cartilage which is avascular. The cells are widely scattered, rather than closely packed, and there is considerable intercellular material, or matrix. In contrast to epithelium, connective tissues do not occur on free surfaces, such as the surfaces of a body cavity or the external surface of the body. The general functions of connective tissues are protection, support, and the binding together of various organs.

The intercellular substance in a connective tissue largely determines the tissue's qualities. These substances are nonliving and may consist of fluid, semifluid, or mucoid (mucuslike) material. In cartilage, the intercellular material is firm but pliable. In bone, it is considerably harder and not pliable. The cells of connective tissue produce the intercellular substances. The cells may also store fat, ingest bacteria and cell debris, form anticoagulants, or give rise to antibodies that protect against disease.

CLASSIFICATION

Connective tissue may be classified in several ways. We will classify them as follows:

I. Embryonic connective tissue
  A. Mesenchyme
  B. Mucous connective tissue
II. Adult connective tissue
  A. Connective tissue proper
    1. Loose (areolar) connective tissue
    2. Adipose tissue
    3. Dense (collagenous) connective tissue
    4. Elastic connective tissue
    5. Reticular connective tissue
  B. Cartilage
    1. Hyaline cartilage
    2. Fibrocartilage
    3. Elastic cartilage

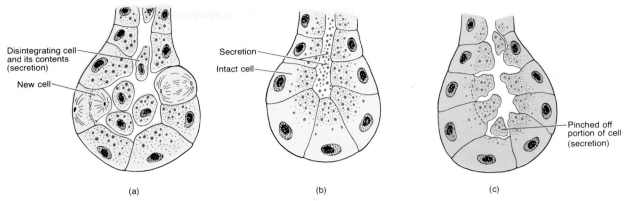

Disintegrating cell
and its contents
(secretion)

New cell

Secretion

Intact cell

Pinched off
portion of cell
(secretion)

(a)   (b)   (c)

FIGURE 4-2 Functional classification of multicellular exocrine glands. (a) Holocrine gland. (b) Merocrine gland. (c) Apocrine gland.

C. Osseous (bone) tissue
D. Vascular (blood) tissue

Each of the connective tissues described in the following sections is illustrated in Exhibit 4-2.

EMBRYONIC CONNECTIVE TISSUE

Connective tissue that is present primarily in the embryo or fetus is called **embryonic connective tissue.** The term *embryo* refers to a developing human from fertilization through the first two months of pregnancy; a *fetus* refers to a developing human from the third month of pregnancy to birth.

One example of embryonic connective tissue found almost exclusively in the embryo is **mesenchyme (MEZ-en-kīm)**—the tissue from which all other connective tissues eventually arise. Mesenchyme may be observed beneath the skin and along the developing bones of the embryo. Some mesenchymal cells are scattered irregularly throughout adult connective tissue, most frequently around blood vessels. Here mesenchymal cells differentiate into fibroblasts that assist in wound healing.

Another kind of embryonic connective tissue is **mucous connective tissue,** found primarily in the fetus. This tissue, also called **Wharton's jelly,** is located in the umbilical cord of the fetus, where it supports the wall of the cord.

ADULT CONNECTIVE TISSUE

Adult connective tissue is connective tissue that exists in the newborn and that does not change after birth. It is subdivided into several kinds.

Connective Tissue Proper

Connective tissue that has a more or less fluid intercellular material and a fibroblast as the typical cell is termed **connective tissue proper.** Five examples of such tissues may be distinguished.

● *Loose (Areolar) Connective Tissue* Loose or areolar (a-RĒ-ō-lar) connective tissue is one of the most widely distributed connective tissues in the body. Structurally, it consists of fibers and several kinds of cells embedded in a semifluid intercellular substance. The term "loose" refers to the loosely woven arrangement of fibers in the intercellular substance. The fibers are neither abundant nor arranged to prevent stretching.

The intercellular substance consists of a viscous material called **hyaluronic acid.** Hyaluronic acid normally facilitates the passage of nutrients from the blood vessels of the connective tissue into adjacent cells and tissues, although the thick consistency of this acid may impede the movement of some drugs. But if an enzyme called **hyaluronidase** is injected into the tissue, hyaluronic acid changes to a watery consistency. This feature is of clinical importance because the reduced viscosity hastens the absorption and diffusion of injected drugs and fluids through the tissue and thus can lessen tension and pain. Some bacteria, white blood cells, and sperm cells produce hyaluronidase.

The three types of fibers embedded between the cells of loose connective tissue are collagenous, elastic, and reticular fibers. **Collagenous (white) fibers** are very tough and resistant to a pulling force, yet are somewhat flexible because they are usually wavy. These fibers often occur in bundles. They are composed of many minute fibers called fibrils lying parallel to one another. The bundle arrangement affords a great deal of strength. Chemically, collagenous fibers consist of the protein collagen. **Elastic (yellow) fibers,** by contrast, are smaller than collagenous fibers and freely branch and rejoin one another. Elastic fibers consist of a protein called elastin. These fibers also provide strength and have great elasticity, up to 50 percent of their length. **Reticular fibers** also consist of collagen, plus some glycoprotein. They are very thin fibers that branch extensively and are not as strong as collagenous fibers. Some authorities believe that reticular fibers are immature collagenous fibers. Like collagenous fibers, reticular fibers provide support and strength and form the *stroma* (framework) of many soft organs.

The cells in loose connective tissue are numerous and varied. Most are **fibroblasts**—large, flat cells with branching processes. If the tissue is injured, the fibroblasts are believed to form collagenous fibers, elastic fibers, and the viscous ground substance. Mature fibroblasts are referred to as **fibrocytes.** The basic distinction between the two is that fibroblasts (or *"blasts"* of any form of cell) are involved in the formation of immature tissue or repair

EXHIBIT 4-2
CONNECTIVE TISSUES

EMBRYONIC

Messenchymal (180×)

Description: Consists of highly branched mesenchymal cells embedded in a fluid substance.

Location: Under skin and along developing bones of embryo; some mesenchymal cells found in adult connective tissue, especially along blood vessels.

Function: Forms all other kinds of connective tissue.

Blood vessels

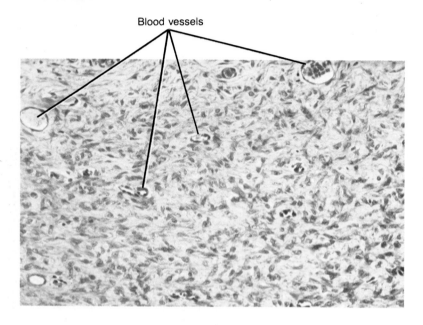

Mucous (320×)

Description: Consists of flattened or spindle-shaped cells embedded in a mucuslike substance containing fine collagenous fibers.

Location: Umbilical cord of fetus.

Function: Support.

Epithelial surface cells of umbilical cord

Collagenous fibers

Nuclei of fiber-forming cells

is located. Specifically, it is in the subcutaneous layer below the skin, around the kidneys, at the base and on the surface of the heart, in the marrow of long bones, as a padding around joints, and behind the eyeball in the orbit. Adipose tissue is a poor conductor of heat and therefore reduces heat loss through the skin. It is also a major energy reserve and generally supports and protects various organs.

● *Dense (Collagenous) Connective Tissue*   Dense (collagenous) connective tissue is characterized by a close packing of fibers and less intercellular substance than in loose connective tissue. The fibers can be either irregularly arranged or regularly arranged. In areas of the body where tensions are exerted in various directions, the fiber bundles are interwoven and without regular orientation. Such a dense connective tissue is referred to as **irregularly arranged** and occurs in sheets. It forms most fasciae, the reticular region of the dermis of the skin, the periosteum of bone, the perichondrium of cartilage, and the *membrane (fibrous) capsules* around organs, such as the kidneys, liver, testes, and lymph nodes.

In other areas of the body, dense connective tissue is adapted for tension in one direction, and the fibers have an orderly, parallel arrangement. Such a dense connective tissue is known as **regularly arranged.** The most common variety of dense regularly arranged connective tissue has a predominance of collagenous (white) fibers arranged in bundles. Fibroblasts are placed in rows between the bundles. The tissue is silvery white, tough, yet somewhat pliable. Because of its great strength, it is the principal component of *tendons,* which attach muscles to bones; many *ligaments* (collagenous ligaments), which hold bones together at joints; and *aponeuroses* (ap'-ō-noo-RŌ-sēz), which are flat bands connecting one muscle with another or with bone.

CLINICAL APPLICATION

Scientists have developed a carbon fiber implant for **reconstructing severely torn ligaments and tendons.** The implant consists of carbon fibers coated with a plastic called polylactic acid. The coated fibers are sewn in and around torn ligaments and tendons to reinforce them and to provide a scaffolding around which the body's own collagenous fibers grow. Within two weeks the polylactic acid is absorbed by the body and the carbon fibers eventually fracture. By this time, the fibers are completely clad in collagen produced by fibroblasts.

● *Elastic Connective Tissue*   Unlike collagenous connective tissue, elastic connective tissue has a predominance of freely branching elastic fibers. These fibers give the tissue a yellowish color. Fibroblasts are present only in the spaces between the fibers. Elastic connective tissue can be stretched and will snap back into shape. It is a component of the cartilages of the larynx, the walls of elastic arteries, the trachea, the bronchial tubes to the lungs, and the lungs themselves. Elastic connective tissue provides stretch and strength, allowing structures to perform their functions efficiently. Yellow elastic ligaments, as contrasted with collagenous ligaments, are composed mostly of elastic fibers; they form the ligamenta flava of the vertebrae (ligaments between successive vertebrae), the suspensory ligament of the penis, and the true vocal cords.

● *Reticular Connective Tissue*   Reticular connective tissue consists of interlacing reticular fibers. It helps to form the stroma of many organs, including the liver, spleen, and lymph nodes. Reticular connective tissue also helps to bind together the cells of smooth muscle tissue. It is especially adapted to providing strength and support.

Cartilage

One type of connective tissue, called cartilage, is capable of enduring considerably more stress than the tissues just discussed. Unlike other connective tissues, cartilage has no blood vessels (except for the perichondrium) or nerves of its own. **Cartilage** consists of a dense network of collagenous fibers and elastic fibers firmly embedded in chondroitin sulfate, a jellylike substance. The cells of mature cartilage, called **chondrocytes** (KON-drō-sīts), occur singly or in groups within spaces called **lacunae** (la-KOO-nē) in the intercellular substance. The surface of cartilage is surrounded by irregularly arranged dense connective tissue called the **perichondrium** (per'-i-KON-drē-um; *chondro* = cartilage, *peri* = around). Three kinds of cartilage are recognized: hyaline cartilage, fibrocartilage, and elastic cartilage (Exhibit 4-2).

● *Growth of Cartilage*   The growth of cartilage follows two basic patterns. In **interstitial (endogenous) growth,** the cartilage increases rapidly in size through the division of existing chondrocytes and continuous deposition of increasing amounts of intercellular matrix by the chondrocytes. The formation of new chondrocytes and their production of new intercellular matrix causes the cartilage to expand from within, thus the term *interstitial growth.* This growth pattern occurs while the cartilage is young and pliable—during childhood and adolescence.

In **appositional (exogenous) growth,** the growth of cartilage occurs because of the activity of the inner chondrogenic layer of the perichondrium. The deeper cells of the perichondrium, the fibroblasts, divide. Some differentiate into chondroblasts (immature cells that develop into specialized cells) and then into chondrocytes. As differentiation occurs, the chondroblasts become surrounded with intercellular matrix and become chondrocytes. As a result, the matrix is deposited on the surface of the cartilage, increasing its size. The new layer of cartilage is added beneath the perichondrium on the surface of the cartilage, causing it to grow in width. Appositional growth starts later than interstitial growth and continues throughout life.

● *Hyaline Cartilage*   This cartilage, also called gristle, appears as a bluish-white, glossy, homogeneous mass. The collagenous fibers, although present, are not visible with ordinary staining techniques, and the prominent chondrocytes are found in lacunae. Hyaline cartilage is the most abundant kind of cartilage in the body. It is found at joints over the ends of long bones (where it is called *articular cartilage*) and forms the *costal cartilages* at the ventral ends of the ribs. Hyaline cartilage also helps to form the nose, larynx, trachea, bronchi, and bronchial tubes leading to the lungs. Most of the embryonic skeleton consists of hyaline cartilage. It affords flexibility and support.

● *Fibrocartilage*   Chondrocytes scattered through many bundles of visible collagenous fibers are found in this type of cartilage. Fibrocartilage is found at the symphysis pubis, the point where the coxal (hip) bones fuse anteriorly at the midline. It is also found in the discs between the vertebrae. This tissue combines strength and rigidity.

● *Elastic Cartilage*   In this tissue, chondrocytes are located in a threadlike network of elastic fibers. Elastic cartilage provides strength and maintains the shape of certain organs—the larynx, the external part of the ear (the pinna), and the auditory tubes (the internal connection between the middle ear cavity and the upper throat).

Osseous Tissue and Vascular Tissue

The details of **osseous tissue** (bone), another kind of connective tissue, are discussed in Chapter 6 as part of the skeletal system. **Vascular tissue,** also known as blood, is treated in Chapter 19 as a component of the cardiovascular system.

MUSCLE TISSUE AND NERVOUS TISSUE

Epithelial and connective tissue can take a variety of forms to provide a variety of body functions. They are all-purpose tissues. By contrast, **muscle tissue** consists of highly modified cells that perform one basic function: contraction (Chapter 10), and **nervous tissue** is specialized to generate and conduct electrical impulses (Chapter 12).

MEMBRANES

The combination of an epithelial layer and an underlying connective tissue layer constitutes an **epithelial membrane.** The principal epithelial membranes of the body are mucous membranes, serous membranes, and the cutaneous membrane, or skin. Another kind of membrane, a **synovial membrane,** does not contain epithelium.

MUCOUS MEMBRANES

A **mucous membrane,** or **mucosa,** lines a body cavity that opens directly to the exterior. Mucous membranes line the entire digestive, respiratory, excretory, and reproductive tracts (see Figure 24-10). The surface tissue of a mucous membrane may vary in type—it is stratified squamous epithelium in the esophagus and simple columnar epithelium in the intestine.

The epithelial layer of a mucous membrane secretes mucus, which prevents the cavities from drying out. It also traps dust in the respiratory passageways and lubricates food as it moves through the digestive tract. In addition, the epithelial layer is responsible for the secretion of digestive enzymes and the absorption of food.

The connective tissue layer of a mucous membrane is called the *lamina propria*. It binds the epithelium to the underlying structures and allows some flexibility of the membrane. It also holds the blood vessels in place, protects underlying muscles from abrasion or puncture, provides the epithelium covering it with oxygen and nutrients, and removes wastes.

SEROUS MEMBRANES

A **serous membrane,** or **serosa,** lines a body cavity that does not open directly to the exterior, and it covers the organs that lie within the cavity. Serous membranes consist of thin layers of loose connective tissue covered by a layer of mesothelium and they are in the form of invaginated double-walled sacs. The part attached to the cavity wall is called the *parietal* (pa-RĪ-i-tal) *portion*. The part that covers the organs inside these cavities is the *visceral portion*. The serous membrane lining the thoracic cavity and covering the lungs is called the *pleura* (see Figure 1-7b). The membrane lining the heart cavity and covering the heart is the *pericardium* (*cardio* = heart). The serous membrane lining the abdominal cavity and covering the abdominal organs and some pelvic organs is called the *peritoneum*.

The epithelial layer of a serous membrane secretes a lubricating fluid that allows the organs to glide easily against one another or against the walls of the cavities. The connective tissue layer of the serous membrane consists of a relatively thin layer of loose connective tissue.

CUTANEOUS MEMBRANE

The **cutaneous membrane,** or skin, constitutes an organ of the integumentary system and is discussed in the next chapter.

SYNOVIAL MEMBRANES

Synovial membranes line the cavities of the joints (see Figure 9-1a). Like serous membranes, they line structures that do not open to the exterior. Unlike mucous, serous, and cutaneous membranes, they do not contain epithelium and are therefore not epithelial membranes. They are composed of loose connective tissue with elastic fibers and varying amounts of fat. Synovial membranes secrete *synovial fluid,* which lubricates the ends of bones as they move at joints and nourishes the articular cartilage covering the bones that form the joints.

TISSUE INFLAMMATION: AN ATTEMPT TO RESTORE HOMEOSTASIS

The tissues of the body perform various roles in attempting to maintain homeostasis. Among these functions are protection, support, filtration, absorption, secretion, movement, transportation, and defense against disease. Damage to a tissue is a stress because it interferes with these vital functions. When cells are injured, many parts of the body work to overcome the stress and return the body to homeostasis. They do this by contributing to the inflammatory response and to the repair process.

When cells are damaged, the injury sets off an **inflammatory response.** The injury, which may be viewed as a form of stress, can have various causes. It could result from mechanical trauma, such as a clean knife incision during surgery. Bacteria that give off toxic chemicals could enter through the nose, pores in the skin, or burns, or by way of a splinter or nail. Cells can be injured if the blood supply is cut off, causing the cells to "starve."

SYMPTOMS

Inflammation is usually characterized by four fundamental symptoms: *redness, pain, heat,* and *swelling.* A fifth symptom can be the *loss of function* of the injured area. Whether loss of function occurs depends on the site and extent of the injury. The inflammatory response serves a protective and defensive role. It is an attempt to neutralize and destroy toxic agents at the site of injury and to prevent their spread to other organs. Thus the inflammatory response is an attempt to restore tissue homeostasis.

The immediate inflammatory response to tissue injury consists of a complicated sequence of physiological and anatomical adjustments. Various body components are involved in the initial response: blood vessels; intercellular fluid mixed with parts of injured cells, called the *exudate* (EKS-yoo-dāt); the cellular components of blood; and the surrounding epithelial and connective tissues. Other factors that affect the inflammatory response are the individual's age and general state of health. Healing processes of all types exert a great demand on the body's store of nutrients. Thus nutrition plays an essential role in healing.

STAGES

The inflammatory response is one of the body's internal systems of defense. The response of a tissue to a rusty nail wound is similar to other inflammatory responses, such as the sore throat that results from bacterial or viral invasion. These are the basic stages of response.

1. Vasodilation and increased permeability of blood vessels. Immediately following tissue damage, there is vasodilation and increased permeability of blood vessels in the area of the injury. **Vasodilation** is an increase in diameter in the blood vessels. **Increased permeability** means that substances normally retained in blood are permitted to pass from the blood vessels. Vasodilation allows more blood to go to the damaged area, and increased permeability permits defensive substances in the blood to enter the injured area. Such defensive substances include white blood cells and clot-forming chemicals. The increased blood supply also removes toxic products and dead cells, preventing them from complicating the injury. These toxic substances include waste products released by invading microorganisms. As the blood clots around the site of activity, the microbe and its toxins are prevented from spreading to other parts of the body.

Vasodilation and increased permeability are caused by the release of certain chemicals by damaged cells in response to injury. One such substance is *histamine* (Figure 4-3a). It is present in many tissues of the body, especially in mast cells in connective tissue, circulating basophils (one type of white blood cell), and blood platelets. Histamine is released in direct response to any injured cells containing it. Neutrophils, another type of white blood cell attracted to the site of injury, can also produce chemicals that cause the release of histamine. *Serotonin* (*5-HT*) also produced by mast cells, assumes a role in

vasodilation. Other substances that play a role in vasodilation and increased permeability of blood vessels are *kinins.* These are chemicals that attract neutrophils to the injured area. Inflammation also results in an increased synthesis of *prostaglandins* (*PG*), especially of the E series. Prostaglandins are potent vasodilators by themselves and they also intensify the effects of histamine, serotonin, and kinins. Prostaglandins also bring about increased permeability of blood vessels.

In addition to the responses in the area of injury, the body may also respond by increasing the metabolic rate and quickening the heartbeat so that more blood circulates to the injured area per minute. Within minutes after the injury, the quickened metabolism and circulation and, especially, the dilation and increased permeability of capillaries produce heat, redness, and swelling. The heat results from the large amount of warm blood that accumulates in the area and, to a certain extent, from the heat energy produced by the metabolic reactions. The large amounts of blood in the area are also responsible for the redness.

CLINICAL APPLICATION

The increased permeability of the capillary walls allows quantities of fluid to move out of the blood and into the intercellular spaces of the tissue. Since the fluid moves into the intercellular spaces faster than it can be drained off, it accumulates in the tissue, causing it to swell. The swelling is called **edema.**

Pain, whether immediate or delayed, is a cardinal symptom of inflammation. It can result from an injury to nerve fibers or from an irritation caused by the release of toxic chemicals from microorganisms. Pain may also be due to increased pressure from edema. Prostaglandins intensify and prolong the pain associated with inflammation. However, kinins affect some nerve endings, causing much of the pain associated with inflammation.

2. Phagocyte migration. Generally, within an hour after the inflammatory process is initiated, phagocytes appear on the scene. Phagocytes include neutrophils (microphages) and monocytes (develop into macrophages); both are types of white blood cells. As the flow of blood decreases, neutrophils begin to stick to the inner surface of the endothelium (lining) of blood vessels. This is called **margination.** Then, the neutrophils begin to squeeze through the wall of the blood vessel to reach the damaged area. This migration, which resembles amoeboid movement, is called **diapedesis** (di'-a-pe-DĒ-sis). It can take as little as two minutes (Figure 4-3b). The movement of neutrophils depends on **chemotaxis** (kē'-mō-TAK-sis; *chemo* = chemicals, *taxis* = arrangement), the attraction of neutrophils by certain chemicals. Neutrophils are attracted by microbes, kinins, and other neutrophils. A steady stream of neutrophils is ensured by the production and release of additional cells from bone marrow. This is brought about by a substance called **leucocytosis-promoting factor,** which is released from inflamed tissues. Neutrophils attempt to destroy the invading microbes by phagocytosis (Figure 4-4).

As the inflammatory response continues, monocytes follow the neutrophils into the infected area. Once in the tissue, monocytes become transformed into **wandering macrophages.** They are so named because they leave the blood and migrate through tissues to infected areas. Other macrophages, called **fixed macrophages** or **histiocytes** (HIS-tē-ō-sīts), are found in the liver (stellate reticuloendothelial cells), lungs (alveolar macrophages), brain (microglia), spleen, lymph nodes, subcutaneous tissue, and bone marrow. Since fixed macrophages can be either reticular (forming a supporting network) or endothelial (lining a sinus), they are referred to as cells of the **reticuloendothelial system (RES).** Wandering macrophages augment the phagocytic activ-

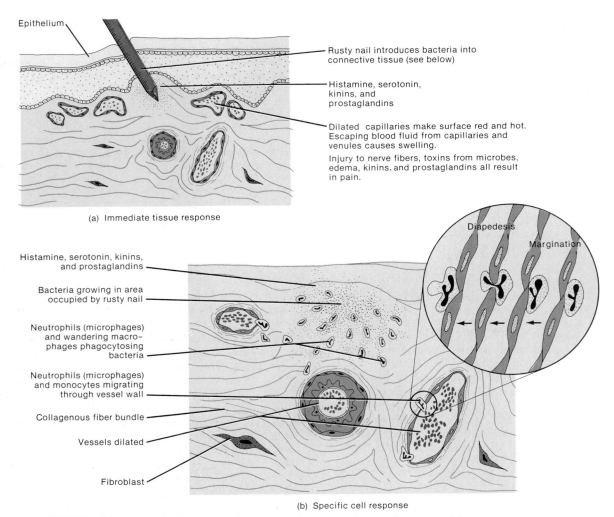

Epithelium

Rusty nail introduces bacteria into connective tissue (see below)

Histamine, serotonin, kinins, and prostaglandins

Dilated capillaries make surface red and hot. Escaping blood fluid from capillaries and venules causes swelling.

Injury to nerve fibers, toxins from microbes, edema, kinins, and prostaglandins all result in pain.

(a) Immediate tissue response

Histamine, serotonin, kinins, and prostaglandins

Bacteria growing in area occupied by rusty nail

Neutrophils (microphages) and wandering macrophages phagocytosing bacteria

Neutrophils (microphages) and monocytes migrating through vessel wall

Collagenous fiber bundle

Vessels dilated

Fibroblast

Diapedesis

Margination

(b) Specific cell response

FIGURE 4-3 Tissue and cell response to an injury. (a) Immediate tissue response. (b) Specific cell response.

Pseudopodium   White blood cell

Pseudopodium

Microbe

(a)   (b)

FIGURE 4-4 Phagocytosis. (a) Color-enhanced photomicrograph of an earlier stage of phagocytosis in which a human white blood cell (neutrophil) is engulfing a microbe. (b) Color-enhanced photomicrograph of a later stage of phagocytosis in which the engulfed microbe is being destroyed. (Courtesy, Abbott Laboratories.)

ity of fixed macrophages. In addition, fixed macrophages mobilize under the stimulus of inflammation and, as wandering macrophages, also migrate to the infected area. The neutrophils predominate in the early stages of infection, but tend to die off rapidly. Macrophages enter the picture during a later stage of the infection. They are several times more phagocytic than neutrophils and large enough to engulf tissue that has been destroyed, neutrophils that have been destroyed, and invading microbes.

3. Release of nutrients. Nutrients stored in the body are used to support the defensive cells. They are also used in the increased metabolic reactions of the cells under attack.

4. Fibrin formation. The blood contains a soluble protein called **fibrinogen** (fī-BRIN-ō-jen). The increased permeability of capillaries causes leakage of fibrinogen to tissues. Fibrinogen is then converted to an insoluble, thick network called **fibrin** (FĪ-brin), which localizes and traps the invading organisms, preventing their spread. This network eventually forms a fibrin clot that prevents hemorrhage and isolates the infected area.

5. Pus formation. In all but very mild inflammations, **pyogenesis** (*pyo* = pus; *genesis* = to produce) occurs. **Pus** is a thick fluid that contains living, as well as nonliving, white blood cells plus debris from other dead tissue.

CLINICAL APPLICATION

If the pus cannot drain out of the body, an abscess develops. An **abscess** is simply an excessive accumulation of pus in a confined space. Common examples are pimples and boils. When inflamed tissue is shed many times, it produces an open sore, called an **ulcer,** on the surface of an organ or tissue. Ulcers may result from a prolonged inflammatory response to a continuously injured tissue. For instance, overproduction of digestive acids in the stomach may cause a steady erosion of the epithelial tissue lining the stomach. People with poor circulation are susceptible to ulcers in the tissues of their legs. The ulcers develop when the tissues are continuously damaged by a shortage of oxygen and nutrients.

TISSUE REPAIR

Tissue repair, the process by which tissues replace dead or damaged cells, begins during the active phase of inflammation, but it can be completed only after all harmful substances have been neutralized or removed from the site of the injury. New cells originate by cell duplication from the **stroma,** the supporting connective tissue, or from the **parenchyma,** cells which form the organ's functioning part. The epithelial cells that secrete and absorb are the parenchymal cells of the intestine, for example. The restoration of an injured organ or tissue to normal structure and function depends entirely on which type of cell—parenchymal or stromal—is active in the repair. If only parenchymal elements accomplish the repair, a perfect or near-perfect reconstruction of the injured tissue may occur. However, if the fibroblast cells of the stroma are active in the repair, the tissue will be replaced with new connective tissue called *scar tissue.* This condition is known as **fibrosis.** Since scar tissue is not specialized to perform the functions of the parenchymal tissue, the function of the tissue is impaired.

CLINICAL APPLICATION

The scar tissue formed by fibrosis can cause the abnormal joining of tissues called **adhesions.** They are common in the abdomen and may occur around a site of previous inflammation such as an inflamed appendix, or they can follow surgery. Such adhesions make subsequent surgery more difficult.

The cardinal factor in tissue repair lies in the capacity of parenchymal tissue to regenerate. This capacity, in turn, depends on the ability of the parenchymal cells to replicate quickly.

REPAIR PROCESS

If injury to a tissue is slight, repair may sometimes be accomplished with the drainage and reabsorption of pus, followed by parenchymal regeneration. When the area of skin loss is great, fluid moves out of the capillaries and the area becomes dry. Fibrin seals the open tissue by hardening into a **scab.**

When tissue and cell damage are extensive and severe, as in large, open wounds, both the connective tissue stroma and the parenchymal cells are active in repair. This repair involves the rapid cell division of many fibroblasts, the manufacture of new collagenous fibers to provide strength, and an increase by cell division of the number of small blood vessels in the area. All these processes create an actively growing, connective tissue called **granulation tissue.** This new granulation tissue forms across a wound or surgical incision to provide a framework (stroma). The framework supports the epithelial cells that migrate into the open area and fill it. The newly formed granulation tissue also secretes a fluid that kills bacteria.

CONDITIONS AFFECTING REPAIR

Three factors affect tissue repair: nutrition, blood circulation, and age. Nutrition is vital in the healing process since a great demand is placed on the body's store of nutrients. Protein-rich diets are important since most of the cell structure is made from proteins. Vitamins also play a direct role in wound healing. Among the vitamins involved and their roles in wound healing are the following:

1. Vitamin A is essential in the replacement of epithelial tissues, especially in the respiratory tract.

2. The B vitamins—thiamine, nicotinic acid, riboflavin—are needed by many enzyme systems in cells. They are needed especially for the enzymes involved in decomposing glucose to CO_2 and H_2O which is crucial to both heart and nervous tissue. These vitamins may relieve pain in some cases.

3. Vitamin C directly affects the normal production and maintenance of intercellular substances. It is required for the manufacture of cementing elements of connective tissues, especially collagen. Vitamin C also strengthens and promotes the formation of new blood vessels. With vitamin C deficiency, even superficial wounds fail to heal, and the walls of the blood vessels become fragile and are easily ruptured.

4. Vitamin D is necessary for the proper absorption of calcium from the intestine. Calcium gives bones their hardness and is necessary for the healing of fractures.

5. Vitamin E is believed to promote healing of injured tissues and may prevent scarring.

6. Vitamin K assists in the clotting of blood and thus prevents the injured person from bleeding to death.

In tissue repair, proper blood circulation is indispensable. It is the blood that transports oxygen, nutrients, antibodies, and many defensive cells to the site of injury. The blood also plays an important role in the removal of tissue fluid, blood cells that have been depleted of oxygen, bacteria, foreign bodies, and debris. These elements would otherwise interfere with healing.

Generally, tissues heal faster and leave less obvious scars in the young than in the aged. The young body is generally in a much better nutritional state, its tissues have a better blood supply, and the cells of younger people have a faster metabolic rate. Thus cells can duplicate their materials and divide more quickly.

STUDY OUTLINE

Types of Tissues (p. 82)

1. A tissue is a group of similar cells and their intercellular substance specialized for a particular function.
2. Depending on their function and structure, the various tissues of the body are classified into four principal types: epithelial, connective, muscular, and nervous.

Epithelial Tissue (p. 82)

1. Epithelium has many cells, little intracellular material, and no blood vessels (avascular). It is attached to connective tissue by a basement membrane. It can replace itself.
2. The subtypes of epithelium include covering and lining epithelium and glandular epithelium.

Covering and Lining Epithelium

1. Layers are arranged as simple (one layer), stratified (several layers), and pseudostratified (one layer that appears as several); cell shapes include squamous (flat), cuboidal (cubelike), columnar (rectangular), and transitional (variable).
2. Simple squamous epithelium is adapted for diffusion and filtration and is found in lungs and kidneys. Endothelium lines the heart and blood vessels. Mesothelium lines the thoracic and abdominopelvic cavities and covers the organs within them.
3. Simple cuboidal epithelium is adapted for secretion and absorption. It is found covering ovaries, in kidneys and eyes, and lining some glandular ducts.
4. Nonciliated simple columnar epithelium lines most of the digestive tract. Specialized cells containing microvilli perform absorption. Goblet cells perform secretion of mucus. In a few portions of the respiratory tract, the cells are ciliated to move foreign particles trapped in mucus out of the body.
5. Stratified squamous epithelium is protective. It lines the upper digestive tract and vagina and forms the outer layer of skin.
6. Stratified cuboidal epithelium is found in adult sweat glands, portion of urethra, pharynx, and epiglottis.
7. Stratified columnar epithelium protects and secretes. It is found in the male urethra and large excretory ducts.
8. Transitional epithelium lines the urinary bladder and is capable of stretching.
9. Pseudostratified epithelium has only one layer but gives the appearance of many. It lines larger excretory ducts, parts of urethra, auditory tubes, and most upper respiratory structures, where it protects and secretes.

Glandular Epithelium

1. A gland is a single cell or a mass of epithelial cells adapted for secretion.
2. Exocrine glands (sweat, oil, and digestive glands) secrete into ducts or directly onto a free surface.
3. Structural classification includes unicellular and multicellular glands; multicellular glands are further classified as tubular, acinar, tubuloacinar, simple, and compound.
4. Functional classification includes holocrine, merocrine, and apocrine glands.
5. Endocrine glands secrete hormones directly into the blood.

Connective Tissue (p. 90)

1. Connective tissue is the most abundant body tissue. It has few cells, an extensive intercellular substance, and a rich blood supply (vascular), except for cartilage. It does not occur on free surfaces.
2. The intercellular substance determines the tissue's qualities.
3. Connective tissue protects, supports, and binds organs together.
4. Connective tissue is classified into two principal types: embryonic and adult.

Embryonic Connective Tissue

1. Mesenchyme forms all other connective tissues.
2. Mucous connective tissue is found in the umbilical cord of the fetus, where it gives support.

Adult Connective Tissue

1. Adult connective tissue is connective tissue that exists in the newborn and that does not change after birth. It is subdivided into several kinds: connective tissue proper, cartilage, bone tissue, and vascular tissue.
2. Connective tissue proper has a more or less fluid intercellular material, and a typical cell is the fibroblast. Five examples of such tissues may be distinguished.
3. Loose (areolar) connective tissue is one of the most widely distributed connective tissues in the body. Its intercellular substance (hyaluronic acid) contains fibers (collagenous, elastic, and reticular) and various cells (fibroblasts, macrophages, plasma, mast, and melanocytes). Loose connective tissue is found in all mucous membranes, around body organs, and in the subcutaneous layer.
4. Adipose tissue is a form of loose connective tissue in which the cells, called adipocytes, are specialized for fat storage. It is found in the subcutaneous layer and around various organs.
5. Dense (collagenous) connective tissue has a close packing of fibers (regularly or irregularly arranged). It is found as a component of fascia, membranes of organs, tendons, ligaments, and aponeuroses.
6. Elastic connective tissue has a predominance of freely branching elastic fibers that give it a yellow color. It is found in the cartilages of the larynx, elastic arteries, trachea, bronchial tubes, and true vocal cords.

7. Reticular connective tissue consists of interlacing reticular fibers and forms the stroma of the liver, spleen, and lymph nodes.
8. Cartilage has a jellylike matrix containing collagenous and elastic fibers and chondrocytes.
9. The growth of cartilage is accomplished by interstitial growth (from within) and appositional growth (from without).
10. Hyaline cartilage is found in the embryonic skeleton, at the ends of bones, in the nose, and in respiratory structures. It is flexible, allows movement, and provides support.
11. Fibrocartilage connects the pelvic bones and the vertebrae. It provides strength.
12. Elastic cartilage maintains the shape of organs such as the larynx, auditory tubes, and external ear.

Muscle Tissue and Nervous Tissue (p. 98)
1. Muscle tissue performs one major function: contraction.
2. Nervous tissue is specialized to conduct electrical impulses.

Membranes (p. 98)
1. An epithelial membrane is an epithelial layer overlying a connective tissue layer. Examples are mucous, serous, and cutaneous.
2. Mucous membranes line cavities that open to the exterior, such as the digestive tract.
3. Serous membranes (pleura, pericardium, peritoneum) line closed cavities and cover the organs in the cavities. These membranes consist of parietal and visceral portions.
4. The cutaneous membrane is the skin.
5. Synovial membranes line joint cavities and do not contain epithelium.

Tissue Inflammation: An Attempt to Restore Homeostasis (p. 98)
1. Damage to a tissue causes an inflammatory response characterized by redness, pain, heat, and swelling; sometimes loss of function occurs.
2. The inflammatory response is initiated by histamine, serotonin, kinins, and prostaglandins released by damaged tissue. They cause vasodilation and increased permeability of blood vessels.
3. Further cell injury is prevented by phagocytes. These include neutrophils (microphages) and macrophages.
4. The role of fibrin is to isolate the infected area.
5. In most inflammations, pus is produced; if it cannot drain out of the body, an abscess develops.

Tissue Repair (p. 101)
1. Tissue repair is the replacement of damaged or destroyed cells by healthy ones.
2. It begins during the active phase of inflammation and is not completed until after harmful substances in the inflamed area have been neutralized or removed.

Repair Process
1. If the injury is superficial, tissue repair involves pus removal (if pus is present), scab formation, and parenchymal regeneration.
2. If damage is extensive, granulation tissue is involved.

Conditions Affecting Repair
1. Nutrition is important to tissue repair. Various vitamins (A, some B, D, C, E, and K) and a protein-rich diet are needed.
2. Adequate circulation of blood is needed.
3. The tissues of young people repair rapidly and efficiently; the process slows down with aging.

REVIEW QUESTIONS

1. Define a tissue. What are the four basic kinds of human tissue?
2. Distinguish covering and lining epithelium from glandular epithelium. What characteristics are common to all epithelium?
3. Describe the origin and composition of the basement membrane.
4. Describe the various layering arrangements and cell shapes of epithelium.
5. How is epithelium classified? List the various types.
6. For each of the following kinds of epithelium, briefly describe the microscopic appearance, location in the body, and functions: simple squamous, simple cuboidal, simple columnar, stratified squamous, stratified cuboidal, stratified columnar, transitional, and pseudostratified.
7. Define the following terms: endothelium, mesothelium, secretion, absorption, goblet cell, and keratin.
8. What is a gland? Distinguish between endocrine and exocrine glands.
9. Describe the classification of exocrine glands according to structure and according to function and give at least one example of each.
10. Enumerate the ways in which connective tissue differs from epithelium.
11. How are connective tissues classified? List the various types.
12. How are embryonic connective tissue and adult connective tissue distinguished?
13. Describe the following connective tissues with regard to microscopic appearance, location in the body, and function: loose (areolar), adipose, dense (collagenous), elastic, reticular, hyaline cartilage, fibrocartilage, and elastic cartilage.
14. Distinguish between the interstitial and appositional growth of cartilage.
15. Define the following terms: hyaluronic acid, collagenous fiber, elastic fiber, reticular fiber, fibroblast, macrophage, plasma cell, mast cell, melanocyte, adipocyte, chondrocyte, and lacuna.
16. Define the following kinds of membranes: mucous, serous, cutaneous, and synovial. Where is each located in the body? What are their functions?
17. Following are some descriptions of various tissues of the body. For each description, name the tissue described.
    a. An epithelium that permits distention (stretching).
    b. A single layer of flat cells concerned with filtration and absorption.
    c. Forms all other kinds of connective tissue.
    d. Specialized for fat storage.
    e. An epithelium with waterproofing qualities.
    f. Forms the framework of many organs.
    g. Produces perspiration, wax, oil, or digestive enzymes.
    h. Cartilage that shapes the external ear.
    i. Contains goblet cells and lines the intestine.
    j. Most widely distributed connective tissue.
    k. Forms tendons, ligaments, and aponeuroses.
    l. Specialized for the secretion of hormones.
    m. Provides support in the umbilical cord.

n. Lines kidney tubules and is specialized for absorption and secretion.

o. Permits extensibility of lung tissue.

18. Define an inflammation. In what sense is the inflammatory response protective?

19. Describe the principal symptoms associated with inflammation.

20. How are vasodilation and increased permeability of blood vessels initiated during inflammation? Why are they important? What causes edema?

21. Describe the mechanism and importance of phagocytic migration during inflammation.

22. What is the role of fibrin in inflammation?

23. What is an abscess?

24. What is meant by tissue repair?

25. Distinguish between stromal and parenchymal repair.

26. What is the importance of granulation tissue?

27. What conditions affect tissue repair?

● Define the integumentary system.

● Describe the various functions of the skin.

● List the various layers of the epidermis and describe their structure and functions.

● Describe the composition and function of the dermis.

● Explain the basis for skin color.

● Explain the basic pattern of epidermal ridges and grooves.

● Describe the development, distribution, structure, color, growth, and replacement of hair.

● Compare the structure, distribution, and functions of sebaceous (oil), sudoriferous (sweat), and ceruminous glands.

● List the parts of a nail and describe their composition.

● Explain the role of the skin in helping to maintain normal body temperature.

● Describe the causes and effects for the following skin disorders: pruritus, acne, impetigo, systemic lupus erythematosus (SLE), psoriasis, decubitus ulcers, warts, cold sores, sunburn, and skin cancer.

● Define a burn and list the systemic effects of a burn.

● Classify burns into first, second, and third degrees.

● Describe how to estimate the extent of a burn.

● Define medical terminology associated with the integumentary system.

● Explain the actions of selected drugs associated with the integumentary system.

5 The Integumentary System

An aggregation of tissues that performs a specific function is an **organ.** The next higher level of organization is a **system**—a group of organs operating together to perform specialized functions. The skin and its derivatives, such as hair, nails, glands, and several specialized receptors, constitute the **integumentary** (in'-teg-yoo-MEN-tar-ē) **system** of the body.

SKIN

The **skin** or **cutis** is an organ because it consists of tissues structurally joined together to perform specific activities. It is not just a simple thin covering that keeps the body together and gives it protection. The skin is quite complex in structure and performs several functions essential for survival.

The skin is one of the larger organs of the body in terms of surface area. For the average adult, the skin occupies a surface area of approximately 19,355 sq cm (3,000 sq inches). It varies in thickness from 0.5 to 3 mm (0.02 to 0.12 inch). In general, skin is thicker on the dorsal surface of a part of the body than it is on the ventral surface, but this condition is reversed in the hand and the foot. The skin of the palmar surface of the hand and the plantar surface of the foot is thicker than on their dorsal aspects.

FUNCTIONS

The skin covers the body and protects the underlying tissues from bacterial invasion, drying out, and harmful light rays. It helps to maintain body temperature; prevents excessive loss of inorganic and organic materials; receives stimuli from the environment; stores chemical compounds; excretes water, salts, and several organic compounds; and synthesizes the hormone, vitamin D. The role of the skin in receiving various stimuli is considered in detail in Chapter 15.

STRUCTURE

Structurally, the skin consists of two principal parts (Figure 5-1). The outer, thinner portion, which is composed of epithelium, is called the **epidermis.** The epidermis is cemented to the inner, thicker, connective tissue part called the **dermis.** Thick skin has a relatively thick epidermis, whereas thin skin has a relatively thin epidermis. Beneath the dermis is a **subcutaneous layer.** This layer, also called the **superficial fascia** or **hypodermis,** consists of areolar and adipose tissues. Fibers from the dermis extend down into the superficial fascia and anchor the skin to the subcutaneous layer. The superficial fascia, in turn, is firmly attached to underlying tissues and organs.

EPIDERMIS

The **epidermis** is composed of *Langerhans' cells* (probably macrophages that have invaded the epidermis) and stratified squamous epithelium organized in four or five cell layers, depending on its location in the body (see Figure 5-1 and Figure 5-2). Where exposure to friction is greatest, such as the palms and soles, the epidermis has five layers.

In all other parts it has four layers. The names of the five layers from the deepest to the most superficial are as follows.

1. Stratum basale. This single layer of cuboidal to columnar cells is capable of continued cell division. As these cells multiply, they push up toward the surface and become part of the layers to be described next. Their nuclei degenerate, and the cells die. Eventually, the cells are shed from the top layer of the epidermis.

2. Stratum spinosum. This layer of the epidermis contains eight to ten rows of polyhedral (many-sided) cells that fit closely together. The surfaces of these cells may assume a prickly appearance when prepared for microscope examination (*spinosum* = prickly). The stratum basale and stratum spinosum are sometimes collectively referred to as the **stratum germinativum** (jer'-mi-na-TĒ-vum) to indicate the layers where new cells are germinated. The deeper layers of the epidermis of hairless skin contain nerve endings sensitive to touch called *Merkel's (tactile) discs* (see Figure 15-1).

3. Stratum granulosum. The third layer of the epidermis consists of three to five rows of flattened cells that contain darkly staining granules of a substance called *keratohyalin* (ker'-a-tō-HĪ-a-lin). This compound is involved in the first step of keratin formation. *Keratin* is a waterproofing protein found in the top layer of the epidermis. The nuclei of the cells in the stratum granulosum are in various stages of degeneration. As these nuclei break down, the cells are no longer capable of carrying out vital metabolic reactions and die.

4. Stratum lucidum. This layer is quite pronounced in the thick skin of the palms and soles. It consists of several rows of clear, flat, dead cells that contain droplets of a substance called *eleidin* (el-Ē-i-din). The layer is so named because eleidin is translucent (*lucidum* = clear). Eleidin is formed from keratohyalin and is eventually transformed to keratin.

5. Stratum corneum. This layer consists of 25 to 30 rows of flat, dead cells completely filled with keratin. These cells are continuously shed and replaced. The stratum corneum serves as an effective barrier against light and heat waves, bacteria, and many chemicals.

DERMIS

The second principal part of the skin, the **dermis,** is composed of connective tissue containing collagenous and elastic fibers (see Figure 5-1). The dermis is very thick in the palms and soles and very thin in the eyelids, penis, and scrotum. It also tends to be thicker on the dorsal aspects of the body than the ventral and thicker on the lateral aspects of extremities than medial aspects. Numerous blood vessels, nerves, glands, and hair follicles are embedded in the dermis.

The upper region of the dermis, about one-fifth of the thickness of the total layer, is named the **papillary region** or **layer.** Its surface area is greatly increased by small, fingerlike projections called **dermal papillae** (pa-PIL-ē). These structures project into the concavities between ridges in the deep surface of the epidermis and many contain loops of capillaries. Some dermal papillae contain *Meissner's* (MĪS-nerz) *corpuscles,* nerve endings sensitive to touch. The papillary region consists of loose connective tissue containing fine elastic fibers.

The remaining portion of the dermis is called the **reticular region** or **layer.** It consists of dense, irregularly arranged connective tissue containing interlacing bundles of collagenous and coarse elastic fibers. It is named the

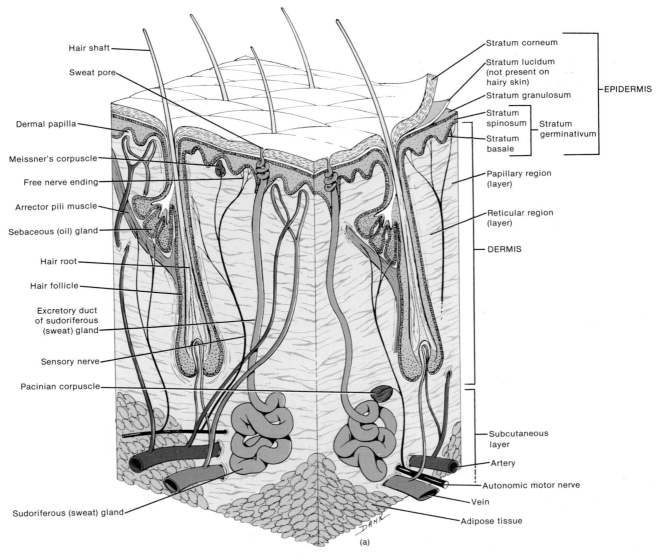

Hair shaft

Sweat pore

Dermal papilla

Meissner's corpuscle

Free nerve ending

Arrector pili muscle

Sebaceous (oil) gland

Hair root

Hair follicle

Excretory duct
of sudoriferous
(sweat) gland

Sensory nerve

Pacinian corpuscle

Sudoriferous (sweat) gland

Stratum corneum

Stratum lucidum
(not present on
hairy skin)

Stratum granulosum

Stratum
spinosum

Stratum
basale

Stratum
germinativum

EPIDERMIS

Papillary region
(layer)

Reticular region
(layer)

DERMIS

Subcutaneous
layer

Artery

Autonomic motor nerve

Vein

Adipose tissue

(a)

FIGURE 5-1 Skin. (a) Structure of the skin and underlying subcutaneous layer. (b) Scanning electron micrograph of the skin and several hairs at a magnification of 260×. (From *Tissues and Organs: A Text-Atlas of Scanning Electron Microscopy,* by Richard K. Kessel and Randy H. Kardon. Copyright © 1979 by Scientific American, Inc.)

Hair shafts

EPIDERMIS

DERMIS

Blood
vessels

(b)

FIGURE 5-2 Photomicrograph of thick skin at a magnification of 50×. The stratum lucidum is not shown here, but is illustrated in Figure 5-1a. (© 1983 by Michael H. Ross. Used by permission.)

reticular layer because the bundles of collagenous fibers interlace in a netlike manner. Spaces between the fibers are occupied by a small quantity of adipose tissue, hair follicles, nerves, oil glands, and the ducts of sweat glands. Varying thicknesses of the reticular region are responsible for differences in the thickness of the skin.

The combination of collagenous and elastic fibers in the reticular region provides the skin with strength, extensibility, and elasticity. (Extensibility is the ability to stretch; elasticity is the ability to return to original shape after extension or contraction.) The ability of the skin to stretch can readily be seen during conditions of pregnancy, obesity, and edema. The small tears that occur during extreme stretching are initially red and remain visible afterward as silvery white streaks called *striae* (STRĪ-ē).

The reticular region is attached to underlying organs, such as bone and muscle, by the subcutaneous layer. The subcutaneous layer also contains nerve endings called *Pacinian* (pa-SIN-ē-an) *corpuscles* that are sensitive to pressure (see Figure 15-1).

CLINICAL APPLICATION
The collagenous fibers in the dermis run in all directions, but in particular regions of the body they tend to run more in one direction than another. The predominant direction of the underlying collagenous fibers is indicated in the skin by **lines of cleavage (tension lines).** The lines are especially evident on the palmar surfaces of the fingers where they run parallel to the long axis of the digit. The characteristic lines for each part of the body are shown in Figure 5-3.

Lines of cleavage are of particular interest to a surgeon because an incision running parallel to the collagen fibers will heal with only a fine scar. An incision made across the rows of fibers disrupts the collagen, and the wound tends to gape open and heal in a broad, thick scar.

SKIN COLOR

The color of the skin is due to melanin, a pigment in the epidermis; carotene, a pigment mostly in the dermis; and blood in the capillaries in the dermis. The amount of **melanin** varies the skin color from pale yellow to black. This pigment is found primarily in the basale and spinosum layers. Melanin is synthesized in cells called **melanocytes** (me-LAN-ō-sīts), located either just beneath or between cells of the stratum basale. Since the number of melanocytes is about the same in all races, differences in skin color are due to the amount of pigment the melanocytes produce and disperse. An inherited inability of an individual in any race to produce melanin results in **albinism** (AL-bi-nizm). The pigment is absent in the hair and eyes as well as the skin. An individual affected with albinism is called an **albino** (al-BĪ-no). In some people, melanin tends to form in patches called **freckles.**

Melanocytes synthesize melanin from the amino acid *tyrosine* in the presence of an enzyme called *tyrosinase*. Exposure to ultraviolet radiation increases the enzymatic activity of melanocytes and leads to increased melanin production. The cell bodies of melanocytes send out long processes between epidermal cells. Upon contact with the processes, epidermal cells take up the melanin by phagocytosis. When the skin is again exposed to ultraviolet radiation, both the amount and the darkness of melanin increase, tanning and further protecting the body against radiation. Thus melanin serves a vital protective function. In mammals, the melanocyte-stimulating hormone (MSH) produced by the anterior pituitary gland causes increased melanin synthesis and dispersion through the epidermis. The exact role of MSH in humans is not clear.

CLINICAL APPLICATION
Overexposure of the skin to the ultraviolet light of the sun may lead to skin cancer. Among the most malignant and lethal skin cancers is **melanoma** (*melano* = dark-colored; *oma* = tumor), cancer of the melanocytes. Fortunately, most skin cancers are operable.

FIGURE 5-3 Lines of cleavage (tension lines) seen in (a) anterior and (b) posterior views.

(a)

(b)

Another pigment, called **carotene** (KAR-o-tēn), is found in the stratum corneum and fatty areas of the dermis in Oriental people. Together carotene and melanin account for the yellowish hue of their skin.

The pink color of Caucasian skin is due to blood in capillaries in the dermis. The redness of the vessels is not heavily masked by pigment. The epidermis has no blood vessels, a characteristic of all epithelia.

EPIDERMAL RIDGES AND GROOVES

The outer surface of the skin of the palms and fingers and soles and toes is marked by a series of ridges and grooves which appear either as fairly straight lines or as a pattern of loops and whorls, as on the tips of the digits.

The **epidermal ridges** develop during the third and fourth fetal months as the epidermis conforms to the contours of the underlying dermal papillae (see Figure 5-1a). Since the ducts of sweat glands open on the summits of the epidermal ridges, fingerprints (or footprints) are left when a smooth object is touched. The ridge pattern, which is genetically determined, is unique for each individual. It does not change throughout life, except to enlarge, and thus can serve as the basis for identification through fingerprints or footprints. The function of the ridges is to increase the grip of the hand or foot by increasing friction.

Epidermal grooves on other parts of the skin divide the surface into a number of diamond-shaped areas. Examine the dorsum of the hand as an example. Note that hairs typically emerge at the points of intersection of the grooves. Note also that the grooves increase in frequency and depth as regions of free joint movement are approached.

CLINICAL APPLICATION

Scientists have now made it possible to improve in many cases the appearance of scars from deep acne, chickenpox, burns, cleft lip, and other disorders and to make age-related wrinkles disappear. The substance that makes this possible is called **Zyderm Collagen Implant.** Collagen is the body's principal structural material. It is found in skin, bone, cartilage, tendons, ligaments, and various viscera and accounts for almost one-third of the total protein of the body.

Zyderm Collagen Implant is prepared from cattle collagen that is suspended in a saline solution containing lidocaine, a local anesthetic. Once injected into the skin, the solution disappears and the collagen, under the influence of body temperature, becomes a stationary fleshlike substance that is incorporated into the skin. The collagen implant becomes colonized by blood vessels and cells and acts as a natural structural framework in the skin.

EPIDERMAL DERIVATIVES

Structures developed from the embryonic epidermis—hair, glands, nails—perform functions that are necessary and sometimes vital. Hair and nails protect the body. The sweat glands help regulate body temperature.

HAIR

Growths of the epidermis variously distributed over the body are **hairs** or **pili.** The primary function of hair is protection. Though the protection is limited, hair guards the scalp from injury and the sun's rays. Eyebrows and eyelashes protect the eyes from foreign particles. Hair in the nostrils and external ear canal protects these structures from insects and dust.

Development and Distribution

During the third and fourth months of fetal life, the epidermis develops downgrowths into the dermis called **hair follicles.** Originating from these follicles are small hairs. By the fifth or six month, the follicles produce delicate hairs called **lanugo** (lan-YOO-gō; *lana* = wool) that covers the fetus. The lanugo is usually shed prior to birth, except in the regions of the eyebrows, eyelids, and scalp. Here they persist and become stronger. Several months after birth, these hairs are shed and are replaced by still coarser ones, while the remainder of the body develops a new hair growth called the **vellus** (fleece). At puberty, coarse hairs develop in the axillary and pubic regions of both sexes and on the face and to a lesser extent on

other parts of the body in the male. The coarse hairs that develop at puberty, plus the hairs of the scalp and eyebrows, are referred to as **terminal hairs.**

Hairs are distributed on nearly all parts of the body. It has been estimated that an average adult has about five million hairs, of which about 100,000 are in the scalp. They are absent from the palms of the hands, the soles of the feet, the dorsal surfaces of the distal phalanges, lips, nipples, clitoris, glans penis, inner surface of the prepuce, inner surfaces of the labia majora and minora, and outer surfaces of the labia minora. Straight hairs are oval or cylindrical in cross section, whereas curly hairs are flattened. Straight hairs are stronger than curly ones.

Structure

Each hair consists of a shaft and a root (Figure 5-4a). The **shaft** is the superficial portion, most of which projects above the surface of the skin. The shaft of coarse hairs consists of three principal parts. The inner **medulla** is composed of rows of polyhedral cells containing granules of eleidin and air spaces. The medulla is poorly developed or not present at all in fine hairs. The second principal part of the shaft is the middle **cortex.** It forms the major part of the shaft and consists of elongated cells that contain pigment granules in dark hair, but mostly air in white hair. The **cuticle of the hair,** the outermost layer, consists of a single layer of thin, flat, scalelike cells that are the most heavily keratinized. They are arranged like shingles on the side of a house, but the free edges of the cuticle cells point upward rather than downward like shingles (Figure 5-4e). The **root** is the portion below the surface that penetrates into the dermis and even into the subcutaneous layer and, like the shaft, contains a medulla, cortex, and cuticle (Figure 5-4a,b).

Surrounding the root is the **hair follicle,** which is made up of an external zone of epithelium (the external root sheath) and an internal zone of epithelium (the internal root sheath). The **external root sheath** is a downward continuation of the basale and spinosum layers of the epidermis. Near the surface it contains all the epidermal layers. As it descends, it does not exhibit the superficial epidermal layers. At the bottom of the hair follicle, the external root sheath contains only the stratum basale. The **internal root sheath** is formed from proliferating cells of the matrix (described shortly) and takes the form of a cellular tubular sheath deep to the external root sheath. The internal root sheath extends only partway up the follicle and consists of (1) an inner layer, the **cuticle of the internal root sheath,** which is a single layer of flattened cells with atrophied nuclei, (2) a middle **granular layer,** which is one to three layers of flattened nucleated cells, and (3) an outer **pallid layer,** which is a single layer of cuboidal cells with flattened nuclei.

The base of each follicle is enlarged into an onion-shaped structure, the **bulb.** This structure contains an indentation, the **papilla of the hair,** filled with loose connective tissue. The papilla of the hair contains many blood vessels and provides nourishment for the growing hair. The bulb also contains a region of cells called the **matrix,** a germinal layer. The cells of the matrix produce new hairs by cell division when older hairs are shed. This replacement occurs within the same follicle.

Sebaceous glands and a bundle of smooth muscle are also associated with hair. Details of the sebaceous glands are discussed shortly. The smooth muscle is called **arrector pili;** it extends from the dermis of the skin to the side of the hair follicle (Figure 5-4a). In its normal position hair is arranged at an angle to the surface of the skin. The arrectores pilorum muscles contract under stresses of fright and cold and pull the hairs into a vertical position. This contraction results in "goosebumps" or "gooseflesh" because the skin around the shaft forms slight elevations.

Around each hair follicle are nerve endings, called *root hair plexuses,* that are sensitive to touch (see Figure 15-1). They respond if a hair shaft is moved.

Color

The color of hair is due primarily to melanin. It is formed by melanocytes distributed in the matrix of the bulb of the follicle. There are only three colors of hair pigment—black, brown, and yellow, due to the pigments black melanin, brown melanin, and pheomelanin, respectively. The many variations in hair color are combinations of different amounts of the three pigments. Graying of hair is the loss of pigment believed to be the result of a progressive inability of the melanocytes to make tyrosinase, the enzyme necessary for the synthesis of melanin. White hair results from air in the medullary shaft.

FIGURE 5-4 Hair. (a) Principal parts of a hair and associated strutures. (b) Longitudinal section of a hair root.

EPIDERMIS

DERMIS

Opening of hair follicle containing sebum from sebaceous (oil) gland

Portion of hair shaft

Hair follicle

Hair follicle

Sebum in hair follicle

Secretory portions of sebaceous (oil) glands

Sebaceous (oil) glands

Site of gland opening into follicle

(c)

(d)

(e)

Growth and Replacement

Hair replacement occurs according to a cyclic pattern, alternating between growing and resting periods. During the growing phase of the cycle, the cells of the matrix are active. They increase in number by cell division and are pushed upward and eventually die, a situation similar to epidermal growth. The product is a hair, which is essentially dead protein tissue. Hair grows about 1 mm (0.04 inch) every 3 days.

FIGURE 5-4 (*Continued*) (c) Photomicrograph of thin skin of the face at a magnification of 60×. (© 1983 by Michael H. Ross. Used by permission.) (d) Photomicrograph of a sebaceous (oil) gland opening into a hair follicle at a magnification of 150×. (© 1983 by Michael H. Ross. Used by permission.) (e) Scanning electron micrograph of the surface of a hair shaft showing the shinglelike cuticular scales at a magnification of 1,000×. (Courtesy of Fisher Scientific Company and S.T.E.M. Laboratories, Inc., Copyright, 1975.)

During the resting phase, the matrix becomes inactive and undergoes atrophy. At this point the root of the hair detaches from the matrix and the hair slowly moves up the follicle. It may remain there for some time until pulled out, shed, or pushed up by a replacing hair.

Either before or after the hair comes out of the follicle, proliferation of cells by the external root sheath forms a new matrix. The new matrix undergoes cell division, forming a new hair that grows up the follicle and replaces the old hair. You might be interested to know that shaving or cutting the hair has no effect on its growth.

The cycle of hair growth varies in different parts of the body. In the scalp, each hair grows steadily and continuously for 2 to 6 years, growth then stops, and after 3 months, the hair is shed. After another 3 months of a resting phase, a new hair starts to grow from the same follicle. Eyebrows, by contrast, have a growing phase of only about 10 weeks. It is for this reason that these hairs do not grow very long.

Normal hair loss in an adult scalp is about 70 to 100 hairs per day. Both the rate of growth and the replacement cycle may be altered by illness, diet, and other factors. For example, high fever, major illness, major surgery, blood loss, or severe emotional stress may increase the rate of shedding. Rapid weight-loss diets involving severe restriction of calories or protein also increase hair loss. An increase in the rate of shedding can also occur for 3 to 4 months after childbirth. Certain drugs and radiation therapy are also factors in increasing hair loss.

The age of onset, degree of thinning, and ultimate hair pattern associated with **common baldness** ("male-pattern" baldness) are determined by male hormones, called androgens, and heredity. Androgens are involved in promoting normal sexual development. Genetic baldness cannot be prevented or helped by local applications of various chemicals, injections, radiation, or any other treatment.

GLANDS

Three kinds of glands associated with the skin are sebaceous glands, sudoriferous glands, and ceruminous glands.

Sebaceous (Oil) Glands

Sebaceous (se-BĀ-shus) or **oil glands,** with few exceptions, are connected to hair follicles (see Figure 5-4a,c, d). The secreting portions of the glands lie in the dermis and those glands associated with hairs open into the necks of hair follicles. Sebaceous glands not associated with hair follicles open directly onto the surface of the skin (lips, glans penis, labia minora, and meibomian glands of the eyelids). Sebaceous glands are simple branched acinar glands. Absent in the palms and soles, they vary in size and shape in other regions of the body. For example, they are small in most areas of the trunk and extremities, but large in the skin of the breasts, face, neck, and upper chest.

The sebaceous glands secrete an oily substance called **sebum** (SĒ-bum), a mixture of fats, cholesterol, proteins, and inorganic salts. Sebum helps keep hair from drying and becoming brittle, forms a protective film that prevents excessive evaporation of water from the skin, and keeps the skin soft and pliable.

CLINICAL APPLICATION
When sebaceous glands of the face become enlarged because of accumulated sebum, **blackheads** develop. Since sebum is nutritive to certain bacteria, **pimples** or **boils** often result. The color of blackheads is due to melanin and oxidized oil, not dirt.

Sudoriferous (Sweat) Glands

Sudoriferous (soo'-dor-IF-er-us; *sudor* = sweat; *ferre* = to bear) or **sweat glands** are distinguished into two principal types on the basis of structure and location. **Apocrine** sweat glands are simple, branched tubular glands. Their distribution is limited primarily to the skin of the axilla, pubic region, and pigmented areas (areolae) of the breasts. The secretory portion of apocrine sweat glands is located in the dermis and the excretory duct opens into hair follicles. Apocrine sweat glands begin to function at puberty and produce a more viscous secretion than the other type of sweat gland.

Eccrine sweat glands are much more common than apocrine sweat glands and are simple, coiled tubular glands. They are distributed throughout the skin except for the margins of the lips, nail beds of the fingers and toes, glans penis, glans clitoris, and eardrums. Eccrine sweat glands are most numerous in the skin of the palms and the soles; their density can be as high as 3,000 per square inch in the palms. The secretory portion of eccrine sweat glands is located in the subcutaneous layer and the excretory duct projects upward through the dermis and epidermis to terminate at a pore at the surface of the epidermis (see Figure 5-1). Eccrine sweat glands function throughout life and produce a secretion that is more watery than that of apocrine sweat glands.

Perspiration, or **sweat,** is the substance produced by sudoriferous glands. It is a mixture of water, salts (mostly NaCl), urea, uric acid, amino acids, ammonia, sugar, lactic acid, and ascorbic acid. Its principal function is to help maintain body temperature. It also helps to eliminate wastes.

Ceruminous Glands

In certain parts of the skin, sudoriferous glands are modified as **ceruminous** (se-ROO-mi-nus) **glands.** Such modified glands are simple, coiled tubular glands present in the external auditory meatus (canal). The secretory portions of ceruminous glands lie in the submucosa, deep to sebaceous glands, and the excretory ducts open either directly onto the surface of the external auditory meatus or into sebaceous ducts. The combined secretion of the ceruminous and sebaceous glands is called **cerumen** (*cera* = wax). Cerumen may accumulate resulting in impacted cerumen (Chapter 17).

NAILS

Hard, keratinized cells of the epidermis are referred to as **nails.** The cells form a clear, solid covering over the dorsal surfaces of the terminal portions of the fingers and toes. Each nail (Figure 5-5) consists of a **nail body,** a **free edge,** and a **nail root.** The nail body is the portion of the nail that is visible, the free edge is the part that projects beyond the distal end of the digit, and the nail root is the portion that is hidden in the nail groove. Most of the nail body is pink because of the underlying vascular tissue. The whitish semilunar area at the proximal end of the body is called the **lunula** (LOO-nyoo-la). It appears whitish because the vascular tissue underneath does not show through.

The fold of skin that extends around the proximal and lateral borders of the nail is known as the **nail fold,** and the epidermis beneath the nail constitutes the **nail bed.** The furrow between the two is the **nail groove.**

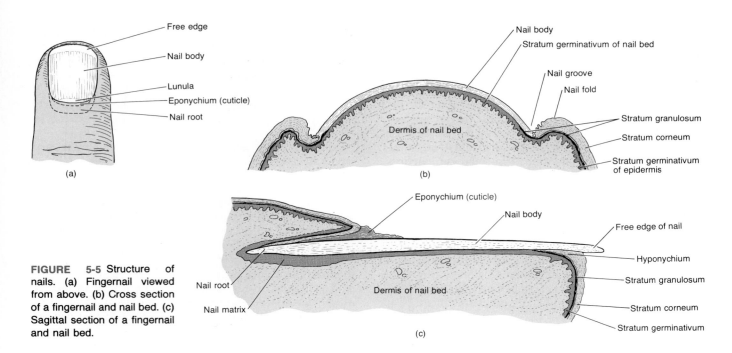

FIGURE 5-5 Structure of nails. (a) Fingernail viewed from above. (b) Cross section of a fingernail and nail bed. (c) Sagittal section of a fingernail and nail bed.

The **eponychium** (ep'-ō-NIK-ē-um) or **cuticle** is a narrow band of epidermis that extends from the margin of the nail wall (lateral border), adhering to it. It occupies the proximal border of the nail and consists of stratum corneum. The thickened area of stratum corneum below the free edge of the nail is referred to as the **hyponychium** (hī'-pō-NIK-ē-um).

The epithelium of the proximal part of the nail bed is known as the **nail matrix.** Its function is to bring about the growth of nails. Essentially, growth occurs by the transformation of superficial cells of the matrix into nail cells. In the process, the outer, harder layer is pushed forward over the stratum germinativum. The average growth in the length of fingernails is about 1 mm (0.04 inch) per week. The growth rate is somewhat slower in toenails.

HOMEOSTASIS

One of the best examples of homeostasis in humans is the maintenance of body temperature by the skin. Humans, like other mammals, are *homeotherms*—warm-blooded organisms. This means that we are able to maintain a remarkably constant body temperature of 37°C (98.6°F) even though the environmental temperature may vary over a broad range.

Suppose you are in an environment where the temperature is 37.8°C (100°F). A sequence of events is set into operation to counteract this above-normal temperature, which may be considered a stress. Sensing devices in the skin called receptors pick up the stimulus—in this case, heat—and activate nerves that send the message to your brain. A temperature-regulating area of the brain then sends nerve impulses to the sudoriferous glands, which produce more perspiration. As the perspiration evaporates from the surface of your skin, it is cooled and your body temperature is maintained. This sequence of events is shown in Figure 5-6.

Note that temperature regulation by the skin involves a feedback system because the output (cooling of the skin) is fed back to the skin receptors and becomes part of a new stimulus-response cycle. In other words, after the sudoriferous glands are activated, the skin receptors keep the brain informed about the external temperature. The brain, in turn, continues to send messages to the sudoriferous glands until the temperature returns to 37°C (98.6°F). Like most of the body's feedback systems, temperature regulation is a negative feedback system—the output, cooling, is the opposite of the original condition, overheating.

Temperature maintenance by perspiration represents only one mechanism by which we maintain normal body temperature. Other mechanisms include adjusting blood flow to the skin, regulating metabolic rate, and regulating skeletal muscle contractions. These other mechanisms are discussed in detail in Chapter 25.

FIGURE 5-6 Role of the skin in maintaining the homeostasis of body temperature.

DISORDERS: HOMEOSTATIC IMBALANCES

Pruritus

Pruritus (*pruire* = to itch), or itching, is probably one of the most common of all dermatological conditions. The causes of pruritus may be placed in three general categories: (1) cutaneous disorders, such as *xerosis* (dry skin), *miliaria* (sweat retention in the skin), and infectious agents (bacteria, fungi, viruses, and body lice); (2) systemic disorders, such as cancer, chronic renal failure, liver diseases, and thyroid diseases; (3) psychogenic factors, as might occur during or just after an episode of emotional stress.

Acne

Acne is an inflammation of sebaceous glands and usually begins at puberty. In fact, a few comedones (blackheads or whiteheads) on the face may be the first signs of approaching puberty. At puberty the sebaceous glands, under the influence of androgens (male hormones) grow in size and increase production of their complex lipid product, sebum. Although testosterone, a male hormone, appears to be the most potent circulating androgen for sebaceous cell stimulation, adrenal and ovarian androgens can stimulate sebaceous secretions as well.

Acne occurs predominantly in sebaceous follicles. The four basic types of acne lesions in order of increasing severity are comedones, papules, pustules, and cysts. The sebaceous follicles are rapidly colonized by microorganisms such as staphylococcal species and *Proprionibacterium acne* bacteria that thrive in the lipid-rich environment of the follicles. When this occurs, the cyst or sac of connective tissue cells can destroy and displace epidermal cells, resulting in permanent scarring. This type of acne, called *cystic acne,* may be treated successfully with a new and powerful drug called Accutane, a synthetic form of vitamin A. Care must be taken to avoid squeezing, pinching, or scratching the lesions.

Impetigo

Impetigo (im'-pe-TĪ-gō) is a superficial skin infection caused by staphylococci or streptococci. It is characterized by isolated pustules that become crusted and rupture. Occurring principally around mouth, nose, and hands, the inflammation is located in the papillary layer of the skin and involves the capillary network and stratum corneum. The disease is most common in children, and epidemics in nurseries (due to staphylococci) may be serious.

Systemic Lupus Erythematosus (SLE)

Systemic lupus erythematosus (er-i'-them-a-TŌ-sus), **SLE,** or **lupus** is an autoimmune, inflammatory disease, occurring mostly in young women in their reproductive years. An *autoimmune disease* is one in which the body attacks its own tissues, failing to differentiate between what is foreign and what is not. In SLE, damage to blood vessel walls results in the release of chemicals that mediate the inflammatory response. The blood vessel damage can be associated with virtually every body system.

The cause of SLE is not known. It is not contagious, and although it is not thought to be hereditary, there seems to be a strong incidence of other connective tissue disorders—especially rheumatoid arthritis (RA) and rheumatic fever—in relatives of SLE victims. The disease may be triggered by medication, such as penicillin, sulfa, or tetracycline, exposure to excessive sunlight, injury, emotional upset, infection, or other stress. These triggering factors, once recognized, are to be avoided by the patient in the future.

Symptoms include low-grade fever, aches, fatigue, photosensitivity, rapid loss of large amounts of scalp hair, and sometimes an eruption across the bridge of the nose and cheeks called a "butterfly rash." Other skin lesions may occur with blistering and ulceration. The erosive nature of some of the SLE skin lesions was thought to resemble the damage inflicted by the bite of a wolf, thus the term *lupus.* The most serious complications of the disease involve inflammation of the kidneys, liver, spleen, lungs, heart, and the central nervous system.

Psoriasis

Psoriasis (sō-RĪ-a-sis) is a chronic, occasionally acute, relapsing skin disease borne by 6 to 8 million people in the United States. In its severe forms, psoriasis is a disabling and disfiguring affliction. It may begin at any age, although it is most severe between ages 10 and 50. It is thought to be hereditary in at least one-third of the patients. Recent studies have traced the complex cause of psoriasis to an abnormally high rate of mitosis in epidermal cells. Triggering factors such as trauma, infections, seasonal and hormonal changes, and emotional stress can initiate and intensify the skin eruptions.

Psoriasis is characterized by distinct, reddish, slightly raised plaques or papules (small, round skin elevations) covered with scales. Itching is seldom severe, and the lesions heal without scarring. Psoriasis ordinarily involves the scalp, the elbows and knees, the back, and the buttocks. Occasionally the disease is generalized.

Decubitus Ulcers

Decubitus (dē-KYOO-be-tus) **ulcers,** also known as **bedsores, pressure sores,** or **trophic ulcers** are caused by a constant deficiency of blood to tissues overlying a bony projection that has been subjected to prolonged pressure against an object such as a bed, cast, or splint. The deficiency results in tissue ulceration. Small breaks in the epidermis become infected, and the sensitive subcutaneous and deeper tissues are damaged. Eventually the tissue is destroyed.

Decubitus ulcers are seen most frequently in patients who are bedridden for long periods of time. The most common areas involved are the skin over the sacrum, heels, ankles, buttocks, and other large bony projections. The chief causes are pressure from infrequent turning of the patient, trauma and maceration of the skin, and malnutrition. Maceration of the skin often follows soaking of bed and clothing by perspiration, urine, or feces.

Warts

Certain viruses (papovaviruses) can cause epithelial skin cells to proliferate into uncontrolled growths called **warts.** Warts are generally benign (noncancerous), and most regress spontaneously. It is possible to spread warts by contact-transfer of the viruses, and sexual contact has been implicated in the transmission of warts on the genitals. After infection, there is an incubation period of several weeks before the appearance of the wart.

Cold Sores (Fever Blisters)

The herpes simplex virus (HSV) has the ability to lay dormant for extended periods of time without expressing any signs or symptoms of disease. Cold sores are caused by type 1 herpes simplex virus, which is transmitted by oral or respiratory routes. During its periods of dormancy, the virus probably infects only a few cells at a time. Although the initial infection usually occurs during infancy, it is for the most part subclinical. The few individuals who express the disease during infancy have characteristic lesions, usually in the oral mucous membrane. These infections subside, but recur as what are called **cold sores (fever blisters).** The appearance of cold sore lesions is associated with stimuli such as exposure to the ultraviolet (UV) radiation from the sun, hormonal changes related to the menstrual cycle, or even emotional upset.

Type 2 herpes simplex virus causes genital herpes infections and is transmitted primarily by sexual contact (see Chapter 28).

Sunburn

Sunburn is injury to the skin as a result of acute, prolonged exposure to the ultraviolet (UV) rays of sunlight. It usually occurs 2 to 8 hours after exposure. Acute redness and pain are maximal in about 12 hours and dissipate approximately 72 to 96 hours later. Sunburn is without doubt one of the most common and painful skin afflictions. The damage to skin cells caused by sunburn is due to inhibition of DNA and RNA synthesis which leads to cell death. There can also be damage to blood vessels as well as other structures in the dermis. Overexposure over a period of years results in a leathery skin texture, wrinkles, skin folds, sagging skin, warty growths called keratoses, freckling, a yellow discoloration due to abnormal elastic tissue, and premature aging of the skin.

The tendency to sunburn is determined by type of skin. There are basically four skin groups. Those who burn and never tan, those who burn but keep some tan, those who burn slightly and develop a good tan, and those who never burn and always tan.

Covering the body remains the best protection against the sun's rays, while a topical sunscreen will help protect uncovered parts. The relative degree of protection provided by a sunscreen-ing agent is discussed in the section on drugs at the end of the chapter.

Extreme and severe sunburn may produce *sunstroke*—a disruption of the body's heat-regulating mechanism. It is characterized by high fever and collapse and sometimes by convulsions, coma, and death (Chapter 25).

Skin Cancer

Excessive sun exposure can result in **skin cancer.** Everyone, regardless of skin pigmentation, is a potential victim of skin cancer if exposure to sunlight is sufficiently intense and continuous. Natural skin pigment can never give complete protection. If you must be in direct sunlight for long periods of time, use a suitable sunscreen. One of the best agents for protection against overexposure to ultraviolet rays of the sun is para-aminobenzoic acid (PABA). The alcohol preparations of PABA are best because the active ingredient binds to the stratum corneum of the skin.

A number of widely prescribed drugs (tetracyclines, sulfa drugs, and others), constituents of foods (such as riboflavins), and saccharin and cyclamates in dietetic drinks are potential photosensitizers. When activated in the body by light, they may produce substances that can damage the tissues in sensitive individuals. A typical response is the appearance of a rash on parts of the body exposed to the sun. Repeated and persistent exposure could produce permanent changes leading to skin cancer.

FIGURE 5-7 The Lund-Browder method for estimating the extent of burns. Relative proportions of various body regions in (a) a young child compared to (b) an adult.

	0 yr.	1 yr.	5 yrs.	10 yrs.	15 yrs.	Adult
A = 1/2 head	9 1/2%	8 1/2%	6 1/2%	5 1/2%	4 1/2%	3 1/2%
B = 1/2 thigh	2 3/4%	3 1/4%	4%	4 1/4%	4 1/2%	4 3/4%
C = 1/2 calf	2 1/2%	2 1/2%	2 3/4%	3%	3 1/4%	3 1/2%

Burns

Tissues may be damaged by thermal (heat), electrical, radioactive, or chemical agents. These agents can destroy the proteins in the exposed cells and cause cell injury or death. Such damage is a **burn.** The injury to tissues directly or indirectly in contact with the damaging agent, such as the skin or the linings of the respiratory and digestive tracts, is the local effect of a burn. Generally, however, the systemic effects of a burn are a greater threat to life than the local effects. The systemic effects of a burn may include (1) a large loss of water, plasma, and plasma proteins, which causes shock; (2) bacterial infection; (3) reduced circulation of blood; and (4) decreased production of urine.

Classification

A burn may extend through the entire thickness of the skin or it may damage or destroy only part of the skin. Clinically, the depth of a burn is determined by its color, the presence or absence of sensation, blister formation, or loss of elasticity. A **first-degree burn** is characterized by mild pain and erythema (redness) and involves only the surface epithelium. Generally, a first-degree burn will heal in about two to three days and may be accompanied by flaking or peeling. A typical sunburn is an example of a first-degree burn.

A **second-degree burn** involves the deeper layers of the epidermis or the upper levels of the dermis. In a superficial second-degree burn, the deeper layers of the epidermis are injured and there is characteristic erythema, blister formation, edema, and pain. Blisters beneath or within the epidermis are called *bullae* (BYOOL-ē), meaning bubbles. Such an injury usually heals within seven to ten days with only mild scarring. In a deep second-degree burn, there is destruction of the epidermis as well as the upper levels of the dermis. Epidermal derivatives, such as hair follicles, sebaceous glands, and sweat glands are usually not injured. If there is no infection, deep second-degree burns heal without grafting in about three to four weeks. Scarring may result.

First- and second-degree burns are collectively referred to as **partial-thickness burns.** A **third-degree burn** or **full-thickness burn** involves destruction of the epidermis, dermis, and the epidermal derivatives. Such burns vary in appearance from marble-white to mahogany colored to charred, dry wounds. There is little if any edema and such a burn is usually not painful to the touch due to destruction of nerve endings. Regeneration is slow and much granulation tissue forms before being covered by epithelium. Even if skin grafting is quickly begun, third-degree burns quickly contract and produce scarring.

The severity of a burn is measured in terms of the amount of surface area affected as well as in terms of the depth of injury. A fairly accurate method for estimating the extent of a burn is to apply the *Lund-Browder method.* This method estimates the extent by measuring the areas affected against the percentage of total surface area for body parts shown in Figure 5-7. For example, if the anterior of the head and neck of an adult is affected, the burn covers 4½ percent of the body surface. Because the proportions of the body change with growth, the percentages vary for different ages. Thus, the extent of burn damage can be made fairly accurately for any age group.

Treatment

A severely burned individual should be moved as quickly as possible to a hospital. Treatment may then include:

1. Cleansing the burn wounds thoroughly.
2. Removing all dead tissue (debridement) so antibacterial agents can directly contact the wound surface and thereby prevent infection.
3. Replacing lost body fluids and electrolytes (ions).
4. Covering wounds with temporary protection as soon as possible. One such covering, *human amniotic membrane,* is now easier to use because of recently developed methods of collection and storage. The membrane is stripped from the placenta ("afterbirth") up to 24 hours after delivery and stored in a saline-penicillin solution for up to 8 weeks. The membrane's desirable permeable properties keep wounds from drying while preventing infection from without. Investigators have found that the membrane is also useful for treating chronic decubitus ulcers. Another covering used as a temporary protection for burns is a newly developed semisynthetic membrane called *Biobrane.* It consists of nylon, Silastic, and a collagen derivative.
5. Removing a thin layer of skin from another part of the burn victim's body and transplanting it to the injured area (skin graft).

MEDICAL TERMINOLOGY

Albinism (*alb* = white; *ism* = condition)   Congenital (existing at birth) absence of pigment from the skin, hair, and parts of the eye.

Anhidrosis (*an* = without; *hidr* = sweating; *osis* = condition)   A rare genetic condition characterized by inability to sweat.

Antiperspirant (*anti* = against; *perspirare* = to breathe through)   An agent that inhibits or prevents perspiration and usually contains aluminum compounds as the active ingredient.

Athlete's foot   A superficial fungus infection of the skin of the foot.

Callus (callosity)   An area of hardened and thickened skin that is usually seen in palms and soles and is due to pressure and friction.

Carbuncle (*carbunculus* = little coal)   A hard, round, deep, painful inflammation of the subcutaneous tissue that causes necrosis (death) and pus formation (abscess).

Comedo (*comedo* = to eat up)   A collection of sebaceous material and dead cells in the hair follicle and excretory duct of the sebaceous gland. Usually found over the face, chest, and back, and more commonly during adolescence. Also called blackhead or whitehead.

Corn   A painful conical thickening of the skin found principally over toe joints and between the toes. It may be hard or soft, depending on the location. Hard corns are usually found over toe joints, and soft corns are usually found between the fourth and fifth toes.

Cyst (*cyst* = sac containing fluid)   A sac with a distinct connective tissue wall, containing a fluid or other material.

Deodorant (*de* = from; *odorare* = perfume)   An agent that masks offensive odors; usually contains aluminum compounds as the active ingredient. Body odor is not caused by perspiration itself as much as by the activities of bacteria that grow in perspiration.

Dermabrasion (*derm* = skin)   Removal of acne, scars, tattoos, or nevi (see below) by sandpaper or a high-speed brush.

Dermatome (*tomy* = cut)   An instrument for excising areas of skin to be used for grafting.

Detritus (*deterere* = to rub away)   Particulate matter produced by or remaining after the wearing away or disintegration of a substance or tissue; scales, crusts, and loosened skin.

Eczema (*ekzein* = to boil out)   An acute or chronic superficial inflammation of the skin, characterized by redness, oozing, crusting, and scaling. Also called chronic dermatitis.

Electrolysis   Method of hair removal in which the hair bulb is destroyed by an electric current so that the hair cannot regrow.

Erythema (*erythema* = redness)   Redness of the skin caused by engorgement of capillaries in lower layers of the skin. Erythema occurs with any skin injury, infection, or inflammation.

Furuncle   A boil; an abscess resulting from infection of a hair follicle.

Hypodermic (*hypo* = under)   Relating to the area beneath the skin. Also called subcutaneous.

Intradermal (*intra* = within)   Within the skin. Also called intracutaneous.

Keratosis (*kera* = horn)   Formation of a hardened growth of tissue.

Nevus   A round, pigmented, flat, or raised skin area varying in color from yellow-brown to black. It may be present at birth or develop later. Also called mole or birthmark.

Nodule (*nodulus* = little knot)   A large cluster of cells raised above the skin but extending deep into the tissues.

Papule   A small, round skin elevation varying in size from a pinpoint to that of a split pea. One example is a pimple.

Polyp   A tumor on a stem found especially on mucous membranes.

Pustule   A small, round elevation of the skin containing pus.

Subcutaneous (*sub* = under)   Beneath the skin. Also called hypodermic.

Topical   Pertaining to a definite area; local. Also, of a medication, applied to the surface rather than ingested or injected.

DRUGS ASSOCIATED WITH THE INTEGUMENTARY SYSTEM

In the sections on drugs associated with the various systems of the body, we have included old, firmly established drugs and new, recently approved medications. Drugs are referred to by generic and nonproprietary names and in some cases by brand names. The generic name indicates the general category of the drug. It usually describes the use or action of the drug, such as keratolytic agent. The nonproprietary name is the common name for the drug as approved by the U.S. Food and Drug Administration (FDA). Nonproprietary names appear in italic type. A brand name is a name assigned to a drug by a manufacturer and is a registered trademark. A specific drug may be manufactured by only one company, or it may be produced by several companies under different brand names. Brand names are given in parentheses following the nonproprietary names, for example, *triamcinolone* (Aristocort, Kenalog).

Disorders of the skin are treated primarily with liquid lotions, liniments, semisolid ointments, pastes, water-miscible bases, or wetting agents. Generally, the skin acts as a barrier to most drugs that are applied to it, and this has the advantage of minimizing the absorption of potentially toxic drugs.

Keratolytic Agents

Keratolytic agents are keratin-dissolving drugs that help speed the rate of peeling of keratin-containing cells. They are used to treat acne, athelete's foot, corns, and calluses. Examples are *salicylic acid, tretinoin,* and *benzoyl peroxide.*

Depilatories

Depilatories are substances that remove superfluous hair. They dissolve the protein in the hair shaft, turning it into a gelatinous mass which can be wiped away. Since the hair root is not affected, regrowth of the hair occurs. Examples of depilatories include calcium hydroxide, strontium hydroxide, and *calcium thioglycollate* (Neet, Nair).

Caustics

Caustics are substances that destroy excessive growths of keratin-containing cells, such as warts. Examples are *glacial acetic acid* and *trichloroacetic acid.*

Astringents

Astringents cause a slight protein-coagulating (tightening) effect on the skin and are used to treat poison ivy, athlete's foot, and acute eczema. An example is *aluminum salts.*

Dusting Powders

Dusting powders absorb moisture from the skin and are used to reduce friction between adjacent skin areas that rub together. Examples include *corn starch, talc,* and *magnesium oxide.*

Protective Dressings

Protective dressings are drugs used to seal small wounds, protect the skin from external irritants, and prevent diaper rash, bed sores, and chapping. Examples include *colloidion* and *dimethicone.*

Corticosteroids

Corticosteroids are antiinflammatory drugs that are useful in treatment of severe sunburn and certain dermatoses. Examples include *hydrocortisone* (Cortaid), *triamcinolone* (Aristocort, Kenalog), *prednisone, dexamethasone* (Decadron), and *betamethasone* (Celestone, Valisone).

Since 0.5 percent hydrocortisone (e.g., Cortaid) is now available over-the-counter (OTC), the harmful effects of topical corticosteroid overuse are likely to become more frequent. One of the primary dangers of continued topical corticosteroid use is masking the symptoms of certain bacterial, viral, and fungal infections. Another is actually worsening the infection, such as occurs in acne, scabies, impetigo, and lice. Topical corticosteroid overuse can also result in thinning and weakening of the epidermis and superficial dermis, especially in the axillae, groin, and face; stimulation of dermal structures causing acne or excessive growth of unwanted hair; and tissue atrophy. A very serious systemic complication is suppression of interactions between the pituitary and suprarenal glands (Chapter 18).

Sunscreening Agents

Presently there are over 50 products available that provide varying degrees of protection from direct ultraviolet light exposure as well as indirect or reflected light exposure from the sun. Such *sunscreening agents* fall into two principal classes: physical reflectors and chemical absorbers. Both classes retard premature aging of the skin and development of skin cancers caused by ultraviolet light. *Physical reflectors* include opaque white and tinted creams containing substances such as titanium dioxide or zinc oxide. Most people, however, prefer chemical absorbers in creams, oils, and lotions, which are more cosmetically attractive than physical reflectors. The most effective ingredient in *chemical absorbers* is para-aminobenzoic acid (PABA) or a related compound.

The FDA has established guidelines to indicate the relative degree of protection provided by a sunscreening agent. The degree of protection is indicated by a numerical rating and is called a *sun protection factor (SPF)*. The SPF is the time required to produce erythema (redness) when a particular sunscreen product is used divided by the time required to produce erythema when the product is not used. For example, if you can remain exposed to the sun with protection for 30 minutes before your skin gets red, but you can remain exposed to the sun without protection for only 5 minutes, then the SPF of the product you are using is 6. In other words, the product allows you to stay in the sun 6 times longer than without protection. SPF ranges from 2 (practically no protection) to 15 or more (super protection).

STUDY OUTLINE

Skin (p. 106)

1. The skin and its derivatives (hair, glands, and nails) constitute the integumentary system.
2. The skin is one of the larger organs of the body. It performs the functions of protection, maintaining body temperature, preventing excessive loss of inorganic and organic materials, receiving stimuli, storage of chemical compounds, synthesis of vitamin D, and excretion of water, salts, and several organic compounds.
3. The principal parts of the skin are the outer epidermis and inner dermis. The dermis overlies the subcutaneous layer.
4. The epidermal layers, from deepest to most superficial, are the stratum basale, spinosum, granulosum, lucidum, and corneum. The basale and spinosum undergo continuous cell division and produce all other layers.
5. The dermis consists of a papillary region and a reticular region. The papillary region is loose connective tissue containing blood vessels, nerves, hair follicles, dermal papillae, and Meissner's corpuscles. The reticular region is dense, irregularly arranged connective tissue containing adipose tissue, hair follicles, nerves, oil glands, and ducts of sweat glands.
6. Lines of cleavage indicate the direction of collagenous fiber bundles in the dermis and are considered during surgery.
7. The color of skin is due to melanin, carotene, and blood in capillaries in the dermis.
8. Epidermal ridges increase friction for better grasping ability and provide the basis for fingerprints and footprints.

Epidermal Derivatives (p. 110)

1. Epidermal derivatives are structures developed from the embryonic epidermis.
2. Among the epidermal derivatives are hair, skin glands (sebaceous, sudoriferous, and ceruminous), and nails.

Hair

1. Hairs are epidermal growths that function in protection.
2. Hair consists of a shaft above the surface, a root that penetrates the dermis and subcutaneous layer, and a hair follicle.
3. Associated with hairs are sebaceous glands, arrectores pilorum muscles, and root hair plexuses.
4. Hair color is due to combinations of various amounts of the three hair pigments, black melanin, brown melanin, and pheomelanin (yellow). Graying is due to the loss of melanin.
5. New hairs develop from cell division of the matrix in the bulb; hair replacement and growth occurs in a cyclic pattern. "Male-pattern" baldness is caused by androgens and heredity.

Glands

1. Sebaceous (oil) glands are usually connected to hair follicles; they are absent in the palms and soles. Sebaceous glands produce sebum which moistens hairs and waterproofs the skin. Enlarged sebaceous glands may produce blackheads, pimples, and boils.
2. Sudoriferous (sweat) glands are distinguished into apocrine and eccrine. Apocrine sweat glands are limited in distribution to the skin of the axilla, pubis, and areolae; their ducts open into hair follicles. Eccrine sweat glands have an extensive distribution; their ducts terminate at pores at the surface of the epidermis. Sudoriferous glands produce perspiration, which carries small amounts of wastes to the surface and assists in maintaining body temperature.
3. Ceruminous glands are modified sudoriferous glands that secrete cerumen. They are found in the external auditory meatus.

Nails

1. Nails are hard, keratinized epidermal cells over the dorsal surfaces of the terminal portions of the fingers and toes.
2. The principal parts of a nail are the body, free edge, root, lunula, eponychium, hyponychium, and matrix. Cell division of the matrix cells produces new nails.

Homeostasis (p. 114)

1. One of the functions of the skin is the maintenance of the normal body temperature of 37°C (98.6°F).
2. If environmental temperature is high, skin receptors sense the stimulus (heat) and generate impulses that are transmitted to the brain. The brain then causes the sweat glands to produce perspiration. As the perspiration evaporates, the skin is cooled.
3. The skin-cooling response is a negative feedback mechanism.
4. Temperature maintenance is also accomplished by adjusting blood flow to the skin, regulating metabolic rate, and regulating skeletal muscle contractions.

Disorders: Homeostatic Imbalances (p. 115)

1. Pruritus or itching is a common skin problem that may be related to skin disorders, systemic diseases, or psychogenic factors.
2. Acne is an inflammation of sebaceous glands.
3. Impetigo is a superficial skin infection caused by bacteria and characterized by pustules that become crusted and rupture.
4. Systemic lupus erythematosus (SLE) is an autoimmune disease of connective tissue.
5. Psoriasis is a chronic skin disease characterized by reddish, raised plaques or papules.
6. Decubitus ulcers are caused by a chronic deficiency of blood to tissues subjected to prolonged pressure.
7. Warts are uncontrolled growths of epithelial skin cells caused by a virus. Most warts are benign.
8. Cold sores (fever blisters) are lesions caused by type 1 herpes simplex virus. The dormant infection is triggered by certain stimuli.
9. Sunburn is a skin injury resulting from prolonged exposure to the ultraviolet (UV) rays of sunlight.
10. Skin cancer can be caused by excessive exposure to sunlight.
11. Tissue damage that destroys protein is called a burn. Depending on the depth of damage, skin burns are classified

as first-degree and second-degree (partial-thickness) and third-degree (full-thickness). One method employed for determining the extent of a burn is the Lund-Browder method.

Burn treatment may include cleansing the wound, removing dead tissue, replacing lost body fluids, and covering wounds with temporary protection, and skin grafting.

REVIEW QUESTIONS

1. Define an organ. In what respect is the skin an organ? What is the integumentary system?
2. List the principal functions of the skin.
3. Compare the structure of epidermis and dermis. What is the subcutaneous layer?
4. List and describe the epidermal layers from the deepest outward. What is the importance of each layer?
5. Contrast the structural differences between the papillary and reticular regions of the dermis.
6. What are lines of cleavage? What is their importance during surgery?
7. Explain the factors that produce skin color. What is an albino?
8. Describe how melanin is synthesized and distributed to epidermal cells.
9. How are epidermal ridges formed? Why are they important?
10. List the receptors in the epidermis, dermis, and subcutaneous layer, and indicate the location and role of each.
11. Describe the development and distribution of hair.
12. Describe the structure of a hair. How are hairs moistened? What produces "goosebumps" or "gooseflesh"?
13. Explain the basis for hair color. Why does hair turn gray?
14. Describe how hair grows and is replaced.
15. Contrast the locations and functions of sebaceous glands, sudoriferous glands, and ceruminous glands. What are the names and chemical components of the secretions of each?
16. Distinguish between apocrine and eccrine sweat glands.
17. From what layer of the skin do nails form? Describe the principal parts of a nail.
18. Explain with a labeled diagram how the skin helps maintain normal body temperature.
19. What is a feedback system? A negative feedback system? Relate these definitions to the maintenance of normal body temperature.
20. Define each of the following disorders of the integumentary system: pruritus, acne, impetigo, systemic lupus erythematosus (SLE), psoriasis, decubitus ulcers, warts, cold sores, sunburn, and skin cancer.
21. Define a burn. Classify burns according to degree.
22. Explain how the Lund-Browder method is used to estimate the extent of a burn.
23. How are burns treated?
24. Refer to the glossary of key medical terms associated with the integumentary system. Be sure that you can define each term.
25. Describe the actions of drugs associated with the integumentary system.

UNIT II

Principles of Support and Movement

This unit considers two primary themes—support and movement. You will study the various ways in which the body is supported and the different movements it can perform. Both support and movement are made possible by the cooperative effort of bones, joints, and muscles.

Osteocyte

Area of
bone formation

Osteoblasts

Osteogenic
layer

Periosteum

Fibrous
layer

FIGURE 6-2 Photomicrograph of compact bone showing the
osteogenic and fibrous layers of the periosteum at a magnification of 300×. (Courtesy of Donald I. Patt, from *Comparative
Vertebrate Histology,* by Donald I. Patt and Gail R. Patt, Harper & Row, Publishers, Inc., New York, 1969.)

nutrients can reach the osteocytes and wastes can be removed. Each Haversian canal, with its surrounding lamellae, lacunae, osteocytes, and canaliculi, is called an **Haversian system** or **osteon.** Haversian systems are characteristic of adult bone. The areas between Haversian
systems contain **interstitial lamellae.** These also possess
lacunae with osteocytes and canaliculi, but their lamellae
are usually not connected to the Haversian systems.

SPONGY BONE

In contrast to compact bone, spongy bone does not contain true Haversian systems. It consists of an irregular
latticework of thin plates of bone called **trabeculae** (tra-
BEK-yoo-lē). The spaces between the trabeculae of some
bones are filled with red marrow. The cells of red marrow
are responsible for producing blood cells. Within the trabeculae lie lacunae, which contain the osteocytes. Blood
vessels from the periosteum penetrate through to the
spongy bone, and osteocytes in the trabeculae are nour-
ished directly from the blood circulating through the marrow cavities.

Most people think of all bone as a very hard, white
material. Yet the bones of an infant are not hard at all,
and a child's bones are generally more pliable than those
of an adult. The final shape and hardness of adult bones
require many years to develop and depend on a complex
series of chemical changes. Let us now see how bones
are formed and how they grow.

OSSIFICATION

The process by which bone forms in the body is called
ossification or **osteogenesis.** The "skeleton" of a human
embryo is composed of either fibrous membranes or hya-

line cartilage. Both are shaped like bones and provide
the medium for ossification. Ossification begins around
the sixth or seventh week of embryonic life and continues
throughout adulthood. Two kinds of bone formation occur. The first is called intramembranous (in'-tra-MEM-
bra-nus) ossification. This term refers to the formation
of bone directly on or within the fibrous membranes (*intra*
= within; *membranous* = membrane). The second kind,
endochondral (en'-dō-KON-dral) ossification, refers to
the formation of bone in cartilage (*endo* = within; *chondro*
= cartilage). These two kinds of ossification do *not* lead
to differences in the structure of mature bones. They simply indicate difference methods of bone formation. Both
mechanisms involve the replacement of a preexisting connective tissue with bone.

The first stage in the development of bone is the migration of mesenchymal cells (embryonic connective tissue
cells) into the area where bone formation is about to
begin. These cells increase in number and size. In some
skeletal structures where capillaries are lacking they become chondroblasts; in others where capillaries are present some become osteoblasts. The **chondroblasts** will be
responsible for cartilage formation. The osteoblasts will
form bone tissue by intramembranous or endochondral
ossification.

INTRAMEMBRANOUS OSSIFICATION

Of the two types of bone formation, the simpler and more
direct is **intramembranous ossification.** The flat bones of
the roof of the skull, parts of the mandible (lower jawbone), and probably part of the clavicles are formed in
this way. The essentials of this process are as follows.

Osteoblasts formed from mesenchymal cells cluster in
the fibrous membrane. The site of such a cluster is called

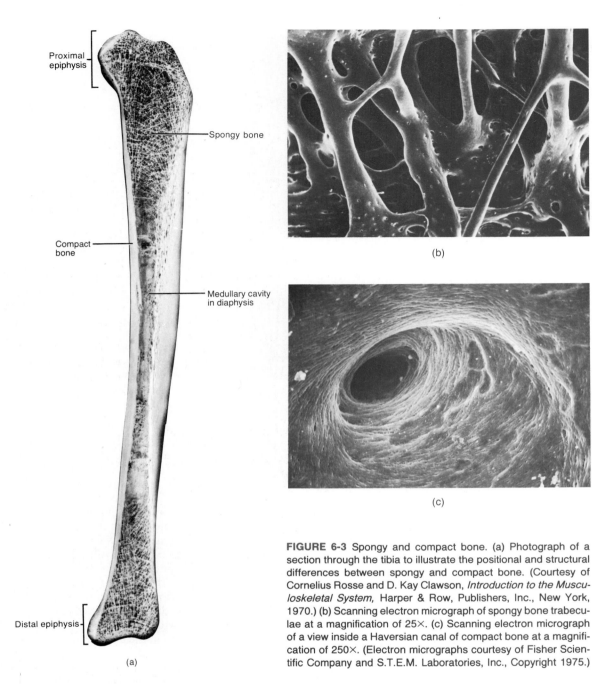

Proximal epiphysis

Spongy bone

Compact bone

Medullary cavity in diaphysis

Distal epiphysis

(a)

(b)

(c)

FIGURE 6-3 Spongy and compact bone. (a) Photograph of a section through the tibia to illustrate the positional and structural differences between spongy and compact bone. (Courtesy of Cornelius Rosse and D. Kay Clawson, *Introduction to the Musculoskeletal System,* Harper & Row, Publishers, Inc., New York, 1970.) (b) Scanning electron micrograph of spongy bone trabeculae at a magnification of 25×. (c) Scanning electron micrograph of a view inside a Haversian canal of compact bone at a magnification of 250×. (Electron micrographs courtesy of Fisher Scientific Company and S.T.E.M. Laboratories, Inc., Copyright 1975.)

a *center of ossification.* The osteoblasts then secrete intercellular substances partly composed of collagenous fibers that form a framework, or matrix, in which calcium salts are quickly deposited. The deposition of calcium salts is called *calcification.* When a cluster of osteoblasts is completely surrounded by the calcified matrix, it is called a *trabecula.* As trabeculae form in nearby ossification centers, they fuse into the open latticework characteristic of spongy bone. With the formation of successive layers of bone, some osteoblasts become trapped in the lacunae. The entrapped osteoblasts lose their ability to form bone and are called osteocytes. The spaces between the trabeculae fill with red marrow. The original connective tissue that surrounds the growing mass of bone then becomes the periosteum. The ossified area has now become true

spongy bone. Eventually, the surface layers of the spongy bone will be reconstructed into compact bone. Much of this newly formed bone will be destroyed and reformed so the bone may reach its final adult size and shape.

ENDOCHONDRAL OSSIFICATION

The replacement of cartilage by bone is called **endochondral** or **intracartilaginous ossification.** Most bones of the body, including parts of the skull, are formed in this way. Since this type of ossification is best observed in a long bone, we will investigate the tibia, or shinbone (Figure 6-4).

Early in embryonic life, a cartilage model or template of the future bone is laid down. This model is covered

FIGURE 6-4 Endochondral ossification of the tibia. (a) Cartilage model. (b) Collar formation. (c) Development of primary ossification center. (d) Entrance of blood vessels. (e) Marrow-cavity formation. (f) Thickening and lengthening of collar. (g) Formation of secondary ossification centers. (h) Remains of cartilage as articular cartilage and epiphyseal plate. (i) Formation of epiphyseal lines.

by a membrane called the *perichondrium* (per-i-KON-drē-um). Midway along the shaft of this model a blood vessel penetrates the perichondrium, stimulating the cells in the internal layer of the perichondrium to enlarge and become osteoblasts. The cells begin to form a collar of compact bone around the middle of the diaphysis of the cartilage model. Once the perichondrium starts to form bone, it is called the *periosteum.* Simultaneously with the appearance of the bone collar and the penetration of blood vessels, changes occur in the cartilage in the center of the diaphysis. In this area, called the **primary**

ossification center, cartilage cells hypertrophy (increase in size)—probably because they accumulate glycogen for energy and produce enzymes to catalyze future chemical reactions. When the hypertrophied cells burst, there is a change in extracellular pH to a more alkaline pH causing the intercellular substance to become *calcified,* that is, minerals are deposited within it. Once the cartilage becomes calcified, nutritive materials required by the cartilage cells can no longer diffuse through the intercellular substance and this may cause the cartilage cells to die. Then the intercellular substance begins to degenerate,

Marrow spaces

Zone of reserve cartilage

Zone of proliferating cartilage

Zone of hypertropic cartilage

Zone of calcified matrix

Bone

FIGURE 6-5 Epiphyseal plate. Photomicrograph of the epiphysis of a long bone showing the various zones of the epiphyseal plate at a magnification of 160×. (© 1983 by Michael H. Ross. Used by permission.)

leaving large cavities in the cartilage model. The blood vessels grow along the spaces where cartilage cells were previously located and enlarge the cavities further. Gradually, these spaces in the middle of the shaft join with each other, and the marrow cavity is formed.

As these developmental changes are occurring, the osteoblasts of the periosteum deposit successive layers of bone on the outer surface so that the collar thickens, becoming thickest in the diaphysis. The cartilage model continues to grow at its ends, steadily increasing in length. Eventually, blood vessels enter the epiphyses and **secondary ossification centers** appear in the epiphyses and also lay down spongy bone. In the tibia, one secondary ossification center develops in the proximal epiphysis soon after birth. The other center develops in the distal epiphysis during the child's second year.

After the two secondary ossification centers have formed, bone tissue has completely replaced cartilage, except in two regions. Cartilage continues to cover the articular surfaces of the epiphyses, where it is called **articular cartilage.** It also remains as a plate between the epiphysis and diaphysis, where it is called the **epiphyseal plate.**

The epiphyseal (ep'-i-FIZ-ē-al) plate consists of four zones (Figure 6-5). The *zone of reserve cartilage* is adjacent to the epiphysis of the bone. It consists of small chondrocytes that are scattered irregularly throughout the intercellular matrix. The cells of this zone do not function in bone growth. The zone of reserve cartilage functions in anchoring the epiphyseal plate to the bone of the epiphysis. Its blood vessels also provide nutrients for the other zones of the epiphyseal plate.

The second zone, the *zone of proliferating cartilage,* consists of slightly larger chondrocytes that are arranged like stacks of coins. The function of this zone is to make new chondrocytes by cell division to replace those that die at the diaphyseal surface of the epiphyseal plate.

The third zone, the *zone of hypertrophic* (hī-per-TRŌF-ik) *cartilage,* consists of even larger chondrocytes also arranged in columns. The cells are in various stages of maturation, with the more mature cells closer to the diaphysis. The lengthwise expansion of the epiphyseal plate is the result of cellular proliferation of the zone of proliferating cartilage and maturation of the cells in the zone of hypertrophic cartilage. Near the diaphyseal end of the bone, the intercellular matrix of the cells of the zone of hypertrophic cartilage becomes calcified and dies.

The fourth zone, the *zone of calcified matrix,* is only a few cells thick and consists mostly of dead cells because the intercellular matrix around them has calcified. The calcified matrix is taken up by osteoclasts, and the area is invaded by osteoblasts and capillaries from the bone in the diaphysis. These cells lay down bone on the calcified cartilage that persists. As a result, the diaphyseal border of the epiphyseal plate is firmly cemented to the bone of the diaphysis.

The region between the diaphysis and epiphysis of a bone where the calcified matrix is replaced by bone is called the *metaphysis* (me-TAF-i-sis). The activity of the epiphyseal plate is the only mechanism by which the diaphysis can increase in length. Unlike cartilage, which can grow by both interstitial and appositional growth, bone can grow only by appositional growth.

The epiphyseal plate allows the diaphysis of the bone to increase in length until early adulthood. As the child grows, cartilage cells are produced by mitosis on the epiphyseal side of the plate. Cartilage cells are then destroyed and the cartilage is replaced by bone on the diaphyseal side of the plate. In this way, the thickness of the epiphyseal plate remains fairly constant, but the bone on the diaphyseal side increases in length. Growth in diameter occurs along with growth in length. In this process, the bone lining the marrow cavity is destroyed so that the cavity increases in diameter. At the same time, osteoblasts from the periosteum add new osseous tissue around the outer surface of the bone. Initially, diaphyseal and epiphyseal ossification produce only spongy bone. Later, by reconstruction, the outer region of spongy bone is reorganized into compact bone.

CLINICAL APPLICATION

The epiphyseal cartilage cells stop dividing and the cartilage is replaced by bone at about age 18 in females and age 20 in males. The newly formed bony structure is called the **epiphyseal line,** a remnant of the once active epiphyseal plate. With the appearance of the epiphyseal line, appositional bone growth stops. The clavicle is the last bone to stop growing.

Ossification of all bones is usually completed by age 25. Bones undergoing either intramembranous or endochondral ossification are continually remodeled from the time that initial calcification occurs until the final structure appears. **Remodeling** is the replacement of old bone tissue by new bone tissue. Compact bone is formed by the transformation of spongy bone. The diameter of a long bone is increased by the destruction of the bone closest to the marrow cavity and the construction of new bone around the outside of the diaphysis. However, even after bones have reached their adult shapes and sizes, old bone is perpetually destroyed and new osseous tissue is formed in its place.

HOMEOSTASIS

Bone shares with skin the feature of replacing itself throughout adult life. Remodeling takes place at different rates in various body regions. The distal portion of the femur (thighbone) is replaced about every 4 months. By contrast, bone in certain areas of the shaft will not be completely replaced during the individual's life. Remodeling allows worn or injured bone to be removed and replaced with new tissue. It also allows bone to serve as the body's storage area for calcium. Many other tissues in the body need calcium in order to perform their functions. For example, nerve cells need calcium for their activities, muscle needs calcium in order to contract, and blood needs calcium in order to clot. The blood continually trades off calcium with the bones, removing calcium when it and other tissues are not receiving enough of this element and resupplying the bones with dietary calcium to keep them from losing too much bone mass.

The cells believed to be responsible for the resorption (loss of a substance through a physiological or pathological process) of bone tissue are called **osteoclasts** (*clast*

= break). In the healthy adult, a delicate homeostasis is maintained between the action of the osteoclasts in removing calcium and the action of the bone-making osteoblasts in depositing calcium. Should too much new tissue be formed, the bones become abnormally thick and heavy. If too much calcium is deposited in the bone, the surplus may form thick bumps, or spurs, on the bone that interfere with movement at joints. A loss of too much tissue or calcium weakens the bones and allows them to break easily or to become very flexible.

In the process of resorption, it is believed that osteoclasts send out projections that secrete proteolytic enzymes released from lysosomes and several acids (lactic and citric). The enzymes may function by digesting the collagen and other organic substances, while the acids may cause the bone salts to dissolve in solution. It is also presumed that the osteoclastic projections may phagocytose whole fragments of collagen and bone salts.

Normal bone growth in the young and bone replacement in the adult depend on several factors. First, sufficient quantities of calcium and phosphorus, components of the primary salt that makes bone hard, must be included in the diet.

Second, the individual must obtain sufficient amounts of vitamins A, C, and D, substances that are responsible for the proper utilization of calcium and phosphorus by the body. Vitamin D, for example, is required for the absorption of calcium by the kidneys and from the digestive tract into the blood. It also has important effects on bone deposition and resorption.

Third, the body must manufacture the proper amounts of the hormones responsible for bone tissue activity (Chapter 18). Growth hormone (GH), secreted by the pituitary gland, is responsible for the general growth of bones. Too much or too little of these hormones during childhood makes the adult abnormally tall or short. Other hormones specialize in regulating the osteoclasts. Calcitonin (CT), produced by the thyroid gland, inhibits osteoclastic activity, while parathormone (PTH), synthesized by the parathyroid glands, increases osteoclastic activity. And still others, especially the sex hormones, aid osteoblastic activity and thus promote the growth of new bone. The sex hormones act as a double-edged sword. They aid in the growth of new bone, but they also bring about the degeneration of all the cartilage cells in the epiphyseal plates. Because of the sex hormones, the typical adolescent experiences a spurt of growth during puberty, when sex hormone levels start to increase. The individual then quickly completes the growth process as the epiphyseal cartilage disappears. Premature puberty can actually prevent one from reaching an average adult height because of the simultaneous premature degeneration of the plates.

DISORDERS: HOMEOSTATIC IMBALANCES

You have already learned how important certain vitamins, minerals, and hormones are to bone growth and development. Many bone disorders result from deficiencies in vitamins or minerals or from too much or too little of the hormones that regulate bone homeostasis. Infection and tumors which disrupt homeostasis are also responsible for certain bone disorders.

Vitamin Deficiencies

Vitamin D is important to normal bone growth and maintenance. It is essential for the synthesis of a protein that transports the calcium obtained from foods across the lining of the intestine and into the extracellular fluid. When the body lacks this vitamin, it is unable to absorb calcium and phosphorus. A deficiency

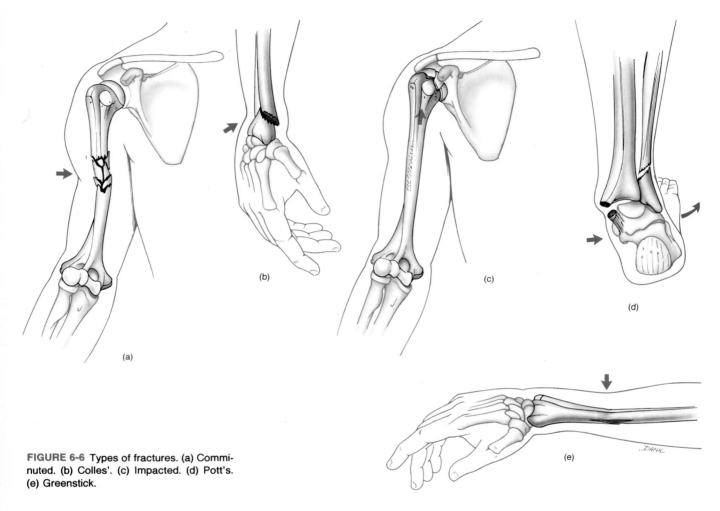

FIGURE 6-6 Types of fractures. (a) Comminuted. (b) Colles'. (c) Impacted. (d) Pott's. (e) Greenstick.

of vitamin D produces rickets in children and osteomalacia in adults.

Rickets

In the condition called **rickets,** there is a deficiency of vitamin D that results in an inability of the body to transport calcium and phosphorus from the digestive tract into the blood for utilization by bones. As a result, epiphyseal cartilage cells cease to degenerate and new cartilage continues to be produced. Epiphyseal cartilage thus becomes wider than normal. At the same time, the soft matrix laid down by the osteoblasts in the diaphysis fails to calcify. As a result, the bones stay soft. When the child walks, the weight of the body causes the bones in the legs to bow. Malformations of the head, chest, and pelvis also occur.

The cure and prevention of rickets consist of adding generous amounts of calcium, phosphorus, and vitamin D to the diet. Exposing the skin to the ultraviolet rays of sunlight also aids the body in manufacturing additional vitamin D.

Osteomalacia

Demineralization caused by vitamin D deficiency is called **osteomalacia** (os'-tē-ō-ma-LA-shē-a; *malacia* = softness). A deficiency of vitamin D in the adult causes the bones to give up excessive amounts of calcium and phosphorus. This loss, called *demineralization,* is especially heavy in the bones of the pelvis, legs, and spine. After the bones demineralize, the weight of the body produces a bowing of the leg bones, a shortening of the backbone, and a flattening of the pelvic bones. Osteomalacia mainly affects women who live on poor cereal diets devoid of

milk, are seldom exposed to the sun, and have repeated pregnancies that deplete the body of calcium. The condition responds to the same treatment as rickets; if the disease is severe enough and threatens life, large doses of vitamin D are given.

Osteomalacia may also result from failure to absorb fat (steatorrhea) because vitamin D is soluble in fats and calcium combines with fats. As a result, vitamin D and calcium remain with the unabsorbed fat and are lost in the feces.

Osteoporosis

Osteoporosis (os'-tē-ō-pō-RŌ-sis) is a bone disorder affecting the middle-aged and elderly, especially white females. Between puberty and the middle years, the sex hormones maintain osseous tissue by stimulating the osteoblasts to form new bone. Women produce smaller amounts of sex hormones after menopause, and both men and women produce smaller amounts during old age. As a result, the osteoblasts become less active and there is a decrease in bone mass. Osteoporosis affects the entire skeletal system, especially the spine, legs, and feet. As the spine collapses and curves, the thorax drops, and the ribs fall on the pelvic rim. This condition leads to gastrointestinal distension and an overall decrease in muscle tone.

Among the factors implicated in bone loss are a decrease in estrogen, calcium deficiency and malabsorption, vitamin D deficiency, loss of musle mass, inactivity, and high-protein diets.

Paget's Disease

Paget's disease is characterized by irregular thickening and softening of the bones. It rarely occurs in individuals under 50. Although the cause, or **etiology** (ē'-tē-OL-ō-jē), of the disease is unknown, some evidence suggests that it might be a slow virus infection. Bone-producing osteoblasts and bone-destroying osteoclasts apparently become uncoordinated. This alters the balance between bone formation and bone destruction. Paget's disease affects the skull, the pelvis, and the bones of the extremities.

Osteomyelitis

The term **osteomyelitis** (os'-tē-ō-mī-i-LĪ-tis) includes all the infectious diseases of bone. These diseases may be localized or widespread and may also involve the periosteum, marrow, and cartilage. Various microorganisms may give rise to bone infection, but most often it is the bacterium *Staphylococcus aureus,* commonly called "staph." These bacteria may reach the bone by various means: the bloodstream, an injury such as a fracture, or an infection such as a sinus infection or a tooth abscess. The infection may destroy extensive areas of bone, spread to nearby joints, and, in rare cases, lead to death by producing abscesses. Antibiotics have been effective in treating the disease and in preventing it from spreading through extensive areas of bone.

Fractures

In simplest terms, a **fracture** is any break in a bone. Usually, the fracture is restored to normal position by manipulation without surgery. This procedure of setting a fracture is called *closed reduction.* In other cases, the fracture must be exposed by surgery before the break is rejoined. This procedure is known as *open reduction.*

Types

Although fractures of bones of the extremities may be classified in several different ways, the following scheme is useful (Figure 6-6).

1. **Partial.** A fracture in which the break across the bone is incomplete.
2. **Complete.** A fracture in which the break across the bone is complete, so that the bone is broken into two pieces.
3. **Closed** or **simple.** A fracture in which the bone does not break through the skin.
4. **Open** or **compound.** A fracture in which the broken ends of the bone protrude through the skin.
5. **Comminuted** (KOM-i-nyoo'-ted). A fracture in which the bone is splintered at the site of impact and smaller fragments of bone are found between the two main fragments.
6. **Greenstick.** A partial fracture in which one side of the bone is broken and the other side bends; occurs only in children.
7. **Spiral.** A fracture in which the bone is usually twisted apart.
8. **Transverse.** A fracture at right angles to the long axis of the bone.
9. **Impacted.** A fracture in which one fragment is firmly driven into the other.
10. **Pott's.** A fracture of the distal end of the fibula, with serious injury of the distal tibial articulation.
11. **Colles'** (KOL-ēz). A fracture of the distal end of the radius in which the distal fragment is displaced posteriorly.
12. **Displaced.** A fracture in which the anatomical alignment of the bone fragments is not preserved.
13. **Nondisplaced.** A fracture in which the anatomical alignment of the bone fragments is preserved.

Fracture Repair

Unlike the skin, which may repair itself within days, or muscle, which may mend in weeks, a bone sometimes requires months to heal. A fractured femur, for example, may take six months to heal. Sufficient calcium to strengthen and harden new bone is deposited only gradually. Also, bone cells grow and reproduce slowly. Moreover, the blood supply to bone is decreased, which helps to explain the difficulty in the healing of an infected bone.

The following steps occur in the repair of a fracture (Figure 6-7).

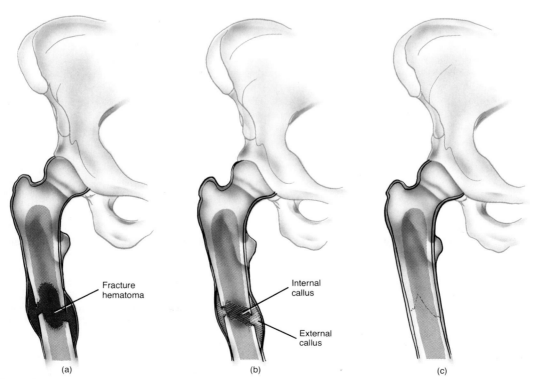

FIGURE 6-7 Fracture repair. (a) Formation of fracture hematoma. (b) Formation of external and internal calli. (c) Completely healed fracture.

1. As a result of the fracture, blood vessels crossing the fracture line are broken. These vessels are found in the periosteum, Haversian systems, and marrow cavity. As blood pours from the torn ends of the vessels, it coagulates and forms a clot in and about the site of the fracture. This clot, called a **fracture hematoma** (hē'-ma-TŌ-ma), usually occurs 6 to 8 hours after the injury. Since the circulation of blood ceases when the fracture hematoma forms, bone cells and periosteal cells at the fracture line die.

2. A growth of new bone tissue—a **callus**—develops in and around the fractured area. It forms a bridge between separated areas of bone. The part of the callus that forms from the osteogenic cells of the torn periosteum and develops around the outside of the fracture is called an *external callus*. The part of the callus that forms from the osteogenic cells of the endosteum and develops between the two ends of bone fragments and between the two marrow cavities is called the *internal callus*.

    Approximately 48 hours after a fracture occurs, the cells that ultimately repair the fracture become actively mitotic. These cells come from the osteogenic layer of the periosteum, the endosteum of the marrow cavity, and the bone marrow. As a result of their accelerated mitotic activity, the cells of the three regions grow toward the fracture. During the first week following the fracture, the cells of the endosteum and bone marrow form new trabeculae in the marrow cavity near the line of fracture. This is the internal callus. During the next few days, osteogenic cells of the periosteum form a collar around each bone fragment. The collar, or external callus, is replaced by trabeculae. The trabeculae of the calli are joined to living and dead portions of the original bone fragments.

3. The final phase of fracture repair is the **remodeling** of the calli. Dead portions of the original fragments are gradually resorbed by osteoclasts. Compact bone replaces spongy bone around the periphery of the fracture. In some cases, the heal-ing is so complete that the fracture line is undetectable, even by x-ray. However, a thickened area on the surface of the bone usually remains as evidence of the fracture site.

CLINICAL APPLICATION

In the past when a fracture failed to unite, the patient could either wait, hoping that nature and time would solve the problem, or choose surgery. Today there is another alternative. The new procedure, called **pulsating electromagnetic fields (PEMFs)**, involves electrotherapy to stimulate bone repair.

Treatment with PEMFs was first used clinically in 1974. Essentially, the fracture is exposed to a weak electrical current generated from coils that are fastened around the cast. Osteoblasts adjacent to the fracture site become more active metabolically in response to the stimulation of the alteration of electrical charges on the surfaces of their membranes. The increased activity apparently causes an acceleration of calcification, vascularization, and endochondral ossification and a subsequent acceleration of fracture repair. It was noted earlier that parathyroid hormone (PTH) increases osteoclastic activity and thus stimulates bone destruction. A recent hypothesis suggests that electricity also heals fractures by keeping PTH from acting on osteoclasts, thus increasing bone formation and repair.

Although the original application of PEMFs was to treat improperly healing fractures, research is now underway to determine if electrical stimulation can be used to regenerate limbs and to stop the growth of tumor cells in humans.

MEDICAL TERMINOLOGY

Achondroplasia (*a* = without; *chondro* = cartilage; *plasia* = growth)   Imperfect ossification within cartilage of long bones during fetal life; also called *fetal rickets*.

Brodie's abscess   Infection in the spongy tissue of a long bone, with a small inflammatory area.

Craniotomy (*cranium* = skull; *tome* = a cutting)   Any surgery that requires cutting through the bones surrounding the brain.

Necrosis (*necros* = death; *osis* = condition)   Death of tissues or organs; in the case of bone, results from deprivation of blood supply resulting from fracture, extensive removal of periosteum in surgery, exposure to radioactive substances, or other causes.

Osteitis (*osteo* = bone)   Inflammation or infection of bone.

Osteoarthritis (*arthro* = joint)   A degenerative condition of bone and also the joint.

Osteoblastoma (*oma* = tumor)   A benign tumor of the osteo blasts.

Osteochondroma (*chondro* = cartilage)   A benign tumor of the bone and cartilage.

Osteoma   A benign bone tumor.

Osteosarcoma (*sarcoma* = connective tissue tumor)   A malignant tumor composed of osseous tissue.

Pott's disease   Inflammation of the backbone, caused by the microorganism that produces tuberculosis.

STUDY OUTLINE

Functions (p. 124)

1. The skeletal system consists of all bones attached at joints and cartilage between joints.
2. The functions of the skeletal system include support, protection, leverage, mineral storage, and blood cell production.

Histology (p. 124)

1. Osseous tissue consists of widely separated cells surrounded by large amounts of intercellular substance. The intercellular substance contains collagenous fibers and abundant hydroxyapatites (mineral salts).
2. Parts of a typical long bone are the diaphysis (shaft), epiphyses (ends), metaphysis, articular cartilage, periosteum, medullary or marrow cavity, and endosteum.
3. Compact (dense) bone consists of Haversian systems with little space between them. Compact bone lies over spongy bone and composes most of the bone tissue of the diaphyses. Functionally, compact bone protects, supports, and resists stress.

4. Spongy (cancellous) bone consists of trabeculae surrounding many red marrow-filled spaces. It forms most of the structure of short, flat, and irregular bones, and the epiphyses of long bones. Functionally, spongy bone stores marrow and provides some support.

Ossification (p. 127)

1. Bone forms by a process called ossification or osteogenesis, which begins when mesenchymal cells become transformed into osteoblasts.
2. The process begins during the sixth or seventh week of embryonic life and continues throughout adulthood. The two types of ossification, intramembranous and endochondral, involve the replacement of a preexisting connective tissue with bone.
3. Intramembranous ossification occurs within fibrous membranes of the embryo and the adult.
4. Endochondral ossification occurs within a cartilage model. The primary ossification center of a long bone is in the diaphysis. Cartilage degenerates, leaving cavities that merge to form the marrow cavity. Osteoblasts lay down bone. Next, ossification occurs in the epiphyses, where bone replaces cartilage, except for the epiphyseal plate.
5. The anatomical zones of the epiphyseal plate are the zones of reserve cartilage, proliferating cartilage, hypertrophic cartilage, and calcified matrix.
6. Because of the activity of the epiphyseal plate, the diaphysis of a bone increases in length by appositional growth.
7. In both types of ossification, spongy bone is laid down first. Compact bone is later reconstructed from spongy bone.

Homeostasis (p. 131)

1. The homeostasis of bone growth and development depends on a balance between bone formation and resorption.
2. Old bone is constantly destroyed by osteoclasts, while new bone is constructed by osteoblasts. This process is called remodeling.
3. Normal growth depends on calcium, phosphorus, and vitamins (A, C, and D) and is controlled by hormones that are responsible for bone mineralization and resorption.

Disorders: Homeostatic Imbalances (p. 131)

1. Rickets is a vitamin D deficiency in children in which the body does not absorb calcium and phosphorus. The bones soften and bend under the body's weight.
2. Osteomalacia is a vitamin D deficiency in adults that leads to demineralization.
3. Osteoporosis is a decrease in the amount and strength of bone tissue due to decreases in hormone output.
4. Paget's disease is the irregular thickening and softening of bones, apparently related to an imbalance between osteoclast and osteoblast activities.
5. Osteomyelitis is a term for the infectious diseases of bones, marrow, and periosteum. It is frequently caused by "staph" bacteria.
6. A fracture is any break in a bone.
7. The types of fractures include: partial, complete, simple, compound, comminuted, greenstick, spiral, transverse, impacted, Pott's, Colles', displaced, and nondisplaced.
8. Fracture repair consists of forming a fracture hematoma, forming a callus, and remodeling.
9. Treatment by pulsating electromagnetic fields (PEMFs) has provided dramatic results in healing fractures that would otherwise not have mended properly. Its application for limb regeneration and stopping the growth of tumor cells is being investigated.

REVIEW QUESTIONS

1. Define the skeletal system. What are its five principal functions?
2. Why is osseous tissue considered a connective tissue? What is its composition?
3. Diagram the parts of a long bone, and list the functions of each part.
4. Distinguish between spongy and compact bone in terms of general appearance, location, and function.
5. Diagram the microscopic appearance of compact bone, and indicate the functions of the various components.
6. How does spongy bone differ histologically from compact bone?
7. What is meant by ossification? Describe the initial events of ossification.
8. Distinguish between the two principal kinds of ossification on the basis of the preexisting connective tissue present.
9. Outline the major events involved in intramembranous and endochondral ossification and explain the principal differences.
10. Describe the histology of the various zones of the epiphyseal plate. How does the plate grow?
11. What is the significance of the epiphyseal line?
12. Define remodeling.
13. How does osteoblast activity in balance with osteoclast activity demonstrate the homeostasis of bone?
14. List the primary factors involved in bone growth and replacement.
15. Distinguish between rickets and osteomalacia. What do the two diseases have in common?
16. What are the principal symptoms of osteoporosis, Paget's disease, and osteomyelitis? What is the etiology of each?
17. What is a fracture? Distinguish several principal kinds.
18. Outline the three basic steps involved in fracture repair.
19. Explain the principal of pulsating electromagnetic fields (PEMFs) in the repair of fractures.
20. Refer to the glossary of medical terminology associated with the skeletal system. Be sure that you can define each term.

- Define the four principal types of bones in the skeleton.
- Describe the various markings on the surfaces of bones.
- Relate the structure of the marking to its function.
- List the components of the axial and appendicular skeleton.
- Identify the bones of the skull and the major markings associated with each.
- Identify the principal sutures and fontanels of the skull.
- Identify the paranasal sinuses of the skull.
- Identify the principal foramina of the skull.
- Identify the bones of the vertebral column and their principal markings.
- List the defining characteristics and curves of each region of the vertebral column.
- Identify the bones of the thorax and their principal markings.
- Contrast herniated (slipped) disc, curvatures, spina bifida, and fractures of the vertebral column as disorders associated with the skeletal system.

7 The Skeletal System: The Axial Skeleton

The skeletal system forms the framework of the body. For this reason, a familiarity with the names, shapes, and positions of individual bones will help you to understand some of the other organ systems. For example, movements such as throwing a ball, typing, and walking require the coordinated use of bones and muscles. To understand how muscles produce different movements, you need to learn the parts of the bones to which the muscles attach. The respiratory system is also highly dependent on bone structure. The bones in the nasal cavity form a series of passageways that help clean, moisten, and warm inhaled air. Furthermore, the bones of the thorax are specially shaped and positioned so the chest can expand during inhalation. Many bones also serve as landmarks to students of anatomy as well as to surgeons. As you will see, bony landmarks can be used to locate the outlines of the lungs and heart, abdominal and pelvic viscera, and structures within the skull. Blood vessels and nerves often run parallel to bones. These structures can be located more easily if the bone is identified first.

We shall study bones by examining the various regions of the body. We shall look at the skull first and see how the bones of the skull relate to each other. We shall then move on to the vertebral column and the chest. This regional approach will allow you to see how all the many bones of the body relate to each other.

TYPES OF BONES

Almost all the bones of the body may be classified into four principal types on the basis of shape: long, short, flat, and irregular. **Long bones** have greater length than width and consist of a diaphysis and two epiphyses. They are slightly curved for strength. A curved bone is structurally designed to absorb the stress of the body weight at several different points so the stress is evenly distributed. If such bones were straight, the weight of the body would be unevenly distributed and the bone would easily fracture. Examples of long bones include bones of the thighs, legs, toes, arms, forearms, and fingers. Figure 6-1a shows the parts of a long bone.

Short bones are somewhat cube-shaped and nearly equal in length and width. Their texture is spongy except at the surface, where there is a thin layer of compact bone. Examples of short bones are the wrist and ankle bones.

Flat bones are generally thin and composed of two more or less parallel plates of compact bone enclosing a layer of spongy bone. The term *diploe* is applied to the spongy bone of the cranial bones. Flat bones afford considerable protection and provide extensive areas for muscle attachment. Examples of flat bones include the cranial bones, which protect the brain, the sternum and ribs, which protect organs in the thorax, and the scapulas.

Irregular bones have complex shapes and cannot be grouped into any of the three categories just described. They also vary in the amount of spongy and compact bone present. Such bones are the vertebrae and certain facial bones.

There are two additional types of bones which are not included in this classification by shape. **Wormian, or sutural (SOO-chur-al), bones** are small clusters of bones between the joints of certain cranial bones (see Figure 7-2e). Their number varies greatly from person to person. **Sesamoid bones** are small bones in tendons where considerable pressure develops, for instance, in the wrist. These, like the Wormian bones, are also variable in number. Two sesamoid bones, the patellas, or kneecaps, are present in all individuals.

SURFACE MARKINGS

The surfaces of bones reveal various structural features adapted to specific functions. These features are called **markings.** Long bones that bear a great deal of weight have large, rounded ends that can form sturdy joints. Other bones have depressions that receive the rounded ends. Rough areas serve as points of attachment for muscles, tendons, and ligaments. Grooves in the surfaces of bones provide for the passage of blood vessels. Openings occur where blood vessels and nerves pass through the bone. Exhibit 7-1 describes the different markings and their functions.

DIVISIONS OF THE SKELETAL SYSTEM

The adult human skeleton usually consists of 206 bones grouped in two principal divisions: the **axial** and the **appendicular.** The longitudinal *axis,* or center, of the human body is a straight line that runs vertically along the body's center of gravity. This imaginary line runs through the head and down to the space between the feet. The midsagittal section is drawn through this line. The axial division of the skeleton consists of the bones that lie around the axis: ribs, breastbone, bones of the skull, and backbone.

The appendicular division contains the bones of the free *appendages,* which are the upper and lower extremities, plus the bones called *girdles,* which connect the free appendages to the axial skeleton.

The 80 bones of the axial division and the 126 bones of the appendicular division are typically grouped as shown in Exhibit 7-2.

Now that you understand how the skeleton is organized into axial and appendicular divisions, refer to Figure 7-1 to see how the two divisions are joined to form the skeleton. The bones of the axial skeleton are shown in yellow. Be certain to locate the following regions of the skeleton: skull, cranium, face, hyoid bone, vertebral column, thorax, shoulder girdle, upper extremity, pelvic girdle, and lower extremity.

SKULL

The **skull,** which contains 22 bones, rests on the superior end of the vertebral column and is composed of two sets of bones: cranial bones and facial bones. The **cranial bones** enclose and protect the brain and the organs of sight, hearing, and balance. The 8 cranial bones are the frontal bone, parietal bones (2), temporal bones (2), occipital bone, sphenoid bone, and ethmoid bone. There are 14 **facial bones:** nasal bones (2), maxillae (2), zygomatic bones (2), mandible, lacrimal bones (2), palatine bones (2), inferior nasal conchae (2), and vomer. Be sure you can locate all the skull bones in the anterior, lateral, median, and posterior views of the skull (Figure 7-2).

EXHIBIT 7-1
BONE MARKINGS

MARKING	DESCRIPTION	EXAMPLE
DEPRESSIONS AND OPENINGS		
Fissure (FISH-ur)	A narrow, cleftlike opening between adjacent parts of bones through which blood vessels or nerves pass.	Superior orbital fissure of the sphenoid bone (Figure 7-2).
Foramen (fō-RĀ-men; *foramen* = hole)	An opening through which blood vessels, nerves, or ligaments pass.	Infraorbital foramen of the maxilla (Figure 7-2).
Meatus (mē-Ā-tus; *meatus* = canal)	A tubelike passageway running within a bone.	External auditory meatus of the temporal bone (Figure 7-2).
Paranasal sinus (*sin* = cavity)	An air-filled cavity within a bone connected to the nasal cavity.	Frontal sinus of the frontal bone (Figure 7-8).
Groove or sulcus (*sulcus* = ditchlike groove)	A furrow or groove that accommodates a soft structure such as a blood vessel, nerve, or tendon.	Intertubercular sulcus of the humerus (Figure 8-4).
Fossa (*fossa* = basinlike depression)	A depression in or on a bone.	Mandibular fossa of the temporal bone (Figure 7-4).
PROCESSES	Any prominent projection.	Mastoid process of the temporal bone (Figure 7-2).
Processes that form joints		
Condyle (KON-dīl; *condylus* = knucklelike process)	A large, articular prominence.	Medial condyle of the femur (Figure 8-10).
Head	A rounded, articular projection supported on the constricted portion (neck) of a bone.	Head of the femur (Figure 8-10).
Facet	A smooth, flat surface.	Articular facet for the tubercle of rib on a vertebra (Figure 7-18).
Processes to which tendons, ligaments, and other connective tissues attach		
Tubercle (TOO-ber-kul; *tube* = knob)	A small, rounded process.	Greater tubercle of the humerus (Figure 8-4).
Tuberosity	A large, rounded, usually roughened process.	Ischial tuberosity of the hipbone (Figure 8-8).
Trochanter (trō-KAN-ter)	A large, blunt projection found only on the femur.	Greater trochanter of the femur (Figure 8-10).
Crest	A prominent border or ridge on a bone.	Iliac crest of the hipbone (Figure 8-7).
Line	A less prominent ridge.	Linea aspera of the femur (Figure 8-10).
Spinous process or spine	A sharp, slender process.	Spinous process of a vertebra (Figure 7-12).
Epicondyle (*epi* = above)	A prominence above a condyle.	Medial epicondyle of the femur (Figure 8-10).

SUTURES

A **suture** (SOO-chur), meaning seam or stitch, is an immovable joint found only between skull bones. Very little connective tissue is found between the bones of the suture. Four prominent skull sutures are:

   1. **Coronal suture** between the frontal bone and the two parietal bones.
   2. **Sagittal suture** between the two parietal bones.
   3. **Lambdoidal** (lam-DOY-dal) **suture** between the parietal bones and the occipital bone.
   4. **Squamosal** (skwa-MŌ-sal) **suture** between the parietal bones and the temporal bones.

Refer to Figures 7-2 and 7-3 for the locations of these sutures. Several other sutures are also shown. Their names are descriptive of the bones they connect. For example, the frontonasal suture is between the frontal bone and the nasal bones. These sutures are indicated in Figures 7-2 to 7-5.

FONTANELS

The "skeleton" of a newly formed embryo consists of cartilage or fibrous membrane structures shaped like bones. Gradually the cartilage or fibrous membrane is replaced by bone. At birth, membrane-filled spaces called **fontanels** (fon'-ta-NELZ; = little fountains), are found between cranial bones (Figure 7-3). These "soft spots" are areas where the bone-making process is not yet complete. They allow the skull to be compressed during birth.

SKULL
Cranium
Face
Hyoid

VERTEBRAL COLUMN

PELVIC GIRDLE

SHOULDER GIRDLE
Clavicle
Scapula
THORAX
Sternum
Ribs
UPPER EXTREMITY
Humerus
Ulna
Radius
Carpals
Metacarpals
Phalanges

VERTEBRAL COLUMN

PELVIC GIRDLE

LOWER EXTREMITY
Femur
Patella
Tibia
Fibula

Tarsals
Metatarsals
Phalanges

(a)

(b)

DANK

FIGURE 7-1 Divisions of the skeletal system. The axial skeleton is indicated in yellow. (a) Anterior view. (b) Posterior view.

Physicians find the fontanels helpful in determining the position of the infant's head prior to delivery. Although an infant may have many fontanels at birth, the form and location of six are fairly constant.

The **anterior (frontal) fontanel** is located between the angles of the two parietal bones and the two segments of the frontal bone. This fontanel is roughly diamond-shaped, and it is the largest of the six fontanels. It usually closes 18 to 24 months after birth.

The **posterior (occipital;** ok-SIP-i-tal) **fontanel** is situated between the two parietal bones and the occipital bone. This diamond-shaped fontanel is considerably

EXHIBIT 7-2
DIVISIONS OF THE SKELETAL SYSTEM

REGIONS OF THE SKELETON	NUMBER OF BONES	REGIONS OF THE SKELETON	NUMBER OF BONES
AXIAL SKELETON		Upper extremities	
Skull		Humerus	2
Cranium	8	Ulna	2
Face	14	Radius	2
Hyoid	1	Carpals	16
Auditory ossicles		Metacarpals	10
(3 in each ear)*	6	Phalanges	28
Vertebral column	26	Pelvic girdle	
Thorax		Coxal, hip, or pelvic bone	2
Sternum	1	Lower extremities	
Ribs	24	Femur	2
	80	Fibula	2
		Tibia	2
APPENDICULAR SKELETON		Patella	2
Pectoral (shoulder) girdles		Tarsals	14
Clavicle	2	Metatarsals	10
Scapula	2	Phalanges	28
			126

* Although the auditory ossicles are not considered part of the axial or appendicular skeleton, but rather as a separate group of bones, they are placed with the axial skeleton for convenience. They are discussed in detail in Chapter 17.

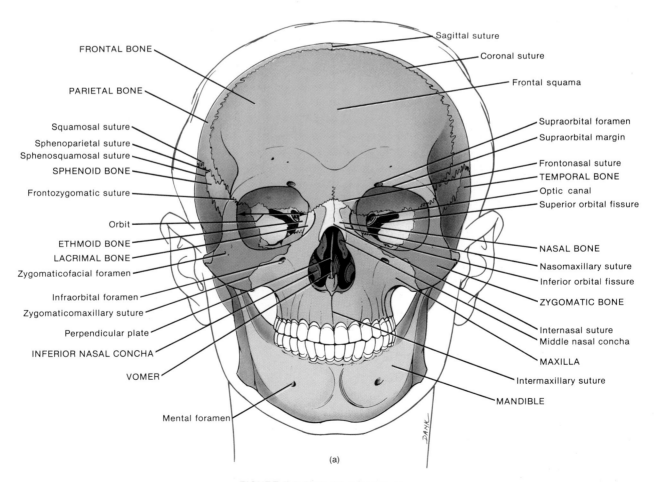

FIGURE 7-2 Skull. (a) Anterior view.

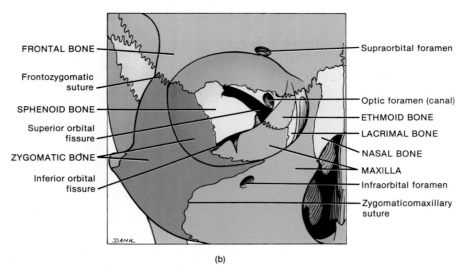

FIGURE 7-2 (*Continued*) Skull. (b) Details of the right orbit in anterior view.

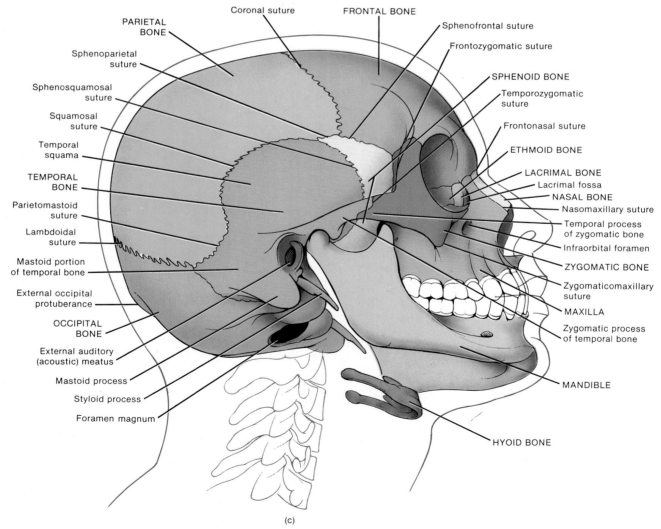

FIGURE 7-2 (*Continued*) Skull. (c) Right lateral view.

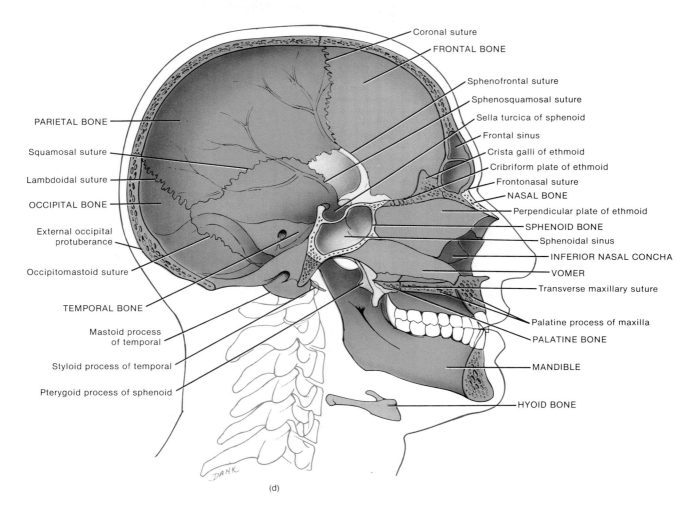

Coronal suture
FRONTAL BONE
Sphenofrontal suture
Sphenosquamosal suture
Sella turcica of sphenoid
Frontal sinus
Crista galli of ethmoid
Cribriform plate of ethmoid
Frontonasal suture
NASAL BONE
Perpendicular plate of ethmoid
SPHENOID BONE
Sphenoidal sinus
INFERIOR NASAL CONCHA
VOMER
Transverse maxillary suture
Palatine process of maxilla
PALATINE BONE
MANDIBLE
HYOID BONE

PARIETAL BONE
Squamosal suture
Lambdoidal suture
OCCIPITAL BONE
External occipital protuberance
Occipitomastoid suture
TEMPORAL BONE
Mastoid process of temporal
Styloid process of temporal
Pterygoid process of sphenoid

DANK

(d)

FIGURE 7-2 (*Continued*) Skull. (d) Median view.

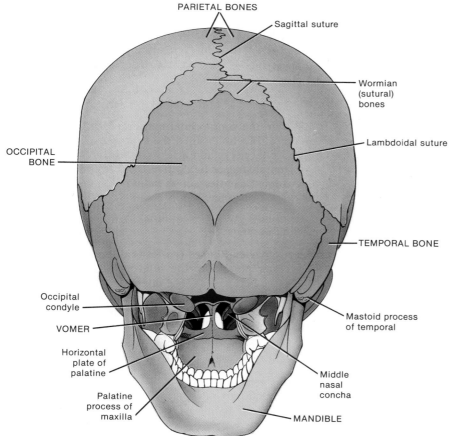

PARIETAL BONES
Sagittal suture
Wormian (sutural) bones
Lambdoidal suture
TEMPORAL BONE
OCCIPITAL BONE
Occipital condyle
VOMER
Horizontal plate of palatine
Palatine process of maxilla
Mastoid process of temporal
Middle nasal concha
MANDIBLE

(e)

FIGURE 7-2 (*Continued*) Skull. (e) Posterior view.

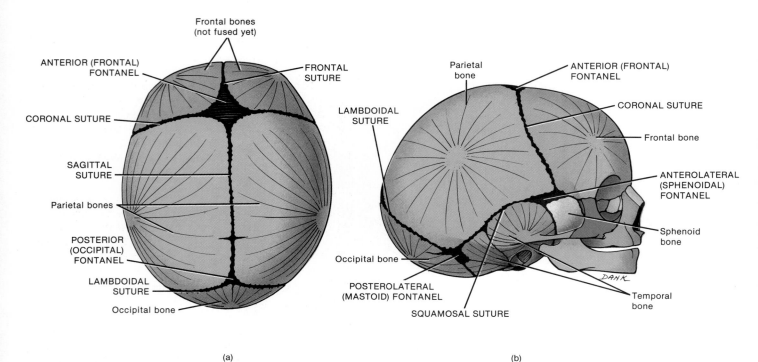

FIGURE 7-3 Fontanels of the skull at birth. (a) Superior view. (b) Right lateral view.

smaller than the anterior fontanel. It generally closes about 2 months after birth.

The **anterolateral (sphenoidal; sfē-NOY-dal) fontanels** are paired. One is located on each side of the skull at the junction of the frontal, parietal, temporal, and sphenoid bones. These fontanels are quite small and irregular in shape. They normally close 3 months after birth.

The **posterolateral (mastoid) fontanels** are also paired. One is situated on each side of the skull at the junction of the parietal, occipital, and temporal bones. These fontanels are irregularly shaped. They begin to close 1 or 2 months after birth, but closure is not generally complete until 12 months.

FRONTAL BONE

The **frontal bone** forms the forehead, the anterior part of the cranium; the roofs of the *orbits* (eye sockets); and most of the anterior part of the cranial floor. Soon after birth the left and right parts of the frontal bone are united by a suture. The suture usually disappears by age 6. If, however, the suture persists throughout life, it is referred to as the **metopic suture.**

If you examine the anterior and lateral views of the skull in Figure 7-2, you will note the **frontal squama** (SKWĀ-ma), or **vertical plate** (*squam* = scale). This scalelike plate, which corresponds to the forehead, gradually slopes down from the coronal suture, then turns abruptly downward.

A thickening of the frontal bone is called the **supraorbital margin.** From this margin the frontal bone extends posteriorly to form the roof of the orbit and part of the floor of the cranial cavity.

CLINICAL APPLICATION

Just above the supraorbital margin is a relatively sharp ridge that overlies the frontal sinus. A blow to the ridge frequently lacerates the skin over it, resulting in bleeding. Bruising of the skin over the ridge causes tissue fluid and blood to accumulate in the surrounding connective tissue and gravitate into the upper eyelid. The resulting swelling and discoloration is called a **"black eye."**

Within the supraorbital margin, slightly medial to its midpoint, is a hole called the **supraorbital foramen.** The supraorbital nerve and artery pass through this foramen. The **frontal sinuses** lie deep to the frontal squama. These mucus-lined cavities act as sound chambers which give the voice resonance.

PARIETAL BONES

The two **parietal bones** (pa-RĪ-i-tal; *paries* = wall) form the greater portion of the sides and roof of the cranial cavity (see Figure 7-2). The internal surfaces of the bones contain many eminences and depressions that accommodate the blood vessels supplying the outer meninx (covering) of the brain called the *dura mater.*

TEMPORAL BONES

The two **temporal bones** form the inferior sides of the cranium and part of the cranial floor. The term *tempora* pertains to the temples.

In the lateral view of the skull in Figure 7-2, notice the **squama** or **squamous portion**—a thin, large, expanded area that forms the anterior and superior part of the temple. Projecting from the inferior portion of the squama is the **zygomatic process,** which articulates with the temporal process of the zygomatic bone. The zygomatic process of the temporal bone and the temporal process of the zygomatic bone constitute the **zygomatic arch.**

At the floor of the cranial cavity, shown in Figure 7-5, is the **petrous portion** of the temporal bone. This portion is triangular and located at the base of the skull between the sphenoid and occipital bones. The petrous portion contains the internal ear, the essential part of the organ of hearing. It also contains the **carotid foramen (canal)** through which the internal carotid artery passes (see Figure 7-4). Posterior to the carotid foramen and anterior to the occipital bone is the **jugular foramen (fossa)** through which the internal jugular vein and the glossopharyngeal (IX) nerve, vagus (X) nerve, and accessory (XI) nerve pass. (As you will see later, the Roman numerals associated with cranial nerves indicate the order in which the nerves arise from the brain, from front to back.)

Between the squamous and petrous portions is a socket called the **mandibular fossa.** Anterior to the mandibular fossa is a rounded eminence, the **articular tubercle.** The mandibular fossa and articular tubercle articulate with the condylar process of the mandible (lower jawbone) to form the temporomandibular joint. The mandibular fossa and articular tubercle are seen best in Figure 7-4.

In the lateral view of the skull in Figure 7-2, you will see the **mastoid portion** of the temporal bone, located posterior and inferior to the external auditory meatus, or ear canal. In the adult, this portion of the bone contains a number of **mastoid air "cells."** These air spaces are separated from the brain only by thin bony partitions.

CLINICAL APPLICATION
If **mastoiditis,** the inflammation of these bony cells, occurs, the infection may spread to the brain or its outer covering. The mastoid air cells do not drain as do the paranasal sinuses.

The **mastoid process** is a rounded projection of the temporal bone posterior to the external auditory meatus. It serves as a point of attachment for several neck muscles. Near the posterior border of the mastoid process is the **mastoid foramen** through which a vein (emissary) to the transverse sinus and a small branch of the occipital artery to the dura mater pass. The **external auditory meatus** is the canal in the temporal bone that leads to the middle ear. The **internal acoustic meatus** is superior to the jugular foramen (see Figure 7-5a). It transmits the facial (VII) and vestibulocochlear (VIII) nerves and the internal auditory artery. The **styloid process** projects downward from the undersurface of the temporal bone and serves as a point of attachment for muscles and ligaments of the tongue and neck.

OCCIPITAL BONE

The **occipital** (ok-SIP-i-tal) **bone** forms the posterior part and a prominent portion of the base of the cranium (Figure 7-4).

The **foramen magnum** is a large hole in the inferior part of the bone through which the medulla oblongata and its membranes, the spinal portion of the accessory (XI) nerve, and the vertebral and spinal arteries pass.

The **occipital condyles** are oval processes with convex surfaces, one on either side of the foramen magnum, which articulate with depressions on the first cervical vertebra.

The **external occipital protuberance** is a prominent projection on the posterior surface of the bone just superior to the foramen magnum. You can feel this structure as a definite bump on the back of your head, just above your neck. The protuberance is also visible in Figure 7-2d.

SPHENOID BONE

The **sphenoid** (SFĒ-noyd) **bone** is situated at the middle part of the base of the skull (Figure 7-5). The combining form *spheno* means wedge. This bone is referred to as the keystone of the cranial floor because it articulates with all the other cranial bones. If you view the floor of the cranium from above, you will note that the sphenoid articulates with the temporal bones anteriorly and the occipital bone posteriorly. It lies posterior and slightly superior to the nasal cavities and forms part of the floor and sidewalls of the eye socket. The shape of the sphenoid is frequently described as a bat with outstretched wings.

The **body** of the sphenoid is the cubelike central portion between the ethmoid and occipital bones. It contains the **sphenoidal sinuses,** which drain into the nasal cavity (see Figure 7-8). On the superior surface of the sphenoid body is a depression called the **sella turcica** (SEL-a-TUR-si-ka = Turk's saddle). This depression houses the pituitary gland.

The **greater wings** of the sphenoid are lateral projections from the body and form the anterolateral floor of the cranium. The greater wings also form part of the lateral wall of the skull just anterior to the temporal bone. The **lesser wings** are anterior and superior to the greater wings. They form part of the floor of the cranium and the posterior part of the **orbit,** or eye socket.

Between the body and lesser wing, you can locate the **optic foramen** through which the optic (II) nerve and ophthalmic artery pass. Lateral to the body between the greater and lesser wings is a somewhat triangular slit called the **superior orbital fissure.** It is an opening for the oculomotor (III) nerve, trochlear (IV) nerve, ophthalmic branch of the trigeminal (V) nerve, and abducens (VI) nerve. This fissure may also be seen in the anterior view of the skull in Figure 7-2.

On the inferior part of the sphenoid bone you can see the **pterygoid** (TER-i-goyd) **processes.** These structures project inferiorly from the points where the body and greater wings unite. The pterygoid processes form part of the lateral walls of the nasal cavities.

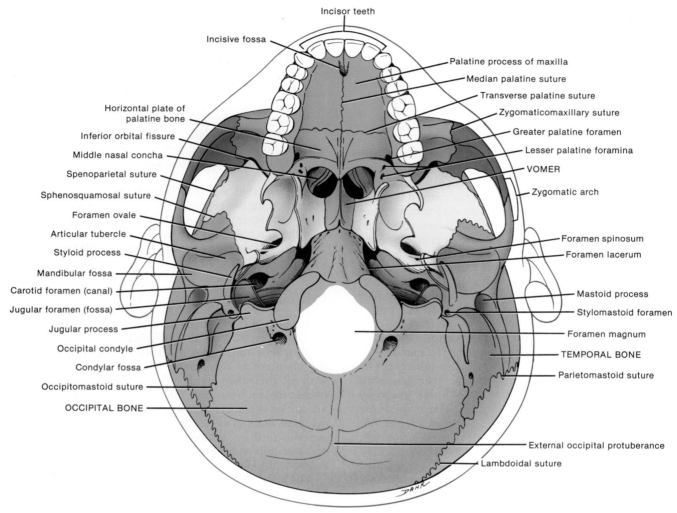

FIGURE 7-4 Skull in inferior view.

ETHMOID BONE

The **ethmoid bone** is a light, spongy bone located in the anterior part of the floor of the cranium between the orbits. It is anterior to the sphenoid and posterior to the nasal bones (Figure 7-6). This bone forms part of the anterior portion of the cranial floor, the medial wall of the orbits, the superior portions of the nasal septum, or partition, and most of the sidewalls of the nasal roof. The ethmoid is the principal supporting structure of the nasal cavities.

Its **lateral masses** or **labyrinths** compose most of the wall between the nasal cavities and the orbits. They contain several air spaces, or "cells," ranging in number from 3 to 18. It is from these "cells" that the bone derives its name (*ethmos* = sieve). The ethmoid "cells" together form the **ethmoidal sinuses.** The sinuses are shown in Figure 7-8. The **perpendicular plate** forms the superior portion of the nasal septum (see Figure 7-7). The **cribriform (horizontal) plate,** lies in the anterior floor of the cranium and forms the roof of the nasal cavity. Projecting upward from the horizontal plate is a triangular process

called the **crista galli** (= cock's comb). This structure serves as a point of attachment for the membranes that cover the brain.

The labyrinths contain two thin, scroll-shaped bones on either side of the nasal septum. These are called the **superior nasal concha** (KONG-ha; *concha* = shell) and the **middle nasal concha.** The conchae allow for the efficient circulation and filtration of inhaled air before it passes into the trachea, the bronchi, and the lungs.

NASAL BONES

The paired **nasal bones** are small, oblong bones that meet at the middle and superior part of the face (see Figures 7-2 and 7-7). Their fusion forms part of the bridge of the nose. The inferior portion of the nose, indeed the major portion, consists of cartilage.

MAXILLAE

The paired maxillary bones unite to form the upper jawbone (Figure 7-7). The **maxillae** (mak-SIL-ē) articulate

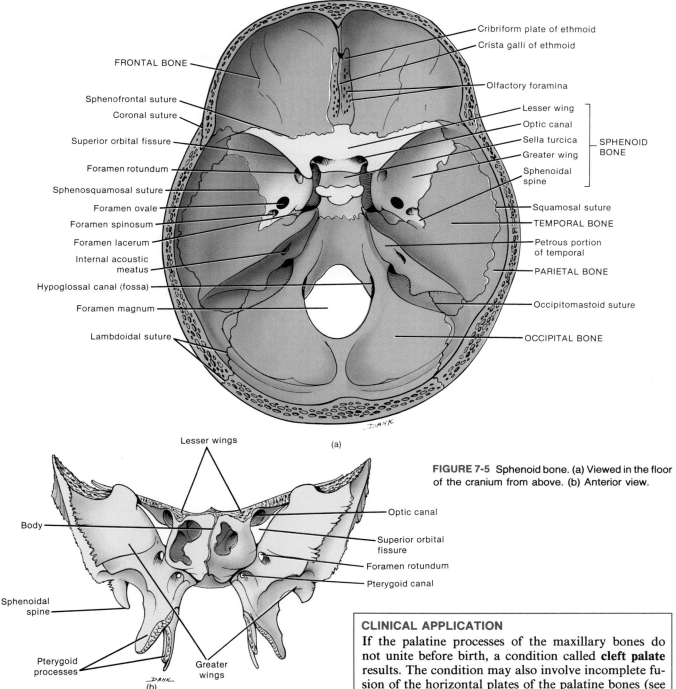

FRONTAL BONE

Sphenofrontal suture
Coronal suture
Superior orbital fissure
Foramen rotundum
Sphenosquamosal suture
Foramen ovale
Foramen spinosum
Foramen lacerum
Internal acoustic meatus
Hypoglossal canal (fossa)
Foramen magnum
Lambdoidal suture

Cribriform plate of ethmoid
Crista galli of ethmoid
Olfactory foramina
Lesser wing
Optic canal
Sella turcica
Greater wing   SPHENOID BONE
Sphenoidal spine
Squamosal suture
TEMPORAL BONE
Petrous portion of temporal
PARIETAL BONE
Occipitomastoid suture
OCCIPITAL BONE

(a)

Lesser wings

Body

Sphenoidal spine

Pterygoid processes

Greater wings

(b)

Optic canal
Superior orbital fissure
Foramen rotundum
Pterygoid canal

FIGURE 7-5 Sphenoid bone. (a) Viewed in the floor of the cranium from above. (b) Anterior view.

with every bone of the face except the mandible, or lower jawbone. They form part of the floors of the orbits, part of the roof of the mouth (most of the hard palate), and part of the lateral walls and floor of the nasal cavities.

Each maxillary bone contains a **maxillary sinus** that empties into the nasal cavity (see Figure 7-8). The **alveolar** (al-VĒ-ō-lar) **process** (*alveolus* = hollow) contains the bony sockets or **alveoli** into which the maxillary (upper) teeth are set. The **palatine process** is a horizontal projection of the maxilla that forms the anterior three-fourths of the hard palate, or anterior portion of the roof of the oral cavity. The two portions of the maxillary bones unite, and the fusion is normally completed before birth.

CLINICAL APPLICATION

If the palatine processes of the maxillary bones do not unite before birth, a condition called **cleft palate** results. The condition may also involve incomplete fusion of the horizontal plates of the palatine bones (see Figure 7-4). Another form of this condition, called **cleft lip,** involves a split in the upper lip. Cleft lip is often associated with cleft palate. Depending on the extent and position of the cleft, speech and swallowing may be affected. A surgical procedure can sometimes improve the condition.

A fissure associated with the maxilla and sphenoid bone is the **inferior orbital fissure.** It is located between the greater wing of the sphenoid and the maxilla (see Figure 7-4). It transmits the maxillary branch of the trigeminal (V) nerve, the infraorbital vessels, and the zygomatic nerve.

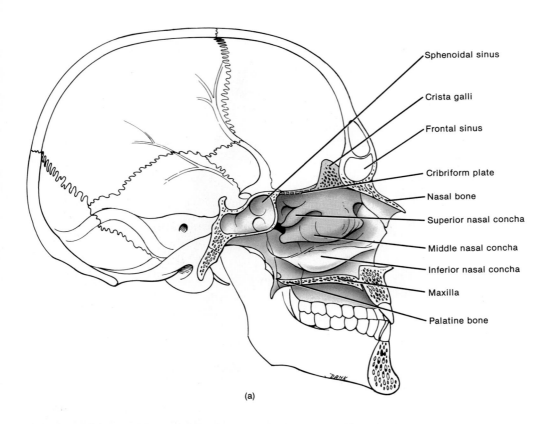

(a)

FIGURE 7-6 Ethmoid bone. (a) Median view showing the ethmoid bone on the inner aspect of the left part of the skull.

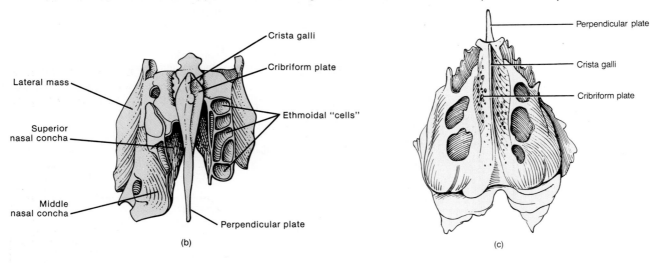

(b)

(c)

FIGURE 7-6 (*Continued*) Ethmoid bone. (b) Anterior view. A frontal section has been made through the left side to expose the ethmoidal "cells." (c) Superior view. (d) Highly diagrammatic representation showing the approximate position of the ethmoid bone in the skull in anterior view.

(d)

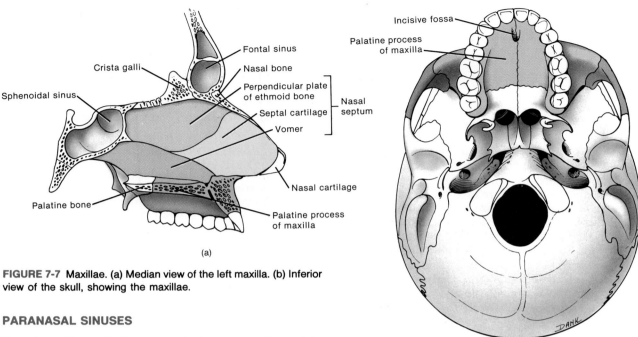

(a)

(b)

FIGURE 7-7 Maxillae. (a) Median view of the left maxilla. (b) Inferior view of the skull, showing the maxillae.

PARANASAL SINUSES

Paired cavities, called **paranasal sinuses,** are located in certain bones near the nasal cavity (Figure 7-8). The paranasal sinuses are lined with mucous membranes that are continuous with the lining of the nasal cavity. Cranial bones containing paranasal sinuses are the frontal bone, the sphenoid, the ethmoid, and the maxillae. (The sinuses were described in the discussion of each of these bones.) Besides producing mucus, the paranasal sinuses lighten the skull bones and serve as resonant chambers for sound.

CLINICAL APPLICATION
Secretions produced by the mucous membranes of the paranasal sinuses drain into the nasal cavity. An inflammation of the membranes due to an allergic reac-

tion or infection is called **sinusitis.** If the membranes swell enough to block drainage into the nasal cavity, fluid pressure builds up in the paranasal sinuses and a sinus headache results.

ZYGOMATIC BONES

The two **zygomatic bones (malars),** commonly referred to as the cheekbones, form the prominences of the cheeks and part of the outer wall and floor of the orbits (Figure 7-2b).

FIGURE 7-8 Paranasal sinuses. (a) Anterior view. (b) Median view.

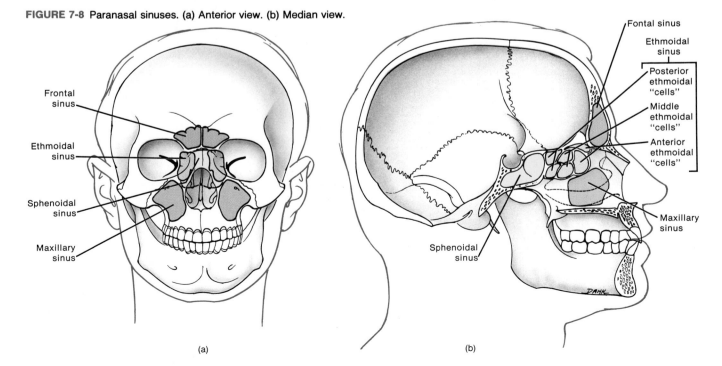

(a)

(b)

The **temporal process** of the zygomatic bone projects posteriorly and articulates with the zygomatic process of the temporal bone. These two processes form the **zygomatic arch** (Figure 7-4).

MANDIBLE

The **mandible** or lower jawbone is the largest, strongest facial bone (Figure 7-9). It is the only movable bone in the skull.

In the lateral view you can see that the mandible consists of a curved, horizontal portion called the **body** and two perpendicular portions called the **rami.** The **angle** of the mandible is the area where each ramus meets the body. Each ramus has a **condylar** (KON-di-lar) **process** that articulates with the mandibular fossa and articular tubercle of the temporal bone to form the temporomandibular (TM) joint. It also has a **coronoid** (KOR-ō-noyd) **process** to which the temporalis muscle attaches. The depression between the coronoid and condylar processes is called the **mandibular notch.** The **alveolar process** is an arch containing the sockets **(alveoli)** for the mandibular (lower) teeth.

CLINICAL APPLICATION

Opening of the mouth very wide, as in a very large yawn, may cause displacement of the condylar process of the mandible from the mandibular fossa of temporal bone, thereby producing a **dislocation of the jaw.** The displacement is usually bilateral and the patient is unable to close the mouth.

LACRIMAL BONES

The paired **lacrimal bones** (LAK-ri-mal; *lacrima* = tear) are thin bones roughly resembling a fingernail in size and shape. They are the smallest bones of the face. These bones are posterior and lateral to the nasal bones in the medial wall of the orbit. They can be seen in the anterior and lateral views of the skull in Figure 7-2. The lacrimal bones form a part of the medial wall of the orbit.

PALATINE BONES

The two **palatine** (PAL-a-tīn) **bones** are L-shaped and form the posterior portion of the hard palate, part of the floor and lateral wall of the nasal cavity, and a small portion of the floors of the orbits. The posterior portion of the hard palate, which separates the nasal cavity from the oral cavity, is formed by the **horizontal plates** of the palatine bones. These can be seen in Figure 7-4.

INFERIOR NASAL CONCHAE

Refer to the views of the skull in Figure 7-2a and 7-6a. The two **inferior nasal conchae** are scroll-like bones that form a part of the lateral wall of the nasal cavity and project into the nasal cavity inferior to the superior and middle nasal conchae of the ethmoid bone. They serve the same function as the superior and middle nasal conchae, that is, they allow for the circulation and filtration of air before it passes into the lungs. The inferior nasal conchae are separate bones and not part of the ethmoid.

VOMER

The **vomer** (= plowshare) is a roughly triangular bone that forms the inferior and posterior part of the nasal septum. It is clearly seen in the anterior view of the skull in Figure 7-2a and the inferior view in Figure 7-4.

The inferior border of the vomer articulates with the cartilage septum that divides the nose into a right and left nostril. Its superior border articulates with the perpendicular plate of the ethmoid bone. Thus the structures that form the **nasal septum,** or partition, are the perpendicular plate of the ethmoid, the septal cartilage, and the vomer (Figure 7-7a). If the vomer is deviated, that is, pushed to one side, the nasal chambers are of unequal size.

CLINICAL APPLICATION

A **deviated nasal septum** is deflected laterally from the midline of the nose. The deviation usually occurs

---

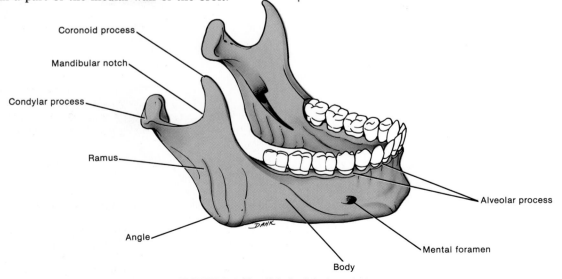

Coronoid process

Mandibular notch

Condylar process

Ramus

Angle

Alveolar process

Mental foramen

Body

FIGURE 7-9 Mandible in right lateral view.

at the junction of bone with the septal cartilage. If the deviation is severe, it may entirely block the nasal passageway. Even though the blockage may not be complete, infection and inflammation develop and cause nasal congestion, blockage of the paranasal sinus openings, and chronic sinusitis.

FORAMINA

Some foramina of the skull were mentioned along with the descriptions of the cranial and facial bones with which they are associated. As preparation for studying other systems of the body, especially the nervous and cardiovascular systems, these foramina, as well as additional ones, and the structures passing through them are listed in Exhibit 7-3. For your convenience and for future reference, the foramina are listed alphabetically.

HYOID BONE

The single **hyoid bone** (*hyoedes* = U-shaped) is a unique component of the axial skeleton because it does not articulate with any other bone. Rather, it is suspended from the styloid process of the temporal bone by ligaments and muscles. The hyoid is located in the neck between the mandible and larynx. It supports the tongue and provides attachment for some of its muscles. Refer to the anterior and lateral views of the skull in Figure 7-2 to see the position of the hyoid bone.

The hyoid consists of a horizontal body and paired projections called the **lesser cornu** (*cornu* = horns) and the **greater cornu** (Figure 7-10). Muscles and ligaments attach to these paired projections.

VERTEBRAL COLUMN

DIVISIONS

The **vertebral column,** or **spine,** together with the sternum and ribs, constitutes the skeleton of the **trunk** of the body. The vertebral column is composed of a series of bones called **vertebrae.** In the average adult, the column measures about 71 cm (28 inches) in length. In effect, the vertebral column is a strong, flexible rod that moves anteriorly, posteriorly, and laterally. It encloses and protects the spinal cord, supports the head, and serves as a point of attachment for the ribs and the muscles of the back.

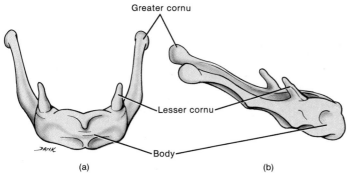

FIGURE 7-10 Hyoid bone. (a) Anterior view. (b) Right lateral view.

Between the vertebrae are openings called **intervertebral foramina.** The nerves that connect the spinal cord to various parts of the body pass through these openings.

The adult vertebral column typically contains 26 vertebrae (Figure 7-11). These are distributed as follows: 7 **cervical vertebrae** (*cervix* = neck) in the neck region; 12 **thoracic vertebrae** posterior to the thoracic cavity; 5 **lumbar vertebrae** (*lumbus* = loin) supporting the lower back; 5 **sacral vertebrae** fused into one bone called the **sacrum;** and usually 4 **coccygeal** (kok-SIJ-ē-al) **vertebrae** fused into one or two bones called the **coccyx** (KOK-six). Prior to the fusion of the sacral and coccygeal vertebrae, the total number of vertebrae is 33.

Between adjacent vertebrae from the axis to the sacrum are fibrocartilaginous **intervertebral discs.** Each disc is composed of an outer fibrous ring consisting of fibrocartilage called the *annulus fibrosus* and an inner soft, pulpy, highly elastic structure called the *nucleus pulposus.* The discs form strong joints, permit various movements of the vertebral column, and absorb vertical shock. Under compression, they flatten, broaden, and bulge from their intervertebral spaces.

CURVES

When viewed from the side, the vertebral column shows four **curves** (Figure 7-11b). From the anterior view, these are alternately convex, meaning they curve toward the viewer, and concave, meaning they curve away from the viewer. The curves of the column, like the curves in a long bone, are important because they increase its strength, help maintain balance in the upright position, absorb shocks from walking, and help protect the column from fracture.

In the fetus, there is only a single anteriorly concave curve. At approximately the third postnatal month, when an infant begins to hold its head erect, the **cervical curve** develops. Later, when the child stands and walks, the **lumbar curve** develops. The cervical and lumbar curves are anteriorly convex. Because they are modifications of the fetal positions, they are called **secondary curves.** The other two curves, the **thoracic curve** and the **sacral curve,** are anteriorly concave. Since they retain the anterior concavity of the fetus, they are referred to as **primary curves.**

TYPICAL VERTEBRA

Although there are variations in size, shape, and detail in the vertebrae in different regions of the column, all the vertebrae are basically similar in structure (Figure 7-12). A typical vertebra consists of the following components.

1. The **body** is the thick, disc-shaped anterior portion that is the weight-bearing part of a vertebra. Its superior and inferior surfaces are roughened for the attachment of intervertebral discs. The anterior and lateral surfaces contain nutrient foramina for blood vessels.
2. The **vertebral arch (neural arch)** extends posteriorly from the body of the vertebra. With the body of the vertebra, it surrounds the spinal cord. It is formed by two short, thick processes, the **pedicles** (PED-i-kuls), which project posteriorly from the body to unite with the laminae. The **laminae** (LAM-

EXHIBIT 7-3
SUMMARY OF FORAMINA OF THE SKULL

FORAMEN	LOCATION	STRUCTURES PASSING THROUGH
Carotid (Figure 7-4)	Petrous portion of temporal.	Internal carotid artery.
Greater palatine (Figure 7-4)	Posterior angle of hard palate.	Greater palatine nerve and greater palatine vessels.
Hypoglossal (Figure 7-5)	Superior to base of occipital condyles.	Hypoglossal (XII) nerve and branch of ascending pharyngeal artery.
Incisive (Figure 7-7)	Posterior to incisor teeth.	Branches of descending palatine vessels and nasopalatine nerve.
Inferior orbital (Figure 7-4)	Between greater wing of sphenoid and maxilla.	Maxillary branch of trigeminal (V) nerve, zygomatic nerve, and infraorbital vessels.
Infraorbital (Figure 7-2a)	Inferior to orbit in maxilla.	Infraorbital nerve and artery.
Jugular (Figure 7-4)	Posterior to carotid canal between petrous portion of temporal and occipital.	Internal jugular vein, glossopharyngeal (IX) nerve, vagus (X) nerve, accessory (XI) nerve, and sigmoid sinus.
Lacerum (Figure 7-5)	Bounded anteriorly by sphenoid, posteriorly by petrous portion of temporal, and medially by the sphenoid and occipital.	Internal carotid artery and branch of ascending pharyngeal artery.
Lacrimal (Figure 7-2c)	Lacrimal bone.	Lacrimal (tear) duct.
Lesser palatine (Figure 7-4)	Posterior to greater palatine foramen.	Lesser palatine nerves and artery.
Magnum (Figure 7-4)	Occipital bone.	Medulla oblongata and its membranes, the accessory (XI) nerve, and the vertebral and spinal arteries and meninges.
Mandibular	Medial surface of ramus of mandible.	Inferior alveolar nerve and vessels.
Mastoid	Posterior border of mastoid process of temporal bone.	Emissary vein to transverse sinus and branch of occipital artery to dura mater.
Mental (Figure 7-9)	Inferior to second premolar tooth in mandible.	Mental nerve and vessels.
Olfactory (Figure 7-5)	Cribriform plate of ethmoid.	Olfactory (I) nerve.
Optic (Figure 7-5)	Between upper and lower portions of small wing of sphenoid.	Optic (II) nerve and ophthalmic artery.
Ovale (Figure 7-5)	Greater wing of sphenoid.	Mandibular branch of trigeminal (V) nerve.
Rotundum (Figure 7-5)	Junction of anterior and medial parts of sphenoid.	Maxillary branch of trigeminal (V) nerve.
Spinosum (Figure 7-5)	Posterior angle of sphenoid.	Middle meningeal vessels.
Stylomastoid (Figure 7-4)	Between styloid and mastoid processes of temporal.	Facial (VII) nerve and stylomastoid artery.
Superior orbital (Figure 7-5)	Between greater and lesser wings of sphenoid.	Oculomotor (III) nerve, trochlear (IV) nerve, ophthalmic branch of trigeminal (V) nerve, and abducens (VI) nerve.
Supraorbital (Figure 7-2a)	Supraorbital margin of orbit.	Supraorbital nerve and artery.
Zygomaticofacial (Figure 7-2a)	Zygomatic bone.	Zygomaticofacial nerve and vessels.

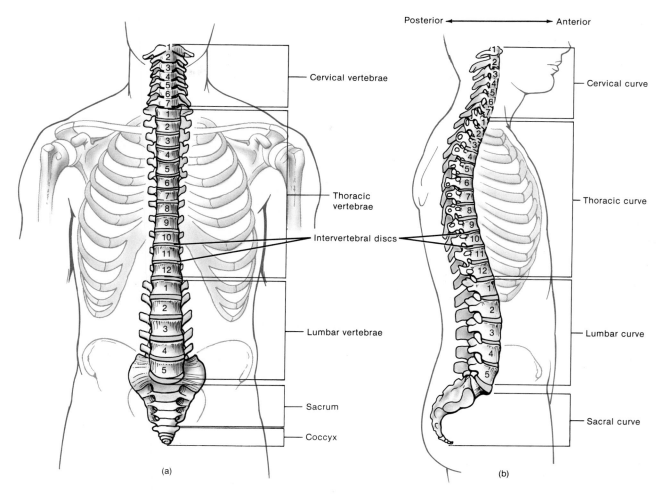

FIGURE 7-11 Vertebral column. (a) Anterior view. (b) Right lateral view.

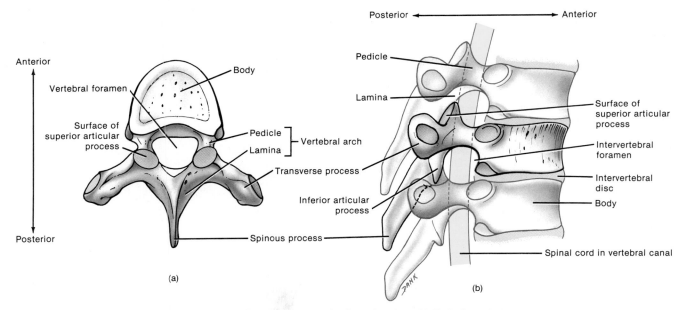

FIGURE 7-12 Typical vertebra. (a) Superior view. (b) Right lateral view.

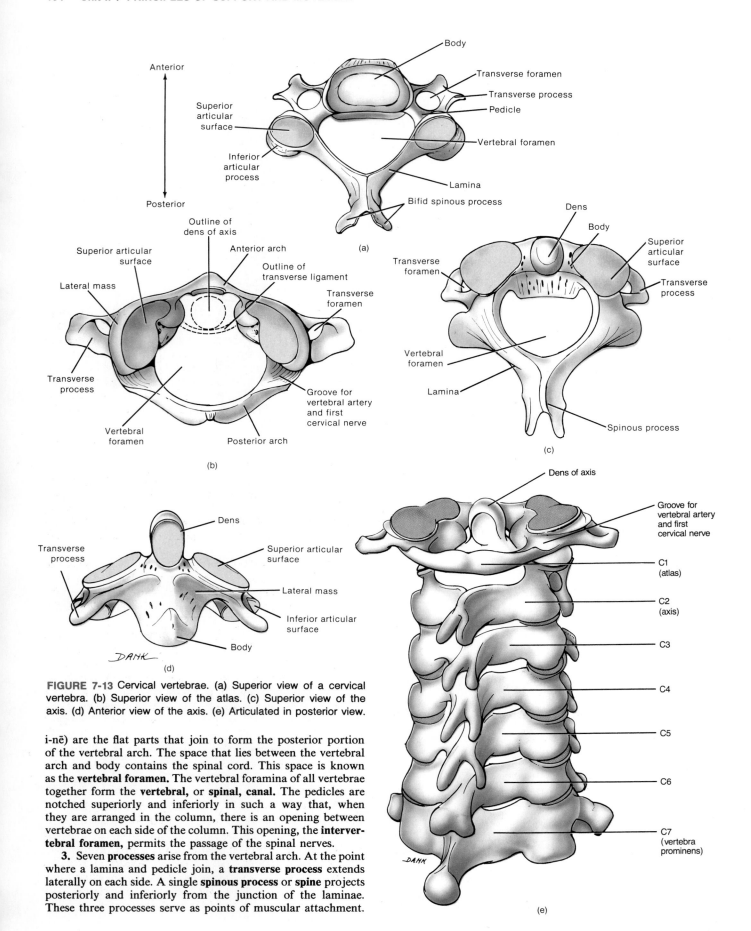

FIGURE 7-13 Cervical vertebrae. (a) Superior view of a cervical vertebra. (b) Superior view of the atlas. (c) Superior view of the axis. (d) Anterior view of the axis. (e) Articulated in posterior view.

i-nē) are the flat parts that join to form the posterior portion of the vertebral arch. The space that lies between the vertebral arch and body contains the spinal cord. This space is known as the **vertebral foramen.** The vertebral foramina of all vertebrae together form the **vertebral,** or **spinal, canal.** The pedicles are notched superiorly and inferiorly in such a way that, when they are arranged in the column, there is an opening between vertebrae on each side of the column. This opening, the **intervertebral foramen,** permits the passage of the spinal nerves.

3. Seven **processes** arise from the vertebral arch. At the point where a lamina and pedicle join, a **transverse process** extends laterally on each side. A single **spinous process** or **spine** projects posteriorly and inferiorly from the junction of the laminae. These three processes serve as points of muscular attachment.

The remaining four processes form joints with other vertebrae. The two **superior articular processes** of a vertebra articulate with the vertebra immediately superior to them. The two **inferior articular processes** of a vertebra articulate with the vertebra inferior to them.

CERVICAL REGION

When viewed from above, it can be seen that the bodies of **cervical vertebrae** are smaller than those of the thoracic vertebrae (Figure 7-13). The arches, however, are larger. The spinous processes of the second through sixth cervical vertebrae are often *bifid,* that is, with a cleft. All cervical vertebrae have three foramina: the vertebral foramen and two transverse foramina. Each cervical transverse process contains a **transverse foramen** through which the vertebral artery and its accompanying vein and nerve fibers pass.

The first two cervical vertebrae differ considerably from the others. The first cervical vertebra (C1), the **atlas,** is named for its support of the head. Essentially, the atlas is a ring of bone with **anterior** and **posterior arches** and large **lateral masses.** It lacks a body and a spinous process. The superior surfaces of the lateral masses, called **superior articular surfaces,** are concave and articulate with the occipital condyles of the occipital bone. This articulation permits the movement seen when nodding the head. The inferior surfaces of the lateral masses, the **inferior articular surfaces,** articulate with the second cervical vertebra. The transverse processes and transverse foramina of the atlas are quite large.

The second cervical vertebra (C2), the **axis,** does have a body. A peglike process called the **dens,** or **odontoid process,** projects up through the ring of the atlas. The dens makes a pivot on which the atlas and head rotate. This arrangement permits side-to-side rotation of the head.

> **CLINICAL APPLICATION**
> In various instances of trauma, the dens of the axis may be driven into the medulla oblongata of the brain, usually with instantly fatal results. This injury is the usual cause of fatality in **whiplash injuries.**

The third through sixth cervical vertebrae (C3–C6) correspond to the structural pattern of the typical cervical vertebra previously described.

The seventh cervical vertebra (C7), called the **vertebra prominens,** is somewhat different. It is marked by a large, nonbifid spinous process that may be seen and felt at the base of the neck (see Figure 7-11).

THORACIC REGION

Viewing a typical **thoracic vertebra** from above, you can see that it is considerably larger and stronger than a vertebra of the cervical region (Figure 7-14). In addition, the spinous process on each vertebra is long, pointed, and directed inferiorly. Thoracic vertebrae also have longer and heavier transverse processes than cervical vertebrae.

Except for the eleventh and twelfth thoracic vertebrae, the transverse processes have **facets** for articulating with the tubercles of the ribs. The bodies of thoracic vertebrae also have whole **facets** or half facets, called **demifacets,** for articulation with the heads of the ribs. The first thoracic vertebra (T1) has, on either side of its body, a superior whole facet and an inferior demifacet. The superior facet articulates with the first rib, and the inferior demifacet, together with the superior demifacet of the second thoracic vertebra (T2), forms a facet for articulation with the second rib. The second through eighth thoracic vertebrae (T2–T8) have two demifacets on each side, a larger superior demifacet and a smaller inferior demifacet. When the vertebrae are articulated, they form whole facets for the heads of the ribs. The ninth thoracic vertebra (T9) has a single superior demifacet on either side of its body. The tenth through twelfth thoracic vertebrae (T10–T12) have whole facets on either side of their bodies.

LUMBAR REGION

The **lumbar vertebrae** (L1–L5) are the largest and strongest in the column (Figure 7-15). Their various projections are short and thick. The superior articular processes are directed medially instead of superiorly. The inferior articular processes are directed laterally instead of inferiorly. The spinous process is quadrilateral in shape, thick and broad, and projects nearly straight posteriorly. The spinous process is well adapted for the attachment of the large back muscles.

SACRUM AND COCCYX

The **sacrum** is a triangular bone formed by the union of five sacral vertebrae. These are indicated in Figure 7-16 as S1 through S5. The sacrum serves as a strong foundation for the pelvic girdle. It is positioned at the posterior portion of the pelvic cavity between the two hipbones.

The concave anterior side of the sacrum faces the pelvic cavity. It is smooth and contains four **transverse lines** that mark the joining of the vertebral bodies. At the ends of these lines are four pairs of **anterior sacral (pelvic) foramina.**

The convex, posterior surface of the sacrum is irregular. It contains a **median sacral crest,** the fused spinous processes of the upper sacral vertebrae; a **lateral sacral crest,** the transverse processes of the sacral vertebrae; and four pairs of **posterior sacral (dorsal) foramina.** These foramina communicate with the anterior sacral foramina through which nerves and blood vessels pass. The **sacral canal** is a continuation of the vertebral canal. The laminae of the fifth sacral vertebra, and sometimes the fourth, fail to meet. This leaves an inferior entrance to the vertebral canal called the **sacral hiatus** (hi-Ā-tus).

The superior border of the sacrum exhibits an anteriorly projecting border, the **sacral promontory** (PROM-on-tō'-rē). It is an obstetrical landmark for measurements of the pelvis. An imaginary line running from the superior surface of the symphysis pubis to the sacral promontory separates the abdominal and pelvic cavities. Laterally, the sacrum has a large **auricular surface** for articulating with the ilium of the hipbone. Its **superior articular processes** articulate with the fifth lumbar vertebra.

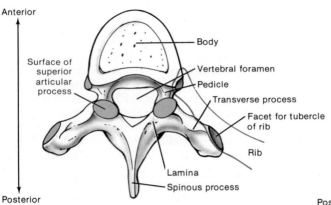

FIGURE 7-14 Thoracic vertebrae. (a) Superior view. (b) Articulated in right lateral view.

The **coccyx** is also triangular in shape and is formed by the fusion of the coccygeal vertebrae, usually the last four. These are indicated in Figure 7-16 as Co1 through Co4. The coccyx articulates superiorly with the sacrum. The coccyx is the most rudimentary part of the column, representing the vestige of a tail.

THORAX

Anatomically, the term **thorax** refers to the chest. The skeletal portion of the thorax is a bony cage formed by the sternum, costal cartilage, ribs, and the bodies of the thoracic vertebrae (Figure 7-17).

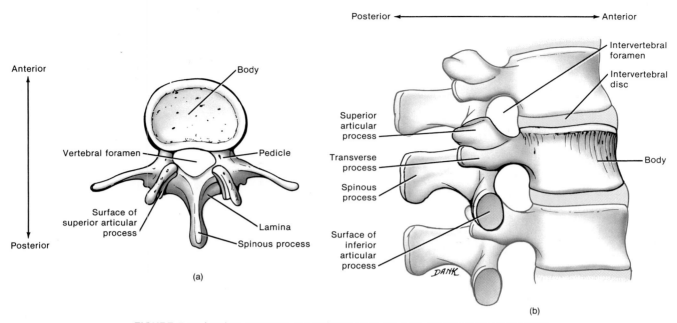

FIGURE 7-15 Lumbar vertebrae. (a) Superior view. (b) Articulated in right lateral view.

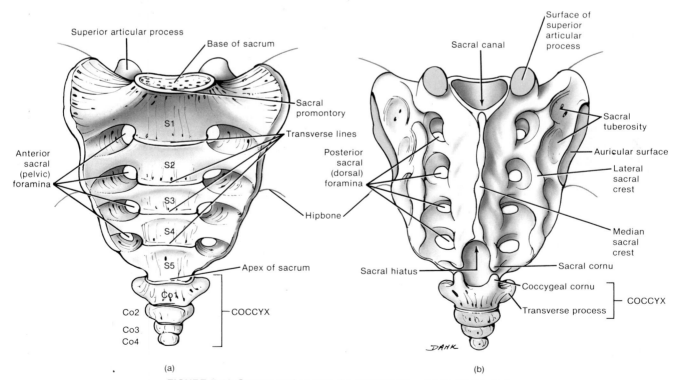

FIGURE 7-16 Sacrum and coccyx. (a) Anterior view. (b) Posterior view.

FIGURE 7-17 Skeleton of the thorax in anterior view.

The thoracic cage is roughly cone-shaped, the narrow portion being superior and the broad portion inferior. It is flattened from front to back. The thoracic cage encloses and protects the organs in the thoracic cavity. It also provides support for the bones of the shoulder girdle and upper extremities.

STERNUM

The **sternum,** or breastbone, is a flat, narrow bone measuring about 15 cm (6 inches) in length. It is located in the median line of the anterior thoracic wall.

The sternum (see Figure 7-17) consists of three basic portions: the **manubrium** (ma-NOO-brē-um), the triangular, superior portion; the **body,** the middle, largest portion; and the **xiphoid** (ZĪ-foyd) **process,** the inferior, smallest portion. The manubrium has a depression on its superior surface called the **jugular (suprasternal) notch.** On each side of the jugular notch are **clavicular notches** that articulate with the medial ends of the clavicles. The manubrium also articulates with the first and second ribs. The body of the sternum articulates directly or indirectly with the second through tenth ribs. The xiphoid process has no ribs attached to it but provides attachment for some abdominal muscles. The xiphoid process consists of hya-

line cartilage during infancy and childhood and does not ossify completely until about age 40. If the hands of a rescuer are mispositioned during cardiopulmonary resuscitation (CPR), there is the danger of fracturing the ossified xiphoid process, separating it from the body, and driving it into the liver.

CLINICAL APPLICATION

Since the sternum possesses red bone marrow throughout life and because it is readily accessible, it is a common site for **marrow biopsy.** Under a local anesthetic, a wide-bore needle is introduced into the marrow cavity of the sternum for aspiration of a sample of red bone marrow. This procedure is called a **sternal puncture.**

The sternum may also be split in the midsaggital plane to allow surgeons access to mediastinal structures such as the thymus gland, heart, and great vessels of the heart.

RIBS

Twelve pairs of **ribs** make up the sides of the thoracic cavity (see Figure 7-17). The ribs increase in length from the first through seventh. Then they decrease in length

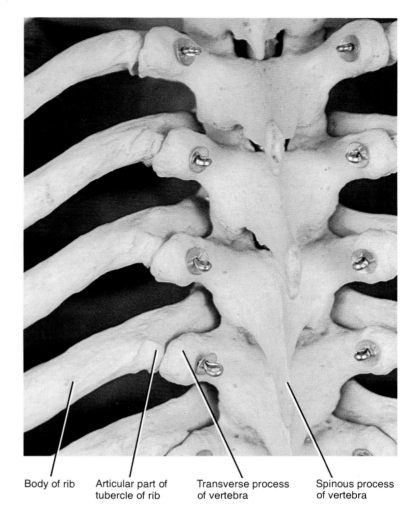

FIGURE 7-18 Typical rib. (a) A left rib viewed from below and behind. (b) Photograph of several ribs articulating with their respective vertebrae in posterior view. (Courtesy of Matt Iacobino.) (c) Photograph of several ribs articulating with their respective vertebrae in anterior view. (Courtesy of Matt Iacobino.)

Body of rib — Articular part of tubercle of rib — Transverse process of vertebra — Spinous process of vertebra

(b)

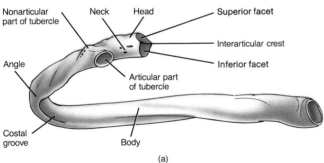

Nonarticular part of tubercle — Neck — Head — Superior facet — Interarticular crest — Inferior facet — Articular part of tubercle — Angle — Costal groove — Body

(a)

Superior facet — Inferior facet — Body of vertebra — Body of rib — Neck of rib — Head of rib

(c)

to the twelfth rib. Each rib articulates posteriorly with its corresponding thoracic vertebra.

The first through seventh ribs have a direct anterior attachment to the sternum by a strip of hyaline cartilage, called **costal cartilage** (*costa* = rib). These ribs are called **true ribs** or **vertebrosternal ribs.** The remaining five pairs of ribs are referred to as **false ribs** because their costal cartilages do not attach directly to the sternum. The cartilages of the eighth, ninth, and tenth ribs attach to each other and then to the cartilage of the seventh rib. These are called **vertebrochondral ribs.** The eleventh and twelfth ribs are designated as **floating ribs** because their anterior ends do not attach even indirectly to the sternum. They attach only posteriorly to the thoracic vertebrae.

Although there is some variation in rib structure, we will examine the parts of a typical (third through ninth) rib when viewed from the right side and from behind (Figure 7-18). The **head** of a typical rib is a projection at the posterior end of the rib. It is wedge-shaped and consists of one or two **facets** that articulate with facets on the bodies of adjacent thoracic vertebrae. The facets on the head of a rib are separated by a horizontal **interarticular crest.** The inferior facet on the head of a rib is larger than the superior facet. The **neck** is a constricted portion just lateral to the head. A knoblike structure on the posterior surface where the neck joins the body is called a **tubercle.** It consists of a **nonarticular part** which affords attachment to the ligament of the tubercle and an **articular part** which articulates with the facet of a transverse process of the inferior of the two vertebrae to which the head of the rib is connected. The **body** or **shaft** is the main part of the rib. A short distance beyond the tubercle, there is an abrupt change in the curvature of the shaft. This point is called the **angle.** The inner surface of the rib has a **costal groove** that protects blood vessels and a small nerve, artery, and vein.

The posterior portion of the rib is connected to a thoracic vertebra by its head and articular part of a tubercle.

The facet of the head fits into a facet on the body of a vertebra, and the articular part of the tubercle articulates with the facet of the transverse process of the vertebra. Each of the second through ninth ribs articulates with the bodies of two adjacent vertebrae. The first, tenth, eleventh, and twelfth ribs articulate with only one vertebra each. On the eleventh and twelfth ribs, there is no articulation between the tubercles and the transverse processes of their corresponding vertebrae.

Spaces between ribs, called **intercostal spaces,** are occupied by intercostal muscles, blood vessels, and nerves.

CLINICAL APPLICATION

Surgical access to the lungs or structures in the mediastinum is commonly undertaken through an intercostal space. Special rib retractors are used to create a wide separation between ribs. The costal cartilages are sufficiently elastic to permit considerable bending.

DISORDERS: HOMEOSTATIC IMBALANCES

Herniated (Slipped) Disc

In their function as shock absorbers, intervertebral discs are subject to compressional forces. The discs between the fourth and fifth lumbar vertebrae and between the fifth lumbar vertebra and sacrum usually are subject to more forces than other discs. If the anterior and posterior ligaments of the discs become injured or weakened, the pressure developed in the nucleus pulposus may be great enough to rupture the surrounding fibrocartilage. If this occurs, the nucleus pulposus may protrude posteriorly or into one of the adjacent vertebral bodies (herniate). This condition is called a **herniated (slipped) disc.**

Most often the nucleus pulposus slips posteriorly toward the spinal cord and spinal nerves. This movement exerts pressure on the spinal nerves, causing considerable, sometimes very acute, pain. If the roots of the sciatic nerve, which passes from the spinal cord to the foot, are pressured, the pain radiates down the back of the thigh, through the calf, and occasionally into the foot. If pressure is exerted on the spinal cord itself, nervous tissue may be destroyed. Traction, bed rest, and analgesia usually relieve the pain. If such treatment is ineffective, surgical decompression of the spinal nerves by laminectomy or removal of some of the nucleus pulposus may be necessary to relieve pain. Removal may involve conventional surgery or use of a proteolytic enzyme called chymopapain. This enzyme is extracted from a papaya plant and is injected into a herniated disc where it dissolves the nucleus pulposus, thus relieving pressure on spinal nerves and pain.

Curvatures

As a result of various conditions, the normal curves of the vertebral column may become exaggerated or the column may acquire a lateral bend. Such conditions are called **curvatures** of the spine.

Scoliosis (skō'-lē-Ō-sis; *scolio* = bent) is a lateral bending of the vertebral column, usually in the thoracic region. This is the most common of the curvatures. It may be congenital, due to an absence of the lateral half of a vertebra (hemivertebra). It may also be acquired from a persistent severe sciatica. Poliomyelitis may cause scoliosis by the paralysis of muscles on one side of the body, which produces a lateral deviation of the trunk toward the unaffected side. Poor posture is also a contributing factor.

Until recently, a child with progressive scoliosis faced either years of treatment with a brace or major corrective surgery. Several research teams are experimenting with using electrical stimulation of muscles to limit the progression of scoliosis and even to reduce curvatures in some cases. The skeletal muscles on the convex side of the spinal curve are electrically stimulated.

Kyphosis (kī-FŌ-sis; *kypho* = hunchback) is an exaggeration of the thoracic curve of the vertebral column. In tuberculosis of the spine, vertebral bodies may partially collapse, causing an acute angular bending of the vertebral column. In the elderly, degeneration of the intervertebral discs leads to kyphosis. Kyphosis may also be caused by rickets and poor posture. The term "round-shouldered" is an expression for mild kyphosis. Electrical muscle stimulation is being studied to assess its effects on kyphosis as well as on scoliosis.

Lordosis (lor-DŌ-sis; *lordo* = swayback) is an exaggeration of the lumbar curve of the vertebral column. It may result from increased weight of abdominal contents as in pregnancy or extreme obesity. Other causes include poor posture, rickets, and tuberculosis of the spine.

Spina Bifida

Spina bifida (SPĪ-na BIF-i-da) is a congenital defect of the vertebral column in which laminae fail to unite at the midline. The lumbar vertebrae are involved in about 50 percent of all cases. In less serious cases, the defect is small and the area is covered with skin. Only a dimple or tuft of hair may mark the site. Symptoms are mild, with perhaps intermittent urinary problems. An operation is not ordinarily required. Larger defects in the vertebral arches with protrusion of the membranes around the spinal cord or spinal cord tissue produce serious problems, such as partial or complete paralysis, partial or complete loss of urinary bladder control, and the absence of reflexes.

Fractures of the Vertebral Column

Fractures of the vertebral column most commonly involve T12, L1, and L2. They usually result from a flexion-compression type of injury such as might be sustained in landing on the feet or buttocks after a fall from a height or having a heavy weight fall on the shoulders. The forceful compression wedges the involved vertebrae. If, in addition to compression, there is forceful forward movement, one vertebra may displace forward on its adjacent vertebra below with either dislocation or fracture of the articular facets between the two (fracture dislocation) and with rupture of the interspinous ligaments.

The cervical vertebrae may be fractured or, more commonly, dislodged by a fall on the head with acute flexion of the neck, as might happen on diving into shallow water. Dislocation may even result from the sudden forward jerk which may occur when an automobile or airplane crashes ("whiplash"). The relatively horizontal intervertebral facets of the cervical vertebrae allow dislocation to take place without their being fractured, whereas the relatively vertical thoracic and lumbar intervertebral facets nearly always fracture in forward dislocation of the thoracolumbar region. Spinal nerve damage may occur as a result of fractures of the vertebral column.

STUDY OUTLINE

Types of Bones (p. 138)

1. On the basis of shape, bones are classified as long, short, flat, or irregular.
2. Wormian or sutural bones are found between the sutures of certain cranial bones. Sesamoid bones develop in tendons or ligaments.

Surface Markings (p. 138)

1. Markings are areas on the surfaces of bones.
2. Each marking is structured for a specific function—joint formation, muscle attachment, or passage of nerves and blood vessels.
3. Terms that describe markings include fissure, foramen, meatus, fossa, process, condyle, head, facet, tuberosity, crest, and spine.

Divisions of the Skeletal System (p. 138)

1. The axial skeleton consists of bones arranged along the longitudinal axis. The parts of the axial skeleton are the skull, hyoid bone, auditory ossicles, vertebral column, sternum, and ribs.
2. The appendicular skeleton consists of the bones of the girdles and the upper and lower extremities. The parts of the appendicular skeleton are the shoulder girdles, the bones of the upper extremities, the pelvic girdle, and the bones of the lower extremities.

Skull (p. 138)

1. The skull consists of the cranium and the face. It is composed of 22 bones.
2. Sutures are immovable joints between bones of the skull. Examples are coronal, sagittal, lambdoidal, and squamosal sutures.
3. Fontanels are membrane-filled spaces between the cranial bones of fetuses and infants. The major fontanels are the anterior, posterior, anterolaterals, and posterolaterals.
4. The 8 cranial bones include the frontal, parietal (2), temporal (2), occipital, sphenoid, and ethmoid.
5. The 14 facial bones are the nasal (2), maxillae (2), zygomatic (2), mandible, lacrimal (2), palatine (2), inferior nasal conchae (2), and vomer.

6. Paranasal sinuses are cavities in bones of the skull that communicate with the nasal cavity. They are lined by mucous membranes. The cranial bones containing the paranasal sinuses are the frontal, sphenoid, ethmoid, and maxilla.
7. The foramina of the skull bones provide passages for nerves and blood vessels.

Hyoid Bone (p. 151)

1. The hyoid bone is a U-shaped bone that does not articulate with any other bone.
2. It supports the tongue and provides attachment for some of its muscles.

Vertebral Column (p. 151)

1. The vertebral column, the sternum, and the ribs constitute the skeleton of the trunk.
2. The bones of the adult vertebral column are the cervical vertebrae (7), thoracic vertebrae (12), lumbar vertebrae (5), the sacrum (5, fused) and the coccyx (4, fused).
3. The vertebral column contains primary curves (thoracic and sacral) and secondary curves (cervical and lumbar). These curves give strength, support, and balance.
4. The vertebra are similar in structure, each consisting of a body, vertebral arch, and seven processes. Vertebra in the different regions of the column vary in size, shape, and detail.

Thorax (p. 156)

1. The thoracic skeleton consists of the sternum, the ribs and costal cartilages, and the thoracic vertebrae.
2. The thorax protects vital organs in the chest area.

Disorders: Homeostatic Imbalances (p. 160)

1. Protrusion of the nucleus pulposus of an intervertebral disc posteriorly or into an adjacent vertebral body is called a herniated (slipped) disc.
2. Exaggeration of a normal curve of the vertebral column is called a curvature. Examples include scoliosis, kyphosis, and lordosis.
3. The imperfect union of the vertebral laminae at the midline, a congenital defect, is referred to as spina bifida.
4. Fractures of the vertebral column most often involve T12, L1, and L2.

REVIEW QUESTIONS

1. What are the four principal types of bones? Give an example of each. Distinguish between a sutural and a sesamoid bone.
2. What are surface markings? Describe and give an example of each.
3. Distinguish between the axial and appendicular skeletons. What subdivisions and bones are contained in each?
4. What are the bones that compose the skull? The cranium? The face?
5. Define a suture. What are the four prominent sutures of the skull? Where are they located?
6. What is a fontanel? Describe the location of the six fairly constant fontanels.
7. What is a paranasal sinus? What cranial bones contain paranasal sinuses?
8. Define the following: mastoiditis, cleft palate, cleft lip, sinusitis, dislocation of the jaw, and deviated nasal septum.

9. What is the hyoid bone? In what respect is it unique? What is its function?

10. What bones form the skeleton of the trunk? Distinguish between the number of nonfused vertebrae found in the adult vertebral column and that of a child.

11. What are the normal curves in the vertebral column? How are primary and secondary curves differentiated? What are the functions of the curves?

12. What are the principal distinguishing characteristics of the bones of the various regions of the vertebral column?

13. What bones form the skeleton of the thorax? What are the functions of the thoracic skeleton?

14. How are ribs classified on the basis of their attachment to the sternum?

15. What is a herniated (slipped) disc? Why does it cause pain? How is it treated?

16. What is a curvature? Describe the symptoms and causes of scoliosis, kyphosis, and lordosis.

17. Define spina bifida.

18. What are some common causes of fractures of the vertebral column?

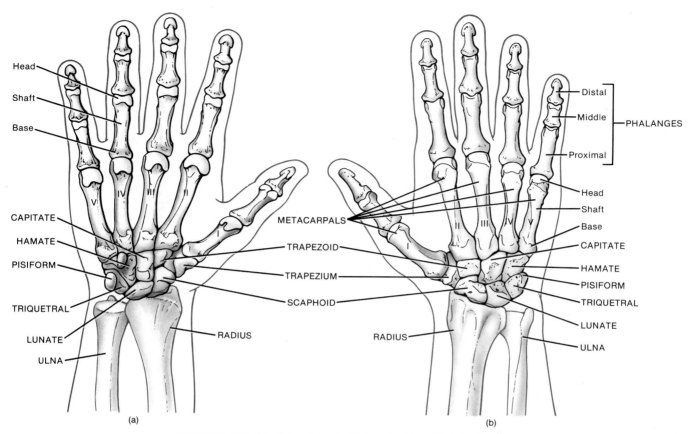

FIGURE 8-6 Right wrist and hand. (a) Anterior view. (b) Posterior view.

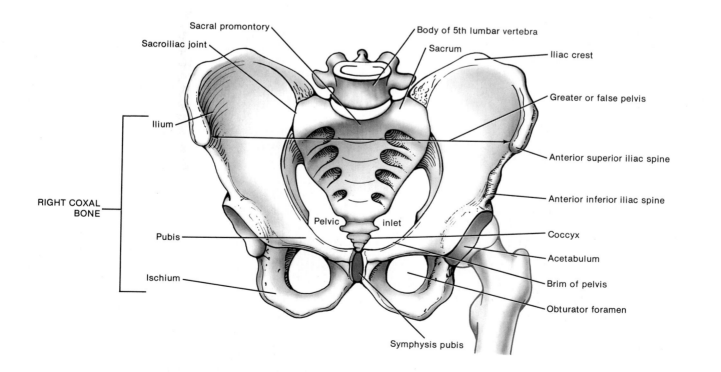

FIGURE 8-7 Pelvic girdle in anterior view.

(a)

(b)

(c)

FIGURE 8-8 Right coxal bone. (a) Lateral view. (b) Medial view. The lines of fusion of the ilium, ischium, and pubis are not actually visible in an adult bone. (c) Three divisions of the coxal bone.

The **greater** or **false pelvis** represents the expanded portion situated superior to the brim of the pelvis. The greater pelvis consists laterally of the superior portions of the ilia and posteriorly of the superior portion of the sacrum. There is no bony component in the anterior aspect of the greater pelvis. Rather, the front is formed by the walls of the abdomen.

The **lesser** or **true pelvis** is inferior and posterior to the brim of the pelvis. It is formed by the inferior portions of the ilia and the sacrum, the coccyx, and the pubes. The lesser pelvis contains a superior opening called the **pelvic inlet** and an inferior opening called the **pelvic outlet.**

CLINICAL APPLICATION
Pelvimetry is the measurement of the size of the inlet and outlet of the birth canal. Measurement of the pelvic cavity is important to the physician, because the fetus must pass through the narrower opening of the lesser pelvis at birth.

united to each other anteriorly at the symphysis (SIM-fi-sis) pubis. They unite posteriorly to the sacrum.

Together with the sacrum and coccyx, the two bones of the pelvic girdle form the basinlike structure called the **pelvis.** The pelvis is divided into a greater pelvis and a lesser pelvis by an oblique plane that passes through sacral promontory (posterior), iliopectineal lines (laterally), and symphysis pubis (anteriorly). The circumference of this oblique plane is called the **brim of the pelvis.**

Each of the two **coxal bones (os coxae)** of a newborn consists of three components: a superior **ilium,** an inferior and anterior **pubis,** and an inferior and posterior **ischium** (Figure 8-8). Eventually, the three separate bones fuse into one. The area of fusion is a deep, lateral fossa called the **acetabulum** (as'-e-TAB-yoo-lum). Although the adult coxae are both single bones, it is common to discuss the bones as if they still consisted of three portions.

The ilium is the largest of the three subdivisions of the coxal bone. Its superior border, the **iliac crest,** ends anteriorly in the **anterior superior iliac spine.** Posteriorly, the iliac crest ends in the **posterior superior iliac spine.** The spines serve as points of attachment for muscles of the abdominal wall. Slightly inferior to the posterior inferior iliac spine is the **greater sciatic (sī-AT-ik) notch.** The internal surface of the ilium seen from the medial side is the **iliac fossa.** It is a concavity where the iliacus muscle attaches. Posterior to this fossa is the auricular surface, which articulates with the sacrum.

The ischium is the inferior, posterior portion of the coxal bone. It contains a prominent **ischial spine,** a **lesser sciatic notch** below the spine, and an **ischial tuberosity.** The rest of the ischium, the **ramus,** joins with the pubis and together they surround the **obturator (OB-too-rā'-ter) foramen.**

The pubis is the anterior and inferior part of the coxal bone. It consists of a **superior ramus,** an **inferior ramus,** and a **body** that contributes to the formation of the symphysis pubis.

The **symphysis pubis** is the joint between the two coxal bones (see Figure 8-7). It consists of fibrocartilage. The **acetabulum** is the fossa formed by the ilium, ischium, and pubis. It is the socket for the head of the femur. Two-fifths of the acetabulum is formed by the ilium, two-fifths by the ischium, and one-fifth by the pubis.

LOWER EXTREMITIES

The **lower extremities** are composed of 60 bones (Figure 8-9). Each extremity includes the femur in the thigh, patella (kneecap), fibula and tibia in the leg, tarsals (ankle bones), metasarsals, and phalanges in the toes.

FEMUR

The **femur,** or thighbone, is the longest and heaviest bone in the body (Figure 8-10). Its proximal end articulates with the coxal bone. Its distal end articulates with the tibia. The shaft of the femur bows medially so that it approaches the femur of the opposite thigh. As a result of this convergence, the knee joints are brought nearer to the body's line of gravity. The degree of convergence is greater in the female because the female pelvis is broader.

The proximal end of the femur consists of a rounded **head** that articulates with the acetabulum of the coxal bone. The **neck** of the femur is a constricted region distal to the head. A fairly common fracture in the elderly occurs at the neck of the femur. Apparently the neck becomes so weak that it fails to support the body. The **greater trochanter (trō-KAN-ter)** and **lesser trochanter** are projections that serve as points of attachment for some of the thigh and buttock muscles.

The shaft of the femur contains a rough vertical ridge on its posterior surface called the **linea aspera.** This ridge serves for the attachment of several thigh muscles.

The distal end of the femur is expanded and includes the **medial condyle** and **lateral condyle.** These articulate with the tibia. Superior to the condyles are the **medial epicondyle** and **lateral epicondyle.** A depressed area between the condyles on the posterior surface is called the **intercondylar (in'-ter-KON-di-lar) fossa.** The **patellar surface** is located between the condyles on the anterior surface.

PATELLA

The **patella,** or kneecap, is a small, triangular bone anterior to the knee joint (Figure 8-11). It is a sesamoid bone that develops in the tendon of the quadriceps femoris muscle. The broad superior end of the patella is called the **base.** The pointed inferior end is the **apex.** The posterior surface contains two **articular facets,** one for the medial condyle and the other for the lateral condyle of the femur.

TIBIA AND FIBULA

The **tibia,** or shinbone, is the larger, medial bone of the leg (Figure 8-12). It bears the major portion of the weight of the leg. The tibia articulates at its proximal end with the femur and fibula and at its distal end with the fibula of the leg and talus of the ankle.

The proximal end of the tibia is expanded into a **lateral condyle** and a **medial condyle.** These articulate with the condyles of the femur. The inferior surface of the lateral condyle articulates with the head of the fibula. The slightly concave condyles are separated by an upward projection called the **intercondylar eminence.** The **tibial tuberosity** on the anterior surface is a point of attachment for the patellar ligament.

The medial surface of the distal end of the tibia forms the **medial malleolus (mal-LĒ-ō-lus).** This structure articulates with the talus bone of the ankle and forms the prominence that can be felt on the medial surface of your ankle. The **fibular notch** articulates with the fibula.

The **fibula** is parallel and lateral to the tibia. It is considerably smaller than the tibia. The **head** of the fibula, the proximal end, articulates with the inferior surface of the lateral condyle of the tibia below the level of the knee joint. The distal end has a projection called the **lateral malleolus,** which articulates with the talus bone of the ankle. This forms the prominence on the lateral surface of the ankle. The inferior portion of the fibula also articulates with the tibia at the fibular notch. A fracture of the lower end of the fibula with injury to the tibial articulation is called a **Pott's fracture.**

TARSUS, METATARSUS, AND PHALANGES

The **tarsus** is a collective designation for the seven bones of the ankle called **tarsals** (Figure 8-13). The term *tarsos* pertains to a broad, flat surface. The **talus** and **calcaneus (kal-KĀ-ne-us)** are located on the posterior part of the foot. The anterior part contains the **cuboid, navicular,** and three **cuneiform bones** called the first (medial), second (intermediate), and third (lateral) cuneiform. The talus, the uppermost tarsal bone, is the only bone of the foot that articulates with the fibula and tibia. It is surrounded on one side by the medial malleolus of the tibia and on the other side by the lateral malleolus of the fibula. During walking, the talus initially bears the entire weight of the

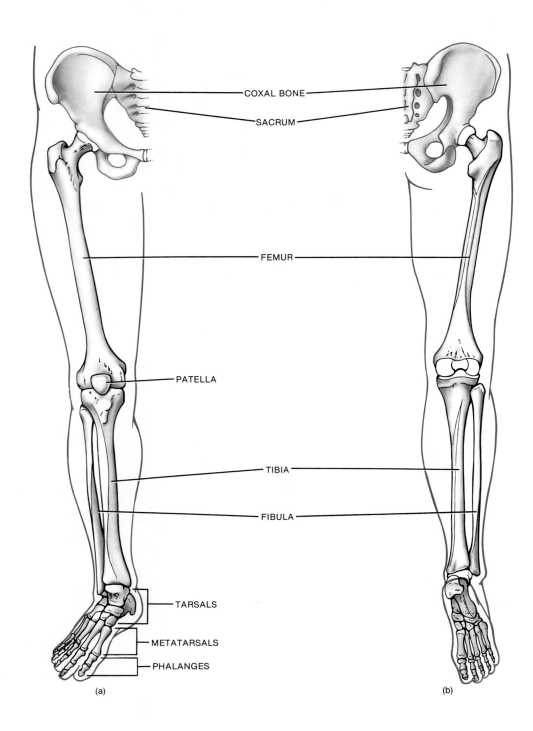

FIGURE 8-9 Right pelvic girdle and lower extremity. (a) Anterior view. (b) Posterior view.

extremity. About half the weight is then transmitted to the calcaneus. The remainder is transmitted to the other tarsal bones. The calcaneus, or heel bone, is the largest and strongest tarsal bone.

The **metatarsus** consists of five metatarsal bones numbered I to V from the medial to lateral position. Like the metacarpals of the palm of the hand, each metatarsal consists of a proximal **base,** a **shaft,** and a distal **head.**

The metatarsals articulate proximally with the first, second, and third cuneiform bones and with the cuboid. Distally, they articulate with the proximal row of phalanges. The first metatarsal is thicker than the others because it bears more weight.

The **phalanges** of the foot resemble those of the hand both in number and arrangement. Each also consists of a proximal **base,** a **shaft,** and a distal **head.** The great

EXHIBIT 8-1
COMPARISON OF TYPICAL MALE AND FEMALE PELVIS

POINT OF COMPARISON	MALE	FEMALE
General structure	Heavy and thick.	Light and thin.
Joint surfaces	Large.	Small.
Muscle attachments	Well marked.	Rather indistinct.
Greater pelvis	Deep.	Shallow.
Pelvic inlet	Heart-shaped.	Larger and more oval.
Pelvic outlet	Comparatively small.	Comparatively large.
First piece of sacrum	Superior surface of the body spans nearly half the width of sacrum.	Superior surface of the body spans about a third the width of sacrum.
Sacrum	Long, narrow, with smooth concavity.	Short, wide, flat, curving forward in lower part.
Auricular surface	Extends well down the third piece of the sacrum.	Extends only to upper border of third piece of the sacrum.
Pubic arch	Less than 90° angle.	Greater than 90° angle.
Inferior ramus of pubis	Presents strong everted surface for attachment of the crus of the penis.	Everted surface not present.
Pubic symphysis	More deep.	Less deep.
Ischial spine	Turned inward more.	Turned inward less.
Ischial tuberosity	Turned inward.	Turned outward.
Ilium	More vertical.	Less vertical.
Iliac fossa	Deep.	Shallow.
Iliac crest	More curved.	Less curved.
Anterior superior iliac spine	Closer.	Wider apart.
Acetabulum	Large.	Small.
Obturator foramen	Round.	Oval.
Greater sciatic notch	Narrow.	Wide.

STUDY OUTLINE

Pectoral (Shoulder) Girdles (p. 164)
1. Each pectoral or shoulder girdle consists of a clavicle and scapula.
2. Each attaches an upper extremity to the trunk.

Upper Extremities (p. 164)
The bones of each upper extremity include the humerus, ulna, radius, carpals, metacarpals, and phalanges.

Pelvic Girdle (p. 167)
1. The pelvic girdle consists of two coxal bones or hipbones.
2. It attaches the lower extremities to the trunk at the sacrum.
3. Each coxal bone consists of three fused components—ilium, pubis, and ischium.

Lower Extremities (p. 171)
1. The bones of each lower extremity include the femur, tibia, fibula, tarsals, metatarsals, and phalanges.
2. The bones of the foot are arranged in two arches, the longitudinal arch and the transverse arch, to provide support and leverage.

Male and Female Skeletons (p. 176)
1. Male bones are generally larger and heavier than female bones and have more prominent markings for muscle attachment.
2. The female pelvis is adapted for pregnancy and childbirth. Differences in pelvic structure are listed in Exhibit 8-1.

REVIEW QUESTIONS

1. What is the pectoral (shoulder) girdle? Why is it important?
2. What are the bones of the upper extremity? What is a Colles' fracture?
3. What is the pelvic girdle? Why is it important?
4. What are the bones of the lower extremity? What is a Pott's fracture?
5. What is pelvimetry? What is its clinical importance?
6. In what ways do the upper extremity and lower extremity differ structurally?
7. Describe the structure of the longitudinal and transverse arches of the foot. What is the function of an arch?
8. How do flatfeet, clawfeet, and bunions arise?
9. What are the principal structural differences between typical male and female skeletons? Use Exhibit 8-1 as a guide in formulating your response.

FIGURE 9-2 (*Continued*) Angular movements at synovial joints.

(a)

FIGURE 9-3 Rotation and circumduction. (a) Rotation at the atlantoaxial joint (left) and rotation of the humerus (right). (b) Circumduction of the humerus at the shoulder joint. (© 1983 by Gerard J. Tortora.)

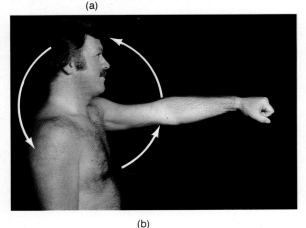

(b)

at gliding joints generally because ligaments or adjacent bones restrict the range of movement. Examples are the joints between carpal bones, tarsal bones, the sternum and clavicle, and the scapula and clavicle.

Hinge

A **hinge** or **ginglymus** (JIN-gli-mus) **joint** is one in which the convex surface of one bone fits into the concave surface of another bone. Movement is primarily in a single plane, and the joint is therefore known as *monaxial,* or *uniaxial* (Figure 9-5b). The motion is similar to that of a hinged door. Movement is usually flexion and extension. Examples of hinge joints are the elbow, ankle, and interphalangeal joints. The movement allowed by a hinge joint is illustrated by flexion and extension at the elbow (see Figure 9-2c).

Pivot

In a **pivot** or **trochoid** (TRŌ-koyd) **joint,** a rounded, pointed, or conical surface of one bone articulates within a ring formed partly by bone and partly by a ligament.

The primary movement permitted is rotation, and the joint is therefore *monaxial* (Figure 9-5c). Examples include the joint between the atlas and axis (atlantoaxial) and between the proximal ends of the radius and ulna. Movement at a pivot joint is illustrated by supination and pronation of the palms and rotation of the head from side to side (see Figure 9-3a).

Ellipsoidal

In an **ellipsoidal** or **condyloid** (KON-di-loyd) **joint,** an oval-shaped condyle of one bone fits into an elliptical cavity of another bone. Since the joint permits side-to-side and back-and-forth movements, it is *biaxial* (Figure 9-5d). The joint at the wrist between the radius and carpals is ellipsoidal. The movement permitted by such a joint is illustrated when you flex and extend and abduct and adduct the wrist (see Figure 9-3d,k).

Saddle

In a **saddle** or **sellaris** (sel-A-ris) **joint,** the articular surfaces of both bones are saddle-shaped, that is, concave

FIGURE 9-4 Special movements. (a) Inversion. (b) Eversion. (c) Retraction. (d) Protraction. (e) Pronation. (f) Supination. (g) Elevation. (h) Depression. (© 1983 by Gerard J. Tortora. Courtesy of Lynne Tortora and James Borghesi.)

in one direction and convex in the other. Essentially, the saddle joint is a modified ellipsoidal joint in which the movement is somewhat freer. Movements at a saddle joint are side to side and back and forth. Thus the joint is *biaxial* (Figure 9-5e). The joint between the trapezium and metacarpal of the thumb is an example of a saddle joint.

Ball-and-Socket

A **ball-and-socket** or **spheroid** (SFĒ-royd) **joint** consists of a ball-like surface of one bone fitted into a cuplike depression of another bone. Such a joint permits *triaxial* movement, or movement in three planes of motion: flexion-extension, abduction-adduction, and rotation (Figure

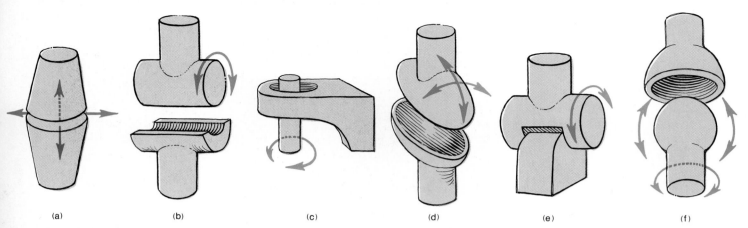

FIGURE 9-5 Simplified representations of subtypes of synovial joints. (a) Gliding. (b) Hinge. (c) Pivot. (d) Ellipsoidal. (e) Saddle. (f) Ball-and-socket.

9-5f). Examples of ball-and-socket joints are the shoulder joint and hip joint. The range of movement at a ball-and-socket joint is illustrated by circumduction of the arm (Figure 9-3b).

SUMMARY OF JOINTS

The summary of joints presented in Exhibit 9-1 is based on the anatomy of the joints. If we rearrange the types of joints into a classification based on movement, we arrive at the following:

Synarthroses: immovable joints
1. **Suture**
2. **Synchondrosis**
3. **Gomphosis**

Amphiarthroses: slightly movable joints
1. **Syndesmosis**
2. **Symphysis**

Diarthroses: freely movable joints
1. **Gliding**
2. **Hinge**
3. **Pivot**
4. **Ellipsoidal**
5. **Saddle**
6. **Ball-and-socket**

SELECTED ARTICULATIONS OF THE BODY

We will now examine in some detail selected articulations of the body. In order to simplify your learning efforts, a series of exhibits has been prepared. Each exhibit considers a specific articulation and contains (1) a definition, that is, a description of the bones that form the joint; (2) the type of joint (its structural classification); and (3) its anatomical components—a description of the major connecting ligaments, articular disc, articular capsule, and other distinguishing features of the joint. Each exhibit also refers you to an illustration of the joint.

EXHIBIT 9-1
JOINTS

TYPE	DESCRIPTION	MOVEMENT	EXAMPLES
FIBROUS	No joint cavity; bones held together by a thin layer of fibrous tissue or dense fibrous tissue.		
Suture	Found only between bones of the skull; articulating bones separated by a thin layer of fibrous tissue.	None (synarthrotic).	Lambdoidal suture between occipital and parietal bones.
Syndesmosis	Articulating bones united by dense fibrous tissue.	Slight (amphiarthrotic).	Distal ends of tibia and fibula.
Gomphosis	Cone-shaped peg fits into a socket; articulating bones separated by periodontal ligament.	None (synarthrotic).	Roots of teeth in alveolar processes.
CARTILAGINOUS	No joint cavity; articulating bones united by cartilage.		
Synchondrosis	Connecting material is hyaline cartilage.	None (synarthrotic).	Temporary joint between the diaphysis and epiphyses of a long bone and permanent joint between first rib and sternum.
Symphysis	Connecting material is a broad, flat disc of fibrocartilage.	Slight (amphiarthrotic).	Intervertebral joints and symphysis pubis.
SYNOVIAL	Joint cavity and articular cartilage present; articular capsule composed of an outer fibrous capsule and an inner synovial membrane; may contain accessory ligaments, articular discs (menisci), and bursae.	Freely movable (diarthrotic).	
Gliding	Articulating surfaces usually flat.	Biaxial (flexion-extension, abduction-adduction).	Intercarpal and intertarsal joints.
Hinge	Spoollike surface fits into a concave surface.	Monaxial (flexion-extension).	Elbow, ankle, and interphalangeal joints.
Pivot	Rounded, pointed, or concave surface fits into a ring formed partly by bone and partly by a ligament.	Monaxial (rotation).	Atlantoaxial and radioulnar joints.
Ellipsoidal	Oval-shaped condyle fits into an elliptical cavity.	Biaxial (flexion-extension, abduction-adduction).	Radiocarpal joint.
Saddle	Articular surfaces concave in one direction and convex in opposite direction.	Biaxial (flexion-extension, abduction-adduction).	Carpometacarpal joint of thumb.
Ball-and-socket	Ball-like surface fits into a cuplike depression.	Triaxial (flexion-extension, abduction-adduction, rotation).	Shoulder and hip joints.

EXHIBIT 9-2
HUMEROSCAPULAR (SHOULDER) JOINT (Figure 9-6)

DEFINITION	Joint formed by the head of the humerus and the glenoid cavity of the scapula.
TYPE OF JOINT	Synovial joint, ball-and-socket (spheroid) type.

ANATOMICAL COMPONENTS

1. **Articular capsule.** Loose sac that completely envelops the joint, extending from the circumference of the glenoid cavity to the anatomical neck of the humerus.
2. **Coracohumeral ligament.** Strong, broad ligament that extends from the coracoid process of the scapula to the greater tubercle of the humerus.
3. **Glenohumeral ligaments.** Three thickenings of the articular capsule over the ventral surface of the joint.
4. **Transverse humeral ligament.** Narrow sheet extending from the greater tubercle to the lesser tubercle of the humerus.
5. **Glenoid labrum.** Narrow rim of fibrocartilage around the edge of the glenoid cavity.
6. Among the bursae associated with the shoulder joint are:
   a. **Subscapular bursa** between the tendon of the subscapularis muscle and the underlying joint capsule.
   b. **Subdeltoid bursa** between the deltoid muscle and joint capsule.
   c. **Subacromial bursa** between the acromion and joint capsule.
   d. **Subcoracoid bursa** either lies between the coracoid process and joint capsule or appears as an extension from the subacromial bursa.

CLINICAL APPLICATION

The strength and stability of the shoulder joint are not provided by the shape of the articulating bones or its ligaments. Instead, deep muscles of the shoulder and their tendons—subscapularis, supraspinatus, infraspinatus, and teres minor—strengthen and stabilize the shoulder joint (see Exhibit 11-14 and Figure 11-15). The muscles and their tendons are so arranged as to form a nearly complete encirclement of the joint. This arrangement is referred to as the **rotator (musculotendinous) cuff** and is a common site of injury to baseball pitchers, especially tearing of the supraspinatus muscle.

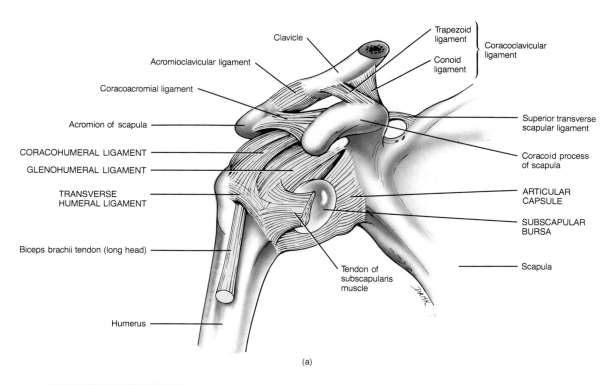

Clavicle

Trapezoid ligament

Conoid ligament

Coracoclavicular ligament

Acromioclavicular ligament

Coracoacromial ligament

Acromion of scapula

CORACOHUMERAL LIGAMENT

GLENOHUMERAL LIGAMENT

TRANSVERSE HUMERAL LIGAMENT

Biceps brachii tendon (long head)

Humerus

Superior transverse scapular ligament

Coracoid process of scapula

ARTICULAR CAPSULE

SUBSCAPULAR BURSA

Scapula

Tendon of subscapularis muscle

(a)

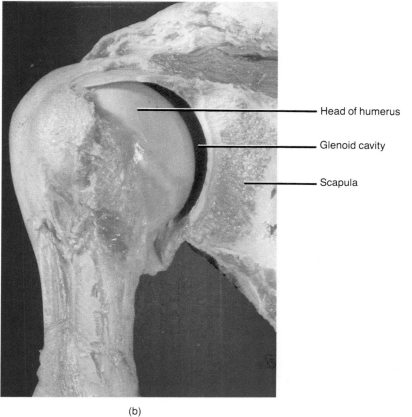

Head of humerus

Glenoid cavity

Scapula

(b)

FIGURE 9-6 Humeroscapular joint. (a) Diagram of anterior view. (b) Photograph of anterior view. (Courtesy of C. Yokochi and J. W. Rohen, *Photographic Anatomy of the Human Body*, 2d ed., 1979, IGAKU-SHOIN, Ltd., Tokyo, New York.)

EXHIBIT 9-3
COXAL (HIP) JOINT (Figure 9-7)

DEFINITION	Joint formed by the head of the femur and the acetabulum of the coxal bone.
TYPE OF JOINT	Synovial, ball-and-socket (spheroid) type.

ANATOMICAL COMPONENTS

1. **Articular capsule.** Extends from the rim of the acetabulum to the neck of the femur. One of the strongest ligaments of the body, the capsule consists of circular and longitudinal fibers. The circular fibers, called the **zona orbicularis**, form a collar around the neck of the femur. The longitudinal fibers are reinforced by accessory ligaments known as the iliofemoral ligament, the pubofemoral ligament, and the ischiofemoral ligament.
2. **Iliofemoral ligament.** Thickened portion of the articular capsule that extends from the anterior inferior iliac spine of the coxal bone to the intertrochanteric line of the femur.
3. **Pubofemoral ligament.** Thickened portion of the articular capsule that extends from the pubic part of the rim of the acetabulum to the neck of the femur.
4. **Ischiofemoral ligament.** Thickened portion of the articular capsule that extends from the ischial wall of the acetabulum to the neck of the femur.
5. **Ligament of the head of the femur (capitate ligament).** Flat, triangular band that extends from the fossa of the acetabulum to the head of the femur.
6. **Acetabular labrum.** Fibrocartilage rim attached to the margin of the acetabulum.
7. **Transverse ligament of the acetabulum.** Strong ligament which crosses over the acetabular notch, converting it to a foramen. It supports part of the acetabular labrum and is connected with the ligament of the head of the femur and the articular capsule.

CLINICAL APPLICATION

Dislocation of the hip among adults is quite rare because of (1) the stability of the ball-and-socket joint, (2) the strong, tough, articular capsule, (3) the strength of the intracapsular ligaments, and (4) the extensive musculature over the joint.

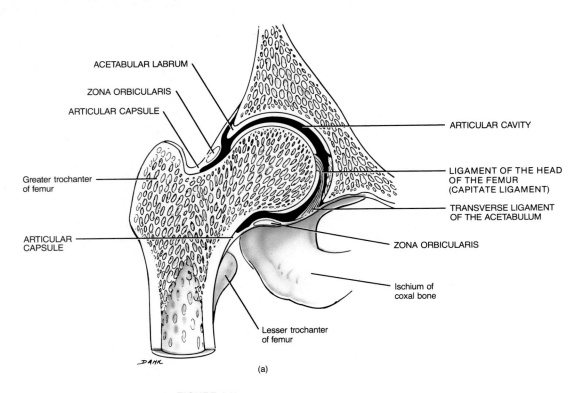

FIGURE 9-7 Coxal (hip) joint. (a) Frontal section.

- List the characteristics and functions of muscle tissue.

- Compare the location, microscopic appearance, nervous control, and functions of the three kinds of muscle tissue.

- Define fascia, epimysium, perimysium, endomysium, tendons, and aponeuroses and list their modes of attachment to muscles.

- Explain the relationship of blood vessels and nerves to skeletal muscles.

- Identify the histological characteristics of skeletal muscle tissue.

- List the principal events associated with the sliding-filament theory.

- Discuss the physiological importance of the motor unit.

- Identify the source of energy for muscular contraction.

- Define the all-or-none principle of muscular contraction.

- Describe different types of normal contractions performed by skeletal muscles.

- Describe the phases of contraction in a typical myogram of a twitch contraction.

- Compare fast (white) muscle with slow (red) muscle.

- Contrast cardiac muscle tissue with smooth muscle tissue.

- Compare oxygen debt, fatigue, and heat production as examples of muscle homeostasis.

- Define such common muscular disorders as fibrosis, fibrositis, "charleyhorse," muscular dystrophy, and myasthenia gravis.

- Compare spasms, cramps, convulsions, fibrillation, and tics as abnormal muscular contractions.

- Define medical terminology associated with the muscular system.

- Explain the actions of selected drugs associated with muscle tissue.

10 Muscle Tissue

Although bones and joints provide leverage and form the framework of the body, they are not capable of moving the body by themselves. Motion is an essential body function that results from the contraction and relaxation of muscles.

Muscle tissue constitutes about 40 to 50 percent of the total body weight and is composed of highly specialized cells.

CHARACTERISTICS

Muscle tissue has four principal characteristics that assume key roles in maintaining homeostasis.

1. **Excitability** is the ability of muscle tissue to receive and respond to stimuli. A stimulus is a change in the internal or external environment strong enough to initiate a nerve impulse (action potential).
2. **Contractility** is the ability to shorten and thicken, or contract, when a sufficient stimulus is received.
3. **Extensibility** is the ability of muscle tissue to stretch. Many skeletal muscles are arranged in opposing pairs. While one is contracting, the other is relaxed and is undergoing extension.
4. **Elasticity** is the ability of muscle to return to its original shape after contraction or extension.

FUNCTIONS

Through contraction, muscle performs three important functions:

1. Motion.
2. Maintenance of posture.
3. Heat production.

Motion is obvious in movements involving the whole body, such as walking and running, and in localized movements, such as grasping a pencil or nodding the head. All of these movements rely on the integrated functioning of the bones, joints, and muscles attached to the bones. Less noticeable kinds of motion produced by muscles are the beating of the heart, the churning of food in the stomach, the pushing of food through the intestines, the contraction of the gallbladder to release bile, and the contraction of the urinary bladder to expel urine.

In addition to the movement function, muscle tissue also enables the body to maintain posture. The contraction of skeletal muscles holds the body in stationary positions, such as standing and sitting.

The third function of muscle tissue is heat production. Skeletal muscle contractions produce heat and are thereby important in maintaining normal body temperature.

TYPES

Types of muscle tissue are categorized by location, microscopic structure, and nervous control.

Skeletal muscle tissue, which is named for its location, is attached to bones. It is **striated** muscle tissue because striations, or bandlike structures, are visible when the tissue is examined under a microscope. It is a **voluntary** muscle tissue because it can be made to contract by conscious control.

Visceral muscle tissue is located in the walls of hollow internal structures, such as blood vessels, the stomach, and the intestines. It is **smooth** or **nonstriated,** that is, it lacks the bandlike structures. It is **involuntary** muscle tissue because its contraction is usually not under conscious control.

Cardiac muscle tissue forms the walls of the heart. It is striated and involuntary.

Thus all muscle tissues are classified in the following way: (1) skeletal, striated, voluntary muscle, (2) visceral, smooth, involuntary muscle, and (3) cardiac, striated, involuntary muscle.

SKELETAL MUSCLE TISSUE

To understand the fundamental mechanisms of muscle movement, you will need some knowledge of its connective tissue components, its nerve and blood supply, and its histology.

FASCIA

The term **fascia** (FASH-ē-a) is applied to a sheet or broad band of fibrous connective tissue beneath the skin or around muscles and other organs of the body. Fasciae may be divided into three types: superficial, deep, and subserous.

The **superficial fascia,** or **subcutaneous layer,** is immediately deep to the skin. It covers the entire body and varies in thickness in different regions. On the back or dorsum of the hand it is quite thin, whereas over the inferior abdominal wall it is thick. The superficial fascia is composed of adipose tissue and loose connective tissue. The outer layer usually contains fat and varies considerably in thickness, while the inner layer is thin and elastic. Between the two layers are found arteries, veins, lymphatics, nerves, the mammary glands, and the facial muscles. Superficial fascia has a number of important functions. (1) It serves as a storehouse for water and particularly for fat. Much of the fat of an overweight person is in the superficial fascia. (2) It forms a layer of insulation protecting the body from loss of heat. (3) It provides mechanical protection from blows. (4) It provides a pathway for nerves and vessels.

The **deep fascia** is by far the most extensive of the three types. It is a dense connective tissue composed of a superficial layer (external investing layer) and a deep layer (internal investing layer). Unlike the superficial fascia, it does not contain fat. The deep fascia lines the body wall and extremities and holds muscles together, separating them into functioning groups. Functionally, deep fascia allows free movement of muscles, carries nerves and blood vessels, fills spaces between muscles, and sometimes provides the origin for muscles.

The **subserous (visceral) fascia** is located between the internal investing layer of deep fascia and a serous membrane. It is composed of loose connective tissue. It forms the fibrous layer of serous membranes, covers and supports the viscera, and attaches the parietal layer of serous membranes to the internal surface of the body wall.

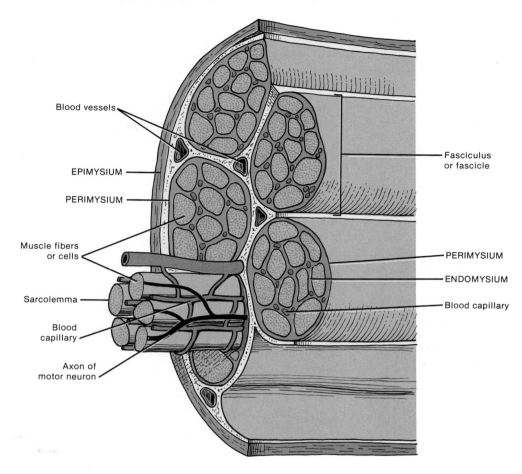

Blood vessels

EPIMYSIUM

PERIMYSIUM

Muscle fibers
or cells

Sarcolemma

Blood
capillary

Axon of
motor neuron

Fasciculus
or fascicle

PERIMYSIUM

ENDOMYSIUM

Blood capillary

FIGURE 10-1 Relationships of connective tissue to skeletal muscle. Shown is a cross section and longitudinal section of a skeletal muscle indicating the relative positions of the epimysium, perimysium and endomysium. Compare this figure with the photomicrograph in Figure 10-2b.

CONNECTIVE TISSUE COMPONENTS

Skeletal muscles are further protected, strengthened, and attached to other structures by several connective tissue components (Figure 10-1). The entire muscle is usually wrapped with a substantial quantity of fibrous connective tissue called the **epimysium** (ep'-i-MĪZ-ē-um). The epimysium is an extension of deep fascia. When the muscle is cut in cross section, invaginations of the epimysium are seen to divide the muscle into bundles of fibers (cells) called **fasciculi** (fa-SIK-yoo-lī) or **fascicles** (FAS-i-kuls). These invaginations of the epimysium are called the **perimysium** (per'-i-MĪZ-ē-um). Perimysium, like epimysium, is an extension of deep fascia. In turn, invaginations of the perimysium, called **endomysium** (en'-dō-MĪZ-ē-um), penetrate into the interior of each fascicle and separate the muscle cells. Endomysium is also an extension of deep fascia.

The epimysium, perimysium, and endomysium are all continuous with the connective tissue that attaches the muscle to another structure, such as bone or other muscle. All three elements may be extended beyond the muscle cells as a **tendon**—a cord of connective tissue that attaches a muscle to the periosteum of a bone. When the connective tissue elements extend as a broad, flat layer, the tendon is called an **aponeurosis.** This structure also attaches to the coverings of a bone or another muscle. When a muscle contracts, the tendon and its corresponding bone or muscle are pulled toward the contracting muscle. In this way skeletal muscles produce movement. Certain tendons, especially those of the wrist and ankle, are enclosed by tubes of fibrous connective tissue called **tendon sheaths.** They are lined by a synovial membrane that permits the tendon to slide easily within the sheath. The sheaths also prevent the tendons from slipping out of place.

CLINICAL APPLICATION

Tendinitis or **tenosynovitis** (ten'-ō-sin-ō-VĪ-tis) frequently occurs as inflammation involving the tendon sheaths and synovial membrane surrounding certain joints. The wrists, shoulders, elbows (tennis elbow), finger joints (trigger finger), ankles, and associated tendons are most often affected. The affected sheaths may become visibly swollen because of fluid accumulation or they may remain dry. Local tenderness is variable, and there may be disabling pain with movement of the body part. The condition often follows some form of trauma, strain, or excessive exercise.

NERVE AND BLOOD SUPPLY

Skeletal muscles are well supplied with nerves and blood vessels. This innervation and vascularization is directly related to contraction, the chief characteristic of muscle. For a skeletal muscle cell to contract, it must first be stimulated by an impulse from a nerve cell. Muscle contraction also requires a good deal of energy and therefore large amounts of nutrients and oxygen. Moreover, the waste products of these energy-producing reactions must be eliminated. Thus muscle action depends on the blood supply.

Generally, an artery and one or two veins accompany each nerve that penetrates a skeletal muscle. The larger branches of the blood vessel accompany the nerve branches through the connective tissue of the muscle.

Microscopic blood vessels called capillaries are arranged in the endomysium. Each muscle cell is thus in close contact with one or more capillaries. Each skeletal muscle cell usually makes contact with a portion of a nerve cell.

HISTOLOGY

When a typical skeletal muscle is teased apart and viewed microscopically, it can be seen to consist of many elongated, cylindrical cells called **muscle fibers** (Figure 10-2a,b). These fibers lie parallel to one another and range from 10 to 100 μm in diameter. Some fibers may reach lengths of 30 cm (12 inches) or more. Each muscle fiber is enveloped by a plasma membrane called the **sarcolemma** (*sarco* = flesh; *lemma* = sheath). The sarcolemma surrounds a quantity of cytoplasm called **sarcoplasm.** Within the sacroplasm of a muscle fiber and lying close to the sarcolemma are many nuclei and a number of mitochondria. Skeletal muscle fibers are thus multinucleate. Also within a muscle fiber is the **sarcoplasmic reticulum** (sar'-kō-PLAZ-mik rē-TIK-yoo-lum), a network of membrane-enclosed tubules comparable to smooth endoplasmic reticulum (Figure 10-2c). Running transversely through the fiber and perpendicularly to the sarcoplasmic reticulum are **T tubules (transverse tubules).** The tubules are extensions of the sarcolemma that open to the outside of the fiber. A **triad** consists of a T tubule and the segments of sarcoplasmic reticulum on either side.

A highly magnified view of skeletal muscle fibers reveals threadlike structures, about 1 or 2 μm in diameter,

Capillary — Nucleus of muscle fiber — Striations

Sarcolemma — Muscle fiber

(a)

FIGURE 10-2 Histology of skeletal muscle tissue. (a) Photomicrograph of several muscle fibers in longitudinal section at a magnification of 800×. (© 1983 by Michael H. Ross. Used by permission.) (b) Photomicrograph of several muscle fibers in cross section at a magnification of 500×. (© 1983 by Michael H. Ross. Used by permission.)

Perimysium — Sarcolemma

Fasciculus or fascicle of muscle fibers

Muscle fiber

Endomysium surrounding muscle fibers

(b)

called **myofibrils** (Figure 10-2c,d,e). The prefix *myo* means muscle. The myofibrils, ranging in number from several hundred to several thousand, run longitudinally through the muscle fiber and consist of two kinds of even smaller structures called **myofilaments.** The **thin myofilaments** are about 6 nm in diameter. The **thick myofilaments** are about 16 nm in diameter.

The myofilaments of a myofibril do not extend the entire length of a muscle fiber—they are stacked in compartments called **sarcomeres.** Sarcomeres are separated from one another by narrow zones of dense material called **Z lines.** Each sarcomere is about 2.6 μm long. In a relaxed muscle fiber, that is, one that is not contracting, the thin and thick myofilaments overlap and form a dark, dense band called the **anisotropic band,** or **A band** (Figure 10-2d). Each A band is about 1.6 μm long. A light-colored,

less dense area called the **isotropic band,** or **I band,** is composed of thin myofilaments only. The I bands are about 1 μm long. This combination of alternating dark and light bands gives the muscle fiber its striped appearance. A narrow **H zone** contains thick myofilaments only. The H zone is about 0.5 μm long. In the center of the H zone is the **M line,** a series of fine threads that appear to connect the middle parts of the adjacent thick filaments.

The thin myofilaments are composed mostly of the protein **actin.** The actin molecules are arranged in a double-stranded coil that gives the thin myofilaments their characteristic shape (Figure 10-3a). Besides actin, the thin myofilaments contain two other protein molecules, **tropomyosin** and **troponin.** Together they are referred to as a **tropomyosin-troponin complex.**

The thick myofilaments are composed mostly of the protein **myosin.** A myosin molecule is shaped like a rod with a round head. The rods form the long axis of the thick myofilaments, and the heads form projections called

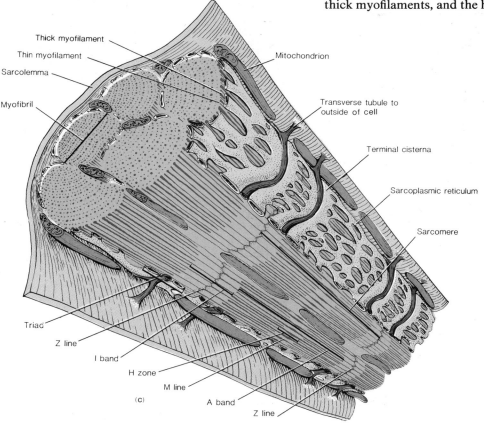

FIGURE 10-2 (*Continued*) Histology of skeletal muscle tissue. (c) Enlarged aspect of several myofibrils of a muscle fiber based on an electron micrograph. (d) Enlarged aspect of a sarcomere showing thin and thick myofilaments.

(e)

FIGURE 10-2 (*Continued*) Histology of skeletal muscle tissue. (e) Electron micrograph of several sarcomeres at a magnification of 35,000×. (Courtesy of D. E. Kelly, from *Introduction to the Musculoskeletal System,* by Cornelius Rosse and D. Kay Clawson, Harper & Row, Publishers, Inc., New York, 1970.)

(a)

(b)

(c)

FIGURE 10-3 Detailed structure of myofilaments. (a) Thin myofilament. (b) Thick myofilament. (c) Cross section of several thin and thick myofilaments showing the arrangement of cross bridges. Note that each thick myofilament is surrounded by six thin myofilaments. (Parts (a) and (b) have been redrawn by permission from Arthur W. Ham, *Histology,* 7th ed., Philadelphia: J. B. Lippincott Company, 1974.)

cross bridges (Figure 10-3b). The cross bridges are arranged in pairs and seem to spiral around the main axis. The relationship between thin and thick myofilaments is shown in Figure 10-3c.

CONTRACTION

SLIDING-FILAMENT THEORY

During muscle contraction, the thin myofilaments slide inward toward the H zone. The sarcomere shortens, but the lengths of the thin and thick myofilaments do not change. The cross bridges of the thick myofilaments connect with portions of actin of the thin myofilaments. The myosin cross bridges move like the oars of a boat on the surface of the thin myofilaments, and the thin and thick myofilaments slide past each other (Figure 10-4). As the thin myofilaments move past the thick myofilaments, the H zone narrows and even disappears when the thin myofilaments meet at the center of the sarcomere (Figure 10-5). In fact, the cross bridges may pull the thin myofilaments of each sarcomere so far inward that

FIGURE 10-4 Movement of thin myofilaments past thick myofilaments. (a) Attachment of myosin cross bridges to actin of thin myofilaments. (b) Mechanism of movement of myosin cross bridges resulting in sliding of thin myofilaments toward H zone.

their ends overlap. As the thin myofilaments slide inward, the Z lines are drawn toward the A band and the sarcomere is shortened. The sliding of myofilaments and shortening of sarcomeres causes the shortening of the muscle fibers. All these events associated with the movement of myofilaments are known as the **sliding-filament theory** of muscle contraction.

MOTOR UNIT

For a skeletal muscle fiber to contract, a stimulus must be applied to it. Such a stimulus is normally transmitted by nerve cells, or **neurons.** A neuron has a threadlike process called a fiber, or axon, that may run 91 cm (3 ft) or more to a muscle. A bundle of such fibers from many different neurons composes a nerve. A neuron that transmits a stimulus to muscle tissue is called a **motor neuron.**

Upon entering a skeletal muscle, the axon of the motor neuron branches into fine endings that come into close approximation at grooves on the muscle membrane. The portion of the muscle cell membrane directly under the termination of the axon is called a **motor end plate** (Figure 10-6). The area of contact between neuron and muscular fiber is called a **neuromuscular junction,** or **myoneural junction.** When a nerve impulse reaches the terminal branches of the nerve fiber, small vesicles in the branches, called synaptic vesicles, release a chemical called **acetylcholine** (as'-ē-til-KŌ-lēn) or **ACh.** The ACh transmits the nerve impulse from the neuron, across the myoneural junction, to the motor end plate, thus initiating contraction. This mechanism is described in detail in Chapter 12.

A motor neuron, together with all the muscle cells it stimulates, is referred to as a **motor unit.** A single motor neuron may innervate about 150 muscle fibers, depending

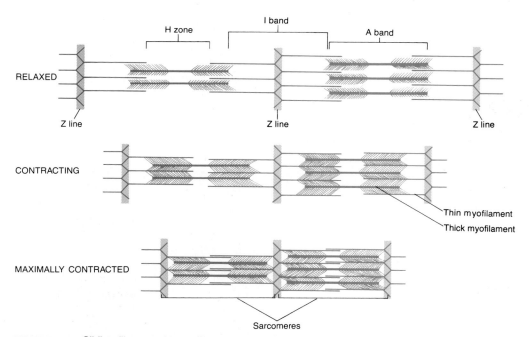

FIGURE 10-5 Sliding-filament theory of muscle contraction. Shown are the positions of the various parts of two sarcomeres in relaxed, contracting, and maximally contracted states. Note the movement of the thin myofilaments and the relative size of the H zone.

Muscle fibers

Motor nerve

Axon terminations

Branches of axon

(a)

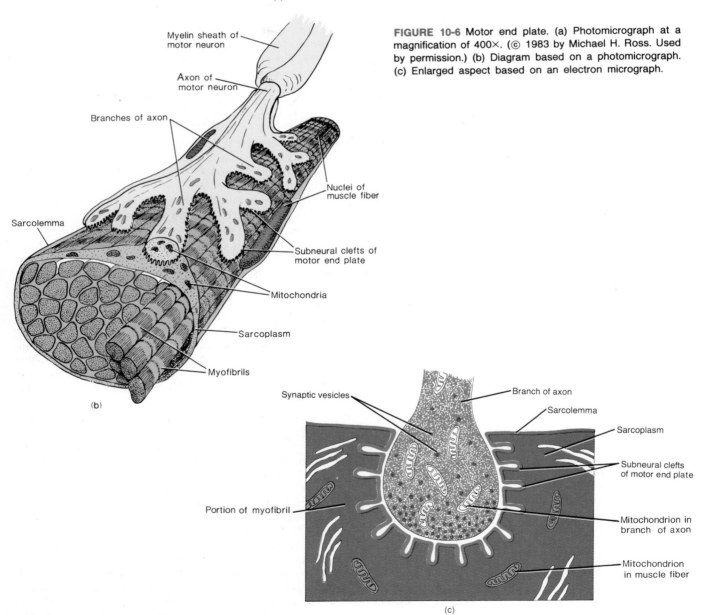

Myelin sheath of motor neuron

Axon of motor neuron

Branches of axon

Sarcolemma

Nuclei of muscle fiber

Subneural clefts of motor end plate

Mitochondria

Sarcoplasm

Myofibrils

(b)

FIGURE 10-6 Motor end plate. (a) Photomicrograph at a magnification of 400×. (© 1983 by Michael H. Ross. Used by permission.) (b) Diagram based on a photomicrograph. (c) Enlarged aspect based on an electron micrograph.

Synaptic vesicles

Branch of axon

Sarcolemma

Sarcoplasm

Subneural clefts of motor end plate

Portion of myofibril

Mitochondrion in branch of axon

Mitochondrion in muscle fiber

(c)

Sarcolemma   Sarcoplasm

Mitochondria in
branch of axon

Synaptic
vesicles

Subneural cleft of
motor end plate

Mitochondria in
muscle fiber

Portion of
myofibril   (d)

FIGURE 10-6 (*Continued*) Motor end plate. (d) Electron micrograph at a magnification of 30,000×. (Courtesy of Cornelius Rosse and D. Kay Clawson, from *Introduction to the Musculoskeletal System,* Harper & Row, Publishers, New York, 1970.)

on the region of the body. This means that stimulation of one neuron will tend to cause the simultaneous contraction of about 150 muscle fibers. In addition, all the muscle fibers of a motor unit that are sufficiently stimulated will contract and relax together. Muscles that control precise movements, such as the extrinsic eye muscles, have fewer than 10 muscle fibers to each motor unit. Muscles of the body that are responsible for gross movements, such as the biceps and gastrocnemius, may have as many as 500 muscle fibers in each motor unit.

Stimulation of a motor neuron produces a contraction in all the muscle fibers in a particular motor unit. Accordingly, the total tension in a muscle can be varied by adjusting the number of motor units that are activated. The process of increasing the number of active motor units is called **recruitment** and is determined by the needs of the body at a given time. The various motor neurons to a given muscle fire asynchronously, that is, while some are excited, others are inhibited. This means that while some motor units are active, others are inactive; all the motor units are not contracting at the same time. Asynchronous firing of motor neurons prevents fatigue while maintaining contraction by allowing a brief rest for the inactive units. The alternating motor units relieve one another so smoothly that the contraction can be sustained for long periods. It also helps to maintain a nearly constant tension in a muscle and is one factor responsible for producing smooth movements during a muscle contraction, rather than a series of jerky movements.

PHYSIOLOGY

When a muscle fiber is relaxed, the concentration of calcium ions (Ca^{2+}) in the sarcoplasm is low; these ions are stored in the sarcoplasmic reticulum. Moreover, molecules of ATP are attached to the myosin cross bridges. The cross bridges are prevented from combining with actin of the thin myofilaments by the binding of the tropomyosin-troponin complex to actin and the binding of ATP to the myosin cross bridges. In other words, the muscle fiber remains relaxed as long as there are few calcium ions in the sarcoplasm, the tropomyosin-troponin complex is attached to actin, and ATP is attached to the myosin cross bridges.

When a nerve impulse reaches the nerve ending, a small amount of calcium enters the nerve ending, causing the synaptic vesicles to release acetylcholine. It diffuses across the myoneural junction and combines with receptor sites on the motor end plate. Acetylcholine alters the motor end plate and initiates an impulse that spreads from the motor end plate over the surface of the sarcolemma and into the T tubules. When the impulse is conveyed from the T tubules to the sarcoplasmic reticulum, the reticulum releases the calcium ions from storage into the sarcoplasm surrounding the myofilaments. The calcium ions move to the myosin cross bridges and activate the myosin. In the presence of calcium ions, myosin acts as an enzyme that catalyzes the breakdown of ATP into ADP + P. With the breakdown of ATP the myosin cross bridges are free to react with actin. Calcium ions also bind to the tropomyosin-troponin complex and permit it to split from the thin myofilament, leaving a free receptor site of actin that can attach to the myosin cross bridge. The energy released from the breakdown of ATP is used for the attachment and movement of the myosin cross bridges and thus the sliding of the myofilaments. As the thin myofilaments slide past the thick myofilaments, the Z lines are drawn toward each other and the sarcomere shortens—the muscle fibers contract and the muscle contracts.

What happens when a muscle fiber goes from a contracted state back to a relaxed state? Acetylcholine is rapidly destroyed by an enzyme called *acetylcholinesterase* (*AChE*) which is found on the membrane of muscle cells. The absence of ACh inhibits nerve impulse conduction from the axon terminals to the motor end plate. After the nerve impulse ends, the calcium ions are actively transported back into the sarcoplasmic reticulum for storage. This is accomplished by a transport protein called *calsequestrin* and involves an expenditure of some ATP. With the removal of calcium ions from the sarcoplasm, the enzymatic activity of myosin stops. The ADP is resynthesized into ATP, which again binds to the myosin cross bridges, and the tropomyosin-troponin complex is reattached to the actin of the thin myofilaments, so that the myosin cross bridges separate from the actin. Since the myosin cross bridges are broken, the thin myofilaments slip back to their relaxed position. The sarcomeres are thereby returned to their resting lengths and the muscle fiber resumes its resting state. ATP is also required to detach the cross bridges.

Exhibit 10-1 shows a summary of the events associated with the contraction and relaxation of a muscle fiber.

ENERGY

Contraction of a muscle requires energy. When a nerve impulse stimulates a muscle fiber, ATP, in the presence of ATPase (activated myosin), breaks down into ADP + P and energy is released. As far as we know, ATP is always the immediate source of energy for muscle contraction.

Like the other cells of the body, muscle cells synthesize ATP as follows:

$$ADP + P + E \text{ (energy)} \rightarrow ATP$$

The energy for replenishing ATP is derived from the breakdown of digested foods. However, unlike most other cells of the body, muscle fibers alternate between great activity and virtual inactivity. A resting muscle needs little energy and produces much more ATP than it can use. When a muscle is contracting, its energy requirements are high and the synthesis of ATP is accelerated. If the exercise is strenuous, ATP is used up even faster than it can be manufactured. Thus muscles must be able to build up a reserve supply of energy. They do this in two ways. First, the muscle fiber stores any excess ATP on the thick myofilaments. Second, when the fiber runs out of storage space for the ATP molecules, it combines the remainder of the ATP with a substance called *creatine*.

EXHIBIT 10-1

SUMMARY OF EVENTS INVOLVED IN CONTRACTION AND RELAXATION OF A SKELETAL MUSCLE FIBER

1. A nerve impulse causes synaptic vesicles in motor axon terminals to release acetylcholine.

2. Acetylcholine diffuses across the myoneural junction and initiates an impulse that spreads from the motor end plate over the surface of the sarcolemma.

3. The impulse enters the T tubules and sarcoplasmic reticulum and stimulates the sarcoplasmic reticulum to release calcium ions from storage into the sarcoplasm.

4. Calcium ions activate myosin, which breaks down ATP. Calcium ions also bind the tropomyosin-troponin complex to permit the complex to split from thin myofilaments.

5. Free receptor sites on actin of thin myofilaments attach to myosin cross bridges, using the energy from ATP breakdown, and thin myofilaments slide past thick myofilaments.

6. The sliding draws the Z lines toward each other, the sarcomere shortens, the muscle fibers contract, and the muscle contracts.

7. Acetylcholine is inactivated by acetylcholinesterase, thus inhibiting nerve impulse conduction from axon terminals to the motor end plate.

8. Once the nerve impulse is inhibited, calcium ions are actively transported back into the sarcoplasmic reticulum by a transport protein called calsequestrin, using energy from ATP breakdown.

9. The low calcium concentration in sarcoplasm stops the enzymatic activity of myosin, ADP is resynthesized to ATP, ATP binds to the myosin cross bridges, the tropomyosin-troponin complex is reattached to actin, myosin cross bridges separate from actin, and the thin myofilaments return to their relaxed position.

10. Sarcomeres return to their resting lengths, muscle fibers relax, and the muscle relaxes.

Creatine, which is produced in the liver, can accept a high-energy phosphate from ATP to become the high-energy compound *creatine phosphate*.

$$ATP + creatine \rightarrow creatine\ phosphate + ADP$$

This reaction is *anaerobic*, that is, it takes place in the absence of oxygen. (An *aerobic* reaction is one that requires oxygen.)

Creatine phosphate is produced when the muscle fibers are resting. During strenuous contraction, the reaction reverses itself. This reaction, also anaerobic, is shown in the following equation:

$$ADP + creatine\ phosphate \rightarrow ATP + creatine$$

CLINICAL APPLICATION

Following death, certain chemical changes occur in muscle tissue that affect the status of the muscles. Due to a lack of ATP, cross bridges of the thick myofilaments remain attached to the thin myofilaments, thus preventing relaxation. The resulting condition, in which muscles are in a state of partial contraction, is called **rigor mortis** (rigidity of death). The time elapsing between death and the onset of rigor mortis varies greatly among individuals. Those who have had long, wasting illnesses undergo rigor mortis more quickly.

ALL-OR-NONE PRINCIPLE

According to the **all-or-none principle,** individual muscle fibers of a motor unit will contract to their fullest extent or will not contract at all, providing conditions remain constant. In other words, *muscle fibers* do not partly contract. The principle does not mean that the entire muscle must be either fully relaxed or fully contracted because, of the many fibers that comprise the entire muscle, some are contracting and some are relaxing. Thus, the muscle as a whole can have graded contractions. The strength of contraction may be decreased by fatigue, lack of nutrients, or lack of oxygen. The weakest stimulus from a neuron that can initiate a contraction is called a **threshold** or **liminal stimulus.** A stimulus of lesser intensity, or one that cannot initiate contraction, is referred to as a **subthreshold** or **subliminal stimulus.**

KINDS

The various skeletal muscles are capable of producing different kinds of contractions, depending on the stimulation frequency.

Twitch

The **twitch contraction** is a rapid, jerky response to a single stimulus. Twitch contractions can be artifically brought about in muscles of an animal. A recording of a twitch contraction is the classic way to illustrate the different phases of one single contraction. Figure 10-7 is a graph of a twitch contraction. This kind of record of a contraction is called a **myogram.** Note that a brief period exists between application of the stimulus and the beginning of contraction: the **latent period.** In frog muscle, it lasts about 10 milliseconds (msec; 10 msec = 0.01 sec) (0.01 second). The second phase, the **contraction period,** lasts about 40 msec (0.04 sec) and is indicated by the

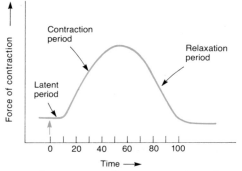

FIGURE 10-7 Myogram of a twitch contraction. The red arrow indicates the point at which the stimulus is applied.

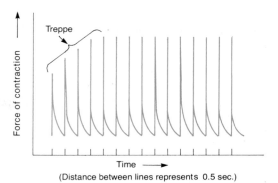

FIGURE 10-8 Myogram of treppe.

upward tracing. The third phase, the **relaxation period,** lasts about 50 msec (0.05 sec) and is indicated by the downward tracing. The duration of these periods depends on the muscle. The latent period, contraction period, and relaxation period for muscles that move the eyes are very short. Muscles that move the leg undergo longer periods.

If additional stimuli are applied to the muscle after the initial stimulus, other responses may be noted. For example, if two stimuli are applied one immediately after the other, the muscle will respond to the first stimulus but not to the second. When a muscle fiber receives enough stimulation to contract, it temporarily loses its irritability and cannot contract again until its responsiveness is regained. This period of lost irritability is the **refractory period.** Its duration also varies with the muscle involved. Skeletal muscle has a short refractory period of 5 msec (0.005 sec). Cardiac muscle has a long refractory period of 300 msec (0.30 sec).

Treppe

Treppe is the condition in which a skeletal muscle contracts more forcefully in response to the same strength of stimulus after it has contracted several times. It is demonstrated by stimulating an isolated muscle with a series of stimuli at the same frequency and voltage. Suppose a series of liminal stimuli are introduced into a muscle. Time must be allowed for the muscle to undergo its latent period, contract, and relax, but this takes only 0.1 sec in frog muscle. If stimuli are repeated at intervals of 0.5 sec, the first few tracings on the myogram will show an increasing height with each contraction. This is treppe—the staircase phenomenon (Figure 10-8). It is the principle athletes use when warming up. After the first few stimuli, the muscle reaches its peak of performance and undergoes its strongest contraction. Treppe is thought to result from an increase in the effectiveness of calcium ions in relieving tropomyosin-troponin inhibition after several contractions.

Tetanus

When two stimuli are applied and the second is delayed until the refractory period is over, the skeletal muscle will respond to both stimuli. In fact, if the second stimulus is applied after the refractory period, but before the muscle has finished relaxing, the second contraction will be

FIGURE 10-9 Myogram of summation of twitches. The second stimulus (long red arrow) is applied before the muscle has finished relaxing. The second contraction is even stronger than the first. The broken line represents the continuation of a twitch contraction.

stronger than the first. This phenomenon is **summation of twitches** (Figure 10-9).

If a frog muscle is stimulated at a rate of 20 to 30 stimuli per second, the muscle can only partly relax between stimuli. As a result, the muscle maintains a sustained contraction called **incomplete (unfused) tetanus** (Figure 10-10a). Stimulation at an increased rate (35 to 50 stimuli per second) results in **complete (fused) tetanus,** a sustained contraction that lacks even partial relaxation (Figure 10-10b). Essentially, both kinds of tetanus result from the addition of calcium ions released from the sarcoplasmic reticulum by the second stimulus to the calcium ions still in the sarcoplasm from the first stimulus. This causes the rapid succession of separate twitches. Relaxation is either partial or does not occur at all. Voluntary contractions, such as contraction of the biceps brachii muscle in order to flex the forearm, are tetanic contractions.

Isotonic and Isometric

Isotonic (*iso* = equal; *tonos* = tension) **contractions** are probably familiar to you. As the contraction occurs, the muscle shortens and pulls on another structure, such as a bone, to produce movement. During such a contraction, the tension remains constant and energy is expended.

In an **isometric contraction,** there is a minimal shortening of the muscle. It remains nearly the same length,

FIGURE 10-10 Myograms of (a) incomplete and (b) complete tetanus.

but the *tension* on the muscle increases greatly. Although isometric contractions do not result in body movement, energy is still expended. You can demonstrate such a contraction by carrying your books with your arm extended. The weight of the books pulls the arm downward, stretching the shoulder and arm muscles. The isometric contraction of the shoulder and arm muscles counteracts the stretch. The two forces—contraction and stretching—applied in opposite directions create the tension.

MUSCLE TONE

A muscle may be in a state of partial contraction even though contraction of the muscle fibers is always complete. At any given time, some cells in a muscle are contracted while others are relaxed. This contraction tightens a muscle, but there may not be enough fibers contracting at the time to produce movement. Asynchronous firing allows the contraction to be sustained for long periods.

A sustained partial contraction of portions of a skeletal muscle in response to activation of stretch receptors results in **muscle tone.** Tone is essential for maintaining posture. For example, when the muscles in the back of the neck are in tonic contraction, they keep the head in the anatomical position and prevent it from slumping forward onto the chest, but they do not apply enough force to pull the head back into hyperextension. The degree of tone in a skeletal muscle is monitored by receptors in the muscle called **muscle spindles.** They provide feedback information on tone to the brain so that adjustments can be made (Chapter 15).

CLINICAL APPLICATION

The term **flaccid** (FLAK-sid) is applied to muscles with less than normal tone. Such a loss of tone may be the result of damage or disease of the nerve that conducts a constant flow of impulses to the muscle. If the muscle does not receive impulses for an extended period of time, it may progress from flaccidity to **atrophy** (AT-rō-fē), which is a state of wasting away. Individual muscle cells decrease in size due to a progressive loss of myofibrils. Muscles may also become flaccid and atrophied if they are not used. Bedridden individuals and people with casts may experience atrophy because the flow of impulses to the inactive muscle is greatly reduced. If the nerve supply to a muscle is cut, it will undergo complete atrophy. In about six months to two years, the muscle will be one-quarter its original size and the muscle fibers will be replaced by fibrous tissue. The transition to fibrous tissue, when complete, cannot be reversed.

Muscular **hypertrophy** (hī-PER-trō-fē) is the reverse of atrophy. It refers to an increase in the diameters of muscle fibers due to the production of more myofibrils, mitochondria, sarcoplasmic reticulum, and nutrients. Hypertrophic muscles are capable of more forceful contractions. Weak muscular activity does not produce significant hypertrophy. It results from very forceful muscular activity or repetitive muscular activity at moderate levels.

FAST (WHITE) AND SLOW (RED) MUSCLE

The duration of contraction of various muscles of the body varies with the functions of the muscles in maintaining homeostasis. For example, the duration of contraction of eye muscles is less than $\frac{1}{100}$ sec, whereas the duration of contraction of the gastrocnemius muscle on the posterior aspect of the leg is about $\frac{1}{30}$ sec. Eye movements must be extremely rapid in order to fix the eyes on different objects very quickly. Movements of the legs during walking or running need not be as rapid as eye movements.

Muscles of the eye contain a predominance of **fast muscle fibers.** Such fibers have a more extensive sarcoplasmic reticulum for the rapid release and uptake of calcium ions needed for rapid contractions. Muscles like the gastrocnemius have a predominance of **slow muscle fibers.** These fibers are smaller and more aerobic; have more blood capillaries and more mitochondria; and have a large amount of myoglobin in the sarcoplasm. **Myoglobin** is a substance similar to the hemoglobin in red blood cells that can store oxygen until needed by the mitochondria. Since slow muscle has a reddish tint due to myoglobin and the large amount of red blood cells in capillaries, it is also known as **red muscle.** Fast muscle lacks myoglobin and has fewer capillaries and is thus also known as **white muscle.**

CARDIAC MUSCLE TISSUE

The principal constituent of the heart wall is **cardiac muscle tissue.** Although it is striated in appearance like skeletal muscle, it is involuntary. The fibers of cardiac muscle tissue are roughly quadrangular and usually have only a single centrally located nucleus (Figure 10-11). Skeletal muscle fibers contain several nuclei that are peripherally located. The thin sarcolemma of cardiac muscle fibers is similar to that of skeletal muscle, but the sarcoplasm is more abundant and the mitochondria are larger and more numerous. Cardiac muscle fibers have the same arrangement of actin and myosin and the same bands, zones, and lines as skeletal muscle fibers. Myofilaments in cardiac muscle fibers are not arranged in discrete myofibrils as in skeletal muscle. The T tubules of mammalian cardiac muscle are larger than those of skeletal muscle and are located at the Z lines rather than at the A-I band junctions as in skeletal muscle fibers. The sarcoplasmic reticulum of cardiac muscle is less well-developed than that in skeletal muscle.

While the groups of skeletal muscle fibers are arranged in a parallel fashion, those of cardiac muscle branch freely with other fibers to form two separate networks. The muscular walls and septum of the upper chambers of the heart (atria) compose one network. The muscular walls and septum of the lower chambers of the heart (ventricles) compose the other network. When a single fiber of either network is stimulated, all the fibers in the network become stimulated as well. Thus each network contracts as a functional unit. The fibers of each network were once thought to be fused together into a multinucleated mass called a syncytium. But it is now known that each fiber in a network is separated from the next fiber by an irregular transverse thickening of the sarco-

FIGURE 10-11 Histology of cardiac muscle tissue. (a) Photomicrograph of several muscle fibers in longitudinal section at a magnification of 400× (© 1983 by Michael H. Ross. Used by permission.) (b) Photomicrograph of several muscle fibers in cross section at a magnification of 640×. (© 1983 by Michael H. Ross. Used by permission.)

Thin myofilament

Thick myofilament

Sarcolemma

Mitochondrion

T tubule to outside of cell

Sarcoplasmic reticulum

Sarcomere

Z line

I band

T tubule

A band

H zone

Z line

(c)

Separation of intercalated discs

Sarcoplasmic reticulum

A band

H zone

I band

Z line

Sarcomere

Z line

Nucleus

Mitochondrion

Thin myofilament

Thick myofilament

Intercalated discs

(d)

FIGURE 10-11 (*Continued*) (c) Diagram based on an electron micrograph showing several myofibrils. **(d)** Diagram based on an electron micrograph showing intercalated discs and related structures.

lemma called an **intercalated** (in-TER-ka-lāt-ed) **disc.** These discs strengthen the cardiac muscle tissue and aid in impulse conduction.

Under normal conditions, cardiac muscle tissue contracts and relaxes rapidly, continuously, and rhythmically about 72 times a minute without stopping while a person is at rest. This is a major physiological difference between cardiac and skeletal muscle tissue. Another difference is the source of stimulation. Skeletal muscle tissue ordinarily contracts only when stimulated by a nerve impulse. In contrast, cardiac muscle tissue can contract without nerve stimulation. Its source of stimulation is a conducting tissue of specialized muscle within the heart. Nerve stimula-

tion merely causes the conducting tissue to increase or decrease its rate of discharge. Cardiac muscle tissue also has an extra long refractory period.

SMOOTH MUSCLE TISSUE

Like cardiac muscle tissue, **smooth muscle tissue** is usually involuntary. However, it is nonstriated. A single fiber of smooth muscle tissue is about 5 to 10 μm in diameter and 30 to 200 μm long. It is spindle-shaped, and within the fiber is a single, oval, centrally located nucleus (Figure 10-12). Smooth muscle cells contain actin and myosin filaments, but because the filaments are not as orderly as in skeletal and cardiac muscle tissue, the well-differentiated striations and sarcomeres do not occur.

Two kinds of smooth muscle tissue, visceral and multiunit, are recognized. The more common type is called **visceral muscle tissue.** It is found in wrap-around sheets that form part of the walls of the hollow viscera such as the stomach, intestines, uterus, and urinary bladder. The terms *smooth muscle tissue* and *visceral muscle tissue* are sometimes used interchangeably. The fibers in visceral muscle tissue are tightly bound together to form a continuous network. When a neuron stimulates one fiber, the impulse travels over the other fibers so that contraction

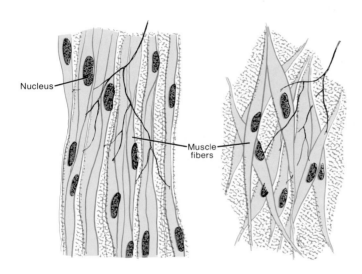

(a)

FIGURE 10-12 Histology of smooth muscle tissue. (a) Diagram of visceral smooth muscle tissue (left) and multiunit smooth muscle tissue (right). (b) Photomicrograph of several muscle fibers in longitudinal section at a magnification of 840×. (© 1983 by Michael H. Ross. Used by permission.) (c) Photomicrograph of several muscle fibers in cross section at a magnification of 840×. (© 1983 by Michael H. Ross. Used by permission.)

(b)

(c)

occurs in a wave over many adjacent fibers. Whereas skeletal muscle cells contract as individual units, visceral muscle cells contract in sequence as the impulse spreads from one cell to another.

The second kind of smooth muscle tissue, **multiunit smooth muscle tissue,** consists of individual fibers each with its own motor-nerve endings. Whereas stimulation of a single visceral muscle fiber causes contraction of many adjacent fibers, stimulation of a single multiunit fiber causes contraction of only that fiber. In this respect, multiunit muscle tissue is like skeletal muscle tissue. Multiunit smooth muscle tissue is found in the walls of blood vessels, in the arrector pili muscles that attach to hair follicles, and in the intrinsic muscles of the eye, such as the iris.

Both kinds of smooth muscle tissue contract and relax more slowly than skeletal muscle tissue. This characteristic is probably due to the arrangement of the thin and thick myofilaments of smooth muscle. Also, both kinds of smooth muscle can maintain a forceful contraction longer than skeletal muscle. Finally, unlike skeletal muscle fibers, smooth muscle fibers can stretch without developing tension. Thus, the smooth muscle in the wall of hollow organs such as the stomach, intestines, and urinary bladder can stretch as the viscera distend, while the pressure within them remains the same.

HOMEOSTASIS

Muscle tissue has a vital role in maintaining the body's homeostasis. Two examples are the relationship of muscle tissue to oxygen and to heat production.

OXYGEN DEBT

The energy required to convert ADP into ATP is produced by the breakdown of digested foods. The primary nutrient that usually supplies this energy is the sugar glucose. The reaction proceeds as follows:

Glucose $\rightarrow$ pyruvic acid + E (energy)

When the skeletal muscle is at rest, the breakdown of glucose is slow enough for the blood to supply sufficient oxygen to participate in the complete catabolism of pyruvic acid. The waste products of this aerobic reaction are carbon dioxide and water.

Pyruvic acid + $O_2 \rightarrow CO_2 + H_2O$ + E (energy)

When a skeletal muscle is contracting and ATP production increases, the breakdown of glucose occurs too rapidly for the blood to supply oxygen for the pyruvic acid to be completely catabolized. The pyruvic acid is only partly catabolized to lactic acid. The conversion of pyruvic acid to lactic acid proceeds *anaerobically*. About 80 percent of the lactic acid diffuses from the skeletal muscles and is transported to the liver. Some of the lactic acid accumulates in the muscle tissue, however. Physiologists suspect that this acid is responsible for the feeling of muscle fatigue. Ultimately, the lactic acid in the skeletal muscle, as well as that which has diffused into the blood, must be catabolized to CO_2 and H_2O. And for this conversion additional oxygen is needed—the **oxygen debt.** When vigorous activity is over, labored breathing continues in order to pay back the debt.

If a skeletal muscle or group of skeletal muscles is continuously stimulated for an extended period of time, the strength of contraction becomes progressively weaker until the muscle no longer responds. This condition is called **fatigue.** It results partly from the diminished availability of oxygen and partly from the toxic effects of lactic acid and carbon dioxide accumulated during exercise. The significant factors that contribute to muscle fatigue are:

1. Excessive activity, resulting in the accumulation of toxic products.
2. Malnutrition, resulting in insufficient supplies of glucose and, therefore, ATP.
3. Cardiovascular disturbances that impair the delivery of useful substances to muscles and the removal of waste products from muscles.
4. Respiratory disturbances that interfere with the oxygen supply and increase the oxygen debt.

HEAT PRODUCTION

The production of heat by skeletal muscle is an important homeostatic mechanism for maintaining normal body temperature. Of the total energy released during muscular contraction, only 20 to 30 percent is used for mechanical work (contraction). The rest is released as heat, which is utilized to help maintain a normal body temperature.

Heat production by muscles may be divided into two phases: (1) *initial heat,* which is produced by the contraction and relaxation of a muscle, and (2) *recovery heat,* which is produced after relaxation. Initial heat is independent of O_2 and is associated with ATP breakdown. Recovery heat is associated with ATP restoration. It includes the anaerobic breakdown of glucose to pyruvic acid and pyruvic acid to lactic acid. It also includes the aerobic breakdown of pyruvic acid to CO_2 and H_2O and the aerobic conversion of lactic acid to CO_2 and H_2O.

DISORDERS: HOMEOSTATIC IMBALANCES

Disorders of the muscular system are related to disruptions of homeostasis. The disorders may involve a lack of nutrients, the accumulation of toxic products, disease, injury, disuse, or faulty nervous connections (innervations).

Fibrosis
The formation of fibrous connective tissue in locations where it normally does not exist is called **fibrosis**. Skeletal and cardiac muscle fibers cannot undergo mitosis, and dead muscle fibers are normally replaced with fibrous connective tissue. Fibrosis, then, is often a consequence of muscle injury or degeneration.

Fibrositis
Fibrositis is an inflammation of fibrous tissue. If it occurs in the lumbar region, it is termed **lumbago** (lum-BĀ-gō). Fibrositis is a common condition characterized by pain, stiffness, or soreness of fibrous tissue, especially in the muscle coverings. It is not destructive or progressive. It may persist for years or sponta-

neously disappear. Attacks of fibrositis may follow an injury, repeated muscular strain, or prolonged muscular tension.

"Charleyhorse"

Fibromyositis refers to a group of symptoms that include pain, tenderness, and stiffness of the joints, muscles, or adjacent structures. These symptoms ordinarily occur in various combinations. When the thigh is involved, it is usually called **"charleyhorse."** It results from a contusion (bruising) and tearing of muscle fibers that produce a hematoma (collection of blood). "Charleyhorse" is characterized by the sudden onset of pain and is aggravated by motion. In some cases, local muscle spasms are noted. The condition is relieved with heat, massage, and rest and completely disappears, although occasionally it may become chronic or recur at frequent intervals.

Muscular Dystrophy

The term **muscular dystrophy** (*dystrophy* = degeneration) applies to a number of inherited myopathies, or muscle-destroying diseases. The disease is characterized by degeneration of the individual muscle cells, which leads to a progressive atrophy of the skeletal muscle. Usually the voluntary skeletal muscles are weakened equally on both sides of the body, whereas the internal muscles, such as the diaphragm, are not affected. Histologically, the changes that occur include the variation in muscle fiber size, degeneration of fibers, and deposition of fat.

The cause of muscular dystrophy has been variously attributed to a genetic defect, faulty metabolism of potassium, protein deficiency, and inability of the body to utilize creatine.

There is no specific drug therapy for this disorder. Treatment involves attempts to prolong ambulation by muscle-strengthening exercises, corrective surgical measures, and appropriate braces. Patients are encouraged to keep physically active.

A simple and reliable screening test for a wide variety of skeletal muscle disorders has been developed. Increased levels of an enzyme called *creatine phosphokinase* (*CPK*) in the blood are typical not only of the dystrophies but of other wasting diseases as well. The measurement of CPK, the most consistently elevated enzyme in cases of dystrophy, provides a diagnostic test even before clinical symptoms appear. Inflammatory muscle diseases can also be detected with this screening test.

Myasthenia Gravis

Myasthenia (mī'-as-THĒ-nē-a) **gravis** is a weakness of the skeletal muscles. It is caused by an abnormality at the neuromuscular junction that prevents the muscle fibers from contracting. Recall that motor neurons stimulate the skeletal muscle fibers to contract by releasing acetylcholine. Myasthenia gravis is believed to be due to antibodies directed against ACh receptors of the motor end plate. The antibodies bind to the receptors, leaving few available to ACh (see Chapter 12). As the disease progresses, more neuromuscular junctions become affected. The muscle becomes increasingly weaker and may eventually cease to function altogether.

Myasthenia gravis is more common in females, occurring most frequently between the ages of 20 and 50. The muscles of the face and neck are most apt to be involved. Initial symptoms include a weakness of the eye muscles and difficulty in swallowing. Later, the individual has difficulty chewing and talking. Eventually, the muscles of the limbs may become involved. Death may result from paralysis of the respiratory muscles, but usually the disorder does not progress to this stage.

Anticholinesterase drugs have been the primary treatment for the disease. More recently, steroid drugs, such as prednisone, have been used with great success. Immunosuppressant drugs are also used to decrease the production of antibodies that interfere with normal muscle contraction. Another recent treatment involves *plasmapheresis,* a procedure that separates blood cells from the plasma that contains the unwanted antibodies. The blood cells are then mixed with a plasma substitute and pumped back into the individual. Preliminary results indicate dramatic improvement in all treated individuals.

Abnormal Contractions

One kind of abnormal contraction of a muscle is **spasm:** a sudden, involuntary contraction of short duration. A **cramp** is a painful spasmodic contraction of a muscle. It is an involuntary, complete tetanic contraction. **Convulsions** are violent, involuntary tetanic contractions of an entire group of muscles. Convulsions occur when motor neurons are stimulated by fever, poisons, hysteria, or changes in body chemistry due to withdrawal of certain drugs. The stimulated neurons send many bursts of seemingly disordered impulses to the muscle fibers. **Fibrillation** is the uncoordinated contraction of individual muscle fibers preventing the smooth contraction of the muscle. A **tic** is a spasmodic twitching made involuntarily by muscles that are ordinarily under voluntary control. Twitching of the eyelid and face muscles are examples. In general, tics are of psychological origin.

MEDICAL TERMINOLOGY

Electromyography or **EMG** (*electro* = electricity; *myo* = muscle; *graph* = to write)   The recording and study of the electrical changes that occur in muscle tissue.

Gangrene (*gangraena* = an eating sore)   Death of a soft tissue, such as muscle, that results from interruption of its blood supply. It is caused by various species of *Clostridium,* bacteria that live anaerobically in the soil.

Myalgia (*algia* = painful condition)   Pain in or associated with muscles.

Myology (*logos* = study of)   Study of muscles.

Myoma (*oma* = tumor)   A tumor consisting of muscle tissue.

Myomalacia (*malaco* = soft)   Softening of a muscle.

Myopathy (*pathos* = disease)   Any disease of muscle tissue.

Myosclerosis (*scler* = hard)   Hardening of a muscle.

Myositis (*itis* = inflammation of)   Inflammation of muscle fibers (cells).

Myospasm   Spasm of a muscle.

Myotonia (*tonia* = tension)   Increased muscular excitability and contractility with decreased power of relaxation; tonic spasm of the muscle.

Paralysis (*para* = beyond; *lyein* = to loosen)   Loss or impairment of motor (muscular) function resulting from a lesion of nervous or muscular origin.

Shin splints   A soreness or pain along the shinbone (tibia) due to straining the flexor digitorum muscle. The condition is often caused by walking or running up and down hills or by vigorous activity of the legs following a period of relative inactivity.

Trichinosis   A myositis caused by the parasitic worm *Trichinella spiralis,* which may be found in the muscles of humans, rats, and pigs. People contract the disease by eating insufficiently cooked infected pork.

Volkmann's contracture (*contra* = against)   Permanent contraction of a muscle due to replacement of destroyed muscle cells with fibrous tissue that lacks ability to stretch. Destruction of muscle cells may occur from interference with circulation caused by a tight bandage, a piece of elastic, or a cast.

Wryneck or torticollis (*tortus* = twisted; *collam* = neck) Spasmodic contraction of several of the superficial and deep muscles of the neck; produces twisting of the neck and an unnatural position of the head.

DRUGS ASSOCIATED WITH THE MUSCULAR SYSTEM

Drugs associated with the muscular system are usually concerned with preventing or controlling abnormal muscular contractions. Specific drug therapy is lacking for the dystrophic diseases.

Skeletal Muscle Relaxants
Skeletal muscle spasms like "charleyhorse" or torticollis can be treated with analgesics for pain or with skeletal muscle relaxants like *methocarbamol* (Robaxin), *diazepam* (Valium), *chlorzoxazone* (Paraflex), *cyclobenzaprine HCl* (Flexeril), and *carisoprodol* (Soma). *Dantrolene* (Dantrium) is used to treat a rare, though deadly, muscle disorder known as malignant hyperpyrexia.

STUDY OUTLINE

Characteristics (p. 204)
1. Excitability is the property of receiving and responding to stimuli.
2. Contractility is the ability to shorten and thicken, contract.
3. Extensibility is the ability to be stretched or extended.
4. Elasticity is the ability to return to original shape after contraction or extension.

Functions (p. 204)
1. Through contraction, muscle tissue performs the three important functions of motion, maintenance of posture, and heat production.

Types (p. 204)
1. Skeletal muscle tissue is attached to bones. It is striated and voluntary.
2. Visceral muscle tissue is located in viscera. It is nonstriated (smooth) and involuntary.
3. Cardiac muscle tissue forms the walls of the heart. It is striated and involuntary.

Skeletal Muscle Tissue (p. 204)
1. The term fascia is applied to a sheet or broad band of fibrous connective tissue underneath the skin or around muscles and organs of the body. There are three types of fascia: superficial, deep, and subserous.
2. Connective tissue components are epimysium, covering the entire muscle; perimysium, covering fasciculi; and endomysium, covering fibers.
3. Tendons and aponeuroses are extensions of connective tissue beyond the muscle cells to attach the muscle to bone or other muscle.
4. Nerves convey impulses for muscular contraction.
5. Blood provides nutrients and oxygen for contraction.
6. Skeletal muscle consists of fibers covered by a sarcolemma. The fibers contain sarcoplasm, nuclei, sarcoplasmic reticulum, and T tubules.
7. Each fiber contains myofibrils that consist of thin and thick myofilaments. The myofilaments are compartmentalized into sarcomeres.
8. Thin myofilaments are composed of actin, tropomyosin, and troponin; thick myofilaments consist of myosin.

Contraction (p. 208)
Sliding-Filament Theory
1. A nerve impulse travels over the sarcolemma and enters the T tubules and sarcoplasmic reticulum.
2. The nerve impulse leads to the release of calcium ions from the sarcoplasmic reticulum, triggering the contractile process.
3. Actual contraction is brought about when the thin myofilaments of a sarcomere slide toward each other.

Motor Unit
1. A motor neuron transmits the stimulus to a skeletal muscle for contraction.
2. The region of the sarcolemma under the terminal branch of an axon of a motor neuron that is specialized to receive the nerve impulse is the motor end plate.
3. The area of contact between a motor neuron and muscle fiber is a neuromuscular, or myoneural, junction.
4. A motor neuron and the muscle fibers it stimulates form a motor unit.

Physiology
1. When a nerve impulse reaches the motor end plate, the neuron releases acetylcholine, which transmits the impulse to the motor end plate and then into the T tubules and sarcoplasmic reticulum.
2. This releases calcium ions that activate myosin, catalyzing the breakdown of ATP, and bind tropomyosin-troponin complex, so that actin of thin myofilaments can attach to myosin cross bridges of thick myofilaments.
3. The energy released from the breakdown of ATP causes the sliding of the myofilaments.

Energy
1. The only direct source of energy for muscle contraction is ATP.
2. A reserve supply of ATP may be built up by storage of excess in thick myofilaments or by anaerobic combination with creatine to form creatine phosphate, which breaks down to produce ATP when muscles contract strenuously.

All-or-None Principle
1. Muscle fibers of a motor unit contract to their fullest extent or not at all.
2. The weakest stimulus capable of causing contraction is a liminal, or threshold, stimulus.
3. A stimulus not capable of inducing contraction is a subliminal, or subthreshold, stimulus.

Kinds
1. The various kinds of contractions are twitch, treppe, tetanus, isotonic, and isometric.
2. A record of a contraction is called a myogram. The refractory period is the time when a muscle has temporarily lost excit-

ability. Skeletal muscles have a short refractory period. Cardiac muscle has a long refractory period.

3. Summation of twitches is the increased strength of a contraction resulting from the application of a second stimulus before the muscle has completely relaxed after a previous stimulus.

Muscle Tone

1. A sustained partial contraction of portions of a skeletal muscle results in muscle tone.
2. Tone is essential for maintaining posture.
3. Flaccidity is a condition of less than normal tone. Atrophy is a wasting away or decrease in size; hypertrophy is an enlargement or overgrowth.

Fast (White) and Slow (Red) Muscle

1. The duration of contraction of various muscles of the body varies with the functions of the muscles in maintaining homeostasis.
2. Fast or white muscles have an extensive sarcoplasmic reticulum.
3. Slow or red muscles have smaller fibers, more blood capillaries, and a large amount of myoglobin.

Cardiac Muscle Tissue (p. 215)

1. This muscle is found only in the heart. It is striated and involuntary.
2. The cells are quadrangular and usually contain a single centrally placed nucleus.
3. Compared to skeletal muscle tissue, cardiac muscle tissue has more sarcoplasm, more mitochondria, less well-developed sarcoplasmic reticulum, and larger T tubules located at Z lines rather than at A-I band junctions. Myofilaments are not arranged in discrete myofibrils.
4. The fibers branch freely to form two continuous networks, each of which contracts as a functional unit.

5. Intercalated discs provide strength and aid impulse conduction.

Smooth Muscle Tissue (p. 217)

1. Smooth muscle is nonstriated and involuntary.
2. Visceral smooth muscle is found in the walls of viscera. The fibers are arranged in a network.
3. Multiunit smooth muscle is found in blood vessels and the eye. The fibers operate singly rather than as a unit.

Homeostasis (p. 218)

1. Oxygen debt is the amount of O_2 needed to convert accumulated lactic acid into CO_2 and H_2O. It occurs during strenuous exercise and is paid back by continuing to breathe rapidly after exercising. Unit it is paid back, the homeostasis between muscular activity and oxygen requirements is not restored.
2. Muscle fatigue results from diminished availability of oxygen and toxic effects of carbon dioxide and lactic acid built up during exercise.
3. The heat given off during muscular contraction maintains the homeostasis of body temperature.

Disorders: Homeostatic Imbalances (p. 218)

1. Fibrosis is the formation of fibrous tissue where it normally does not exist; it frequently occurs in damaged muscle tissue.
2. Fibrositis is an inflammation of fibrous tissue. If it occurs in the lumbar region, it is called lumbago.
3. "Charleyhorse" refers to pain, tenderness, and stiffness of joints, muscles, and related structures in the thigh.
4. Muscular dystrophy is a hereditary disease of muscles characterized by degeneration of individual muscle cells.
5. Myasthenia gravis is a disease characterized by great muscular weakness and fatigability resulting from improper neuromuscular transmission.
6. Abnormal contractions include spasms, cramps, convulsions, fibrillations, and tics.

REVIEW QUESTIONS

1. How is the skeletal system related to the muscular system? What are the three basic functions of the muscular system?
2. What are the four characteristics of muscle tissue?
3. How can the three types of muscle tissue be distinguished?
4. What is fascia? What are the three different types of fascia and where are they found in the body?
5. Define epimysium, perimysium, endomysium, tendon, and aponeurosis. Describe the nerve and blood supply to a skeletal muscle.
6. Discuss the microscopic structure of skeletal muscle.
7. In considering the contraction of skeletal muscle, describe the following: motor unit, role of calcium, sources of energy, and sliding-filament theory.
8. What is the all-or-none principle? Relate it to a liminal and subliminal stimulus.
9. Define each of the following contractions and state the importance of each: twitch, treppe, tetanus, isotonic, and isometric.
10. What is a myogram?
11. Describe the latent period, contraction period, and relaxation period of muscle contraction. Construct a diagram to illustrate your answer.
12. Define the refractory period. How does it differ between skeletal and cardiac muscle? What is summation?
13. What is muscle tone? Why is it important? Distinguish

between atrophy and hypertrophy.
14. Compare fast (white) and slow (red) muscle with respect to structure and function.
15. Compare cardiac and skeletal muscle with regard to microscopic structure, functions, and locations.
16. Compare cardiac and smooth muscle with regard to microscopic structure, functions, and locations.
17. Discuss each of the following as examples of muscle homeostasis: oxygen debt, fatigue, and heat production.
18. What do you think might be the relationship between shivering (uncontrolled muscular contractions) and body temperature? Can you relate sweating (cooling of the skin) after strenuous exercise to the homeostasis of body temperature?
19. Define fibrosis and fibrositis.
20. What is a "charleyhorse"?
21. What is muscular dystrophy?
22. What is myasthenia gravis? In this disease, why do the muscles not contract normally?
23. Define each of the following abnormal muscular contractions: spasm, cramp, convulsion, fibrillation, and tic.
24. Refer to the glossary of medical terminology associated with the muscular system. Be sure that you can define each term.
25. Describe the actions of selected drugs associated with muscle tissue.

EXHIBIT 11-2
MUSCLES OF FACIAL EXPRESSION (Figure 11-4)

MUSCLE	ORIGIN	INSERTION	ACTION	INNERVATION
Epicranius (*epi* = over; *crani* = skull)	This muscle is divisible into two portions: the frontalis over the frontal bone and the occipitalis over the occipital bone. The two muscles are united by a strong aponeurosis, the galea aponeurotica, which covers the superior and lateral surfaces of the skull.			
Frontalis (*front* = forehead)	Galea aponeurotica.	Skin superior to supraorbital line.	Draws scalp forward, raises eyebrows, and wrinkles forehead horizontally.	Facial (VII) nerve.
Occipitalis (*occipito* = base of skull)	Occipital bone and mastoid process of temporal bone.	Galea aponeurotica.	Draws scalp backward.	Facial (VII) nerve.
Orbicularis oris (*orb* = circular; *or* = mouth)	Muscle fibers surrounding opening of mouth.	Skin at corner of mouth.	Closes lips, compresses lips against teeth, protrudes lips, and shapes lips during speech.	Facial (VII) nerve.
Zygomaticus major (*zygomatic* = cheek bone; *major* = greater)	Zygomatic bone.	Skin at angle of mouth and orbicularis oris.	Draws angle of mouth upward and outward as in smiling or laughing.	Facial (VII) nerve.
Levator labii superioris (*levator* = raises or elevates; *labii* = lip; *superioris* = upper)	Superior to infraorbital foramen of maxilla.	Skin at angle of mouth and orbicularis oris.	Elevates (raises) upper lip.	Facial (VII) nerve.
Depressor labii inferioris (*depressor* = depresses or lowers; *inferioris* = lower)	Mandible.	Skin of lower lip.	Depresses (lowers) lower lip.	Facial (VII) nerve.
Buccinator (*bucc* = cheek)	Alveolar processes of maxilla and mandible and pterygomandibular raphe (fibrous band extending from the pterygoid hamulus to the mandible).	Orbicularis oris.	Major cheek muscle; compresses cheek as in blowing air out of mouth and causes cheeks to cave in, producing the action of sucking.	Facial (VII) nerve.
Mentalis (*mentum* = chin)	Mandible.	Skin of chin.	Elevates and protrudes lower lip and pulls skin of chin up as in pouting.	Facial (VII) nerve.
Platysma (*platy* = flat, broad)	Fascia over deltoid and pectoralis major muscles.	Mandible, muscles around angle of mouth, and skin of lower face.	Draws outer part of lower lip downward and backward as in pouting; depresses mandible.	Facial (VII) nerve.
Risorius (*risor* = laughter)	Fascia over parotid (salivary) gland.	Skin at angle of mouth.	Draws angle of mouth laterally as in tenseness.	Facial (VII) nerve.
Orbicularis oculi (*ocul* = eye)	Medial wall of orbit.	Circular path around orbit.	Closes eye.	Facial (VII) nerve.
Corrugator supercilii (*corrugo* = to wrinkle; *supercilium* = eyebrow)	Medial end of superciliary arch of frontal bone.	Skin of eyebrow.	Draws eyebrow downward as in frowning.	Facial (VII) nerve.
Levator palpebrae superioris (*palpebrae* = eyelids) (see Figure 11-6b)	Roof of orbit (lesser wing of sphenoid bone).	Skin of upper eyelid.	Elevates upper eyelid.	Oculomotor (III) nerve.

FIGURE 11-4 Muscles of facial expression. (a) Anterior superficial view. (b) Anterior deep view.

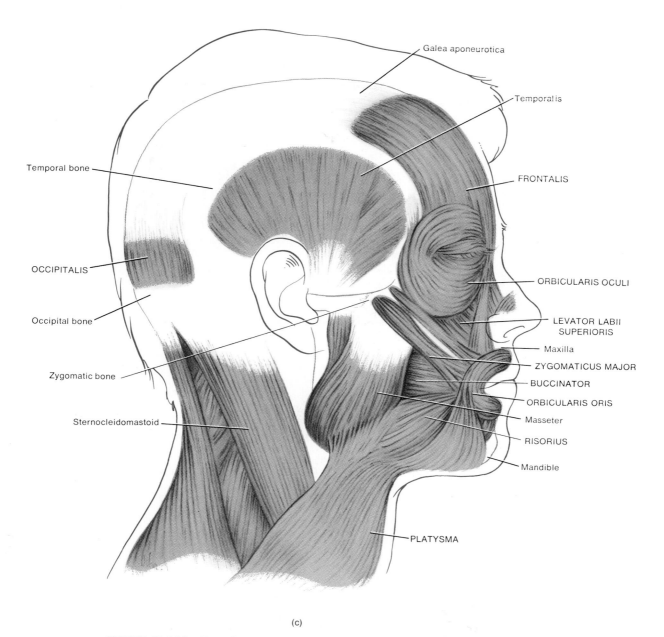

(c)

FIGURE 11-4 (*Continued*) Muscles of facial expression. (c) Right lateral superficial view.

EXHIBIT 11-3
MUSCLES THAT MOVE THE LOWER JAW (Figure 11-5)

MUSCLE	ORIGIN	INSERTION	ACTION	INNERVATION
Masseter (*maseter* = chewer)	Maxilla and zygomatic arch.	Angle and ramus of mandible.	Elevates mandible as in closing mouth and protracts (protrudes) mandible.	Mandibular branch of trigeminal (V) nerve.
Temporalis (*tempora* = temples)	Temporal bone.	Coronoid process of mandible.	Elevates and retracts mandible.	Temporal nerve from mandibular division of trigeminal (V) nerve.
Medial pterygoid (*medial* = closer to midline; *pterygoid* = like a wing; pterygoid plate of sphenoid bone)	Medial surface of lateral pterygoid plate of sphenoid; maxilla.	Angle and ramus of mandible.	Elevates and protracts mandible and moves mandible from side to side.	Mandibular branch of trigeminal (V) nerve.
Lateral pterygoid (*lateral* = farther from midline)	Greater wing and lateral surface of lateral pterygoid plate of sphenoid.	Condyle of mandible; temporomandibular articulation.	Protracts mandible, opens mouth, and moves mandible from side to side.	Mandibular branch of trigeminal (V) nerve.

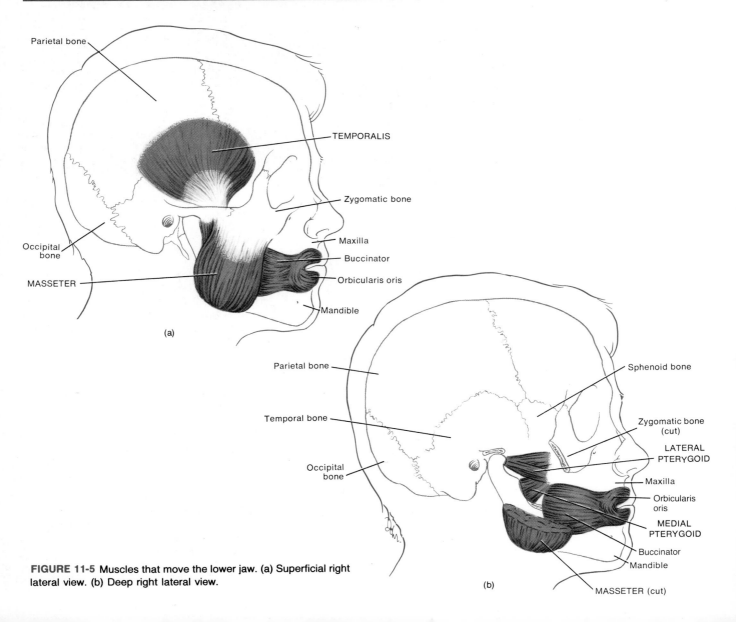

FIGURE 11-5 Muscles that move the lower jaw. (a) Superficial right lateral view. (b) Deep right lateral view.

EXHIBIT 11-4

MUSCLES THAT MOVE THE EYEBALLS—EXTRINSIC MUSCLES* (Figure 11-6)

MUSCLE	ORIGIN	INSERTION	ACTION	INNERVATION
Superior rectus (*superior* = above; *rectus* = in this case, muscle fibers running parallel to long axis of eyeball)	Tendinous ring attached to bony orbit around optic foramen.	Superior and central part of eyeball.	Rolls eyeball upward.	Oculomotor (III) nerve.
Inferior rectus (*inferior* = below)	Same as above.	Inferior and central part of eyeball.	Rolls eyeball downward.	Oculomotor (III) nerve.
Lateral rectus	Same as above.	Lateral side of eyeball.	Rolls eyeball laterally.	Abducens (VI) nerve.
Medial rectus	Same as above.	Medial side of eyeball.	Rolls eyeball medially.	Oculomotor (III) nerve.
Superior oblique (*oblique* = in this case, muscle fibers running diagonally to long axis of eyeball)	Same as above.	Eyeball between superior and lateral recti.	Rotates eyeball on its axis; directs cornea downward and laterally; note that it moves through a ring of fibrocartilaginous tissue called the trochlea (*trochlea* = pulley).	Trochlear (IV) nerve.
Inferior oblique	Maxilla (front of orbital cavity).	Eyeball between inferior and lateral recti.	Rotates eyeball on its axis; directs cornea upward and laterally.	Oculomotor (III) nerve.

* Muscles situated on the outside of the eyeball.

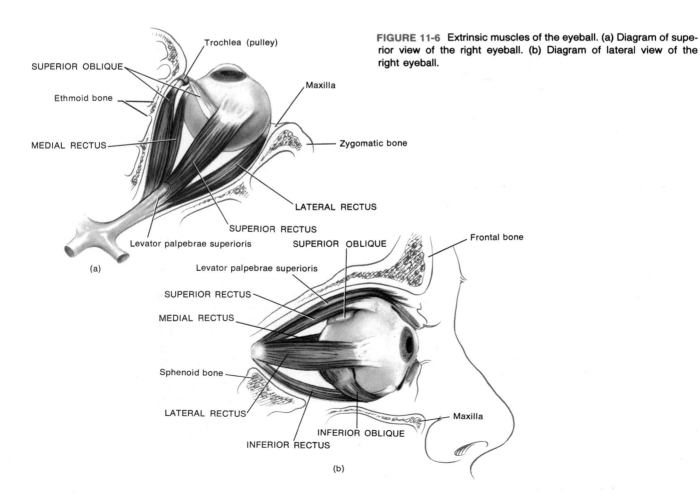

FIGURE 11-6 Extrinsic muscles of the eyeball. (a) Diagram of superior view of the right eyeball. (b) Diagram of lateral view of the right eyeball.

Trochlea

Superior oblique muscle

Superior rectus muscle

Levator palpebrae superioris muscle

Lateral rectus muscle

Optic nerve

(c)

Superior rectus muscle

Lateral rectus muscle

Inferior rectus muscle

Superior oblique muscle

Trochlea

Medial rectus muscle

Inferior oblique muscle

(d)

Figure 11-6 (*Continued*) Extrinsic muscles of the eyeball. (c) Photograph of superior view of the right eyeball. (d) Photograph of the right eye muscles extended. (Photographs courtesy of C. Yokochi and J. W. Rohen, *Photographic Anatomy of the Human Body,* 2d ed., 1979, IGAKU-SHOIN, Ltd., Tokyo, New York.)

EXHIBIT 11-5

MUSCLES THAT MOVE THE TONGUE (Figure 11-7)

MUSCLE	ORIGIN	INSERTION	ACTION	INNERVATION
Genioglossus (*geneion* = chin; *glossus* = tongue)	Mandible.	Undersurface of tongue and hyoid bone.	Depresses tongue and thrusts it forward (protraction).	Hypoglossal (XII) nerve.
Styloglossus (*stylo* = stake or pole; styloid process of temporal bone)	Styloid process of temporal bone.	Side and undersurface of tongue.	Elevates tongue and draws it backward (retraction).	Hypoglossal (XII) nerve.
Palatoglossus (*palato* = palate)	Anterior surface of soft palate.	Side of tongue.	Elevates tongue and draws soft palate down on tongue.	Hypoglossal (XII) nerve.
Hyoglossus	Body of hyoid bone.	Side of tongue.	Depresses tongue and draws down its sides.	Hypoglossal (XII) nerve.

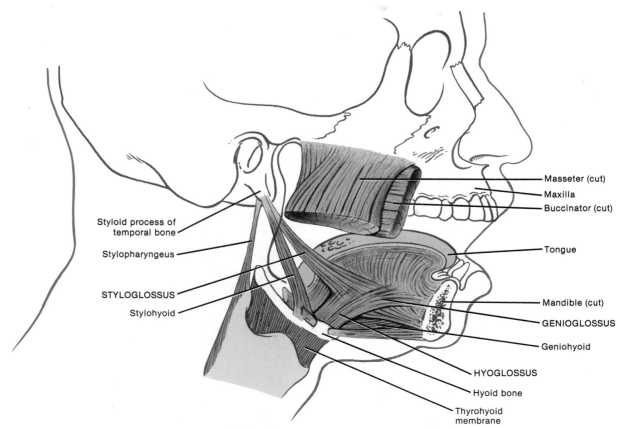

FIGURE 11-7 Muscles that move the tongue viewed from the right side.

EXHIBIT 11-6

MUSCLES OF THE PHARYNX (Figure 11-8)

MUSCLE	ORIGIN	INSERTION	ACTION	INNERVATION
Inferior constrictor (*inferior* = below; *constrictor* = decreases diameter of a lumen)	Cricoid and thyroid cartilages of larynx.	Posterior median raphe of pharynx.	Constricts inferior portion of pharynx to propel a bolus into esophagus.	Pharyngeal plexus.
Middle constrictor	Greater and lesser cornu of hyoid bone and stylohyoid ligament.	Posterior median raphe of pharynx.	Constricts middle portion of pharynx to propel a bolus into esophagus.	Pharyngeal plexus.
Superior constrictor (*superior* = above)	Pterygoid process, pterygomandibular raphe, and mylohyoid line of mandible.	Posterior median raphe of pharynx.	Constricts superior portion of pharynx to propel a bolus into esophagus.	Pharyngeal plexus.
Stylopharyngeus (*stylo* = stake or pole; styloid process of temporal bone; *pharyngo* = pharynx) (see also Figure 11-7)	Medial side of base of styloid process.	Lateral aspects of pharynx and thyroid cartilage.	Elevates larynx and dilates pharynx to help bolus descend.	Glossopharyngeal (IX) nerve.
Salpingopharyngeus (*salping* = pertaining to the auditory or uterine tube)	Inferior portion of auditory tube.	Posterior fibers of palatopharyngeus muscle.	Elevates superior portion of lateral wall of pharynx during swallowing and opens orifice of auditory tube.	Pharyngeal plexus.
Palatopharyngeus (*palato* = palate)	Soft palate.	Posterior border of thyroid cartilage and lateral and posterior wall of pharynx.	Elevates larynx and pharynx and helps close nasopharynx during swallowing.	Pharyngeal plexus.

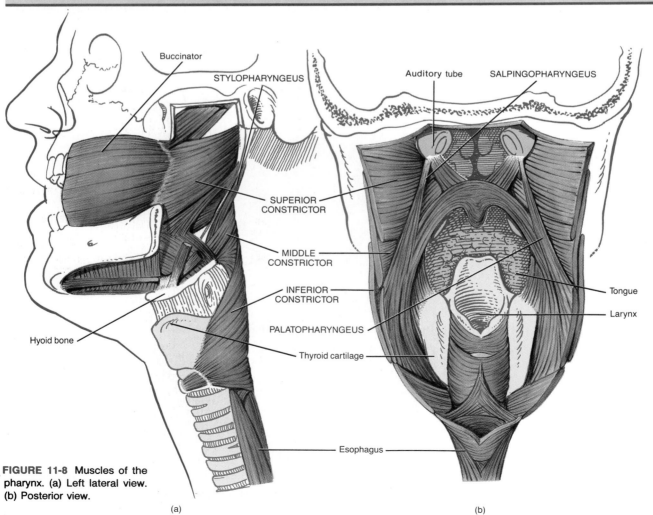

FIGURE 11-8 Muscles of the pharynx. (a) Left lateral view. (b) Posterior view.

(a)　　　　　(b)

EXHIBIT 11-7

MUSCLES OF THE LARYNX (Figure 11-9)

MUSCLE	ORIGIN	INSERTION	ACTION	INNERVATION
EXTRINSIC				
Omohyoid (*omo* = relationship to the shoulder; *hyoedes* = U-shaped; pertaining to hyoid bone)	Superior border of scapula and superior transverse ligament.	Body of hyoid bone.	Depresses hyoid bone.	Branches of ansa cervicalis nerve (C1–C3).
Sternohyoid (*sterno* = sternum)	Medial end of clavicle and manubrium of sternum.	Body of hyoid bone.	Depresses hyoid bone.	Branches of ansa cervicalis nerve (C1–C3).
Sternothyroid (*thyro* = thyroid gland)	Manubrium of sternum.	Thyroid cartilage of larynx.	Depresses thyroid cartilage.	Branches of ansa cervicalis nerve (C1–C3).
Thyrohyoid	Thyroid cartilage of larynx.	Greater cornu of hyoid bone.	Elevates thyroid cartilage and depresses hyoid bone.	Cervical nerves C1–C2 and descending hypoglossal (XII) nerve.
Stylopharyngeus	See Exhibit 11-6.			
Palatopharyngeus	See Exhibit 11-6.			
Inferior constrictor	See Exhibit 11-6.			
Middle constrictor	See Exhibit 11-6.			
INTRINSIC				
Cricothyroid (*crico* = cricoid cartilage of larynx)	Anterior and lateral portion of cricoid cartilage of larynx.	Anterior border of inferior cornu of thyroid cartilage of larynx and posterior part of inferior border of lamina of thyroid cartilage.	Produces tension and elongation of vocal folds.	External laryngeal branch of vagus (X) nerve.

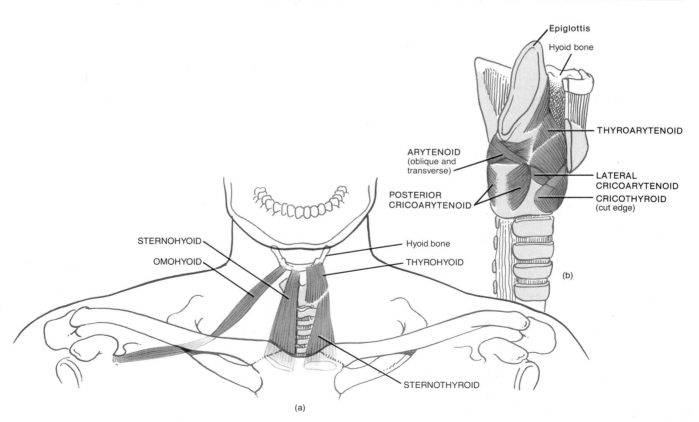

FIGURE 11-9 Muscles of the larynx. (a) Anterior view. (b) Right posterolateral view.

EXHIBIT 11-8
MUSCLES THAT MOVE THE HEAD

MUSCLE	ORIGIN	INSERTION	ACTION	INNERVATION
Sternocleidomastoid (*sternum* = breastbone; *cleido* = clavicle; *mastoid* = mastoid process of temporal bone) (see Figure 11-14)	Sternum and clavicle.	Mastoid process of temporal bone.	Contraction of both muscles flex the cervical part of the vertebral column, draw the head forward, and elevate chin; contraction of one muscle rotates face toward side opposite contracting muscle.	Accessory (XI) nerve; cervical nerves C2–C3.
Semispinalis capitis (*semi* = half; *spine* = spinous process; *caput* = head) (see Figure 11-18)	Articular process of seventh cervical vertebra and transverse processes of first six thoracic vertebrae.	Occipital bone.	Both muscles extend head; contraction of one muscle rotates face toward same side as contracting muscle.	Dorsal rami of spinal nerves.
Splenius capitis (*splenion* = bandage) (see Figure 11-18)	Ligamentum nuchae and spines of seventh cervical vertebra and first four thoracic vertebrae.	Occipital bone and mastoid process of temporal bone.	Both muscles extend head; contraction of one rotates it to same side as contracting muscle.	Dorsal rami of middle and lower cervical nerves.
Longissimus capitis (*longissimus* = longest) (see Figure 11-18)	Transverse processes of last four cervical vertebrae.	Mastoid process of temporal bone.	Extends head and rotates face toward side opposite contracting muscle.	Dorsal rami of middle and lower cervical nerves.

EXHIBIT 11-9
MUSCLES THAT ACT ON THE ANTERIOR ABDOMINAL WALL (Figure 11-10)

MUSCLE	ORIGIN	INSERTION	ACTION	INNERVATION
Rectus abdominis (*rectus* = fibers parallel to midline; *abdomino* = belly)	Pubic crest and symphysis pubis.	Cartilage of fifth to seventh ribs and xiphoid process.	Flexes vertebral column.	Branches of thoracic nerves T7–T12.
External oblique (*external* = closer to surface; *oblique* = fibers diagonal to midline)	Lower eight ribs.	Iliac crest and linea alba (midline aponeurosis).	Contraction of both compresses abdomen; contraction of one side alone bends vertebral column laterally.	Branches of thoracic nerves T7–T12 and iliohypogastric nerve.
Internal oblique (*internal* = farther from surface)	Iliac crest, inguinal ligament, and thoracolumbar fascia.	Cartilage of last three or four ribs.	Compresses abdomen; contraction of one side alone bends vertebral column laterally.	Branches of thoracic nerves T8–T12, iliohypogastric, and ilioinguinal nerves.
Transversus abdominis (*transverse* = fibers perpendicular to midline)	Iliac crest, inguinal ligament, lumbar fascia, and cartilages of last six ribs.	Xiphoid process, linea alba, and pubis.	Compresses abdomen.	Branches of thoracic nerves T8–T12, iliohypogastric, and ilioinguinal nerves.

FIGURE 11-10 Muscles of the anterior abdominal wall. (a) Superficial view. (b) Deep view.

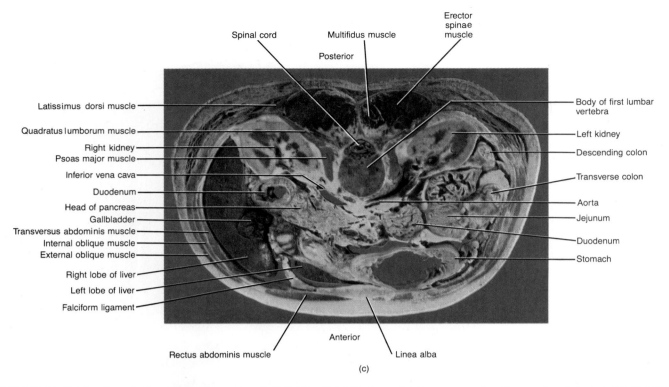

Spinal cord   Multifidus muscle   Erector spinae muscle

Posterior

Latissimus dorsi muscle

Quadratus lumborum muscle

Right kidney
Psoas major muscle

Inferior vena cava

Duodenum

Head of pancreas

Gallbladder

Transversus abdominis muscle
Internal oblique muscle
External oblique muscle

Right lobe of liver

Left lobe of liver

Falciform ligament

Body of first lumbar vertebra

Left kidney

Descending colon

Transverse colon

Aorta

Jejunum

Duodenum

Stomach

Anterior

Rectus abdominis muscle   Linea alba

(c)

FIGURE 11-10 (*Continued*) Muscles of the anterior abdominal wall. (c) Photograph of a cross section through the abdomen showing the musculature and related viscera. (Courtesy of Stephen A. Kieffer and E. Robert Heitzman, *An Atlas of Cross-Sectional Anatomy*, Harper & Row, Publishers, Inc., New York, 1979.)

EXHIBIT 11-10
MUSCLES USED IN BREATHING (Figure 11-11)

MUSCLE	ORIGIN	INSERTION	ACTION	INNERVATION
Diaphragm (*dia* = across; *phragma* = wall)	Xiphoid process, costal cartilages of last six ribs, and lumbar vertebrae.	Central tendon.	Forms floor of thoracic cavity; pulls central tendon downward during inspiration and thus increases vertical length of thorax.	Phrenic nerve.
External intercostals (*inter* = between; *costa* = rib)	Inferior border of rib above.	Superior border of rib below.	Elevate ribs during inspiration and thus increase lateral and anteroposterior dimensions of thorax.	Intercostal nerves.
Internal intercostals	Superior border of rib below.	Inferior border of rib above.	Draw adjacent ribs together during forced expiration and thus decrease lateral and anteroposterior dimensions of thorax.	Intercostal nerves.

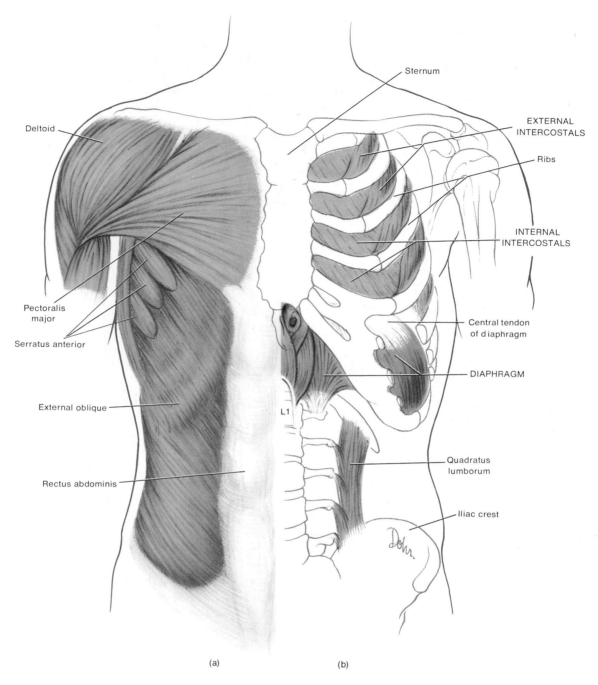

FIGURE 11-11 Muscles used in breathing. (a) Diagram of superficial view. (b) Diagram of deep view.

(c)

FIGURE 11-11 (*Continued*) Muscles used in breathing. (c) Photograph of superior view of diaphragm. (Courtesy of C. Yokochi and J. W. Rohen, *Photographic Anatomy of the Human Body,* 2d ed., 1979, IGAKU-SHOIN, Ltd., Tokyo, New York.)

EXHIBIT 11-11

MUSCLES OF THE PELVIC FLOOR* (Figure 11-12)

MUSCLE	ORIGIN	INSERTION	ACTION	INNERVATION
Levator ani (*levator* = raises; *ani* = anus)	This muscle is divisible into two parts, the pubococcygeus muscle and the iliococcygeus muscle. It forms the funnel-shaped floor of the pelvic cavity and supports the pelvic structures. It contains openings for the anal canal and urethra in both sexes and the vagina in the female.			
Pubococcygeus (*pubo* = pubis; *coccygeus* = coccyx)	Pubis.	Coccyx, urethra, anal canal, and central tendon of perineum.	Supports and slightly raises pelvic floor, resists increased intraabdominal pressure, and draws anus toward pubis and constricts it.	Sacral nerves S3–S4 or S4 and perineal branch of pudendal nerve.
Iliococcygeus (*ilio* = ilium)	Ischial spine.	Coccyx.	Supports and slightly raises pelvic floor, resists increased intraabdominal pressure, and draws anus toward pubis and constricts it.	Sacral nerves S3–S4 or S4 and perineal branch of pudendal nerve.
Coccygeus	Ischial spine.	Lower sacrum and upper coccyx.	Supports and slightly raises pelvic floor, resists intraabdominal pressure, and pulls coccyx forward following defecation or parturition.	Sacral nerve S3 or S4.

* The muscles of the pelvic floor, together with the fasciae covering their internal and external surfaces, are collectively referred to as the **pelvic diaphragm.**

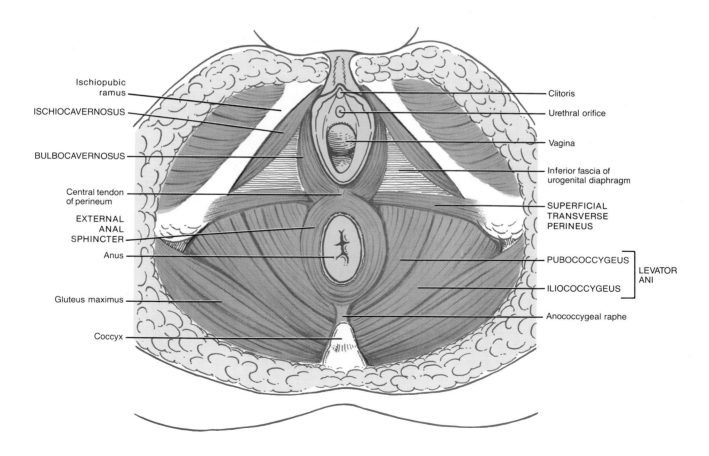

FIGURE 11-12 Muscles of the pelvic floor seen in the female perineum.

EXHIBIT 11-12
MUSCLES OF THE PERINEUM* (Figure 11-13)

MUSCLE	ORIGIN	INSERTION	ACTION	INNERVATION
Superficial transverse perineus (*superficial* = closer to surface; *transverse* = across; *perineus* = perineum)	Ischial tuberosity.	Central tendon of perineum.	Helps to stabilize the central tendon of the perineum.	Perineal branch of pudendal nerve.
Bulbocavernosus (*bulbus* = bulb; *caverna* = hollow place)	Central tendon of perineum.	Inferior fascia of urogenital diaphragm, corpus spongiosum of penis, and deep fascia on dorsum of penis in male; pubic arch and root and dorsum of clitoris in female.	Helps expel last drops of urine during micturition, helps propel semen along urethra, and assists in erection of the penis in male; decreases vaginal orifice and assists in erection of clitoris in female.	Perineal branch of pudendal nerve.
Ischiocavernosus (*ischion* = hip)	Ischial tuberosity and ischial and pubic rami.	Corpus cavernosum of penis in male and clitoris in female.	Maintains erection of penis in male and clitoris in female.	Perineal branch of pudendal nerve.
Deep transverse perineus † (*deep* = farther from surface)	Ischial rami.	Central tendon of perineum.	Helps eject last drops of urine and semen in male and urine in female.	Perineal branch of pudendal nerve.
Urethral sphincter † (*sphincter* = circular muscle that decreases size of an opening; *urethrae* = urethra)	Ischial and pubic rami.	Median raphe in male and vaginal wall in female.	Helps eject last drops of urine and semen in male and urine in female.	Perineal branch of pudendal nerve.
External anal sphincter	Anococcygeal raphe.	Central tendon of perineum.	Keeps anal canal and orifice closed.	Sacral nerve S4 and inferior rectal branch of pudendal nerve.

* The **perineum** is the entire outlet of the pelvis. It is a diamond-shaped area at the lower end of the trunk between the thighs and buttocks. It is bordered anteriorly by the symphysis pubis, laterally by the ischial tuberosities, and posteriorly by the coccyx. A transverse line drawn between the ischial tuberosities divides the perineum into an anterior **urogenital triangle** that contains the external genitals and a posterior **anal triangle** that contains the anus.

† The deep transverse perineus, the urethral sphincter, and a fibrous membrane constitute the **urogenital diaphragm.** It surrounds the urogenital ducts and helps to strengthen the pelvic floor.

FIGURE 11-13 Muscles of the male perineum.

EXHIBIT 11-13

MUSCLES THAT MOVE THE PECTORAL (SHOULDER) GIRDLE (Figure 11-14)

MUSCLE	ORIGIN	INSERTION	ACTION	INNERVATION
Subclavius (*sub* = under; *clavius* = clavicle)	First rib.	Clavicle.	Depresses clavicle.	Nerve to subclavius.
Pectoralis minor (*pectus* = breast, chest, thorax; *minor* = lesser)	Third through fifth ribs.	Coracoid process of scapula.	Depresses scapula, rotates shoulder joint anteriorly, and elevates third through fifth ribs during forced inspiration when scapula is fixed.	Medial pectoral nerve.
Serratus anterior (*serratus* = saw-toothed; *anterior* = front)	Upper eight or nine ribs.	Vertebral border and inferior angle of scapula.	Rotates scapula laterally and elevates ribs when scapula is fixed.	Long thoracic nerve.
Trapezius (*trapezoides* = trapezoid-shaped)	Occipital bone, ligamentum nuchae, and spines of seventh cervical and all thoracic vertebrae.	Clavicle and acromion and spine of scapula.	Elevates clavicle, adducts scapula, elevates or depresses scapula, and extends head.	Accessory (XI) nerve and cervical nerves C3–C4.
Levator scapulae (*levator* = raises; *scapulae* = scapula)	Upper four or five cervical vertebrae.	Vertebral border of scapula.	Elevates scapula.	Dorsal scapular nerve and cervical nerves C3–C5.
Rhomboideus major (*rhomboides* = rhomboid or diamond-shaped)	Spines of second to fifth thoracic vertebrae.	Vertebral border of scapula.	Adducts scapula and slightly rotates it upward.	Dorsal scapular nerve.
Rhomboideus minor	Spines of seventh cervical and first thoracic vertebrae.	Superior angle of scapula.	Adducts scapula.	Dorsal scapular nerve.

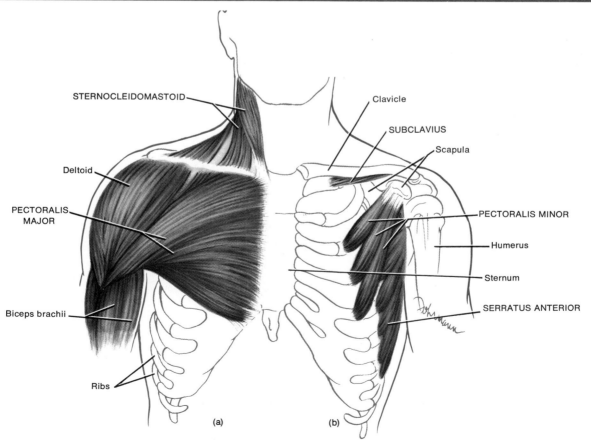

FIGURE 11-14 Muscles that move the pectoral (shoulder) girdle. (a) Anterior superficial view. (b) Anterior deep view.

EXHIBIT 11-14
MUSCLES THAT MOVE THE ARM (Figure 11-15)

MUSCLE	ORIGIN	INSERTION	ACTION	INNERVATION
Pectoralis major (see Figure 11-14a)	Clavicle, sternum, cartilages of second to sixth ribs.	Greater tubercle of humerus.	Flexes, adducts, and rotates arm medially.	Medial and lateral pectoral nerve.
Deltoid (*delta* = triangular)	Clavicle and acromion and spine of scapula.	Deltoid tuberosity of humerus.	Abducts arm.	Axillary nerve.
Subscapularis (*sub* = below; *scapularis* = scapula)	Subscapular fossa of scapula.	Lesser tubercle of humerus.	Rotates arm medially.	Upper and lower subscapular nerves.
Supraspinatus (*supra* = above; *spinatus* = spine of scapula)	Fossa superior to spine of scapula.	Greater tubercle of humerus.	Assists deltoid muscle in abducting arm.	Suprascapular nerve.

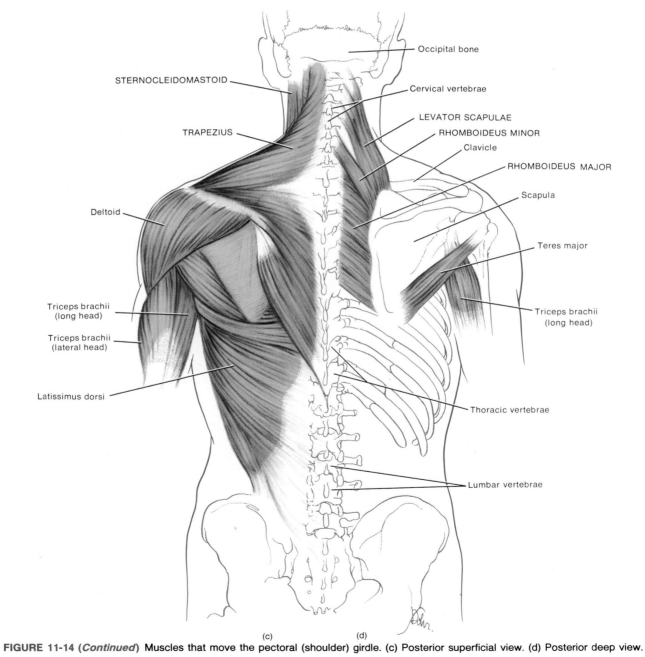

FIGURE 11-14 (*Continued*) Muscles that move the pectoral (shoulder) girdle. (c) Posterior superficial view. (d) Posterior deep view.

MUSCLE	ORIGIN	INSERTION	ACTION	INNERVATION
Infraspinatus (*infra* = below)	Fossa inferior to spine of scapula.	Greater tubercle of humerus.	Rotates arm laterally.	Suprascapular nerve.
Latissimus dorsi (*latissimus* = widest; *dorsum* = back)	Spines of lower six thoracic vertebrae, lumbar vertebrae, crests of sacrum and ilium, lower four ribs.	Intertubercular groove of humerus.	Extends, adducts, and rotates arm medially; draws shoulder downward and backward.	Thoracodorsal nerve.
Teres major (*teres* = long and round)	Inferior angle of scapula.	Distal to lesser tubercle of humerus.	Extends arm and draws it down; assists in adduction and medial rotation of arm.	Lower subscapular nerve.
Teres minor	Lateral border of scapula.	Greater tubercle of humerus.	Rotates arm laterally.	Axillary nerve.

(a) (b)

FIGURE 11-15 Muscles that move the arm. (a) Posterior superficial view. (b) Posterior deep view.

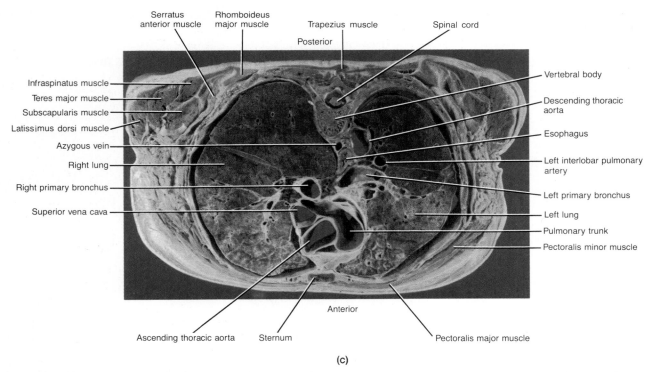

Serratus anterior muscle
Rhomboideus major muscle
Trapezius muscle
Spinal cord
Posterior
Infraspinatus muscle
Teres major muscle
Subscapularis muscle
Latissimus dorsi muscle
Azygous vein
Right lung
Right primary bronchus
Superior vena cava
Vertebral body
Descending thoracic aorta
Esophagus
Left interlobar pulmonary artery
Left primary bronchus
Left lung
Pulmonary trunk
Pectoralis minor muscle
Ascending thoracic aorta
Sternum
Anterior
Pectoralis major muscle

(c)

FIGURE 11-15 (*Continued*) Muscles that move the arm. (c) Photograph of a cross section of the thorax showing the musculature and related viscera. (Courtesy of Stephen A. Kieffer and E. Robert Heitzman, *An Atlas of Cross-Sectional Anatomy,* Harper & Row, Publishers, Inc., New York, 1979.)

UNIT III

Control Systems of the Human Body

This unit will show you the significance of the nerve impulse in making rapid adjustments for maintaining homeostasis. You will learn how the nervous system detects changes in the environment, decides on a course of action, and responds to the change. We will also investigate the role of hormones in maintaining long-term homeostasis.

12 Nervous Tissue

The **nervous system** is the body's control center and communications network. In humans, the nervous system serves three broad functions. First, it senses changes within the body and in the outside environment. Second, it interprets the changes. Third, it responds to the interpretation by initiating action in the form of muscular contractions or glandular secretions.

Through sensation, integration, and response the nervous system represents the body's most rapid means of maintaining homeostasis. Its split-second reactions, carried out by nerve impulses, can normally make the adjustments necessary to keep the body functioning efficiently. As you will see later, the nervous system shares the maintenance of homeostasis with the endocrine system. Although the adjustments made by hormones secreted by endocrine glands are slower than those made by nerve impulses, they are no less effective.

ORGANIZATION

The nervous system may be divided into two principal divisions, the central nervous system and the peripheral nervous system, and several subdivisions (Figure 12-1).

The **central nervous system (CNS)** is the control center for the entire system and consists of the brain and spinal cord. All body sensations must be relayed from receptors to the central nervous system if they are to be interpreted and acted on. All the nerve impulses that stimulate muscles to contract and glands to secrete must also pass from the central nervous system.

The various nerve processes that connect the brain and spinal cord with receptors, muscles, and glands constitute the **peripheral (pe-RIF-er-al) nervous system (PNS).** The peripheral nervous system may be divided into an afferent system and an efferent system. The **affer-**ent (AF-er-ent) **system** consists of nerve cells that convey information from receptors in the periphery of the body to the central nervous system. These nerve cells, called *afferent (sensory) neurons,* are the first cells to pick up incoming information. The **efferent** (EF-er-ent) **system** consists of nerve cells that convey information from the central nervous system to muscles and glands. These nerve cells are called *efferent (motor) neurons.*

The efferent system is subdivided into a somatic nervous system and an autonomic nervous system. The **somatic nervous system** or **SNS** (*soma* = body) consists of efferent neurons that conduct impulses from the central nervous system to skeletal muscle tissue. Since the somatic nervous system produces movement only in skeletal muscle tissue, it is under conscious control and therefore voluntary. The **autonomic nervous system** or **ANS** (*auto* = self; *nomos* = law), by contrast, contains efferent neurons that convey impulses from the central nervous system to smooth muscle tissue, cardiac muscle tissue, and glands. Since it produces responses only in involuntary muscles and glands, it is usually considered to be involuntary.

With few exceptions, the viscera receive nerve fibers from the two divisions of the autonomic nervous system: the **sympathetic division** and the **parasympathetic division.** In general, the fibers of one division stimulate or increase an organ's activity, while the fibers from the other inhibit or decrease activity (see Chapter 16).

HISTOLOGY

Despite the organizational complexity of the nervous system, it consists of only two principal kinds of cells: neurons and neuroglia. Neurons make up the nervous tissue that forms the structural and functional portion of the

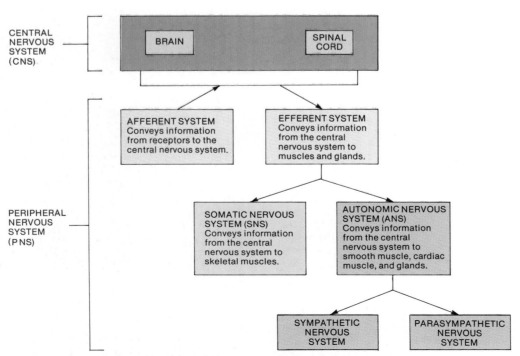

FIGURE 12-1 Organization of the nervous system.

EXHIBIT 12-1
NEUROGLIA OF CENTRAL NERVOUS SYSTEM

TYPE	DESCRIPTION	MICROSCOPIC APPEARANCE	FUNCTION
Astrocytes (*astro* = star; *cyte* = cell)	Star-shaped cells with numerous processes. *Protoplasmic astrocytes* are found in the gray matter of the CNS, and *fibrous astrocytes* are found in the white matter of the CNS.	Astrocyte Process / Blood vessel / Fiber of neuron	Twine around nerve cells to form supporting network in brain and spinal cord; attach neurons to their blood vessels.
Oligodendrocytes (*oligo* = few; *dendro* = tree)	Resemble astrocytes in some ways, but processes are fewer and shorter.	Neuron / Oligodendrocytes	Give support by forming semi-rigid connective tissue rows between neurons in brain and spinal cord; produce a myelin sheath around axons of neurons of central nervous system.
Microglia (*micro* = small; *glia* = glue)	Small cells with few processes; derived from monocytes; normally stationary, but may migrate to site of injury; also called brain macrophages.	Microglial cell / Microbes / Phagocytic vacuole	Engulf and destroy microbes and cellular debris; may migrate to area of injured nervous tissue and function as small macrophages.
Ependyma (*ependyma* = upper garment)	Epithelial cells arranged in a single layer and ranging in shape from squamous to columnar; many are ciliated.	Cilia / Ependyma cells	Form a continuous epithelial lining for the ventricles of the brain (spaces that form and circulate cerebrospinal fluid) and the central canal of the spinal cord.

system. Neurons are highly specialized for impulse conduction and for all special functions attributed to the nervous system: thinking, controlling muscle activity, regulating glands. Neuroglia serve as a special supporting and protective component of the nervous system.

NEUROGLIA

The cells of the nervous system that perform the functions of support and protection are called **neuroglia** (noo-ROG-lē-a; *neuro* = nerve; *glia* = glue) or **glial cells.** About 50 percent of all brain cells are neuroglial cells. Many of the glial cells form a supporting network by twining around nerve cells or lining certain structures in the brain and spinal cord. Others bind nervous tissue to supporting structures and attach the neurons to their blood vessels. A few types of glial cells also serve specialized protective functions. For example, many nerve fibers are coated with a thick, fatty sheath produced by one type of neuroglia.

Certain small glial cells are phagocytic; they protect the central nervous system from disease by engulfing invading microbes and clearing away debris. Neuroglia are of clinical interest because they are a common source of tumors of the nervous system. Exhibit 12-1 lists the neuroglial cells and summarizes their functions.

NEURONS

Nerve cells, called **neurons,** are responsible for conducting impulses from one part of the body to another. They are the structural and functional units of the nervous system.

Structure

A neuron consists of three distinct portions: (1) cell body, (2) dendrites, and (3) axon (Figure 12-2a). The **cell body, soma,** or **perikaryon** (per'-i-KAR-ē-on), contains a well-

Labels on figure (a):
- Cytoplasm
- Neurofibril
- Nissl bodies
- Nucleus
- Nucleolus
- Mitochondrion
- Cell body or perikaryon
- Axon hillock
- Dendrites
- Axon collateral
- Axon or axis cylinder
- Myelin sheath of Schwann cell
- Neurilemma of Schwann cell
- Nucleus of Schwann cell
- Node of Ranvier
- Telodendria
- Synaptic knobs (end feet)

Labels on figure (b):
- Neurofibril
- Myelin sheath
- Nucleus of Schwann cell
- Neurilemma of Schwann cell
- Axon or axis cylinder
- Node of Ranvier
- Axolemma

Labels on figure (c):
- Processes
- Cell body or perikaryon

(a)

(b)

(c)

FIGURE 12-2 Structure of a neuron. (a) An entire multipolar neuron. Arrows indicate the direction in which nerve impulses travel. (b) Sections through a myelinated fiber. (c) Photomicrograph of a multipolar neuron at a magnification of 640×. (Courtesy of Biophoto Associates.)

defined nucleus and nucleolus surrounded by a granular cytoplasm. Within the cytoplasm are typical organelles such as lysosomes, mitochondria, and Golgi complexes. Many neurons also contain cytoplasmic inclusions such as *lipofuscin* pigment that occurs as clumps of yellowish-brown granules. Lipofuscin may be a by-product of lyso-somal activity. Although its significance is unknown, lipo-

fuscin is related to aging; the amount of pigment increases with age. Also located in the cytoplasm are structures characteristic of neurons: Nissl bodies and neurofibrils. *Nissl bodies* are orderly arrangements of granular (rough) endoplasmic reticulum whose function is protein synthesis. Newly synthesized proteins pass from the perikaryon into the neuronal processes, mainly the axon, at the rate of about 1 mm (0.04 inch) per day. These proteins replace those lost during metabolism and are used for growth of neurons and regeneration of peripheral nerve fibers. *Neurofibrils* are long, thin fibrils composed of microtubules. They may assume a function in support and the transportation of nutrients. Mature neurons do not contain a mitotic apparatus. The significance of this is noted shortly.

The cytoplasmic processes of neurons generally depend on the direction in which they conduct impulses. There are two kinds: dendrites and axons. **Dendrites** (*dendro* = tree) are highly branched, thick extensions of the cytoplasm of the cell body. They typically contain Nissl bodies, mitochondria, and other cytoplasmic organelles. A neuron usually has several main dendrites. Their function is to conduct an impulse toward the cell body.

The second type of cytoplasmic process, called an **axon (axis cylinder),** is a single, highly specialized, long, thin process that conducts impulses away from the cell body to another neuron or tissue. It usually originates from the cell body as a small conical elevation called the *axon hillock.* An axon contains mitochondria and neurofibrils but no Nissl bodies; thus it does not carry on protein synthesis. Its cytoplasm, called *axoplasm,* is surrounded by a plasma membrane known as the *axolemma* (*lemma* = sheath or husk). Axons vary in length from a few millimeters (1 mm = 0.04 inch) in the brain to a meter (3.28 ft) or more between the spinal cord and toes. Along the course of an axon, there may be side branches called *axon collaterals.* The axon and its collaterals terminate by branching into many fine filaments called *telodendria.* The distal ends of telodendria are expanded into bulblike structures called *synaptic knobs* (*end feet*), which are important in nerve impulse conduction. They contain membrane-enclosed sacs called *synaptic vesicles* that store chemicals that determine whether impulse conduction occurs or not.

The cell body of a neuron is essential for the synthesis of many substances that sustain the life of the nerve cell. Neurons have two types of intracellular systems for transporting synthesized materials from the cell body. The slower one, called **axoplasmic flow,** conveys axoplasm in one direction only—from the cell body toward nerve fiber terminals. This mechanism may occur by protoplasmic streaming and supplies new axoplasm for developing or regenerating axons and renews axoplasm in growing and mature axons. The faster type of intracellular transport is called **axonal transport.** It conveys materials in both directions—away from the cell body and toward the cell body—possibly along tracks formed by microtubules and filaments. Axonal transport moves various organelles and materials that form the membranes of the axolemma, synaptic knobs, and synaptic vesicles. Materials returning to the cell body are degraded or recycled.

CLINICAL APPLICATION
The route taken by materials back to the cell body by axonal transport is the route by which the **herpes virus** and **rabies virus** make their way back to nerve cell bodies where they multiply and cause their damage. The toxin produced by the **tetanus bacterium** uses

the same route to reach the central nervous system. In fact, the time delay between the release of the toxin and the first appearance of symptoms is in part due to the time required for movement of the toxin by axonal transport.

The term **nerve fiber** is applied to an axon and its sheaths. Figure 12-2b shows two sections of a nerve fiber of the peripheral nervous system. Many axons, especially large, peripheral axons, are surrounded by a multilayered white, phospholipid, segmented covering called the **myelin sheath.** Axons containing such a covering are *myelinated,* while those without it are *unmyelinated.* The function of the myelin sheath is to increase the speed of nerve impulse conduction and to insulate and maintain the axon. Myelin is responsible for the color of the white matter in the nerves, brain, and spinal cord.

The myelin sheath of axons of the peripheral nervous system is produced by flattened cells, called **Schwann cells,** located along the axons. In the formation of a sheath, a developing Schwann cell encircles the axon until its ends meet and overlap (Figure 12-3). The cell then winds around the axon many times and, as it does so, the cytoplasm and nucleus are pushed to the outside layer. The inner portion, consisting of several layers of Schwann cell membrane, is the myelin sheath. The peripheral nucleated cytoplasmic layer of the Schwann cell (the outer layer that encloses the sheath) is called the **neurilemma (sheath of Schwann).**

The neurilemma is found only around fibers of the peripheral nervous system. Its function is to assist in the regeneration of injured axons. Between the segments of the myelin sheath are unmyelinated gaps called **nodes of Ranvier** (ron-VĒ-ā). Unmyelinated fibers are also enclosed by Schwann cells, but without multiple wrappings.

Nerve fibers of the central nervous system may also be myelinated or unmyelinated. Myelination of central nervous system axons is accomplished by oligodendrocytes in somewhat the same manner that Schwann cells myelinate peripheral nervous system axons (see Exhibit 12-1). Myelinated axons of the central nervous system also contain nodes of Ranvier, but they are not so numerous.

Myelin sheaths are first laid down during the later part of the fetal development and during the first year postnatally. The amount of myelin increases from birth to maturity. Since myelination is still in progress during infancy, an infant's responses to stimuli are not as rapid or coordinated as those of an older child or an adult.

Classification

The different neurons in the body may be classified by structure and function.

The structural classification is based on the number of processes extending from the cell body. **Multipolar neurons** have several dendrites and one axon. Most neurons in the brain and spinal cord are of this type. **Bipolar neurons** have one dendrite and one axon and are found in the retina of the eye, the inner ear, and the olfactory area. **Unipolar neurons** have only one process extending from the cell body. The single process divides into a cen-

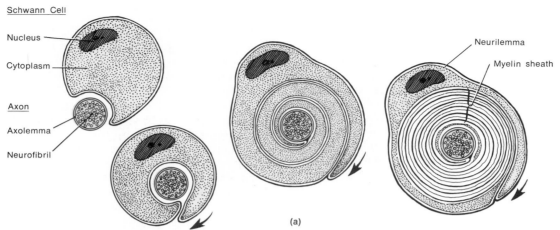

FIGURE 12-3 Myelin sheath. (a) Stages in the formation of a myelin sheath by a Schwann cell.

(b)

FIGURE 12-3 (*Continued*) Myelin sheath. (b) Electron micrographs of a myelinated axon (above) and several unmyelinated axons (below). (© Biology Media, Schulz, 1980, Photo Researchers.)

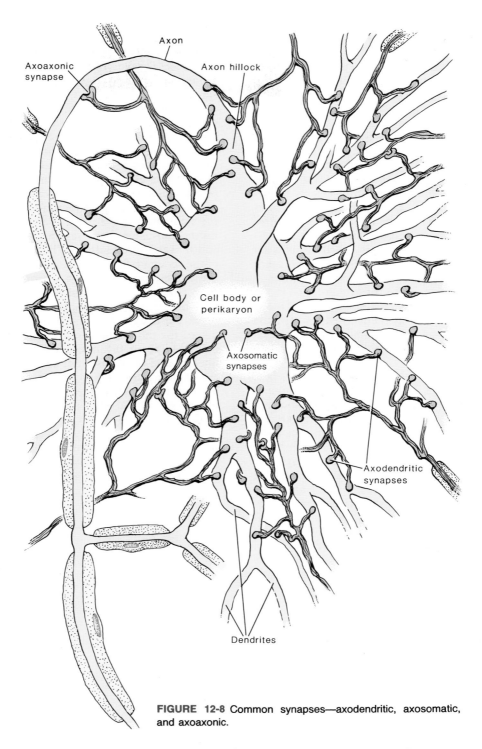

FIGURE 12-8 Common synapses—axodendritic, axosomatic, and axoaxonic.

nication. And synapses are the sites of action for most drugs that affect the brain, including therapeutic and addictive substances. Figure 12-7 shows that within a synapse is a minute gap, about 200 Å across, called the **synaptic cleft.** A **presynaptic neuron** is a neuron located before a synapse. A **postsynaptic neuron** is located after a synapse.

Impulses are conducted from a neuron to a muscle cell across an area of contact called a **neuromuscular (myoneural) junction.** The area of contact between a neuron and glandular cells is known as a **neuroglandular junction.** Together, neuromuscular and neuroglandular junctions are known as **neuroeffector junctions.**

The telodendrium of an axon terminates in **synaptic knobs (end feet).** The synaptic knobs of a presynaptic neuron commonly synapse with the dendrites, cell body, or axon hillock of a postsynaptic neuron. Accordingly, synapses may be classified as *axodendritic, axosomatic,* and *axoaxonic* (Figure 12-8). The synaptic knobs from a single presynaptic neuron may synapse with several postsynaptic neurons. Such an arrangement, called **divergence,** permits a single presynaptic neuron to influence several postsynaptic neurons or several muscle or gland cells at the same time (see Figure 12-11a). In another arrangement, called **convergence,** the synaptic knobs of several presynaptic neurons synapse with a single postsynaptic neuron (see Figure 12-11b). This arrangement permits stimulation or inhibition of the postsynaptic neuron.

At a synapse there is only **one-way impulse conduction**—from a presynaptic axon to a postsynaptic dendrite, cell body, or axon hillock. Impulses must move forward over their pathway. They cannot back up into another presynaptic neuron. Such a mechanism is crucial in preventing impulse conduction along improper pathways.

Whether an impulse is conducted across a synapse, neuromuscular junction, or neuroglandular junction depends on the presence of chemicals called **transmitter substances (neurotransmitters).** These chemicals are made by the neuron, usually from amino acids. Following its production and transportation to the synaptic knobs, the transmitter substance is stored in the knobs in small membrane-enclosed sacs called **synaptic vesicles** (Figure 12-7). Each of the thousands of synaptic vesicles present may contain between 10,000 and 100,000 transmitter molecules.

When a nerve impulse arrives at a synaptic knob of a presynaptic neuron, it is believed that a small amount of calcium ions leaks into the knob, attracts synaptic vesicles to the plasma membrane, and helps to liberate the transmitter molecules from the vesicles. Some scientists believe that the synaptic vesicles fuse with the plasma membrane of the presynaptic neuron, form openings, and release the transmitter through the openings into the synaptic cleft. Other scientists think that the transmitter leaves through small channels to enter the synaptic cleft. In either case, the transmitter enters the synaptic cleft and, depending on the chemical nature of the transmitter and the interaction of the transmitter with receptors of the postsynaptic plasma membrane, several things can happen.

Excitatory Transmission

An **excitatory transmitter-receptor interaction** is one that can lower (make less negative) the postsynaptic neuron's membrane potential so that a new impulse can be generated across the synapse. If the potential is lowered enough, the membrane becomes depolarized, the potential inside the membrane becomes positive and the potential outside becomes negative, and a nerve impulse is initiated. Generally, the release of a transmitter by a single presynaptic knob is not sufficient to develop an action potential in a postsynaptic neuron. However, its release does bring the resting membrane potential closer to threshold level as Na^+ ions move into the cell. This change from the resting membrane potential level in the direction of the threshold level is called the **excitatory postsynaptic potential (EPSP)** (Figure 12-9a). The EPSP is always less negative than the resting membrane potential of the neuron but more negative than its threshold level. Even though the release of a transmitter by a single presynaptic knob is not sufficient to initiate an action potential in a postsynaptic neuron, the postsynaptic neuron does become

more excitable to impulses from presynaptic neurons, so that the possibility for generating an impulse is greater. This effect is called **facilitation.** In other words, the postsynaptic neuron is prepared for subsequent stimuli that can trigger an impulse.

The EPSP lasts only a few milliseconds. If several presynaptic knobs release their transmitter at the same time, however, the combined effect may initiate a nerve impulse—this effect is **summation.** If hundreds of presynaptic knobs release their transmitter simultaneously, they increase the chance of initiating a nerve impulse in the postsynaptic neuron. The greater the summation, the greater the probability an impulse will be initiated. When the summation is the result of the accumulation of transmitter substance from several presynaptic knobs, it is called **spatial summation.** When summation is the result of the accumulation of transmitter substance from a single presynaptic knob firing two or more times in rapid succession, it is called **temporal summation.** Since the EPSP lasts only about 15 msec, the second firing must follow quickly if temporal summation is to occur. The time required for the impulse to cross a synapse—the **synaptic delay**—is about 0.5 msec. This delay is caused by the liberation of the transmitter substance, its passage across the synapse, stimulation of the postsynaptic neuron to become more permeable to Na^+ ions, and the inward movement of Na^+ ions that initiates the nerve impulse in the postsynaptic neuron.

Excitatory transmitter-receptor interactions are thought to occur by two different mechanisms. In the first mechanism, chemical transmitters bind to proteins

(a)

(b)

FIGURE 12-9 Comparison between (a) EPSP and (b) IPSP.

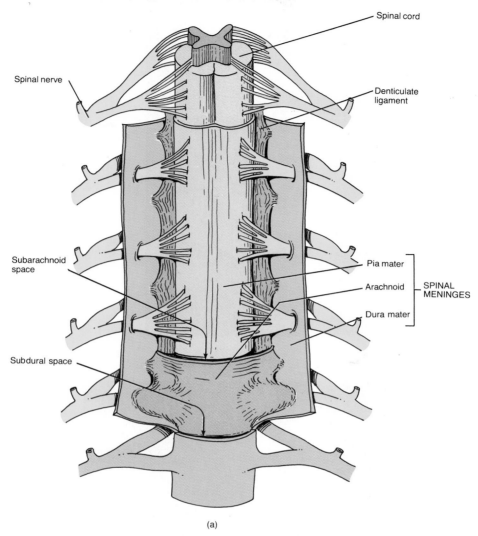

(a)

FIGURE 13-2 Spinal meninges. (a) Location of the spinal meninges as seen in a cross section of the spinal cord.

tral and dorsal nerve roots on either side. The ligaments protect the spinal cord against shock and sudden displacement.

CLINICAL APPLICATION

Inflammation of the meninges is known as **meningitis.** If only the dura mater becomes inflamed, the condition is *pachymeningitis.* Inflammation of the arachnoid and pia mater is *leptomeningitis,* the most common form of meningitis.

Cerebrospinal fluid is removed from the subarachnoid space in the inferior lumbar region of the spinal cord by a **spinal (lumbar) puncture** or **tap.** The procedure is normally performed between the third and fourth or fourth and fifth lumbar vertebrae. The spinous process of the fourth lumbar vertebra is easily located. A line drawn across the highest points of the iliac crests will pass through the spinous process. A lumbar puncture is below the spinal cord and thus poses little danger to it. If the patient lies on one side, drawing the knees and chest together, the vertebrae separate slightly so that a needle can be conveniently inserted. In its course the needle pierces the skin, superficial fascia, supraspinous ligament, interspinous ligament, epidural space, and arachnoid, to enter the subarachnoid space. Lumbar punctures are used to withdraw fluid for diagnostic purposes and to introduce antibiotics (as in the case of meningitis) and radiopaque dyes.

STRUCTURE IN CROSS SECTION

The spinal cord consists of both gray and white matter. Figure 13-3 shows that the gray matter forms an H-shaped area within the white matter. The gray matter consists primarily of nerve cell bodies and unmyelinated axons and dendrites of association and motor neurons. The white matter consists of bundles of myelinated axons of motor and sensory neurons.

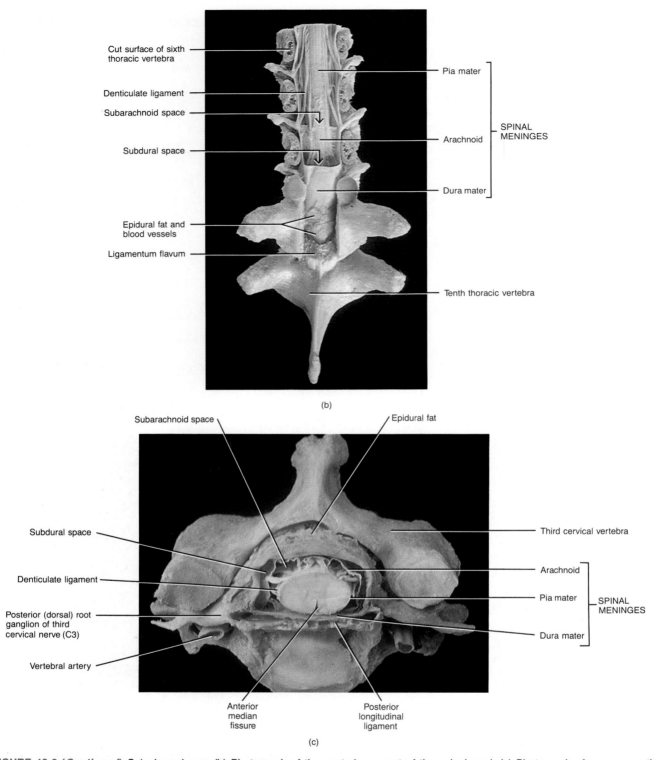

Cut surface of sixth
thoracic vertebra

Denticulate ligament

Subarachnoid space

Subdural space

Epidural fat and
blood vessels

Ligamentum flavum

Pia mater

Arachnoid

SPINAL
MENINGES

Dura mater

Tenth thoracic vertebra

(b)

Subarachnoid space

Epidural fat

Subdural space

Third cervical vertebra

Denticulate ligament

Arachnoid

Posterior (dorsal) root
ganglion of third
cervical nerve (C3)

Pia mater

SPINAL
MENINGES

Vertebral artery

Dura mater

Anterior
median
fissure

Posterior
longitudinal
ligament

(c)

FIGURE 13-2 (*Continued*) Spinal meninges. (b) Photograph of the posterior aspect of the spinal cord. (c) Photograph of a cross section through the spinal cord between the second and third thoracic vertebrae. (Courtesy of N. Gluhbegovic and T. H. Williams, *The Human Brain: A Photographic Guide,* Harper & Row, New York, 1980.)

The cross bar of the H is formed by the **gray commissure** (KOM-mi-shur). In the center of the gray commissure is a small space called the **central canal.** This canal runs the length of the spinal cord and is continuous with the fourth ventricle of the medulla. It contains cerebrospinal fluid. Anterior to the gray commissure is the **anterior (ventral) white commissure,** which connects the white matter of the right and left sides of the spinal cord.

The upright portions of the H are further subdivided into regions. Those closer to the front of the cord are called **anterior (ventral) gray horns.** They represent the motor part of the gray matter. The regions closer to the back of the cord are referred to as **posterior (dorsal) gray horns.** They represent the sensory part of the gray matter. The regions between the anterior and posterior gray horns are intermediate **lateral gray horns.** The lateral gray horns

(a)

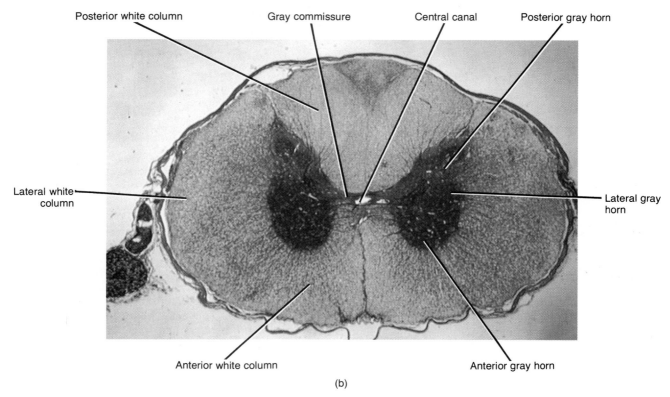

(b)

FIGURE 13-3 Spinal cord. (a) The organization of gray and white matter in the spinal cord as seen in cross section. The front of the figure has been sectioned at a lower level than the back so that you can see what is inside the posterior root ganglion, posterior root of the spinal nerve, anterior root of the spinal nerve, and the spinal nerve. (b) Photograph of the spinal cord at the seventh cervical segment. (Courtesy of Victor B. Eichler, Wichita State University.)

are present in the thoracic, upper lumbar, and sacral segments of the cord.

The gray matter of the cord also contains several nuclei that serve as relay stations for impulses and origins for certain nerves. Nuclei are clusters of nerve cell bodies and dendrites in the spinal cord and brain.

The white matter, like the gray matter, is organized into regions. The anterior and posterior gray horns divide the white matter on each side into three broad areas:

anterior (ventral) white columns, posterior (dorsal) white columns, and **lateral white columns.** Each column (or *funiculus*) in turn consists of distinct bundles of myelinated fibers that run the length of the cord. These bundles are called **tracts** or **fasciculi** (fa-SIK-yoo-lī). The longer **ascending tracts** consist of sensory axons that conduct impulses which enter the spinal cord upward to the brain. The longer **descending tracts** consist of motor axons that conduct impulses from the brain downward into the spinal

EXHIBIT 13-1
SELECTED ASCENDING AND DESCENDING TRACTS OF SPINAL CORD

TRACT	LOCATION (WHITE COLUMN)	ORIGIN	TERMINATION	FUNCTION
ASCENDING TRACTS				
Anterior (ventral) spinothalamic	Anterior (ventral) column.	Posterior (dorsal) gray horn on one side of cord, but crosses to opposite side of brain.	Thalamus; impulses eventually conveyed to cerebral cortex.	Conveys sensations for touch and pressure from one side of body to opposite side of thalamus. Eventually sensations reach cerebral cortex.
Lateral spinothalamic	Lateral column.	Posterior (dorsal) gray horn on one side of cord, but crosses to opposite side of brain.	Thalamus; impulses eventually conveyed to cerebral cortex.	Conveys sensations for pain and temperature from one side of body to opposite side of thalamus. Eventually sensations reach cerebral cortex.
Fasciculus gracilis and **fasciculus cuneatus**	Posterior (dorsal) column.	Axons of afferent neurons from periphery that enter posterior (dorsal) column on one side of cord and rise to same side of brain.	Nucleus gracilis and nucleus cuneatus of medulla; impulses eventually conveyed to cerebral cortex.	Convey sensations from one side of body to same side of medulla for touch; two-point discrimination (ability to distinguish that two points on skin are touched even though close together); proprioception (awareness of precise position of body parts and their direction of movement); stereognosis (ability to recognize size, shape, and texture of object); weight discrimination (ability to assess weight of an object); and vibration. Eventually sensations may reach cerebral cortex.
Posterior (dorsal) spinocerebellar	Posterior (dorsal) portion of lateral column.	Posterior (dorsal) gray horn on one side of cord and rises to same side of brain.	Cerebellum.	Conveys sensations from one side of body to same side of cerebellum for subconscious proprioception.
Anterior (ventral) spinocerebellar	Anterior (ventral) portion of lateral column.	Posterior (dorsal) gray horn on one side of cord; contains both crossed and uncrossed fibers.	Cerebellum.	Conveys sensations from both sides of body to cerebellum for subconscious proprioception.

cord where they synapse with other neurons whose axons pass out to muscles and glands. Thus the ascending tracts are sensory tracts and the descending tracts are motor tracts. Still other short tracts contain ascending or descending axons that convey impulses from one level of the cord to another.

FUNCTIONS

A major function of the spinal cord is to convey sensory impulses from the periphery to the brain and to conduct motor impulses from the brain to the periphery. This is accomplished within spinal tracts. A second principal function is to provide a means of integrating reflexes. Both functions are essential to maintaining homeostasis.

Spinal Tracts

The vital function of conveying sensory and motor information to and from the brain is carried out by the ascending and descending tracts of the cord. The names of the tracts indicate the white column (funiculus) in which the tract travels, where the cell bodies of the tract originate, and where the axons of the tract terminate. Since the origin and termination are specified, the direction of impulse conduction is also indicated by the name. For example, the anterior spinothalamic tract is located in the *anterior* white column, it originates in the *spinal* cord, and it terminates in the *thalamus* (a region of the brain). It is an ascending (sensory) tract since it conveys impulses from the cord upward to the brain.

The principal ascending and descending tracts are listed in Exhibit 13-1 and shown in Figure 13-4.

Reflex Center

The second principal function of the spinal cord is to provide reflexes. Spinal nerves are the paths of communication between the spinal cord tracts and the periphery. Figure 13-3 reveals that each pair of spinal nerves is connected to a segment of the cord by two points of attachment called roots. The **posterior** or **dorsal (sensory) root**

TRACT	LOCATION (WHITE COLUMN)	ORIGIN	TERMINATION	FUNCTION
DESCENDING TRACTS				
Lateral corticospinal	Lateral column.	Cerebral cortex on one side of brain, but crosses in base of medulla to opposite side of cord.	Anterior (ventral) gray horn.	Conveys motor impulses from one side of cortex to anterior gray horn of opposite side. Eventually impulses reach skeletal muscles on opposite side of body that coordinate precise, discrete movements.
Anterior (ventral) corticospinal	Anterior (ventral) column.	Cerebral cortex on one side of brain, uncrossed in medulla, but crosses to opposite side of cord.	Anterior (ventral) gray horn.	Conveys motor impulses from one side of cortex to anterior gray horn of opposite side. Eventually impulses reach skeletal muscles on opposite side of body that coordinate precise, discrete movements.
Rubrospinal	Lateral column.	Midbrain (red nucleus) on one side of brain, but crosses to opposite side of cord.	Anterior (ventral) gray horn.	Conveys motor impulses from one side of midbrain to skeletal muscles on opposite side of body that are concerned with muscle tone and posture.
Tectospinal	Anterior (ventral) column.	Midbrain on one side of brain, but crosses to opposite side of cord.	Anterior (ventral) gray horn.	Conveys motor impulses from one side of midbrain to skeletal muscles on opposite side of body that control movements of head in response to auditory, visual, and cutaneous stimuli.
Vestibulospinal	Anterior (ventral) column.	Medulla on one side of brain and descends to same side of cord.	Anterior (ventral) gray horn.	Conveys motor impulses from one side of medulla to skeletal muscles on same side of body that regulate body tone in response to movements of head (equilibrium).

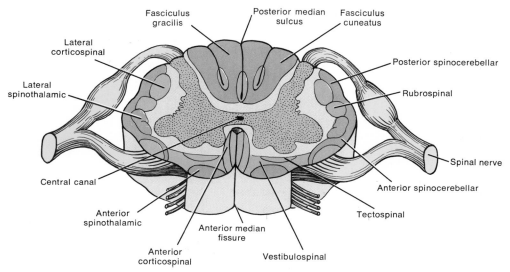

FIGURE 13-4 Selected tracts of the spinal cord. Ascending (sensory) tracts are indicated in pink; descending (motor) tracts are shown in blue.

contains sensory nerve fibers only and conducts impulses from the periphery to the spinal cord. These fibers extend into the posterior (dorsal) gray horn. Each dorsal root also has a swelling, the **posterior** or **dorsal (sensory) root ganglion,** which contains the cell bodies of the sensory neurons from the periphery. The other point of attachment of a spinal nerve to the cord is the **anterior** or **ventral (motor) root.** It contains motor nerve fibers only and conducts impulses from the spinal cord to the periphery.

The cell bodies of the motor neurons are located in the gray matter of the cord. If the motor impulse supplies a skeletal muscle, the cell bodies are located in the anterior (ventral) gray horn. If, however, the impulse supplies smooth muscle, cardiac muscle, or a gland through the autonomic nervous system, the cell bodies are located in the lateral gray horn.

Reflex Arc and Homeostasis

The path an impulse follows from its origin in the dendrites or cell body of a neuron in one part of the body to its termination elsewhere in the body is called a **conduction pathway.** All conduction pathways consist of circuits of neurons. One pathway is known as a **reflex arc,** the functional unit of the nervous system. A reflex arc contains two or more neurons over which impulses are conducted from a receptor to the brain or spinal cord and then to an effector. The basic components of a reflex arc are as follows (see Figure 13-5).

1. **Receptor.** The distal end of a dendrite or a sensory structure associated with the distal end of a dendrite. Its role in the reflex arc is to respond to a change in the internal or external environment by initiating a nerve impulse in a sensory neuron.
2. **Sensory neuron.** Passes the impulse from the receptor to its axonal termination in the central nervous system.
3. **Center.** A region, usually in the central nervous system, where an incoming sensory impulse generates an outgoing motor impulse. In the center, the impulse may be inhibited, transmitted, or rerouted. In the center of some reflex arcs, the sensory

neuron directly generates the impulse in the motor neuron. The center may also contain an association neuron between the sensory neuron and the motor neuron leading to a muscle or a gland.
4. **Motor neuron.** Transmits the impulse generated by the sensory or association neuron in the center to the organ of the body that will respond.
5. **Effector.** The organ of the body, either muscle or gland, that responds to the motor impulse. This response is called a **reflex action** or **reflex.**

Reflexes are fast responses to changes in the internal or external environment that allow the body to maintain homeostasis. Reflexes are associated not only with skeletal muscle contraction, but also with body functions such as heart rate, respiration, digestion, urination, and defecation. Reflexes carried out by the spinal cord alone are called **spinal reflexes.** Reflexes that result in the contraction of skeletal muscles are known as **somatic reflexes.** Those that cause the contraction of smooth or cardiac muscle or secretion by glands are **visceral (autonomic) reflexes.** Our concern at this point is to examine a few somatic spinal reflexes: the stretch reflex, the flexor reflex, and the crossed extensor reflex.

● *Stretch Reflex*   The **stretch reflex** is based on a two-neuron or *monosynaptic reflex arc.* Only two neurons are involved, and there is only one synapse in the pathway (Figure 13-5). This reflex results in the contraction of a muscle when it is stretched. Slight stretching of a muscle stimulates receptors in the muscle called *muscle spindles.* The spindles monitor changes in the length of the muscle. Once the spindle is stimulated, an impulse is sent along a sensory neuron to the spinal cord. The sensory neuron lies in the posterior root of a spinal nerve and synapses with a motor neuron in the anterior gray horn. The sensory neuron generates an impulse at the synapse that is transmitted along the motor neuron. The motor neuron lies in the anterior root of the spinal nerve and terminates in a skeletal muscle. Once the impulse reaches the stretched muscle, it contracts. Thus the stretch is counteracted by contraction.

FIGURE 13-5 Stretch reflex. A stretch reflex is monosynaptic. Since only two neurons are involved, there is only one synapse in the pathway. Why is the reflex arc shown referred to as an ipsilateral reflex arc?

FIGURE 14-2 Brain and meninges. (a) Brain and meninges seen in sagittal section. Arrows indicate the direction of flow of cerebrospinal fluid.

break open and the released enzymes bring about self-destruction of brain cells. Occasionally during childbirth the oxygen supply from the mother's blood is interrupted before the baby leaves the birth canal and can breathe. Often such babies are stillborn or suffer permanent brain damage that may result in mental retardation, epilepsy, and paralysis.

Blood supplying the brain also contains glucose, the principal source of energy for brain cells. Because carbohydrate storage in the brain is limited, the supply of glu-

Superior sagittal sinus

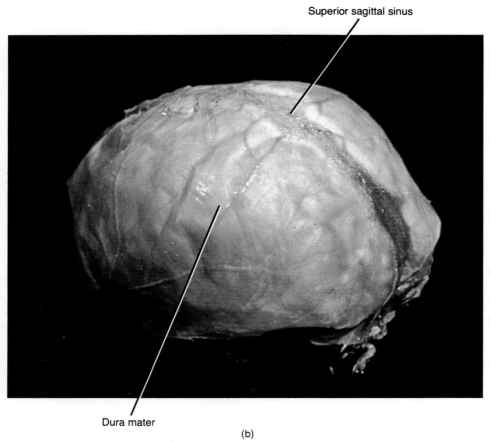

Dura mater

(b)

FIGURE 14-2 (*Continued*) Brain and meninges. (b) Photograph of a brain with dura mater. Note the superior sagittal sinus. (Courtesy of J. W. Eads.) (c) Frontal section through the brain showing the relationship of the superior sagittal sinus to the arachnoid villi.

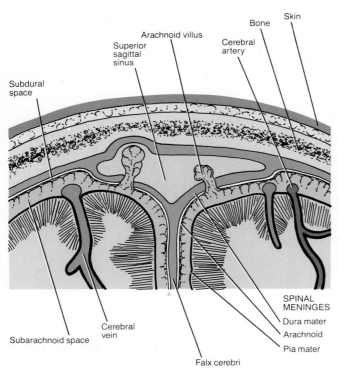

(c)

cose must be continuous. If blood entering the brain has a low glucose level, mental confusion, dizziness, convulsions, and loss of consciousness may occur.

Both carbon dioxide and oxygen have potent effects on cerebral blood flow. Carbon dioxide increases cerebral blood flow by combining with water to form carbonic acid (H_2CO_3), which breaks down into hydrogen ions (H^+) and bicarbonate ions (HCO_3^-). The H^+ ions then cause vasodilation of cerebral vessels and increased blood flow. Dilation is almost directly proportional to an increase in H^+ ion concentration. A decrease in oxygen in the blood also causes vasodilation and increased cerebral blood flow. Other factors related to blood flow and pressure are discussed in detail in Chapter 21.

Glucose, oxygen, and certain ions pass rapidly from the circulating blood into brain cells. Other substances, such as creatinine, urea, chloride, insulin, and sucrose, enter quite slowly. Still other substances—proteins and most antibiotics—do not pass at all from the blood into brain cells. The differential rates of passage of certain materials from the blood into the brain suggest a concept called the **blood–brain barrier (BBB).** Electron micrograph studies of the capillaries of the brain reveal that they differ structurally from other capillaries. Brain capillaries are constructed of more densely packed cells and are surrounded by large numbers of neuroglial cells and a continuous basement membrane. These features form a barrier to the passage of certain materials. Substances that cross the barrier are either very small molecules or require the assistance of a carrier molecule to cross by active transport. The blood–brain barrier functions as a selective barrier to protect brain cells from harmful substances. An injury to the brain due to trauma, inflammation, or toxins causes a breakdown of the blood–brain barrier, permitting the passage of normally restricted substances into brain tissue.

CLINICAL APPLICATION

Various **drugs** differ with respect to their passage through the blood–brain barrier. The antibiotics *chloramphenicol* (Chloromycetin), *tetracycline* (Achromycin V), and *sulfonamides* (Sonilyn) cross easily. *Penicillin* (Bicillin) crosses in only very small amounts.

Theopental sodium (Pentothal), a general anesthetic, rapidly crosses the blood–brain barrier following intravenous injection. *Atropine* (Atropisol), an antispasmodic, also quickly enters the brain. *Anisotropine methylbromide* (Valpin), another antispasmodic, and *phenylbutazone* (Azolid), an antiinflammatory drug, cannot cross the blood–brain barrier.

BRAIN STEM

Medulla Oblongata

The **medulla oblongata** (me-DULL-la ob'-long-GA-ta), or simply **medulla,** is a continuation of the upper portion of the spinal cord and forms the inferior part of the brain stem (Figure 14-3). Its position in relation to the other parts of the brain may be noted in Figure 14-1. It lies just superior to the level of the foramen magnum and extends upward to the inferior portion of the pons varolii. The medulla measures 3 cm (about 1 inch) in length.

The medulla contains all ascending and descending tracts that communicate between the spinal cord and various parts of the brain. These tracts constitute the white matter of the medulla. Some tracts cross as they pass through the medulla. Let us see how this crossing occurs and what it means.

On the ventral side of the medulla are two roughly triangular structures called **pyramids** (Figures 14-3 and 14-4). The pyramids are composed of the largest motor tracts that pass from the outer region of the cerebrum

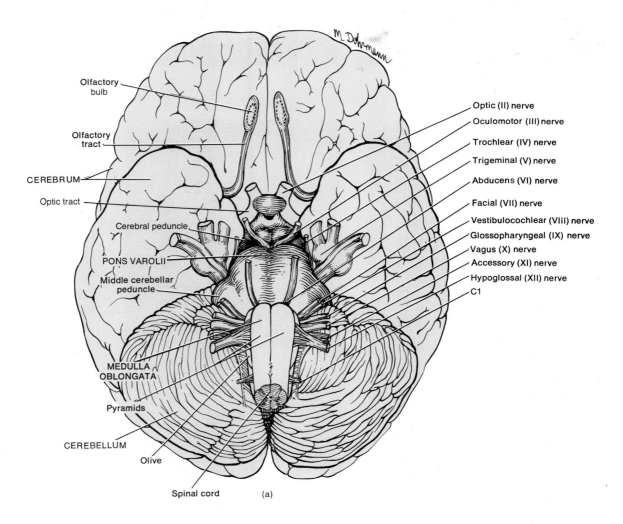

FIGURE 14-3 Brain stem. (a) Diagram of the ventral surface of the brain showing the structure of the brain stem in relation to the cranial nerves and associated structures.

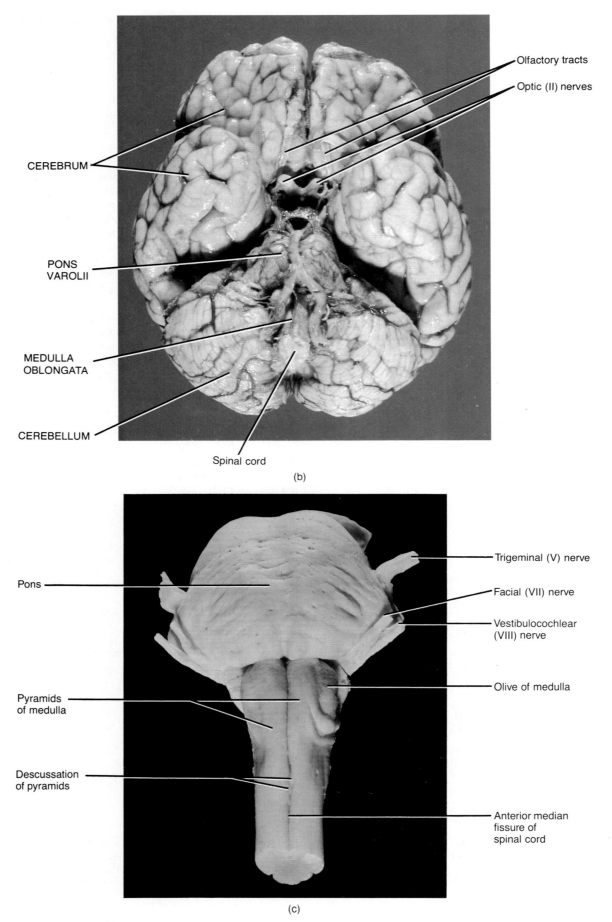

Olfactory tracts

Optic (II) nerves

CEREBRUM

PONS
VAROLII

MEDULLA
OBLONGATA

CEREBELLUM

Spinal cord

(b)

Pons

Trigeminal (V) nerve

Facial (VII) nerve

Vestibulocochlear
(VIII) nerve

Pyramids
of medulla

Olive of medulla

Descussation
of pyramids

Anterior median
fissure of
spinal cord

(c)

FIGURE 14-3 (*Continued*) Brain stem. (b) Photograph of the ventral surface of the brain showing the brain stem in relation to associated structures. (Courtesy of Martin Rotker, Taurus Photos.) (c) Photograph of the ventral surface of the brain stem. (Courtesy of N. Gluhbegovic and T. H. Williams, *The Human Brain: A Photographic Guide,* Harper & Row, New York, 1980.)

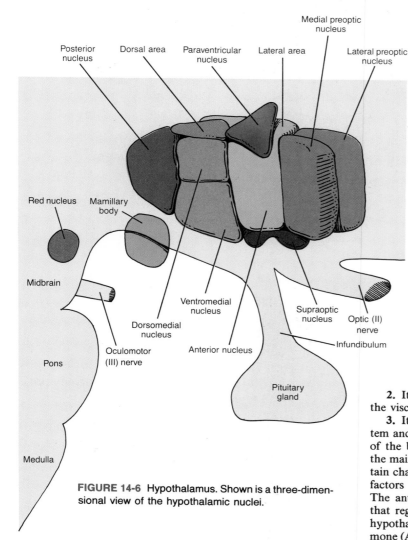

FIGURE 14-6 Hypothalamus. Shown is a three-dimensional view of the hypothalamic nuclei.

stasis. Although differentiation of the hypothalamic nuclei is far from precise, it is possible to identify certain nuclei. Some of these nuclei are more readily identified in lower animals and are more distinct in fetuses than adults. Also, within a given nucleus there may be several kinds of cells that can be differentiated histologically. The localization of function, with a few exceptions, is not specific to the individual nuclei; certain functions tend to overlap nuclear boundaries. For this reason, functions are attributed to regions rather than specific nuclei. Since the hypothalamic nuclei are useful landmarks in understanding subsequent discussion of functions, a three-dimensional view of the hypothalamic nuclei is shown in Figure 14-6.

The chief functions of the hypothalamus are as follows.

1. It controls and integrates the autonomic nervous system, which stimulates smooth muscle, regulates the rate of contraction of cardiac muscle, and controls the secretions of many glands. This is accomplished by axons of neurons whose dendrites and cell bodies are in hypothalamic nuclei. The axons form tracts from the hypothalamus to sympathetic and parasympathetic centers in the brain stem and spinal cord. Through the autonomic nervous system, the hypothalamus is the main regulator of visceral activities. It regulates heart rate, movement of food through the digestive tract, and contraction of the urinary bladder.

2. It is involved in the reception of sensory impulses from the viscera.

3. It is the principal intermediary between the nervous system and the endocrine system—the two major control systems of the body. The hypothalamus lies just above the pituitary, the main endocrine gland. When the hypothalamus detects certain changes in the body, it releases chemicals called regulating factors that stimulate or inhibit the anterior pituitary gland. The anterior pituitary then releases or holds back hormones that regulate various physiological activities of the body. The hypothalamus also produces two hormones, antidiuretic hormone (ADH) and oxytocin, which are transported to and stored in the posterior pituitary gland. The hormones are released from storage when needed by the body.

4. It is the center for the mind-over-body phenomenon. When the cerebral cortex interprets strong emotions, it often sends impulses along the tracts that connect the cortex with the hypothalamus. The hypothalamus then directs impulses via the autonomic nervous system and also releases chemicals that stimulate the anterior pituitary gland. The result can be a wide range of changes in body activities. For instance, when you panic, impulses leave the hypothalamus to stimulate your heart to beat faster. Likewise, continued psychological stress can produce long-term abnormalities in body function that result in serious illness. These so-called psychosomatic disorders are definitely real.

5. It is associated with feelings of rage and aggression.

6. It controls normal body temperature. Certain cells of the hypothalamus serve as a thermostat. If blood flowing through the hypothalamus is above normal temperature, the hypothalamus directs impulses along the autonomic nervous system to stimulate activities that promote heat loss. Heat can be lost through relaxation of the smooth muscle in the blood vessels, causing vasodilation of cutaneous vessels and increased heat loss from the skin. Heat loss also occurs by sweating. Conversely, if the temperature of the blood is below normal, the hypothalamus generates impulses that promote heat retention. Heat can be retained through the constriction of cutaneous blood vessels, cessation of sweating, and shivering.

(a)

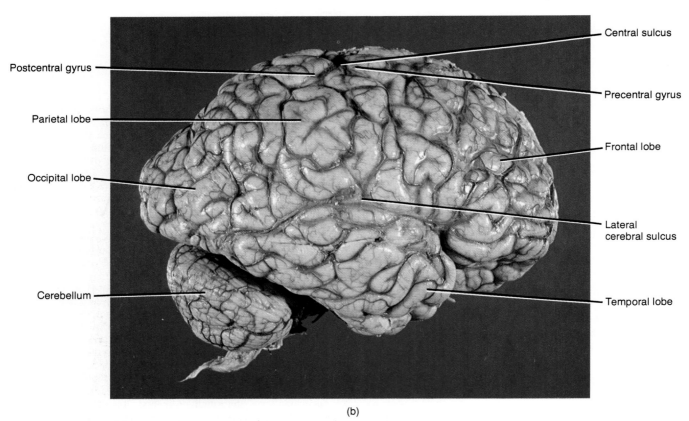

(b)

FIGURE 14-7 Lobes and fissures of the cerebrum. (a) Diagram of right lateral view. Since the insula cannot be seen externally, it has been projected to the surface. It can be seen in Figure 14-5a, b. (b) Photograph of right lateral view. (Courtesy of Martin Rotker, Taurus Photos.)

7. It regulates food intake through two centers. The **feeding center** is stimulated by hunger sensations from an empty stomach. When sufficient food has been ingested, the **satiety** (sa-TĪ-e-tē) **center** is stimulated and sends out impulses that inhibit the feeding center.

8. It contains a **thirst center.** Certain cells in the hypothalamus are stimulated when the extracellular fluid volume is reduced. The stimulated cells produce the sensation of thirst.

9. It is one of the centers that maintains the waking state and sleep patterns.

10. It exhibits properties of a self-sustained oscillator and, as such, acts as a pacemaker to drive many biological rhythms.

CEREBRUM

Supported on the brain stem and forming the bulk of the brain is the **cerebrum** (see Figure 14-1). The surface of the cerebrum is composed of gray matter 2 to 4 mm (0.08 to 0.16 inch) thick and is referred to as the **cerebral**

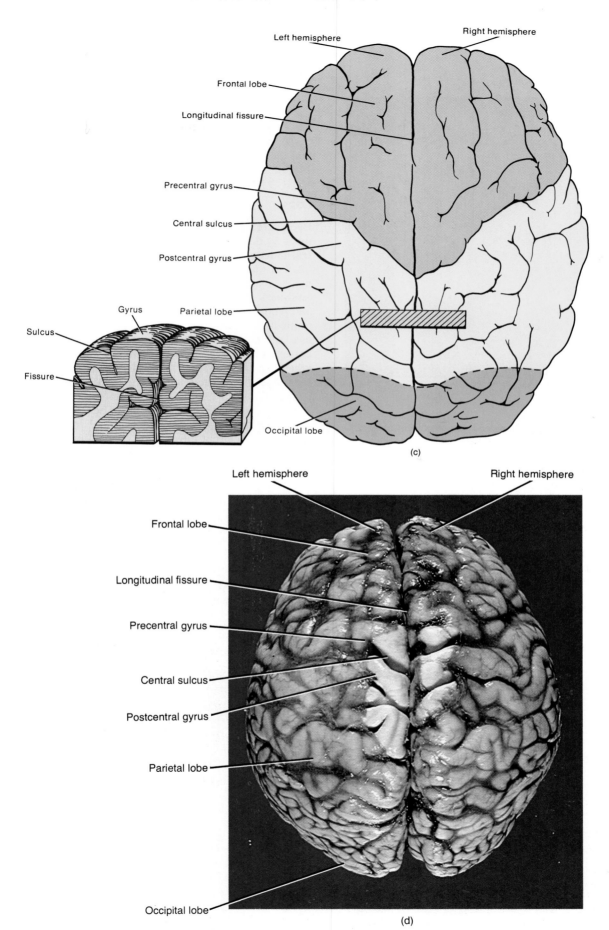

Left hemisphere

Right hemisphere

Frontal lobe

Longitudinal fissure

Precentral gyrus

Central sulcus

Postcentral gyrus

Parietal lobe

Occipital lobe

Gyrus

Sulcus

Fissure

(c)

Left hemisphere

Right hemisphere

Frontal lobe

Longitudinal fissure

Precentral gyrus

Central sulcus

Postcentral gyrus

Parietal lobe

Occipital lobe

(d)

FIGURE 14-7 (*Continued*) Lobes and fissures of the cerebrum. (c) Diagram of superior view. The insert to the left indicates the relative differences among a gyrus, sulcus, and fissure. (d) Photograph of superior view. (Courtesy of Martin Rotker, Taurus Photos.)

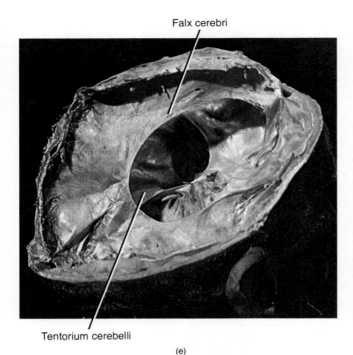

Falx cerebri

Tentorium cerebelli

(e)

FIGURE 14-7 (*Continued*) Lobes and fissures of the cerebrum. (e) Photograph of the skull showing the tentorium cerebelli and falx cerebri. (Courtesy of J. W. Eads.)

cortex (*cortex* = rind or bark). The cortex, containing roughly 15 billion cells, consists of six layers of nerve cell bodies in most areas. Beneath the cortex lies the cerebral white matter.

During embryonic development, when there is a rapid increase in brain size, the gray matter of the cortex enlarges out of proportion to the underlying white matter. As a result, the cortical region rolls and folds upon itself. The folds are called **gyri** (JĪ-rī) or **convolutions** (Figure 14-7a–d). The deep grooves between folds are referred to as **fissures;** the shallow grooves between folds are **sulci** (SUL-sī). The most prominent fissure, the **longitudinal fissure,** nearly separates the cerebrum into right and left

halves, or **hemispheres** (Figure 14-7c,d). The hemispheres, however, are connected internally by a large bundle of transverse fibers composed of white matter called the **corpus callosum** (kal-LŌ-sum; *corpus* = body; *callosus* = hard). Between the hemispheres is an extension of the cranial dura mater called the **falx (FALKS) cerebri** (Figure 14-7e).

Lobes

Each cerebral hemisphere is further subdivided into four lobes by deep sulci or fissures (Figure 14-7a–d). The **central sulcus** separates the **frontal lobe** from the **parietal lobe.** A major gyrus, the **precentral gyrus,** is located immediately anterior to the central sulcus. The gyrus is a landmark for the primary motor area of the cerebral cortex. Another major gyrus, the **postcentral gyrus,** is located immediately posterior to the central sulcus. This gyrus is a landmark for the general sensory area of the cerebral cortex. The **lateral cerebral sulcus** separates the **frontal lobe** from the **temporal lobe.** The **parietooccipital sulcus** separates the **parietal lobe** from the **occipital lobe.** Another prominent fissure, the **transverse fissure,** separates the cerebrum from the cerebellum. The frontal lobe, parietal lobe, temporal lobe, and occipital lobe are named after the bones that cover them. A fifth part of the cerebrum, the **insula,** lies deep within the lateral cerebral fissure, under the parietal, frontal, and temporal lobes. It cannot be seen in an external view of the brain (Figure 14-7a).

White Matter

The white matter underlying the cortex consists of myelinated axons running in three principal directions (Figure 14-8).

1. **Association fibers** connect and transmit impulses between gyri in the same hemisphere.
2. **Commissural fibers** transmit impulses from the gyri in one cerebral hemisphere to the corresponding gyri in the opposite cerebral hemisphere. Three important groups of commis-

Association fibers

Cerebral cortex

Mamillary body

Commissural and projection fibers

Corpus callosum

Anterior commissure

Commissural fibers

FIGURE 14-8 White matter tracts of the left cerebral hemisphere seen in sagittal section. (Courtesy of N. Gluhbegovic and T. H. Williams, *The Human Brain: A Photographic Guide,* Harper & Row, New York, 1980.)

FIGURE 14-9 Basal ganglia. (a) In this diagram of the right lateral view of the cerebrum, the basal ganglia have been projected to the surface. Refer to Figure 14-5a, b to note the positions of the basal ganglia in the frontal section of the cerebrum. (b) Photograph of the medial surface of the left cerebral hemisphere showing portions of the basal ganglia. (Courtesy of N. Gluhbegovic and T. H. Williams, *The Human Brain: A Photographic Guide,* Harper & Row, New York, 1980.)

sural fibers are the *corpus callosum, anterior commissure,* and *posterior commissure.*

3. Projection fibers form ascending and descending tracts that transmit impulses from the cerebrum to other parts of the brain and spinal cord.

Basal Ganglia (Cerebral Nuclei)

The **basal ganglia,** or **cerebral nuclei,** are paired masses of gray matter in each cerebral hemisphere (Figures 14-5 and 14-9). The largest of the basal ganglia of each hemisphere is the **corpus striatum** (strī-Ā-tum; *corpus* = body; *striatus* = striped). It consists of the **caudate** (*cauda* = tail) **nucleus** and the **lentiform** (*lenticula* = shaped like a lentil or lens) **nucleus.** The lentiform nucleus, in turn, is subdivided into a lateral portion called the **putamen** (pu-TĀ-men; *putamen* = shell) and a medial portion called the **globus pallidus** (*globus* = ball; *pallid* = pale).

The portion of the *internal capsule* passing between the lentiform nucleus and the caudate nucleus and between the lentiform nucleus and thalamus is sometimes considered part of the corpus striatum. The internal capsule is made up of a group of sensory and motor white matter tracts that connect the cerebral cortex with the brain stem and spinal cord.

Other structures frequently considered part of the basal ganglia are the claustrum and amygdaloid nucleus. The **claustrum** (KLAWS-trum) is a thin sheet of gray matter lateral to the putamen. The **amygdaloid** (a-MIG-da-loyd; *amygda* = almond) **nucleus** is located at the tail end of the caudate nucleus. Some authorities also consider the **substantia nigra,** the **subthalamic nucleus,** and the **red nucleus** to be part of the basal ganglia. The substantia nigra is a large nucleus in the midbrain. The subthalamic nucleus lies against the internal capsule. Its major connection is with the globus pallidus.

The basal ganglia are interconnected by many fibers. They are also connected to the cerebral cortex, thalamus, and hypothalamus. The caudate nucleus and the putamen control large subconscious movements of the skeletal muscles, such as swinging the arms while walking. Such gross movements are also consciously controlled by the cerebral cortex. The globus pallidus is concerned with the regulation of muscle tone required for specific body movements.

CLINICAL APPLICATION

Damage to the basal ganglia results in abnormal body movements, such as uncontrollable shaking, called **tremor,** and **involuntary movements of skeletal muscles.** Moreover, destruction of a substantial portion of the caudate nucleus almost totally **paralyzes** the side of the body opposite to the damage. The caudate nucleus is an area often affected by a stroke.

A lesion in the subthalamic nucleus results in a motor disturbance on the opposite side of the body called **hemiballismus** (*hemi* = half; *ballismos* = jumping), which is characterized by involuntary movements occurring suddenly with great force and rapidity. The movements are purposeless and generally of the withdrawal type, although they may be jerky. The spontaneous movements affect the proximal portions of the extremities most severely, especially the arms.

Limbic System

Certain components of the cerebral hemispheres and diencephalon constitute the **limbic** (*limbus* = border) **system.** Among its components are the following regions of gray matter.

1. **Limbic lobe.** Formed by two gyri of the cerebral hemisphere: the cingulate gyrus and the hippocampal gyrus.
2. **Hippocampus.** An extension of the hippocampal gyrus that extends into the floor of the lateral ventricle.
3. **Amygdaloid nucleus.** Located at the tail end of the caudate nucleus.
4. **Mammillary bodies of the hypothalamus.** Two round masses close to the midline near the cerebral peduncles.
5. **Anterior nucleus of the thalamus.** Located in the floor of the lateral ventricle.

The limbic system is a wishbone-shaped group of structures that encircles the brain stem and functions in the emotional aspects of behavior related to survival. The hippocampus, together with portions of the cerebrum, also functions in memory. Although behavior is a function of the entire nervous system, the limbic system controls most of its involuntary aspects. Experiments on the limbic system of monkeys and other animals indicate that the amygdaloid nucleus assumes a major role in controlling the overall pattern of behavior.

Other experiments have shown that the limbic system is associated with pleasure and pain. When certain areas of the limbic system of the hypothalamus, thalamus, and midbrain are stimulated, the reactions of experimental animals indicate they are experiencing intense punishment. When other areas are stimulated, the animals' reactions indicate they are experiencing extreme pleasure. In

still other studies, stimulation of the perifornical nuclei of the hypothalamus results in a behavioral pattern called *rage.* The animal assumes a defensive posture—extending its claws, raising its tail, hissing, spitting, growling, and opening its eyes wide. Stimulating other areas of the limbic system results in an opposite behavioral pattern: docility, tameness, and affection. Because the limbic system assumes a primary function in emotions such as pain, pleasure, anger, rage, fear, sorrow, sexual feelings, docility, and affection, it is sometimes called the "visceral" or "emotional" brain.

CLINICAL APPLICATION

Brain injuries are commonly associated with head injuries and result from displacement and distortion of neuronal tissue at the moment of impact. The various degrees of brain injury are described by the following terms.

1. **Concussion.** An abrupt but temporary loss of consciousness following a blow to the head or a sudden stopping of a moving head. A concussion produces no visible bruising of the brain.
2. **Contusion.** A visible bruising of the brain due to trauma and blood leaking from microscopic vessels. The pia mater is stripped from the brain over the injured area and may be torn, allowing blood to enter the subarachnoid space. A contusion usually results in an extended loss of consciousness, ranging from several minutes to many hours.
3. **Laceration.** Tearing of the brain, usually from a skull fracture or gunshot wound. A laceration results in rupture of large blood vessels with bleeding into the brain and subarachnoid space. Consequences include cerebral hematoma, edema, and increased intracranial pressure.

Functional Areas of Cerebral Cortex

The functions of the cerebrum are numerous and complex. In a general way, the cerebral cortex is divided into sensory, motor, and association areas. The **sensory areas** interpret sensory impulses, the **motor areas** control muscular movement, and the **association areas** are concerned with emotional and intellectual processes.

● *Sensory Areas*   The **general sensory area** or **somesthetic** (sō'-mes-THET-ik; *soma* = body; *aisthesis* = perception) **area** is located directly posterior to the central sulcus of the cerebrum in the postcentral gyrus of the parietal lobe. It extends from the longitudinal fissure on the top of the cerebrum to the lateral cerebral sulcus. In Figure 14-10 the general sensory area is designated by the areas numbered 1, 2, and 3.*

The general sensory area receives sensations from cutaneous, muscular, and visceral receptors in various parts of the body. Each point of the area receives sensations from specific parts of the body, and essentially the entire

---

* These numbers, as well as most of the others shown, are based on K. Brodmann's cytoarchitectural map of the cerebral cortex. His map, first published in 1909, is an attempt to correlate structure and function.

FIGURE 14-10 Functional areas of the cerebrum. This lateral view indicates the sensory and motor areas of the right hemisphere. Although Broca's area is in the left hemisphere of most people, it is shown here to indicate its location.

body is spatially represented in it. The size of the portion of the sensory area receiving stimuli from body parts is not dependent on the size of the part, but on the number of receptors the part contains. For example, a larger portion of the sensory area receives impulses from the lips than from the thorax (see Figure 15-3). The major function of the general sensory area is to localize exactly the points of the body where the sensations originate. The thalamus is capable of localizing sensations in a general way, that is, it receives sensations from large areas of the body, but cannot distinguish between specific areas of stimulation. This ability is reserved to the general sensory area of the cortex.

Posterior to the general sensory area is the **somesthetic association area.** It corresponds to the areas numbered 5 and 7 in Figure 14-10. The somesthetic association area receives input from the thalamus, other lower portions of the brain, and the general sensory area. Its role is to integrate and interpret sensations. This area permits you to determine the exact shape and texture of an object without looking at it, to determine the orientation of one object to another as they are felt, and to sense the relationship of one body part to another. Another role of the somesthetic association area is the storage of memories of past sensory experiences. Thus you can compare sensations with previous experiences.

Other sensory areas of the cortex include:

1. Primary visual area (area 17). Located on the medial surface of the occipital lobe and occasionally extends around to the lateral surface. It receives sensory impulses from the eyes and interprets shape and color.

2. Visual association area (areas 18 and 19). Located in the occipital lobe. It receives sensory signals from the primary visual area and the thalamus. It relates present to past visual experiences with recognition and evaluation of what is seen.

3. Primary auditory area (areas 41 and 42). Located in the superior part of the temporal lobe near the lateral cerebral sulcus. It interprets the basic characteristics of sound such as pitch and rhythm.

4. Auditory association area (area 22). Inferior to the primary auditory area in the temporal cortex. It determines if a sound is speech, music, or noise. It also interprets the meaning of speech by translating words into thoughts.

5. Primary gustatory area (area 43). Located at the base of the postcentral gyrus above the lateral cerebral sulcus in the parietal cortex. It interprets sensations related to taste.

6. Primary olfactory area. Located in the temporal lobe on the medial aspect. It interprets sensations related to smell.

7. Gnostic (NOS-tik; *gnosis* = knowledge) **area** (areas 5, 7, 39, and 40). This *common integrative area* is located between the somesthetic, visual, and auditory association areas. The gnostic area receives impulses from these areas, as well as from the taste and smell areas, the thalamus, and lower portions of the brain stem. It integrates sensory interpretations from the association areas and impulses from other areas so that a common thought can be formed from the various sensory inputs. It then transmits signals to other parts of the brain to cause the appropriate response to the sensory signal.

CLINICAL APPLICATION

In Chapter 2, we discussed the principle and clinical applications of a recently developed type of radioisotope scanning called **positron emission tomography (PET).** At that point it was indicated that PET is being used in the diagnosis of an assortment of diseases. It is also being used by scientists to probe the healthy brain. By detecting and recording changes in glucose

Brain activity:

—— Highest

—— Intermediate

—— Lowest

(a)

(b)

Orientation of brain

Front

Left   Right

Back

FIGURE 14-11 Positron emission tomography (PET) scans showing brain activity in response to visual and auditory stimulation. (a) Response of the visual areas of the cerebral cortex (areas between arrows) when the eyes are closed, open to white light, and open to a complex scene. (b) Response of auditory areas of the cerebral cortex (areas at horizontal arrows) and the frontal lobe (areas at vertical arrow) in response to language and music, language alone, and music alone. Note that when the auditory stimulation consists of both language and music, the auditory cortex on both sides of the brain is active; when the auditory stimulation is language only, there is a predominant left-sided activation of the auditory cortex; and when the auditory stimulation is music only, there is a predominant right-sided activation of the auditory cortex. Frontal lobe activation probably results from a high level of mentation and the promise that all subjects participating in the experiment would be paid in proportion to how much they could remember about the auditory stimulation for a subsequent test. (Courtesy of Dr. Michael E. Phelps, Division of Biophysics, Department of Radiological Sciences, UCLA School of Medicine, Los Angeles, California.)

metabolism (consumption), it is possible to identify which specific areas of the brain are involved in specific sensory and motor activities. The PET scans in Figure 14-11 indicate brain activity in response to certain visual and auditory stimulation.

● *Motor Areas*   The **primary motor area** (area 4) is located in the precentral gyrus of the frontal lobe (see Figure 14-10). Like the general sensory area, the primary motor area consists of regions that control specific muscles or groups of muscles (see Figure 15-7). Stimulation of a specific point of the primary motor area results in a muscular contraction, usually on the opposite side of the body.

The **premotor area** (area 6) is anterior to the primary motor area. It is concerned with learned motor activities of a complex and sequential nature. It generates impulses that cause a specific group of muscles to contract in a

specific sequence. An example of this is writing. Thus the premotor area controls skilled movements.

The **frontal eye field area** (area 8) in the frontal cortex is sometimes included in the premotor area. This area controls voluntary scanning movements of the eyes— searching for a word in a dictionary, for instance.

The **language areas** are also significant parts of the motor cortex. The translation of speech or written words into thoughts involves sensory areas—primary auditory, auditory association, primary visual, visual association, and gnostic—as we just described. The translation of thoughts into speech involves **Broca's** (BRŌ-kaz) **area** or the **motor speech area** (area 44), located in the frontal lobe just superior to the lateral cerebral sulcus. From this area, a sequence of signals is sent to the premotor regions that control the muscles of the larynx, pharynx, and mouth. The impulses from the premotor area to the muscles result in specific, coordinated contractions that enable you to speak. Simultaneously, impulses are sent from Broca's area to the primary motor area. From here, impulses reach your breathing muscles to regulate the proper flow of air past the vocal cords. The coordinated contractions of your speech and breathing muscles enable you to translate your thoughts into speech.

CLINICAL APPLICATION

Broca's area and other language areas are located in the left cerebral hemisphere of most individuals regardless of whether they are left-handed or right-handed. Injury to the sensory or motor speech areas results in **aphasia** (a-FĀ-zē-a; *a* = without; *phasis* = speech), an inability to speak; **agraphia** (*a* = without; *graph* = write), an inability to write; **word deafness,** an inability to understand spoken words; or **word blindness,** an inability to understand written words.

● *Association Areas*   The **association areas** of the cerebrum are made up of association tracts that connect motor and sensory areas (see Figure 14-8). The association region of the cortex occupies the greater portion of the lateral surfaces of the occipital, parietal, and temporal lobes and the frontal lobes anterior to the motor areas. The association areas are concerned with memory, emotions, reasoning, will, judgment, personality traits, and intelligence.

Brain Waves

Brain cells can generate electrical activity as a result of literally millions of action potentials of individual neurons. These electrical potentials are called **brain waves** and indicate activity of the cerebral cortex. Brain waves pass through the skull easily and can be detected by sensors called electrodes. A record of such waves is called an **electroencephalogram (EEG).** An EEG is obtained by placing electrodes on the head and amplifying the waves with an electroencephalograph. As indicated in Figure 14-12, four kinds of waves are produced by normal individuals.

1. Alpha waves. These rhythmic waves occur at a frequency of about 10 to 12 cycles per second. (The unit commonly used to express frequency is the Hertz; 1 Hz = 1 cycle per second.) They are found in the EEGs of nearly all normal individuals when awake and in the resting state. These waves disappear entirely during sleep.

2. Beta waves. The frequency of these waves is between 15 and 60 Hz. Beta waves generally appear when the nervous system is active, that is, during periods of sensory input and mental activity.

3. Theta waves. These waves have frequencies of 5 to 8 Hz. Theta waves normally occur in children and in adults experiencing emotional stress.

4. Delta waves. The frequency of these waves is 1 to 5 Hz. Delta waves occur during sleep. They are normal in an awake infant. When produced by an awake adult, they indicate brain damage.

CLINICAL APPLICATION

Distinct **EEG patterns** appear in certain abnormalities. In fact, the EEG is used clinically in the diagnosis of epilepsy, infectious diseases, tumors, trauma, and hematomas. Electroencephalograms also furnish information regarding sleep and wakefulness.

BRAIN LATERALIZATION (SPLIT-BRAIN CONCEPT)

A distinctive characteristic of the human brain is the allocation of functions to the two cerebral hemispheres. At a glance the brain appears to have perfect bilateral symmetry, like most other organs of the body. It might therefore be expected that the two hemispheres of the brain would also be functionally equivalent. Actually many of the more specialized functions are found in only one hemisphere or the other. In fact, careful examination indicates that even the apparent anatomical symmetry is an illusion, since brain asymmetries have been detected by CT scans. In these images a peculiar departure from bilateral symmetry is observed. In right-handed people the right frontal lobe is usually wider than the left, but the left occipital and parietal lobes are wider than the right. The inner surfaces of the skull itself bulge at the right front and the left rear to accommodate the protuberances.

The distribution of the more specialized functions is profoundly asymmetrical. For example, the left hemisphere is more important for right-hand control, spoken and written language, numerical and scientific skills, and reasoning. The right hemisphere is more important for left-hand control, musical and artistic awareness, space and pattern perception, insight, imagination, and generating mental images of sight, sound, touch, taste, and smell in order to compare relationships.

In everyday life this lateralization of function can seldom be detected because information is readily passed between the hemispheres through several commissures, including the corpus callosum.

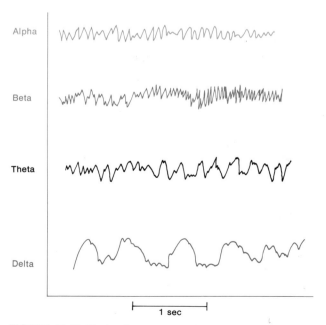

Alpha

Beta

Theta

Delta

|— 1 sec —|

FIGURE 14-12 Kinds of waves recorded in an electroencephalogram (EEG).

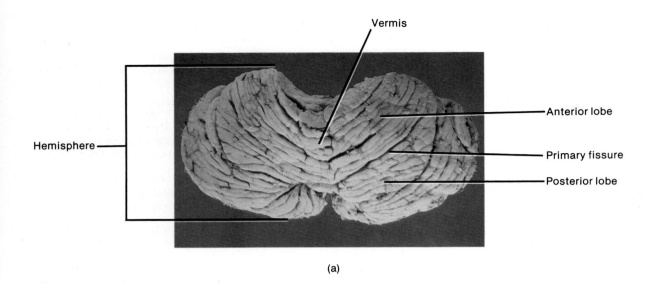

Vermis

Hemisphere

Anterior lobe

Primary fissure

Posterior lobe

(a)

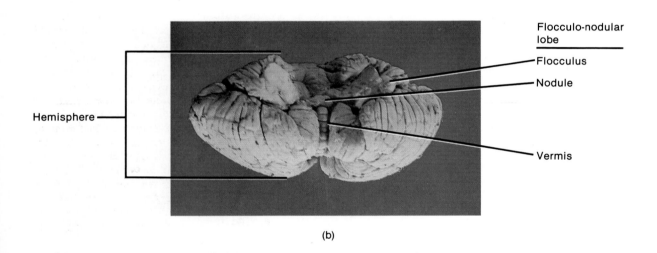

Hemisphere

Flocculo-nodular lobe

Flocculus

Nodule

Vermis

(b)

Folia

Gray matter

Arbor vitae
(White matter)

Pons

Medulla

(c)

FIGURE 14-13 Cerebellum. (a) Superior view. (b) Posteroinferior view. (c) Viewed in sagittal section. (Courtesy of M. L. Barr, *The Human Nervous System,* 2nd ed., Harper & Row, New York, 1974.)

One of the most recent findings is that different emotional reactions result from damage to the right and left sides of the brain. Lesions in most areas on the left side are accompanied by the feelings of loss that might be expected as a result of any serious injury, and the patient is disturbed and depressed. Damage in much of the right hemisphere sometimes leaves the patient unconcerned with his condition.

A common cerebral dominance and also one of the most puzzling is the phenomenon of handedness. In the human population, no more than 9 percent are left-handed. This considerable bias toward right-handedness may represent a unique specialization of the human brain. Studies indicate that the distribution of brain asymmetries in left-handed people is different from that in right-handed people.

Scientists are testing the possibility that faulty connections between the hemispheres may be a cause of dyslexia and that other peculiarities in communication between the hemispheres may sometimes be clues to psychiatric symptoms and may also be involved in creativity and invention.

CEREBELLUM

The **cerebellum** is the second-largest portion of the brain (almost ⅛ of the brain's mass) and occupies the inferior and posterior aspects of the cranial cavity. Specifically, it is posterior to the medulla and pons and below the occipital lobes of the cerebrum. It is separated from the cerebrum by the **transverse fissure** (see Figure 14-1) and by an extension of the cranial dura mater called the **tentorium** (*tentorium* = tent) **cerebelli** (see Figure 14-7e).

The cerebellum is shaped somewhat like a butterfly. The central constricted area is the **vermis,** which means worm-shaped, and the lateral "wings" or lobes are referred to as **hemispheres** (Figure 14-13). Each hemisphere consists of lobes that are separated by deep and distinct fissures. The **anterior lobe** and **posterior lobe** are concerned with subconscious movements of skeletal muscles. The **flocculonodular lobe** is concerned with the sense of equilibrium (Chapter 17). Between the hemispheres is another extension of the cranial dura mater: the **falx cerebelli.** It passes only a short distance between the cerebellar hemispheres.

The surface of the cerebellum, called the **cortex,** consists of gray matter in a series of slender, parallel ridges called **folia.** They are less prominent than the gyri of the cerebral cortex. Beneath the gray matter are **white matter tracts (arbor vitae)** that resemble branches of a tree. Deep within the white matter are masses of gray matter, the **cerebellar nuclei.**

The cerebellum is attached to the brain stem by three paired bundles of fibers called **cerebellar peduncles** (see Figure 14-3d,e). **Inferior cerebellar peduncles** connect the cerebellum with the medulla at the base of the brain stem and with the spinal cord. **Middle cerebellar peduncles** connect the cerebellum with the pons. **Superior cerebellar peduncles** connect the cerebellum with the midbrain.

The cerebellum is a motor area of the brain concerned with certain subconscious movements in the skeletal muscles. These movements are required for coordination, maintenance of posture, and balance. The cerebellar peduncles are the fiber tracts that allow the cerebellum to perform its functions.

Let us now see how the cerebellum produces coordinated movement. Motor areas of the cerebral cortex voluntarily initiate muscle contraction. Once the movement has begun, the sensory areas of the cortex receive impulses from nerves in the joints. The impulses provide information about the extent of muscle contraction and the amount of joint movement. The term *proprioception* is applied to this sense of the position of one body part relative to another. The cerebral cortex uses the proprioceptive sensations to determine which muscles are required to contract next and with what strength they are to contract in order to continue moving in the desired direction. Then a pattern of impulses is generated by the cerebral cortex along tracts to the pons and midbrain, which relay the impulses over the middle and superior cerebellar peduncles to the cerebellum. The cerebellum then generates subconscious motor impulses along the inferior cerebellar peduncles to the medulla and spinal cord. The impulses pass downward along the spinal cord and out the nerves that stimulate the prime movers and synergists to contract and that inhibit the contraction of the antagonists. The result is a smooth, coordinated movement. A well-functioning cerebellum is essential for delicate movements such as playing the piano.

The cerebellum also transmits impulses that control postural muscles, that is, it maintains normal muscle tone. The cerebellum also maintains body equilibrium. The inner ear contains structures that sense balance. Information such as whether the body is leaning to the left or right is transmitted from the inner ear to the cerebellum. The cerebellum then discharges impulses that cause the contraction of the muscles necessary for maintaining equilibrium.

There is some evidence that the cerebellum may play a role in a person's emotional development, modulating sensations of anger and pleasure.

CLINICAL APPLICATION
Damage to the cerebellum through trauma or disease is characterized by certain symptoms involving skeletal muscles on the same side of the body as the damage. The effects are ipsilateral because of a double-crossing of tracts within the cerebellum. There may be lack of muscle coordination, called *ataxia* (*a* = without; *taxis* = order). Blindfolded people with ataxia cannot touch the tip of their nose with a finger because they cannot coordinate movement with their sense of where a body part is located. Another sign of ataxia is a change in the speech pattern due to a lack of coordination of speech muscles. Cerebellar damage may also result in *disturbances of gait,* in which the subject staggers or cannot coordinate normal walking movements, and *severe dizziness.*

TRANSMITTER SUBSTANCES IN THE BRAIN

There are over 40 different substances that are either known or suspected transmitter substances in the brain. These substances can facilitate, excite, or inhibit postsynaptic neurons. They establish the lines of communication between brain cells and are localized in certain parts of the brain. In Chapter 12, we noted that **acetylcholine (ACh)** is released at some neuromuscular junctions, at some neuroglandular junctions, and at synapses between certain brain and spinal cord cells. In most parts of the body, ACh leads to excitation. **Norepinephrine (NE),** another transmitter that leads to excitation, is released at some neuromuscular junctions and some neuroglandular junctions. It is also concentrated in a group of neurons in the brain stem near the fourth ventricle called the *locus coeruleus* (LŌ-kus sē-ROO-lē-us). This area, which literally means blue place, projects axons into the hypothalamus, cerebellum, and cerebral cortex. In these places, NE has been implicated in maintaining arousal (awakening from deep sleep), dreaming, and the regulation of mood. Neurons containing the transmitter **dopamine (DA)** are clustered in the midbrain in the substantia nigra. DA leads to excitation. Some axons projecting from the substantia nigra terminate in the cerebral cortex, where DA is thought to be involved in emotional responses. Other axons project to the corpus striatum (basal ganglia), where the DA is involved in gross subconscious movements of skeletal muscles. The degeneration of axons in the basal ganglia results in Parkinson's disease, which is discussed at the end of this chapter. **Serotonin (5-HT)** is a transmitter that leads to excitation and is concentrated in the neurons in a part of the brain stem called the *raphe nucleus*. Axons projecting from the nucleus terminate in the hypothalamus, thalamus, and other regions of the brain. Serotonin is thought to be involved in inducing sleep, sensory perception, and temperature regulation. The amino acids **glutamic acid** and **aspartic acid** are also believed to lead to excitation in the brain.

The most common transmitter in the brain leading to inhibition is **gamma aminobutyric acid (GABA).** It has been implicated as a likely target for antianxiety drugs such as diazempan (Valium). These drugs enhance the action of GABA. Another transmitter that leads to inhibition in the spinal cord is **glycine.**

In recent years, another group of chemical messengers in the brain has been identified. These are known as **peptides** and consist of chains of amino acids that occur naturally in the brain. In 1975 the first peptides, referred to as **enkephalins** (en-KEF-a-lins; en = without; keph = head), were discovered. These chemicals are similar in structure to morphine, the painkiller derived from the opium poppy. Enkephalins are concentrated in the thalamus, in parts of the limbic system, and in those spinal cord pathways that relay impulses for pain. It has been suggested that enkephalins are the body's natural painkillers. They do this by inhibiting impulses in the pain pathway and by binding to the same receptors in the brain as morphine.

Other naturally occurring peptides, called **endorphins** (en·DOR-fins), were subsequently isolated from the pituitary gland. Like the enkephalins, they have morphinelike properties that suppress pain. Endorphins have also been linked to memory and learning; sexual activity; control of body temperature; regulation of hormones that affect the onset of puberty, sexual drive, and reproduction; and mental illnesses such as depression and schizophrenia. One of the better studied endorphins is *beta-endorphin,* or *β-endorphin.* A peptide that is believed to work with endorphins is known as *substance P.* It is found in sensory nerves, spinal cord pathways, and parts of the brain associated with pain. When substance P is released by neurons, it conducts pain-related nerve impulses from peripheral pain receptors into the central nervous system. It is now suspected that endorphins may exert their analgesic effects by suppressing the release of substance P. Substance P has also been shown to counter the effects of certain nerve-damaging chemicals, prompting speculation that it might prove useful as a treatment for nerve degeneration.

A brain peptide called **dynorphin** (*dynamis* = power), discovered in 1979, is 200 times more powerful in action than morphine and 50 times more powerful than *β*-endorphin. Its exact functions have yet to be determined. However, the chemical might be a factor in controlling pain and registering emotions.

As research into brain peptides continues, a long list of substances is being drawn up. Interestingly enough, many of these brain peptides are also found in other parts of the body, where they serve as hormones or other regulators of physiological responses. Although we will be discussing many of them in subsequent chapters, a few will be mentioned here to give you some idea of the diversity of the peptides.

    1. Angiotensin. This is a substance that can raise blood pressure and is found in the bloodstream as well. It is produced by a kidney enzyme called renin. Some evidence suggests that angiotensin in the brain may be part of the brain's mechanism for regulating its own blood pressure.

    2. Cholecystokinin (ko'-lē-sis'-tō-KĪN-in). This peptide hormone (also called pancreozymin) is produced by the lining of the small intestine and causes the pancreas to release digestive juice and the gallbladder to release bile. Although the significance of cholecystokinin in the brain is unclear, there are some suggestions that it may be related to the control of feeding.

    3. Neurotensin. This substance helps to regulate blood glucose level by its action on glucose and glucagon. It is also possible that neurotensin may be involved in pain pathways.

    4. Regulating factors. These are chemicals produced by the hypothalamus that regulate the release of hormones by the pituitary gland.

CRANIAL NERVES

Of the 12 pairs of **cranial nerves,** 10 originate from the brain stem, but all leave the skull through foramina of the skull (see Figure 14-3a). The cranial nerves are designated with Roman numerals and with names. The Roman numerals indicate the order in which the nerves arise from the brain (front to back). The names indicate the distribution or function. Cranial nerves that contain both sensory and motor fibers are termed *mixed nerves.* Other cranial nerves contain sensory fibers only. The cell bodies of sensory fibers are located in ganglia outside the brain. The cell bodies of motor fibers lie in nuclei within the brain.

STUDENT OBJECTIVES

● Define a sensation and list the four prerequisites necessary for its transmission.

● Define projection, adaptation, afterimage, and modality as characteristics of sensations.

● Classify receptors on the basis of location, stimulus detected, and simplicity or complexity.

● Describe the distribution of cutaneous receptors.

● List the location and function of the receptors for tactile sensations (touch, pressure, vibration), thermoreceptive sensations (heat and cold), pain, and proprioception.

● Distinguish somatic, visceral, referred, and phantom pain.

● Describe how acupuncture is believed to relieve pain.

● Discuss the origin, neuronal components, and destination of the posterior column and spinothalamic pathways.

● Explain the neural pathways for pain and temperature; crude touch and pressure; and fine touch, proprioception, and vibration.

● Contrast the roles of the cerebellar tracts in conveying sensations.

● Compare the course of the pyramidal and extrapyramidal motor pathways.

● List the functions of the lateral corticospinal, anterior corticospinal, corticobulbar, rubrospinal, tectospinal, and vestibulospinal tracts.

● Compare integrative functions such as memory, sleep, and wakefulness.

15 The Sensory, Motor, and Integrative Systems

N

ow that you have examined the structure of the nervous system and its activities, we will see how its different parts cooperate in performing its three essential functions: (1) receiving sensory information; (2) transmitting motor impulses that result in movement or secretion; and (3) integration, an activity that deals with memory, sleep, and emotions.

SENSATIONS

Your ability to sense stimuli is vital to your survival. If pain could not be sensed, burns would be common. An inflamed appendix or stomach ulcer would progress unnoticed. A lack of sight would increase the risk of injury from unseen obstacles, a loss of smell would allow a harmful gas to be inhaled, a loss of hearing would prevent recognition of automobile horns, and a lack of taste would allow toxic substances to be ingested. In short, if you could not "sense" your environment and make the necessary homeostatic adjustments, you could not survive on your own.

DEFINITION

In its broadest context, **sensation** refers to a state of awareness of external or internal conditions of the body. **Perception** refers to the conscious registration of a sensory stimulus. For a sensation to occur, four prerequisites must be fulfilled.

1. A **stimulus,** or change in the environment, capable of initiating a response by the nervous system must be present.
2. A **receptor** or **sense organ** must pick up the stimulus and convert it to a nerve impulse. A sense receptor or sense organ may be viewed as specialized nervous tissue that is extremely sensitive to internal or external conditions.
3. The impulse must be **conducted** along a nervous pathway from the receptor or sense organ to the brain.
4. A region of the brain must **translate** the impulse into a sensation.

Receptors are capable of converting a specific stimulus into a nerve impulse. The stimulus may be light, heat, pressure, mechanical energy, or chemical energy. Each stimulus is capable of causing the membrane of the receptor to depolarize. This depolarization is called a **generator (receptor) potential.**

The generator potential is a graded response; within limits, the magnitude increases with stimulus strength and frequency. When the generator potential reaches the threshold level, it initiates an action potential (nerve impulse). Once initiated, the action potential is propagated along the nerve fiber. Whereas a generator potential is a local, graded response, an action potential obeys the all-or-none principle. The function of a generator potential is to initiate an action potential by transducing (converting) a stimulus into a nerve impulse.

A receptor may be quite simple. It may consist of the dendrites of a single neuron in the skin that are sensitive to pain stimuli. Or it may be contained in a complex organ such as the eye. Regardless of complexity, all sense receptors contain the dendrites of sensory neurons. The dendrites occur either alone or in close association with specialized cells of other tissues.

Receptors are at the same time very excitable and very specialized. Except for pain receptors, each has a low threshold of response to its specific stimulus and a high threshold to all others.

Once a stimulus is received by a receptor and converted into an impulse, the impulse is conducted along an afferent pathway that enters either the spinal cord or the brain. Many sensory impulses are conducted to the sensory areas of the cerebral cortex. It is in this region that stimuli produce conscious sensations. Sensory impulses that terminate in the spinal cord or brain stem can initiate motor activities, but typically do not produce conscious sensations. The thalamus detects pain sensations but cannot distinguish the intensity or location from which they arise. This is a function of the cerebrum.

CHARACTERISTICS

Most conscious sensations or perceptions occur in the cortical regions of the brain. In other words, you see, hear, and feel in the brain. You seem to see with your eyes, hear with your ears, and feel pain in an injured part of your body only because the cortex interprets the sensation as coming from the stimulated sense receptor. The term **projection** describes this process by which the brain refers sensations to their point of stimulation.

A second characteristic of many sensations is **adaptation.** The perception of a sensation may disappear even though a stimulus is still being applied. When you get into a tub of hot water, you might feel a burning sensation. But soon the sensation decreases to one of comfortable warmth, even though the stimulus (hot water) is still present. Other examples of adaptation include placing a ring on your finger, putting on your shoes or hat, and sitting on a chair. Initially, you are conscious of the sensations involved, but they are lost soon thereafter.

Sensations may also be characterized by **afterimages,** that is, some sensations persist even though the stimulus has been removed. This phenomenon is the reverse of adaptation. One common example of afterimage occurs when you look at a bright light and then look away or close your eyes. You still see the light for several seconds or minutes afterward.

Another characteristic of sensations is **modality:** the specific sensation felt. The sensation may be one of pain, pressure, touch, body position, equilibrium, hearing, vision, smell, or taste. In other words, the distinct property by which one sensation may be distinguished from another is its modality.

CLASSIFICATION OF RECEPTORS

Location

One convenient method of classifying receptors is by their location. **Exteroceptors** (eks'-ter-ō-SEP-tors) provide information about the external environment. They are sensitive to stimuli outside the body and transmit sensations of hearing, sight, smell, taste, touch, pressure, temperature, and pain. Exteroceptors are located near the surface of the body.

Visceroceptors (vis'-er-ō-SEP-tors), or **enteroceptors,** provide information about the internal environment. These sensations arise from within the body and may be felt as pain, pressure, fatigue, hunger, thirst, and nausea. Visceroceptors are located in blood vessels and viscera.

Proprioceptors (prō'-prē-ō-SEP-tors) provide information about body position and movement. Such sensations give us information about muscle tension, the position and tension of our joints, and equilibrium. These receptors are located in muscles, tendons, joints, and the internal ear.

Stimulus Detected

Another method of classifying receptors is by the type of stimuli they detect. **Mechanoreceptors** detect mechanical deformation of the receptor itself or in adjacent cells. Stimuli so detected include those related to touch, pressure, vibration, proprioception, hearing, equilibrium, and blood pressure. **Thermoreceptors** detect changes in temperature. **Nociceptors** detect pain, usually as a result of physical or chemical damage to tissues. **Electromagnetic (photo) receptors** detect light on the retina of the eye. **Chemoreceptors** detect taste in the mouth, smell in the nose, and chemicals in body fluids, such as oxygen, carbon dioxide, water, and glucose.

Simplicity or Complexity

As will be described shortly, receptors may also be classified according to the simplicity or complexity of their structure and the neural pathway involved. **Simple receptors** and neural pathways are associated with **general senses.** The receptors for general sensations are numerous and widespread. Examples include cutaneous sensations such as touch, pressure, vibration, heat, cold, and pain. **Complex receptors** and neural pathways are associated with **special senses.** The receptors for each special sense are found in only one or two specific areas of the body. Among the special senses are smell, taste, sight, and hearing.

GENERAL SENSES

CUTANEOUS SENSATIONS

Cutaneous sensations include tactile sensations (touch, pressure, vibration), thermoreceptive sensations (cold and heat), and pain. The receptors for these sensations are in the skin, connective tissue, and the ends of the gastrointestinal tract.

The cutaneous receptors are distributed over the body surface in such a way that certain parts of the body are densely populated with receptors and other parts contain only a few. Areas of the body that have few cutaneous receptors are insensitive; those containing many are very sensitive.

This type of distribution can be demonstrated in the skin by using the *two-point discrimination test* for touch. A compass is applied to the skin, and the distance in millimeters between the two points of the compass is var-

ied. The subject then indicates when two points are felt and when only one is felt. The compass may be placed on the tip of the tongue, an area where receptors are very densely packed. At a distance of 1.4 mm (0.06 inch), the points are able to stimulate two different receptors, and the subject feels touched by two objects. If the distance is less than 1.4 mm, the subject feels only one point, even though both points are touching the tongue, because the points are so close together that they reach only one receptor. If the compass is placed on the back of the neck, the subject feels two distinctly different points only if the distance between them is 36.2 mm (1.43 inch) or greater, because the receptors are few in number and far apart. The results of this test indicate that the more sensitive the area, the closer the compass points may be placed and still be felt separately. The following order for these receptors, from greatest sensitivity to least, has been established: tip of tongue, tip of finger, side of nose, back of hand, and back of neck.

Cutaneous receptors have simple structures. They consist of the dendrites of sensory neurons that may or may not be enclosed in a capsule of epithelial or connective tissue. Impulses generated by cutaneous receptors pass along somatic afferent neurons in spinal and cranial nerves, through the thalamus, to the general sensory area of the parietal lobe of the cortex.

Tactile Sensations

Even though the **tactile sensations** are divided into separate sensations of touch, pressure, and vibration, they are all detected by the same types of receptors, that is, receptors subject to deformation.

● *Touch*   **Touch sensations** generally result from stimulation of tactile receptors in the skin or tissues immediately beneath the skin. The term *light touch* refers to the ability to recognize exactly what point of the body is touched. *Crude touch* refers to the ability to perceive that something has touched the skin, although its exact location, shape, size, or texture cannot be determined.

Tactile receptors for touch include root hair plexuses, free nerve endings, Merkel's discs, Meissner's corpuscles, and end organs of Ruffini (Figure 15-1). **Root hair plexuses** are dendrites arranged in networks around the roots of hairs. They are not surrounded by supportive or protective structures. If a hair shaft is moved, the dendrites are stimulated. Root hair plexuses detect movements mainly on the surface of the body.

Other receptors that are not surrounded by supportive or protective structures are called **free (naked) nerve endings.** Free nerve endings are found everywhere in the skin and many other tissues.

Merkel's (MER-kelz) **discs** are receptors for touch that consist of disclike formations of dendrites attached to deeper layers of epidermal cells. They are distributed in many of the same locations as Meissner's corpuscles.

Meissner's (MĪS-nerz) **corpuscles** are egg-shaped receptors containing a mass of dendrites enclosed by connective tissue. They are located in the dermal papillae of the skin. Meissner's corpuscles are most numerous in the fingertips, palms of the hand, and soles of the feet.

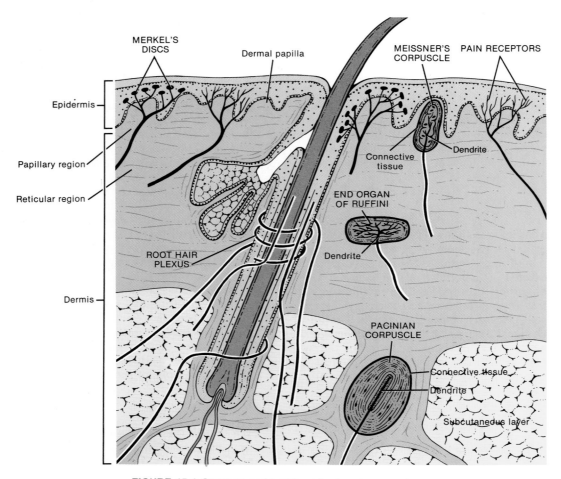

FIGURE 15-1 Structure and location of cutaneous receptors.

They are also abundant in the eyelids, tip of the tongue, lips, nipples, clitoris, and tip of penis.

End organs of Ruffini are embedded deeply in the dermis and in deeper tissues of the body. They detect heavy and continuous touch sensations.

● *Pressure*   **Pressure sensations** generally result from stimulation of tactile receptors in deeper tissues and are longer lasting and have less variation in intensity than touch sensations. Moreover, pressure is felt over a larger area than touch.

Pressure receptors are free nerve endings, end organs of Ruffini, and Pacinian corpuscles. **Pacinian (pa-SIN-ē-an) corpuscles** (Figure 15-1) are oval structures composed of a capsule resembling an onion and consist of connective tissue layers enclosing dendrites. Pacinian corpuscles are located in the subcutaneous tissue under the skin, the deep subcutaneous tissues that lie under mucous membranes, in serous membranes, around joints and tendons, in the perimysium of muscles, in the mammary glands, in the external genitalia of both sexes, and in certain viscera.

● *Vibration*   **Vibration sensations** result from rapidly repetitive sensory signals from tactile receptors.

The receptors for vibration sensations are Meissner's corpuscles and Pacinian corpuscles. Whereas Meissner's corpuscles detect low frequency vibration, Pacinian corpuscles detect higher frequency vibration.

Thermoreceptive Sensations

The **thermoreceptive sensations** are heat and cold. The **thermoreceptive receptors** are not known, but might be free nerve endings.

Pain Sensations

The receptors for **pain** are simply the branching ends of the dendrites of certain sensory neurons (Figure 15-1). Pain receptors are found in practically every tissue of the body. They may be stimulated by any type of stimulus. When stimuli for other sensations, such as touch, pressure, heat, and cold, reach a certain threshold, they stimulate the sensation of pain as well. Excessive stimulation of a sense organ causes pain. Additional stimuli for pain receptors include excessive distension or dilation of a structure, prolonged muscular contractions, muscle spasms, inadequate blood flow to an organ, or the presence of certain chemical substances. Pain receptors, because of their sensitivity to all stimuli, perform a protective function by identifying changes that may endanger the body. Pain receptors adapt only slightly or not at all.

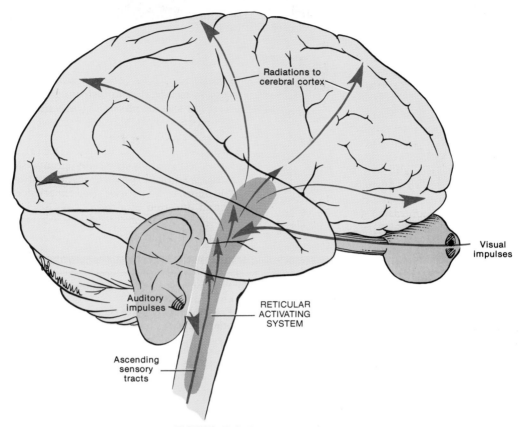

FIGURE 15-9 Reticular formation.

system with the spinal cord that is composed of many circuits. Impulses from the activated RAS are transmitted down the spinal cord and then to skeletal muscles. Muscle activation causes proprioceptors to return impulses that activate the RAS. The two feedback systems maintain activation of the RAS, which in turn maintains activation of the cerebral cortex. The result is a state of wakefulness called *consciousness*. The RAS is the physical basis of consciousness, the brain's chief watchguard. It continuously sifts and selects, forwarding only the essential, unusual, or dangerous to the conscious mind. Since humans experience different levels of consciousness (alertness, attentiveness, relaxation, nonattentiveness), it is assumed that the level of consciousness depends on the number of feedback circuits operating at the time.

CLINICAL APPLICATION

Consciousness may be altered by various factors. Amphetamines probably activate the RAS to produce a state of wakefulness and alertness. Meditation produces a lack of consciousness. Anesthetics produce a state of consciousness called anesthesia. Damage to the nervous system, as well as disease, can produce a lack of consciousness called coma. And drugs such as LSD can alter consciousness also.

If this theory of wakefulness is accepted, how then does sleep occur if the activating feedback systems are in continual operation? One explanation is that the feed-back system slows tremendously or is inhibited. Inactivation of the RAS produces a state known as *sleep.*

Just as there are different levels of consciousness, there are different levels of sleep. Normal sleep consists of two types: nonrapid eye movement sleep (NREM) and rapid eye movement sleep (REM).

NREM sleep consists of four stages. Each has been identified by EEG recordings:

Stage 1. The person is relaxing with eyes closed. During this time, respirations are regular, pulse is even, and the person has fleeting thoughts. If awakened, the person will frequently say he has not been sleeping. Alpha waves appear on the EEG.

Stage 2. It is a little harder to awaken the person. Fragments of dreams may be experienced, and the eyes may slowly roll from side to side. The EEG shows *sleep spindles*—sudden, short bursts of sharply pointed alpha waves that occur at 14 to 16 Hz (cycles) per second.

Stage 3. The person is very relaxed. Body temperature begins to fall and blood pressure decreases. It is difficult to awaken the person and the *EEG* shows a mixture of sleep spindles and delta waves. This stage occurs about 20 minutes after falling asleep.

Stage 4. Deep sleep occurs. The person is very relaxed and responds slowly if awakened. Bed-wetting and sleep-walking may occur during this stage. The EEG is dominated by delta waves.

In a typical 7- or 8-hour sleep period, a person goes from stages 1 to 4 of NREM sleep. Then the person ascends to stages 3 and 2 and then to REM sleep within 50 to 90 minutes.

In **REM sleep,** the EEG readings are similar to those of stage 1 of NREM sleep. There are significant physiological differences, however. During REM sleep, respirations and pulse rate increase and are irregular. Blood pressure also fluctuates considerably. It is during REM sleep that most dreaming occurs. Following REM sleep, the person descends again to stages 3 and 4 of NREM sleep.

REM and NREM sleep alternate throughout the night with approximately 90-minute intervals between REM periods. This cycle repeats itself from 3 to 5 times during the entire sleep period. The REM periods start out lasting from 5 to 10 minutes and gradually lengthen until the final one lasts about 50 minutes. In a normal night's sleep, REM totals 90 to 120 minutes. As much as 50 percent of an infant's sleep is REM as contrasted with 20 percent

for adults. Most sedatives significantly decrease REM sleep.

CLINICAL APPLICATION

A **polysomograph** (*poly* = many; *somnus* = sleep; *graph* = to write) is an instrument that uses electrodes to record several physiological variables during sleep. Among these variables are brain activity recorded as an electroencephalogram (EEG), eye movements recorded as an electrooculogram (EOG), and muscle tonus recorded as an electromyogram (EMG). These recordings indicate precisely when patients fall asleep, how many wake periods they experience, and the quality and duration of sleep.

STUDY OUTLINE

Sensations (p. 346)
Definition
1. Sensation is a state of awareness of external and internal conditions of the body.
2. The prerequisites for sensation are reception of a stimulus, conversion of the stimulus into a nerve impulse by a receptor, conduction of the impulse to the brain, and translation of the impulse into a sensation by a region of the brain.

Characteristics
1. Projection occurs when the brain refers a sensation to the point of stimulation.
2. Adaptation is the loss of sensation even though the stimulus is still applied.
3. An afterimage is the persistence of a sensation even though the stimulus is removed.
4. Modality is the property by which one sensation is distinguished from another.

Classification of Receptors
1. According to location, receptors are classified as exteroceptors, visceroceptors, and proprioceptors.
2. On the basis of type of stimulus detected, receptors are classified as mechanoreceptors, thermoreceptors, nociceptors, electromagnetic receptors, and chemoreceptors.
3. In terms of simplicity or complexity, simple receptors are associated with general senses and complex receptors are associated with special senses.

General Senses (p. 347)
Cutaneous Sensations
1. Cutaneous sensations include tactile sensations (touch, pressure, vibration), thermoreceptive sensations (heat and cold), and pain. Receptors for these sensations are located in the skin, connective tissues, and the ends of the gastrointestinal tract.
2. Receptors for touch are root hair plexuses, free nerve endings, Merkel's discs, Meissner's corpuscles, and end organs of Ruffini. Receptors for pressure are free nerve endings, end organs of Ruffini, and Pacinian corpuscles. Receptors for vibration are Meissner's corpuscles and Pacinian corpuscles.
3. Pain receptors are located in nearly every body tissue.
4. Two kinds of pain recognized in the parietal lobe of the cortex are somatic and visceral.
5. Referred pain is felt in the skin near or away from the organ sending pain impulses.

6. Phantom pain is the sensation of pain in a limb that has been amputated.
7. Pain impulses may be inhibited by drugs, surgery, and acupuncture.

Proprioceptive Sensations
1. Receptors located in muscles, tendons, and joints convey impulses related to muscle tone, movement of body parts, and body position.
2. The receptors include joint kinesthetic receptors, muscle spindles, and tendon organs.

Levels of Sensation
1. Sensory fibers terminating in the lower brain stem bring about far more complex motor reactions than simple spinal reflexes.
2. When sensory impulses reach the lower brain stem, they cause subconscious motor reactions.
3. Sensory impulses that reach the thalamus can be localized crudely in the body.
4. When sensory impulses reach the cerebral cortex, we experience precise localization.

Sensory Pathways (p. 351)
1. Sensory information from all parts of the body terminates in a specific area of the somatosensory cortex.
2. In the posterior column pathway and the spinothalamic pathway there are first-order, second-order, and third-order neurons.
3. The neural pathway for pain and temperature is the lateral spinothalamic pathway.
4. The neural pathway for crude touch and pressure is the anterior spinothalamic pathway.
5. The neural pathway for light touch, proprioception, and vibration is the posterior column pathway.
6. The pathways to the cerebellum are the anterior and posterior spinocerebellar tracts.

Motor Pathways (p. 353)
1. The muscles of all parts of the body are controlled by a specific area of the motor cortex.
2. Voluntary motor impulses are conveyed from the brain through the spinal cord along the pyramidal pathways and the extrapyramidal pathways.
3. Pyramidal pathways include the lateral corticospinal, anterior corticospinal, and corticobulbar tracts.
4. Major extrapyramidal tracts are the rubrospinal, tectospinal, and vestibulospinal tracts.

Integrative Functions (p. 356)

1. Memory is defined as the ability to recall thoughts; it consists of activated and long-term components.
2. Sleep and wakefulness are integrative functions that are controlled by the reticular activating system (RAS).
3. Nonrapid eye movement (NREM) sleep consists of four stages identified by EEG recordings.
4. Most dreaming occurs during rapid eye movement (REM) sleep.

REVIEW QUESTIONS

1. Define a sensation and a sense receptor. What prerequisites are necessary for the perception of a sensation?
2. Describe the following characteristics of a sensation: projection, adaptation, afterimage, modality.
3. Name some examples of adaptation not discussed in the text.
4. Classify receptors on the basis of location, stimulus detected, and simplicity or complexity.
5. Distinguish between a general sense and a special sense.
6. What is a cutaneous sensation? Distinguish tactile, thermoreceptive, and pain sensations.
7. How are cutaneous receptors distributed over the body? Relate your response to the two-point discrimination test.
8. For each of the following cutaneous sensations, describe the receptor involved in terms of structure, function, and location: touch, pressure, vibration, and pain.
9. How do cutaneous sensations help maintain homeostasis?
10. Why are pain receptors important? Differentiate somatic pain, visceral pain, referred pain, and phantom pain.
11. Why is the concept of referred pain useful to the physician in diagnosing internal disorders?
12. What is acupuncture? Describe the "gate control" theory of acupuncture.
13. How is acupuncture currently being used in the United States?
14. What is the proprioceptive sense? Where are the receptors for this sense located?
15. Relate proprioception to the maintenance of homeostasis.
16. Describe how various parts of the body are represented in the somatosensory cortex.
17. Distinguish between the posterior column and spinothalamic pathways.
18. What is the sensory pathway for pain and temperature? How does it function?
19. Which pathway controls crude touch and pressure? How does it function?
20. Which pathway is responsible for light touch, proprioception, and vibration? How does it function?
21. Describe how various parts of the body are represented in the motor cortex.
22. Which pathways control voluntary motor impulses from the brain through the spinal cord? How are they distinguished?
23. Define memory. What are the two kinds of memory?
24. Describe how sleep and wakefulness are related to the reticular activating system.
25. What are the four stages of nonrapid eye movement (NREM) sleep? How is NREM sleep distinguished from rapid eye movement (REM) sleep?

16 The Autonomic Nervous System

The portion of the nervous system that regulates the activities of smooth muscle, cardiac muscle, and glands is the **autonomic nervous system (ANS).** Structurally, the system consists of visceral efferent neurons organized into nerves, ganglia, and plexuses. Functionally, it usually operates without conscious control. The system was originally named *autonomic* because physiologists thought it functioned with no control from the central nervous system, that it was autonomous or self-governing. It is now known that the autonomic system is neither structurally nor functionally independent of the central nervous system. It is regulated by centers in the brain, in particular by the cerebral cortex, hypothalamus, and medulla oblongata. However, the old terminology has been retained, and since the autonomic nervous system does differ from the somatic nervous system in some ways, the two are separated for convenience of study.

SOMATIC EFFERENT AND AUTONOMIC NERVOUS SYSTEMS

Whereas the somatic efferent nervous system produces conscious movement in skeletal muscles, the autonomic nervous system (visceral efferent nervous system) regulates visceral activities, and it generally does so involuntarily and automatically. Examples of visceral activities regulated by the autonomic nervous system are changes in the size of the pupil, accommodation for near vision, dilation and constriction of blood vessels, adjustment of the rate and force of the heartbeat, movements of the gastrointestinal tract, and secretion by most glands. These activities usually lie beyond conscious control. They are automatic.

The autonomic nervous system is entirely motor. All its axons are efferent fibers, which transmit impulses from the central nervous system to visceral effectors. Autonomic fibers are called **visceral efferent fibers. Visceral effectors** include cardiac muscle, smooth muscle, and glandular epithelium. This does not mean there are no afferent (sensory) impulses from visceral effectors, however. Impulses that give rise to visceral sensations pass over visceral afferent neurons that have cell bodies located in the posterior (dorsal) root ganglia of spinal nerves.

Some functions of these afferent neurons were described with the cranial and spinal nerves. The hypothalamus, which largely controls the autonomic nervous system, also receives impulses from the visceral sensory fibers.

The autonomic nervous system consists of two principal divisions: the **sympathetic** and the **parasympathetic.** Many organs innervated by the autonomic nervous system receive visceral efferent neurons from both components of the autonomic system—one set from the sympathetic division, another from the parasympathetic division. In general, impulses transmitted by the fibers of one division stimulate the organ to start or increase activity, whereas impulses from the other division decrease the organ's activity. Organs that receive impulses from both sympathetic and parasympathetic fibers are said to have *dual innervation*. In the somatic efferent nervous system, only one kind of motor neuron innervates an organ, which is always a skeletal muscle. When the somatic neurons stimulate the cells of the skeletal muscle, the muscle becomes active. When the neuron ceases to stimulate the muscle, contraction stops altogether.

STRUCTURE OF THE AUTONOMIC NERVOUS SYSTEM

VISCERAL EFFERENT PATHWAYS

Autonomic visceral efferent pathways always consist of two neurons. One extends from the central nervous system to a ganglion. The other extends directly from the ganglion to the effector (muscle or gland).

The first of the visceral efferent neurons in an autonomic pathway is called a **preganglionic neuron** (Figure 16-1). Its cell body is in the brain or spinal cord. Its myelinated axon, called a **preganglionic fiber,** passes out of the central nervous system as part of a cranial or spinal nerve. At some point, the fiber separates from the nerve and courses to an autonomic ganglion, where it synapses with the dendrites or cell body of the postganglionic neuron, the second neuron in the visceral efferent pathway.

The **postganglionic neuron** lies entirely outside the central nervous system. Its cell body and dendrites (if it has dendrites) are located in the autonomic ganglion,

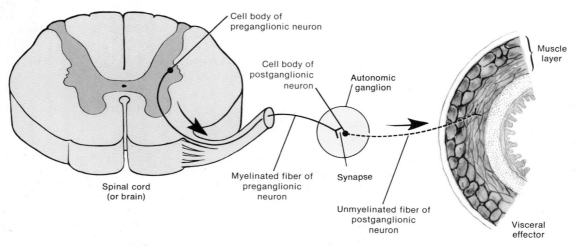

FIGURE 16-1 Relationship between preganglionic and postganglionic neurons.

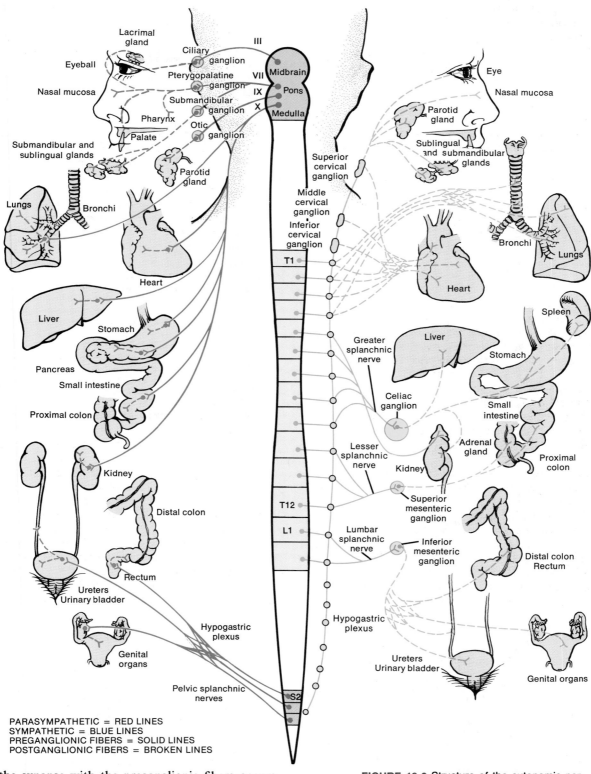

Lacrimal gland
Eyeball
Nasal mucosa
Ciliary ganglion
III
Pterygopalatine ganglion
VII
Midbrain
Pons
IX
Submandibular ganglion
X
Medulla
Pharynx
Otic ganglion
Palate
Submandibular and sublingual glands
Parotid gland
Lungs
Bronchi
Heart
Liver
Stomach
Pancreas
Small intestine
Proximal colon
Kidney
Distal colon
Rectum
Ureters
Urinary bladder
Hypogastric plexus
Genital organs
Pelvic splanchnic nerves

Superior cervical ganglion
Middle cervical ganglion
Inferior cervical ganglion
T1
Greater splanchnic nerve
Celiac ganglion
Lesser splanchnic nerve
Kidney
T12
Lumbar splanchnic nerve
L1
Superior mesenteric ganglion
Inferior mesenteric ganglion
Hypogastric plexus
S2

Eye
Nasal mucosa
Parotid gland
Sublingual and submandibular glands
Bronchi
Lungs
Heart
Spleen
Liver
Stomach
Adrenal gland
Small intestine
Proximal colon
Distal colon
Rectum
Ureters
Urinary bladder
Genital organs

PARASYMPATHETIC = RED LINES
SYMPATHETIC = BLUE LINES
PREGANGLIONIC FIBERS = SOLID LINES
POSTGANGLIONIC FIBERS = BROKEN LINES

where the synapse with the preganglionic fibers occurs. The axon of a postganglionic neuron, called a **postganglionic fiber,** is unmyelinated and terminates in a visceral effector.

Thus preganglionic neurons convey efferent impulses from the central nervous system to autonomic ganglia. Postganglionic neurons relay the impulses from autonomic ganglia to visceral effectors.

FIGURE 16-2 Structure of the autonomic nervous system. Although the parasympathetic division is shown only on the left side of the figure and the sympathetic division is shown only on the right side, keep in mind that each division is actually on both sides (bilateral symmetry).

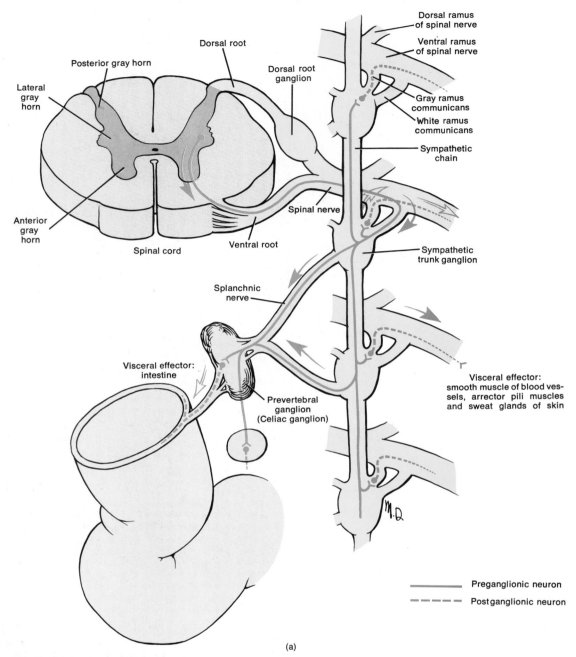

FIGURE 16-3 Ganglia and rami communicantes of the sympathetic division of the autonomic nervous system. (a) Diagram.

Preganglionic Neurons

In the sympathetic division, the preganglionic neurons have their cell bodies in the lateral gray horns of the twelve thoracic segments and first two lumbar segments of the spinal cord (Figure 16-2). It is for this reason that the sympathetic division is also called the **thoracolumbar** (thō'-ra-kō-LUM-bar) **division** and the fibers of the sympathetic preganglionic neurons are known as the **thoracolumbar outflow.**

The cell bodies of the preganglionic neurons of the parasympathetic division are located in the nuclei of cranial nerves III, VII, IX, and X in the brain stem and in the lateral gray horns of the second through fourth sacral segments of the spinal cord. Hence the parasympathetic division is also known as the **craniosacral division,** and the fibers of the parasympathetic preganglionic neurons are referred to as the **craniosacral outflow.**

Autonomic Ganglia

Autonomic pathways include **autonomic ganglia,** where synapses between visceral efferent neurons occur. Autonomic ganglia differ from posterior root ganglia. The latter contain cell bodies of sensory neurons and no synapses occur in them.

Cut edge of dura mater

Cervical enlargement

Posterior (dorsal) root ganglion
of first thoracic nerve (T1)

Sympathetic trunk

Dorsal ramus of
second thoracic nerve (T2)

White ramus
communicans

Gray ramus
communicans

(b)

FIGURE 16-3 (*Continued*) Ganglia and rami communicantes of the sympathetic division of the autonomic nervous system. (b) Photograph of the posterior aspect of the lower cervical and upper thoracic segments of the spinal cord. (Courtesy of N. Gluhbegovic and T. H. Williams, *The Human Brain: A Photographic Guide,* Harper & Row, New York, 1980.)

The autonomic ganglia may be divided into three general groups. The **sympathetic trunk** or **vertebral chain ganglia** are a series of ganglia that lie in a vertical row on either side of the vertebral column, extending from the base of the skull to the coccyx (Figure 16-3). They are also known as **paravertebral** or **lateral ganglia.** They receive preganglionic fibers only from the thoracolumbar (sympathetic) division (Figure 16-2).

The second kind of autonomic ganglion also belongs to the sympathetic division. It is called a **prevertebral** or **collateral ganglion** (Figure 16-3). The ganglia of this group lie anterior to the spinal column and close to the large abdominal arteries from which their names are derived. Examples of prevertebral ganglia so named are the celiac ganglion, on either side of the celiac artery just below the diaphragm; the superior mesenteric ganglion, near the beginning of the superior mesenteric artery in the upper abdomen; and the inferior mesenteric ganglion, located near the beginning of the inferior mesenteric artery in the middle of the abdomen (Figure 16-2). Prevertebral ganglia receive preganglionic fibers from the thoracolumbar (sympathetic) division.

The third kind of autonomic ganglion belongs to the parasympathetic division and is called a **terminal** or **intramural ganglion.** The ganglia of this group are located at the end of a visceral efferent pathway very close to visceral effectors or within the walls of visceral effectors. Terminal ganglia receive preganglionic fibers from the craniosacral (parasympathetic) division. The pregan-

glionic fibers do not pass through sympathetic trunk ganglia (Figure 16-2).

In addition to autonomic ganglia, the autonomic nervous system contains **autonomic plexuses.** Slender nerve fibers from ganglia containing postganglionic nerve cell bodies arranged in a branching network constitute an autonomic plexus.

Postganglionic Neurons

Axons from preganglionic neurons of the sympathetic division pass to ganglia of the sympathetic trunk. They can either synapse in the sympathetic chain ganglia with postganglionic sympathetics or they can continue, without synapsing, through the chain ganglia to end at a prevertebral ganglion where synapses with the postganglionic sympathetics can take place. Each sympathetic preganglionic fiber synapses with several postganglionic fibers in the ganglion, and the postganglionic fibers pass to several visceral effectors. Upon exiting their ganglia, the postsynaptic fibers innervate their visceral effectors.

Axons from preganglionic neurons of the parasympathetic division pass to terminal ganglia near or within a visceral effector. In the ganglion, the presynaptic neuron usually synapses with only four or five postsynaptic neurons to a single visceral effector. Upon exiting their ganglia, the postsynaptic fibers supply their visceral effectors.

With this background in mind, we can now examine some specific structural features of the sympathetic and parasympathetic divisions of the autonomic nervous system.

SYMPATHETIC DIVISION

The preganglionic fibers of the sympathetic division have their cell bodies located in the lateral gray horn of the spinal cord in the thoracic and first two lumbar segments (Figure 16-2). The preganglionic fibers are myelinated and leave the spinal cord through the ventral root of a spinal nerve along with the somatic efferent fibers at the same segmental levels. After exiting through the intervertebral foramina, the preganglionic sympathetic fibers enter a white ramus to pass to the nearest sympathetic trunk ganglion on the same side. Collectively, the white rami are called the **white rami communicantes** (kō-myoo-ne-KAN-tēz). Their name indicates that they contain myelinated fibers. Only thoracic and upper lumbar nerves have white rami communicantes. The white rami communicantes connect the ventral ramus of the spinal nerve with the ganglia of the sympathetic trunk.

The paired sympathetic trunks are situated anterolaterally to the spinal cord, one on either side. Each consists of a series of ganglia arranged more or less segmentally. The divisions of the sympathetic trunk are named on the basis of location. Typically, there are 22 ganglia in each chain: 3 cervical, 11 thoracic, 4 lumbar, and 4 sacral. Although the trunk extends downward from the neck, thorax, and abdomen to the coccyx, it receives preganglionic fibers only from the thoracic and lumbar segments of the spinal cord (Figure 16-2).

The cervical portion of each sympathetic trunk is located in the neck anterior to the prevertebral muscles. It is subdivided into a superior, middle, and inferior ganglion (Figure 16-2). The *superior cervical ganglion* is posterior to the internal carotid artery and anterior to the transverse processes of the second cervical vertebra. Postganglionic fibers leaving the ganglion serve the head, where they are distributed to the sweat glands, the smooth muscle of the eye and blood vessels of the face, the nasal mucosa, and the submandibular, sublingual, and parotid salivary glands. Gray rami communicantes from the ganglion also pass to the upper two to four cervical spinal nerves. The *middle cervical ganglion* is situated near the sixth cervical vertebra at the level of the cricoid cartilage. Postganglionic fibers from it innervate the heart. The *inferior cervical ganglion* is located near the first rib, anterior to the transverse processes of the seventh cervical vertebra. Its postganglionic fibers also supply the heart.

The thoracic portion of each sympathetic trunk usually consists of 11 segmentally arranged ganglia, lying ventral to the necks of the corresponding ribs. This portion of the sympathetic trunk receives most of the sympathetic preganglionic fibers. Postganglionic fibers from the thoracic sympathetic trunk innervate the heart, lungs, bronchi, and other thoracic viscera.

The lumbar portion of each sympathetic trunk is found on either side of the corresponding lumbar vertebrae. The sacral portion of the sympathetic trunk lies in the pelvic cavity on the medial side of the sacral foramina. Postganglionic fibers from the lumbar and sacral sympathetic

chain ganglia are distributed with the respective spinal nerves via gray rami, or they may join the hypogastric plexus via direct visceral branches.

When a preganglionic fiber of a white ramus communicans enters the sympathetic trunk, it may terminate (synapse) in several ways. Some fibers synapse in the first ganglion at the level of entry. Others pass up or down the sympathetic trunk for a variable distance to form the fibers on which the ganglia are strung. These fibers, known as **sympathetic chains** (Figure 16-3), may not synapse until they reach a ganglion in the cervical or sacral area. Most rejoin the spinal nerves before supplying peripheral visceral effectors such as sweat glands and the smooth muscle in blood vessels and around hair follicles in the extremities. The **gray ramus communicans** (kō-MYOO-ne-kanz) is the structure containing the postganglionic fibers that connect the ganglion of the sympathetic trunk to the spinal nerve (Figure 16-3). The fibers are unmyelinated. All spinal nerves have gray rami communicantes. Gray rami communicantes outnumber the white rami, since there is a gray ramus leading to each of the 31 pairs of spinal nerves.

In most cases, a sympathetic preganglionic fiber terminates by synapsing with a large number of postganglionic cell bodies in a ganglion, usually 20 or more. Often the postganglionic fibers then terminate in widely separated organs of the body. Thus an impulse that starts in a single preganglionic neuron may reach several visceral effectors. For this reason, most sympathetic responses have widespread effects on the body.

Some preganglionic fibers pass through the sympathetic trunk without terminating in the trunk. Beyond the trunk, they form nerves known as **splanchnic** (SPLANK-nik) **nerves** (Figure 16-2). After passing through the trunk of ganglia, the splanchnic nerves terminate in the *celiac* (SĒ-lē-ak) or *solar plexus*. In the plexus, the preganglionic fibers synapse in ganglia with postganglionic cell bodies. These ganglia are prevertebral ganglia. The greater splanchnic nerve passes to the celiac ganglion of the celiac plexus. From here, postganglionic fibers are distributed to the stomach, spleen, liver, kidney, and small intestine. The lesser splanchnic nerve passes through the celiac plexus to the superior mesenteric ganglion of the superior mesenteric plexus. Postganglionic fibers from this ganglion innervate the small intestine and colon. The lowest splanchnic nerve, not always present, enters the renal plexus. Postganglionics supply the renal artery and ureter. The lumbar splanchnic nerve enters the inferior mesenteric plexus. In the plexus, the preganglionic fibers synapse with postganglionic fibers in the inferior mesenteric ganglion. These fibers pass through the hypogastric plexus and supply the distal colon and rectum, urinary bladder, and genital organs. As noted earlier, the postganglionic fibers leaving the prevertebral ganglia follow the course of various arteries to abdominal and pelvic visceral effectors.

PARASYMPATHETIC DIVISION

The preganglionic cell bodies of the parasympathetic division are found in nuclei in the brain stem and the lateral gray horn of the second through fourth sacral segments of the spinal cord (Figure 16-2). Their fibers emerge as part of a cranial nerve or as part of the ventral root of a spinal nerve. The **cranial parasympathetic outflow** consists of preganglionic fibers that leave the brain stem by way of the oculomotor (III) nerves, facial (VII) nerves, glossopharyngeal (IX) nerves, and vagus (X) nerves. The **sacral parasympathetic outflow** consists of preganglionic fibers that leave the ventral roots of the second through fourth sacral nerves. The preganglionic fibers of both the cranial and sacral outflows end in terminal ganglia, where they synapse with postganglionic neurons. We will first look at the cranial outflow.

The cranial outflow has five components: four pairs of ganglia and the plexuses associated with the vagus nerve. The four pairs of cranial parasympathetic ganglia innervate structures in the head and are located close to the organs they innervate. The *ciliary ganglion* is near the back of an orbit lateral to each optic nerve. Preganglionic fibers pass with the oculomotor (III) nerve to the ciliary ganglion. Postganglionic fibers from the ganglion innervate smooth muscle cells in the eyeball. Each *pterygopalatine* (ter'-i-gō-PAL-a-tin) *ganglion* is situated lateral to a sphenopalatine foramen. It receives preganglionic fibers from the facial (VII) nerve and transmits postganglionic fibers to the nasal mucosa, palate, pharynx, and lacrimal gland. Each *submandibular ganglion* is found near the duct of a submandibular salivary gland. It receives preganglionic fibers from the facial (VII) nerve and transmits postganglionic fibers that innervate the submandibular and sublingual salivary glands. The *otic ganglia* are situated just below each foramen ovale. The otic ganglion receives preganglionic fibers from the glossopharyngeal (IX) nerve and transmits postganglionic fibers that innervate the parotid salivary gland. Ganglia associated with the cranial outflow are classified as terminal ganglia. Since the terminal ganglia are close to their visceral effectors, postganglionic parasympathetic fibers are short. Postganglionic sympathetic fibers are relatively long.

The last component of the cranial outflow consists of the preganglionic fibers that leave the brain via the vagus (X) nerves. This component has the most extensive distribution of the parasympathetic fibers. It provides about 80 percent of the craniosacral outflow. Each vagus (X) nerve enters into the formation of several plexuses in the thorax and abdomen. As it passes through the thorax, it sends fibers to the *superficial cardiac plexus* in the arch of the aorta and the *deep cardiac plexus* anterior to the branching of the trachea. These plexuses contain terminal ganglia, and the postganglionic parasympathetic fibers emerging from them supply the heart. Also in the thorax is the *pulmonary plexus,* in front of and behind the roots of the lungs and within the lungs themselves. It receives preganglionic fibers from the vagus and transmits postganglionic parasympathetic fibers to the lungs and bronchi. Other plexuses associated with the vagus (X) nerve are described in later chapters in conjunction with the appropriate thoracic, abdominal, and pelvic viscera. Postganglionic fibers from these plexuses innervate viscera such as the liver, pancreas, stomach, kidneys, small intestine, and part of the colon.

The sacral parasympathetic outflow consists of pregan-

EXHIBIT 16-1
STRUCTURAL FEATURES OF SYMPATHETIC AND PARASYMPATHETIC DIVISIONS

SYMPATHETIC	PARASYMPATHETIC
Forms thoracolumbar outflow.	Forms craniosacral outflow.
Contains sympathetic trunk and prevertebral ganglia.	Contains terminal ganglia.
Ganglia are close to the CNS and distant from visceral effectors.	Ganglia are near or within visceral effectors.
Each preganglionic fiber synapses with many postganglionic neurons that pass to many visceral effectors.	Each preganglionic fiber usually synapses with four or five postganglionic neurons that pass to a single visceral effector.
Distributed throughout the body, including the skin.	Distribution limited primarily to head and viscera of thorax, abdomen, and pelvis.

glionic fibers from the ventral roots of the second through fourth sacral nerves. Collectively, they form the *pelvic splanchnic nerves.* They pass into the hypogastric plexus. From ganglia in the plexus, parasympathetic postganglionic fibers are distributed to the colon, ureters, urinary bladder, and reproductive organs.

The salient structural features of the sympathetic and parasympathetic divisions are compared in Exhibit 16-1.

PHYSIOLOGY

TRANSMITTER SUBSTANCES

Autonomic fibers, like other axons of the nervous system, release transmitter substances at synapses as well as at points of contact with visceral effectors. These latter points are called **neuroeffector junctions.** Neuroeffector junctions may be either neuromuscular or neuroglandular junctions. On the basis of the transmitter substance produced, autonomic fibers may be classified as either cholinergic or adrenergic.

Cholinergic (kō'-lin-ER-jik) **fibers** release **acetylcholine (ACh)** and include the following: (1) all sympathetic and parasympathetic preganglionic axons, (2) all parasympathetic postganglionic axons, and (3) some sympathetic postganglionic axons. The cholinergic sympathetic postganglionic axons include those to sweat glands and blood vessels in skeletal muscles, to the skin, and to the external genitalia. Since acetylcholine is quickly inactivated by the enzyme **acetylcholinesterase (AChE),** the effects of cholinergic fibers are short-lived and local.

Adrenergic (ad'-ren-ER-jik) **fibers** produce **norepinephrine (NE).** Most sympathetic postganglionic axons are adrenergic. Since norepinephrine is inactivated much more slowly by **catechol-o-methyltransferase (COMT)** or **monoamine oxidase (MAO)** than acetylcholine is by acetylcholinesterase, and since norepinephrine may enter the bloodstream, the effects of sympathetic stimulation are longer lasting and more widespread than parasympathetic stimulation.

EXHIBIT 16-2
ACTIVITIES OF AUTONOMIC NERVOUS SYSTEM

VISCERAL EFFECTOR	EFFECT OF SYMPATHETIC STIMULATION	EFFECT OF PARASYMPATHETIC STIMULATION
Eye		
Iris	Contraction of dilator muscle that results in dilation of pupil.	Contraction of sphincter muscle that results in constriction of pupil.
Ciliary muscle	No innervation.	Contraction that results in lens accommodation for near vision.
Glands		
Sweat	Stimulates secretion.	No innervation.
Lacrimal (tear)	Vasoconstriction, which inhibits secretion.	Normal or excessive secretion.
Salivary	Vasoconstriction, which decreases secretion.	Stimulates secretion and vasodilation.
Gastric	Vasoconstriction, which inhibits secretion.	Stimulates secretion.
Intestinal	Vasoconstriction, which inhibits secretion.	Stimulates secretion.
Adrenal medulla	Promotes epinephrine and norepinephrine secretion.	No innervation.
Adrenal cortex	Promotes glucocorticoid secretion.	No innervation.
Lungs (bronchial tubes)	Dilation.	Constriction.
Heart	Increases rate and strength of contraction; dilates coronary vessels that supply blood to heart muscle cells.	Decreases rate and strength of contraction; constricts coronary vessels.
Blood vessels		
Skin	Constriction.	No innervation for most.
Skeletal muscle	Dilation.	No innervation.
Visceral organs (except heart and lungs)	Constriction.	No innervation for most.
Liver	Promotes glycogenolysis; decreases bile secretion.	Promotes glycogenesis; increases bile secretion.
Gallbladder	Relaxation.	Contraction.
Stomach	Decreases motility.	Increases motility.
Intestines	Decreases motility.	Increases motility.
Gastrointestinal sphincters	Increases tone.	Relaxation.
Kidney	Constriction of blood vessels that results in decreased urine volume.	No innervation.
Pancreas	Inhibits secretion.	Promotes secretion.
Spleen	Contraction and discharge of stored blood into general circulation.	No innervation.
Urinary bladder	Relaxation of muscular wall; increases tone in internal sphincter.	Contraction of muscular wall; relaxation of internal sphincter.
Arrector pili of hair follicles	Contraction that results in erection of hairs.	No innervation.
Uterus	Inhibits contraction if nonpregnant; stimulates contraction if pregnant.	Minimal effect.
Sex organs	In male, vasoconstriction of ductus deferens, seminal vesicle, prostate; results in ejaculation. In female, reverse uterine peristalsis.	Vasodilation and erection in both sexes; secretion in female.

ACTIVITIES

Most visceral effectors have dual innervation, that is, they receive fibers from both the sympathetic and the parasympathetic divisions. In these cases, impulses from one division stimulate the organ's activities, whereas impulses from the other division inhibit the organ's activities. The stimulating division may be either the sympathetic or the parasympathetic, depending on the organ. For example, sympathetic impulses increase heart activity, whereas parasympathetic impulses decrease it. On the other hand, parasympathetic impulses increase digestive activities, whereas sympathetic impulses inhibit them. The actions of the two systems are carefully integrated to help maintain homeostasis. A summary of the activities of the autonomic nervous system is presented in Exhibit 16-2.

The parasympathetic division is primarily concerned with activities that restore and conserve body energy. It is a **rest-repose system.** Under normal body conditions, for instance, parasympathetic impulses to the digestive glands and the smooth muscle of the digestive system dominate over sympathetic impulses. Thus energy-supplying food can be digested and absorbed by the body.

The sympathetic division, by contrast, is primarily concerned with processes involving the expenditure of energy. When the body is in homeostasis, the main function of the sympathetic division is to counteract the parasympathetic effects just enough to carry out normal processes requiring energy. During extreme stress, however, the sympathetic dominates the parasympathetic. When people are confronted with a stress condition, for example, their bodies become alert and they sometimes perform feats of unusual strength. Fear stimulates the sympathetic division.

Activation of the sympathetic division sets into operation a series of physiological responses collectively called the **fight-or-flight response.** It produces the following effects.

1. The pupils of the eyes dilate.
2. The heart rate increases.
3. The blood vessels of the skin and viscera constrict.
4. The remainder of the blood vessels dilate. This reaction causes a rise in blood pressure and a faster flow of blood into the dilated blood vessels of skeletal muscles, cardiac muscle, lungs, and brain—organs involved in fighting off danger.
5. Rapid breathing occurs as the bronchioles dilate to allow faster movement of air in and out of the lungs.
6. Blood sugar level rises as liver glycogen is converted to glucose to supply the body's additional energy needs.
7. The medulla of the adrenal gland is stimulated to produce epinephrine and norepinephrine, hormones that intensify and prolong the sympathetic effects noted above.
8. Processes that are not essential for meeting the stress situation are inhibited. For example, muscular movements of the gastrointestinal tract and digestive secretions are slowed down or even stopped.

CLINICAL APPLICATION

If the sympathetic trunk is cut on one side, the sympathetic supply to that side of the head is removed and the result is **Horner's syndrome,** in which the patient exhibits (on the affected side): ptosis (drooping of the upper eyelid), slight elevation of the lower eyelid, narrowing of the palpebral fissure, enophthalmos (the eye appears sunken), a constricted pupil, anhydrosis (lack of sweating), and flushing of the skin.

VISCERAL AUTONOMIC REFLEXES

A **visceral autonomic reflex** adjusts the activity of a visceral effector. In other words, it results in the contraction of smooth or cardiac muscle or secretion by a gland. Such reflexes assume a key role in activities such as regulating heart action, blood pressure, respiration, digestion, defecation, and urinary bladder functions.

A visceral autonomic reflex arc consists of the following components.

1. **Receptor.** The receptor is the distal end of an afferent neuron in an exteroceptor or enteroceptor.
2. **Afferent neuron.** This neuron, either a somatic afferent or visceral afferent neuron, conducts the sensory impulse to the spinal cord or brain.

3. **Association neurons.** These neurons are found in the central nervous system.
4. **Visceral efferent preganglionic neuron.** In the thoracic and abdominal regions, this neuron is in the lateral gray horn of the spinal cord. The axon passes through the ventral root of the spinal nerve, the spinal nerve, and the white ramus communicans. It then enters a sympathetic trunk or prevertebral ganglion, where it synapses with a postganglionic neuron. In the cranial and sacral regions, the visceral efferent preganglionic axon leaves the central nervous system and passes to a terminal ganglion, where it synapses with a postganglionic neuron. The role of the visceral efferent preganglionic neuron is to convey a motor impulse from the brain or spinal cord to an autonomic ganglion.
5. **Visceral efferent postganglionic neuron.** This neuron conducts a motor impulse from a visceral efferent preganglionic neuron to the visceral effector.
6. **Visceral effector.** A visceral effector is smooth muscle, cardiac muscle, or a gland.

The basic difference between a somatic reflex arc and a visceral autonomic reflex arc is that in a somatic reflex arc, only one efferent neuron is involved. In a visceral autonomic reflex arc, two efferent neurons are involved.

Visceral sensations do not always reach the cerebral cortex. Most remain at subconscious levels. Under normal conditions, you are not aware of muscular contractions of the digestive organs, heartbeat, changes in the diameter of blood vessels, and pupil dilation and constriction. When your body is making adjustments in such visceral activities, they are handled by visceral reflex arcs whose centers are in the spinal cord or lower regions of the brain. Among such centers are the cardiac, respiratory, vasomotor, swallowing, and vomiting centers in the medulla and the temperature control center in the hypothalamus. Stimuli delivered by somatic or visceral afferent neurons synapse in these centers, and the returning motor impulses conducted by visceral efferent neurons bring about an adjustment in the visceral effector without conscious recognition. The impulses are interpreted and acted on subconsciously. Some visceral sensations do give rise to conscious recognition: hunger, nausea, and fullness of the urinary bladder and rectum.

CONTROL BY HIGHER CENTERS

The autonomic nervous system is not a separate nervous system. Axons from many parts of the central nervous system are connected to both the sympathetic and the parasympathetic divisions of the autonomic nervous system and thus exert considerable control over it. Autonomic centers in the cerebral cortex are connected to autonomic centers of the thalamus, for example. These, in turn, are connected to the hypothalamus. In this hierarchy of command, the thalamus sorts incoming impulses before they reach the cerebral cortex. The cerebral cortex then turns over control and integration of visceral activities to the hypothalamus. It is at the level of the hypothalamus that the major control and integration of the autonomic nervous system is exerted.

The hypothalamus is connected to both the sympathetic and the parasympathetic divisions of the autonomic nervous system. The posterior and lateral portions of the hypothalamus appear to control the sympathetic division.

When these areas are stimulated, there is an increase in visceral activities—an increase in heart rate, a rise in blood pressure due to vasoconstriction of blood vessels, an increase in the rate and depth of respiration, dilation of the pupils, and inhibition of the digestive tract. On the other hand, the anterior and medial portions of the hypothalamus seem to control the parasympathetic division. Stimulation of these areas results in a decrease in heart rate, lowering of blood pressure, constriction of the pupils, and increased motility of the digestive tract.

Control of the autonomic nervous system by the cerebral cortex occurs primarily during emotional stress. In extreme anxiety, the cerebral cortex can stimulate the hypothalamus. This stimulation, in turn, increases heart rate and blood pressure. If the cortex is stimulated by an extremely unpleasant sight, the stimulation causes vasodilation of blood vessels, a lowering of blood pressure, and fainting.

Evidence of even more direct control of visceral responses is provided by data gathered from studies of biofeedback and meditation.

BIOFEEDBACK

In the simplest terms, **biofeedback** is a process in which people get constant signals, or feedback, about visceral body functions such as blood pressure, heart rate, and muscle tension. By using special monitoring devices, they can control these visceral functions consciously.

Suppose you are connected to a monitor that informs you of your heartbeat by means of lights. A red light indicates a fast heart beat, an amber light a normal rate, and a green light a slow rate. When you see the red light flash, you know your heart is beating too fast. You have been informed of a visceral response. This is the biofeedback. According to some researchers, you can be taught to slow down the heart rate by thinking of something pleasant and thus relaxing the body. The green light flashes and your reward is a slower heart rate. In similar experiments, some individuals have learned to control heart rhythm.

Researchers estimate that between 6 and 8 percent of the American population suffers from migraine headaches. Moreover, no effective treatment has been developed that does not have significant side effects and serious risks. One approach to alleviating migraine headaches is biofeedback.* The first clue that biofeedback could be used in this way came when a patient in the voluntary control laboratory of the Menninger Foundation demonstrated that with a 5°C (10°F) rise in hand skin temperature (as a result of vasodilation and increased blood flow) she could spontaneously recover from a migraine headache.

In a study conducted at the Menninger Foundation, subjects suffering from migraine headaches received instructions in the use of a monitor that registers the skin temperature of the right index finger. Subjects were also

* Much of the following discussion of the use of biofeedback for the treatment of migraine headaches is based upon the information provided by Dr. Joseph D. Sargent of the Menninger Foundation, Topeka, Kansas.

given a typewritten sheet containing two sets of phrases. The first set was designed to help them relax the entire body. The second set was designed to bring about an increased flow of blood in the hands. The subjects practiced raising their skin temperature at home for 5 to 15 minutes a day. When skin temperature increased, the monitor emitted a high-pitched sound. In time, the monitor was abandoned.

Once the subjects learned how to vasodilate their blood vessels, the migraine headaches lessened. Since migraine headaches are believed to involve a distension of blood vessels in the head, the shunting of blood from head to hands relieved the distension and thus the pain.

Other experiments have shown that biofeedback can be applied to childbirth. Women were given monitors hooked up to their fingers and arms to measure electrical conductivity of the skin and skeletal muscle tension. Both conductivity and tension increase with nervousness and make labor difficult. Muscle tension was recorded as a sirenlike sound that became louder with nervousness. Skin conductivity was recorded as a crackling noise that also increased with nervousness. The monitors kept the women informed of their nervousness. This was the biofeedback. Having pleasant thoughts reduced the sound levels. The reward was less nervousness. The results of the study indicate that the women needed less medication during labor and labor time itself was shortened.

There is no way to determine where biofeedback will lead. Perhaps the outstanding contribution of biofeedback research has been to demonstrate that the autonomic nervous system is not autonomous. Visceral responses can be controlled. Strong supporters point out that possible applications of biofeedback in medicine are legion. They envision its use in lowering blood pressure in patients with hypertension, altering heart rates and rhythms, relieving pain from migraine headaches, making delivery easier, and controlling anxiety related to a host of illnesses that may be linked to stress. One researcher concludes that "an important trend is beginning to take place in the areas of psychosomatic disorders and medicine. This is the increasing involvement of the patient in his own treatment. The traditional doctor-patient relationship is giving way slowly to a shared responsibility."

MEDITATION

Yoga, which literally means union, is defined as a higher consciousness achieved through a fully rested and relaxed body and a fully awake and relaxed mind. One widely practiced technique for achieving higher consciousness is called **transcendental meditation (TM).** One sits in a comfortable position with the eyes closed and concentrates on a suitable sound or thought.

Research indicates that transcendental meditation can alter physiological responses. Oxygen consumption decreases drastically along with carbon dioxide elimination. Subjects have experienced a reduction in metabolic rate and blood pressure. Researchers have also observed a decrease in heart rate, an increase in the intensity of alpha brain waves, a sharp decrease in the amount of lactic acid in the blood, and an increase in the skin's electrical resistance. These last four responses are characteristic of a highly relaxed state of mind. Alpha waves are found in the EEGs of almost all individuals in a resting, but awake state; they disappear during sleep.

These responses have been called an **integrated response**—essentially, a hypometabolic state due to inactivation of the sympathetic division of the autonomic nervous system. The response is the exact opposite of the fight-or-flight response, which is a hyperactive state of the sympathetic division. The existence of the integrated response suggests that the central nervous system does exert some control over the autonomic nervous system.

STUDY OUTLINE

Somatic Efferent and Autonomic Nervous Systems (p. 362)
1. The autonomic nervous system, or visceral efferent nervous system, regulates visceral activities, that is, activities of smooth muscle, cardiac muscle, and glands.
2. It usually operates without conscious control.
3. It is regulated by centers in the brain, in particular by the cerebral cortex, the hypothalamus, and the medulla oblongata.
4. The somatic efferent nervous system produces conscious movement in skeletal muscles.

Structure of the Autonomic Nervous System (p. 362)
1. The autonomic nervous system consists of visceral efferent neurons organized into nerves, ganglia, and plexuses.
2. It is entirely motor. All autonomic axons are efferent fibers.
3. Efferent neurons are preganglionic (with myelinated axons) and postganglionic (with unmyelinated axons).
4. The autonomic system consists of two principal divisions: sympathetic (thoracolumbar) and parasympathetic (craniosacral).
5. Autonomic ganglia are classified as sympathetic trunk ganglia (on sides of spinal column), prevertebral ganglia (anterior to spinal column), and terminal ganglia (near or inside visceral effectors).

Physiology (p. 367)
1. Autonomic fibers release chemical transmitters at synapses. On the basis of the transmitter produced, these fibers may be classified as cholinergic or adrenergic.
2. Cholinergic fibers release acetylcholine (ACh). Adrenergic fibers produce norepinephrine (NE).
3. Sympathetic responses are widespread and, in general, concerned with energy expenditure. Parasympathetic responses are restricted and are typically concerned with energy restoration and conservation.

Visceral Autonomic Reflexes (p. 369)
1. A visceral autonomic reflex adjusts the activity of a visceral effector.
2. A visceral autonomic reflex arc consists of a receptor, afferent neuron, association neuron, visceral efferent preganglionic neuron, visceral efferent postganglionic neuron, and visceral effector.

Control by Higher Centers (p. 369)
1. The hypothalamus controls and integrates the autonomic nervous system. It is connected to both the sympathetic and the parasympathetic divisions.
2. Biofeedback is a process in which people learn to monitor visceral functions and to control them consciously. It has been used to control heart rate, to alleviate migraine headaches, and to make childbirth easier.
3. Yoga is a higher consciousness achieved through a fully rested and relaxed body and a fully awake and relaxed mind.
4. Transcendental meditation (TM) produces the following physiological responses: decreased oxygen consumption and carbon dioxide elimination, reduced metabolic rate, decrease in heart rate, increase in the intensity of alpha brain waves, a sharp decrease in the amount of lactic acid in the blood, and an increase in the skin's electrical resistance.

REVIEW QUESTIONS

1. What are the principal components of the autonomic nervous system? What is its general function? Why is it called involuntary?
2. What is the principal anatomical difference between the voluntary nervous system and the autonomic nervous system?
3. Relate the role of visceral efferent fibers and visceral effectors to the autonomic nervous system.
4. Distinguish between preganglionic neurons and postganglionic neurons with respect to location and function.
5. What is an autonomic ganglion? Describe the location and function of the three types of autonomic ganglia. Define white and gray rami communicantes.
6. On what basis are the sympathetic and parasympathetic divisions of the autonomic nervous system differentiated anatomically and functionally?
7. Discuss the distinction between cholinergic and adrenergic fibers of the autonomic nervous system.
8. Give examples of the antagonistic effects of the sympathetic and parasympathetic divisions of the autonomic nervous system.
9. Summarize the principal functional differences between the voluntary nervous system and the autonomic nervous system.
10. Give the *sympathetic response* in a fear situation for each of the following body parts: hair follicles, sweat glands, iris of eye, lungs, spleen, adrenal glands, kidneys, urinary bladder, liver, heart, and blood vessels of the viscera, skeletal muscles, and skin.
11. Define a visceral autonomic reflex and give three examples.
12. Describe a complete visceral autonomic reflex in proper sequence.
13. Describe how the hypothalamus controls and integrates the autonomic nervous system.
14. Define biofeedback. Explain how it could be useful.
15. What is transcendental meditation (TM)? How is the integrated response related to the autonomic nervous system?

Each of the four primary tastes is caused by a different response to different chemicals. Certain regions of the tongue react more strongly than others to certain taste sensations. Although the tip of the tongue reacts to all four primary taste sensations, it is highly sensitive to sweet and salty substances. The posterior portion of the tongue is highly sensitive to bitter substances. The lateral edges of the tongue are more sensitive to sour substances (see Figure 24-5).

ADAPTATION AND TASTE THRESHOLDS

Adaptation to taste occurs rapidly. Complete adaptation can occur in 1 to 5 minutes of continuous stimulation. As with odors, receptor adaptation alone cannot account for the speed of complete adaptation. Although taste receptors have a period of rapid adaptation during the first 2 to 3 seconds after contact with food, adaptation is slow thereafter. Adaptation of receptors to smell contributes to taste adaptation, but still does not account for its speed. As with adaptation to odors, adaptation to taste involves a psychological adaptation in the central nervous system.

The threshold for taste varies for each of the primary tastes. The threshold for bitter substances, as measured by quinine, is lowest. This may have a protective function. The threshold for sour substances, as measured by hydrochloric acid, is somewhat higher. The thresholds for salt substances, as measured by sodium chloride, and sweet substances, as measured by sucrose, are about the same and are higher than both bitter and sour substances.

GUSTATORY PATHWAY

The cranial nerves that supply afferent fibers to taste buds are the facial (VII) nerve and trigeminal (V) nerve which supply the anterior two-thirds of the tongue; the glossopharyngeal (IX), which supplies the posterior one-third of the tongue; and the vagus (X), which supplies the epiglottis. Taste impulses are conveyed from the gustatory cells in taste buds along the nerves to the medulla and then to the thalamus. They terminate in the primary gustatory area in the parietal lobe of the cerebral cortex.

VISUAL SENSATIONS

The structures related to vision are the eyeball, the optic (II) nerve, the brain, and a number of accessory structures.

ACCESSORY STRUCTURES OF EYE

Among the **accessory structures** are the eyebrows, eyelids, eyelashes, and the lacrimal apparatus (Figure 17-3). The **eyebrows** form a transverse arch at the junction of the upper eyelid and forehead. Structurally, they resemble the hairy scalp. The skin of the eyebrows is richly supplied with sebaceous glands. The hairs are generally coarse and directed laterally. Deep to the skin of the eyebrows are the fibers of the orbicularis oculi muscles. The eyebrows help to protect the eyeballs from foreign objects, perspiration, and the direct rays of the sun.

The upper and lower **eyelids,** or **palpebrae** (PAL-pe-brē), have several important roles. They shade the eyes during sleep, protect the eyes from excessive light and foreign objects, and spread lubricating secretions over the eyeballs. The upper eyelid is more movable than the lower and contains in its superior region a special levator muscle known as the **levator palpebrae superioris.** The space between the upper and lower eyelids that exposes the eyeball is called the **palpebral fissure.** Its angles are known as the **lateral canthus** (KAN-thus), which is narrower and closer to the temporal bone, and the **medial canthus,** which is broader and nearer the nasal bone. In the medial canthus there is a small reddish elevation, the **lacrimal caruncle** (KAR-ung-kul), containing sebaceous and sudoriferous glands. A whitish material secreted by the caruncle collects in the medial canthus.

From superficial to deep, each eyelid consists of epidermis, dermis, subcutaneous areolar connective tissue, fibers of the orbicularis oculi muscle, a tarsal plate, tarsal glands, and a conjunctiva. The **tarsal plate** is a thick fold of connective tissue that forms much of the inner wall of each eyelid and gives form and support to the eyelids. Embedded in grooves on the deep surface of each tarsal plate is a row of elongated tarsal glands known as **Meibomian** (mī-BŌ-mē-an) **glands.** These are modified sebaceous glands, and their oily secretion helps keep the eyelids from adhering to each other. Infection of the Meibomian glands produces a tumor or cyst on the eyelid called a **chalazion** (ka-LĀ-zē-on). The **conjunctiva** (kon'-junk-TĪ-va) is a thin mucous membrane. It is called the **palpebral conjunctiva** when it lines the inner aspect of the eyelids. It is called the **bulbar** or **ocular conjunctiva** when it is reflected from the eyelids onto the anterior surface of the eyeball to the periphery of the cornea.

Projecting from the border of each eyelid, anterior to the Meibomian glands, is a row of short, thick hairs, the **eyelashes.** In the upper lid, they are long and turn upward; in the lower lid, they are short and turn downward. Sebaceous glands at the base of the hair follicles of the eyelashes called **glands of Zeis** (ZĪS) pour a lubricating fluid into the follicles. Infection of these glands is called a **sty.**

The **lacrimal apparatus** is a term used for a group of structures that manufactures and drains away tears. These structures are the lacrimal glands, the excretory lacrimal ducts, the lacrimal canals, the lacrimal sacs, and the nasolacrimal ducts. A **lacrimal gland** is a compound tubuloacinar gland located at the superior lateral portion of each orbit. Each is about the size and shape of an almond. Leading from the lacrimal glands are 6 to 12 **excretory**

Lacrimal sac
Punctum lacrimale
Lacrimal gland
Excretory lacrimal duct
Eyelashes
Superior and inferior lacrimal canals
Nasolacrimal duct
Inferior nasal concha

Eyebrow
Superior palpebra
Lateral canthus
Inferior palpebra
Lacrimal caruncle
Medial canthus

(a)

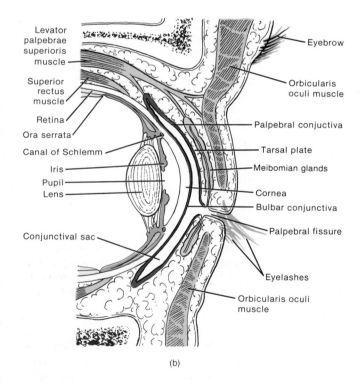

Levator palpebrae superioris muscle
Superior rectus muscle
Retina
Ora serrata
Canal of Schlemm
Iris
Pupil
Lens
Conjunctival sac

Eyebrow
Orbicularis oculi muscle
Palpebral conjuctiva
Tarsal plate
Meibomian glands
Cornea
Bulbar conjunctiva
Palpebral fissure
Eyelashes
Orbicularis oculi muscle

(b)

FIGURE 17-3 Accessory structures of the eye. (a) Anterior view. (b) Sagittal section of the eyelids and anterior portion of the eyeball.

lacrimal ducts that empty lacrimal fluid, or tears, onto the surface of the conjunctiva of the upper lid. From here the lacrimal fluid passes medially and enters two small openings called **puncta lacrimalia** that appear as two small pores, one in each papilla of the eyelid, at the medial canthus of the eye. The lacrimal secretion then passes into two ducts, the **lacrimal canals,** and is next conveyed into the lacrimal sac. The lacrimal canals are located in the lacrimal grooves of the lacrimal bones. The **lacrimal sac** is the superior expanded portion of the **nasolacrimal duct,** a canal that transports the lacrimal secretion into the inferior meatus of the nose.

The **lacrimal secretion** is a watery solution containing salts, some mucus, and a bactericidal enzyme called **lysozyme.** It cleans, lubricates, and moistens the eyeball. After being secreted by the lacrimal glands, it is spread over the surface of the eyeball by the blinking of the eyelids. Usually, 1 ml per day is produced. Normally, the secretion is carried away by evaporation or by passing into the lacrimal canals and then into the nasal cavities as fast as it is produced. If, however, an irritating substance makes contact with the conjunctiva, the lacrimal glands are stimulated to oversecrete. Tears then accumulate more rapidly than they can be carried away. This is a protective mechanism since the tears dilute and wash away the irritating substance. "Watery" eyes also occur when an inflammation of the nasal mucosa, such as a cold, obstructs the nasolacrimal ducts so that drainage of tears is blocked.

(c)

FIGURE 17-3 (*Continued*) Accessory structures of the eye. (c) Surface anatomy of the eye. (© 1982 by Gerard J. Tortora. Courtesy of Lynne Tortora.)

1. **Pupil.** Opening of center of iris of eyeball for light transmission.
2. **Iris.** Circular, pigmented muscular membrane behind cornea.
3. **Sclera.** "White" of eye, a coat of fibrous tissue that covers entire eyeball except for cornea.
4. **Conjunctiva.** Membrane that covers exposed surface of eyeball and lines eyelids.
5. **Palpebrae (eyelids).** Folds of skin and muscle lined by conjunctiva.
6. **Palpebral fissure.** Space between eyelids when they are open.
7. **Medial canthus.** Site of union of upper and lower eyelids near nose.
8. **Lateral canthus.** Site of union of upper and lower eyelids away from nose.
9. **Lacrimal caruncle.** Fleshy, yellowish projection of medial canthus that contains modified sweat and sebaceous glands.
10. **Eyelashes.** Hairs on margins of eyelids, usually arranged in two or three rows.
11. **Eyebrows.** Several rows of hairs superior to upper eyelids.

STRUCTURE OF EYEBALL

The adult **eyeball** measures about 2.5 cm (1 inch) in diameter. Of its total surface area, only the anterior one-sixth is exposed. The remainder is recessed and protected by the orbit into which it fits. Anatomically, the eyeball can be divided into three layers: fibrous tunic, vascular tunic, and retina or nervous tunic (Figure 17-4).

Fibrous Tunic

The **fibrous tunic** is the outer coat of the eyeball. It can be divided into two regions: the posterior portion is the sclera and the anterior portion is the cornea. The **sclera** (SKLE-ra), the "white of the eye," is a white coat of fibrous tissue that covers all the eyeball except the anterior colored portion. The sclera gives shape to the eyeball and protects its inner parts. Its posterior surface is pierced by the optic (II) nerve. The **cornea** (KOR-nē-a) is a nonvascular, nervous, transparent fibrous coat that covers the iris, the colored part of the eye. The cornea's outer

surface is covered by an epithelial layer continuous with the epithelium of the bulbar conjunctiva. At the junction of the sclera and cornea is a venous sinus known as the **canal of Schlemm.** (Surface features of the eye are shown in Figure 17-3c.)

> **CLINICAL APPLICATION**
>
> Each year, about 10,000 **corneal transplants (keratoplasty)** are performed in the United States. Corneal transplants are considered to be the most successful type of transplantation. The surgical procedure is performed under an operating microscope at magnifications of between 6× and 40×. A general or local anesthesia is used. The defective cornea, usually 7 to 8 mm (about 0.3 inch) in diameter, is excised with microcorneal scissors and replaced with a donor cornea of similar diameter. The transplanted cornea is sewn into place with nylon sutures. Patients usually remain hospitalized for 3 to 5 days and are later fitted with hard or soft contact lenses or eyeglasses.

(a)

(b)

FIGURE 17-4 Structure of the eyeball. (a) Diagram of gross structure in transverse section. (b) Photograph of midsagittal section. (Courtesy of C. Yokochi and J. W. Rohen, *Photographic Anatomy of the Human Body,* 2nd ed., 1979, IGAKU-SHOIN, Tokyo, New York.)

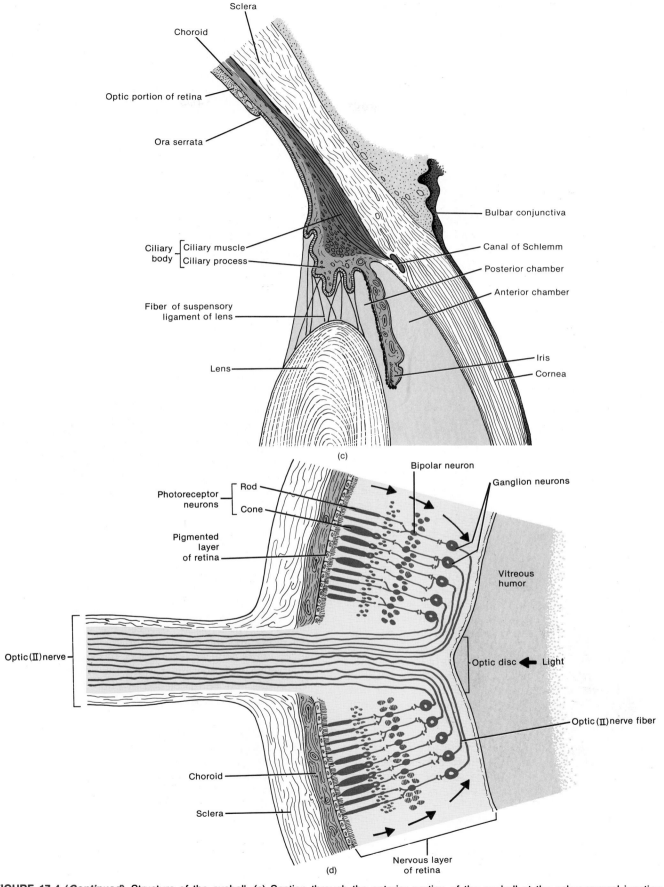

Sclera

Choroid

Optic portion of retina

Ora serrata

Ciliary body { Ciliary muscle / Ciliary process }

Fiber of suspensory ligament of lens

Lens

Bulbar conjunctiva

Canal of Schlemm

Posterior chamber

Anterior chamber

Iris

Cornea

(c)

Bipolar neuron

Ganglion neurons

Photoreceptor neurons { Rod / Cone }

Pigmented layer of retina

Vitreous humor

Optic (II) nerve

Optic disc ← Light

Optic (II) nerve fiber

Choroid

Sclera

Nervous layer of retina

(d)

FIGURE 17-4 (*Continued*) Structure of the eyeball. (c) Section through the anterior portion of the eyeball at the sclerocorneal junction. (d) Diagram of the microscopic structure of the retina, exaggerated for emphasis.

Ganglion neurons

Inner nuclear layer

Outer nuclear layer

Rods and cones

Choroid

Inner plexiform layer

Outer plexiform layer

Pigmented epithelium

(e)

FIGURE 17-4 (Continued) Structure of the eyeball. (e) Photomicrograph of a portion of the retina at a magnification of 300×. The outer nuclear layer contains nuclei of photoreceptor neurons; the outer plexiform layer is where photoreceptor and bipolar neurons synapse; the inner nuclear layer contains the nuclei of bipolar neurons; and the inner plexiform layer is where bipolar and ganglion neurons synapse. (© 1983 by Michael H. Ross. Used by permission.)

Vascular Tunic

The **vascular tunic** is the middle layer of the eyeball and is composed of three portions: the choroid, the ciliary body, and the iris. Collectively, these three structures are called the **uvea** (YOO-vē-a). The **choroid** (KŌ-royd), the posterior portion of the vascular tunic, is a thin, dark brown membrane that lines most of the internal surface of the sclera. It contains numerous blood vessels and a large amount of pigment. The choroid absorbs light rays so they are not reflected back out of the eyeball. Through its blood supply, it nourishes the retina. Like the sclera, it is pierced by the optic (II) nerve at the back of the eyeball.

In the anterior portion of the vascular tunic, the choroid becomes the **ciliary** (SIL-ē-ar'-ē) **body.** It is the thickest portion of the vascular tunic. It extends from the **ora serrata** (Ō-ra ser-RĀ-ta) of the retina (inner tunic) to a point just behind the sclerocorneal junction. The ora serrata is simply the jagged margin of the retina. The ciliary body consists of the ciliary processes and ciliary muscle. The **ciliary processes** consist of protrusions or folds on the internal surface of the ciliary body that secrete aqueous humor. The **ciliary muscle** is a smooth muscle that alters the shape of the lens for near or far vision.

The **iris** is the third portion of the vascular tunic. It consists of circular and radial smooth muscle fibers arranged to form a doughnut-shaped structure. The black hole in the center of the iris is the **pupil,** the area through which light enters the eyeball. The iris is suspended between the cornea and the lens and is attached at its outer margin to the ciliary process. A principal function of the iris is to regulate the amount of light entering the eyeball. When the eye is stimulated by bright light, the circular muscles of the iris contract and decrease the size of the pupil. When the eye must adjust to dim light, the radial muscles of the iris contract and increase the pupil's size.

Retina (Nervous Tunic)

The third and inner coat of the eye, the **retina (nervous tunic),** lies only in the posterior portion of the eye. Its primary function is image formation. It consists of an inner nervous tissue layer and an outer pigmented layer. The retina covers the choroid. At the edge of the ciliary body, it ends in a scalloped border called the ora serrata. This is where the nervous layer or visual portion of the retina ends. The pigmented layer extends anteriorly over the back of the ciliary body and the iris as the nonvisual portion of the retina.

CLINICAL APPLICATION
Detachment of the retina may occur in trauma, such as a blow to the head. The tear in the retina causes blindness in the corresponding field of vision. The actual detachment occurs between the sensory part of the retina and the underlying pigmented layer. Fluid accumulates between these layers, forcing the thin, pliable retina to billow out toward the vitreous humor. The retina may be reattached by photocoagulation by laser beam, cryosurgery, or scleral resection.

The nervous layer contains three zones of neurons. These three zones, named in the order in which they conduct impulses, are the **photoreceptor neurons,** the **bipolar neurons,** and the **ganglion neurons.** The dendrites of the photoreceptor neurons are called rods and cones because of their shapes. They are visual receptors highly specialized for stimulation by light rays. Functionally, rods and cones develop generator potentials. **Rods** are specialized for vision in dim light. They also allow us to discriminate between different shades of dark and light and permit us to see shapes and movement. **Cones** are specialized for color vision and sharpness of vision (*visual acuity*). They are stimulated only by bright light. This is why we cannot see color by moonlight. It is estimated that there are 7 million cones and somewhere between 10 and 20 times as many rods. Cones are most densely concentrated in the **central fovea,** a small depression in the center of the macula lutea. The **macula lutea** (MAK-yoo-la LOO-tē-a), or yellow spot, is in the exact center of the posterior portion of the retina, corresponding to the visual axis of the eye. The fovea is the area of sharpest vision because of the high concentration of cones. Rods are absent from the fovea and macula and increase in density toward the periphery of the retina.

CLINICAL APPLICATION
In a disease called **senile macular degeneration (SMD),** new blood vessels grow over the macula lutea. The effect ranges from distorted vision to blindness. SMD accounts for nearly all new cases of blindness in people over 65. Its cause is unknown. Laser beam treatment has been used effectively in arresting blood vessel proliferation and restoring normal vision in over 80 percent of cases. Treatment, however, must be obtained soon after the onset of symptoms. In order to determine if SMD exists, a simple test can be performed without any assistance or special instruments. Stare, one eye at a time (covering the other with your hand), at any long straight line, such as a door frame. If the line appears bent or twisted, or if a black spot appears, to either eye, a physician should be informed immediately.

When information has passed through the photoreceptor neurons, it is conducted across synapses to the bipolar neurons in the intermediate zone of the nervous layer of the retina. From here it is passed to the ganglion neurons. These cells transmit their signals through optic (II) nerve fibers to the brain in the form of nerve impulses.

The axons of the ganglion neurons extend posteriorly to a small area of the retina called the **optic disc (blind spot).** This region contains openings through which the axons of the ganglion neurons exit as the optic (II) nerve. Since it contains no rods or cones, an image striking it cannot be perceived. Thus it is called the blind spot.

Lens

In addition to the fibrous tunic, vascular tunic, and retina, the eyeball itself contains the lens, just behind the pupil and iris. The **lens** is constructed of numerous layers of protein fibers arranged like the layers of an onion. Normally, the lens is perfectly transparent. It is enclosed by a clear connective tissue capsule and held in position by the **suspensory ligament.** A loss of transparency of the lens is known as a **cataract.**

Interior

The interior of the eyeball is a large cavity divided into two smaller ones by the lens. The **anterior cavity,** the division anterior to the lens, is further divided into the **anterior chamber,** which lies behind the cornea and in front of the iris, and the **posterior chamber,** which lies behind the iris and in front of the suspensory ligament

and lens. The anterior cavity is filled with a watery fluid, similar to cerebrospinal fluid, called the **aqueous humor.** The fluid is believed to be secreted into the posterior chamber by choroid plexuses of the ciliary processes of the ciliary bodies behind the iris. Once the fluid is formed, it permeates the posterior chamber and then passes forward between the iris and the lens, through the pupil, into the anterior chamber. From the anterior chamber, the aqueous humor, which is continually produced, is drained off into the **canal of Schlemm** and then into the blood. The anterior chamber thus serves a function similar to the subarachnoid space around the brain and spinal cord. The canal of Schlemm is analogous to a venous sinus of the dura mater. The pressure in the eye, called **intraocular pressure,** is produced mainly by the aqueous humor. The intraocular pressure keeps the retina smoothly applied to the choroid so the retina will form clear images. Normal intraocular pressure (about 24 mm Hg) is maintained by drainage of the aqueous humor through the canal of Schlemm. Excessive intraocular pressure, called **glaucoma** (glow-KŌ-ma), results in degeneration of the retina and blindness. Besides maintaining intraocular pressure, the aqueous humor is also the principal link between the circulatory system and the lens and cornea. Neither the lens nor the cornea has blood vessels.

The second, and larger, cavity of the eyeball is the **posterior cavity.** It lies between the lens and the retina and contains a jellylike substance called the **vitreous humor.** This substance contributes to intraocular pressure, helps to prevent the eyeball from collapsing, and holds the retina flush against the internal portions of the eyeball. The vitreous humor, unlike the aqueous humor, does not undergo constant replacement. It is formed during embryonic life and is not replaced thereafter.

PHYSIOLOGY OF VISION

Before light can reach the rods and cones of the retina, it must pass through the cornea, aqueous humor, pupil, lens, and vitreous humor. For vision to occur, light reaching the rods and cones must form an image on the retina. The resulting nerve impulses must then be conducted to the visual areas of the cerebral cortex. In discussing the physiology of vision, let us first consider retinal image formation.

Retinal Image Formation

The formation of an image on the retina requires four basic processes, all concerned with focusing light rays: (1) refraction of light rays, (2) accommodation of the lens, (3) constriction of the pupil, and (4) convergence of the eyes. Accommodation and pupil size are functions of the smooth muscle cells of the ciliary muscle and the dilator and sphincter muscles of the iris. They are termed *intrinsic eye muscles* since they are inside the eyeball. Convergence is a function of the voluntary muscles attached to the outside of the eyeball called the *extrinsic eye muscles.*

● *Refraction of Light Rays* When light rays traveling through a transparent medium (such as air) pass into a

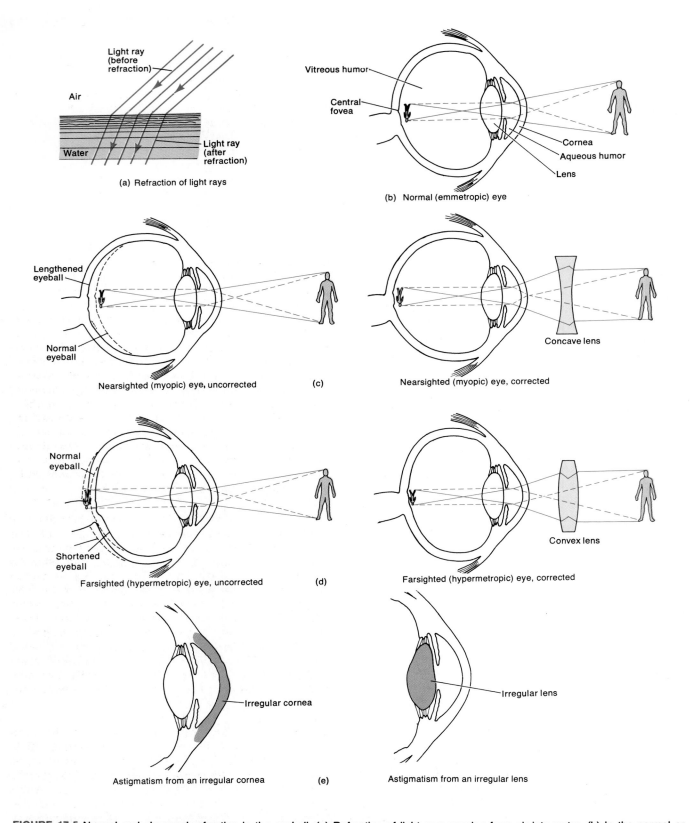

FIGURE 17-5 Normal and abnormal refraction in the eyeball. (a) Refraction of light rays passing from air into water. **(b)** In the normal or emmetropic eye, light rays from an object are bent sufficiently by the four refracting media and converged on the central fovea. A clear image is formed. **(c)** In the nearsighted or myopic eye, the image is focused in front of the retina. The condition may result from an elongated eyeball or thickened lens. Correction is by use of a concave lens which diverges entering light rays so that they have to travel further through the eyeball and are focused exactly on the retina. **(d)** In the farsighted or hypermetropic eye, the image is focused behind the retina. The condition results from a shortened eyeball or a thin lens. Correction is by a convex lens which converges entering light rays so that they focus exactly on the retina. **(e)** An astigmatism is an irregular curvature of the cornea (shown on left) or lens (shown on right). As a result, horizontal and vertical rays are focused at two different points on the retina, and the image is not focused on the area of sharpest vision of the retina. This results in blurred or distorted vision. Suitable glasses correct the refraction of an astigmatic eye.

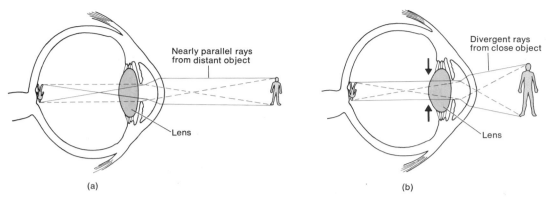

FIGURE 17-6 Accommodation. (a) For objects 6 m (20 ft) or more away. (b) For objects nearer than 6 m.

second transparent medium with a different density (such as water), they bend at the surface of the two media. This is **refraction** (Figure 17-5a). The eye has four such media of refraction: cornea, aqueous humor, lens, and vitreous humor. Light rays entering the eye from the air are refracted at the following points: (1) the anterior surface of the cornea as they pass from the lighter air into the denser cornea, (2) the posterior surface of the cornea as they pass into the less dense aqueous humor, (3) the anterior surface of the lens as they pass from the aqueous humor into the denser lens, and (4) the posterior surface of the lens as they pass from the lens into the less dense vitreous humor.

The degree of refraction that takes place at each surface in the eye is very precise. When an object is 6 m (20 ft) or more away from the viewer, the light rays reflected from the object are nearly parallel to one another. The parallel rays must be bent sufficiently to fall exactly on the central fovea, where vision is sharpest. Light rays that are reflected from near objects are divergent rather than parallel. As a result, they must be refracted toward each other to a greater extent. This change in refraction is brought about by the lens of the eye (Figure 17-5b and 17-6).

● *Accommodation of the Lens*   If the surface of a lens curves outward, as in a convex lens, the lens will refract incoming rays toward each other so they eventually intersect. The greater the curve, the more acutely it bends the rays toward each other. Conversely, when the surface of a lens curves inward, as in a concave lens, the rays bend away from each other. The lens of the eye is biconvex. Furthermore, it has the unique ability to change the focusing power of the eye by becoming moderately curved at one moment and greatly curved the next. When the eye is focusing on a close object, the lens curves greatly in order to bend the rays toward the central fovea. This increase in the curvature of the lens is called **accommodation** (Figure 17-6). In near vision, the ciliary muscle contracts, pulling the ciliary process and choroid forward toward the lens. This action releases the tension on the lens and suspensory ligament. Because of its elasticity, the lens shortens, thickens, and bulges. In far vision, the ciliary muscle is relaxed and the lens is flatter. With aging, the lens loses elasticity and, therefore, its ability to accommodate.

CLINICAL APPLICATION

The normal eye, known as an **emmetropic** (em'-e-TROP-ik) **eye,** can sufficiently refract light rays from an object 6 m (20 ft) away to focus a clear image on the retina. Many individuals, however, do not have this ability because of abnormalities related to improper refraction. Among these abnormalities are **myopia** (mī-Ō-pē-a) or nearsightedness, **hypermetropia** (hī'-per-mē-TRO-pē-a) or farsightedness, and **astigmatism** (a-STIG-ma-tizm) or irregularities in the surface of the lens or cornea. The conditions are illustrated and explained in Figure 17-5c–e. The inability to focus on nearby objects due to loss of elasticity of the lens with age is called **presbyopia** (prez-bē-OP-ē-a).

● *Constriction of Pupil*   The circular muscle fibers of the iris also assume a function in the formation of clear retinal images. Part of the accommodation mechanism consists of the contraction of the dilator and sphincter muscles of the iris to constrict the pupil. **Constriction of the pupil** means narrowing the diameter of the hole through which light enters the eye. This action occurs simultaneously with accommodation of the lens and prevents light rays from entering the eye through the periphery of the lens. Light rays entering at the periphery would not be brought to focus on the retina and would result in blurred vision. The pupil, as noted earlier, also constricts in bright light to protect the retina from sudden or intense stimulation.

● *Convergence*   Because of the alignment of the orbits in the skull, many animals see a set of objects off to the left through one eye and an entirely different set off to the right through the other. This characteristic doubles their field of vision and allows them to detect predators behind them. In humans, both eyes focus on only one set of objects—a characteristic called **single binocular vision.**

Single binocular vision occurs when light rays from an object are directed toward corresponding points on the two retinas. When we stare straight ahead at a distant object, the incoming light rays are aimed directly at both pupils and are refracted to identical spots on the retinas of both eyes. But as we move closer to the object, our eyes must rotate medially for the light rays from the object to hit the same points on both retinas. The term **convergence** refers to this medial movement of the two eyeballs so they are both directed toward the object being viewed. The nearer the object, the greater the degree of convergence necessary to maintain single binocular vision. Convergence is brought about by the coordinated action of the extrinsic eye muscles.

● *Inverted Image* Images are focused upside down on the retina. They also undergo mirror reversal, that is, light reflected from the right side of an object hits the left side of the retina and vice versa. Note in Figure 17-5b that reflected light from the top of the object crosses light from the bottom of the object and strikes the retina below the central fovea. Reflected light from the bottom of the object crosses light from the top of the object and strikes the retina above the central fovea. The reason we do not see an inverted world is that the brain learns early in life to coordinate visual images with the exact locations of objects. The brain stores memories of reaching and touching objects and automatically turns visual images right-side-up and right-side-around.

Stimulation of Photoreceptors

● *Excitation of Rods* After an image is formed on the retina by refraction, accommodation, constriction of the pupil, and convergence, light impulses must be converted into nerve impulses. The initial step is the development of generator potentials by rods and cones. Rods contain a reddish-purple photosensitive pigment called **rhodopsin (visual purple).** This substance consists of the protein **scotopsin** plus **retinene (visual yellow),** a derivative of vitamin A. When light rays strike a rod, rhodopsin rapidly breaks down into its components, neither of which can absorb light. The chemical breakdown stimulates generator potential development by the rods (Figure 17-7).

Rhodopsin is highly light sensitive—even the light rays from the moon or a candle will break down some of it and thereby allow us to see. The rods, then, are specialized for night vision. However, they are of only limited help for daylight vision. In bright light, the rhodopsin is broken down faster than it can be manufactured. In dim light, production is able to keep pace with a slower rate of breakdown. These characteristics of rhodopsin are responsible for the experience of having to adjust to a dark room after walking in from the sunshine. The normal period of adjustment is the time it takes for the completely dissociated rhodopsin to reform. **Night blindness,** which is also referred to as **nyctalopia** (nik'-ta-LŌ-pē-a), is the lack of normal night vision following the adjustment period. It is most often caused by vitamin A deficiency.

● *Excitation of Cones* Cones are the receptors for bright light and color. The photosensitive pigments in cones are almost the same as those in rods. Both contain retinene, but the protein portion in cones is **photopsin,** which is slightly different from the scotopsin in rods. Unlike rhodopsin, the photosensitive pigments of the cones require bright light for their breakdown and they reform quickly. It is believed that there are three types of cones and that each contains a different combination of retinene

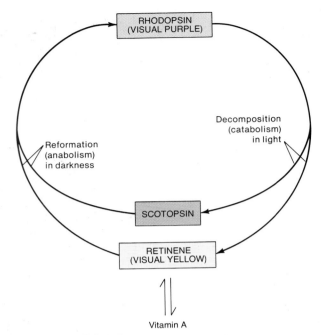

FIGURE 17-7 Rhodopsin cycle.

and photopsin. Each combination has a different maximum absorption of light of a different wavelength so that each responds best to light of a given color. One type of cone responds best to red light, the second to green light, and the third to blue light. Just as an artist can obtain almost any color by mixing colors, it is believed that cones can perceive any color by differential stimulation (Figure 17-8). Stimulation of a cone by two or more colors may produce any combination of colors.

CLINICAL APPLICATION

If a single group of color receptive cones is missing from the retina, an individual cannot distinguish some colors from others and is said to be **color-blind.** The most common type is **red-green color blindness** in which the cones best receptive to red light and to green light are missing. As a result, the person cannot distinguish between red and green. Color blindness is an inherited condition that affects males far more frequently than females. The inheritance of the condition is discussed in Chapter 29 and illustrated in Figure 29-22.

● *Light and Dark Adaptation* If an individual is exposed to bright light for an extended period of time, a considerable amount of the photosensitive pigment in both rods and cones is broken down into retinene and opsin (scotopsin or photopsin). In addition, most of the retinene in both rods and cones is converted to vitamin A. As a result, the concentration of the photosensitive pigment in both is reduced considerably and the sensitivity of the eye to light is greatly reduced. This is called **light adaptation.**

Conversely, if an individual is exposed to darkness for an extended period of time, a considerable amount of retinene and opsins in the rods and cones is converted

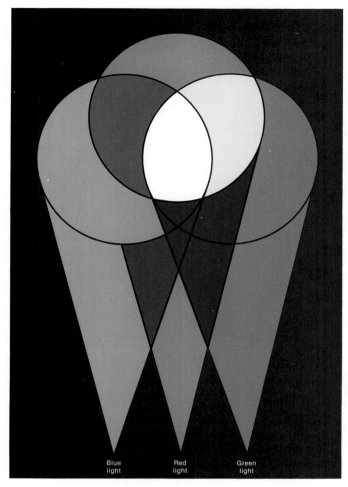

Blue
light   Red   Green
light   light

FIGURE 17-8 Differential stimulation of cones. Stimulation of a cone by two or more colors can produce any combination of colors.

into the photosensitive pigments. Also, large amounts of vitamin A are converted into retinene, which is then converted into photosensitive pigments upon combination with opsins. As a result, the concentration of photosensitive pigments in rods and cones is increased considerably and the sensitivity of the eye to light is greatly increased. This is called **dark adaptation.**

Visual Pathway

From the rods and cones, information is transmitted through bipolar neurons to ganglion cells. Ganglion cells initiate nerve impulses. The cell bodies of the ganglion cells lie in the retina, and their axons leave the eyeball via the **optic (II) nerve** (Figure 17-9). The axons pass through the **optic chiasma** (kī-AZ-ma), a crossing point of the optic (II) nerves. Some fibers cross to the opposite side. Others remain uncrossed. On passing through the optic chiasma, the fibers, now part of the **optic tract,** enter the brain and terminate in the lateral geniculate nucleus of the thalamus. Here the fibers synapse with third-order neurons whose axons pass to the visual areas located in the occipital lobes of the cerebral cortex.

Analysis of the afferent pathway to the brain reveals that the visual field of each eye is divided into two regions:

the **nasal (medial) half** and the **temporal (lateral) half.** For each eye, light rays from an object in the nasal half of the visual field fall on the temporal half of the retina. Light rays from an object in the temporal half of the visual field fall on the nasal half of the retina (Figure 17-9). Also, light rays from objects at the top of the visual field of each eye fall on the inferior portion of the retina, and light rays from objects at the bottom of the visual field fall on the superior portion of the retina. In the optic chiasma, nerve fibers from the nasal halves of the retinas cross and continue on to the lateral geniculate nuclei of the thalamus; nerve fibers from the temporal halves of the retinas do not cross, but continue directly on to the lateral geniculate nuclei. As a result, the primary visual area of the cerebral cortex of the right occipital lobe interprets visual sensations from the left side of an object via impulses from the temporal half of the retina of the right eye and the nasal half of the retina of the left eye. The primary visual area of the cerebral cortex of the left occipital lobe interprets visual sensations from the right side of an object via impulses from the nasal half of the right eye and the temporal half of the left eye.

CLINICAL APPLICATION

A **scotoma** (skō-TŌ-ma) or **blind spot** in the field of vision, other than the normal blind spot (optic disc), may indicate a brain tumor along one of the afferent pathways. For instance, a symptom of a tumor in the right optic tract might be an inability to see the left side of a normal field of vision without moving the eyeball.

AUDITORY SENSATIONS AND EQUILIBRIUM

In addition to containing receptors for sound waves, the ear also contains receptors for equilibrium. Anatomically, the ear is divided into three principal regions: the external (outer) ear, the middle ear, and the internal (inner) ear.

EXTERNAL OR OUTER EAR

The **external (outer) ear** is structurally designed to collect sound waves and direct them inward (Figure 17-10a,b). It consists of the pinna, the external auditory canal, and the tympanic membrane, also called the eardrum.

The **pinna (auricle)** is a trumpet-shaped flap of elastic cartilage covered by thick skin. The rim of the pinna is called the **helix;** the inferior portion is the **lobule.** The pinna is attached to the head by ligaments and muscles. Additional surface anatomy features of the external ear are shown in Figure 17-10a.

The **external auditory canal** or **meatus** is a tube about 2.5 cm (1 inch) in length that lies in the external auditory meatus of the temporal bone. It leads from the pinna to the eardrum. The walls of the canal consist of bone lined with cartilage that is continuous with the cartilage of the pinna. The cartilage in the external auditory canal is covered with thin, highly sensitive skin. Near the exterior opening, the canal contains a few hairs and specialized sebaceous glands called **ceruminous** (se-ROO-mi-nus)

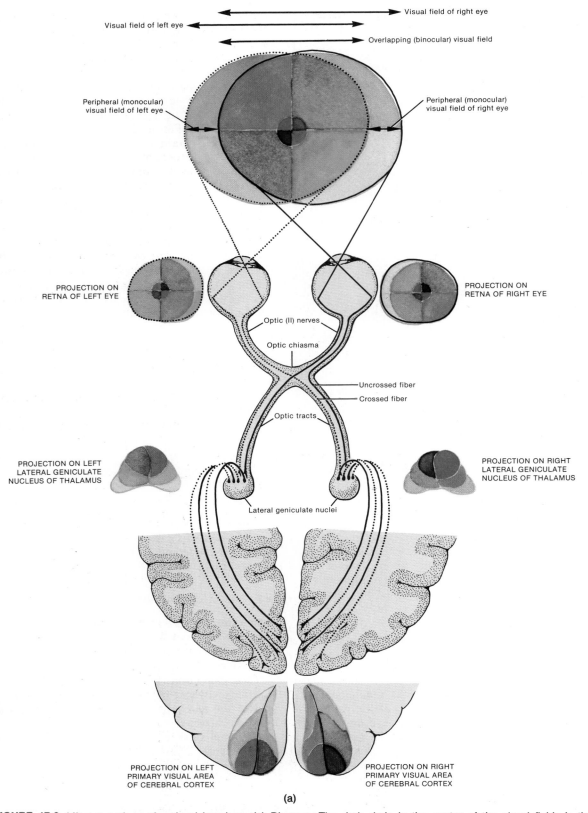

FIGURE 17-9 Afferent pathway for visual impulses. (a) Diagram. The dark circle in the center of the visual fields is the macula lutea. The center of the macula lutea is the central fovea, the area of sharpest vision.

glands, which secrete **cerumen** (earwax). The combination of hairs and cerumen (se-ROO-min) helps to prevent foreign objects from entering the ear.

The **tympanic** (tim-PAN-ik) **membrane,** or **eardrum,** is a thin, semitransparent partition of fibrous connective tissue between the external auditory canal and the middle ear. Its external surface is concave and covered with skin. Its internal surface is convex and covered with a mucous membrane.

CLINICAL APPLICATION
Rupture of the tympanic membrane may result from foreign bodies, pressure, or infection. Severe bleeding and escape of cerebrospinal fluid through a ruptured tympanic membrane may occur following a severe blow to the head and indicate skull fracture.

MIDDLE EAR

The **middle ear,** also called the **tympanic cavity,** is a small, epithelial-lined, air-filled cavity hollowed out of the temporal bone (Figure 17-10b–d). The cavity is separated from the external ear by the eardrum and from the internal ear by a thin bony partition that contains two small openings: the oval window and the round window.

The posterior wall of the cavity communicates with the mastoid air cells of the temporal bone through a chamber called the **tympanic antrum.** This anatomical fact explains why a middle ear infection may spread to the temporal bone, causing mastoiditis, or even to the brain.

The anterior wall of the cavity contains an opening that leads directly into the **auditory tube.** The auditory tube connects the middle ear with the nasopharynx of the throat. Through this passageway, infections may travel from the throat and nose to the ear. The function of the tube is to equalize air pressure on both sides of the tympanic membrane. Abrupt changes in external or internal air pressure might otherwise cause the eardrum to rupture. During swallowing and yawning, the tube opens to allow atmospheric air to enter or leave the middle ear until the internal pressure equals the external pressure. Any sudden pressure changes against the eardrum may be equalized by deliberately swallowing.

Extending across the middle ear are three exceedingly small bones called **auditory ossicles** (OS-si-kuls). The bones, named for their shape, are the malleus, incus, and stapes, commonly called the hammer, anvil, and stirrup, respectively. They are connected by synovial joints. The "handle" of the **malleus** is attached to the internal surface of the tympanic membrane. Its head articulates with the body of the incus. The **incus** is the intermediate bone in the series and articulates with the head of the stapes. The base or footplate of the **stapes** fits into a small opening in the thin bony partition between the middle and inner ear. The opening is called the **fenestra vestibuli** (fe-NES-tra ves-TIB-yoo-lī), or **oval window.** Directly below the oval window is another opening, the **fenestra cochlea** (fe-NES-tra KŌK-lē-a), or **round window.** This opening is enclosed by a membrane called the secondary tympanic membrane. The auditory ossicles are attached to the tympanic cavity by means of ligaments.

(b)

FIGURE 17-9 (Continued) Afferent pathway for visual impulses. (b) Photograph as seen from the ventral aspect of the brain. (Courtesy of N. Gluhbegovic and T. H. Williams, *The Human Brain: A Photographic Guide,* Harper & Row, New York, 1980.)

(a)

FIGURE 17-10 Structure of the auditory apparatus. (a) Surface anatomy of the right ear seen in lateral view. (© 1982 by Gerard J. Tortora. Courtesy of James Borghesi.) (b) Divisions of the right ear into external, middle, and internal portions seen in a frontal section through the right side of the skull. (c) Details of the middle ear and bony labyrinth of the internal ear. (d) Auditory ossicles of the middle ear.

1. **Pinna.** Portion of external ear not contained in head, also called auricle or trumpet.
2. **Tragus.** Cartilaginous projection anterior to external opening to ear.
3. **Antitragus.** Cartilaginous projection opposite tragus.
4. **Concha.** Hollow of auricle.
5. **Helix.** Superior and posterior free margin of auricle.
6. **Antihelix.** Semicircular ridge posterior and superior to concha.
7. **Triangular fossa.** Depression in superior portion of antihelix.
8. **Lobule.** Inferior portion of pinna devoid of cartilage.
9. **External auditory meatus.** Canal extending from external ear to eardrum.

(b)

(c)

(d)

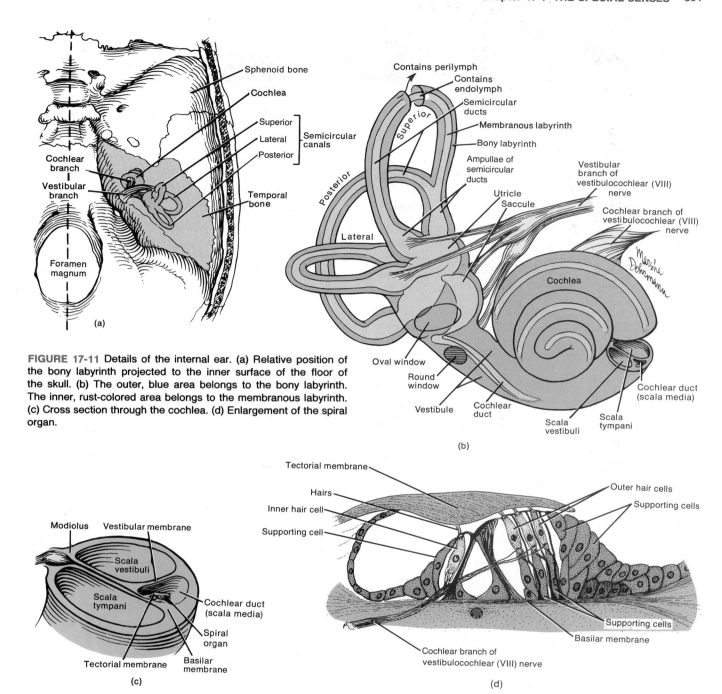

FIGURE 17-11 Details of the internal ear. (a) Relative position of the bony labyrinth projected to the inner surface of the floor of the skull. (b) The outer, blue area belongs to the bony labyrinth. The inner, rust-colored area belongs to the membranous labyrinth. (c) Cross section through the cochlea. (d) Enlargement of the spiral organ.

INTERNAL OR INNER EAR

The **internal (inner) ear** is also called the **labyrinth** (LAB-i-rinth) because of its complicated series of canals (Figure 17-11). Structurally, it consists of two main divisions: a bony labyrinth and a membranous labyrinth that fits in the bony labyrinth. The **bony labyrinth** is a series of cavities in the petrous portion of the temporal bone. It can be divided into three areas named on the basis of shape: the vestibule, cochlea, and semicircular canals. The bony labyrinth is lined with periosteum and contains a fluid called **perilymph.** This fluid surrounds the **membranous labyrinth,** a series of sacs and tubes lying inside and having the same general form as the bony labyrinth. The membra-

nous labyrinth is lined with epithelium and contains a fluid called **endolymph.**

The **vestibule** constitutes the oval central portion of the bony labyrinth. The membranous labyrinth in the vestibule consists of two sacs called the **utricle** and **saccule.** These sacs are connected to each other by a small duct.

Projecting upward and posteriorly from the vestibule are the three bony **semicircular canals.** Each is arranged at approximately right angles to the other two. On the basis of their positions, they are called the superior, posterior, and lateral canals. One end of each canal enlarges into a swelling called the **ampulla** (am-POOL-la). The portions of the membranous labyrinth that lie inside the bony semicircular canals are called the **semicircular ducts (membranous semicircular canals).** These structures are almost identical in shape to the semicircular canals and communicate with the utricle of the vestibule.

Lying in front of the vestibule is the **cochlea** (KŌK-lē-a), so designated because of its resemblance to a snail's shell. The cochlea consists of a bony spiral canal that makes about 2¾ turns around a central bony core called the **modiolus.** A cross section through the cochlea shows the canal is divided into three separate channels by partitions that together have the shape of the letter Y. The stem of the Y is a bony shelf that protrudes into the canal; the wings of the Y are composed mainly of membranous labyrinth. The channel above the bony partition is the **scala vestibuli;** the channel below is the **scala tympani.** The cochlea adjoins the wall of the vestibule, into which the scala vestibuli opens. The scala tympani terminates at the round window. The perilymph of the vestibule is continuous with that of the scala vestibuli. The third channel (between the wings of the Y) is the membranous labyrinth: the **cochlear duct (scala media).** The cochlear duct is separated from the scala vestibuli by the **vestibular membrane.** It is separated from the scala tympani by the **basilar membrane.**

Resting on the basilar membrane is the **spiral organ,** the organ of hearing. The spiral organ is a series of epithelial cells on the inner surface of the basilar membrane. It consists of a number of supporting cells and hair cells,

which are the receptors for auditory sensations. The inner hair cells are medially placed in a single row and extend the entire length of the cochlea. The outer hair cells are arranged in several rows throughout the cochlea. The hair cells have long hairlike processes at their free ends that extend into the endolymph of the cochlear duct. The basal ends of the hair cells are in contact with fibers of the cochlear branch of the vestibulocochlear (VIII) nerve. Projecting over and in contact with the hair cells of the spiral organ is the **tectorial membrane,** a delicate and flexible gelatinous membrane.

SOUND WAVES

Sound waves result from the alternate compression and decompression of air molecules. They originate from a vibrating object, much the same way that waves travel over the surface of water. The sounds heard most acutely by human ears are those from sources that vibrate at frequencies between 1,000 and 4,000 cycles per second (Hz). The entire range extends from 20 to 20,000 Hz.

The frequency of vibration of the source of the sound is directly related to pitch. The greater the vibration, the higher the pitch. Also, the greater the force of the vibration, the louder the sound. Sound intensities (loudness) are usually measured in units called **decibels (dB).** Decibels measure the relative intensity of sounds; 1 dB represents an increase in intensity of 1.26 times. In the

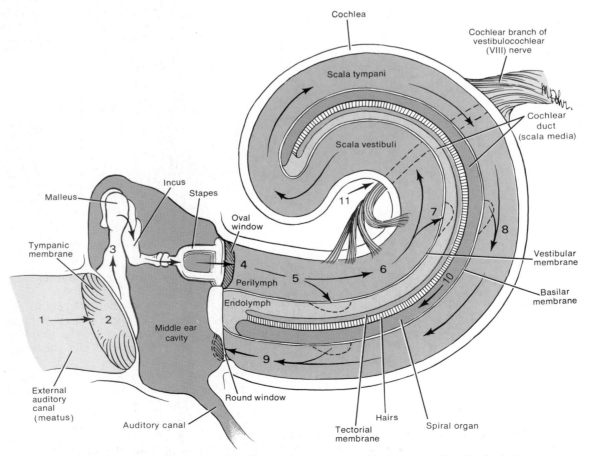

FIGURE 17-12 Physiology of hearing. The numbers correspond to the events listed in the text.

- Describe the relationship between the endocrine system and the nervous system in maintaining homeostasis.
- Discuss the functions of the endocrine system in maintaining homeostasis.
- Define an endocrine gland and list the endocrine glands of the body.
- Distinguish between water-soluble and lipid-soluble hormones.
- Explain the mechanism of hormonal action.
- Identify the role of prostaglandins in hormonal action.
- Describe the control of hormonal secretions via feedback cycles and explain several examples.
- Describe the structural and functional division of the pituitary gland into the adenohypophysis and the neurohypophysis.
- Discuss how the pituitary gland and hypothalamus are structurally and functionally related.
- List the hormones of the adenohypophysis, their principal actions, and their associated hypothalamic regulating factors.
- Describe the release of hormones stored in the neurohypophysis and their principal actions.
- Discuss the symptoms of pituitary dwarfism, giantism, acromegaly, and diabetes insipidus as pituitary gland disorders.
- Describe the location and histology of the thyroid gland.
- Explain the synthesis, storage, and release of thyroid hormones.
- Discuss the principal actions and control of thyroid gland hormones.
- Discuss the symptoms of cretinism, myxedema, exophthalmic goiter, and simple goiter as thyroid gland disorders.
- Describe the location and histology of the parathyroid glands.
- Explain the principal actions and control of the parathyroid hormone.
- Discuss the symptoms of tetany and osteitis fibrosa cystica as parathyroid gland disorders.
- Describe the location and histology of the adrenal (suprarenal) glands and explain their subdivision into cortical and medullary portions.
- Distinguish the effects of adrenal cortical mineralocorticoids, glucocorticoids, and gonadocorticoids on physiological activities, and explain how the hormones are controlled.
- Discuss the symptoms of aldosteronism, Addison's disease, Cushing's syndrome, and adrenogenital syndrome as adrenal cortical disorders.
- Identify the function of the adrenal medullary secretions as supplements of sympathetic responses.
- Discuss the symptoms of pheochromocytomas as an adrenal medullary disorder.
- Describe the location and histology of the pancreas.
- Compare the principal actions of the pancreatic hormones and describe how they are controlled.
- Discuss the symptoms of diabetes mellitus and hyperinsulinism as endocrine disorders of the pancreas.
- Explain why the ovaries and testes are classified as endocrine glands.
- Describe the location and histology of the pineal gland and the possible functions of its hormones.
- Describe the location and histology of the thymus gland and the function of its hormones in immunity.
- Explain why portions of the gastrointestinal tract, placenta, kidneys, and skin are considered as endocrine structures.
- Define the general adaptation syndrome and compare homeostatic responses and stress responses.
- Identify the body reactions during the alarm, resistance, and exhaustion stages of stress.
- Define medical terminology associated with the endocrine system.

18 The Endocrine System

The nervous system controls homeostasis through electrical impulses delivered over neurons. The body's other control system, the **endocrine system,** affects bodily activities by releasing chemical messengers, called hormones, into the bloodstream. Whereas the nervous system causes muscles to contract and glands to secrete, the endocrine system brings about changes in the metabolic activities of body tissues. The release of nerve impulses is much more rapid than that of hormones. Also, the effects of nervous system stimulation are generally brief compared to the effects of endocrine stimulation.

Obviously, the body could not function if the two great control systems were to pull in opposite directions. The nervous and endocrine systems coordinate their activities like an interlocking supersystem. Certain parts of the nervous system stimulate or inhibit the release of hormones, and hormones, in turn, are quite capable of stimulating or inhibiting the flow of nerve impulses.

Although the effects of hormones are many and varied, their actions can be categorized into four broad areas:

1. They help to control the internal environment by regulating its chemical composition and volume.

2. They respond to marked changes in environmental conditions to help the body cope with emergency demands such as infection, trauma, emotional stress, dehydration, starvation, hemorrhage, and temperature extremes.

3. They assume a role in the smooth, sequential integration of growth and development.

4. They contribute to the basic processes of reproduction, including gamete production, fertilization, nourishment of the embryo and fetus, delivery, and nourishment of the newborn.

ENDOCRINE GLANDS

The endocrine glands make up the **endocrine system.** The body contains two kinds of glands: exocrine and endocrine. **Exocrine glands** secrete their products onto a free surface or into ducts. The ducts carry the secretions into body cavities, into the lumens of various organs, or to the body's surface. Exocrine glands include sweat, sebaceous, mucous, and digestive glands. **Endocrine** (*endo* = within) **glands** by contrast, secrete their products (hormones) into the extracellular space around the secretory cells, rather than into ducts. The secretion then passes into capillaries to be transported in the blood. The endocrine glands of the body include the pituitary (hypophysis), thyroid, parathyroids, adrenals (suprarenals), pancreas, ovaries, testes, pineal (epiphysis cerebri), thymus, kidneys, stomach, small intestine, and placenta. (As you will see, some of these glands are mixed, that is, both endocrine and exocrine. The pancreas is an example). The location of many organs of the endocrine system is illustrated in Figure 18-1.

CHEMISTRY OF HORMONES

The secretions of endocrine glands are called **hormones** (*hormone* = set in motion). Although hormones are chemically diverse, they can be grouped into two principal classes on the basis of their solubility: water-soluble and lipid-soluble. **Water-soluble** hormones include protein hormones or derivatives of proteins (glycoproteins) or amino acids (catecholamines). Examples are hormones of the pituitary, parathyroids, pancreas, and adrenal medulla. These hormones are synthesized in the manner described for protein synthesis in Chapter 3. **Lipid-soluble** hormones include steroid hormones, lipids synthesized from cholesterol, and thyroid hormones. Examples are hormones of the thyroid, adrenal cortex, ovaries, and testes. These hormones are produced in mitochondria and agranular endoplasmic reticulum.

As you will learn in Chapter 29, all tissues and organs of the body are derived from three embryonic tissues called the endoderm, mesoderm, and ectoderm. Endocrine glands that originate from the endoderm (e.g., parathyroids and pancreas) or ectoderm (e.g., adrenal medulla and pituitary) generally secrete water-soluble hormones. Endocrine glands of mesodermal origin (e.g., adrenal cortex, ovaries, testes) generally secrete lipid-soluble hormones.

The one thing all hormones have in common—whether water-soluble or lipid-soluble—is the function of maintaining homeostasis by changing the physiological activities of cells.

MECHANISM OF HORMONAL ACTION

The general effects of most hormones are fairly well known. However, the manner in which specific hormones affect different cells of the body is just starting to become clear. The cells that respond to the effects of a hormone are called **target cells.** It appears as though hormones exert their effects on target cells in two quite different ways—one involves interaction with plasma membrane receptors and the other involves activation of genes.

INTERACTION WITH PLASMA MEMBRANE RECEPTORS

One mechanism of hormone action involves an interaction at the cell surface between the hormone and a receptor site on the plasma membrane. A hormone released from an endocrine gland circulates in the blood, reaches a target cell, and brings a specific message to that cell. The hormone is called the **first messenger.** To give the cell its message, the hormone must attach to a specific receptor site (integral protein) on the plasma membrane. This attachment can alter cell function by two different pathways.

In one pathway there is an increase in the synthesis of **cyclic adenosine-3′, 5′-monophosphate (cyclic AMP)** (see Figure 2-16). Cyclic AMP is synthesized from ATP, the main energy-storing chemical in cells. This synthesis requires an enzyme present in the plasma membrane called *adenyl cyclase.* When the first messenger attaches to its receptors, there is an activation of adenyl cyclase, resulting in the conversion of ATP into cyclic AMP in the cell (Figure 18-2a). Cyclic AMP then diffuses throughout the cell, binds to an intracellular receptor, and acts as a **second messenger,** altering cell function according to the message indicated by the hormone. Cyclic AMP can activate the appropriate cellular enzymes to get a specific biochemical response. Calcium ions may sometimes be involved as second messengers along with cyclic normal height is reached. This condition is called **pituitary dwarfism.** Other organs of the body also fail to grow, and the pituitary dwarf is childlike in many physical respects. Treatment requires administration of GH during childhood before the epiphyseal plates close.

Hypersecretion of GH during childhood results in **giantism,** an abnormal increase in the length of long bones. Hypersecretion during adulthood is called **acromegaly** (ak'-rō-MEG-a-lē). Acromegaly cannot produce further lengthening of the long bones because the epiphyseal plates are already closed. Instead, the bones of the hands, feet, cheeks, and jaws thicken. Other tissues also grow. The eyelids, lips, tongue, and nose enlarge and the skin thickens and furrows, especially on the forehead and soles of the feet.

Thyroid-Stimulating Hormone (TSH)

Thyroid-stimulating hormone (TSH) is also called **thyrotropin** (thī-rō-TRŌ-pin). It stimulates the synthesis and secretion of the hormones produced by the thyroid gland. Secretion is controlled by a regulating factor produced by the hypothalamus called **thyrotropin releasing factor (TRF).** Release of TRF depends on blood levels of thyroxine and the body's metabolic rate among other factors, and operates according to a negative feedback system.

Adrenocorticotropic Hormone (ACTH)

Adrenocorticotropic hormone (ACTH) is also called **adrenocorticotropin** (ad-rē'-nō-kor'-ti-kō-TRŌ-pin). Its tropic function is to control the production and secretion of certain adrenal cortex hormones. Secretion of ACTH is governed by a regulating factor produced by the hypothalamus called **corticotropin releasing factor (CRF).** Release of CRF depends on a number of stimuli and hormones and operates as a negative feedback system.

Follicle-Stimulating Hormone (FSH)

In the female, **follicle-stimulating hormone (FSH)** is transported from the adenohypophysis by the blood to the ovaries, where it initiates the development of ova each month. FSH also stimulates cells in the ovaries to secrete estrogens, or female sex hormones. In the male, FSH stimulates the testes to initiate sperm production. Secretion of FSH is subject to a regulating factor produced by the hypothalamus called **gonadotropin releasing factor (GnRF).** GnRF is released in response to estrogens and possibly progesterone in the female and testosterone in the male and involves a negative feedback system.

Luteinizing Hormone (LH)

In the female, **luteinizing** (LOO-tē-in'-īz-ing) **hormone (LH),** together with estrogens, stimulates the ovary to release an ovum (ovulation) and prepares the uterus for implantation of a fertilized ovum. It also stimulates formation of the corpus luteum in the ovary, which secretes progesterone (another female sex hormone) and readies

the mammary glands for milk secretion. In the male, LH stimulates the interstitial endocrinocytes in the testes to develop and secrete large amounts of testosterone. Secretion of LH, like that of FSH, is controlled by GnRF. Release of GnRF is governed by a negative feedback system involving estrogens, progesterone, and testosterone. Recent evidence suggests that the placenta, formed during pregnancy, may be an extrahypothalamic source of GnRF. The GnRF it secretes has been designated as **placental luteotropic releasing factor (pLRF).** Its role in pregnancy will be discussed in Chapter 29.

Prolactin (PRL)

Prolactin (PRL) or the **lactogenic hormone,** together with other hormones, initiates and maintains milk secretion by the mammary glands. The actual ejection of milk by the mammary glands is controlled by a hormone stored in the posterior lobe called oxytocin. Together, milk secretion and ejection are referred to as lactation. PRL acts directly on tissues. By itself, it has little effect; it requires preparation by estrogens, progesterone, corticosteroids, growth hormone, thyroxine, and insulin. When the mammary glands have been primed by these hormones, PRL brings about milk secretion.

PRL has both an inhibitory and an excitatory negative control system. During menstrual cycles, **prolactin inhibiting factor (PIF),** a regulating factor from the hypothalamus, inhibits the release of PRL from the anterior pituitary. As the levels of estrogens and progesterone fall during the late secretory phase of the menstrual cycle, the secretion of PIF diminishes and the blood level of PRL rises. However, its rising level does not last long enough to have much effect on the breasts, which may be tender because of the presence of PRL just before menstruation. As the menstrual cycle starts up again and the level of estrogens again rises, PIF is again secreted and the PRL level drops.

PRL levels rise during pregnancy. Apparently a regulating factor from the hypothalamus, called **prolactin releasing factor (PRF),** stimulates PRL secretion after long periods of inhibition. PRL levels fall after delivery and rise again during breast feeding. Sucking reduces hypothalamic secretion of PIF. Mechanical stimulation of the nonlactating female breast brings about increased secretion of PRL; stimulation of the male breast does not.

Melanocyte-Stimulating Hormone (MSH)

The **melanocyte-stimulating hormone (MSH)** increases skin pigmentation by stimulating the dispersion of melanin granules in melanocytes. In the absence of the hormone, the skin may be pallid. An excess of MSH may cause darkening of the skin. Secretion of MSH is stimulated by a hypothalamic regulating factor called **melanocyte-stimulating hormone releasing factor (MRF).** It is inhibited by a **melanocyte-stimulating hormone inhibiting factor (MIF).**

A summary of anterior pituitary hormones, their principal actions, and associated hypothalamic regulating factors is presented in Exhibit 18-2.

EXHIBIT 18-2

ANTERIOR PITUITARY HORMONES, PRINCIPAL ACTIONS, AND ASSOCIATED HYPOTHALAMIC REGULATING FACTORS

HORMONE	PRINCIPAL ACTIONS	ASSOCIATED HYPOTHALAMIC REGULATING FACTORS
Growth hormone (GH)	Growth of body cells; protein anabolism.	Growth hormone releasing factor (GHRF); Growth hormone inhibiting factor (GHIF).
Thyroid-stimulating hormone (TSH)	Controls secretion of hormones by thyroid gland.	Thyrotropin releasing factor (TRF).
Adrenocorticotropic hormone (ACTH)	Controls secretion of some hormones by adrenal cortex.	Corticotropin releasing factor (CRF).
Follicle-stimulating hormone (FSH)	In female, initiates development of ova and induces ovarian secretion of estrogens. In male, stimulates testes to produce sperm.	Gonadotropin releasing factor (GnRF).
Luteinizing hormone (LH)	In female, together with estrogens, stimulates ovulation and formation of progesterone-producing corpus luteum, prepares uterus for implantation, and readies mammary glands to secrete milk. In male, stimulates interstitial cells in testes to develop and produce testosterone.	Gonadotropin releasing factor (GnRF).
Prolactin (PRL)	Together with other hormones, initiates and maintains milk secretion by the mammary glands.	Prolactin inhibiting factor (PIF); Prolactin releasing factor (PRF).
Melanocyte-stimulating hormone (MSH)	Stimulates dispersion of melanin granules in melanocytes.	Melanocyte-stimulating hormone releasing factor (MRF); Melanocyte-stimulating hormone inhibiting factor (MIF).

NEUROHYPOPHYSIS

In a strict sense, the posterior lobe, or **neurohypophysis,** is not an endocrine gland, since it does not synthesize hormones. The posterior lobe consists of cells called **pituicytes** (pi-TOO-i-sītz), which are similar in appearance to the neuroglia of the nervous system. It also contains axon terminations of secretory neurons of the hypothalamus (Figure 18-5). Such neurons are called **neurosecretory cells.** The cell bodies of the neurons originate in nuclei in the hypothalamus. The fibers project from the hypothalamus, form the **hypothalamic-hypophyseal tract,** and terminate on blood capillaries in the neurohypophysis. The cell bodies of the neurosecretory cells produce two hormones: **oxytocin (OT)** and **antidiuretic hormone (ADH).** OT is produced primarily in the paraventricular nucleus and ADH is synthesized primarily in the supraoptic nucleus.

Following their production, the hormones are transported in the neuron fibers by a carrier protein called *neurophysin* into the neurohypophysis and stored in the axon terminals. Later, when the hypothalamus is properly stimulated, it sends impulses over the neurosecretory cells. The impulses cause the release of the hormones from the axon terminals into the blood.

The blood supply to the neurohypophysis is from the *inferior hypophyseal arteries,* derived from the internal carotid arteries. In the neurohypophysis, the inferior hypophyseal arteries form a plexus of capillaries called the *plexus of the infundibular process.* From this plexus hormones stored in the neurohypophysis pass into the *posterior hypophyseal veins* for distribution to tissue cells.

Oxytocin (OT)

Oxytocin (ok'-sē-TŌ-sin) or **OT** stimulates the contraction of the smooth muscle cells in the pregnant uterus and the contractile cells around the ducts of the mammary glands. It is released in large quantities just prior to giving birth (Figure 18-6). When labor begins, the cervix of the uterus is distended. This distension initiates afferent impulses to the neurosecretory cells of the paraventricular nucleus in the hypothalamus that stimulate the secretion of OT. The OT is transported by neurophysin to the neurohypophysis. The impulses also cause the neurohypophysis to release OT into the blood. It is then carried by the blood to the uterus to reinforce uterine contractions. As the contractions become more forceful, the resulting afferent impulses stimulate the synthesis of more OT. Then, a positive feedback cycle is established. Note also that the afferent part of the cycle is neural, whereas the efferent part is hormonal.

OT affects milk ejection. Milk formed by the glandular cells of the breasts is stored until the baby begins active sucking. For about 30 seconds to 1 minute after nursing begins, the baby receives no milk. During this latent period, nerve impulses from the nipple are transmitted to the hypothalamus. According to a mechanism similar to that involved in forming and releasing OT for uterine muscle contractions, OT flows from the neurohypophysis

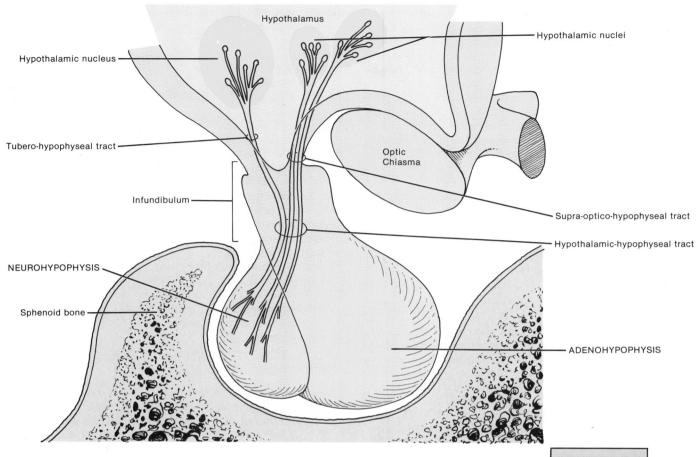

FIGURE 18-5 Hypothalamic-hypophyseal tract.

via the blood to the mammary glands, where it stimulates smooth muscle cells to contract and eject milk.

OT is inhibited by progesterone, but works together with estrogens. Estrogens and progesterone inhibit, via PIF, the release of PRL. PRL accumulates in the pituitary gland during gestation. The sucking stimulation that produces the release of OT also inhibits the release of PIF.

Antidiuretic Hormone (ADH)

An **antidiuretic** is any chemical substance that prevents excessive urine production. The principal physiological activity of **antidiuretic hormone (ADH)** is its effect on urine volume. ADH causes the kidneys to remove water from newly forming urine and return it to the bloodstream, thus decreasing urine volume. This involves an increase in the permeability of plasma membranes of kidney water-reabsorbing cells so that more water passes from urine into kidney cells. In the absence of ADH, urine output may be increased tenfold.

ADH can also raise blood pressure by bringing about constriction of arterioles. This effect is noted if there is a severe loss of blood volume due to hemorrhage.

The amount of ADH normally secreted varies with the body's needs. When the body is dehydrated, the concentration of water in the blood falls below normal limits as the salt-to-water ratio changes. Receptors in the hypothalamus called osmoreceptors detect the low water con-

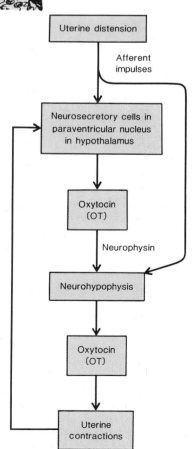

FIGURE 18-6 Regulation of the secretion of oxytocin (OT) during labor.

centration in the plasma and stimulate the neurosecretory cells of the supraoptic nucleus in the hypothalamus to produce ADH. The hormone is transported by neurophysin to the neurohypophysis. It is then released into the bloodstream and transported to the kidneys. The kidneys respond by decreasing urine output, and water is conserved. ADH also decreases the rate at which perspiration is produced during dehydration. By contrast, if the blood contains a higher than normal water concentration, the receptors detect the increase and hormone secretion is stopped. The kidneys can then release large quantities of urine, and the volume of body fluid is brought down to normal.

Secretion of ADH can also be altered by a number of other conditions. Pain, stress, trauma, anxiety, acetylcholine (ACh), nicotine, and drugs, such as morphine, tranquilizers, and some anesthetics stimulate secretion of the hormone. Alcohol inhibits secretion and thereby increases urine output. This may be why thirst is one symptom of a hangover.

CLINICAL APPLICATION

The principal abnormality associated with dysfunction of the neurohypophysis is **diabetes insipidus** (in-SIP-i-dus). Diabetes means overflow and insipidus means tasteless. This disorder should not be confused with diabetes mellitus (*meli* = honey), a disorder of the pancreas characterized by sugar in the urine. Diabetes insipidus is the result of a hyposecretion of ADH, usually caused by damage to the neurohypophysis or the supraoptic nucleus in the hypothalamus. Symptoms include excretion of large amounts of urine and subsequent thirst. Diabetes insipidus is treated by administering ADH.

A summary of posterior pituitary hormones, their principal actions, and control of secretion is presented in Exhibit 18-3.

THYROID

The **thyroid gland** is located just below the larynx. The right and left **lateral lobes** lie one on either side of the trachea. The lobes are connected by a mass of tissue called an **isthmus** (IS-mus) that lies in front of the trachea, just below the cricoid cartilage (Figure 18-7). The **pyramidal lobe,** when present, extends upward from the isthmus. The gland weighs about 25 g (almost 1 oz) and has a rich blood supply, receiving about 80 to 120 ml of blood per minute.

Histologically, the thyroid gland is composed of spherical sacs called **thyroid follicles** (Figure 18-8). The walls of each follicle consist of cells that reach the surface of the lumen of the follicle (**follicular cells**) and cells that do not reach the lumen (**parafollicular** or **C cells**). When the cells are inactive, they tend to be cuboidal, but when actively secreting hormones, they become more columnar. The follicular cells manufacture **thyroxine** (thī-ROK-sēn) or T_4, since it contains four atoms of iodine, and **triiodothyronine** (trī-ī'-od-ō-THĪ-rō-nēn) or T_3, since it contains three atoms of iodine. Together these hormones are referred to as the **thyroid hormones.** Thyroxine is normally secreted in greater quantity than triiodothyronine, but triiodothyronine is 3 to 4 times more potent. Moreover, in peripheral tissues, especially the liver and lungs, about one-third of triiodothyronine is converted to thyroxine. Both hormones are similar functionally. The parafollicular cells produce **calcitonin** (kal-si-TŌ-nin) or **CT.**

FORMATION, STORAGE, AND RELEASE OF THYROID HORMONES

One of the thyroid gland's unique features is its ability to store hormones and release them in a steady flow over a long period of time. The first stage in the synthesis of thyroid hormones is the active transport of iodine ions from the blood into the follicle cells. Under normal conditions, the iodide concentration in the cells is about 40 times that of the blood; during periods of maximal activity, the concentration can increase to over 300 times that of the blood. In the follicle cells, the iodide is oxidized to iodine. Through a series of enzymatically controlled reactions, the iodine combines with the amino acid tyrosine to form the thyroid hormones. This combination occurs within a large glycoprotein molecule, called **thyroglobulin (TGB)**, which is secreted by the follicle cells into the follicle.

The thyroid hormones are an integral part of TGB and, as such, are stored (sometimes for months) until needed. The entire complex in the follicle, consisting of TGB and the stored hormones, constitutes the **thyroid**

EXHIBIT 18-3

SUMMARY OF POSTERIOR PITUITARY HORMONES, PRINCIPAL ACTIONS, AND CONTROL OF SECRETION

HORMONE	PRINCIPAL ACTIONS	CONTROL OF SECRETION
Oxytocin (OT)	Stimulates contraction of smooth muscle cells of pregnant uterus during labor and stimulates contraction of contractile cells of mammary glands for milk ejection.	Neurosecretory cells of hypothalamus secrete OT in response to uterine distension and stimulation of nipples.
Antidiuretic hormone (ADH)	Principal effect is to decrease urine volume; also raise blood pressure by constricting arteries during severe hemorrhage.	Neurosecretory cells of hypothalamus secrete ADH in response to low water concentration of the blood, pain, stress, trauma, anxiety, acetylcholine (ACh), nicotine, morphine, and tranquilizers; alcohol inhibits secretion.

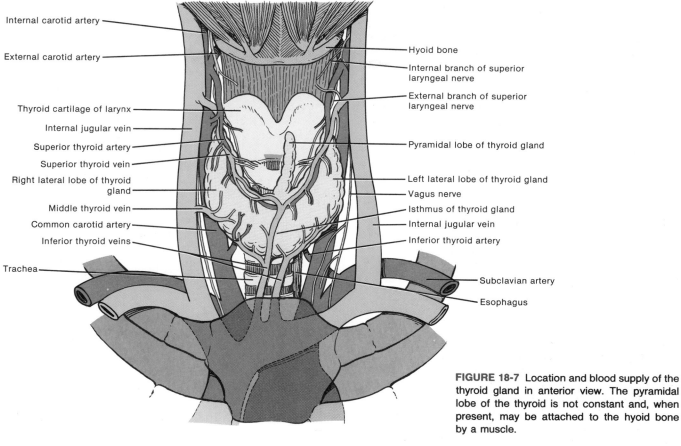

Internal carotid artery

External carotid artery

Thyroid cartilage of larynx

Internal jugular vein

Superior thyroid artery

Superior thyroid vein

Right lateral lobe of thyroid gland

Middle thyroid vein

Common carotid artery

Inferior thyroid veins

Trachea

Hyoid bone

Internal branch of superior laryngeal nerve

External branch of superior laryngeal nerve

Pyramidal lobe of thyroid gland

Left lateral lobe of thyroid gland

Vagus nerve

Isthmus of thyroid gland

Internal jugular vein

Inferior thyroid artery

Subclavian artery

Esophagus

FIGURE 18-7 Location and blood supply of the thyroid gland in anterior view. The pyramidal lobe of the thyroid is not constant and, when present, may be attached to the hyoid bone by a muscle.

Epithelium of follicle

Thyroid follicle

Thyroglobulin

FIGURE 18-8 Histology of the thyroid gland. Photomicrograph at a magnification of 230×. (© 1983 by Michael H. Ross. Used by permission.)

colloid. Before the thyroid hormones can be released into the blood, droplets of colloid are absorbed into the follicular cells by pinocytosis and the thyroid hormones are split from TGB. Following their diffusion into the blood, most of the thyroid hormones combine with plasma proteins, mainly **thyroxine-binding globulin (TBG).** Thyroid hormones combined with plasma proteins are referred to as **protein-bound iodine (PBI).**

FUNCTION AND CONTROL OF THYROID HORMONES

The thyroid hormones have three principal effects on the body: (1) regulation of metabolism, (2) regulation of growth and development, and (3) regulation of the activity of the nervous system. With respect to the regulation of metabolism, the thyroid hormones stimulate virtually all aspects of carbohydrate and lipid catabolism in most

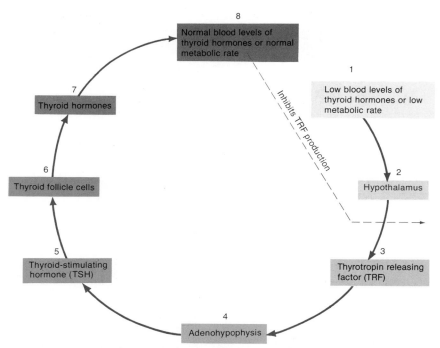

FIGURE 18-9 Regulation of the secretion of thyroid hormones.

cells of the body. They also increase the rate of protein synthesis. Since their overall effect is to increase catabolism, they increase the basal metabolic rate. The energy produced raises body temperature as heat is given off and this phenomenon is called the **calorigenic effect.**

The thyroid hormones help to regulate tissue growth and development, especially in children. They work with GH to accelerate body growth, particularly the growth of nervous tissue. Deficiency of the hormones during fetal development can result in fewer and smaller neurons, defective myelination of axons, and mental retardation. During the early years of life, deficiency of the hormones results in small stature and poor development of certain organs.

Finally, the thyroid hormones increase the reactivity of the nervous system. This results in increased blood flow, increased and more forceful heartbeats, increased blood pressure, increased motility of the gastrointestinal tract, and increased nervousness.

The secretion of thyroid hormones is stimulated by several factors (Figure 18-9). If thyroid hormone levels in the blood fall below normal or the metabolic rate decreases, chemical sensors in the hypothalamus detect the change in blood chemistry and stimulate the hypothalamus to secrete a regulating factor called thyrotropin releasing factor (TRF). TRF stimulates the adenohypophysis to secrete thyroid-stimulating hormone (TSH). Then TSH stimulates the thyroid to release thyroid hormones until the metabolic rate returns to normal. Conditions that increase the body's need for energy—a cold environment, high altitude, pregnancy—also trigger this feedback system and increase the secretion of thyroid hormones.

Thyroid activity can be inhibited by a number of other factors. When large amounts of certain sex hormones (estrogens and androgens) are circulating in the blood, for example, TSH secretion diminishes. Aging slows down

the activities of most glands, and thyroid production may decrease.

eight times more frequently in females than in males. Its symptoms are abolished by the administration of thyroxine.

Hypersecretion of thyroid hormones gives rise to **exophthalmic** (ek'-sof-THAL-mik) **goiter** (GOY-ter). This disease, like myxedema, is also more frequent in females. One of its primary symptoms is an enlarged thyroid, called a **goiter,** which may be two to three times its original size. Two other symptoms are an edema behind the eye, which causes the eye to protrude **(exophthalmos),** and an abnormally high metabolic rate. The high metabolic rate produces a range of effects that are generally opposite to those of myxedema—increased pulse, high body temperature, and moist, flushed skin. The person loses weight and is usually full of "nervous" energy. The thyroid hormones also increase the responsiveness of the nervous system, causing the person to become irritable and exhibit tremors of the extended fingers. Hyperthyroidism is usually treated by administering drugs that suppress thyroid hormone synthesis or by surgically removing part of the gland.

Goiter is a symptom of many thyroid disorders. It may also occur if the gland does not receive enough iodine to produce sufficient thyroxine for the body's needs. The follicular cells then enlarge in a futile attempt to produce more thyroid hormones and they secrete large quantities of thyroglobulin (TGB). This condition is called **simple goiter.** Simple goiter is most often caused by a lower than average amount of iodine in the diet. It may also develop if iodine intake is not increased during certain conditions that put a high demand on the body for thyroxine, such as frequent exposure to cold and high fat and protein diets.

CALCITONIN (CT)

The hormone produced by the parafollicular cells of the thyroid gland is **calcitonin** (kal-si-TŌ-nin) or **CT.** It is involved in the homeostasis of blood calcium level. CT lowers the amount of calcium and phosphate in the blood by inhibiting bone breakdown and accelerating the absorption of calcium by the bones. It appears to exert its effect in lowering calcium and phosphate blood levels by inhibiting osteoclastic action and the parathyroid hormone (PTH). If CT is administered to a person with a normal level of blood calcium, it causes **hypocalcemia** (low blood calcium level). Hypocalcemia is also a complication of magnesium deficiency. If CT is given to a person with **hypercalcemia** (high blood calcium level), the level returns to normal. It is suspected that the blood calcium level directly controls the secretion of CT according to a negative feedback system that does not involve the pituitary gland (Figure 18-10).

A summary of hormones produced by the thyroid gland, their principal actions, and control of secretion is presented in Exhibit 18-4.

PARATHYROIDS

Typically embedded on the posterior surfaces of the lateral lobes of the thyroid are small, round masses of tissue called the **parathyroid glands.** Usually, two parathyroids, superior and inferior, are attached to each lateral thyroid lobe (Figure 18-11). They measure about 3 to 6 mm (0.1 to 0.3 inch) in length, 2 to 5 mm (0.07 to 0.2 inch) in width, and 0.5 to 2 mm (0.02 to 0.07 inch) in thickness.

Histologically, the parathyroids contain two kinds of epithelial cells (Figure 18-12). The more numerous cells, called **principal** or **chief cells,** are believed to be the major synthesizer of **parathyroid hormone (PTH).** Some researchers believe that the other kind of cell, called an **oxyphil cell,** synthesizes a reserve capacity of hormone.

PARATHYROID HORMONE (PTH)

Parathyroid hormone (PTH) or **parathormone** controls the homeostasis of ions in the blood, especially calcium and phosphate ions. If adequate amounts of vitamin D are present, PTH increases the rate of calcium, some magnesium, and phosphate absorption from the gastrointestinal tract into the blood. It also leads to the activation of vitamin D. Moreover, PTH increases the number and activity of osteoclasts, or bone-destroying cells. As a result, bone tissue is broken down and calcium and phosphate are released into the blood. PTH produces two

EXHIBIT 18-4
SUMMARY OF THYROID GLAND HORMONES, PRINCIPAL ACTIONS, AND CONTROL OF SECRETION

HORMONE	PRINCIPAL ACTIONS	CONTROL OF SECRETION
Thyroid hormones **Thyroxine (T4)**	Regulates metabolism, growth and development, and activity of nervous system.	Thyrotropin releasing factor (TRF) is released from hypothalamus in response to low thyroid hormone levels, low metabolic rate, cold, pregnancy, and high altitudes; TRF secretion is inhibited in response to high thyroid hormone levels, high metabolic rate, high levels of estrogens and androgens, and aging.
Triiodothyronine (T3)	Same as above.	Same as above.
Calcitonin (CT)	Lowers blood levels of calcium by accelerating calcium absorption by bones.	High blood calcium levels stimulate secretion; low levels inhibit secretion.

FIGURE 18-10 Regulation of the secretion of the parathyroid hormone (PTH) and calcitonin (CT).

changes in the kidneys. It increases the rate at which the kidneys remove calcium and magnesium from urine that is being formed and returns them to the blood. It accelerates the transportation of phosphate from the blood into the urine for elimination. More phosphate is lost through the urine than is gained from the bones.

The overall effect of PTH with respect to ions, then, is to decrease blood phosphate level and increase blood calcium level. As far as blood calcium level is concerned, PTH and CT are antagonists.

PTH secretion is not controlled by the pituitary gland. When the calcium level of the blood falls, more PTH is released (see Figure 18-10). Conversely, when the calcium level of the blood rises, less PTH (and more CT) is secreted. This is another example of a negative feedback control system that does not involve the pituitary gland.

CLINICAL APPLICATION

A normal amount of calcium in the extracellular fluid is necessary to maintain the resting state of neurons. A deficiency of calcium caused by *hypoparathyroidism*

---

causes neurons to depolarize without the usual stimulus. As a result, nervous impulses increase and result in muscle twitches, spasms, and convulsions. This condition is called **tetany.** The effects of hypocalcemic tetany are observed in the *Trousseau* (troo-SŌ) and *Chvostek* (VOS-tek) *signs.* Trousseau sign is observed when the binding of a blood pressure cuff around the arm produces contraction of the fingers and inability to open the hand. The Chvostek sign is a contracture of the facial muscles elicited by tapping the facial nerves at the angle of the jaw. Hypoparathyroidism results from surgical removal of the parathyroids or from parathyroid damage caused by parathyroid disease, infection, hemorrhage, or mechanical injury.

Hyperparathyroidism causes demineralization of bone. This condition is called **osteitis fibrosa cystica** because the areas of destroyed bone tissue are replaced by cavities that fill with fibrous tissue. The bones thus become deformed and are highly susceptible to fracture. Hyperparathyroidism is usually caused by a tumor in the parathyroids.

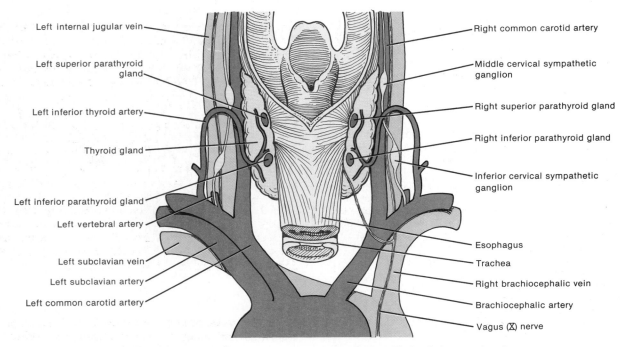

FIGURE 18-11 Location and blood supply of the parathyroid glands in posterior view.

Acini

Islet of
Langerhans

FIGURE 18-18 Histology of the pancreas. Photomicrograph at a magnification of 600×. (© 1983 by Michael H. Ross. Used by permission.)

18-19). It accelerates the transport of glucose from the blood into cells, especially skeletal muscle cells. It also accelerates the conversion of glucose into glycogen (glycogenesis). Insulin also decreases glycogenolysis and gluconeogenesis, stimulates the conversion of glucose or other nutrients into fatty acids (lipogenesis), and helps stimulate protein synthesis.

The regulation of insulin secretion, like that of glucagon secretion, is directly determined by the level of sugar in the blood and is based on a negative feedback system.

Other hormones can indirectly affect insulin production, however. For instance, GH raises blood glucose level, and the rise in glucose level triggers insulin secretion. ACTH, by stimulating the secretion of glucocorticoids, brings about hyperglycemia and also indirectly stimulates the release of insulin. Gastrointestinal hormones like stomach and intestinal gastrin, secretin, cholecystokinin (CCK), and gastric inhibitory peptide (GIP) also stimulate insulin secretion. GHIF (somatostatin) inhibits the secretion of insulin.

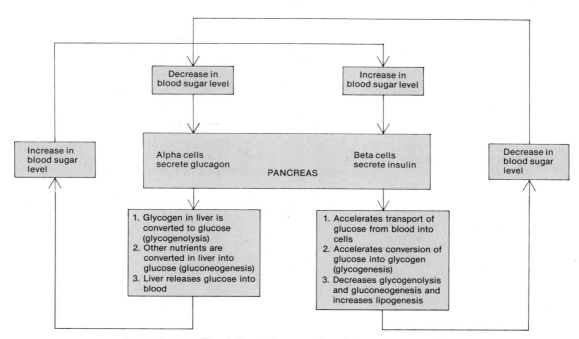

FIGURE 18-19 Regulation of the secretion of glucagon and insulin.

CLINICAL APPLICATION

An absolute or relative deficiency of insulin results in a number of clinical symptoms referred to as **diabetes mellitus** (MEL-it-us). Research has led many to conclude that diabetes mellitus is not a single hereditary disease, but rather a heterogeneous group of hereditary diseases, all of which ultimately lead to an elevation of glucose in the blood (hyperglycemia) and excretion of glucose in the urine as hyperglycemia increases. Diabetes mellitus is also characterized by the three "polys": an inability to reabsorb water, resulting in increased urine production (*polyuria*); excessive thirst (*polydipsia*); and excessive eating (*polyphagia*).

Two major types of diabetes have been distinguished: the maturity-onset type and the juvenile-onset type. *Maturity-onset diabetes* is much more common, representing more than 90 percent of all cases. It most often occurs in people who are over 40 and overweight. Clinical symptoms are mild, and the high glucose levels in the blood can usually be controlled by diet alone. Many maturity-onset diabetics have a sufficiency or even a surplus of insulin in the blood. For these individuals, diabetes arises not from a shortage of insulin, but probably from defects in the molecular machinery that mediates the action of insulin on its target cells. Maturity-onset diabetes is therefore called *non-insulin-dependent diabetes*.

Juvenile-onset diabetes develops in people younger than age 20. Its onset is more abrupt and it is generally more severe than maturity-onset diabetes. It is usually preceded by a virus infection like measles or mumps. The disease is characterized by a marked decline in the number of beta cells in the pancreas leading to insufficient production of insulin and an elevation of glucose in the blood. It is known as *insulin-dependent diabetes*. The deficiency of insulin accelerates the breakdown of the body's reserve of fat resulting in the production of organic acids called ketones. This causes a form of acidosis called *ketosis*, which lowers the pH of the blood and can result in death. The catabolism of stored fats and proteins also causes weight loss. As lipids are transported by the blood from storage depots to hungry cells, lipid particles are deposited on the walls of blood vessels. The deposition leads to atherosclerosis and a multitude of cardiovascular problems.

Some cases of juvenile-onset diabetes may be explained primarily on a genetic basis and others on an environmental basis. Still other cases appear to rise from a complex interaction between genetic background and environment.

Hyperinsulinism is much rarer than hyposecretion and is generally the result of a malignant tumor in an islet. The principal symptom is a decreased blood glucose level, which stimulates the secretion of epinephrine, glucagon, and GH. As a consequence, anxiety, sweating, tremor, increased heart rate, and weakness occur. Moreover, brain cells do not have enough glucose to function efficiently. This condition leads to mental disorientation, convulsions, unconsciousness, shock, and eventual death as the vital centers in the medulla are affected.

A summary of the hormones produced by the pancreas, their principal actions, and control of secretion is presented in Exhibit 18-7.

OVARIES AND TESTES

The female gonads, called the **ovaries,** are paired oval bodies located in the pelvic cavity. The ovaries produce female sex hormones called **estrogens** and **progesterone.** These hormones are responsible for the development and maintenance of the female sexual characteristics. Along with the gonadotropic hormones of the pituitary, the sex hormones also regulate the menstrual cycle, maintain pregnancy, and prepare the mammary glands for lactation. The ovaries (and placenta) also produce a hormone called **relaxin** which relaxes the symphysis pubis and helps dilate the uterine cervix toward the end of pregnancy.

The male has two oval glands, called **testes,** that lie in the scrotum. The testes produce **testosterone,** the primary male sex hormone, that stimulates the development and maintenance of the male sexual characteristics. The testes also produce the hormone **inhibin** that inhibits secretion of FSH. The detailed structure of the ovaries and testes and the specific roles of gonadotropic hormones and sex hormones will be discussed in Chapter 28.

EXHIBIT 18-7

SUMMARY OF HORMONES PRODUCED BY THE PANCREAS, PRINCIPAL ACTIONS, AND CONTROL OF SECRETION

HORMONE	PRINCIPAL ACTIONS	CONTROL OF SECRETION
Glucagon	Raises blood sugar level by accelerating conversion of glucogen into glucose in liver (glycogenolysis) and conversion of other nutrients into glucose in the liver (gluconeogenesis) and releasing glucose into blood.	Decreased blood level of glucose, exercise, and largely protein meals stimulate glucagon secretion; GHIF (somatostatin) inhibits glucagon secretion.
Insulin	Lowers blood sugar level by accelerating transport of glucose into cells, converting glucose into glycogen (glycogenesis), and decreasing glycogenolysis and gluconeogenesis; also increases lipogenesis and stimulates protein synthesis.	Increased blood level of glucose, GH, ACTH, and gastrointestinal hormones stimulate insulin secretion; GHIF (somatostatin) inhibits insulin secretion.

PINEAL (EPIPHYSIS CEREBRI)

The endocrine gland attached to the roof of the third ventricle is known as the **pineal gland** (because of its resemblance to a pine cone) or **epiphysis cerebri** (see Figure 18-1). The gland is about 5 to 8 mm (0.2 to 0.3 inch) long and 5 mm wide. It weighs about 0.2 g. It is covered by a capsule formed by the pia mater and consists of masses of **neuroglial cells** and parenchymal secretory cells called **pinealocytes.** Around the cells are scattered preganglionic sympathetic fibers. The pineal gland starts to calcify at about the time of puberty. Such calcium deposits are referred to as **brain sand.** Contrary to a once widely held belief, there is no evidence that the pineal atrophies with age and that the presence of brain sand is an indication of atrophy. In fact, the presence of brain sand may indicate increased secretory activity.

Although many anatomical facts concerning the pineal gland have been known for years, its physiology is still somewhat obscure. One hormone secreted by the pineal gland is **melatonin,** which appears to inhibit reproductive activities by inhibiting gonadotropic hormones. Some evidence also exists that the pineal secretes a second hormone called **adrenoglomerulotropin** (a-drē'-nō-glō-mer'-yoo-lō-TRŌ-pin). This hormone may stimulate the adrenal cortex to secrete aldosterone. Other substances found in the pineal gland include norepinephrine (NE), serotonin, histamine, GnRF, and GABA.

THYMUS

Usually a bilobed lymphatic organ, the **thymus gland** is located in the superior mediastinum, posterior to the sternum and between the lungs (Figure 18-20). The two **thymic lobes** are held in close proximity by an enveloping layer of connective tissue. Each lobe is enclosed by a fibrous connective tissue **capsule.** The capsule gives off extensions into the lobes called **trabeculae** (tra-BEK-yoo-lē), which divide the lobes into **lobules** (Figure 18-21a).

Each lobule consists of a deeply staining peripheral **cortex** and a lighter staining central **medulla.** The cortex is composed almost entirely of small, medium, and large tightly packed lymphocytes held in place by reticular tissue fibers (Figure 18-21b). Since the reticular tissue differs in origin and structure from that usually found in other lymphatic organs, it is referred to as **epithelioreticular** supporting tissue. The medulla consists mostly of epithelial cells and more widely scattered lymphocytes, and its reticulum is more cellular than fibrous. In addition, the medulla contains characteristic **thymic corpuscles,** concentric layers of epithelial cells. Their significance is unknown.

The thymus gland is conspicuous in the infant, and it reaches its maximum size of about 40 grams during puberty. After puberty, most of the thymic tissue is replaced by fat and connective tissue. By the time the person reaches maturity, the gland has atrophied substantially.

The function of the thymus gland is related to immunity. Lymphoid tissue of the body consists primarily of lymphocytes that may be distinguished into two kinds: B cells and T cells. Both are derived originally in the embryo from lymphocytic stem cells in bone marrow. Before migrating to their positions in lymphoid tissue, the descendants of the stem cells follow two distinct pathways. About half of them migrate to the thymus gland, where they are processed to become thymus-dependent lymphocytes, or **T cells.** The thymus gland confers on some of them the ability to destroy antigens (foreign microbes and substances) directly. The remaining stem cells are processed in some as yet undetermined area of the body, possibly bone marrow, the fetal liver and spleen, or gut-associated lymphoid tissue, and are known as **B cells.** These cells, under the influence of hormones produced by the thymus gland, called **thymosin, thymic humoral factor (THF), thymic factor (TF),** and **thymopoietin,** differentiate into plasma cells. Plasma cells, in turn, produce antibodies against antigens. Thymosin is also secreted by macrophages. The details of the roles of B cells and T cells in immunity are presented in Chapter 22.

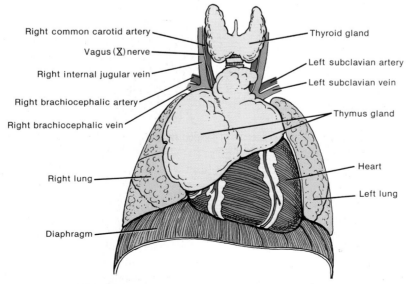

FIGURE 18-20 Location of the thymus gland in a young child.

FIGURE 18-21 Histology of the thymus gland. (a) Photomicrograph of several lobules at a magnification of 40×. (© 1983 by Michael H. Ross. Used by permission.) (b) Photomicrograph of an enlarged aspect of thymic corpuscles at a magnification of 600×. (© 1983 by Michael H. Ross. Used by permission.)

OTHER ENDOCRINE TISSUES

Before leaving our discussion of hormones, it should be noted that body tissues other than endocrine glands also secrete hormones. The gastrointestinal tract synthesizes several hormones that regulate digestion in the stomach and small intestine. Among these hormones are **stomach gastrin, enteric gastrin, secretin, cholecystokinin (CCK),** **enterocrinin,** and **gastric inhibitory peptide (GIP).** The actions of these hormones and their control of secretion are summarized in Exhibit 24-1.

The placenta produces **human chorionic gonadotropin (HCG), estrogens, progesterone, relaxin,** and **human chorionic somatomammotropin (HCS),** all of which are related to pregnancy (Chapter 29).

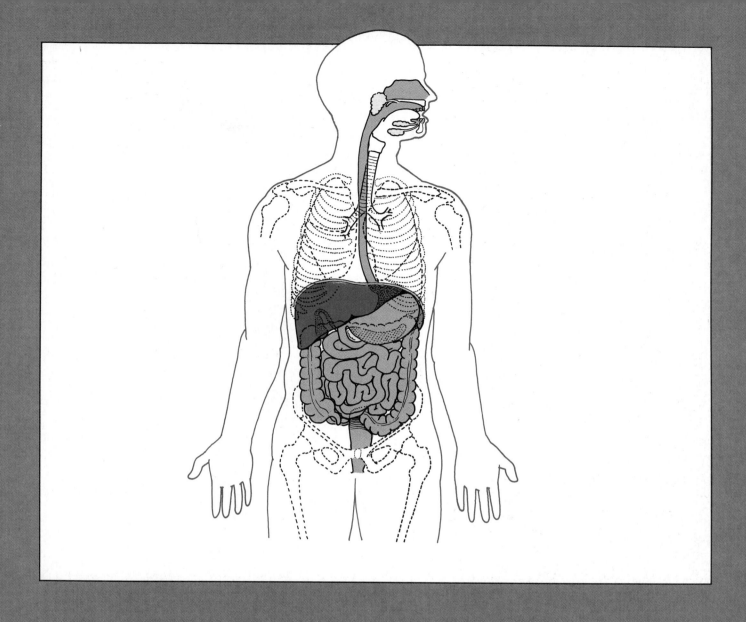

UNIT IV

Maintenance of the Human Body

This unit explains how the body maintains homeostasis on a day-to-day basis. In these chapters you will be studying the interrelations among the cardiovascular, lymphatic, respiratory, digestive, and urinary systems. You will also learn about metabolism, fluid and electrolyte balance, and acid–base dynamics.

(b)

FIGURE 19-2 (*Continued*) Blood cells. (b) Scanning electron micrographs of erythrocytes (left), leucocytes (center), and a platelet (right) at a magnification of 10,000×. (Courtesy of Fisher Scientific Company and S.T.E.M. Laboratories, Inc., Copyright 1975.)

blood cells because hemoglobin molecules are small and, if they were free in plasma, they would leak through the endothelial membranes of blood vessels and be lost in the urine. The biconcave shape of a red blood cell has a much greater surface area than, say, a sphere or a cube. The erythrocyte thus presents the maximum surface area for the diffusion of gas molecules that pass through the membrane to combine with hemoglobin.

CLINICAL APPLICATION

Clinical trials have been performed to test the effectiveness of a **blood substitute** called *Fluosol-DA.* It is a slippery, white liquid that has a very high solubility of oxygen. Since its only function is to transport oxygen, it is actually a hemoglobin substitute. It may eventually turn out to be useful in providing oxygen to tissues which are oxygen-deficient as a result of carbon monoxide poisoning, sickle cell anemia, strokes, heart attacks, and burns. Its use is limited by the fact that it must be kept frozen until just before use and recipients must receive oxygen by mask. Nevertheless, patients would be free of the risks of transfusion reactions and posttransfusion hepatitis.

FIGURE 19-3 Diagram of a hemoglobin molecule. The globin (protein) portions of the molecule are indicated in green and blue and the four heme (iron-containing) portions are in the center of each globin molecule.

● *Life Span and Number* The cell membrane of a red blood cell becomes fragile and the cell is nonfunctional in about 120 days. A healthy male has about 5.4 million red blood cells per cubic millimeter (mm³) of blood, and a healthy female has about 4.8 million. The higher value in the male is because of his higher rate of metabolism. To maintain normal quantities of erythrocytes, the body must produce new mature cells at the astonishing rate of 2 million per second. In the adult, production takes place in the red bone marrow in the spongy bone of the cranium, ribs, sternum, bodies of vertebrae, and proximal epiphyses of the humerus and femur.

● *Production* The process by which erythrocytes are formed is called **erythropoiesis** (e-rith'-rō-poy-Ē-sis). Erythropoiesis starts with the transformation of a hemocytoblast into a rubriblast (Figure 19-2a). The *rubriblast* (proerythroblast) gives rise to a *prorubricyte* (early erythroblast), which then develops into a *rubricyte* (intermediate erythroblast), the first cell in the sequence that begins to synthesize hemoglobin. The rubricyte next develops into a *metarubricyte* (late erythroblast). In the metarubricyte, hemoglobin synthesis is at a maximum. In the next stage, the metarubricyte develops into a *reticulocyte* and the nucleus is lost by extrusion. The reticulocyte in turn becomes an *erythrocyte,* or mature red blood cell. Once the erythrocyte is formed, it leaves the marrow and enters the bloodstream. Aged erythrocytes are destroyed by reticuloendothelial cells in the liver and spleen. The hemoglobin molecules are split apart, the iron is reused, and the rest of the molecule is converted into other substances for reuse or elimination.

Normally erythropoiesis and red cell destruction proceed at the same pace. But if the body suddenly needs more erythrocytes or if erythropoiesis is not keeping up with red blood cell destruction, a homeostatic mechanism steps up erythrocyte production. The mechanism is triggered by the reduced supply of oxygen for body cells that results from a reduced number of erythrocytes. If certain kidney cells become oxygen-deficient, they release an enzyme called **renal erythropoietic factor** that converts a plasma protein into the hormone **erythropoietin.** This hormone circulates through the blood to the red bone marrow, where it stimulates more hemocytoblasts to develop into red blood cells. Erythropoietin is also produced by other body tissues, particularly the liver.

Cellular oxygen deficiency, called **hypoxia** (hī-POKS-ē-a), may occur if you do not breathe in enough oxygen. This deficiency commonly occurs at high altitudes, where the air contains less oxygen. Oxygen deficiency may also occur because of anemia. The term **anemia** means that the number of functional red blood cells or their hemoglobin content is below normal. Consequently, the erythrocytes are unable to transport enough oxygen from the lungs to the cells. Anemia has many causes; lack of iron, lack of certain amino acids, and lack of vitamin B_{12} are but a few. Iron is needed for the oxygen-carrying part of the hemoglobin molecule. The amino acids are needed for the protein, or globin, part. Vitamin B_{12} helps the red bone marrow to produce erythrocytes. This vitamin is obtained from meat, especially liver, but it cannot be absorbed by the lining of the small intestine without the help of another substance—**intrinsic factor** produced by the mucosal cells of the stomach. Intrinsic factor facilitates the absorption of vitamin B_{12}. Once absorbed, it is stored in the liver. (Several types of anemia are described later in the chapter.)

CLINICAL APPLICATION
The rate of erythropoiesis is measured by a procedure called a **reticulocyte** (re-TIK-yoo-lō-sīt) **count.** Some reticulocytes are normally released into the bloodstream before they become mature red blood cells. If the number of reticulocytes in a sample of blood is less than 0.5 percent of the number of mature red blood cells in the sample, erythropoiesis is occurring too slowly. A low reticulocyte count might confirm a diagnosis of nutritional or pernicious anemia. Or it might indicate a kidney disease that prevents the kidney cells from producing erythropoietin. If the reticulocytes number more than 1.5 percent of the mature red blood cells, erythropoiesis is abnormally rapid. Any number of problems may be responsible for a high reticulocyte count: anemia, oxygen deficiency, and uncontrolled red blood cell production caused by a cancer in the bone marrow. A high count may also indicate that treatment for a condition causing anemia has been effective and the bone marrow is making up for lost time.

Another test with important clinical applications is the hematocrit. **Hematocrit** is the percentage of blood that is made up of red blood cells. It is determined by centrifuging blood and noting the ratio of red blood cells to whole blood. The average hematocrit for males is 47 percent. This means that in 100 ml of blood there are 47 ml of cells and 53 ml of plasma. The average hematocrit for females is 42 percent. Anemic blood may have a hematocrit of 15 percent; polycythemic blood (an abnormal increase in the number of functional erythrocytes) may have a hematocrit of 65 percent. Athletes, however, may have higher than normal hematocrits, reflecting constant physical activity rather than a pathological condition.

Leucocytes

● *Structure and Types* Unlike red blood cells, **leucocytes** (LOO-kō-sīts), or **white blood cells** (**wbc**'s) have nuclei and do not contain hemoglobin (see Figure 19-2). Leucocytes fall into two major groups. The first group is the **granular leucocytes.** They develop from red bone marrow, have granules in the cytoplasm, and possess lobed nuclei. The three kinds of granular leucocytes are **neutrophils** or **polymorphs** (10 to 12 μm in diameter), **eosinophils** (10 to 12 μm in diameter), and **basophils** (8 to 10 μm in diameter).

The second principal group of leucocytes is the **agranular leucocytes.** They develop from lymphoid and myeloid tissue, no cytoplasmic granules can be seen under a light microscope, and their nuclei are usually spherical. The two kinds of agranular leucocytes are **lymphocytes** (7 to 15 μm in diameter) and **monocytes** (14 to 19 μm in diameter).

● *Functions* The skin and mucous membranes of the body are continuously exposed to microbes and their toxins. Some of these microbes are capable of invading deeper tissues to cause disease. The general function of leucocytes is to combat these microbes once they enter the body by phagocytosis or antibody production. Neutrophils and monocytes are actively **phagocytotic**—they can ingest bacteria and dispose of dead matter (see Figure 4-4). Neutrophils (NOO-trō-fils) are the most active leucocytes in response to tissue destruction by bacteria. In addition to carrying on phagocytosis, they release the enzyme lysozyme, which destroys certain bacteria. Apparently monocytes (MON-ō-sīts) take longer to reach the site of infection than do neutrophils, but once they arrive, they do so in larger numbers and destroy more microbes. Monocytes that have migrated to infected tissues are called **wandering macrophages.** They clean up cellular debris following an infection.

A number of different chemicals in inflamed tissue cause phagocytes to migrate toward the tissue. This phenomenon is called **chemotaxis.** Among the substances that provide stimuli for chemotaxis are toxins and degenerative products of damaged tissues.

Most leucocytes possess, to some degree, the ability to crawl through the minute spaces between the cells that form the walls of capillaries and through connective and epithelial tissue. This movement, like that of amoebas, is called **diapedesis** (dī'-a-pe-DĒ-sis). First, part of the cell membrane stretches out like an arm. Then the cytoplasm and nucleus flow into the projection. Finally, the rest of the membrane snaps up into place. Another projection is made, and so on, until the cell has crawled to its destination (see Figure 4-3b).

Eosinophils (ē'-ō-SIN-ō-fils) are believed to combat the irritants that cause allergies. Eosinophils leave the capillaries, enter the tissue fluid, and produce antihistamines that destroy antigen-antibody complexes. Eosinophils might also be phagocytic.

Basophils (BĀ-sō-fils) are also believed to be involved in allergic reactions. Basophils leave the capillaries, enter the tissues, and become the mast cells of the tissues, liberating heparin, histamine, and serotonin.

Lymphocytes (LIM-fō-sīts) are involved in the production of antibodies. **Antibodies** (AN-ti-bod'-ēz) are special proteins that inactivate antigens. An **antigen** (AN-ti-jen) is any substance that will stimulate the production of antibodies and is capable of reacting specifically with the antibody. Most antigens are proteins, and most are not synthesized by the body. Many of the proteins that make up the cell structures and enzymes of bacteria are antigens. The toxins released by bacteria are also antigens. When antigens enter the body, they react chemically with substances in the lymphocytes and stimulate some lymphocytes, called B cells, to become **plasma cells** (Figure 19-4). The plasma cells then produce antibodies, globulin-type proteins that attach to antigens much as enzymes attach to substrates. Like enzymes, a specific antibody will generally attach only to a certain antigen. However, unlike enzymes, which enhance the reactivity of the substrate, antibodies "cover" their antigens so the antigens cannot come in contact with other chemicals in the body. In this way, bacterial poisons can be sealed up and rendered harmless. The bacteria themselves are destroyed by the antibodies. This process is called the **antigen-antibody response.** Phagocytes in tissues destroy the antigen-antibody complexes. The antigen-antibody response helps us to combat infection and gives us immunity to some diseases. It is also responsible for blood types, allergies, and the body's rejection of organs transplanted from an individual with a different genetic makeup.

CLINICAL APPLICATION
An increase in the number of white cells present in the blood indicates a state of inflammation or infection. Because each type of white cell plays a different role, determining which types have increased aids in diagnosis of the condition. A **differential count** is the number of each kind of white cell in 100 white blood cells. A normal differential count falls within the following percentages.

Neutrophils	60–70%
Eosinophils	2–4%
Basophils	0.5–1%
Lymphocytes	20–25%
Monocytes	3–8%
	100%

Particular attention is paid to the neutrophils in a differential count. More often than not, a high neutrophil count indicates damage by invading bacteria. An increase in the number of monocytes generally indicates a chronic infection. Eosinophils and basophils are elevated during allergic reactions. High lymphocyte counts indicate antigen-antibody reactions.

FIGURE 19-4 Antigen-antibody response. An antigen entering the body stimulates a B cell to develop into an antibody-producing plasma cell. The antibodies attach to the antigen, cover it, and render it harmless.

● *Life Span and Number*   Foreign bacteria exist everywhere in the environment and have continuous access to the body through the mouth, nose, and pores of the skin. Furthermore, many cells, especially those of epithelial tissue, age and die, and their remains must be disposed of daily. Even when the body is healthy, the leucocytes actively ingest bacteria and debris. However, a leucocyte can phagocytose only a certain number of substances before they interfere with the leucocyte's normal metabolic activities and bring on its death. Consequently, the life span of most leucocytes is very short. In a healthy body, some white blood cells will live only a few days. During a period of infection they may live only a few hours.

Leucocytes are far less numerous than red blood cells, averaging from 5,000 to 9,000 cells per cubic millimeter of blood. Red blood cells, therefore, outnumber white blood cells about 700 to 1. The term **leucocytosis** (loo'-kō-sī-TŌ-sis) refers to an increase in the number of white blood cells. If the increase exceeds 10,000, a pathological condition is usually indicated. An abnormally low level of white blood cells (below 5,000/mm³) is termed **leucopenia** (loo-kō-PĒ-nē-a).

● *Production*   Granular leucocytes are produced in red bone marrow (myeloid tissue); agranular leucocytes are produced in both myeloid and lymphoid tissue. The developmental sequences for the five types of leucocytes are shown in Figure 19-2a.

Thrombocytes

● *Structure*   In addition to the immature cell types that develop into erythrocytes and leucocytes, hemocytoblasts differentiate into still another kind of cell, called a megakaryoblast (see Figure 19-2). Megakaryoblasts are transformed into megakaryocytes, large cells that shed fragments of cytoplasm. Each fragment becomes enclosed by a piece of the cell membrane and is called a **thrombocyte** (THROM-bō-sīt) or **platelet**. Platelets are disc-shaped cells without a nucleus. They average from 2 to 4 μm in diameter.

● *Function*   Platelets prevent fluid loss by initiating a chain of reactions that results in blood clotting. This mechanism is described shortly.

● *Life Span and Number*   Like the other formed elements of the blood, platelets have a short life, probably only 5 to 9 days. Between 250,000 and 400,000 platelets appear in each cubic millimeter of blood.

● *Production*   Platelets are produced in red bone marrow according to the developmental sequence shown in Figure 19-2a.

A summary of the formed elements in blood is presented in Exhibit 19-1.

CLINICAL APPLICATION

The most commonly ordered hematology test is the **complete blood count (CBC)**. It generally includes determination of hemoglobin, hematocrit, red blood cell count, white blood cell count, differential count, and comments about red blood cell, white blood cell, and platelet morphology.

PLASMA

When the formed elements are removed from blood, a straw-colored liquid called **plasma** is left. Exhibit 19-2 outlines the chemical composition of plasma. Note that about 7 percent of the solutes are proteins. Some of these proteins are also found elsewhere in the body, but in blood they are called *plasma proteins*. Albumins, which constitute 55 percent of plasma proteins, are largely responsible for blood's viscosity. The concentration of albumins is about four times higher in plasma than in interstitial fluid. Along with the electrolytes, albumins also help

---

EXHIBIT 19-1

SUMMARY OF THE FORMED ELEMENTS IN BLOOD

FORMED ELEMENT	NUMBER	DIAMETER (in μm)	LIFE SPAN	FUNCTION
Erythrocyte (red blood cell)	4.8 million/mm³ in females 5.4 million/mm³ in males	7.7	120 days	Transport oxygen and carbon dioxide.
Leucocyte (white blood cell)	5,000–9,000/mm³		Few hours to a few days	
Granular				
Neutrophil	60–70% of total	10–12		Phagocytosis.
Eosinophil	2–4% of total	10–12		Combat allergens.
Basophil	0.5–1% of total	8–10		Combat allergens.
Agranular				
Lymphocyte	20–25% of total	7–15		Immunity (antigen-antibody reactions).
Monocyte	3–8% of total	14–19		Phagocytosis.
Thrombocyte (platelet)	250,000–400,000/ mm³	2–4	5–9 days	Blood clotting.

EXHIBIT 19-2

CHEMICAL COMPOSITION AND DESCRIPTION OF SUBSTANCES IN PLASMA

CONSTITUENT	DESCRIPTION
WATER	Liquid portion of blood; constitutes about 91.5 percent of plasma. Ninety percent of water derived from absorption from digestive tract; 10 percent from cellular respiration. Acts as solvent and suspending medium for solid components of blood and absorbs heat.
SOLUTES Proteins	Constitute about 8.5 percent of plasma.
Albumins	Smallest plasma proteins. Produced by liver and provide blood with viscosity, a factor related to maintenance and regulation of blood pressure. Also exert considerable osmotic pressure to maintain water balance between blood and tissues and regulate blood volume.
Globulins	Protein group to which antibodies belong. Gamma globulins attack measles, hepatitis, and polio viruses and tetanus bacterium.
Fibrinogen	Produced by liver. Plays essential role in clotting.
Nonprotein nitrogen (NPN) substances	Contain nitrogen but are not proteins. Include urea, uric acid, creatine, creatinine, and ammonium salts. Represent breakdown products of protein metabolism and are carried by blood to organs of excretion.
Food substances	Products of digestion passed into blood for distribution to all body cells. Include amino acids (from proteins), glucose (from carbohydrates), and fatty acids and glycerol (from fats).
Regulatory substances	Enzymes, produced by body cells, to catalyze chemical reactions. Hormones, produced by endocrine glands, to regulate growth and development in body.
Respiratory gases	Oxygen and carbon dioxide. These gases are more closely associated with hemoglobin or red blood cells than plasma itself.
Electrolytes	Inorganic salts of plasma. Cations include Na^+, K^+, Ca^{2+}, Mg^{2+}; anions include Cl^-, PO_4^{3-}, SO_4^{2-}, HCO_3^-. Help maintain osmotic pressure, normal pH, physiological balance between tissues and blood.

to regulate blood volume by preventing the water in the blood from diffusing into the interstitial fluid. Recall that water moves by osmosis from an area of high water (low solute) concentration to an area of low water (high solute) concentration. Globulins, which comprise 38 percent of plasma proteins, are antibody proteins released by plasma cells. Gamma globulin is especially well known because it is able to form an antigen-antibody complex with the proteins of the hepatitis and measles viruses and the tetanus bacterium, among others. Fibrinogen makes up about 7 percent of plasma proteins and takes part in the blood-clotting mechanism along with the platelets.

HEMOSTASIS

Hemostasis refers to the stopping of bleeding. When blood vessels are damaged or ruptured, three basic mechanisms operate to prevent blood loss: (1) vascular spasm, (2) platelet plug formation, and (3) blood coagulation (clotting). Although these mechanisms are useful for preventing hemorrhage in smaller blood vessels, extensive hemorrhage usually requires intervention of some type.

VASCULAR SPASM

When a blood vessel is damaged, the smooth muscle in its wall contracts immediately. Such a **vascular spasm** reduces blood loss for up to 30 minutes, during which time the other hemostatic mechanisms can go into operation. The spasm is probably caused by damage to the vessel wall and from reflexes initiated by pain receptors.

PLATELET PLUG FORMATION

When platelets come into contact with parts of a damaged blood vessel, such as collagen or endothelium, their characteristics change drastically. They begin to enlarge and their shapes become even more irregular. They also become sticky and begin to adhere to the collagen fibers. Their synthesis of ADP and enzymes causes the formation of substances that activate more platelets, causing them to stick to the original platelets. The accumulation and attachment of large numbers of platelets form a **platelet plug.** The plug is very effective in preventing blood loss in a small vessel. Although the platelet plug is initially loose, it becomes quite tight when reinforced by fibrin threads formed during coagulation.

COAGULATION

Normally, blood maintains its liquid state as long as it remains in the vessels. If it is drawn from the body, however, it thickens and forms a gel. Eventually, the gel separates from the liquid. The straw-colored liquid, called **serum,** is simply plasma minus its clotting proteins. The gel is called a **clot** and consists of a network of insoluble fibers in which the cellular components of blood are trapped.

The process of clotting is called **coagulation.** If the blood clots too easily, the result can be thrombosis—clotting in an unbroken blood vessel. If the blood takes too long to clot, a hemorrhage can result.

Clotting involves various chemicals known as **coagulation factors.** In plasma, these factors are called *plasma coagulation factors.* A few *platelet coagulation factors* are released by platelets. One coagulation factor is released by damaged body tissues. Coagulation factors are listed in Exhibit 19-3. *thromboplastin?*

Clotting is a complex process, but it can be described basically as a sequence of three stages.

Stage 1. Formation of a substance called thromboplastin.
Stage 2. Conversion of prothrombin, a plasma protein, into thrombin, an enzyme. This stage requires the presence of thromboplastin and several other plasma coagulation factors.
Stage 3. Conversion of fibrinogen, another plasma protein, into insoluble fibrin. The reaction is catalyzed by thrombin.

Thromboplastin also from platelets

EXHIBIT 19-3
COAGULATION FACTORS

FACTOR	COMMENTS
PLASMA COAGULATION FACTORS	
I: Fibrinogen	Synthesized in liver. Important factor in stage 3 of clotting, in which it is converted to fibrin. Plasma minus fibrinogen is called serum.
II: Prothrombin	Synthesized in liver; formation requires vitamin K. Important in stage 2 of clotting, in which it is converted to thrombin.
III: Thromboplastin	In extrinsic pathway it is known as extrinsic thromboplastin and is formed from tissue thromboplastin. In intrinsic pathway it is called intrinsic thromboplastin and is formed from platelet disintegration. Formation of thromboplastin signifies end of stage 1 of clotting.
IV: Calcium ions	Apparently involved in all three stages of clotting. Removal of calcium or its binding in plasma prevents coagulation.
V: Proaccelerin or labile factor	Synthesized in liver. Required for stages 1 and 2 of both extrinsic and intrinsic pathways.
VI	No longer used in coagulation theory. Number has not been reassigned.
VII: Serum prothrombin conversion accelerator (SPCA) or stable factor	Synthesized in liver; formation requires vitamin K. Required in stage 1 of extrinsic pathway.
VIII: Antihemophilic factor	Synthesized by liver. Required for stage 1 of intrinsic pathway. Deficiency causes classic hemophilia A.
IX: Christmas factor or plasma thromboplastin component (PTC)	Synthesized by liver; formation requires vitamin K. Required for stage 1 of intrinsic pathway. Deficiency causes hemophilia B.
X: Stuart factor or Stuart-Prower factor	Synthesized by liver; formation requires vitamin K. Required for stages 1 and 2 of extrinsic and intrinsic pathways. Deficiency results in nosebleeds, bleeding into joints, or bleeding into soft tissues.
XI: Plasma thromboplastin antecedent (PTA)	Synthesized by liver. Required for stage 1 of intrinsic pathway. Deficiency results in hemophilia C.
XII: Hageman factor	Required for stage 1 of intrinsic pathway. Known to be activated by contact with glass and may assume role in initiating coagulation outside body.
XIII: Fibrin stabilizing factor (FSF)	Required for stage 3 of clotting.
PLATELET COAGULATION FACTORS	
Pf$_1$: Platelet factor 1 or platelet accelerator	Essentially same as plasma coagulation factor V.
Pf$_2$: Platelet factor 2 or thrombin accelerator	Accelerates formation of thrombin in stage 1 of intrinsic pathway and conversion of fibrinogen to fibrin.
Pf$_3$: Platelet factor 3 or platelet thromboplastic factor	Required for stage 1 of intrinsic pathway.
Pf$_4$: Platelet factor 4	Binds heparin, an anticoagulant, during clotting.

Fibrin forms the threads of the clot. (Cigarette smoke contains at least two substances that interfere with fibrin formation.)

Depending on whether thromboplastin is released by damaged tissues or formed by platelet disintegration, blood clotting may proceed along one of two routes: the extrinsic pathway or the intrinsic pathway.

Extrinsic Pathway

The **extrinsic pathway** of blood clotting begins when a blood vessel is ruptured (Figure 19-5). The damaged tissues surrounding the blood vessel or the ruptured area of the blood vessel itself release a lipoprotein called *tissue thromboplastin*. Tissue thromboplastin, in reaction with plasma coagulation factors IV, V, VII, and X, forms *extrinsic thromboplastin*. This process is stage 1 of the extrinsic pathway. In stage 2 prothrombin is converted to thrombin. This conversion requires extrinsic thromboplastin and several plasma coagulating factors such as IV, V, VII, and X. In stage 3 fibrinogen, which is soluble, is converted to fibrin, which is insoluble, by the thrombin formed in stage 2. This reaction requires plasma coagulation factors IV and XIII. Except for a few initial steps of the extrinsic pathway, all other stages require calcium ions (Ca^{2+}).

Since type AB blood can theoretically receive all other blood types, it is called the **universal recipient** (Figure 19-8). Type O blood will not agglutinate with any of the other types. Theoretically, it may be given to individuals whose blood type is O, A, B, and AB. Since type O blood has no agglutinogens to serve as antigens in another person's body, it is known as the **universal donor.** Type O persons requiring blood may receive only type O blood. In practice, only matching blood types are used for transfusions.

CLINICAL APPLICATION

Scientists recently converted type B blood into type O blood by using an enzyme called alpha galactosidase. The enzyme is isolated from green coffee beans and has the ability to remove a sugar from the surface of red blood cells, thus **changing type B into type O blood.** Preliminary tests on animals and a few human volunteers showed no adverse reactions. It is hoped that further testing will confirm the safety of the converted blood and that other conversions can be made so that blood banks could juggle their supplies as needed.

Exhibit 19-4 summarizes the interreactions of the four blood types.

Rh

The **Rh system** of blood classification is so named because it was first worked out in the blood of the *Rhesus* monkey. Like the ABO grouping, the Rh system is based on agglutinogens that lie on the surfaces of erythrocytes. Individuals whose erythrocytes have the Rh agglutinogens are designated *Rh⁺*. Those who lack Rh agglutinogens are designated *Rh⁻*. It is estimated that 85 percent of whites and 88 percent of blacks in the United States are Rh⁺, whereas 15 percent of whites and 12 percent of blacks are Rh⁻.

Under normal circumstances, human plasma does not contain anti-Rh agglutinins. However, if an Rh⁻ person receives Rh⁺ blood, the body starts to make anti-Rh agglutinins that will remain in the blood. If a second transfusion of Rh⁺ blood is given later, the previously formed anti-Rh agglutinins will react against the donated blood and a severe reaction may occur.

CLINICAL APPLICATION

One of the most common problems with Rh incompatibility arises from pregnancy (Figure 19-9). During pregnancy, some of the fetus's blood may leak from

---

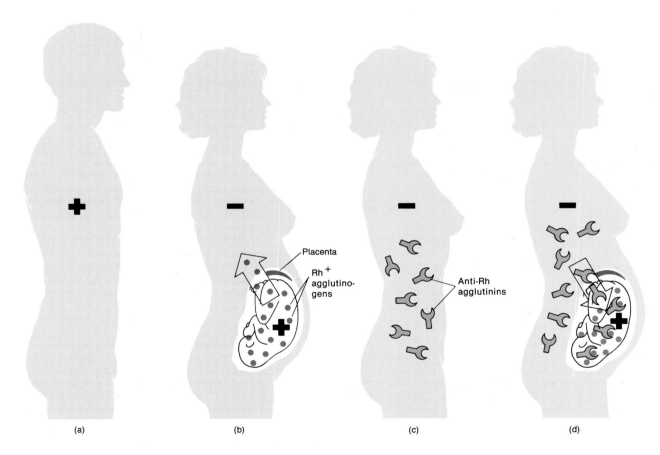

FIGURE 19-9 Development of erythroblastosis fetalis. (a) Rh⁺ father. (b) Rh⁻ mother and Rh⁺ fetus. If an Rh⁻ female is impregnated by an Rh⁺ male and the fetus is Rh⁺, fetal Rh⁺ agglutinogens may enter the maternal blood via the placenta during delivery. (c) Upon exposure to the fetal Rh⁺ agglutinogens, the mother will make anti-Rh agglutinins. (d) If the female becomes pregnant again, her anti-Rh agglutinins will cross the placenta into the fetal blood. If the fetus is Rh⁺, erythroblastosis fetalis will result.

the placenta (afterbirth) into the mother's bloodstream. If the fetus is Rh^+ and the woman is Rh^-, she, upon exposure to the Rh^+ fetal cells, will make anti-Rh agglutinins. If the woman becomes pregnant again, her anti-Rh agglutinins will cross the placenta and make their way into the bloodstream of the baby. If the fetus is Rh^-, no problem will occur, since Rh^- blood does not have the Rh agglutinogen. If the fetus is Rh^+, an antigen-antibody response called **hemolysis** (*lysis* = dissolve) may occur in the fetal blood. Hemolysis means a swelling and subsequent rupture of erythrocytes resulting in the liberation of hemoglobin. The hemolysis brought on by fetal-maternal incompatibility is called **erythroblastosis fetalis** or **hemolytic disease of newborn.** When a baby is born with this condition, all the blood is slowly removed and replaced with agglutinin-free blood. It is even possible to transfuse blood into the unborn child if the disease is suspected. More important, though, is the fact that the disorder can be prevented with an injection of an anti-Rh gamma$_2$-globulin agglutinin preparation, administered to Rh^- mothers right after delivery or abortion. These agglutinins tie up the fetal agglutinogens so the mother cannot respond to the foreign agglutinogens by producing agglutinins. Thus the fetus of the next pregnancy is protected. In the case of an Rh^+ mother and an Rh^- child, there are no complications, since the fetus cannot make agglutinins.

INTERSTITIAL FLUID AND LYMPH

Whole blood does not flow into the tissue spaces; it remains in closed vessels. However, certain constituents of the plasma do move through the capillary walls, and once they move out of the blood, they are called interstitial fluid. Interstitial fluid and lymph are basically the same. The major difference between the two is location. When the fluid bathes the cells, it is called **interstitial fluid, intercellular fluid,** or **tissue fluid.** When it flows through the lymphatic vessels, it is called **lymph.**

Both fluids are similar in composition to plasma. The principal chemical difference is that they contain less protein than plasma because the larger protein molecules are not easily filtered through the cells that form the walls of the capillaries. The transfer of materials between blood and interstitial fluid occurs by osmosis, diffusion, and filtration across the cells that make up the capillary walls.

Interstitial fluid and lymph also differ from plasma in that they contain variable numbers of leucocytes. Leucocytes can enter the tissue fluid by diapedesis, and the lymphoid tissue itself is one of the sites of agranular leucocyte production. Like plasma, interstitial fluid and lymph lack erythrocytes and platelets.

Other substances, especially organic molecules, in interstitial fluid and lymph vary in relation to the location of the sample analyzed. The lymph vessels that drain the organs of the digestive tract, for example, contain a great deal of lipid absorbed from food.

DISORDERS: HOMEOSTATIC IMBALANCES

Anemia

Anemia is a sign, not a diagnosis. Many kinds of anemia exist, all characterized by insufficient erythrocytes or hemoglobin. These conditions lead to fatigue and intolerance to cold, both of which are related to lack of oxygen needed for energy and heat production, and to paleness, which is due to low hemoglobin content.

Nutritional Anemia

Nutritional anemia arises from an inadequate diet, one that provides insufficient amounts of iron, the necessary amino acids, or vitamin B_{12}.

Pernicious Anemia

Pernicious anemia is the insufficient production of erythrocytes resulting from an inability of the body to produce intrinsic factor.

Hemorrhagic Anemia

An excessive loss of erythrocytes through bleeding is called **hemorrhagic anemia.** Common causes are large wounds, stomach ulcers, and heavy menstrual bleeding. If bleeding is extraordinarily heavy, the anemia is termed acute. Excessive blood loss can be fatal. Slow, prolonged bleeding is apt to produce a chronic anemia; the chief symptom is fatigue.

Hemolytic Anemia

If erythrocyte cell membranes rupture prematurely, the cells remain as "ghosts" and their hemoglobin pours out into the plasma. A characteristic sign of this condition, called **hemolytic anemia,** is distortion in the shape of erythrocytes that are pro-

gressing toward hemolysis. There may also be a sharp increase in the number of reticulocytes, since the destruction of red blood cells stimulates erythropoiesis.

The premature destruction of red blood cells may result from inherent defects, such as hemoglobin defects, abnormal red blood cell enzymes, or defects of the red blood cell membrane. Agents that may cause hemolytic anemia are parasites, toxins, and antibodies from incompatible blood (Rh^- mother and Rh^+ fetus, for instance). *Erythroblastosis fetalis* is an example of a hemolytic anemia.

The term **thalassemia** (thal'-a-SĒ-mē-a) represents a group of hereditary hemolytic anemias, resulting from a defect in the synthesis of hemoglobin, which produces extremely thin and fragile erythrocytes. It occurs primarily in populations from countries bordering the Mediterranean Sea. Treatment generally consists of blood transfusions.

Aplastic Anemia

Destruction or inhibition of the red bone marrow results in **aplastic anemia.** Typically, the marrow is replaced by fatty tissue, fibrous tissue, or tumor cells. Toxins, gamma radiation, and certain medications are causes. Many of the medications inhibit the enzymes involved in hemopoiesis. Bone marrow transplants can now be done with a reasonable hope of success in patients with aplastic anemia. They are done early, before the victim is sensitized by transfusions. Immunosuppressive drugs are given for 4 days before the transplant and with decreasing frequency for 100 days afterward.

Sickle Cell Anemia

The erythrocytes of a person with **sickle cell anemia** manufac-

- Identify the location of the heart in the mediastinum.
- Distinguish between the structure and location of the fibrous and serous pericardium.
- Contrast the structure of the epicardium, myocardium, and endocardium.
- Identify the chambers, great vessels, and valves of the heart.
- Explain the initiation and conduction of nerve impulses through the electrical conduction system of the heart.
- Discuss the route of blood in coronary (cardiac) circulation.
- Compare angina pectoris and myocardial infarction as abnormalities of coronary (cardiac) circulation.
- Explain the pressure changes associated with blood flow through the heart.
- Define systole and diastole as the two principal events of the cardiac cycle.
- Relate the events of the cardiac cycle to time.
- Contrast the sounds of the heart and their clinical significance.
- Discuss the surface anatomy features of the heart.

- Define cardiac output and explain what determines it.
- Explain what is meant by cardiac reserve.
- Define Starling's law of the heart.
- Contrast the effects of sympathetic and parasympathetic stimulation of the heart.
- Define the role of pressoreceptors in reflex pathways in controlling heart rate.
- Explain how chemicals, temperature, emotions, and age affect heart rate.
- Define circulatory shock and explain the homeostatic mechanisms that compensate for it.
- List the risk factors involved in heart disease.
- Explain why inadequate blood supply, anatomical disorders, and malfunctions of conduction are primary reasons for heart disease.
- Define congestive heart failure (CHF).
- Explain the principle and diagnostic importance of cardiac catheterization.
- Define medical terminology associated with the heart.

20 The Cardiovascular System: The Heart

The heart is the center of the cardiovascular system. It is a hollow, muscular organ that weighs about 342 grams (11 oz) and beats over 100,000 times a day to pump blood through over 60,000 miles of blood vessels. The blood vessels form a network of tubes that carry blood from the heart to the tissues of the body and then return it to the heart. The anatomical features of the heart that we will consider are its location, covering, wall and chambers, great vessels, valves, and conduction system. Functional aspects of the heart to be considered are the electrocardiogram, cardiac cycle (heartbeat), cardiac output, and circulatory shock.

LOCATION

The **heart** is situated obliquely between the lungs in the mediastinum. About two-thirds of its mass lies to the left of the body's midline (Figures 20-1a and 20-10). The heart is shaped like a blunt cone about the size of your closed fist—12 cm (5 inches) long, 9 cm (3½ inches) wide at its broadest point, and 6 cm (2½ inches) thick.

Its pointed end, the *apex*, is formed by the tip of the left ventricle, projects inferiorly, anteriorly, and to the left and lies superior to the central depression of the diaphragm. Anteriorly, the apex is in the fifth intercostal space—about 7.5 to 8 cm (3 inches) from the midline of the body.

The *left border* is formed almost entirely by the left ventricle, although the left atrium forms part of the upper end of the border. The left border of the heart is indicated approximately as a curved line drawn from the left fifth costochondral junction to the left second costochondral junction.

The *superior border,* where the great vessels enter and leave the heart, is formed by both atria. It is represented by a line joining the second left intercostal space to the third right costal cartilage.

The *base* of the heart projects superiorly, posteriorly, and to the right. It is formed by the atria, mostly the left atrium. It lies opposite the fifth to ninth thoracic vertebrae. Anteriorly, it lies just inferior to the second rib.

The *right border* is formed by the right atrium and corresponds to a curved line drawn from the xiphisternal articulation to about the middle of the right third costal cartilage.

The *inferior border* is formed by the right ventricle and slightly by the left ventricle. It is represented by a line passing from the inferior end of the right border through the xiphisternal joint to the apex of the heart.

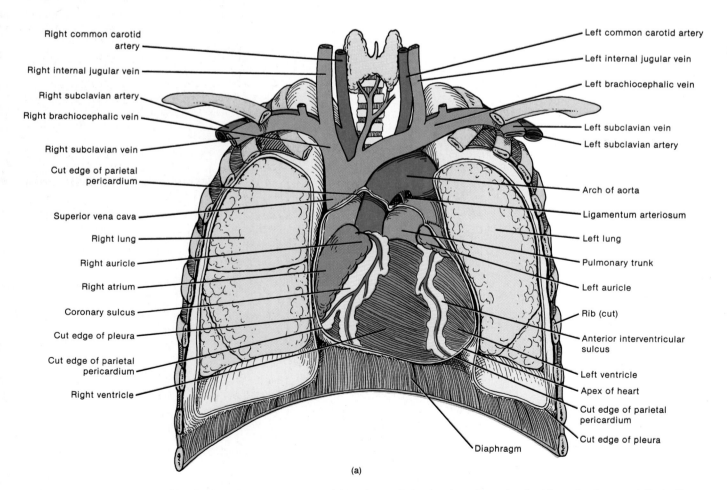

Right common carotid artery
Right internal jugular vein
Right subclavian artery
Right brachiocephalic vein
Right subclavian vein
Cut edge of parietal pericardium
Superior vena cava
Right lung
Right auricle
Right atrium
Coronary sulcus
Cut edge of pleura
Cut edge of parietal pericardium
Right ventricle

Left common carotid artery
Left internal jugular vein
Left brachiocephalic vein
Left subclavian vein
Left subclavian artery
Arch of aorta
Ligamentum arteriosum
Left lung
Pulmonary trunk
Left auricle
Rib (cut)
Anterior interventricular sulcus
Left ventricle
Apex of heart
Cut edge of parietal pericardium
Cut edge of pleura
Diaphragm

(a)

FIGURE 20-1 Heart. (a) Position of the heart and associated blood vessels in the thoracic cavity. In this and subsequent illustrations, vessels that carry oxygenated blood are colored red; vessels that carry deoxygenated blood are blue.

The *sternocostal* (*anterior*) *surface* is formed mainly by the right ventricle and right atrium, whereas the *diaphragmatic* (*inferior*) *surface* is formed by the left and right ventricles, mostly the left.

PARIETAL PERICARDIUM (PERICARDIAL SAC)

The heart is enclosed in a loose-fitting serous membrane called the **parietal pericardium** or **pericardial sac** (Figure 20-1a, b). It consists of two layers: the fibrous layer and the serous layer. The *fibrous layer* (*fibrous pericardium*) is the outer layer and consists of a tough, fibrous connective tissue. It is attached to the large blood vessels entering and leaving the heart, to the diaphragm, and to the inside of the sternal wall of the thorax. It also adheres to the parietal pleurae. The fibrous pericardium prevents overdistension of the heart, surrounds it with a tough protective membrane, and anchors it in the mediastinum.

The inner layer of the parietal pericardium, known as the *serous layer* (*serous pericardium*), is thinner and more delicate. It is continuous with the visceral pericardium (the outer layer of the wall of the heart) at the base of the heart and around the large blood vessels.

An inflammation of the parietal pericardium is known as **pericarditis.**

HEART WALL

The wall of the heart (Figure 20-1b) is divided into three portions: the epicardium (external layer), the myocardium (middle layer), and the endocardium (inner layer). The **epicardium (visceral pericardium)** is the thin, transparent outer layer of the wall. It is composed of serous tissue and mesothelium. Between the serous pericardium and the epicardium is a potential space called the *pericardial cavity.* The cavity contains a watery fluid, known as *pericardial fluid,* which prevents friction between the membranes as the heart moves.

The **myocardium,** which is cardiac muscle tissue, constitutes the bulk of the heart. Cardiac muscle fibers are involuntary, striated, and branched, and the tissue is arranged in interlacing bundles of fibers. The myocardium is responsible for the contraction of the heart.

The **endocardium** is a thin layer of endothelium overlying a thin layer of connective tissue pierced by tiny blood vessels and bundles of smooth muscle. It lines the inside of the myocardium and covers the valves of the heart and the tendons that hold them open. It is continuous with the endothelial lining of the large blood vessels of the heart.

Inflammation of the epicardium, myocardium, and endocardium are referred to as **epicarditis, myocarditis,** and **endocarditis,** respectively.

CHAMBERS OF THE HEART

The interior of the heart is divided into four cavities, or **chambers,** which receive the circulating blood (Figure 20-2d–f). The two upper chambers are called the right and left **atria.** Each atrium has an appendage called an *auricle* (AWR-i-kul; *auris* = ear), so named because its shape resembles a dog's ear. The auricle increases the atrium's surface area. The lining of the atria is smooth, except for the anterior atrial walls and the lining of the auricles, which contain projecting muscle bundles that are parallel to one another and resemble the teeth of a comb: the *musculi pectinati* (MUS-kyoo-lī pek-ti-NA-tē). These bundles give the lining of the auricles a ridged appearance.

The atria are separated by a partition called the *interatrial septum.* A prominent feature of this septum is an oval depression, the *fossa ovalis,* which corresponds to the site of the foramen ovale, an opening in the interatrial septum of the fetal heart. The fossa ovalis faces the opening of the inferior vena cava and is located in the septal wall of the right atrium.

The two lower chambers are the right and left **ventricles.** They are separated by an *interventricular septum.*

The muscle tissue of the atria and ventricles is separated by connective tissue that also forms the valves. This "cardiac skeleton" effectively divides the myocardium

PARIETAL PERICARDIUM (PERICARDIAL SAC)
Fibrous pericardium
Serous pericardium
Pericardial cavity

HEART WALL
VISCERAL PERICARDIUM (EPICARDIUM)
MYOCARDIUM (HEART MUSCLE)
ENDOCARDIUM
Trabeculae

(b)

FIGURE 20-1 (*Continued*) Heart. (b) Structure of the parietal pericardium and heart wall.

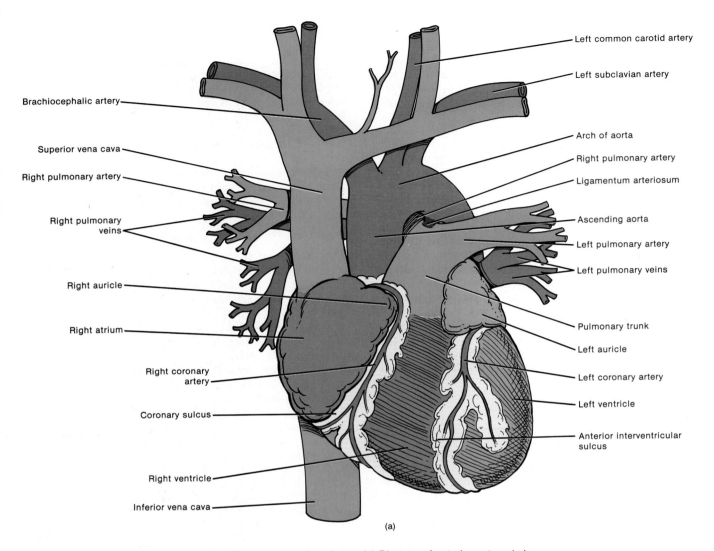

Brachiocephalic artery

Superior vena cava

Right pulmonary artery

Right pulmonary veins

Right auricle

Right atrium

Right coronary artery

Coronary sulcus

Right ventricle

Inferior vena cava

Left common carotid artery

Left subclavian artery

Arch of aorta

Right pulmonary artery

Ligamentum arteriosum

Ascending aorta

Left pulmonary artery

Left pulmonary veins

Pulmonary trunk

Left auricle

Left coronary artery

Left ventricle

Anterior interventricular sulcus

(a)

FIGURE 20-2 Structure of the heart. (a) Diagram of anterior external view.

into two separate muscle masses. Externally, a groove known as the *coronary sulcus* (SUL-kus) separates the atria from the ventricles. It encircles the heart and houses the coronary sinus and circumflex branch of the left coronary artery. The *anterior interventricular sulcus* and *posterior interventricular sulcus* separate the right and left ventricles externally. The sulci contain coronary blood vessels and a variable amount of fat (Figure 20-2a–c).

GREAT VESSELS OF THE HEART

The right atrium receives blood from all parts of the body except the lungs. It receives the blood through three veins. The **superior vena cava** brings blood from parts of the body superior to the heart; the **inferior vena cava,** brings blood from parts of the body inferior to the heart; and the **coronary sinus** drains blood from most of the vessels supplying the walls of the heart (Figure 20-2c). The right atrium then delivers the blood into the right ventricle, which pumps it into the **pulmonary trunk.** The pulmonary trunk divides into a **right** and **left pulmonary artery,** each of which carries blood to the lungs. In the

lungs, the blood releases its carbon dioxide and takes on oxygen. It returns to the heart via four **pulmonary veins** that empty into the left atrium. The blood then passes into the left ventricle, which pumps the blood into the **ascending aorta.** From here the blood is passed into the **coronary arteries, arch of the aorta, thoracic aorta, and abdominal aorta.** These blood vessels transport the blood to all body parts except the lungs.

The sizes of the four chambers vary according to function (Figure 20-2d–f). The right atrium, which must collect blood coming from almost all parts of the body, is slightly larger than the left atrium, which receives blood only from the lungs. The thickness of the chamber walls varies too. The atria are thin-walled because they need only enough cardiac muscle tissue to deliver the blood into the ventricles with the aid of a reduced pressure created by the expanding ventricles. The right ventricle has a thicker layer of myocardium than the atria, since it must send blood to the lungs and around back to the left atrium. The left ventricle has the thickest walls, since it must pump blood at high pressure through literally thousands of miles of vessels in the head, trunk, and extremities.

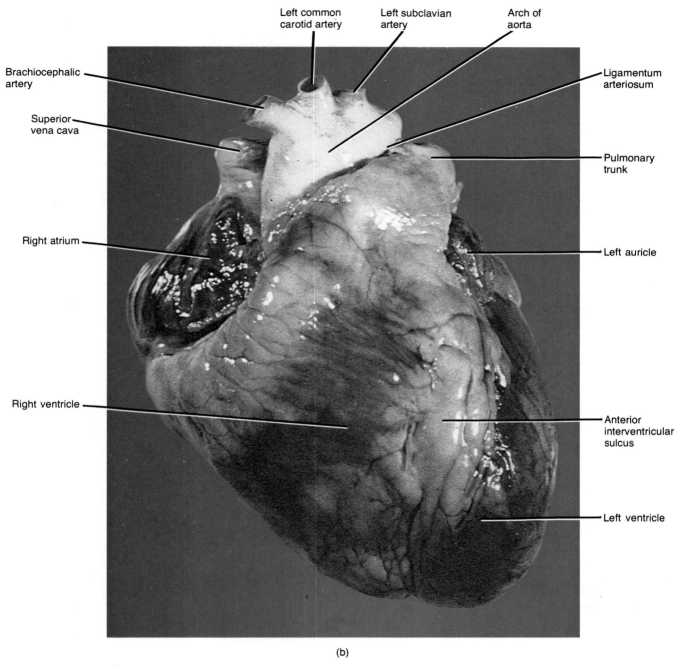

Left common
carotid artery

Left subclavian
artery

Arch of
aorta

Brachiocephalic
artery

Ligamentum
arteriosum

Superior
vena cava

Pulmonary
trunk

Right atrium

Left auricle

Right ventricle

Anterior
interventricular
sulcus

Left ventricle

(b)

FIGURE 20-2 (*Continued*) Structure of the heart. (b) Photograph of anterior external view. (Courtesy of C. Yokochi and J. W. Rohen, *Photographic Anatomy of the Human Body,* 2nd ed., 1979, IGAKU-SHOIN, Ltd., Tokyo, New York.)

VALVES OF THE HEART

As each chamber of the heart contracts, it pushes a portion of blood into a ventricle or out of the heart through an artery. In order to keep the blood from flowing backward, the heart has **valves.**

ATRIOVENTRICULAR (AV) VALVES

Atrioventricular (AV) valves lie between the atria and ventricles (Figure 20-2d). The right atrioventricular valve

between the right atrium and right ventricle, is also called the **tricuspid valve** because it consists of three flaps, or cusps. These flaps are fibrous tissues that grow out of the walls of the heart and are covered with endocardium. The pointed ends of the cusps project into the ventricle. Cords called *chordae tendineae* (KOR-dē TEN-di-nē) connect the pointed ends to small conical projections— the *papillary muscles* (muscular columns)—located on the inner surface of the ventricles. The irregular surface of ridges and folds of the myocardium in the ventricles is known as the *trabeculae carneae* (tra-BEK-yoo-lē

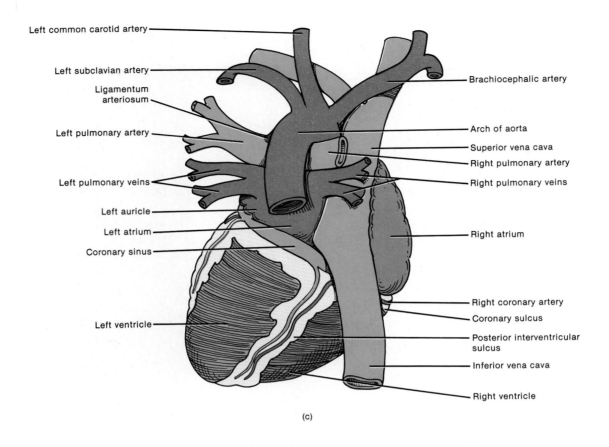

Left common carotid artery

Left subclavian artery

Ligamentum arteriosum

Left pulmonary artery

Left pulmonary veins

Left auricle

Left atrium

Coronary sinus

Left ventricle

Brachiocephalic artery

Arch of aorta

Superior vena cava

Right pulmonary artery

Right pulmonary veins

Right atrium

Right coronary artery

Coronary sulcus

Posterior interventricular sulcus

Inferior vena cava

Right ventricle

(c)

Brachiocephalic artery

Right pulmonary artery

Ascending aorta

Superior vena cava

Right pulmonary vein

Pulmonary trunk

Pulmonary semilunar valve

Right pulmonary vein

Fossa ovalis

Right atrium

Valve of coronary sinus

Valve of inferior vena cava

Tricuspid valve

Chordae tendineae

Right ventricle

Trabeculae carneae

Inferior vena cava

Left common carotid artery

Left subclavian artery

Arch of aorta

Ligamentum arteriosum

Left pulmonary artery

Left pulmonary veins

Left atrium

Aortic semilunar valve

Bicuspid valve

Left ventricle

Interventricular septum

Papillary muscle

Descending thoracic aorta

(d)

FIGURE 20-2 (*Continued*) Structure of the heart. (c) Diagram of posterior external view. (d) Diagram of anterior internal view.

(e)

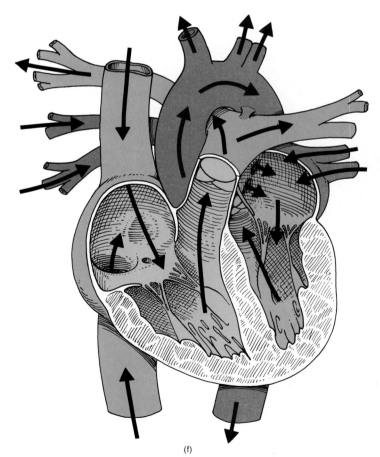

(f)

FIGURE 20-2 (*Continued*) Structure of the heart. (e) Photograph of a portion of anterior internal view. (Courtesy of J. W. Eads.) (f) Path of blood through the heart.

KAR-nē). The chordae tendineae and their muscles keep the flaps pointing in the direction of the blood flow. As the atrium relaxes and the ventricle pumps the blood out of the heart, any blood driven back toward the atrium is pushed between the flaps and the ventricle walls (Figure 20-3). This action drives the cusps upward until their edges meet and close the opening. At the same time, contraction of the papillary muscles helps prevent the valve from swinging upward into the atrium.

The atrioventricular valve between the left atrium and left ventricle is called the **bicuspid (mitral) valve.** It is also known as the left atrioventricular (left AV) valve. It has two cusps that work in the same way as the cusps of the tricuspid valve. Its cusps are also attached by way of the chordae tendineae to papillary muscles.

SEMILUNAR VALVES

Both arteries that leave the heart have a valve that prevents blood from flowing back into the heart. These are the **semilunar valves** (Figure 20-3). The **pulmonary semilunar valve** lies in the opening where the pulmonary trunk leaves the right ventricle. The **aortic semilunar valve** is situated at the opening between the left ventricle and the aorta.

Both valves consist of three semilunar (half-moon or crescent-shaped) cusps. Each cusp is attached by its convex margin to the artery wall. The free borders of the cusps curve outward and project into the opening inside the blood vessel. Like the atrioventricular valves, the semilunar valves permit blood to flow in only one direction—in this case, from the ventricles into the arteries.

(a)

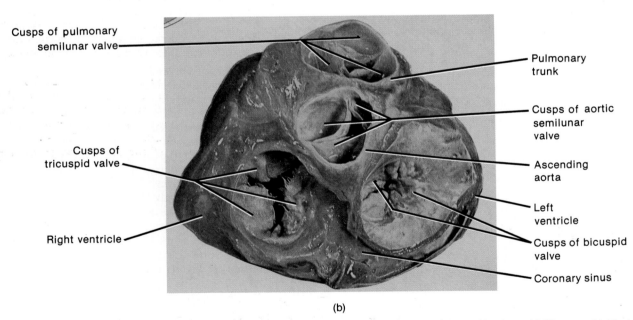

(b)

FIGURE 20-3 Valves of the heart. The atria have been removed to expose the tricuspid and bicuspid valves. (a) Diagram. (b) Photograph. (Courtesy of C. Yokochi and J. W. Rohen, *Photographic Anatomy of the Human Body*, 2nd ed., 1979, IGAKU-SHOIN, Ltd., Tokyo, New York.)

CLINICAL APPLICATION

The development of **artificial heart valves** made from synthetic materials has made it possible for cardiac surgeons to repair defective natural valves to alleviate many valvular problems (Figure 20-4). In some valvular disorders, a heart valve does not close properly, causing blood to backflow.

Even more amazing than artificial heart valves is the development of a permanently implanted **artificial heart.** On December 2, 1982, Dr. Barney B. Clark, a 61-year-old retired dentist, made medical history by becoming the first human to receive such an artificial heart. The 7½-hour operation was performed by Dr. William DeVries, a surgeon at the University of Utah Medical Center.

The artificial heart, known as Jarvik-7 after its inventor, Dr. Robert Jarvik, consists of an aluminum base and a pair of rigid plastic chambers that serve as ventricles (Figure 20-5). Each chamber contains a flexible diaphragm driven by compressed air that forces blood past mechanical valves into the major blood vessels. The surgery essentially consisted of removing the ventricles from Dr. Clark's heart, but leaving the atria intact. Then Dacron connectors were sutured onto the atria, aorta, and pulmonary trunk. The artificial heart was then snapped into position via the connectors.

The external power source for the artificial heart is a 375-pound unit that includes two compressors and a back-up compressor that force compressed air through six-foot-long air tubes into the artificial heart. These tubes enter through the abdomen. The unit also contains a three-hour supply of pressurized air should there be a power failure, a dehumidifier that removes moisture from the compressed air, and mechanisms that control air pressure and heart rate. The unit is always within six feet of the patient, the length of the air tubes. The artificial heart is designed to pump sufficient blood to sustain only moderate activity.

On the 112th day after the implantation, Dr. Clark died from massive circulatory collapse, shock, and failure of all of his natural organs. This was probably due to complications of antibiotic therapy to treat pneumonia and the damage that his body suffered just prior to implantation of the artificial heart. While in his body, the artificial heart beat almost 13 million times for 2,688 hours. Except for a defective biscuspid valve that had to be replaced, the heart showed no evidence of clots, infection, or physical deterioration. Dr. Clark's courage and determination have made him one of the outstanding pioneers in the advancement of medical research; his efforts will help scientists to develop even more successful artificial hearts.

CONDUCTION SYSTEM

The heart is innervated by the autonomic nervous system (Chapter 16), but the autonomic neurons only increase or decrease the time it takes to complete a cardiac cycle, that is, they do not initiate contraction. The chamber walls can go on contracting and relaxing, contracting and relaxing, without any direct stimulus from the nervous system. This action is possible because the heart has an intrinsic regulating system called the **conduction system.** The conduction system is composed of specialized muscle tissue that generates and distributes the electrical impulses which stimulate the cardiac muscle fibers to contract. These tissues are the sinoatrial (sinuatrial) node, the atrioventricular (AV) node, the atrioventricular (AV) bundle, the bundle branches, and the Purkinje fibers. The cells of the conduction system develop during embryological life from certain cardiac muscle cells. These cells lose their ability to contract and become specialized for impulse transmission.

A **node** of the conducting system is a compact mass of conducting cells. The **sinoatrial (sinuatrial) node,** also known as the **SA node** or **pacemaker,** is located in the right atrial wall inferior to the opening of the superior vena cava (Figure 20-6a). The SA node initiates each cardiac cycle, and thereby sets the basic pace for the heart rate—hence its common name, pacemaker. The rate set by the SA node may be altered by nervous impulses from the autonomic nervous system or by certain blood-borne chemicals such as thyroid hormones and epinephrine.

Once an action potential is initiated by the SA node, the impulse spreads out over both atria, causing them to contract and at the same time depolarizing the **atrioventricular (AV) node.** Because of its location near the inferior portion of the interatrial septum, the AV node is one of the last portions of the atria to be depolarized.

Artificial atrioventricular valve

FIGURE 20-4 Photograph of a superior view of an artificial atrioventricular valve. (Courtesy of John W. Eads.)

(a)   (b)

FIGURE 20-5 Artificial heart. (a) Photograph of an artificial heart prior to implantation. (b) Photograph of an implanted artificial heart. (Courtesy Nelson, University of Utah.)

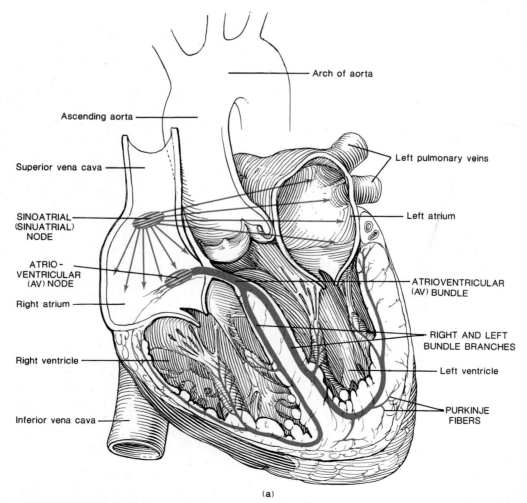

(a)

FIGURE 20-6 Conduction system of the heart. (a) Location of the nodes and bundles of the conduction system.

(b)

(c)

FIGURE 20-6 (*Continued*) Conduction system of the heart. (b) Normal electrocardiogram of a single heartbeat, enlarged for emphasis. (c) Normal electrocardiogram of several heartbeats.

From the AV node, a tract of conducting fibers called the **atrioventricular (AV) bundle** runs through the cardiac skeleton to the top of the interventricular septum. It then continues down both sides of the septum as the **right** and **left bundle branches.** The atrioventricular bundle distributes the action potential over the medial surfaces of the ventricles. Actual contraction of the ventricles is stimulated by the **Purkinje (pur-KIN-jē) fibers** that emerge from the bundle branches and pass into the cells of the myocardium.

ELECTROCARDIOGRAM

Impulse transmission through the conduction system generates electrical currents that may be detected on the body's surface. A recording of the electrical changes that accompany the cardiac cycle is called an **electrocardiogram (ECG or EKG).** The instrument used to record the changes is an *electrocardiograph.*

Each portion of the cardiac cycle produces a different electrical impulse. These impulses are transmitted from

the electrodes to a recording pen that graphs the impulses as a series of up-and-down waves called *deflection waves.* In a typical record (Figure 20-6b, c), three clearly recognizable waves accompany each cardiac cycle. The first, called the **P wave,** is a small upward wave. It indicates atrial depolarization—the spread of an impulse from the SA node through the muscle of the two atria. A fraction of a second after the P wave begins, the atria contract. The second wave, called the **QRS wave (complex),** begins as a downward deflection, continues as a large, upright, triangular wave, and ends as a downward wave at its base. This deflection represents ventricular depolarization, that is, the spread of the electrical impulse through the ventricles. The third recognizable deflection is a dome-shaped **T wave.** This wave indicates ventricular repolarization. There is no deflection to show atrial repolarization because the stronger QRS wave masks this event.

In reading an electrocardiogram, it is important to note the size of the deflection waves and certain time intervals. Enlargement of the P wave, for example, indicates enlargement of the atrium, as in mitral stenosis.

(a)

In this condition, the mitral valve narrows, blood backs up into the atrium, and there is expansion of the atrial wall.

The *P-R interval* is measured from the beginning of the P wave to the beginning of the R wave. It represents the conduction time from the beginning of atrial excitation to the beginning of ventricular excitation. The P-R interval is the time required for an impulse to travel through the atria and atrioventricular node to the remaining conducting tissues. The lengthening of this interval, as in atherosclerotic heart disease and rheumatic fever, occurs because the heart tissue covered by the P-R interval, namely the atria and atrioventricular node, is scarred or inflamed. Thus the impulse must travel at a slower rate and the interval is lengthened. The normal P-R interval covers no more than 0.2 sec.

An enlarged Q wave may indicate a myocardial infarction (heart attack). An enlarged R wave generally indicates enlarged ventricles.

The *S-T segment* begins at the end of the S wave and terminates at the beginning of the T wave. It represents the time between the end of the spread of the impulse through the ventricles and repolarization of the ventricles. The S-T segment is elevated in acute myocardial infarction and depressed when the heart muscle receives insufficient oxygen.

The T wave represents ventricular repolarization. It is flat when the heart muscle is receiving insufficient oxygen, as in atherosclerotic heart disease. It may be elevated when the body's potassium level is increased.

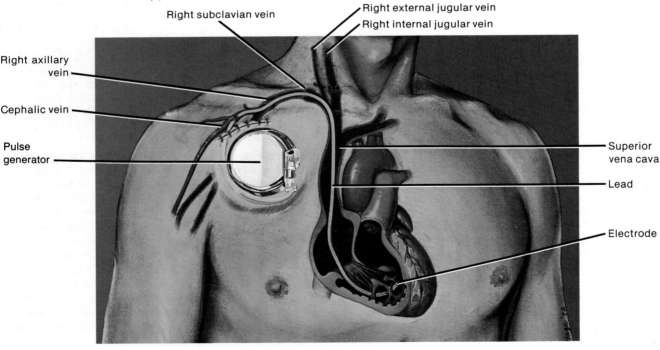

(b)

FIGURE 20-7 Artificial pacemaker. (a) Photograph of a pacemaker prior to implantation. (Courtesy of Medtronic, Inc.) (b) Diagram of an implanted artificial pacemaker. In this method of implantation, the lead is inserted into the cephalic vein under fluoroscopic guidance, following an incision inferior to the clavicle. The lead is then guided into the axillary and subclavian veins, superior vena cava, right atrium, and finally into the right ventricle to make contact with the ventricular myocardium. The pulse generator is placed in a subcutaneous pocket. (Courtesy of Medtronic, Inc.)

The ECG is useful in diagnosing abnormal cardiac rhythms and conduction patterns and following the course of recovery from a heart attack. It can also detect the presence of fetal life or determine the presence of several fetuses. For some people, ECG monitoring may be done by using a *Holter monitor.* This is an ambulatory system in which the monitoring hardware is worn by an individual while going about everyday routines. The Holter monitor is used especially to detect rhythm disorders in the conduction system. It is also used to correlate rhythm disorders and symptoms and to follow the effectiveness of drugs.

CLINICAL APPLICATION

When major elements of the conduction system are disrupted, heart block of varying degrees may result. Complete heart block results most commonly from a disturbance in AV conduction. The ventricles fail to receive atrial impulses, causing the ventricles and atria to beat independently of each other. In patients with heart block, normal heart rate can be restored and maintained with an **artificial pacemaker** (Figure 20-7). A pacemaker is a device that sends out small electrical charges that stimulate the heart. It consists of three basic parts: a *pulse generator,* which contains the battery cells; a *lead,* which is a flexible wire connected to the pulse generator; and an *electrode,* which delivers the charge to the heart.

Pulse generators are of two types. Demand (ventricular-inhibited) generators sense when the heart is sending out its own electrical charges and therefore will not send an electrical charge to the heart. But when the heart fails to beat, the demand generator senses the lack of activity and immediately sends an electrical charge to the heart. Demand generators are used for patients who still have some natural heart activity. The other pulse generator is called a fixed-rate (asynchronous) generator. This unit sends a series of electrical charges to the heart at a steady rate, usually 80 beats per minute. Unlike demand generators, fixed-rate generators supply all the stimulation, since the natural activity of the heart is too slow.

Pacemaker leads are referred to as endocardial and myocardial. Endocardial leads are passed through a vein so the tip, called the electrode, makes contact with the endocardium. This is the type shown in Figure 20-7b. Myocardial leads are designed so that the electrode attaches directly to the outside of the heart.

BLOOD SUPPLY

The wall of the heart, like any other tissue, including large blood vessels, has its own blood vessels. Nutrients could not possibly diffuse through all the layers of cells that make up the heart tissue. The flow of blood through the numerous vessels that pierce the myocardium is called the **coronary (cardiac) circulation** (Figure 20-8).

The vessels that serve the myocardium include the *left coronary artery,* which originates as a branch of the ascending aorta. This artery runs under the left atrium

and divides into the anterior interventricular and circumflex branches. The *anterior interventricular branch* follows the anterior interventricular sulcus and supplies oxygenated blood to the walls of both ventricles. The *circumflex branch* distributes oxygenated blood to the walls of the left ventricle and left atrium.

The *right coronary artery* also originates as a branch of the ascending aorta. It runs under the right atrium and divides into the posterior interventricular and marginal branches. The *posterior interventricular branch* follows the posterior interventricular sulcus and supplies the walls of the two ventricles with oxygenated blood. The *marginal branch* transports oxygenated blood to the myocardium of the right ventricle and right atrium. The left ventricle receives the most abundant blood supply because of the enormous work it must do.

As blood passes through the atrial system of the heart, it delivers oxygen and nutrients and collects carbon dioxide and wastes. Most of the deoxygenated blood, which carries the carbon dioxide and wastes, is collected by a large vein, the *coronary sinus,* which empties into the right atrium. A vascular sinus is a vein with a thin wall that has no smooth muscle to alter its diameter. The principal tributaries of the coronary sinus are the *great cardiac vein,* which drains the anterior aspect of the heart, and the *middle cardiac vein,* which drains the posterior aspect of the heart.

CLINICAL APPLICATION

Most heart problems result from faulty coronary circulation. If a reduced oxygen supply weakens the cells but does not actually kill them, the condition is called **ischemia** (is-KĒ-mē-a). **Angina pectoris** (an-JĪ-na PEK-tō-ris), meaning "chest pain," is ischemia of the myocardium. (Remember that pain impulses originating from most visceral muscles are referred to an area on the surface of the body.) Angina pectoris occurs when coronary circulation is somewhat reduced for some reason. Stress, which produces constriction of vessel walls, is a common cause. Equally common is strenous exercise after a heavy meal. When any quantity of food enters the stomach, the body increases blood flow to the digestive tract. As a consequence, some blood is diverted away from other organs, including the heart. Exercise, however, increases heart muscle activity and thus the heart's need for oxygen. Doing heavy work while food is in the stomach can therefore lead to oxygen deficiency in the myocardium. Agina pectoris weakens the heart muscle, but it does not produce a full-scale heart attack. The simple remedy of taking nitroglycerin, a drug that dilates coronary vessels and thereby increased blood flow to the area, brings coronary circulation back to normal and stops the pain of angina.

A much more serious problem is **myocardial infarction** (in-FARK-shun), commonly called a "coronary" or "heart attack." **Infarction** means the death of an area of tissue because of an interrupted blood supply. Myocardial infarction may result from a thrombus or embolus in one of the coronary arteries. The tissue distal to the obstruction dies and is replaced by noncon-

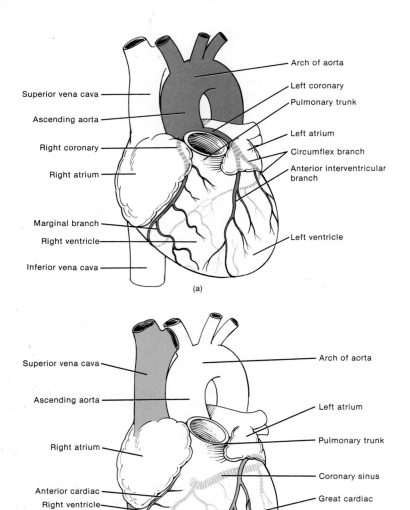

Superior vena cava

Ascending aorta

Right coronary

Right atrium

Marginal branch

Right ventricle

Inferior vena cava

Arch of aorta

Left coronary

Pulmonary trunk

Left atrium

Circumflex branch

Anterior interventricular branch

Left ventricle

(a)

Superior vena cava

Ascending aorta

Right atrium

Anterior cardiac

Right ventricle

Small cardiac

Inferior vena cava

Arch of aorta

Left atrium

Pulmonary trunk

Coronary sinus

Great cardiac

Left ventricle

Middle cardiac

(b)

FIGURE 20-8 Coronary (cardiac) circulation. (a) Anterior view of arterial distribution. (b) Anterior view of venous drainage.

tractile scar tissue. Thus, the heart muscle loses at least some of its strength. The aftereffects depend partly on the size and location of the infarcted, or dead, area.

Diagnostic tests have been developed that enable physicians to diagnose heart attacks within hours. During a heart attack, certain enzymes are released into the bloodstream and are the strongest indicators that a heart attack has occurred. One test detects creatine kinase MB isoenzyme (CK-MB), which is found only in heart cells. The test detects nearly all the CK-MB released after a heart attack and gives physicians a more accurate measure of the extent of the heart damage. Another test measures lactic dehydrogenase-1 (LDH-1), an enzyme liberated during a myocardial infarction. This test will yield results in an hour. Other enzymes that are always elevated after a myocardial infarction are serum glutamic oxaloacetic transaminase (SGOT) and creatine phosphokinase (CPK).

CARDIAC CYCLE

Before we examine the events of the cardiac cycle, it will be useful to first look at how blood flows through the heart.

BLOOD FLOW THROUGH THE HEART

Two phenomena control the movement of blood through the heart: the opening and closing of the valves and the contraction and relaxation of the myocardium. Both these activities occur without direct stimulation from the nervous system. The valves are controlled by pressure changes in each heart chamber. The contraction of the cardiac muscle is stimulated by its conduction system.

Blood flows from an area of higher pressure to an area of lower pressure. The pressure developed in a heart chamber is related primarily to the chamber's size and the volume of blood it contains. The greater the volume of blood, the higher the pressure. Figure 20-9b indicates pressure changes associated with the left side of the heart during the cardiac cycle. Although the pressures in the right side of the heart are somewhat lower, the same results are achieved.

The pressure in the atria is called **atrial pressure.** When the atria are relaxed, the pressure in them steadily increases because blood flows into them continuously from their vessels. When the ventricles contract, ventricular pressure drops below atrial pressure and the atrioventricular valves open. Atrial blood drains into the ventricles, which are empty and consequently have a lower pressure. When atrial pressure builds up, the atria

- Contrast the structure and function of arteries, arterioles, capillaries, venules, and veins.

- Explain why blood circulates between different regions of the cardiovascular system.

- Relate the importance of cardiac output, blood volume, and peripheral resistance to blood pressure.

- Explain the role of the vasomotor center in controlling blood pressure.

- Contrast the roles of pressoreceptors and chemoreceptors in regulating blood pressure.

- Describe the effects of epinephrine, norepinephrine (NE), antidiuretic hormone (ADH), angiotensin II, histamine, and kinins on blood pressure.

- Define autoregulation and explain its importance.

- Discuss the movement of fluids through capillaries.

- Explain how velocity of blood flow, skeletal muscle contractions, valves in veins, and breathing assist in the return of venous blood to the heart.

- Define a blood reservoir and explain its importance.

- Explain the effects of exercise on the cardiovascular system.

- Define pulse and identify the arteries where pulse may be felt.

- Define blood pressure (BP).

- Explain one clinical method for recording systolic and diastolic pressure.

- Contrast the clinical significance of systolic, diastolic, and pulse pressures.

- Compare systemic, hepatic portal, pulmonary, fetal, and cerebral circulation.

- Identify the principal arteries and veins of systemic circulation.

- Trace the route of blood involved in hepatic portal circulation and explain its importance.

- Identify the major blood vessels of pulmonary circulation.

- Contrast fetal and adult circulation.

- Describe the importance of cerebral circulation.

- Explain the fate of fetal circulation structures once postnatal circulation is established.

- List the causes and symptoms of aneurysms, coronary artery disease, atherosclerosis, coronary artery spasm, and hypertension.

- Define medical terminology associated with the cardiovascular system.

- Explain the actions of selected drugs on the cardiovascular system.

21 The Cardiovascular System: Vessels and Routes

The blood vessels form a network of tubes that carry blood away from the heart, transport it to the tissues of the body, and then return it to the heart. **Arteries** are the vessels that carry blood from the heart to the tissues. Large, elastic arteries leave the heart and divide into medium-sized, muscular vessels that head toward the various regions of the body. The medium-sized arteries divide into small arteries, which, in turn, divide into still smaller arteries called **arterioles.** As the arterioles enter a tissue, they branch into countless microscopic vessels called **capillaries.** Through the walls of the capillaries, substances are exchanged between the blood and body tissues. Before leaving the tissue, groups of capillaries reunite to form small veins called **venules.** These, in turn, merge to form progressively larger tubes called veins. **Veins,** then, are blood vessels that convey blood from the tissues back to the heart. Since blood vessels require oxygen and nutrients just like other tissues of the body, they also have blood vessels in their own walls called **vasa vasorum.**

ARTERIES

Arteries have walls constructed of three coats or tunics and a hollow core, called a *lumen,* through which the blood flows (Figures 21-1 and 21-4). The inner coat of an arterial wall is the **tunica interna (intima).** It is composed of a lining of endothelium (simple squamous epithelium) that is in contact with the blood and a layer of elastic tissue called the internal elastic membrane. The middle coat, or **tunica media,** is usually the thickest layer. It consists of elastic fibers and smooth muscle. The outer coat, the **tunica externa (adventitia),** is composed principally of elastic and collagenous fibers. An external elastic membrane may separate the tunica media from the tunica externa.

As a result of the structure of the middle coat especially, arteries have two major properties: elasticity and contractility. When the ventricles of the heart contract and eject blood into the large arteries, the arteries expand to contain the extra blood. Then, as the ventricles relax, the elastic recoil of the arteries forces the blood onward. The contractility of an artery comes from its smooth muscle. The smooth muscle is arranged longitudinally and in rings around the lumen somewhat like a doughnut and is innervated by sympathetic branches of the autonomic nervous system. When there is sympathetic stimulation, the smooth muscle contracts, squeezes the wall around the lumen, and narrows the vessel. Such a decrease in the size of the lumen is called **vasoconstriction.** Conversely, when sympathetic stimulation is removed, the smooth muscle fibers relax and the size of the arterial lumen increases. This increase is called **vasodilation** and is often due to the inhibition of vasoconstriction.

The contractility of arteries also serves a function in stopping bleeding. This is called vascular spasm, one of the three mechanisms involved in hemostasis (Chapter 19). The blood flowing through an artery is under a great deal of pressure. Thus great quantities of blood can be quickly lost from a broken artery. When an artery is cut, its wall constricts so that blood does not escape quite so rapidly. However, there is a limit to how much vasoconstriction can help.

ELASTIC ARTERIES

Large arteries are referred to as **elastic** or **conducting arteries.** They include the aorta and brachiocephalic, common carotid, subclavian, vertebral, and common iliac arteries. The wall of elastic arteries is relatively thin in proportion to their diameter and their tunica media contains more elastic fibers and less smooth muscle. As the heart alternately contracts and relaxes, the rate of blood flow tends to be intermittent. When the heart contracts and forces blood into the aorta, the wall of the elastic arteries stretches. During relaxation of the heart, the wall of the elastic arteries recoils, moving the blood forward in a more continuous flow. Elastic arteries are called conducting arteries because they conduct blood from the heart to medium-sized arteries.

MUSCULAR ARTERIES

Medium-sized arteries are called **muscular** or **distributing arteries.** They include the axillary, brachial, radial, intercostal, splenic, mesenteric, femoral, popliteal, and tibial arteries. Their tunica media contains more smooth muscle than elastic fibers and they are capable of vasoconstriction and vasodilation to adjust the volume of blood to suit the needs of the structure supplied. The wall of muscular arteries is relatively thick, mainly due to the large amounts of smooth muscle. Muscular arteries are called distributing arteries because they distribute blood to various parts of the body.

ANASTOMOSIS

Most parts of the body receive branches from more than one artery. In such areas the distal ends of the vessels unite. The junction of two or more vessels supplying the same body region is called an **anastomosis** (a-nas-tō-MŌ-sis). Anastomoses may also occur between the origins of veins and between arterioles and venules. Anastomoses between arteries provide alternate routes by which blood can reach a tissue or organ. Thus if a vessel is occluded by disease, injury, or surgery, circulation to a part of the body is not necessarily stopped. The alternate route of blood to a body part through an anastomosis is known as **collateral circulation.** An alternate blood route may also be from nonanastomosing vessels that supply the same region of the body.

Arteries that do not anastomose are known as **end arteries.** Occlusion of an end artery interrupts the blood supply to a whole segment of an organ, producing necrosis (death) of that segment.

> **CLINICAL APPLICATION**
> A procedure that provides physicians with a much clearer look at diseased arteries is called **digital subtraction angiography (DSA).** The procedure employs a computer technique that compares an x-ray image

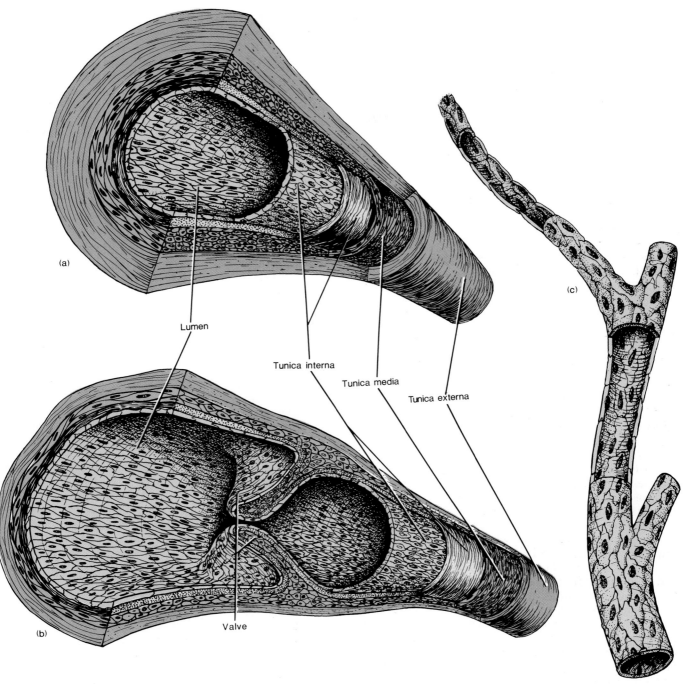

Lumen

Tunica interna

Tunica media

Tunica externa

Valve

(a)

(b)

(c)

FIGURE 21-1 Comparative structure of (a) an artery, (b) a vein, and (c) a capillary. The relative size of the capillary is enlarged for emphasis.

of the same region of the body before and after a contrast substance has been introduced intravenously. Any tissue or blood vessels that show up in the first image can be subtracted (erased) from the second image, leaving an unobstructed view of an artery. DSA figuratively lifts an artery out of the body so that it can be studied in isolation. DSA helps in diagnosing lesions in the carotid arteries leading to the brain, a potential cause of strokes, and in evaluating patients before surgery and after coronary bypass and certain transplant operations.

ARTERIOLES

An **arteriole** is a small artery that delivers blood to capillaries. Arterioles closer to the arteries from which they branch have a tunica interna like that of arteries, a tunica media composed of smooth muscle and very few elastic fibers, and a tunica externa composed mostly of elastic and collagenous fibers (see Figure 21-4c). As arterioles get smaller in size, the tunics change character so that arterioles closest to capillaries consist of little more than a layer of endothelium surrounded by a few scattered smooth muscle cells (Figure 21-2).

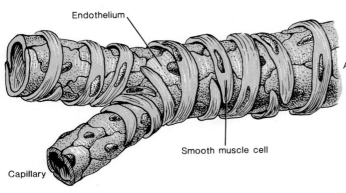

FIGURE 21-2 Structure of an arteriole.

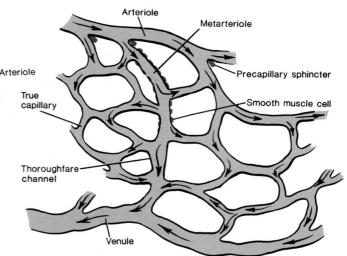

FIGURE 21-3 Details of a capillary network.

Arterioles play a key role in regulating blood flow from arteries into capillaries. The smooth muscle of arterioles, like that of arteries, is subject to vasoconstriction and vasodilation. During vasoconstriction, blood flow into capillaries is restricted; during vasodilation, the flow is significantly increased. The relationship of arterioles to blood flow will be considered in detail later in the chapter.

CAPILLARIES

Capillaries are microscopic vessels that usually connect arterioles and venules (Figure 21-1). They are found near almost every cell in the body. The distribution of capillaries in the body varies with the activity of the tissue. For example, in places where activity is higher, such as muscles, liver, kidneys, lungs, and nervous system, there are rich capillary supplies. In areas where activity is lower, such as tendons and ligaments, the capillary supply is not as extensive. The epidermis, cornea of the eye, and cartilage are devoid of capillaries.

The primary function of capillaries is to permit the exchange of nutrients and wastes between the blood and tissue cells. The structure of the capillaries is admirably suited to this purpose. Capillary walls are composed of only a single layer of cells (endothelium). They have no tunica media or tunica externa. Thus a substance in the blood must pass through the plasma membrane of just one cell to reach tissue cells. This vital exchange of materials occurs only through capillary walls—the thick walls of arteries and veins present too great a barrier.

Although in some places in the body capillaries pass directly from arterioles to venules, in other places they form extensive branching networks. These networks increase the surface area for diffusion and thereby allow a rapid exchange of large quantities of materials. In most tissues, blood normally flows through only a small portion of the capillary network when metabolic needs are low. But, when a tissue becomes active, the entire capillary network fills with blood.

The flow of blood through capillaries is regulated by vessels with smooth muscle in their walls. A **metarteriole** (*met* = beyond) is a vessel that emerges from an arteriole, transverses the capillary network, and empties into a venule (Figure 21-3). The proximal portions of metarterioles are surrounded by scattered smooth muscle cells whose

contraction and relaxation help to regulate the amount and force of blood. The distal portion of a metarteriole has no smooth muscle cells and is called a *thoroughfare channel.* It serves as a low-resistance channel that increases blood flow. **True capillaries** emerge from arterioles or metarterioles and are not on the direct flow route from arteriole to venule. At their sites of origin, there is a ring of smooth muscle called a **precapillary sphincter** that controls the flow of blood entering a true capillary. The various factors that regulate the contraction of smooth muscle cells and precapillary sphincters are discussed later.

Some capillaries of the body, such as those found in muscle tissue and other locations, are referred to as **continuous capillaries.** These capillaries are so named because the cytoplasm of the endothelial cells is continuous when viewed in cross section through a microscope; the cytoplasm appears as an uninterrupted ring, except for the endothelial junction. Other capillaries of the body are referred to as **fenestrated capillaries.** They differ from continuous capillaries in that their endothelial cells have numerous fenestrae (pores) where the cytoplasm is absent. The fenestrae range from 700 to 1000 Å in diameter and are closed by a thin diaphragm, except in the capillaries in the kidneys where they are assumed to be open. Fenestrated capillaries are also found in the villi of the small intestine, choroid plexuses of the ventricles in the brain, ciliary processes of the eyes, and endocrine glands.

Microscopic blood vessels in certain parts of the body, such as the liver, are termed **sinusoids.** They are wider than capillaries and more tortuous. Also, instead of the usual endothelial lining, sinusoids are lined largely by phagocytic cells. In the liver, such cells are called *stellate reticuloendothelial cells.* Like capillaries, sinusoids convey blood from arterioles to venules. Other regions containing sinusoids include the spleen, adenohypophysis, parathyroid glands, and adrenal cortex.

VENULES

When several capillaries unite, they form small veins called **venules** (see Figure 21-4c). Venules collect blood

FIGURE 21-4 Histology of blood vessels. (a) Photomicrograph of an artery and its accompanying vein at a magnification of 50×. (b) Photomicrograph of a portion of the wall of an artery and its accompanying vein at a magnification of 250×.

from capillaries and drain it into veins. The venules closest to the capillaries consist of a tunica interna of endothelium and a tunica externa of connective tissue. As the venules approach the veins, they also contain the tunica media characteristic of veins.

VEINS

Veins are composed of essentially the same three coats as arteries, but they have considerably less elastic tissue and smooth muscle and contain more white fibrous tissue

(Figures 21-1 and 21-4). However, they are still distensible enough to adapt to variations in the volume and pressure of blood passing through them.

By the time the blood leaves the capillaries and moves into the veins, it has lost a great deal of pressure. The difference in pressure can be observed in the blood flow from a cut vessel; blood leaves a cut vein in an even flow rather than in the rapid spurts characteristic of arteries. Most of the structural differences between arteries and veins reflect this pressure difference. For example, veins do not need walls as strong as those of arteries.

(c)

FIGURE 21-4 (*Continued*) Histology of blood vessels. (c) Photomicrograph of an arteriole and a venule at a magnification of 800×. (© 1983 by Michael H. Ross. Used by permission.)

The low pressure in veins, however, has its disadvantages. When you stand, the pressure pushing blood up the veins in your lower extremities is barely enough to balance the force of gravity pushing it back down. For this reason, many veins, especially those in the limbs, contain valves that prevent backflow. Normal valves ensure the flow of blood toward the heart.

CLINICAL APPLICATION

In people with weak venous valves, large quantities of blood are forced by gravity back down into distal parts of the vein. This pressure overloads the vein and pushes the walls outward. After repeated overloading, the walls lose their elasticity and become stretched and flabby. A vein damaged in this way is called a **varicose vein.** Varicose veins may be caused by prolonged standing and pregnancy. Because a varicosed wall is not able to exert a firm resistance against the blood, blood tends to accumulate in the pouched-out area of the vein, causing it to swell and forcing fluid into the surrounding tissue. Veins close to the surface of the legs are highly susceptible to varicosities. Veins that lie deeper are not as vulnerable because surrounding skeletal muscles prevent their walls from overstretching.

Varicosities are also common in the veins that lie in the wall of the anal canal. These varicosities are called **hemorrhoids** (HEM-o-royds). Hemorrhoids may be caused by constipation. Repeated straining during defecation forces blood down into the superior hemorrhoidal plexus, increasing pressure in these veins. Constipation is related to low-fiber diets, especially in

---

North America. One hypothesis links increased intra-abdominal pressure, caused by straining during evacuation of firm feces, directly to hemorrhoids or varicose veins. A new development in the treatment of hemorrhoids is *cryosurgery* (*cryo* = cold), in which external hemorrhoids are destroyed by freezing them with a solution of nitrous oxide or liquid nitrogen.

A **vascular (venous) sinus** is a vein with a thin endothelial wall that has no smooth muscle to alter its diameter. Surrounding tissue replaces the tunica media and tunica externa to provide support. Intracranial vascular sinuses, which are supported by the dura mater, return cerebrospinal fluid and deoxygenated blood from the brain to the heart. Another example of a vascular sinus is the coronary sinus of the heart.

PHYSIOLOGY OF CIRCULATION

BLOOD FLOW AND BLOOD PRESSURE

Blood flows through its system of closed vessels because of different pressures in various parts of the system. It always flows from regions of higher pressure to regions of lower pressure. The mean (average) pressure in the aorta is about 100 mm Hg (mercury). This pressure continually decreases rapidly through the arterial system and more slowly through the venous system (Figure 21-5). Because of the continuous drop in pressure, blood flows from the aorta (100 mm Hg) to the arteries (100–40 mm Hg) to arterioles (40–25 mm Hg) to capillaries (25–12 mm Hg) to venules (12–8 mm Hg) to veins (10–5 mm

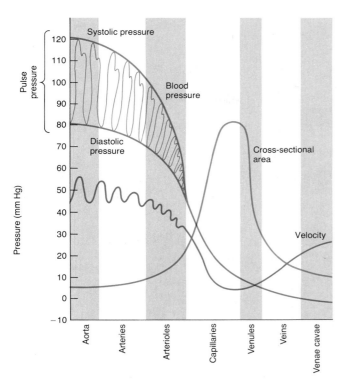

FIGURE 21-5 Relationship between blood pressure, velocity of blood flow, and total cross-sectional area in various blood vessels of the cardiovascular system. The velocity curve (green) and cross-sectional area curve (blue) are not related to the pressures shown in millimeters of mercury (mm Hg).

Hg) to the venae cavae (2 mm Hg). The pressure in the right atrium is 0 mm Hg.

Other mechanisms aid the flow of blood also. When blood leaves the capillaries, it enters the venules and veins, which are larger in diameter and thereby offer less resistance to flow. Contraction of skeletal muscles around the veins also helps drive blood toward the heart.

Factors That Affect Arterial Blood Pressure

We have defined blood pressure as the pressure exerted by blood on the wall of any blood vessel. In clinical use, however, the term refers to the pressure in arteries. Here we will identify several factors that influence arterial blood pressure.

● *Cardiac Output* (*CO*)   Cardiac output (CO), the amount of blood ejected by the left ventricle into the aorta each minute, is the principal determinant of blood pressure. As noted in Chapter 20, CO is calculated by multiplying stroke volume by heart rate. In a normal, resting adult, it is about 5.25 liters/min (70 ml × 75 beats/min). Blood pressure varies directly with CO. If CO is increased by any increase in stroke volume or heart rate, then blood pressure increases. A decrease in CO causes a decrease in blood pressure.

● *Blood Volume*   Blood pressure is directly proportional to the volume of blood in the cardiovascular system. The normal volume of blood in a human body is about 5

liters (5 qt). Any decrease in this volume, as from hemorrhage, decreases the amount of blood that is circulated through the arteries each minute. As a result, blood pressure drops. Conversely, anything that increases blood volume, such as high salt intake and therefore water retention, increases blood pressure.

● *Peripheral Resistance*   **Peripheral resistance** refers to the resistance (impedance) to blood flow by the force of friction between blood and the walls of blood vessels. It is related to the viscosity of blood and blood vessel diameter.

The viscosity of blood is a function of the number of red blood cells and amount of plasma proteins and also the ratio of red blood cells and solutes to fluid. Any condition that increases the viscosity of blood, such as dehydration or an unusually high number of red blood cells, increases blood pressure. A depletion of plasma proteins or red blood cells, as a result of anemia or hemorrhage, decreases blood viscosity and blood pressure.

The smaller the diameter of a vessel, the more resistance it offers the blood. A major function of arterioles is to control peripheral resistance and, therefore, blood pressure, by changing their diameters. The center for this regulation is the vasomotor center in the medulla, which will be described shortly.

The factors that affect arterial blood pressure are summarized in Figure 21-6.

Control of Blood Pressure

In Chapter 20 we considered the increase and decrease of heart rate and force of contraction by the cardioacceleratory center (CAC) and the cardioinhibitory center (CIC). We also took a look at how certain chemicals (epinephrine, potassium, sodium, and calcium), temperature, emotions, sex, and age affect heart rate. By their effect on heart rate and force of contraction, these factors also control blood pressure. Any increase in heart rate and force of contraction increases blood pressure. Conversely, any decrease will lower blood pressure.

At this point we will examine those factors that help to regulate blood pressure by acting on the blood vessels themselves.

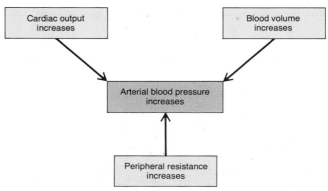

FIGURE 21-6 Summary of factors that affect arterial blood pressure.

● *Vasomotor Center*   Within the medulla is a cluster of neurons referred to as the **vasomotor** (*vas* = vessel, *motor* = movement) **center.** The function of this center is to control the diameter of blood vessels, especially arterioles. It continually sends impulses to the smooth muscle in arteriole walls that result in a moderate state of vasoconstriction at all times. This state of tonic contraction, called *vasomotor tone,* assumes a role in maintaining peripheral resistance and blood pressure. The vasomotor center brings about vasoconstriction by increasing the number of sympathetic impulses above normal. It causes vasodilation by decreasing the number of sympathetic impulses below normal. In other words, in this case, the sympathetic division of the ANS can bring about both vasoconstriction and vasodilation.

The vasomotor center is affected by several factors, all of which influence blood pressure.

● *Pressoreceptors*   It will be recalled that stimulated **pressoreceptors** in the carotid sinus and aorta send impulses to the cardiac center that result in increased or decreased cardiac output to help regulate blood pressure (see Figure 20-12). This reflex not only acts on the heart, but it also involves arterioles. For example, if there is an increase in blood pressure, the pressoreceptors stimulate the cardioinhibitory center and inhibit the cardioacceleratory center. The result is a decrease in cardiac output and a decrease in blood pressure. The pressoreceptors also send impulses to the vasomotor center. In response, the vasomotor center decreases sympathetic stimulation to the arterioles. The result is vasodilation and a decrease in blood pressure (Figure 21-7). If there is a decrease in blood pressure, the pressoreceptors inhibit the cardioinhibitory center and stimulate the cardioacceleratory center. The result is an increase in cardiac output and an increase in blood pressure. Also, the pressoreceptors send impulses to the vasomotor center that increase sympathetic stimulation to arterioles. The result is vasoconstriction and an increase in blood pressure.

● *Chemoreceptors*   Receptors sensitive to chemicals in the blood are called **chemoreceptors.** Chemoreceptors in the carotid sinus and aorta are called *carotid* and *aortic bodies,* respectively. They are sensitive to arterial blood levels of oxygen, carbon dioxide, and hydrogen ions. Given a deficiency of oxygen (hypoxia), a decrease in hydrogen ion concentration, or an excess of carbon dioxide (hypercapnia), the chemoreceptors are stimulated and send impulses to the vasomotor center. In response, the vasomotor center increases sympathetic stimulation to arterioles. This brings about vasoconstriction and an increase in blood pressure. As you will see in Chapter 23, chemoreceptors also stimulate the medullary rhythmicity area in response to these chemical stimuli to adjust the rate of respiration.

● *Control by Higher Brain Centers*   In response to strong emotions, **higher brain centers,** such as the cerebral cortex, can have a significant influence on blood pressure. For example, during periods of intense anger, the cerebral cortex stimulates the vasomotor center to fire sympathetic impulses to arterioles. This causes vasoconstriction and

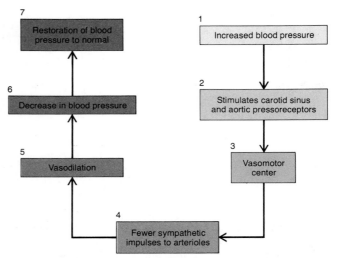

FIGURE 21-7 Vasomotor center control of blood pressure by stimulation of pressoreceptors.

an increase in blood pressure. When a person is emotionally upset, impulses from higher brain centers cause a decrease in sympathetic stimulation by the vasomotor center. This produces vasodilation and a decrease in blood pressure. A frequent result is fainting because blood flow to the brain is diminished.

● *Chemicals*   Several **chemicals** affect blood pressure by causing vasoconstriction. Epinephrine and norepinephrine (NE), produced by the adrenal medulla, increase the rate and force of heart contractions and bring about vasoconstriction of abdominal and cutaneous arterioles. They also bring about dilation of cardiac and skeletal arterioles. Antidiuretic hormone (ADH), produced by the hypothalamus and released from the neurohypophysis, causes vasoconstriction if there is a severe loss of blood due to hemorrhage. Angiotensin II helps to raise blood pressure by stimulating secretion of aldosterone (increases sodium ion concentration and water reabsorption) and causing vasoconstriction due to release of renin. Histamine produced by mast cells and kinins found in plasma are vasodilators that assume key functions during the inflammatory response.

● *Autoregulation (Local Control)*   **Autoregulation** refers to a local, automatic adjustment of blood flow in a given region of the body in response to the particular needs of the tissue. In most body tissues, oxygen is the principal stimulus for autoregulation. A suggested mechanism for autoregulation is as follows. In response to low oxygen supplies, the cells in the immediate area produce and release *vasodilator substances.* Such substances are thought to include potassium ions, hydrogen ions, carbon dioxide, lactic acid, and adenosine. Once released, the vasodilator substances produce a local dilation of arterioles and relaxation of precapillary sphincters. The result is an increased flow of blood into the tissue, which restores oxygen levels to normal. The autoregulation mechanism is important in meeting the nutritional demands of active tissues, such as muscle tissue, where the demand might increase as much as tenfold.

Tissue cells

Interstitial fluid

FIGURE 21-8 Hydrostatic and osmotic pressure forces involved in moving fluid out of plasma (filtration) and into plasma (reabsorption).

Key:
BHP = Blood hydrostatic pressure
IFHP = Interstitial fluid hydrostatic pressure
BOP = Blood osmotic pressure
IFOP = Interstitial fluid osmotic pressure

Capillary Exchange

For reasons to be discussed shortly, you will see that, although blood pressure decreases consistently from the aorta to the venae cavae, the velocity of blood decreases as it flows through the aorta, arterioles, and capillaries and then increases as it passes into venules and veins. The velocity of blood flow in capillaries is the slowest in the cardiovascular system, and this is important to allow for the exchange of materials between blood and body tissues.

Blood usually does not flow in a continuous manner through capillary networks. Rather, it flows intermittently because of contraction and relaxation of the smooth muscle cells of metarterioles and the precapillary sphincters of true capillaries. The intermittent contraction and relaxation, which may occur 5 to 10 times per minute, is called **vasomotion.** The most important factor in controlling vasomotion is the oxygen concentration in tissues, a mechanism similar to autoregulation in arterioles.

Diffusion is the principal mechanism by which substances are exchanged between capillary blood and body cells. Most substances are exchanged by this process through endothelial junctions of continuous capillaries and fenestrae of fenestrated capillaries. Some larger molecules are exchanged by pinocytosis.

The movement of water and dissolved substances, except proteins, through capillaries is greatly affected by hydrostatic pressure and osmotic pressure. Blood pressure in capillaries is called **blood hydrostatic pressure (BHP).** It tends to move fluid out of a capillary and averages 25 mm Hg at the arterial end of a capillary and 10 mm Hg at the venous end (Figure 21-8). **Interstitial fluid hydrostatic pressure (IFHP)** is the pressure of interstitial fluid against cells of a tissue and the endothelial cells of capillaries. It also tends to move fluid into capillaries and averages −6.3 mm Hg at both the arterial and venous ends of capillaries. **Blood osmotic pressure (BOP)** is due

principally to the presence of large amounts of plasma proteins and tends to move fluid into capillaries by osmosis. It averages 28 mm Hg at both ends of capillaries. **Interstitial fluid osmotic pressure (IFOP),** due to small amounts of proteins in interstitial fluid, tends to move fluid out of capillaries. It averages 5.0 mm Hg at both ends of capillaries.

Whether fluids leave or enter capillaries depends on how the pressures relate to each other. If the forces that tend to move fluid out of capillaries are greater than the forces that tend to move fluid into capillaries, the fluid will move from capillaries into tissue spaces. If, on the other hand, the inward force is greater, then fluid will move from tissue spaces into capillaries. The term **effective filtration pressure (Peff)** is used to show the direction of fluid movement. It is calculated as follows:

$$Peff = (BHP + IFOP) - (IFHP + BOP)$$

Substituting measured values at the arterial end of a capillary,

$$Peff = (25 + 5) - (-6.3 + 28)$$
$$= (30) - (21.7)$$
$$= 8.3 \text{ mm Hg}$$

Substituting measured values at the venous end of a capillary,

$$Peff = (10 + 5) - (-6.3 + 28)$$
$$= (15) - (21.7)$$
$$= -6.7 \text{ mm Hg}$$

Thus, at the arterial end of a capillary, there is a *net outward force* (8.3 mm Hg), and fluid moves out of the capillary (filtered) into tissue spaces. At the venous end of a capillary, the negative value (−6.7 mm Hg) represents a *net inward force,* and fluid moves into the capillary (reabsorbed) from tissue spaces.

Not all the fluid filtered at one end of the capillary is reabsorbed at the other end. When we discuss fluid dynamics in more detail in Chapter 27, you will see that some of the filtered fluid and any proteins that escape from blood into interstitial fluid are returned by the lymphatic system to the cardiovascular system. Under normal conditions there is a state of near equilibrium at the arterial and venous ends of a capillary in which filtered fluid and absorbed fluid plus that returned to the lymphatic system are nearly equal. This near equilibrium is known as **Starling's law of the capillaries.**

Factors That Aid Venous Return

The establishment of a pressure gradient is the primary reason why blood flows. But a number of other factors help the blood to return through the veins: an increased rate of flow, contractions of the skeletal muscles, valves, and breathing.

● *Velocity of Blood Flow*   The velocity of blood flow is inversely related to the cross-sectional area of the blood vessels. Blood flows most rapidly where the cross-sectional area is least (see Figure 21-5). Each time an artery branches, the total cross-sectional area of all the branches is greater than that of the original vessel. When branches combine, the resulting cross-sectional area is less than that of the original branches. The cross-sectional area of the aorta is 2.5 cm², and the velocity of the blood there is 40 cm/sec. Capillaries have a cross-sectional area of 2,500 cm², and the velocity is less than 0.1 cm/sec. In the venae cavae, the cross-sectional area is 8 cm² and the velocity is 5 to 20 cm/sec. The area and velocity for the different types of blood vessels are listed in Exhibit 21-1.

Thus, the velocity of blood decreases as it flows from the aorta to arteries to arterioles to capillaries. The velocity of blood in the capillaries is the slowest in the cardiovascular system. Thus there is adequate exchange (diffusion) time between the capillaries and adjacent tissues. As blood vessels leave capillaries and approach the heart, their cross-sectional area decreases. Therefore, the velocity of blood increases as it flows from capillaries to venules to veins to the heart.

CLINICAL APPLICATION
Blood flow can be measured to diagnose certain circulatory disorders. **Circulation time** is the time required for blood to pass from the right atrium, through pulmonary circulation, back to the left ventricle, through systemic circulation down to the foot, and back again to the right atrium. Such a trip usually takes about 1 minute and requires about 28 heartbeats.

● *Skeletal Muscle Contractions and Valves*   The combination of skeletal muscle contractions and valves in veins is important in returning venous blood to the heart. Many veins, especially those in the extremities, contain valves. When skeletal muscles contract, they tighten around the veins running through them and the valves open. This pressure drives the blood toward the heart—the action is called **milking** (Figure 21-9). When the muscles relax,

EXHIBIT 21-1

RELATIONSHIP BETWEEN BLOOD VELOCITY AND CROSS-SECTIONAL AREA

VESSEL	CROSS-SECTIONAL AREA (cm²)	VELOCITY (cm/sec)
Aorta	2.5	40
Arteries	20	10–40
Arterioles	40	0.1
Capillaries	2,500	less than 0.1
Venules	250	0.3
Veins	80	0.3–5.0
Venae cavae	8	5–20

the valves close to prevent the backflow of blood away from the heart. Individuals who are immobilized through injury or disease cannot take advantage of these contractions. As a result, the return of venous blood to the heart is slower and the heart has to work harder. For this reason, periodic massage is helpful.

● *Breathing*   Another factor that is important in maintaining venous circulation is breathing. During inspiration, the diaphragm moves downward. This causes a decrease in pressure in the chest cavity and an increase in pressure in the abdominal cavity. Once the pressure difference is established, blood is squeezed from the abdominal veins into the thoracic veins. When the pressures reverse during expiration, blood in the veins is prevented from backflowing by the valves.

BLOOD RESERVOIRS

The volume of blood in various parts of the cardiovascular system varies considerably. Veins, venules, and venous sinuses contain about 59 percent of the blood in the system, arteries about 13 percent, pulmonary vessels about 12 percent, the heart about 9 percent, and arterioles and capillaries about 7 percent. Since systemic veins contain so much of the blood, they are referred to as **blood reservoirs.** They serve as storage depots for blood, which can be moved out of them quickly to other parts of the body if the need arises. When there is increased muscular activity, the vasomotor center sends increasing sympathetic impulses to veins that serve as blood reservoirs. The result is vasoconstriction. This permits the distribution of blood from venous reservoirs to skeletal muscles, where it is needed most. A similar mechanism operates in cases of hemorrhage when blood volume and pressure decrease. Vasoconstriction of veins in venous reservoirs helps to compensate for the blood loss. Among the principal blood reservoirs are the veins of the abdominal organs (especially the liver and spleen) and the veins of the skin.

CLINICAL APPLICATION
Of the various types of **exercise,** some are more effective than others for improving the health of the cardio-

FIGURE 21-9 Role of skeletal muscle contractions and venous valves in returning blood to the heart. (a) When skeletal muscles contract, the valves open, and blood is forced toward the heart. (b) When skeletal muscles relax, the valves close to prevent the backflowing of blood from the heart. (c) Photograph of a one-way valve in a vein. (Courtesy of J. W. Eads.)

vascular system because they involve movements of large body muscles. The sustained movements of large muscles increases the flow of blood to the heart. Brisk walking, running, bicycling, cross-country skiing, and swimming are examples of sustained exercises.

Sustained exercise increases the oxygen demand of the muscles, and whether the demand is met depends primarily on the adequacy of cardiac output. After several weeks of training, the healthy individual increases cardiac output and thereby increases the rate of oxygen delivery to the tissues.

Another benefit of sustained exercise is the reduced oxygen requirement of skeletal muscles. After conditioning, an individual will have a decrease in skeletal muscle blood flow and lactic acid production. After conditioning, therefore, the heart is required to pump less blood to the muscles for a given amount of oxygen required by exercise.

Physical conditioning also causes an interesting effect upon systemic blood pressure. After a conditioning period, hypertensive individuals show a reduction in systolic pressure amounting to an average of 13 mm Hg. Blood pressure response to a given work load also is less after training, while lower pressure itself helps reduce myocardial oxygen requirements. So conditioning also may be useful in the management of systemic hypertension.

Additional benefits to be gained from physical conditioning are a reduction in high-density lipoprotein (HDL), a substance that seems to counter the impact of cholesterol in heart disease (discussed shortly). Exercise also helps to control weight and increases the body's ability to dissolve blood clots by increasing fibrinolytic activity. Intense exercise increases levels of endorphins, the body's natural painkillers. This may explain the psychological "high" that runners experience with strenuous training and the "low" they feel when they miss regular workouts.

CHECKING CIRCULATION

PULSE

The alternate expansion and elastic recoil of an artery with each systole of the left ventricle is called the **pulse.** Pulse is strongest in the arteries closest to the heart. It becomes weaker as it passes over the arterial system, and it disappears altogether in the capillaries. The pulse may be felt in any artery that lies near the surface of the body and over a bone or other firm tissue. The radial artery at the wrist is most commonly used. Other arteries that may be used for determining pulse are the:

1. Temporal artery, above and toward the outside of the eye.
2. Facial artery, at the lower jawbone on a line with the corners of the mouth.
3. Common carotid artery, on the side of the neck.
4. Brachial artery, along the inner side of the biceps brachii muscle.
5. Femoral artery, near the pelvic bone.
6. Popliteal artery, behind the knee.
7. Posterior tibial artery, behind the medial malleolus of the tibia.
8. Dorsalis pedis artery, over the instep of the foot.

The pulse rate is the same as the heart rate and averages between 70 and 90 beats per minute in the resting state. The term **tachycardia** (tak'-ē-KAR-dē-a; *tachy* = fast) is applied to a rapid heart or pulse rate (over 100/min). The term **bradycardia** (brad'-ē-KAR-dē-a; *brady* = slow) indicates a slow heart or pulse rate (under 50/min).

Other characteristics of the pulse may give additional information about circulation. For example, the intervals between beats should be equal in length. If a pulse is

FIGURE 21-10 Measurement of blood pressure using a sphygmomanometer and a stethoscope. (© 1982 by Gerard J. Tortora. Courtesy of Geraldine C. Tortora.)

missed at intervals, the pulse is said to be irregular. Also, each pulse beat should be of equal strength. Irregularities in strength may indicate a lack of muscle tone in the heart or arteries.

MEASUREMENT OF BLOOD PRESSURE (BP)

In clinical use, the term **blood pressure (BP)** refers to the pressure in arteries exerted by the left ventricle when it undergoes systole and the pressure remaining in the arteries when the ventricle is in diastole. Blood pressure is usually taken in the left brachial artery, and it is measured by a *sphygmomanometer* (sfig'-mō-ma-NOM-e-ter; *sphygmo* = pulse). A commonly used sphygmomanometer consists of a rubber cuff attached by a rubber tube to a compressible hand pump or bulb (Figure 21-10). Another tube attaches to the cuff and to a column of mercury or pressure dial marked off in millimeters. This column measures the pressure. The cuff is wrapped around the arm over the brachial artery and inflated by squeezing the bulb. The inflation creates a pressure on the artery. The bulb is squeezed until the pressure in the cuff exceeds the pressure in the artery. At this point, the walls of the brachial artery are compressed tightly against each other, and no blood can flow through. Compression of the artery may be evidenced in two ways. First, if a stethoscope is placed over the artery below the cuff, no pulse can be heard. Second, no pulse can be felt by placing the fingers over the radial artery at the wrist.

Next the cuff is deflated gradually until the pressure in the cuff is slightly less than the maximal pressure in the brachial artery. At this point, the artery opens, a spurt of blood passes through, and the pulse may be heard through the stethoscope. As cuff pressure is further reduced, the sound suddenly becomes faint. Finally, the sound disappears altogether. When the first sound is heard, a reading on the mercury column is made. This sound corresponds to **systolic blood pressure**—the force with which blood is pushing against arterial walls during ventricular contraction. The pressure recorded on the mercury column when the sounds suddenly become faint is called **diastolic blood pressure.** It measures the force of blood in arteries during ventricular relaxation. Whereas systolic pressure indicates the force of the left ventricular contraction, diastolic pressure provides information about the resistance of blood vessels.

CLINICAL APPLICATION

The average blood pressure of a young adult male is about 120 mm Hg systolic and 80 mm Hg diastolic, expressed as 120/80. In young adult females, the pressures are 8 to 10 mm Hg less. The difference between systolic and diastolic pressure is called **pulse pressure.** This pressure, which averages 40 mm Hg, provides information about the condition of the arteries. For example, conditions such as atherosclerosis and patent ductus arteriosus greatly increase pulse pressure. The normal ratio of systolic pressure to diastolic pressure to pulse pressure is about 3 : 2 : 1.

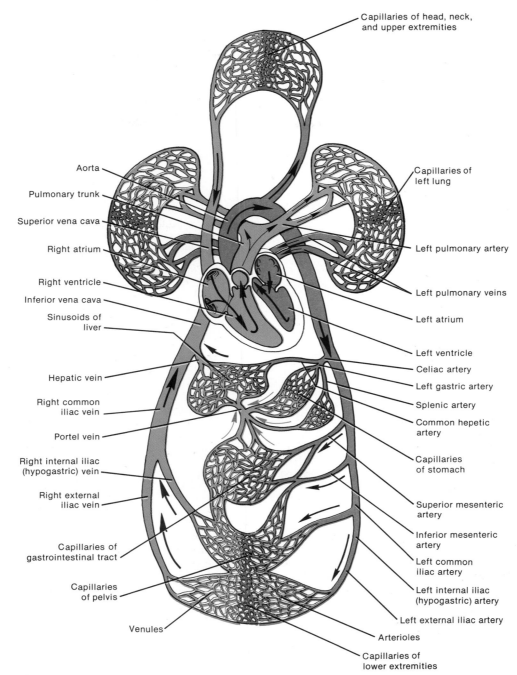

Capillaries of head, neck, and upper extremities

Aorta

Pulmonary trunk

Superior vena cava

Right atrium

Right ventricle

Inferior vena cava

Sinusoids of liver

Hepatic vein

Right common iliac vein

Portel vein

Right internal iliac (hypogastric) vein

Right external iliac vein

Capillaries of gastrointestinal tract

Capillaries of pelvis

Venules

Capillaries of left lung

Left pulmonary artery

Left pulmonary veins

Left atrium

Left ventricle

Celiac artery

Left gastric artery

Splenic artery

Common hepetic artery

Capillaries of stomach

Superior mesenteric artery

Inferior mesenteric artery

Left common iliac artery

Left internal iliac (hypogastric) artery

Left external iliac artery

Arterioles

Capillaries of lower extremities

FIGURE 21-11 Circulatory routes. Systemic circulation is indicated by heavy black arrows, pulmonary circulation by thin black arrows, and hepatic portal circulation by thin colored arrows. Refer to Figure 20-7 for the details of coronary (cardiac) circulation and Figure 21-24 for the details of fetal circulation.

CIRCULATORY ROUTES

The arteries, arterioles, capillaries, venules, and veins are organized into definite routes in order to circulate the blood throughout the body. We can now look at the basic routes the blood takes as it is transported through its vessels.

Figure 21-11 shows a number of basic **circulatory routes** through which the blood travels. **Systemic circulation** includes all the oxygenated blood that leaves the left ventricle through the aorta and the deoxygenated blood that returns to the right atrium after traveling to all the organs including the nutrient arteries to the lungs. Two of the many subdivisions of the systemic circulation

are the **coronary (cardiac) circulation** (see Figure 20-7), which supplies the myocardium of the heart, and the **hepatic portal circulation,** which runs from the digestive tract to the liver. Blood leaving the aorta and traveling through the systemic arteries is a bright red color. As it moves through the capillaries, it loses its oxygen and takes on carbon dioxide, which gives the blood in the systemic veins its dark red color.

When blood returns to the heart from the systemic route, it goes out of the right ventricle through the **pulmonary circulation** to the lungs. In the lungs, it loses its carbon dioxide and takes on oxygen. It is now bright red again. It returns to the left atrium of the heart and reenters the systemic circulation.

EXHIBIT 21-2
AORTA AND ITS BRANCHES (Figure 21-12)

DIVISION OF AORTA	ARTERIAL BRANCH	REGION SUPPLIED
Ascending aorta	Right and left coronary	Heart.
Arch of aorta	Brachiocephalic → Right common carotid	Right side of head and neck.
	Brachiocephalic → Right subclavian	Right upper extremity.
	Left common carotid	Left side of head and neck.
	Left subclavian	Left upper extremity.
Thoracic aorta	Intercostals	Intercostal and chest muscles and pleurae.
	Superior phrenics	Posterior and superior surfaces of diaphragm.
	Bronchials	Bronchi of lungs.
	Esophageals	Esophagus.
Abdominal aorta	Inferior phrenics	Inferior surface of diaphragm.
	Celiac → Common hepatic	Liver.
	Celiac → Left gastric	Stomach and esophagus.
	Celiac → Splenic	Spleen, pancreas, and stomach.
	Superior mesenteric	Small intestine, cecum, and ascending and transverse colons.
	Suprarenals	Adrenal (suprarenal) glands.
	Renals	Kidneys.
	Gonadals → Testicular	Testes.
	Gonadals → Ovarians	Ovaries.
	Inferior mesenteric	Transverse, descending, and sigmoid colons and rectum.
	Common iliacs → External iliacs	Lower extremities.
	Common iliacs → Internal iliacs (hypogastrics)	Uterus, prostate, muscles of buttocks, and urinary bladder.

Another major route—the **fetal circulation**—exists only in the fetus and contains special structures that allow the developing fetus to exchange materials with its mother.

Cerebral circulation (circle of Willis) is discussed in Exhibit 21-4.

SYSTEMIC CIRCULATION

The flow of blood from the left ventricle to all parts of the body and back to the right atrium is called the **systemic circulation.** The purpose of systemic circulation is to carry oxygen and nutrients to body tissues and to remove carbon dioxide and other wastes from the tissues. All systemic arteries branch from the *aorta,* which arises from the left ventricle of the heart.

As the aorta emerges from the left ventricle, it passes upward and deep to the pulmonary artery. At this point, it is called the *ascending aorta.* The ascending aorta gives off two coronary branches to the heart muscle. Then it

turns to the left, forming the *arch of the aorta* before descending to the level of the fourth thoracic vertebra as the *descending aorta.* The descending aorta lies close to the vertebral bodies, passes through the diaphragm, and terminates at the level of the fourth lumbar vertebra by dividing into two *common iliac arteries,* which carry blood to the lower extremities. The section of the descending aorta between the arch of aorta and the diaphragm is referred to as the *thoracic aorta.* The section between the diaphragm and the common iliac arteries is termed the *abdominal aorta.* Each section of the aorta gives off arteries that continue to branch into distributing arteries leading to organs and finally into the arterioles and capillaries that pierce the tissues.

Blood is returned to the heart through the systemic veins. All the veins of the systemic circulation flow into either the *superior* or *inferior venae cavae* or the *coronary sinus.* They in turn empty into the right atrium. The principal arteries and veins of systemic circulation are described and illustrated in Exhibits 21-2 to 21-13 and Figures 21-12 to 21-21.

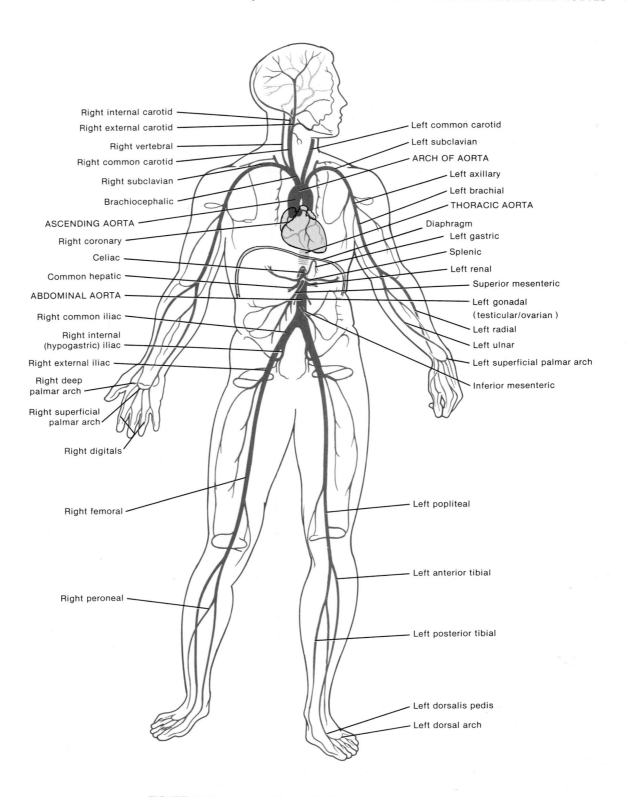

FIGURE 21-12 Aorta and its principal branches in anterior view.

EXHIBIT 21-3
ASCENDING AORTA (Figure 21-13)

BRANCH	DESCRIPTION AND REGION SUPPLIED
Coronary arteries	Right and left branches arise from ascending aorta just superior to aortic semilunar valve. They form crown around heart, giving off branches to atrial and ventricular myocardium.

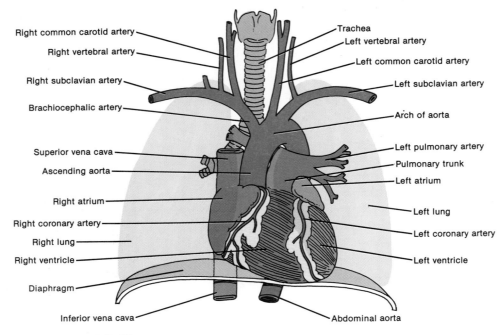

Right common carotid artery
Right vertebral artery
Right subclavian artery
Brachiocephalic artery
Superior vena cava
Ascending aorta
Right atrium
Right coronary artery
Right lung
Right ventricle
Diaphragm
Inferior vena cava

Trachea
Left vertebral artery
Left common carotid artery
Left subclavian artery
Arch of aorta
Left pulmonary artery
Pulmonary trunk
Left atrium
Left lung
Left coronary artery
Left ventricle
Abdominal aorta

FIGURE 21-13 Ascending aorta and its branches in anterior view.

EXHIBIT 21-4
ARCH OF AORTA (Figure 21-14)

BRANCH	DESCRIPTION AND REGION SUPPLIED
Brachiocephalic	**Brachiocephalic artery** is first branch off arch of aorta. It divides to form right subclavian artery and right common carotid artery. **Right subclavian artery** extends from brachiocephalic to first rib and then passes into armpit (axilla) and supplies arm, forearm, and hand. Continuation of right subclavian into axilla is called **axillary artery.*** From here, it continues into arm as **brachial artery.** At bend of elbow, brachial artery divides into medial **ulnar** and lateral **radial arteries.** These vessels pass down to palm, one on each side of forearm. In palm, branches of two arteries anastomose to form two palmar arches—**superficial palmar arch** and **deep palmar arch.** From these arches arise **digital arteries,** which supply fingers and thumb. Before passing into axilla, right subclavian gives off major branch to brain called **vertebral artery.** Right vertebral artery passes through foramina of transverse processes of cervical vertebrae and enters skull through foramen magnum to reach undersurface of brain. Here it unites with left vertebral artery to form **basilar artery.**
Right common carotid	**Right common carotid artery** passes upward in neck. At upper level of larynx, it divides into **right external** and **right internal carotid arteries.** External carotid supplies right side of thyroid gland, tongue, throat, face, ear, scalp, and dura mater. Internal carotid supplies brain, right eye, and right sides of forehead and nose. Anastomoses of left and right internal carotids along with basilar artery form arterial circle at base of brain called **circle of Willis.** From this anastomosis arise arteries supplying brain. Essentially, circle of Willis is formed by union of **anterior cerebral arteries** (branches of internal carotids) and **posterior cerebral arteries** (branches of basilar artery). Posterior cerebral arteries are connected with internal carotids by **posterior communicating arteries.** Anterior cerebral arteries are connected by **anterior communicating arteries.** Circle of Willis equalizes blood pressure to brain and provides alternate routes for blood to brain should arteries become damaged.
Left common carotid	**Left common carotid** is second branch off arch of aorta (see Figure 21-13). Corresponding to right common carotid, it divides into basically same branches with same names, except that arteries are now labeled "left" instead of "right."
Left subclavian	**Left subclavian artery** is third branch off arch of aorta (see Figure 21-13). It distributes blood to left vertebral artery and vessels of left upper extremity. Arteries branching from left subclavian are named like those of right subclavian.

* The right subclavian artery is a good example of the practice of giving the same vessel different names as it passes through different regions.

FIGURE 21-14 Arch of the aorta and its branches. (a) Anterior view of the arteries of the right upper extremity. (b) Right lateral view of the arteries of the neck and head. (c) Arteries of the base of the brain. Note the arteries that comprise the circle of Willis.

EXHIBIT 21-5
THORACIC AORTA (Figure 21-15)

BRANCH	DESCRIPTION AND REGION SUPPLIED
	Thoracic aorta runs from fourth to twelfth thoracic vertebrae. Along its course, it sends off numerous small arteries to viscera and skeletal muscles of the chest. Branches of an artery that supply viscera are called **visceral branches.** Those that supply body wall structures are **parietal branches.**
VISCERAL	
Pericardial	Several minute **pericardial arteries** supply blood to dorsal aspect of pericardium.
Bronchial	One **right** and two **left bronchial arteries** supply the bronchial tubes, areolar tissue of the lungs, bronchial lymph nodes, and esophagus.
Esophageal	Four or five **esophageal arteries** supply the esophagus.
Mediastinal	Numerous small **mediastinal arteries** supply blood to structures in the posterior mediastinum.
PARIETAL	
Posterior intercostal	Nine pairs of **posterior intercostal arteries** supply the intercostal, pectoral, and abdominal muscles; overlying subcutaneous tissue and skin; mammary glands; and vertebral canal and its contents.
Subcostal	The **left** and **right subcostal arteries** have a distribution similar to that of the posterior intercostals.
Superior phrenic	Small **superior phrenic arteries** supply the posterior surface of the diaphragm.

EXHIBIT 21-6
ABDOMINAL AORTA (Figure 21-15)

BRANCH	DESCRIPTION AND REGION SUPPLIED
VISCERAL	
Celiac	**Celiac artery (trunk)** is first visceral aortic branch below diaphragm. It has three branches: (1) **common hepatic artery,** which supplies tissues of liver, (2) **left gastric artery,** which supplies stomach, and (3) **splenic artery,** which supplies spleen, pancreas, and stomach.
Superior mesenteric	**Superior mesenteric artery** distributes blood to small intestine and part of large intestine.
Suprarenals	Right and left **suprarenal arteries** supply blood to adrenal (suprarenal) glands.
Renals	Right and left **renal arteries** carry blood to kidneys.
Gonadals (testiculars and **ovarians**)	Right and left **testicular arteries** extend into scrotum and terminate in testes; right and left **ovarian arteries** are distributed to ovaries.
Inferior mesenteric	**Inferior mesenteric artery** supplies major part of large intestine and rectum.
PARIETAL	
Inferior phrenics	**Inferior phrenic arteries** are distributed to undersurface of diaphragm.
Lumbars	**Lumbar arteries** supply spinal cord and its meninges and muscles and skin of lumbar region of back.
Middle sacral	**Middle sacral artery** supplies sacrum, coccyx, gluteus maximus muscles, and rectum.

FIGURE 21-15 Thoracic and abdominal aorta and their principal branches in anterior view.

EXHIBIT 21-7

ARTERIES OF PELVIS AND LOWER EXTREMITIES (Figure 21-16)

BRANCH	DESCRIPTION AND REGION SUPPLIED
Common iliacs	At about level of fourth lumbar vertebra, abdominal aorta divides into right and left **common iliac arteries.** Each passes downward about 5 cm (2 inches) and gives rise to two branches: internal iliac and external iliac.
Internal iliacs	**Internal iliac** or **hypogastric arteries** form branches that supply psoas major, quadratus lumborum, and medial side of each thigh, urinary bladder, rectum, prostate gland, ductus deferens, uterus, and vagina.
External iliacs	**External iliac arteries** diverge through pelvis, enter thighs, and here become right and left **femoral arteries.** Both femorals send branches back up to genitals and wall of abdomen. Other branches run to muscles of thigh. Femoral continues down medial and posterior side of thigh at back of knee joint, where it becomes **popliteal artery.** Between knee and ankle, popliteal runs down back of leg and is called **posterior tibial artery.** Below knee, **peroneal artery** branches off posterior tibial to supply structures on medial side of fibula and calcaneus. In calf, **anterior tibial artery** branches off popliteal and runs along front of leg. At ankle, it becomes **dorsalis pedis artery.** At ankle, posterior tibial divides into **medial** and **lateral plantar arteries.** These arteries anastomose with dorsalis pedis and supply blood to foot.

FIGURE 21-16 Arteries of the pelvis and right lower extremity. (a) Anterior view. (b) Posterior view.

EXHIBIT 21-8
VEINS OF SYSTEMIC CIRCULATION (Figure 21-17)

VEIN	DESCRIPTION AND REGION DRAINED
	All systemic and cardiac veins return blood to the right atrium of the heart through one of three large vessels. Return flow in coronary circulation is taken up by **cardiac veins,** which empty into the large vein of the heart, the **coronary sinus.** From here, the blood empties into the right atrium of the heart (see Figure 20-7). Return flow in systemic circulation empties into the superior vena cava or inferior vena cava.
Superior vena cava	Veins that empty into the **superior vena cava** are veins of the head and neck, upper extremities, and some from the thorax.
Inferior vena cava	Veins that empty into the **inferior vena cava** are some from the thorax and veins of the abdomen, pelvis, and lower extremities.

CLINICAL APPLICATION

The inferior vena cava is commonly compressed during the later stages of pregnancy due to the enlargement of the uterus. This produces edema of the ankles and feet and temporary varicose veins.

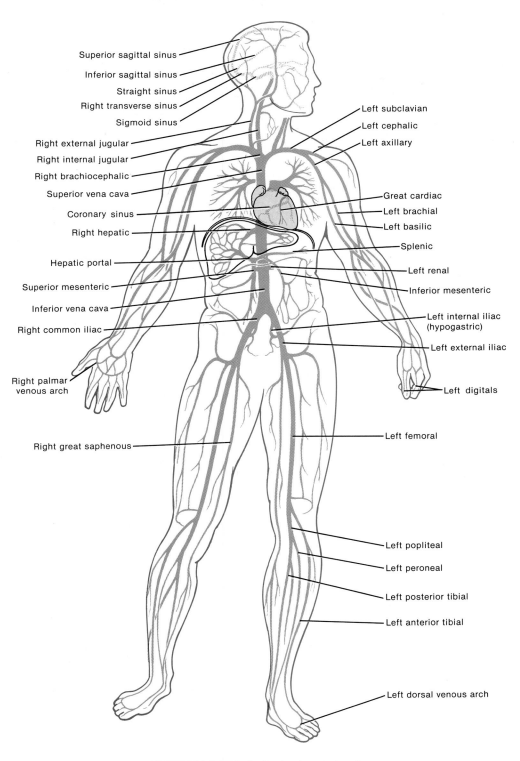

Superior sagittal sinus
Inferior sagittal sinus
Straight sinus
Right transverse sinus
Sigmoid sinus
Right external jugular
Right internal jugular
Right brachiocephalic
Superior vena cava
Coronary sinus
Right hepatic
Hepatic portal
Superior mesenteric
Inferior vena cava
Right common iliac
Right palmar venous arch
Right great saphenous

Left subclavian
Left cephalic
Left axillary
Great cardiac
Left brachial
Left basilic
Splenic
Left renal
Inferior mesenteric
Left internal iliac (hypogastric)
Left external iliac
Left digitals
Left femoral
Left popliteal
Left peroneal
Left posterior tibial
Left anterior tibial
Left dorsal venous arch

FIGURE 21-17 Principal veins in anterior view.

EXHIBIT 21-9

VEINS OF HEAD AND NECK (Figure 21-18)

VEIN	DESCRIPTION AND REGION DRAINED
Internal jugulars	Right and left **internal jugular veins** receive blood from face and neck. They arise as continuation of **sigmoid sinuses** at base of skull. Intracranial vascular sinuses are located between layers of dura mater and receive blood from brain. Other sinuses that drain into internal jugular include **superior sagittal sinus, inferior sagittal sinus, straight sinus,** and **transverse (lateral) sinuses.** Internal jugulars descend on either side of neck and pass behind clavicles, where they join with right and left subclavian veins. Unions of internal jugulars and subclavians form right and left brachiocephalic veins. From here blood flows into superior vena cava.
External jugulars	Right and left **external jugular veins** run down neck along outside of internal jugulars. They drain blood from parotid (salivary) glands, facial muscles, scalp, and other superficial structures into subclavian veins.

CLINICAL APPLICATION

In cases of heart failure, the venous pressure in the right atrium may rise. In such patients the pressure in the column of blood in the external jugular vein rises so that, even with the patient at rest and sitting in a chair, the external jugular vein will be visibly distended. Temporary distention of the vein is often seen in healthy adults when the intrathoracic pressure is raised in coughing and physical exertion

EXHIBIT 21-10

VEINS OF UPPER EXTREMITIES (Figure 21-19)

VEIN	DESCRIPTION AND REGION DRAINED
	Blood from each upper extremity is returned to the heart by deep and superficial veins. Both sets of veins contain valves. **Deep veins** are located deep in the body. They usually accompany arteries, and many have the same names as corresponding arteries. **Superficial veins** are located just below the skin and are often visible. They anastomose extensively with each other and deep veins.
SUPERFICIAL	
Cephalics	**Cephalic vein** of each upper extremity begins in the medial part of **dorsal arch** and winds upward around radial border of forearm. In front of elbow, it is connected to basilic vein by the **median cubital vein.** Just below elbow, cephalic vein unites with **accessory cephalic vein** to form cephalic vein of upper extremity. Ultimately, cephalic vein empties into axillary vein.
Basilics	**Basilic vein** of each upper extremity originates in the ulnar part of dorsal arch. It extends along posterior surface of ulna to point below elbow where it receives **median cubital vein.** If a vein must be punctured for an injection, transfusion, or removal of a blood sample, median cubitals are preferred. The median cubital vein joins the basilic vein to form the axillary vein.
Median antebrachials	**Median antebrachial veins** drain venous plexus on palmar surface of hand, ascend on ulnar side of anterior forearm, and end in median cubital veins.
DEEP	
Radials	**Radial veins** receive **dorsal metacarpal veins.**
Ulnars	**Ulnar veins** receive tributaries from **deep palmar arch.** Radial and ulnar veins unite in bend of elbow to form brachial vein.
Brachials	Located on either side of brachial artery, **brachial veins** join into axillary veins.
Axillaries	**Axillary veins** are a continuation of brachials and basilics. Axillaries end at first rib, where they become subclavians.
Subclavians	Right and left **subclavian veins** unite with internal jugulars to form brachiocephalic veins. Thoracic duct of lymphatic system delivers lymph into left subclavian vein at junction with internal jugular. Right lymphatic duct delivers lymph into right subclavian vein at corresponding junction.

Right subclavian

Right axillary

Right brachial

Right cephalic

Right accessory
cephalic

Right cephalic

Right
digital

Right external jugular

Right internal jugular

Right brachiocephalic

Superior vena cava

FIGURE 21-19 Veins of the right upper extremity in anterior view.

Right basilic

Right median cubital

Right basilic

Right median antebrachial

Right palmar
venous arch

FIGURE 21-18 Veins of the head and neck
in right lateral view.

Superior sagittal sinus

Inferior sagittal sinus

Straight sinus

Right transverse
(lateral) sinus

Right sigmoid sinus

Right vertebral

Right external jugular

Right
axillary

Right internal
jugular

Right subclavian

Right brachiocephalic

Superior
vena cava

EXHIBIT 21-11
VEINS OF THORAX (Figure 21-20)

VEIN	DESCRIPTION AND REGION DRAINED
Brachiocephalic	Right and left **brachiocephalic veins,** formed by union of subclavians and internal jugulars, drain blood from head, neck, upper extremities, mammary glands, and upper thorax. Brachiocephalics unite to form superior vena cava.
Azygos veins	**Azygos veins,** besides collecting blood from thorax, may serve as bypass for inferior vena cava that drains blood from lower body. Several small veins directly link azygos veins with inferior vena cava. And large veins that drain lower extremities and abdomen dump blood into azygos. If inferior vena cava or hepatic portal vein becomes obstructed, azygos veins can return blood from lower body to superior vena cava.
Azygos	**Azygos vein** lies in front of vertebral column, slightly right of midline. It begins as continuation of right ascending lumbar vein. It connects with inferior vena cava, right common iliac, and lumbar veins. Azygos receives blood from **right intercostal veins** that drain chest muscles; from hemiazygos and accessory hemiazygos veins; from several **esophageal, mediastinal,** and **pericardial veins;** and from right **bronchial vein.** Vein ascends to fourth thoracic vertebra, arches over right lung, and empties into superior vena cava.
Hemiazygos	**Hemiazygos vein** is in front of vertebral column and slightly left of midline. It begins as continuation of left ascending lumbar vein. It receives blood from lower four or five **intercostal veins** and some **esophageal** and **mediastinal veins.** At level of ninth thoracic vertebra, it joins azygos vein.
Accessory hemiazygos	**Accessory hemiazygos vein** is also in front and to left of vertebral column. It receives blood from three or four **intercostal veins** and left **bronchial vein.** It joins azygos at level of eighth thoracic vertebra.

EXHIBIT 21-12
VEINS OF THE ABDOMEN AND PELVIS (Figure 21-20)

VEIN	DESCRIPTION AND REGION DRAINED
Inferior vena cava	**Inferior vena cava** is the largest vein of the body. It is formed by union of two common iliac veins that drain lower extremities and abdomen. Inferior vena cava extends upward through abdomen and thorax to right atrium. Numerous small veins enter the inferior vena cava. Most carry return flow from branches of abdominal aorta and names correspond to names of arteries.
Common iliacs	**Common iliac veins** are formed by union of internal (hypogastric) and external iliac veins and represent distal continuation of inferior vena cava at its bifurcation.
Internal iliacs (hypogastrics)	Tributaries of **internal iliac (hypogastric) veins** basically correspond to branches of external iliac arteries. Internal iliacs drain gluteal muscles, medial side of thigh, urinary bladder, rectum, prostate gland, ductus deferens, uterus, and vagina.
External iliacs	**External iliac veins** are continuation of femoral veins and receive blood from lower extremities and inferior part of anterior abdominal wall.
Renals	**Renal veins** drain kidneys.
Gonadals (testiculars and ovarians)	**Testicular veins** drain testes (left testicular vein empties into left renal vein); **ovarian veins** drain ovaries (left ovarian vein empties into left renal vein).
Suprarenals	**Suprarenal veins** drain suprarenal glands (left suprarenal vein empties into left renal vein).
Inferior phrenics	**Inferior phrenic veins** drain diaphragm (left inferior phrenic vein sends tributary to left renal vein).
Hepatics	**Hepatic veins** drain liver.
Lumbars	A series of parallel **lumbar veins** drain blood from both sides of posterior abdominal wall. Lumbars connect at right angles with right and left **ascending lumbar veins,** which form origin of corresponding azygos or hemiazygos vein. Lumbars drain blood into ascending lumbars and then run to inferior vena cava, where they release remainder of flow.

FIGURE 21-20 Veins of the thorax, abdomen, and pelvis in anterior view.

EXHIBIT 21-13
VEINS OF LOWER EXTREMITIES (Figure 21-21)

VEIN	DESCRIPTION AND REGION DRAINED
	Blood from each lower extremity is returned by superficial set and deep set of veins.
SUPERFICIAL VEINS	
Great saphenous	**Great saphenous vein,** longest vein in body, begins at medial end of **dorsal venous arch** of foot. It passes in front of medial malleolus and then upward along medial aspect of leg and thigh. It receives tributaries from superficial tissues and connects with deep veins as well. It empties into femoral vein in groin.

> **CLINICAL APPLICATION**
> The great saphenous vein is very constant in its position anterior to the medial malleolus. It is frequently used for prolonged administration of intravenous fluids. This is particularly important in very young babies and in patients of any age who are in shock and whose veins are collapsed. It and the small saphenous vein are subject to varicosity.

VEIN	DESCRIPTION AND REGION DRAINED
Small saphenous	**Small saphenous vein** begins at lateral end of dorsal venous arch of foot. It passes behind lateral malleolus and ascends under skin of back of leg. It receives blood from foot and posterior portion of leg. It empties into popliteal vein behind knee.
DEEP VEINS	
Posterior tibial	**Posterior tibial vein** is formed by union of **medial** and **lateral plantar veins** behind medial malleolus. It ascends deep in muscle at back of leg, receives blood from **peroneal vein,** and unites with anterior tibial vein just below knee.
Anterior tibial	**Anterior tibial vein** is upward continuation of **dorsalis pedis veins** in foot. It runs between tibia and fibula and unites with posterior tibial to form popliteal vein.
Popliteal	**Popliteal vein,** just behind knee, receives blood from anterior and posterior tibials and small saphenous vein.
Femoral	**Femoral vein** is upward continuation of popliteal just above knee. Femorals run up posterior of thighs and drain deep structures of thighs. After receiving great saphenous veins in groin, they continue as right and left external iliac veins.

FIGURE 21-21 Veins of the pelvis and right lower extremity. (a) Anterior view. (b) Posterior view.

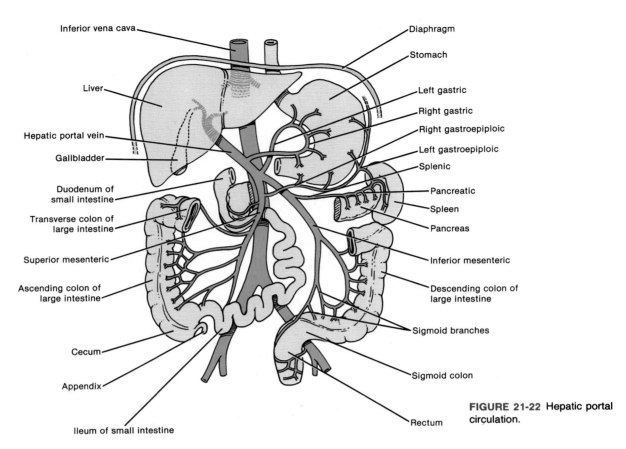

FIGURE 21-22 Hepatic portal circulation.

HEPATIC PORTAL CIRCULATION

Blood enters the liver from two sources. The hepatic artery delivers oxygenated blood from the systemic circulation; the hepatic portal vein delivers deoxygenated blood from the digestive organs. The term **hepatic portal circulation** refers to this flow of venous blood from the digestive organs to the liver before returning to the heart (Figure 21-22). Hepatic portal blood is rich with substances absorbed from the digestive tract. The liver monitors these substances before they pass into the general circulation. For example, the liver stores nutrients such as glucose. It modifies other digested substances so they may be used by cells. It detoxifies harmful substances that have been absorbed by the digestive tract and destroys bacteria by phagocytosis.

The hepatic portal system includes veins that drain blood from the pancreas, spleen, stomach, intestines, and gallbladder and transport it to the portal vein of the liver. The *hepatic portal vein* is formed by the union of the superior mesenteric and splenic veins. The *superior mesenteric vein* drains blood from the small intestine and portions of the large intestine and stomach. The *splenic vein* drains the spleen and receives tributaries from the stomach, pancreas, and portions of the colon. The tributaries from the stomach are the *gastric, pyloric,* and *gastroepiploic veins.* The *pancreatic veins* come from the pancreas, and the *inferior mesenteric veins* come from portions of the colon. Before the hepatic portal vein enters the liver, it receives the *cystic vein* from the gallbladder and other veins. Ultimately, blood leaves the liver through the *hepatic veins,* which enter the inferior vena cava.

PULMONARY CIRCULATION

The flow of deoxygenated blood from the right ventricle to the lungs and the return of oxygenated blood from the lungs to the left atrium is called the **pulmonary circulation** (Figure 21-23). The *pulmonary trunk* emerges from the right ventricle and passes upward, backward, and to the left. It then divides into two branches. The *right pulmonary artery* runs to the right lung; the *left pulmonary artery* goes to the left lung. On entering the lungs, the branches divide and subdivide until ultimately they form capillaries around the alveoli in the lungs. Carbon dioxide is passed from the blood into the alveoli to be breathed out of the lungs. Oxygen breathed in by the lungs is passed from the alveoli into the blood. The capillaries unite, venules and veins are formed, and eventually, two *pulmonary veins* exit from each lung and transport the oxygenated blood to the left atrium. The pulmonary veins are the only postnatal veins that carry oxygenated blood. Contractions of the left ventricle then sends the blood into the systemic circulation.

FETAL CIRCULATION

The circulatory system of a fetus, called **fetal circulation,** differs from an adult's because the lungs, kidneys, and digestive tract of a fetus are nonfunctional. The fetus derives its oxygen and nutrients from the maternal blood and eliminates its carbon dioxide and wastes into the maternal blood (Figure 21-24).

The exchange of materials between fetal and maternal circulation occurs through a structure called the *placenta*

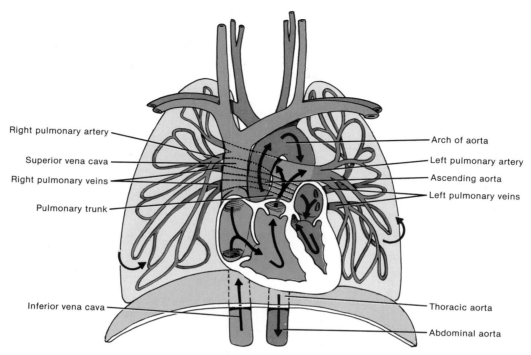

Right pulmonary artery

Superior vena cava

Right pulmonary veins

Pulmonary trunk

Inferior vena cava

Arch of aorta

Left pulmonary artery

Ascending aorta

Left pulmonary veins

Thoracic aorta

Abdominal aorta

FIGURE 21-23 Pulmonary circulation.

(pla-SEN-ta). It is attached to the umbilicus of the fetus by the umbilical (um-BIL-i-kal) cord, and it communicates with the mother through countless small blood vessels that emerge from the uterine wall. The umbilical cord contains blood vessels that branch into capillaries in the placenta. Wastes from the fetal blood diffuse out of the capillaries, into spaces containing maternal blood (intervillous spaces) in the placenta, and finally into the mother's uterine blood vessels. Nutrients travel the opposite route—from the maternal blood vessels to the intervillous spaces to the fetal capillaries. Normally there is no mixing of maternal and fetal blood since all exchanges occur through capillaries.

Blood passes from the fetus to the placenta via two *umbilical arteries*. These branches of the internal iliac (hypogastric) arteries are included in the umbilical cord. At the placenta, the blood picks up oxygen and nutrients and eliminates carbon dioxide and wastes. The oxygenated blood returns from the placenta via a single *umbilical vein*. This vein ascends to the liver of the fetus, where it divides into two branches. Some blood flows through the branch that joins the hepatic portal vein and enters the liver. Although the fetal liver manufactures red blood cells, it does not function in digestion. Therefore, most of the blood flows into the second branch, the *ductus venosus* (DUK-tus ve-NŌ-sus). The ductus venosus eventually passes its blood to the inferior vena cava, bypassing the liver.

In general, circulation through other portions of the fetus is not unlike postnatal circulation. Deoxygenated blood returning from the lower regions is mingled with oxygenated blood from the ductus venosus in the inferior vena cava. This mixed blood then enters the right atrium. The circulation of blood through the upper portion of the fetus is also similar to postnatal flow. Deoxygenated

blood returning from the upper regions of the fetus is collected by the superior vena cava, and it also passes into the right atrium.

Most of the blood does not pass through the right ventricle to the lungs, as it does in postnatal circulation, since the fetal lungs do not operate. In the fetus, an opening called the *foramen ovale* (fō-RĀ-men ō-VAL-ē) exists in the septum between the right and left atria. A valve in the inferior vena cava directs about a third of the blood through the foramen ovale so that it may be sent directly into the systemic circulation. The blood that does descend into the right ventricle is pumped into the pulmonary trunk, but little of this blood actually reaches the lungs. Most blood in the pulmonary trunk is sent through the *ductus arteriosus* (ar-tē-rē-Ō-sus). This small vessel connecting the pulmonary trunk with the aorta enables blood in excess of nutrient requirements to bypass the fetal lungs. The blood in the aorta is carried to all parts of the fetus through its systemic branches. When the common iliac arteries branch into the external and internal iliacs (hypogastric), part of the blood flows into the internal iliacs (hypogastrics). It then goes to the umbilical arteries and back to the placenta for another exchange of materials. The only vessel that carries fully oxygenated blood is the umbilical vein.

At birth, when lung, renal, digestive, and liver functions are established, the special structures of fetal circulation are no longer needed and the following changes occur.

1. The umbilical arteries atrophy to become the *lateral umbilical ligaments*.

2. The umbilical vein becomes the *round ligament* of the liver.

3. The placenta is delivered by the mother as the *"afterbirth."*

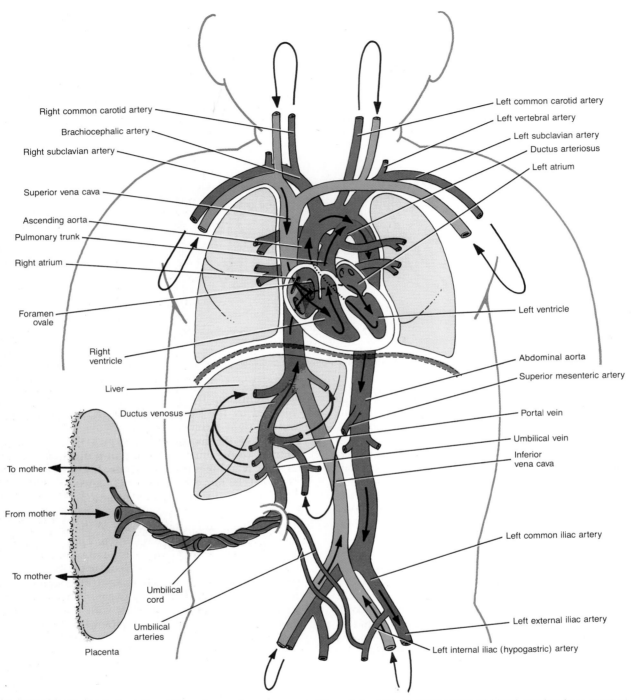

Right common carotid artery
Brachiocephalic artery
Right subclavian artery
Superior vena cava
Ascending aorta
Pulmonary trunk
Right atrium
Foramen ovale
Right ventricle
Liver
Ductus venosus
To mother
From mother
To mother
Placenta
Umbilical cord
Umbilical arteries

Left common carotid artery
Left vertebral artery
Left subclavian artery
Ductus arteriosus
Left atrium
Left ventricle
Abdominal aorta
Superior mesenteric artery
Portal vein
Umbilical vein
Inferior vena cava
Left common iliac artery
Left external iliac artery
Left internal iliac (hypogastric) artery

FIGURE 21-24 Fetal circulation. The various colors represent different degrees of oxygenation of blood ranging from greatest (red) to intermediate (two shades of purple) to least (blue).

4. The ductus venosus becomes the *ligamentum venosum,* a fibrous cord in the liver.

5. The foramen ovale normally closes shortly after birth to become the *fossa ovalis,* a depression in the interatrial septum.

6. The ductus arteriosus closes, atrophies, and becomes the *ligamentum arteriosum.*

Anatomical defects resulting from failure of these changes to occur are described in Chapter 20.

DISORDERS: HOMEOSTATIC IMBALANCES

Aneurysm

A blood-filled sac formed by an outpouching in an arterial or venous wall is called an **aneurysm** (AN-yoo-rizm). Aneurysms may occur in any major blood vessel. A **berry** aneurysm is a small aneurysm, frequently in the cerebral artery. If it ruptures, it may cause a hemorrhage below the dura mater. Hemorrhaging is one cause of a stroke. A **ventricular** aneurysm is a dilation of a ventricle of the heart. An **aortic** aneurysm is a dilation of the aorta. The aorta has a higher incidence of aneurysms than any other artery, probably because of its curved shape, large size, and high pressure. In the thorax, aneurysms usually occur in the ascending or descending aorta but seldom in the arch of the aorta. Symptoms of thoracic aortic aneurysm depend on pressure exerted on adjacent structures. For example, pressure on the inferior (recurrent) laryngeal nerve causes hoarseness and a brassy cough. Pressure on the esophagus may cause dysphagia (difficulty in swallowing). Dyspnea (difficulty in breathing) may follow pressure on the trachea, the root of the lung, or the phrenic nerve.

Coronary Artery Disease (CAD)

In Chapter 20 it was indicated that the common causes of heart disease are related to inadequate coronary blood supply, anatomical disorders, and arrhythmias. **Coronary artery disease (CAD)** is a condition in which the heart muscle receives inadequate blood because of an interruption of blood supply. Depending on the degree of interruption, symptoms can range from a mild chest pain to a full-scale heart attack. The underlying causes of CAD are many and varied. Two of the principal ones are atherosclerosis and coronary artery spasm.

Atherosclerosis

Atherosclerosis (ath'-er-ō-skle-RŌ-sis) is a process in which fatty substances, especially cholesterol and triglycerides (ingested fats), are deposited in the walls of medium-sized and large arteries in response to certain stimuli. It is believed that the first event in atherosclerosis is damage to the endothelial lining of the artery. Contributing factors include high blood pressure (hypertension), carbon monoxide in cigarettes, and a high cholesterol level. Following endothelial damage, platelets aggregate in the area of injury, adhere to the injured surface, and begin releasing their clot-forming chemicals. One of the chemicals increases permeability of the endothelial lining and modifies the cholesterol so that it is taken up by smooth muscle cells in the tunica media. The more cholesterol there is in the blood, the faster the process occurs. Platelets also release a chemical that stimulates the smooth muscle cells to proliferate. The result of enlarged, proliferating muscle cells is the development of an **atherosclerotic plaque** which deforms the arterial wall (Figure 21-25).

The plaque looks like a pearly gray or yellow mound of tissue. As it grows, it may obstruct blood flow in the affected artery and damage the tissues the artery supplies. An additional danger is that the plaque may provide a roughened surface for clot formation, and a thrombus may form. If the clot breaks off and forms an embolus, it may obstruct small arteries, capillaries, and veins quite a distance from the site of formation.

Cholesterol is a major building block of all plasma membranes and a key compound for the synthesis of some hormones and bile salts. Some cholesterol is present in foods (eggs, dairy products, organ meats, beef, pork, and processed luncheon meats); some is synthesized by the liver. As it turns out, foods containing cholesterol are not the main source of cholesterol in the blood. It is the type of fat eaten. Saturated fats (Chapter 2) have their carbon atoms filled with all the hydrogen atoms they can accommodate and are high in cholesterol. They occur mostly in animal foods and contribute significantly to a rise in blood cholesterol levels.

Neither cholesterol nor triglycerides dissolve in water and thus cannot travel in the blood in their unaltered forms. They are made water-soluble by combination with proteins produced by the liver and intestine and the complexes are called **lipoproteins**. The two major classes are called **low-density lipoproteins (LDL)** and **high-density lipoproteins (HDL)**. There is an important difference between the two: LDL seems to pick up cholesterol and deposit it in body cells including, under abnormal conditions, smooth muscle cells in arteries. HDL seems to gather cholesterol from body cells and transports it to the liver for elimination. According to this proposed relationship, a person's susceptibility to CAD from atherosclerosis depends on the relative balance between LDL and HDL. Supposedly, the higher the ratio of HDL to total cholesterol (and to LDL), the less the risk of CAD. Several ways to improve the HDL-LDL ratio are regular sustained exercise, diet (reducing animal fats, especially), and giving up cigarettes.

Diagnosis of atherosclerosis is based upon a stress ECG, angiography, and nuclear cardiology. The *stress ECG* is done while a person is exercising on a treadmill or bicycle in order to determine the presence and extent of CAD that might be missed on a resting ECG. *Angiography* (an-jē-OG-ra-fē) involves injecting a dye directly into the blood and then taking x-rays of the arteries. The technique is also called *arteriography*, and the film is called an *arteriogram* (Figure 21-25c). Angiography of the blood vessels of the heart is called *coronary angiography*. *Nuclear cardiology* involves injection of a radioactive substance into a vein followed by analysis of its distribution in the heart. An example of such a test is a *thallium scan*. As the radioisotope thallium 201 circulates through the coronary arteries, a camera placed over the chest picks up the signals from the radioisotope, relays them to a computer for translation, and a picture of the heart is displayed. An area of poor circulation of thallium in the heart indicates blockage of a coronary artery.

The treatment of CAD varies with the nature and urgency of the symptoms. Among the possible treatments are drug therapy, coronary bypass surgery, and percutaneous transluminal angioplasty (PTA). The two most commonly used drugs are *nitroglycerin* (which dilates blood vessels to increase blood flow to the heart) and *beta blockers* (which block the heart's response to sympathetic stimulation and cut down on the work load of the heart). Antithrombotic drugs that suppress the activity of platelets are being tested on a trial basis. In *coronary bypass surgery,* a vein taken from a leg or a synthetic vein is used as a conduit to literally bypass an area (or areas) of blockage in one or more coronary arteries. *Percutaneous transluminal angioplasty (PTA)* is the balloon catheterization technique to dilate obstructed vessels, described in Chapter 20. Unfortunately, calcified plaques are not crushed by the balloon, and the balloon can damage the arterial wall. Despite these drawbacks, it is hoped that the procedure can be refined so that it will come into general use in the treatment of atherosclerosis.

Coronary Artery Spasm

Atherosclerosis results in a fixed obstruction to blood flow. Recently, it has been learned that obstruction can also be caused by **coronary artery spasm,** a condition in which the smooth muscle of a coronary artery undergoes a sudden contraction, resulting in vasoconstriction. Coronary artery spasm can occur in individuals with or without atherosclerosis and may result in chest pain during rest (variant angina), chest pain during exertion (typical angina), heart attacks, and sudden death. Although the causes of coronary artery spasm are unknown, sev-

(a)

Atheromatous plaque   Obstructed lumen

(b)

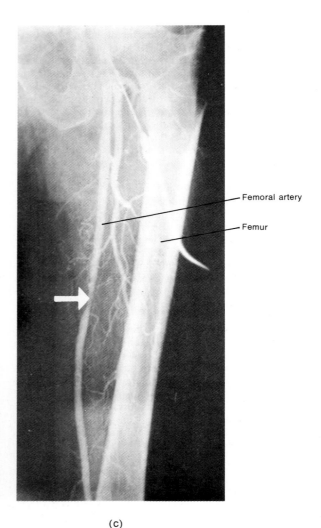

(c)

FIGURE 21-25 Atherosclerosis. (a) Atherosclerotic plaque formation. Compare the size of the normal lumen and the obstructed lumen. (b) Photomicrograph of a cross section of an artery partially obstructed by an atherosclerotic plaque. (Courtesy of National Heart, Lung, and Blood Institute.) (c) Femoral arteriogram showing an atherosclerotic plaque (arrow) in the middle third of the thigh. (Courtesy of Lester W. Paul and John H. Juhl, *The Essentials of Roentgen Interpretation*, 3rd ed., Harper & Row, Publishers, Inc., New York, 1972.)

eral factors are receiving attention. These include smoking, stress, and a vasoconstrictor chemical released by platelets. There is considerable interest in aspirin and other drugs that might inhibit the vasoconstrictor chemical. Coronary artery spasm is treated with long-acting nitrates (nitroglycerin) and calcium-blocking agents (drugs that block calcium transport into muscle cells, a requirement for muscle contraction).

Hypertension

Hypertension, or high blood pressure, is the most common disease affecting the heart and blood vessels. Statistics indicate that hypertension afflicts one out of every five American adults. Although there is some disagreement as to what defines hypertension, a strong consensus has emerged suggesting that a blood pressure of 120/80 is normal and desirable in a healthy adult.

STUDENT OBJECTIVES

- Identify the components and functions of the lymphatic system.

- Describe the structure and origin of lymphatics and contrast them with veins.

- Describe the histological aspects of lymph nodes and explain their functions.

- Trace the general plan of lymph circulation from lymphatics into the thoracic duct or right lymphatic duct.

- Explain the forces responsible for maintaining the circulation of lymph.

- Give the locations and compare the functions of the tonsils, spleen, and thymus gland.

- Explain the difference between nonspecific and specific resistance.

- Discuss the roles of the skin and mucous membranes, antimicrobial substances, phagocytosis, inflammation, and fever as components of nonspecific resistance.

- Define immunity.

- Explain the relationship between an antigen and an antibody.

- Contrast the main characteristics of antigens and antibodies.

- Contrast the role of T cells in cellular immunity and the role of B cells in humoral immunity.

- Discuss the relationship of immunology to cancer.

- Define a monoclonal antibody and explain its importance.

- Describe hypersensitivity, tissue rejection, autoimmune diseases, and acquired immune deficiency syndrome (AIDS) as malfunctions of the immune system.

- Define medical terminology associated with the lymphatic system.

22 The Lymphatic System and Immunity

Lymph, lymphatic vessels, a series of small masses of lymphoid tissue called lymph nodes, and three organs—tonsils, thymus, and spleen—make up the **lymphatic (lim-FAT-ik) system.** The primary function of the lymphatic system is to drain from the tissue spaces protein-containing fluid that escapes from the blood capillaries. Such proteins cannot be directly reabsorbed by the cardiovascular system. Other functions of the lymphatic system are to transport fats from the digestive tract to the blood, to produce lymphocytes, and to develop antibodies.

LYMPHATIC VESSELS

Lymphatic vessels originate as blind-ended tubes that begin in spaces between cells (Figure 22-1). The tubes, called **lymph capillaries,** occur singly or in extensive plexuses. Lymph capillaries originate throughout the body, but not in avascular tissue, the central nervous system, splenic pulp, and bone marrow. They are slightly larger and more permeable than blood capillaries.

Just as blood capillaries converge to form venules and veins, lymph capillaries unite to form larger and larger lymph vessels called **lymphatics** (Figure 22-2). Lymphatics resemble veins in structure, but have thinner walls and more valves and contain lymph nodes at various intervals. Lymphatics of the skin travel in loose subcutaneous tissue and generally follow veins. Lymphatics of the viscera generally follow arteries, forming plexuses around them. Ultimately, lymphatics converge into two main channels—the thoracic duct and the right lymphatic duct. This will be described shortly.

CLINICAL APPLICATION
Lymphangiography (lim-fan'-jē-OG-ra-fē) is the x-ray examination of lymphatic vessels and lymph organs after they are filled with a radiopaque substance. Such an x-ray is called a **lymphangiogram** (lim-FAN-jē-ō-gram). Lymphangiograms are useful in detecting edema and carcinomas and in locating lymph nodes for surgical or radiotherapeutic treatment. A normal lymphangiogram of lymphatic vessels and a few nodes in the upper thighs and pelvis is shown in Figure 22-3.

STRUCTURE OF LYMPH NODES

The oval or bean-shaped structures located along the length of lymphatics are called **lymph nodes.** They range from 1 to 25 mm (0.04 to 1 inch) in length. A lymph node contains a slight depression on one side called a *hilus* (HĪ-lus) or *hilum,* where blood vessels and an efferent lymphatic vessel leave the node (Figure 22-4). Each node is covered by a *capsule* of fibrous connective tissue that extends into the node. The capsular extensions are called *trabeculae* (tra-BEK-yoo-lē). The capsule, trabeculae, and hilus constitute the stroma (framework) of a lymph node. The parenchyma of a lymph node is specialized into two regions. The outer *cortex* contains densely packed lymphocytes arranged in masses called *lymph nodules.* The nodules often contain lighter-staining central areas, the *germinal centers,* where lymphocytes are produced. The inner region of a lymph node is called the *medulla.* In the medulla, the lymphocytes are arranged in strands called *medullary cords.*

The circulation of lymph through a node involves afferent (to convey toward a center) lymphatic vessels, sinuses in the node, and efferent (to convey away from a center) lymphatic vessels. *Afferent lymphatic vessels* enter the convex surface of the node at several points. They contain valves that open toward the node so that the lymph is directed inward. Once inside the node, the lymph enters the sinuses, which are a series of irregular channels. Lymph from the afferent lymphatic vessels enters the *cortical sinuses* under the capsule. From here it circulates to the *medullary sinuses* between the medullary cords. From these sinuses the lymph circulates into a usually single *efferent lymphatic vessel.* This vessel is located at the hilus of the lymph node. The efferent vessel is wider than the afferent vessels and contains valves that open away from the node to convey the lymph out of the node.

Lymph nodes are scattered throughout the body, usually in groups (see Figure 22-2b). Typically, these groups are arranged in two sets: *superficial* and *deep.*

As the lymph circulates through the nodes, it is processed by macrophages. The macrophages are fixed phagocytic cells of the reticuloendothelial system that line the sinuses. They remove bacteria, foreign material, and cell debris from the lymph. Lymph nodes also give rise to lymphocytes and plasma cells. The plasma cells produce antibodies.

Histological features of lymphatics and lymph nodes are shown in Figure 22-5.

CLINICAL APPLICATION
At times the number of microbes entering the nodes in the lymph is so great that the macrophages and

---

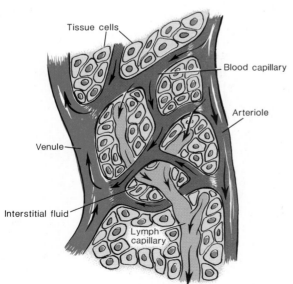

FIGURE 22-1 Relationship of lymph capillaries to tissue cells and blood capillaries.

Tissue cells

Blood capillary

Arteriole

Venule

Interstitial fluid

Lymph capillary

To cardiovascular system

FIGURE 22-9 Relationship of an antigen to antibodies. Since most antigens contain more than one antigenic determinant site, they are referred to as multivalent. Most human antibodies are bivalent; they have two reaction sites that are complementary to the antigenic determinant sites for which they are specific. Each antibody has reaction sites for specific antigenic determinant sites only.

nogenicity (i'-myoo-nō-jen-IS-it-ē), the ability to stimulate the formation of specific antibodies. The second is **reactivity,** the ability of the antigen to react specifically with the produced antibodies. An antigen with both of these characterisitics is called a **complete antigen.**

Characteristics

Chemically, the vast majority of antigens are proteins, nucleoproteins (nucleic acid + protein), lipoproteins (lipid + protein), glycoproteins (carbohydrate + protein), and certain large polysaccharides. In general, they have molecular weights of 10,000 or greater.

The entire microbe, such as a bacterium or virus, or components of microbes may act as an antigen. For example, bacterial structures such as flagella, capsules, and cell walls are also antigenic; bacterial toxins are highly antigenic. Nonmicrobial examples of antigens include pollen, egg white, incompatible blood cells, and transplanted tissues and organs. The myriad of antigens in the environment provides many opportunities for the production of antibodies by the body.

Antibodies do not form against the whole antigen. At specific regions on the surface of the antigen, called **antigenic determinant sites** (Figure 22-9), specific chemical groups of the antigen combine with the antibody. This combination depends upon the size and shape of the determinant site and the manner in which it corresponds to the chemical structure of the antibody. The combination is very much like the lock and key analogy used to describe the combination of enzyme and substrate molecule.

The number of antigenic determinant sites on the surface of an antigen is called *valence.* Most antigens are *multivalent,* that is, they have several antigenic determinant sites. It is believed that in order to induce antibody formation, an antigen must have at least two determinant sites (*bivalent*). If an antigen is chemically broken down, it is possible to separate the determinant site. Its molecular weight, compared to that of the antigen as a whole, is

small. It might be only 200 to 1,000. An isolated determinant site still has the ability to react with an antibody in response to the original antigen (reactivity), but does not have the ability to stimulate the production of antibodies (immunogenicity) when injected into an animal. Its lower molecular weight prevents it from serving as a complete antigen. A determinant site that has reactivity but not immunogenicity is called a **partial antigen** or **hapten** (*haptein* = to grasp). A hapten can stimulate an immune response if it is attached to a larger carrier molecule so that the combined molecule has two determinant sites. Some drugs of low molecular weight, like penicillin, may combine with proteins of high molecular weight in the body. This combination forms two determinant sites, and the complex becomes antigenic. Antibodies formed against the complex are responsible for the allergic reactions to drugs and other chemicals.

As a rule, antigens are foreign substances. They are not usually part of the chemistry of the body. The body's own substances, recognized as "self," do not act as antigens, identified as "nonself," to stimulate antibody production. However, there are certain conditions in which the distinction between self and nonself breaks down, and antibodies which attack the body are produced. These conditions (to be described later) result in what are called autoimmune disease.

Let's now look at some of the characteristics of antibodies.

ANTIBODIES

Definition

An **antibody** is a protein produced by the body in response to the presence of an antigen and is capable of combining specifically with the antigen. This is essentially the complementary definition of an antigen.

The specific fit of the antibody with antigen depends not only on the size and shape of the antigenic determi-

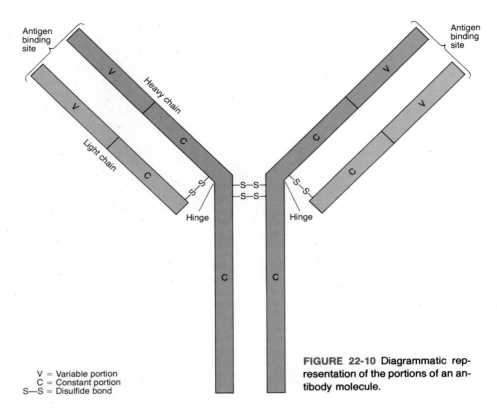

V = Variable portion
C = Constant portion
S—S = Disulfide bond

FIGURE 22-10 Diagrammatic representation of the portions of an antibody molecule.

nant site, but also on the corresponding antibody site, much like the lock and key analogy (see Figure 22-9). An antibody, like an antigen, also has a valence. Whereas most antigens are multivalent, antibodies are bivalent or multivalent. The majority of human antibodies are bivalent.

Antibodies belong to a group of proteins called globulins, and for this reason they are also known as **immunoglobulins** (i'-myoo-nō-GLOB-yoo-lins) or **Ig.** Five different classes of immunoglobulins are known to exist in humans. These are designated as IgG, IgA, IgM, IgD, and IgE. Each has a distinct chemical structure and a specific biological role. IgG antibodies enhance phagocytosis, neutralize toxins, and protect the fetus and newborn. IgA antibodies provide localized protection on mucosal surfaces; IgM antibodies are especially effective against microbes by causing agglutination and lysis. IgD antibodies may be involved in stimulating antibody-producing cells to manufacture antibodies. IgE antibodies are involved in allergic reactions.

Structure

Since antibodies are proteins, they consist of polypeptide chains. Most antibodies contain four pairs of polypeptide chains (Figure 22-10). Two of the chains are identical to each other and are called *heavy (H) chains.* Each consists of more than 400 amino acids. The other two chains, also identical to each other, are called *light (L) chains,* and each consists of 200 amino acids. The antibody consists of identical halves held together by disulfide bonds (S—S). Each half consists of a heavy and a light chain, also held together by a disulfide bond (S—S). Over-

all, the antibody molecule sometimes assumes the shape of a letter Y. At other times, it resembles a letter T. The reasons for these differences in shape will be explained shortly.

Within each H and L chain, there are two distinct regions. The tops of the H and L chains, called the *variable portions,* contain the antigen binding site. The variable portion is different for each kind of antibody, and it is this portion that allows the antibody to recognize and attach specifically to a particular type of antigen. Since most antibodies have two variable portions for attachment of antigens, they are bivalent. The remainder of each polypeptide chain is called the *constant portion.* The constant portion is the same in all antibodies of the same class and is responsible for the type of antigen-antibody reaction that occurs. However, the constant portion differs from one class of antibody to another, and its structure serves as a basis for distinguishing the classes.

In performing its role in immunity, the antibody is believed to behave as a switch. When antibodies are viewed in the electron microscope before combination with antigen, they resemble the letter T. But after combination with antigen, they appear to be smaller and resemble the letter Y. It is possible that the binding of antigen causes a rearrangement of the structure of antibody. If you examine Figure 22-10 again, you will note that there is a hinge area along the H chains; this hinge provides the antibody molecule with flexibility. When antigen and antibody are combined, the rearrangement of antibody might consist of a pivoting movement in which the H chain moves from a T shape into a Y shape. This pivoting could cause exposure of the constant portion, which then functions in a particular antigen-antibody reaction, such as fixing complement.

CELLULAR AND HUMORAL IMMUNITY

The ability of the body to defend itself against invading agents such as bacteria, toxins, viruses, and foreign tissues consists of two closely allied components. One component consists of the formation of specially sensitized lymphocytes that have the capacity to attach to the foreign agent and destroy it. This is called **cellular (cell-mediated) immunity** and is particularly effective against fungi, parasites, intracellular viral infections, cancer cells, and foreign tissue transplants. In the other component, the body produces circulating antibodies that are capable of attacking an invading agent. This is called **humoral immunity** and is particularly effective against bacterial and viral infections.

Cellular immunity and humoral immunity are the product of the body's lymphoid tissue. The bulk of lymphoid tissue is located in the lymph nodes, but it is also found in the spleen and gastrointestinal tract and, to a lesser extent, in bone marrow. The placement of lymphoid tissue is strategically designed to intercept an invading agent before it can spread too extensively into the general circulation.

Formation of T Cells and B Cells

Lymphoid tissue consists primarily of lymphocytes that may be distinguished into two kinds. **T cells** are responsible for cellular immunity. **B cells** develop into specialized cells (plasma cells) that produce antibodies and provide humoral immunity.

Both types of lymphocytes are derived originally in the embryo from lymphocytic stem cells in bone marrow (Figure 22-11). Before migrating to their positions in lymphoid tissues, the descendants of the stem cells follow two distinct pathways. About half of them first migrate

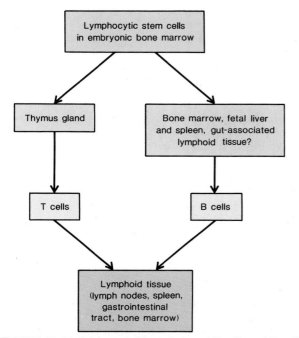

FIGURE 22-11 Origin and differentiation of T cells and B cells.

to the thymus gland, where they are processed to become T cells. Since the processing occurs in the thymus gland, they are called T cells. The thymus gland in some way confers what is called *immunologic competence* on the T cells. This means that they develop the ability to differentiate into cells that perform specific immune reactions. They then leave the thymus gland and become embedded in lymphoid tissue. Immunologic competence is conferred by the thymus gland shortly before birth and for a few months after birth. Removal of the thymus gland from an animal before it has processed the T cells results in a failure to develop cellular immunity. Removal after processing has little effect.

The remaining stem cells are processed in some unknown area of the body, possibly bone marrow, fetal liver and spleen, or gut-associated lymphoid tissue. They become the B cells. They are given this name because in birds they are processed in the bursa of Fabricius, a small pouch of lymphoid tissue attached to the intestine. Once processed, B cells migrate to lymphoid tissue and take up their positions there. Although both T cells and B cells occupy the same lymphoid tissues, they localize in separate areas of the tissues.

T Cells and Cellular Immunity

There are literally thousands of different types of T cells, and each is capable of responding to a specific antigen. At any given time, most T cells are inactive. When an antigen enters the body, only the particular T cell specifically programmed to react with the antigen becomes activated. Such an activated T cell is said to be *sensitized.* Activation occurs when macrophages phagocytose the antigen and present it to the T cell. Sensitized T cells increase in size and divide, each cell giving rise to a *clone,* or population of cells identical to itself (Figure 22-12). Four subpopulations of cells within the clone can be recognized: killer T cells, helper T cells, suppressor T cells, and memory T cells.

Killer T cells leave the lymphoid tissue and migrate to the site of invasion. Here, several things happen. They attach to the invading cell and secrete *cytotoxic substances* that destroy the antigen directly. The cytotoxic substances are probably lysosomal enzymes synthesized by the killer T cells. The cells can also release a protein substance called *transfer factor.* This substance reacts with nonsensitized lymphocytes at the site of invasion, causing them to take on the same characteristics as the sensitized killer T cells. In this way, additional lymphocytes are recruited, and the effect of the sensitized killer T cells is intensified. Killer T cells also secrete a substance called *macrophage chemotactic factor.* This substance attracts macrophages to the site of invasion where they destroy the antigens by phagocytosis. Another substance released by killer T cells, called *macrophage activating factor (MAF),* greatly increases the phagocytic activity of macrophages. In summary, then, killer T cells can destroy antigens directly by releasing cytotoxic substances and indirectly by recruiting additional lymphocytes, attracting macrophages, and intensifying phagocytosis by macrophages. Killer T cells are especially effective against slowly developing bac-

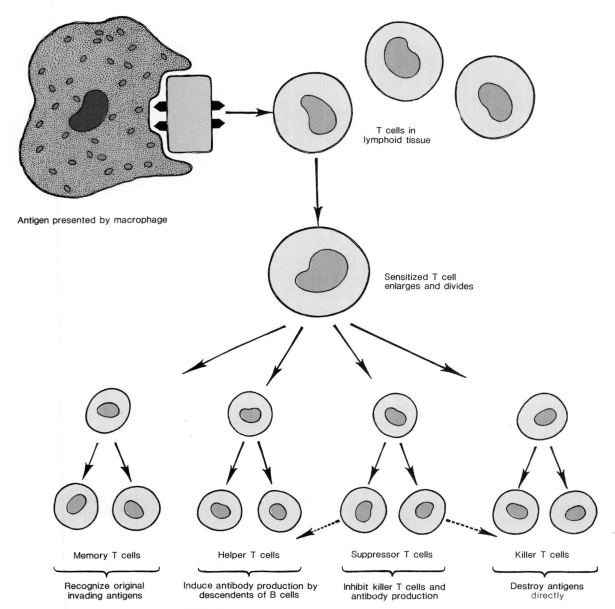

FIGURE 22-12 Role of T cells in cellular immunity.

terial diseases, such as tuberculosis and brucellosis, some viruses, fungi, transplanted cells, and cancer cells.

Helper T cells cooperate with B cells to help amplify antibody production. Although helper T cells do not themselves secrete antibody, antigens first interact with them before inducing antibody production by the descendants of B cells.

Suppressor T cells assume a role in regulating parts of the immune response. They can inhibit secretion of injurious substances by killer T cells and inhibit the development of B cells into antibody-producing plasma cells. The normal ratio of helper T cells to suppressor T cells is 2:1.

Memory T cells are programmed to recognize the original invading antigen. Should the pathogen invade the body at a later date, the memory cells initiate a far swifter reaction than during the first invasion. In fact, the second response is so swift that the pathogens are usually destroyed before any signs or symptoms of the disease occur.

B Cells and Humoral Immunity

The body not only contains thousands of different T cells, it also contains thousands of different B cells, each capable of responding to a specific antigen. Whereas killer T cells leave their reservoirs of lymphoid tissue to meet a foreign antigen, B cells respond differently. They differentiate into cells that produce specific antibodies that then circulate in the lymph and blood to reach the site of invasion. When a foreign antigen has been prepared and presented by macrophages to B cells in lymph nodes, the spleen, or lymphoid tissue in the gastrointestinal tract, B cells

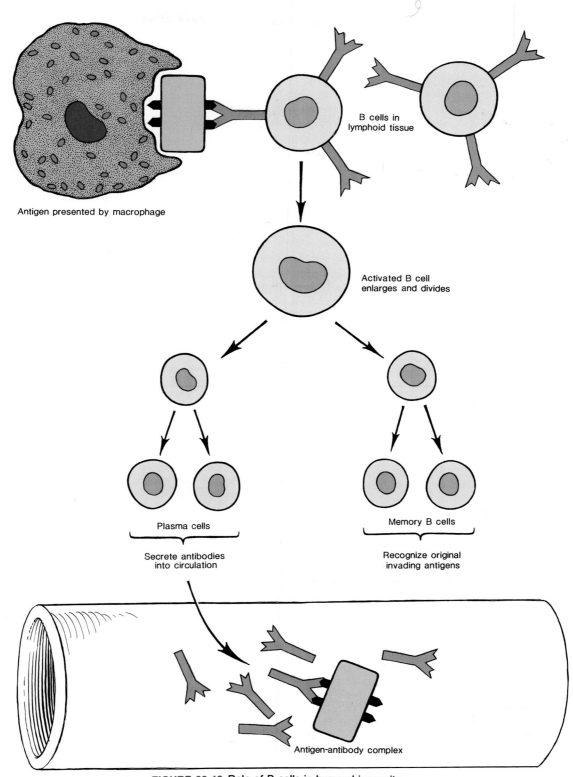

Antigen presented by macrophage

B cells in lymphoid tissue

Activated B cell enlarges and divides

Plasma cells

Secrete antibodies into circulation

Memory B cells

Recognize original invading antigens

Antigen-antibody complex

FIGURE 22-13 Role of B cells in humoral immunity.

specific for the antigen are activated. Some of them enlarge and divide and differentiate into a clone of **plasma cells** under the influence of thymic hormones (Figure 22-13). Plasma cells secrete the antibody. The phenomenal rate of secretion is about 2,000 antibody molecules per second for each cell, and it occurs for several days until the plasma cell dies (in 4 to 5 days). The activated B cells that do not differentiate into plasma cells remain

FIGURE 22-14 Primary and anamnestic (secondary) responses to the injection of an antigen.

as **memory B cells,** ready to respond more rapidly and forcefully should the same antigen appear at a future time.

Different antigens stimulate different B cells to develop into plasma cells and their accompanying memory B cells because the B cells of a particular clone are capable of secreting only one kind of antibody. Each specific antigen activates only those B cells already predetermined to secrete antibody specific to that antigen. The specific antigen selects a specific B cell because that particular B cell has on its surface the antibody molecules that it is capable of producing. The surface antibodies serve as receptor sites with which the antigen can combine.

Once the antigen-antibody complex is formed, the antibody activates complement enzymes for attack and fixes the complement to the surface of the antigen.

CLINICAL APPLICATION

The immune response of the body, whether cellular or humoral, is much more intense after a second or subsequent exposure to an antigen than after the initial exposure. This can be demonstrated by measuring the amount of antibody in serum, called the *antibody titer* (TĪ-ter). After an initial contact with an antigen, there is a period of several days during which no antibody is present. Then there is a slow rise in the antibody titer, followed by a gradual decline. Such a response of the body to the first contact is called the **primary response** (Figure 22-14). During the primary response, in which the body is said to be primed or sensitized, there is proliferation of immunocompetent lymphocytes. When the antigen is contacted again, whether for the second time or the two hundredth, there is an immediate proliferation of immunocompetent lymphocytes, and the antibody titer is far greater than that for the primary response. This accelerated, more intense response is called the **anamnestic** (an'-am-NES-tik; *anamne* = recall) or **secondary response.** The reason for the anamnestic response is that some of the immunocompetent lymphocytes formed during the primary response remain as memory cells. These memory cells not only add to the pool of cells that can respond to the antigen—their response is more intense as well.

Primary and anamnestic responses occur during microbial infection. When you recover from an infection without taking antibiotics, it is usually because of the primary response. If, at a later time, you contact the same microbe, the anamnestic response could be so swift that the microbes are quickly destroyed and you do not exhibit any signs or symptoms.

You have probably guessed by now that the anamnestic response provides the basis for immunization against certain diseases. When you receive the initial immunization, your body is sensitized. Should you encounter the pathogen again as an infecting microbe or "booster dose," your body experiences the anamnestic response. The booster dose literally boosts the antibody titer to a higher level. Since their effects are not permanent, boosters must be given periodically, and immunization schedules are designed to maintain high antibody titers.

IMMUNOLOGY AND CANCER

When a normal cell becomes transformed into a cancer cell, the tumor cell assumes cell surface components called **tumor-specific antigens.** It is believed that the immune system usually recognizes tumor-specific antigens as "nonself" and destroys the cancer cells carrying them. Such an immune response is called **immunologic surveillance.** Most investigators believe that cellular immunity is the basic mechanism involved in tumor destruction. In this regard, it is presumed that sensitized killer T cells react with tumor-specific antigens, initiating lysis of tumor cells. Sensitized macrophages may also be involved.

Despite the mechanism of immunologic surveillance, some cancer cells escape destruction, a phenomenon called **immunologic escape.** One possible explanation is that tumor cells may shed their tumor-specific antigens, thus evading recognition. Another theory is that antibodies produced by plasma cells bind to tumor-specific antigens, preventing recognition by killer T cells. It is reasonable to assume that immunologic techniques may be used in the future to prevent cancer.

MONOCLONAL ANTIBODIES

Scientists have known for many years how to induce laboratory animals (or humans) to produce antibodies. By injecting a particular antigen, antibodies are produced against the antigen by plasma cells (descendants of B cells). However, since the antibodies are produced by many different plasma cells, they vary physically and chemically; they are not pure. What scientists have learned to do is isolate a single B cell and fuse it with a tumor cell that is capable of proliferating endlessly. The resulting hybrid cell is called a **hybridoma cell.** These cells are a long-term source of substantial quantities of pure antibodies called **monoclonal antibodies.** Such antibodies will go after one and only one antigen.

One clinical use for monoclonal antibodies is the diagnosis of allergies and diseases such as hepatitis, rabies, and certain sexually transmitted diseases. Another application is the detection of cancer at an early stage and determining the extent of metastasis. One of the most exciting applications is in the area of treating diseases.

Monoclonal antibodies, either alone or in combination with drugs, are being used in trial studies to treat cancer. This approach has the advantage of destroying diseased tissues only, while sparing healthy tissues, thus overcoming some of the major adverse effects of chemotherapy and radiation treatment. Monoclonal antibodies may also be useful in preparing vaccines to counteract the rejection associated with transplants and to treat autoimmune diseases. One such disease that has been treated with some success is severe combined immunodeficiency (SCID), the most serious of all immune deficiency diseases. SCID victims are born lacking all immune defenses for fighting disease. This is the rare disease that affects the boy who is internationally known as the "boy in a bubble."

DISORDERS: HOMEOSTATIC IMBALANCES

The antigen-antibody response is essential to survival and typically provides a state of immunity. Under certain circumstances, however, it may create problems. Three such problems are hypersensitivity (allergy), tissue rejection, and autoimmunity.

Hypersensitivity (Allergy)

A person who is overly reactive to an antigen is said to be **hypersensitive** or **allergic.** Whenever an allergic reaction occurs, there is tissue injury. The antigens that induce an allergic reaction are called **allergens.** Almost any substance can be an allergen for some individual. Common allergens include certain foods (milk, eggs), antibiotics such as penicillin, cosmetics, chemicals in plants such as poison ivy, pollens, dust, molds, and even microbes.

One type of hypersensitivity is called **anaphylaxis** (an'-a-fi-LAK-sis), the interaction of humoral antibodies (IgE) with mast cells and basophils. Anaphylaxis literally means "against protection." Some anaphylactic reactions are localized and produce *allergic rhinitis* (hay fever), *bronchial asthma, atopic dermatitis* (eczema), and *urticaria* (hives). A severe anaphylactic reaction, called *acute anaphylaxis* (*anaphylactic shock*), may produce life-threatening systemic effects such as circulatory shock and asphyxia.

In response to certain allergens, some people produce IgE antibodies which bind to basophils in the blood and mast cells that are especially numerous in the skin, respiratory system, and endothelium of blood vessels. The antibodies may remain attached to the cells for weeks before a response occurs. This attachment, however, has the effect of sensitizing the cells, that is, they become subject to injury if the allergen enters the body at a later date. When the allergen enters the body again, tissue damage results. The allergen is believed to bind the cell-bound IgE antibodies together, forming a lattice. This stimulates the release of mediators of anaphylaxis stored in granules in the cytoplasm of basophils and mast cells. Among these mediators are histamine, eosinophil chemotactic factor of anaphylaxis (ECFA), platelet-activating factors (PAFs), slow-reacting substance of anaphylaxis (SRS-A), and serotonin. Collectively, these mediators of anaphylaxis increase blood capillary permeability, increase smooth muscle contraction in bronchial tubes, increase gastic secretion of hydrochloric acid, enhance complement reactions, increase mucus secretion, lyse platelets, and inhibit coagulation. Increased capillary permeability results in edema and redness and other symptoms of inflammation seen in anaphylaxis. Constriction of bronchial tubes results in difficulty in breathing. Excessive mucus secretion is responsible for the "runny" nose.

In acute anaphylaxis, large amounts of histamine are released into circulation. As a result, there is flushing of the skin; the respiratory tubes are continuously constricted, resulting in wheezing and shortness of breath; and edema is accelerated. This condition lowers the blood volume even more. If the effects of histamine are not counteracted, death may result, usually from respiratory failure. The effects of histamine may be reversed by administering epinephrine or antihistamines, drugs that inactivate histamine.

Tissue Rejection

Transplantation involves the replacement of an injured or diseased tissue or organ. Usually, the body recognizes the proteins in the transplanted tissue or organ as foreign and produces antibodies against them. This phenomenon is known as **tissue rejection.** Rejection can be somewhat reduced by matching donor and recipient and by administering drugs that inhibit the body's ability to form antibodies.

Types of Transplants

The closer the relationship of donor and recipient, the more successful the transplant. The most successful transplants are **isografts,** transplants in which the donor and recipient have identical genetic backgrounds. They include transplants between identical twins and from one part of the body to another.

An **allograft (homograft)** is a transplant between individuals of the same species, but with different genetic backgrounds. The success of this type of transplant has been moderate. Frequently, it is used as a temporary measure until the damaged or diseased tissue is able to repair itself. Skin transplants from other individuals and blood transfusions might properly be considered allografts. The one organ allograft that has been quite successful is the thymus. Children born without a thymus can now receive the gland from an aborted fetus. The thymus-deficient child cannot produce antibodies and thus cannot reject the transplant. Rejection later on indicates that the child is manufacturing antibodies and no longer needs the organ.

A **xenograft** is a transplant between animals of different species. This type of transplantation is used primarily as a physiological dressing over severe burns. Xenografts are presently restricted to laboratory animals.

Immunosuppressive Therapy

Until recently, **immunosuppressive drugs** suppressed not only the recipient's immune rejection of the donor organ, but the immune response to all antigens as well. This causes patients to become very susceptible to infectious diseases. A new drug called cyclosporin A, derived from a fungus, has largely overcome this problem. It is a selective immunosuppressive drug. It inhibits T cells which are responsible for tissue rejection, but has only a minimal effect on B cells. Thus, rejection is avoided and resistance against disease is still maintained.

Autoimmune Diseases

Under normal conditions, the body's immune mechanism is able to recognize its own tissues and chemicals. It normally does not produce T cells or B cells against its own substances. Such recognition of self is called **immunologic tolerance.** Although the mechanism of tolerance is not completely understood, it is believed that suppressor T cells may inhibit the differentiation of B cells into antibody-producing plasma cells or inhibit helper T cells which cooperate with B cells to amplify antibody production.

At times, however, immunologic tolerance breaks down and the body has difficulty in discriminating between its own antigens and foreign antigens. This loss of immunologic tolerance leads to an **autoimmune disease (autoimmunity).** Such diseases are immunologic responses mediated by antibodies against a person's own tissue antigens. Among human autoimmune diseases are rheumatoid arthritis (RA), systemic lupus erythematosus (SLE), thyroiditis, rheumatic fever, glomerulonephritis, encephalomyelitis, hemolytic and pernicious anemia, myasthenia gravis, and multiple sclerosis (MS).

Acquired Immune Deficiency Syndrome (AIDS)

Never before has science been confronted with an epidemic in which the primary disease only lowers the victim's immunity and then a second unrelated disease produces the symptoms that may result in death. This primary disease, first recognized in 1981, is called **acquired immune deficiency syndrome (AIDS)** and has a mortality rate of nearly 40 percent. One of the principal problems associated with AIDS is that its victims have too few T cells. In addition, the ratio of helper T cells to suppressor T cells, normally 2:1, is reversed. Symptoms of AIDS may develop for months or years and include malaise, a low-grade fever, coughing, shortness of breath, sore throat, muscle aches, weight loss, and enlarged lymph nodes in the neck, axilla, and groin.

The two diseases that most often kill AIDS victims are Kaposi's sarcoma and *Pneumocystis carinii* pneumonia. Kaposi's sarcoma is a deadly form of skin cancer prevalent in equatorial Africa, but previously almost unknown in the United States. *Pneumocystis carinii* pneumonia is a rare form of pneumonia caused by the protozoan, *Pneumocystis carinii.* AIDS victims are also subject to a form of herpes that attacks the central nervous system and a bacterial infection that usually causes tuberculosis in chickens and pigs.

AIDS is found primarily among homosexual males, intravenous drug users, hemophiliacs, and Haitian immigrants. It is also found among infants and other patients who have received blood transfusions. The cause of AIDS is unknown. However, most experts feel that it is caused by an infectious organism. The pattern of AIDS closely resembles the occurrence of hepatitis B (serum hepatitis), a liver disease that commonly strikes homosexual drug addicts using contaminated needles, and sometimes patients getting blood transfusions. Like hepatitis B, AIDS might be transmitted by sexual contact or blood.

MEDICAL TERMINOLOGY

Adenitis (*adeno* = gland; *itis* = inflammation of)   Enlarged, tender, and inflamed lymph nodes resulting from an infection.

Elephantiasis   Great enlargement of a limb (especially lower limbs) and scrotum resulting from obstruction of lymph glands or vessels by a parasitic worm.

Hypersplenism (*hyper* = over)   Abnormal splenic activity involving highly increased blood cell destruction.

Lymphadenectomy (*ectomy* = removal)   Removal of a lymph node.

Lymphadenopathy (*patho* = disease)   Enlarged, sometimes tender lymph glands.

Lymphangioma (*angio* = vessel; *oma* = tumor).   A benign tumor of the lymph vessels.

Lymphangitis   Inflammation of the lymphatic vessels.

Lymphedema (*edema* = swelling)   Accumulation of lymph fluid producing subcutaneous tissue swelling.

Lymphoma   Any tumor composed of lymph tissue. Malignancy of reticuloendothelial cells of lymph nodes is called Hodgkin's disease.

Lymphostasis (*stasis* = halt)   A lymph flow stoppage.

Splenomegaly (*mega* = large)   Enlarged spleen.

STUDY OUTLINE

Lymphatic Vessels (p. 520)

1. The lymphatic system consists of lymph, lymphatic vessels, lymph nodes, and lymph organs.
2. Lymphatic vessels begin as blind-ended lymph capillaries in tissue spaces between cells.
3. Lymph capillaries merge to form larger vessels, called lymphatics, which ultimately converge into the thoracic duct or right lymphatic duct.
4. Lymphatics have thinner walls and more valves than veins.

Structure of Lymph Nodes (p. 520)

1. Lymph nodes are oval structures located along lymphatics.
2. Lymph enters nodes through afferent lymphatic vessels and exits through efferent lymphatic vessels.
3. Lymph passing through the nodes is processed by macrophages.

Lymph Circulation (p. 521)

1. The passage of lymph is from interstitial fluid, to lymph capillaries, to lymphatics, to lymph trunks, to the thoracic duct or right lymphatic trunk, to the subclavian veins.
2. Lymph flows as a result of skeletal muscle contractions and respiratory movements. It is also aided by valves in the lymphatics.

Lymphatic Organs (p. 523)

1. Tonsils are masses of lymphoid tissue embedded in mucous membranes. They include the pharyngeal, palatine, and lingual tonsils.
2. The spleen functions as a lymphatic organ in phagocytosis of bacteria and worn-out cells and production of lymphocytes and plasma cells. It also acts as a reservoir for blood.
3. The thymus gland functions in immunity by processing T cells and stimulating B cells to develop into antibody-producing plasma cells.

Nonspecific Resistance to Disease (p. 526)

1. The ability to ward off disease using a number of defenses is called resistance. Lack of resistance is called susceptibility.
2. Nonspecific resistance refers to a wide variety of body responses against a wide range of pathogens.
3. Nonspecific resistance includes mechanical factors (skin, mucous membranes, lacrimal apparatus, saliva, mucus, cilia, epiglottis, and flow of urine), chemical factors (gastric juice, acid pH of skin, unsaturated fatty acids, and lysozyme), antimicrobial substances (interferon, complement, and properdin), phagocytosis, inflammation, and fever.

Immunity (Specific Resistance to Disease) (p. 530)

1. Specific resistance to disease involves the production of a specific lymphocyte or antibody against a specific antigen and is called immunity.
2. Antigens are chemical substances that, when introduced into the body, stimulate the production of antibodies that react with the antigen.

3. Examples of antigens are microbes, microbial structures, pollen, incompatible blood cells, and transplants.
4. Antigens are characterized by immunogenicity, reactivity, and multivalence.
5. Antibodies are proteins produced in response to antigens.
6. Based on chemistry and structure, antibodies are distinguished into five principal classes, each with specific biological roles (IgG, IgA, IgM, IgD, and IgE).
7. Antibodies consist of heavy and light chains and variable and constant portions.
8. Cellular immunity refers to destruction of antigens by T cells and humoral immunity refers to destruction of antigens by antibodies.
9. T cells are processed in the thymus gland; B cells may be processed in bone marrow, fetal liver and spleen, or gut-associated lymphoid tissue.
10. T cells consist of subpopulations: killer T cells destroy antigens directly; helper T cells help B cells to produce antibodies; suppressor T cells help to regulate the immune response; and memory T cells initiate response to subsequent invasions by the antigen.
11. B cells develop into antibody-producing plasma cells under the influence of thymic hormones; memory B cells recognize the original, invading antigen.
12. The anamnestic response provides the basis for immunization against certain diseases.
13. Cancer cells contain tumor-specific antigens and are frequently destroyed by the body's immune system (immuno-logic surveillance); some cancer cells escape detection and destruction, a phenomenon called immunologic escape.
14. Monoclonal antibodies are pure antibodies produced by fusing a B cell with a tumor cell; they are important in diagnosis, detection of disease, treatment, preparing vaccines, and countering rejection by transplants and autoimmune diseases.

Disorders: Homeostatic Imbalances (p. 537)

1. Hypersensitivity is overreactivity to an antigen. Localized anaphylactic reactions include hay fever, asthma, eczema, and hives; acute anaphylaxis is a severe reaction with systemic effects.
2. Tissue rejection of a transplanted tissue or organ involves antibody production against the proteins (antigens) in the transplant. It may be overcome with immunosuppressive drugs.
3. Autoimmune diseases result when the body does not recognize "self" antigens and produces antibodies against them. Several human autoimmune diseases are rheumatoid arthritis (RA), systemic lupus erythematosus (SLE), rheumatic fever, hemolytic and pernicious anemias, myasthenia gravis, and multiple sclerosis (MS).
4. Acquired immune deficiency syndrome (AIDS) lowers the body's immunity by decreasing the number of T cells and reversing the ratio of helper T cells to suppressor T cells. AIDS victims frequently develop Karposi's sarcoma and *Pneumocystis carinii* pneumonia.

REVIEW QUESTIONS

1. Identify the components and functions of the lymphatic system.
2. How do lymphatic vessels originate? Compare veins and lymphatics with regard to structure.
3. What is a lymphangiogram? What is its diagnostic value?
4. Describe the structure of a lymph node. What functions do lymph nodes serve?
5. Construct a diagram to indicate the route of lymph circulation.
6. List and explain the various factors involved in the maintenance of lymph circulation.
7. Define edema. What are some of its causes?
8. Identify the tonsils by location.
9. Describe the location, gross anatomy, histology, and functions of the spleen. What is a splenectomy?
10. Describe the role of the thymus gland in immunity.
11. Describe the operation of the various mechanical and chemical factors involved in nonspecific resistance.
12. Outline the role of the following antimicrobial substances: interferon (IFN), complement, and properdin.
13. What is phagocytosis? Describe the steps involved in adherence and ingestion.
14. Compare the roles of microphages and macrophages in phagocytosis.
15. Define immunity.
16. Distinguish between an antigen and an antibody.
17. List the various characteristics of antigens.
18. By means of a labeled diagram, describe the structure of an antibody.
19. What are the characteristics of antibodies?
20. Where are T cells and B cells processed? How do they differ in function?
21. Describe the role of T cells in cellular immunity. Be sure to cite the roles of each type of T cell.
22. Describe the role of B cells in humoral immunity.
23. Discuss the importance of the secondary (anamnestic) response of the body to an antigen.
24. Explain how immunology is related to cancer.
25. What are monoclonal antibodies? How are they produced? Why are they clinically important?
26. Define hypersensitivity, anaphylaxis, and anaphylactic shock.
27. Describe the mechanism of anaphylaxis.
28. Why does tissue rejection occur? How is this problem overcome?
29. How are transplants classified?
30. Define an autoimmune disease. Give several examples.
31. Describe the symptoms of acquired immune deficiency syndrome (AIDS). What are the complications of AIDS?
32. Refer to the glossary of medical terminology associated with the lymphatic system. Be sure that you define each term.

● Identify the organs of the respiratory system.

● Compare the structure and function of the external and internal nose.

● Differentiate the three regions of the pharynx and describe their roles in respiration.

● Describe the structure of the larynx and explain its function in respiration and voice production.

● Explain the structure and function of the trachea.

● Contrast tracheostomy and intubation as alternative methods for clearing obstructed air passageways.

● Discuss the composition of the bronchial tree.

● Identify the coverings of the lungs and the division of the lungs into lobes.

● Describe the composition of a lobule of the lung.

● Explain the structure of the alveolar-capillary membrane and its function in the diffusion of respiratory gases.

● List the events involved in inspiration and expiration.

● Explain how compliance and airway resistance relate to breathing.

● Define coughing, sneezing, sighing, yawning, sobbing, crying, laughing, and hiccuping as modified respiratory movements.

● Compare the volumes and capacities of air exchanged during respiration.

● Define the partial pressure of a gas, Boyle's law, Charles' law, Dalton's law, and Henry's law.

● Explain how external and internal respiration occur, basing your explanation on differences in the partial pressure of the respiratory gases.

● Explain how the respiratory gases are transported by the blood.

● Describe how the oxygen-carrying capacity of the blood is affected by pO_2, pCO_2, temperature, and DPG.

● Explain how the respiratory center functions in establishing the basic rhythm of respiration.

● Describe how various factors may modify the rate of respiration.

● List the basic steps involved in cardiopulmonary resuscitation (CPR).

● Explain how the abdominal thrust (Heimlich) maneuver is performed.

● Define bronchogenic carcinoma, nasal polyps, bronchial asthma, emphysema, pneumonia, tuberculosis, infant respiratory distress syndrome (RDS), sudden infant death syndrome (SIDS), coryza (common cold), and influenza (flu) as disorders of the respiratory system.

● Define medical terminology associated with the respiratory system.

● Explain the actions of selected drugs on the respiratory system.

23 The Respiratory System

Cells need a continuous supply of oxygen to carry out the activities that are vital to their survival. Many of these activities release quantities of carbon dioxide. Since an excessive amount of carbon dioxide produces acid conditions that are poisonous to cells, the gas must be eliminated quickly and efficiently. The two systems that supply oxygen and eliminate carbon dioxide are the cardiovascular system and the respiratory system. The **respiratory system** consists of organs that exchange gases between the atmosphere and blood. These organs are the nose, pharynx, larynx, trachea, bronchi, and lungs (Figure 23-1). The cardiovascular system transports the gases in the blood between the lungs and the cells. The term *upper respiratory system* refers to the nose and throat and associated structures. The *lower respiratory system* refers to the remainder of the system.

The overall exchange of gases between the atmosphere, the blood, and the cells is **respiration.** Three basic processes are involved. The first process, **pulmonary ventilation,** or breathing, is the inflow and outflow of air between the atmosphere and the lungs. The second and third processes involve the exchange of gases within the body. **External respiration** is the exchange of gases between the lungs and blood. **Internal respiration** is the exchange of gases between the blood and the cells.

The respiratory and cardiovascular systems participate equally in respiration. Failure of either system has the same effect on the body: disruption of homeostasis and rapid death of cells from oxygen starvation.

ORGANS

NOSE

Anatomy

The **nose** has an external portion and an internal portion inside the skull (Figure 23-2). The external portion consists of a supporting framework of bone and cartilage covered with skin and lined with mucous membrane. The bridge of the nose is formed by the nasal bones, which hold it in a fixed position. Because it has a framework of pliable cartilage, the rest of the external nose is quite flexible. On the undersurface of the external nose are two openings called the **nostrils** or **external nares** (NA-rēz; *sing.,* naris). The surface anatomy of the nose is shown in Figure 23-2a.

The internal portion of the nose is a large cavity in the skull that lies inferior to the cranium and superior to the mouth. Anteriorly, the internal nose merges with the external nose, and posteriorly it communicates with the throat (pharynx) through two openings called the **internal nares (choanae).** Four paranasal sinuses (frontal, sphenoidal, maxillary, and ethmoidal) and the nasolacrimal ducts also open into the internal nose. The lateral walls of the internal nose are formed by the ethmoid, maxillae, and inferior nasal conchae bones. The ethmoid also forms the roof. The floor is formed by the palatine bones and the palatine process of the maxilla, which together comprise the hard palate.

The inside of both the external and internal nose consists of a **nasal cavity,** divided into right and left sides

by a vertical partition called the **nasal septum.** The anterior portion of the septum is made primarily of cartilage. The remainder is formed by the vomer and the perpendicular plate of the ethmoid (see Figure 7-7a). The anterior portion of the nasal cavity, just inside the nostrils, is called the **vestibule.** It is surrounded by cartilage. The upper nasal cavity is surrounded by bone.

Physiology

The interior structures of the nose are specialized for three functions: incoming air is warmed, moistened, and filtered; olfactory stimuli are received; and large hollow resonating chambers are provided for speech sounds.

When air enters the nostrils, it passes first through the vestibule. The vestibule is lined by skin containing coarse hairs that filter out large dust particles. The air then passes into the upper nasal cavity. Three shelves formed by projections of the superior, middle, and inferior nasal conchae extend out of the lateral wall of the cavity. The conchae, almost reaching the septum, subdivide each side of the nasal cavity into a series of groovelike passageways—the **superior, middle,** and **inferior meatuses.** Mucous membrane lines the cavity and its shelves. The olfactory receptors lie in the membrane lining the area superior to the superior nasal conchae and is also called the **olfactory region.** Below the olfactory region, the membrane

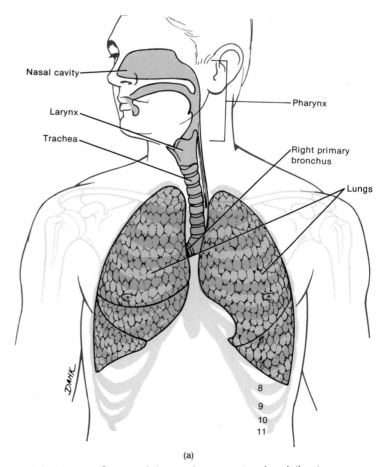

FIGURE 23-1 Organs of the respiratory system in relation to surrounding structures. (a) Diagram.

(b)

FIGURE 23-1 (*Continued*) Organs of the respiratory system in relation to surrounding structures. (b) Photograph. (Courtesy of C. Yokochi and J. W. Rohen, *Photographic Anatomy of the Human Body,* 2nd ed., 1979, IGAKU-SHOIN, Ltd., Tokyo, New York.)

contains pseudostratified ciliated columnar cells with many goblet cells and capillaries. As the air whirls around the conchae and meatuses, it is warmed by the capillaries. Mucus secreted by the goblet cells moistens the air and traps dust particles. Drainage from the lacrimal ducts, and perhaps secretions from the paranasal sinuses, also help moisten the air. The cilia move the mucus-dust packages along the pharynx so they can be eliminated from the body.

CLINICAL APPLICATION

Nosebleed, or **epistaxis** (ep'-i-STAK-sis), is common because of the exposure of the nose to trauma and the extensive blood supply of the nose. Bleeding, either

---

arterial or venous, usually occurs on the anterior part of the septum and can be arrested by firm packing of the external nares. If the point of bleeding is in the posterior region, plugging of both the external and internal nares may be necessary. In extreme emergency, the external carotid artery may have to be ligated (tied) in order to control the hemorrhage.

PHARYNX

The **pharynx** (FAR-inks), or throat, is a somewhat funnel-shaped tube about 13 cm (5 inches) long that starts at the internal nares and extends partway down the neck

1. **Root.** Superior attachment of nose at forehead located between eyes.
2. **Apex.** Tip of nose.
3. **Dorsum nasi.** Rounded anterior border connecting root and apex; in profile, may be straight, convex, concave, or wavy.
4. **Nasofacial angle.** Point at which side of nose blends with tissues of face.
5. **Ala.** Convex flared portio of inferior lateral surface; unites with upper lip.
6. **External naris.** External opening into nose.
7. **Bridge.** Superior portion of dorsum nasi, superficial to nasal bones.

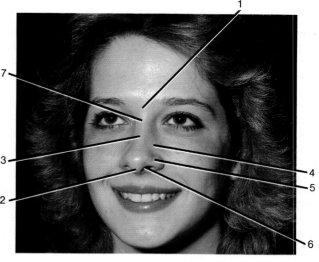

(a)

FIGURE 23-2 Nose. (a) Surface anatomy of the nose in anterior view. (© 1982 by Gerard J. Tortora. Courtesy of Lynne Tortora.)

(b)

FIGURE 23-2 (*Continued*) Nose. (b) Diagram of the left side seen in sagittal section with the nasal septum removed.

(Figure 23-3). It lies just posterior to the nasal cavity and oral cavity and just anterior to the cervical vertebrae. Its wall is composed of skeletal muscles and lined with mucous membrane. The functions of the pharynx are to serve as a passageway for air and food and to provide a resonating chamber for speech sounds.

The uppermost portion of the pharynx, called the **naso-pharynx,** lies posterior to the internal nasal cavity and extends to the plane of the soft palate. There are four openings in its wall: two internal nares and two openings that lead into the auditory tubes. The posterior wall also contains the pharyngeal tonsil, or adenoid. Through the internal nares the nasopharynx exchanges air with the nasal cavities and receives the packages of dust-laden mucus. It is lined with pseudostratified ciliated epithelium, and the cilia move the mucus down toward the mouth. The nasopharynx also exchanges small amounts of air

with the auditory tubes so that the air pressure inside the middle ear equals the pressure of the atmospheric air flowing through the nose and pharynx.

The middle portion of the pharynx, the **oropharynx,** lies posterior to the oral cavity and extends from the soft palate inferior to the level of the hyoid bone. It has only one opening, the *fauces* (FAW-sēz), or opening from the mouth. It is lined by stratified squamous epithelium. This portion of the pharynx is both respiratory and diges-tive in function, since it is a common passageway for both air and food. Two pairs of tonsils, the palatine and lingual tonsils, are found in the oropharynx. The lingual tonsil lies at the base of the tongue (see also Figure 24-5a).

The lowest portion of the pharynx, the **laryngopharynx** (la-rin'-gō-FAR-inks), extends downward from the hyoid bone and becomes continuous with the esophagus (food

FIGURE 23-2 (*Continued*) Nose. (c) Photograph of the left side seen in sagittal section with the nasal septum removed. (Courtesy of C. Yokochi and J. W. Rohen, *Photographic Anatomy of the Human Body,* 2nd ed., 1979, IGAKU-SHOIN, Ltd., Tokyo, New York.)

tube) posteriorly and the larynx (voice box) anteriorly. Like the oropharynx, the laryngopharynx is a respiratory and a digestive pathway and is lined by stratified squamous epithelium.

LARYNX

The **larynx,** or voice box, is a short passageway that connects the pharynx with the trachea. It lies in the midline of the neck anterior to the fourth through sixth cervical vertebrae.

Anatomy

The wall of the larynx is supported by nine pieces of cartilage (Figure 23-4). Three are single and three are paired. The three single pieces are the thyroid cartilage, epiglottic cartilage (epiglottis), and cricoid cartilage. Of the paired cartilages, the arytenoid cartilages are the most important. The paired corniculate and cuneiform cartilages are of lesser significance.

The **thyroid cartilage,** or Adam's apple, consists of two fused plates that form the anterior wall of the larynx and give it its triangular shape. It is larger in males than in females.

The **epiglottis** is a large, leaf-shaped piece of cartilage lying on top of the larynx (see also Figure 23-3). The

"stem" of the epiglottis is attached to the thyroid cartilage, but the "leaf" portion is unattached and free to move up and down like a trapdoor. During swallowing, there is elevation of the larynx. This causes the free edge of the epiglottis to form a lid over the glottis and closure of the glottis. The **glottis** is the space between the vocal folds (true vocal cords) in the larynx. In this way, the larynx is closed off and liquids and foods are routed into the esophagus and kept out of the trachea. If anything but air passes into the larynx, a cough reflex attempts to expel the material.

The **cricoid** (KRĪ-koyd) **cartilage** is a ring of cartilage forming the inferior wall of the larynx. It is attached to the first ring of cartilage of the trachea.

The paired **arytenoid** (ar'-i-TĒ-noyd) **cartilages** are pyramidal in shape and located at the superior border of the cricoid cartilage. They attach to the vocal folds and pharyngeal muscles and by their action can move the vocal cords.

The paired **corniculate** (kor-NIK-yoo-lāt) **cartilages** are cone-shaped. One is located at the apex of each arytenoid cartilage. The paired **cuneiform** (kyoo-NĒ-i-form) **cartilages** are rod-shaped cartilages that connect the epiglottis to the arytenoid cartilages.

The epithelium lining the larynx below the vocal folds is pseudostratified. It consists of ciliated columnar cells, goblet cells, and basal cells, and it helps to trap dust not removed in the upper passages.

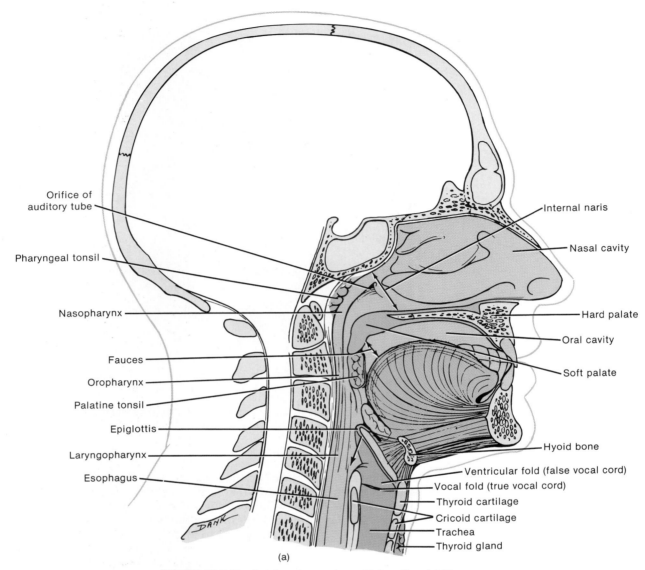

Orifice of
auditory tube
Pharyngeal tonsil
Nasopharynx
Fauces
Oropharynx
Palatine tonsil
Epiglottis
Laryngopharynx
Esophagus

Internal naris
Nasal cavity
Hard palate
Oral cavity
Soft palate
Hyoid bone
Ventricular fold (false vocal cord)
Vocal fold (true vocal cord)
Thyroid cartilage
Cricoid cartilage
Trachea
Thyroid gland

(a)

FIGURE 23-3 Head and neck seen in sagittal section. (a) Diagram.

Voice Production

The mucous membrane of the larynx is arranged into
two pairs of folds—an upper pair called the **ventricular
folds (false vocal cords)** and a lower pair called simply
the **vocal folds (true vocal cords)** (Figure 23-4c). When
the ventricular folds are brought together they function
in holding the breath against pressure in the thoracic
cavity such as might occur when a person exerts a strain
while lifting a heavy weight. The mucous membrane of
the vocal folds is lined by nonkeratinized stratified squamous epithelium. Under the membrane lie bands of elastic
ligaments stretched between pieces of rigid cartilage like
the strings on a guitar. Skeletal muscles of the larynx,
called intrinsic muscles, are attached internally to the
pieces of rigid cartilage and to the vocal folds themselves
(see Figure 11-9). When the muscles contract, they pull
the strings of elastic ligaments tight and stretch the vocal
folds out into the air passageways so that the glottis is
narrowed. If air is directed against the vocal folds, they

vibrate and set up sound waves in the column of air in
the pharynx, nose, and mouth. The greater the pressure
of air, the louder the sound.

Pitch is controlled by the tension on the vocal folds.
If they are pulled taut by the muscles, they vibrate more
rapidly and a higher pitch results. Lower sounds are produced by decreasing the muscular tension on the vocal
folds. Vocal folds are usually thicker and longer in males
than in females, and therefore they vibrate more slowly.
Thus men generally have a lower range of pitch than
women.

Sound originates from the vibration of the vocal folds,
but other structures are necessary for converting the
sound into recognizable speech. The pharynx, mouth, nasal cavity, and paranasal sinuses all act as resonating
chambers that give the voice its human and individual
quality. By constricting and relaxing the muscles in the
walls of the pharynx, we produce the vowel sounds. Muscles of the face, tongue, and lips help us to enunciate
words.

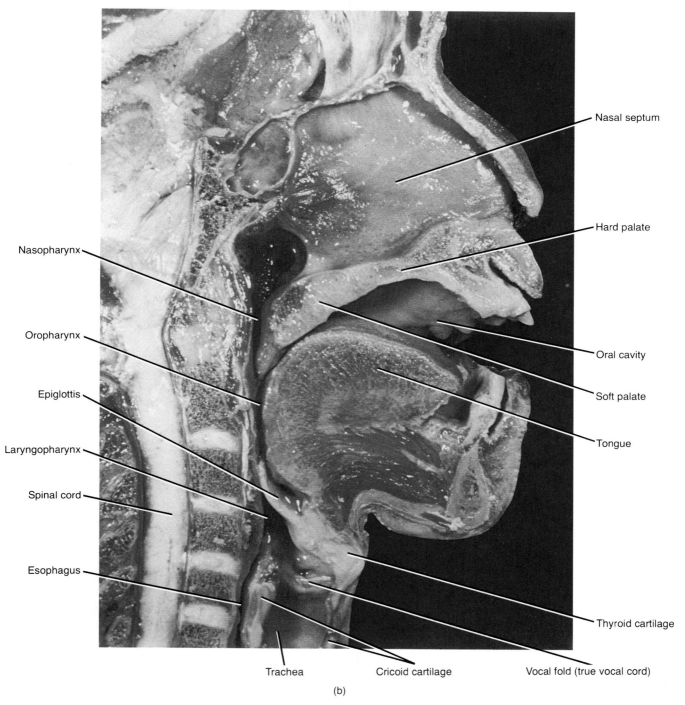

Nasal septum

Hard palate

Nasopharynx

Oropharynx

Epiglottis

Laryngopharynx

Spinal cord

Esophagus

Oral cavity

Soft palate

Tongue

Thyroid cartilage

Trachea

Cricoid cartilage

Vocal fold (true vocal cord)

(b)

FIGURE 23-3 (*Continued*) Head and neck seen in sagittal section. (b) Photograph. (Courtesy of C. Yokochi and J. W. Rohen, *Photographic Anatomy of the Human Body,* 2nd ed., 1979, IGAKU-SHOIN, Ltd., Tokyo, New York.)

CLINICAL APPLICATION

Laryngitis is an inflammation of the larynx that is most often caused by a respiratory infection or irritants such as cigarette smoke. Inflammation of the vocal folds themselves causes hoarseness or loss of voice by interfering with the contraction of the folds or by causing them to swell to the point where they cannot vibrate freely. Many long-term smokers acquire a permanent hoarseness from the damage done by chronic inflammation.

TRACHEA

The **trachea** (TRĀ-kē-a), or windpipe, is a tubular passageway for air about 12 cm (4½ inches) in length and 2.5 cm (1 inch) in diameter. It is located anterior to the esophagus and extends from the larynx to the fifth thoracic vertebra, where it divides into right and left primary bronchi (see Figure 23-6a).

The tracheal epithelium is pseudostratified. It consists of ciliated columnar cells that reach the lumenal surface, goblet cells, and basal cells that do not reach the lumenal

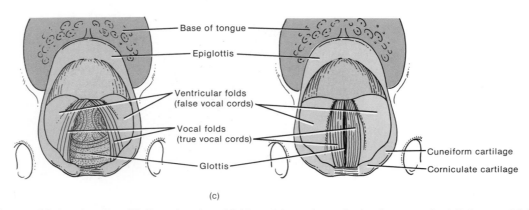

FIGURE 23-4 Larynx. (a) Anterior view. (b) Posterior view. (c) Viewed from above. In the figure on the left the vocal folds (true vocal cords) are relaxed and the glottis is open. In the figure on the right the vocal folds are pulled taut and the glottis is closed.

surface (Figure 23-5). The epithelium provides the same protection against dust as the membrane lining the larynx. The wall of the trachea is composed of smooth muscle and elastic connective tissue. It also consists of a series of 16 to 20 horizontal incomplete rings of hyaline cartilage that look like a series of letter Cs stacked one on top of another. The open parts of the Cs face the esophagus and permit it to expand into the trachea during swallowing. Transverse smooth muscle fibers, called the *trachealis muscle,* attach the open ends of the cartilage rings. The open ends of the rings of cartilage are also attached by elastic connective tissue. The solid parts of the Cs provide a rigid support so the tracheal wall does not collapse inward and obstruct the air passageway.

CLINICAL APPLICATION

At the point where the trachea bifurcates into right and left primary bronchi, there is an internal ridge called the **carina** (ka-RĪ-na). It is formed by a posterior and somewhat inferior projection of the last tracheal cartilage. Widening and distortion of the carina, which can be seen in an examination by bronchoscopy, is a serious prognostic sign, since it usually indicates a carcinoma of the lymph nodes around the bifurcation of the trachea. **Bronchoscopy** is the visual examination of the bronchi through a **bronchoscope,** an illuminated, tubular instrument which can be passed through the trachea into the bronchi.

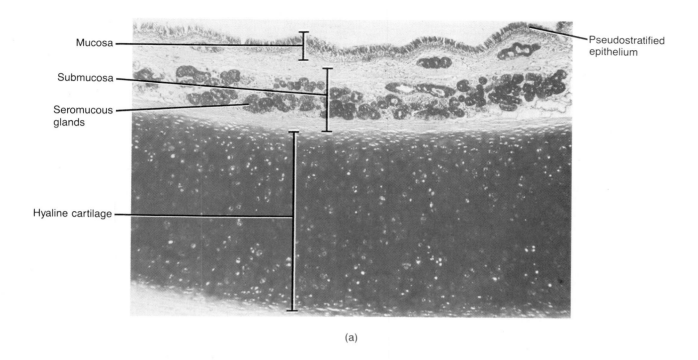

Mucosa

Submucosa

Seromucous glands

Hyaline cartilage

Pseudostratified epithelium

(a)

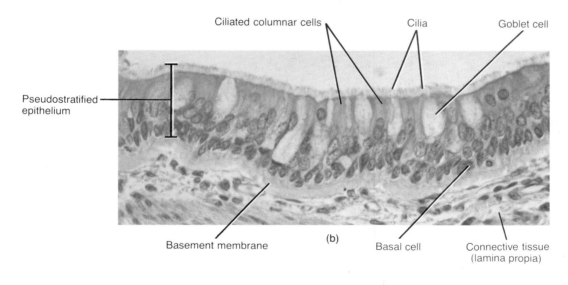

Ciliated columnar cells

Cilia

Goblet cell

Pseudostratified epithelium

Basement membrane

(b)

Basal cell

Connective tissue (lamina propia)

FIGURE 23-5 Histology of the trachea. (a) Photomicrograph of a portion of the tracheal wall at a magnification of 80×. (© 1983 by Michael H. Ross. Used by permission.) (b) Photomicrograph of an enlarged aspect of the tracheal epithelium at a magnification of 600×. (© 1983 by Michael H. Ross. Used by permission.) (c) Scanning electron micrograph of the tracheal epithelium at a magnification of 200×. (Courtesy of Fisher Scientific Company and S.T.E.M. Laboratories, Inc., Copyright, 1975.)

(c)

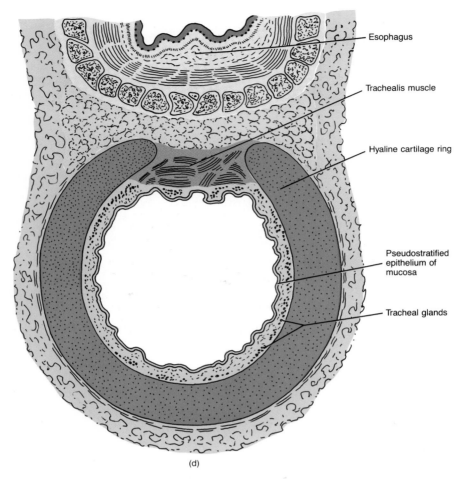

Esophagus

Trachealis muscle

Hyaline cartilage ring

Pseudostratified epithelium of mucosa

Tracheal glands

(d)

FIGURE 23-5 (*Continued*) Histology of the trachea. (d) Diagram of a cross section through the trachea and a portion of the esophagus.

Occasionally the respiratory passageways are unable to protect themselves from obstruction. The rings of cartilage may be accidentally crushed; the mucous membrane may become inflamed and swell so much that it closes off the air space; inflamed membranes secrete a great deal of mucus that may clog the lower respiratory passageways; or a large object may be breathed in (aspirated) while the glottis is open. The passageways must be cleared quickly. If the obstruction is above the level of the chest, a **tracheostomy** (trā-kē-OS-tō-mē) may be performed. An incision is made in the neck and into the part of the trachea below the obstructed area. The patient breathes through a tube inserted through the incision. Another method is **intubation.** A tube is inserted into the mouth or nose and passed down through the larynx and trachea. The firm wall of the tube pushes back any flexible obstruction, and the inside of the tube provides a passageway for air. If mucus is clogging the trachea, it can be suctioned out through the tube.

BRONCHI

The trachea terminates in the chest by dividing at the sternal angle into a **right primary bronchus** (BRON-kus),

which goes to the right lung, and a **left primary bronchus,** which goes to the left lung (Figure 23-6a, b). The right primary bronchus is more vertical, shorter, and wider than the left. As a result, foreign objects in the air passageways are more likely to enter it than the left and frequently lodge in it. Like the trachea, the primary bronchi (BRON-kē) contain incomplete rings of cartilage and are lined by a pseudostratified ciliated epithelium.

Upon entering the lungs, the primary bronchi divide to form smaller bronchi—the **secondary (lobar) bronchi,** one for each lobe of the lung (the right lung has three lobes; the left lung has two). The secondary bronchi continue to branch, forming still smaller bronchi, called **tertiary (segmental) bronchi** which divide into **bronchioles.** Bronchioles, in turn, branch into even smaller tubes called **terminal bronchioles.** This continuous branching from the trachea resembles a tree trunk with its branches and is commonly referred to as the **bronchial tree.**

As the branching becomes more extensive in the bronchial tree, several structural changes may be noted. First, rings of cartilage are replaced by plates of cartilage that finally disappear in the bronchioles. Second, as the cartilage decreases, the amount of smooth muscle increases. Third, the epithelium changes from pseudostratified ciliated to simple cuboidal in the terminal bronchioles.

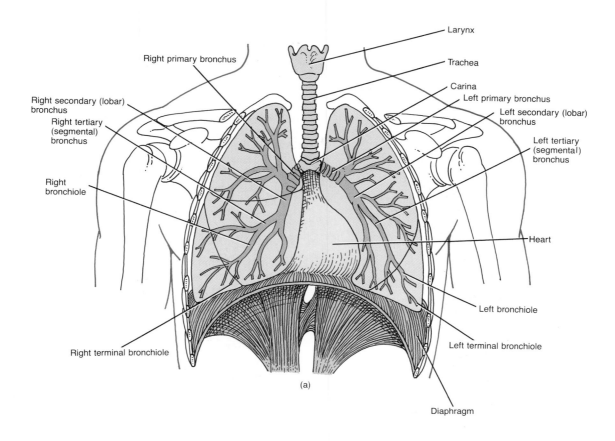

Larynx

Right primary bronchus

Trachea

Carina
Left primary bronchus

Right secondary (lobar)
bronchus

Left secondary (lobar)
bronchus

Right tertiary
(segmental)
bronchus

Left tertiary
(segmental)
bronchus

Right
bronchiole

Heart

Left bronchiole

Right terminal bronchiole

Left terminal bronchiole

(a)

Diaphragm

Glands

Pseudostratified
ciliated epithelium

Smooth muscle

Smooth muscle

Cartilage plates

(b)

FIGURE 23-6 Air passageways to the lungs. (a) Diagram of the bronchial tree in relation to the lungs. (b) Photomicrograph of a cross section of a primary bronchus at a magnification of 100×. (© 1983 by Michael H. Ross. Used by permission.)

(c)

FIGURE 23-6 (*Continued*) Air passageways to the lungs. (c) Anteroposterior bronchogram. (Courtesy of Lester W. Paul and John H. Juhl, *The Essentials of Roentgen Interpretation,* 3rd ed., Harper & Row, Publishers, Inc., New York, 1972.)

CLINICAL APPLICATION

The fact that the walls of the bronchioles contain a great deal of smooth muscle but no cartilage is clinically significant. During an **asthma attack** the muscles go into spasm. Because there is no supporting cartilage, the spasms can close off the air passageways.

Bronchography (brong-KOG-ra-fē) is a technique for examining the bronchial tree. An intratracheal catheter is passed into the mouth or nose, through the glottis, and into the trachea. Then, an opaque contrast medium, usually containing iodine is introduced, by means of gravity, into the trachea and distributed through the bronchial branches. Roentgenograms of the chest in various positions are taken and the developed film, a **bronchogram** (BRONG-kō-gram), provides a picture of the tree (Figure 23-6c).

LUNGS

The **lungs** are paired, cone-shaped organs lying in the thoracic cavity. They are separated from each other by the heart and other structures in the mediastinum (see Figure 20-1a). Two layers of serous membrane, collectively called the **pleural membrane,** enclose and protect each lung. The outer layer is attached to the wall of the thoracic cavity and is called the **parietal pleura.** The inner layer, the **visceral pleura,** covers the lungs themselves. Between the visceral and parietal pleura is a small potential space, the **pleural cavity,** which contains a lubricating fluid secreted by the membranes (see Figure 1-7b). This fluid prevents friction between the membranes and allows them to move easily on one another during breathing.

CLINICAL APPLICATION

In certain conditions, the pleural cavity may fill with air (**pneumothorax**), blood (**hemothorax**), or pus. Air in the pleural cavity, mostly commonly introduced in a surgical opening of the chest, may cause the lung to collapse. Fluid can be drained from the pleural cavity by inserting a needle, usually posteriorly through the seventh intercostal space. The needle is passed along the superior border of the lower rib to avoid damage to the intercostal nerves and blood vessels. Below the seventh intercostal space there is danger of penetrating the diaphragm.

Inflammation of the pleural membrane, or **pleurisy,** causes friction during breathing that can be quite painful when the swollen membranes rub against each other.

Gross Anatomy

The lungs extend from the diaphragm to a point about 1.5 to 2.5 cm (¾ to 1 inch) superior to the clavicles and lie against the ribs anteriorly and posteriorly. The broad inferior portion of the lung, the **base,** is concave and fits over the convex area of the diaphragm (Figure 23-7). The narrow superior portion of the lung is termed the **apex (cupula).** The surface of the lung lying against the ribs, the **costal surface,** is rounded to match the curvature of the ribs. The **mediastinal (medial) surface** of each lung contains a vertical slit, the **hilus,** through which bronchi, pulmonary vessels, and nerves enter and exit. The blood vessels, bronchi, and nerves are held together by the pleura and connective tissue, and they constitute the **root** of the lung. Medially, the left lung also contains a concavity, the **cardiac notch,** in which the heart lies.

The right lung is thicker and broader than the left. It is also somewhat shorter than the left because the diaphragm is higher on the right side to accommodate the liver that lies below it.

Lobes and Fissures

Each lung is divided into lobes by one or more fissures. Both lungs have an **oblique fissure,** which extends downward and forward. The right lung also has a **horizontal fissure.** The oblique fissure in the left lung separates the **superior lobe** from the **inferior lobe.** The upper part of the oblique fissure of the right lung separates the superior lobe from the inferior lobe, whereas the lower part of the oblique fissure separates the inferior lobe from the **middle lobe.** The horizontal fissure of the right lung subdivides the superior lobe, thus forming a middle lobe.

Each lobe receives its own secondary (lobar) bronchus. Thus the right primary bronchus gives rise to three secondary (lobar) bronchi called the **superior, middle,** and **inferior secondary (lobar) bronchi.** The left primary bronchus gives rise to a **superior** and an **inferior secondary (lobar) bronchus.** Within the substance of the lung, the secondary bronchi give rise to the **tertiary (segmental) bronchi,** which are constant in both origin and distribution. The segment of lung tissue that each supplies is called a **bronchopulmonary segment.** Bronchial and pulmonary disorders, such as tumors or abscesses, may be localized in a bronchopulmonary segment and may be surgically removed without seriously disrupting surrounding lung tissue.

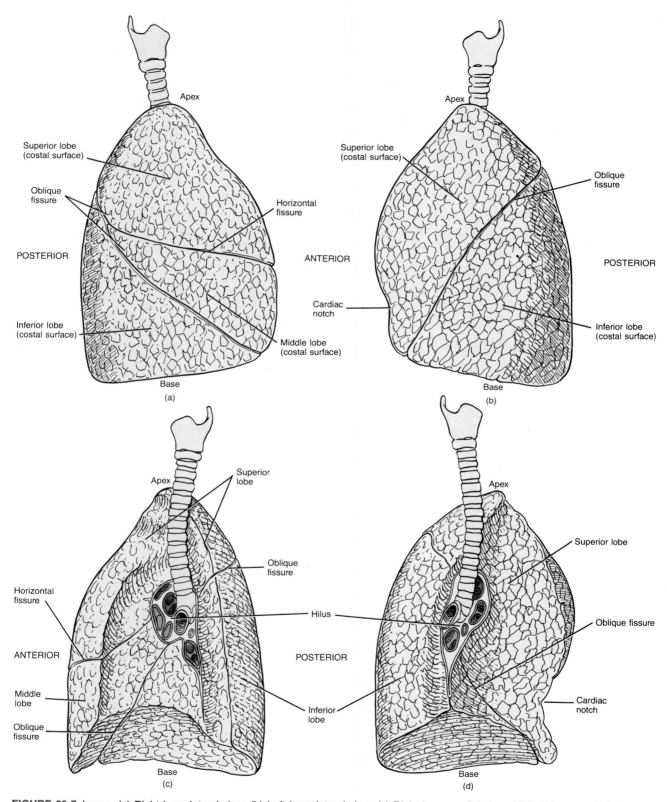

FIGURE 23-7 Lungs. (a) Right lung, lateral view. (b) Left lung, lateral view. (c) Right lung, medial view. (d) Left lung, medial view.

Lobules

Each bronchopulmonary segment of the lungs is broken up into many small compartments called **lobules** (Figure 23-8a). Each lobule is wrapped in elastic connective tissue and contains a lymphatic vessel, an arteriole, a venule, and a branch from a terminal bronchiole. Terminal bronchioles subdivide into microscopic branches called **respi-**

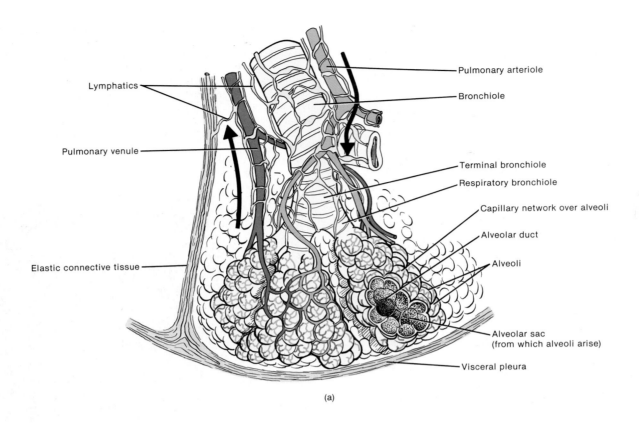

Pulmonary arteriole

Bronchiole

Lymphatics

Pulmonary venule

Terminal bronchiole

Respiratory bronchiole

Capillary network over alveoli

Alveolar duct

Alveoli

Elastic connective tissue

Alveolar sac
(from which alveoli arise)

Visceral pleura

(a)

Terminal
bronchiole

Alveoli

Surface
of lung

Alveolar
sac

Alveolar
duct

Respiratory
bronchioles

(b)

FIGURE 23-8 Histology of the lungs. (a) Diagram of a lobule of the lung. (b) Photomicrograph of a terminal bronchiole, respiratory bronchioles, alveolar duct, alveolar sac, and alveoli at a magnification of 50×. (© 1983 by Michael H. Ross. Used by permission.)

(c)

FIGURE 23-8 (*Continued*) Histology of the lungs. (c) Scanning electron micrograph of alveolar sacs at a magnification of 100×. (Courtesy of Fisher Scientific Company and S.T.E.M. Laboratories, Inc., Copyright, 1975). (d) Scanning electron micrograph of an alveolus showing squamous pulmonary alveolar cells and an alveolar macrophage at a magnification of 3,430×. (Courtesy of Richard K. Kessel and Randy H. Kardon, *Tissues and Organs: A Text-Atlas of Scanning Electron Microscopy*. Copyright © 1979 by Scientific American, Inc.) (e) Diagram of the structure of the alveolar-capillary membrane.

(d)

ratory bronchioles (Figure 23-8a, b). As the respiratory bronchioles become more distal, the epithelial lining changes from cuboidal to squamous. Respiratory bronchioles, in turn, subdivide into several (2 to 11) **alveolar ducts (atria).**

Around the alveolar ducts are numerous alveoli and alveolar sacs. An **alveolus** (al-VĒ-ō-lus) is a cup-shaped outpouching lined by epithelium and supported by a thin elastic basement membrane. **Alveolar sacs** are two or more alveoli that share a common opening (Figure 23-8a–c). The alveolar walls consist of two principal types of epithelial cells: *squamous pulmonary epithelial cells* and *septal cells.* The squamous pulmonary epithelial cells are the larger of the two types of cells and form a continuous lining of the alveolar wall, except for occasional septal cells. Septal cells are much smaller, cuboidal in shape, and are dispersed among the squamous pulmonary epithelial cells. Septal cells produce a phospholipid substance called *surfactant* (sur-FAK-tant), which lowers surface tension. Also found within the alveolar wall are free *alveolar macrophages* (*dust cells*). They are highly phagocytic and serve to remove dust particles or other debris that gain entrance to alveolar spaces. Deep to the layer of squamous pulmonary epithelial cells is an elastic basement membrane. Over the alveoli, the arteriole and venule disperse into a capillary network (Figure 23-8a). The blood capillaries consist of a single layer of endothelial cells and a basement membrane.

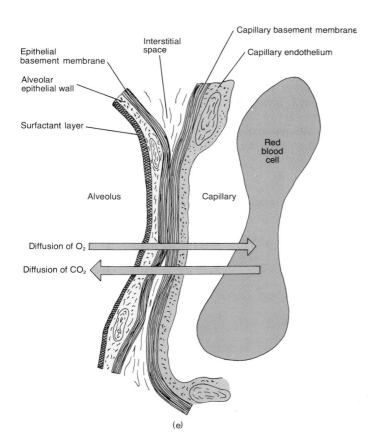

(e)

Alveolar-Capillary (Respiratory) Membrane

The exchange of respiratory gases between the lungs and blood takes place by diffusion across the alveoli and capillary walls. This membrane, through which the respiratory gases move, is collectively known as the **alveolar-capillary (respiratory) membrane** (Figure 23-8e). It consists of:

1. A layer of squamous pulmonary epithelial cells with septal cells and free alveolar macrophages that constitute the alveolar (epithelial) wall.
2. An epithelial basement membrane underneath the alveolar wall.
3. A capillary basement membrane that is often fused to the epithelial basement membrane.
4. The endothelial cells of the capillary.

Despite the large number of layers, the alveolar-capillary membrane averages only 0.5 μm in thickness. This is of considerable importance to the efficient diffusion of respiratory gases. Moreover, it has been estimated that the lungs contain 300 million alveoli, providing an immense surface area of 70 m² (753 ft²) for the exchange of gases.

Blood Supply

The arterial supply of the lungs is derived from the pulmonary trunk. It divides into a left pulmonary artery which enters the left lung and a right pulmonary artery which enters the right lung. The venous return of the oxygenated blood is by way of the pulmonary veins, typically two in number on each side—the right and left superior and inferior pulmonary veins. All four veins drain into the left atrium (see Figure 21-23).

CLINICAL APPLICATION
Pulmonary embolism is a common cause of sickness and death. An embolus may arise in a distant site, such as a leg vein, and pass through the right side of the heart to a lung via a pulmonary artery. The immediate effect of a pulmonary embolism is complete or partial obstruction of the pulmonary arterial blood flow to the lung resulting in dysfunction of the affected lung tissue. A large embolus can produce death in a few minutes.

RESPIRATION

The principal purpose of **respiration** is to supply the cells of the body with oxygen and remove the carbon dioxide produced by cellular activities. The three basic processes of respiration are pulmonary ventilation, external respiration, and internal respiration.

PULMONARY VENTILATION

Pulmonary ventilation (breathing) is the process by which gases are exchanged between the atmosphere and lung alveoli. Air flows between the atmosphere and lungs for the same reason that blood flows through the body—a pressure gradient exists. We breathe in when the pressure inside the lungs is less than the air pressure in the atmosphere. We breathe out when the pressure inside the lungs is greater than the pressure in the atmosphere. Let us examine the mechanics of pulmonary ventilation by first looking at inspiration.

Inspiration

Breathing in is called **inspiration** or **inhalation.** Just before each inspiration, the air pressure inside the lungs equals the pressure of the atmosphere, which is about 760 mm Hg, or 1 atmosphere (atm), at sea level. For air to flow into the lungs, the pressure inside the lungs must become lower than the pressure in the atmosphere. This condition is achieved by increasing the volume of the lungs.

The pressure of a gas in a closed container is inversely proportional to the volume of the container. If the size of a closed container is increased, the pressure of the air inside the container decreases. If the size of the container is decreased, then the pressure inside it increases. This is referred to as **Boyle's law** and may be demonstrated as follows. Suppose we place a gas in a cylinder that has a movable piston and a pressure gauge, and the initial pressure is 1 atm (Figure 23-9). This pressure is created by the gas molecules striking the wall of the container. If the piston is pushed down, the gas is concentrated in a smaller volume. This means that the same number of gas molecules are striking less wall space. The gauge shows that the pressure doubles as the gas is compressed to half its volume. In other words, the same number of molecules in half the space produces twice the pressure. Conversely, if the piston is raised to increase the volume, the pressure decreases. Thus, the volume of a gas varies inversely with pressure (assuming that temperature is constant). Boyle's law applies to the operation of a bicycle pump and the blowing up of a balloon. Differences in pressure force air into our lungs when we inhale and force the air out when we exhale.

In order for inspiration to occur, the lungs must be expanded. This increases lung volume and thus decreases the pressure in the lungs. The first step toward increasing lung volume involves contraction of the principal inspiratory muscles—the diaphragm and external intercostals

Piston

Pressure gauge

Volume = 1
Pressure = 1

Volume = 1/2
Pressure = 2

(a)   (b)

FIGURE 23-9 Boyle's law. The volume of gas varies inversely with the pressure. If the volume is decreased to ¼, what happens to the pressure?

(a)

(b)

FIGURE 23-12 Spirometers. (a) Collins spirometer. This type of spirometer is the one commonly used in college biology laboratories. (Photographed in a biology laboratory at Bergen Community College by Gerard J, Tortora.) (b) Ohio 842 spirometer. This instrument is a highly sophisticated spirometer that utilizes a computerized mechanism for recording results. (Courtesy of Lenny Patti.)

EXCHANGE OF RESPIRATORY GASES

As soon as the lungs fill with air, oxygen diffuses from the alveoli into the blood, through the interstitial fluid, and finally into the cells. Carbon dioxide diffuses in the opposite direction—from the cells, through interstitial fluid to the blood, and to the alveoli. To understand how respiratory gases are exchanged in the body, you need to know a few gas laws.

Charles' Law

According to **Charles' law**, the volume of a gas is directly proportional to its absolute temperature, assuming that the pressure remains constant. Recall the cylinder we used to demonstrate Boyle's law. Suppose now that the gas in the cylinder exerts an initial pressure of 1 atm when the piston is halfway down (Figure 23-14). When the gas is heated, the gas molecules move faster and the

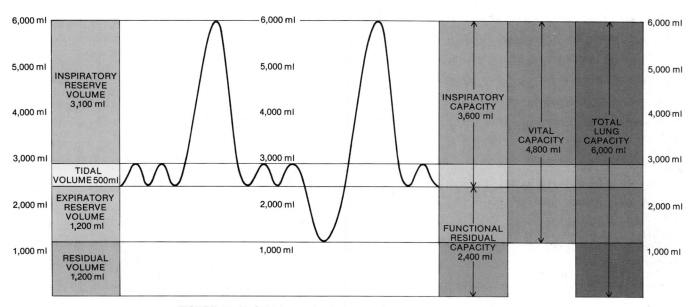

FIGURE 23-13 Spirogram of pulmonary volumes and capacities.

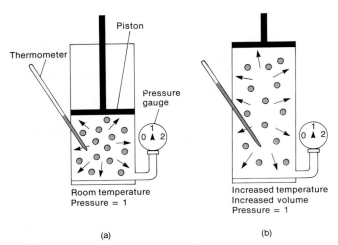

FIGURE 23-14 Charles' law. The volume of a gas is directly proportional to its absolute temperature, assuming that the pressure is constant.

number of collisions within the cylinder increases. Assuming that the piston moves freely (no external pressure is applied), the force of the molecules hitting it moves it upward. As the gas expands, the movement of the piston provides a measure of the increase in volume. As the space in the cylinder increases, the molecules have farther to travel, so the number of collisions decreases as the space increases. The pressure of 1 atm is maintained, and the volume increases in direct proportion to the temperature increase.

Dalton's Law

According to **Dalton's law,** each gas in a mixture of gases exerts its own pressure as if all the other gases were not present. This *partial pressure* is denoted as *p*. The total pressure of the mixture is calculated by simply adding all the partial pressures. Atmospheric air is a mixture of several gases—oxygen, carbon dioxide, nitrogen, water vapor, and a number of other gases that appear in such small quantities that we will ignore them. Atmospheric pressure is the sum of the pressures of all these gases:

Atmospheric pressure = $pO_2 + pCO_2 + pN_2 + pH_2O$
   (760 mm Hg)

We can determine the partial pressure exerted by each component in the mixture by multiplying the percentage of the mixture the particular gas constitutes by the total pressure of the mixture. For example, to find the partial pressure of oxygen in the atmosphere, multiply the percentage of atmospheric air composed of oxygen (21 percent) by the total atmospheric pressure (760 mm Hg):

Atmospheric pO_2 = 21% × 760 mm Hg
         = 159.60 or 160 mm Hg

Since the percentage of CO_2 in the atmosphere is 0.04,

Atmospheric pCO_2 = 0.04% × 760 mm Hg
          = 0.3040 or 0.3 mm Hg

The partial pressures of the respiratory gases and nitrogen in the atmosphere, alveoli, blood, and tissue cells are shown in Exhibit 23-2. These partial pressures are important in determining the movement of oxygen and carbon dioxide between the atmosphere and lungs, the lungs and blood, and the blood and body cells. When a mixture of gases diffuses across a semipermeable membrane, each gas diffuses from the area where its partial pressure is greater to the area where its partial pressure is less. Every gas is on its own and behaves as if the other gases in the mixture did not exist.

The amounts of respiratory gases vary in inspired (atmospheric), alveolar, and expired air (Exhibit 23-3). Inspired (atmospheric) air contains about 21 percent oxygen and 0.04 percent carbon dioxide. Expired air contains less oxygen (about 16 percent) and more carbon dioxide (about 4.5 percent) than inspired air. Compared to alveolar air, expired air contains more oxygen (about 16 percent versus 14 percent) and less carbon dioxide (about 4.5 percent versus 5.5 percent) because some of the expired air is dead air volume that has not participated in gaseous exchange. Expired air is actually a mixture of inspired and alveolar air.

Henry's Law

You have probably noticed that a bottle of soda makes a hissing sound when the top is removed, and bubbles rise to the surface for some time afterward. The gas dissolved in carbonated beverages is carbon dioxide. The ability of a gas to stay in solution depends on its partial pressure and solubility coefficient, that is, its physical or chemical attraction for water. The solubility coefficient of carbon dioxide is high (0.57), that of oxygen is lower (0.024), and that of nitrogen is still lower (0.012). The higher the partial pressure of a gas over a liquid and

EXHIBIT 23-2
PARTIAL PRESSURES (IN mm Hg) OF RESPIRATORY GASES AND NITROGEN IN ATMOSPHERIC AIR, ALVEOLAR AIR, BLOOD, AND TISSUE CELLS

	ATMOSPHERIC AIR (SEA LEVEL)	ALVEOLAR AIR	DEOXYGENATED BLOOD	OXYGENATED BLOOD	TISSUE CELLS
pO_2	160	105	40	105	40
pCO_2	0.3	40	45	40	45
pN_2	597	569	569	569	569

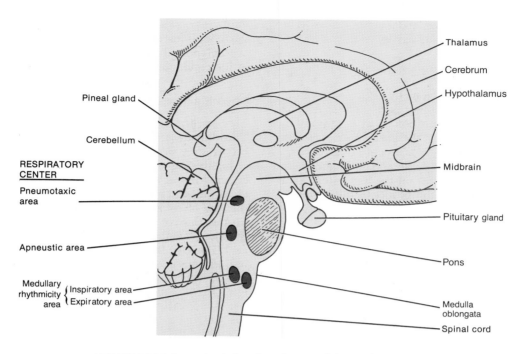

FIGURE 23-21 Approximate location of areas of the respiratory center.

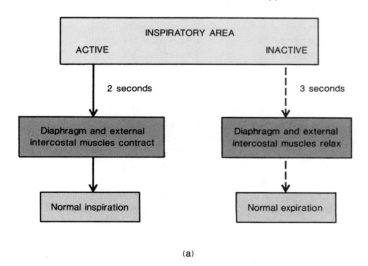

(a)

from the active inspiratory area last for about 2 seconds and travel to the muscles of inspiration. The impulses reach the diaphragm by the phrenic nerves and the external intercostal muscles by the intercostal nerves. When the impulses reach the inspiratory muscles, the muscles contract and inspiration occurs. At the end of 2 seconds, the inspiratory muscles become inactive again, and the cycle repeats itself over and over.

It is believed that the expiratory neurons remain inactive during most normal quiet respiration. During quiet respiration, inspiration is accomplished by active contraction of the inspiratory muscles and expiration results from passive recoil of the lungs and thoracic wall as the inspiratory muscles relax. However, during high levels of ventilation, it is believed that impulses from the inspiratory area activate the expiratory area (Figure 23-22b). Impulses discharged from the expiratory area cause contraction

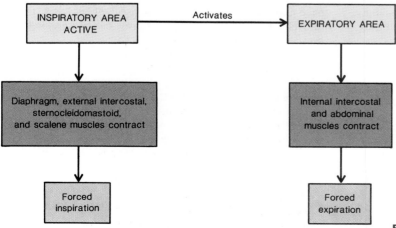

FIGURE 23-22 Proposed role of the medullary rhythmicity area in controlling the basic rhythm of respiration. (a) During normal quiet breathing. (b) During high levels of respiration.

(b)

of the internal intercostals and abdominal muscles that decrease the size of the thoracic cavity.

Pneumotaxic Area

Although the medullary rhythmicity area controls the basic rhythm of respiration, other parts of the nervous system help coordinate the transition between inspiration and expiration. One of these is the **pneumotaxic area** in the upper pons (see Figure 23-21). It continuously transmits impulses to the inspiratory area. The major effect of these impulses is to help turn off the inspiratory area before the lungs become too full of air. In other words, the impulses limit inspiration and thus facilitate expiration.

Apneustic Area

Another part of the nervous system that coordinates the transition between inspiration and expiration is the **apneustic area** in the lower pons (see Figure 23-21). It sends impulses to the inspiratory area that activate it and prolong inspiration, thus inhibiting expiration. This occurs when the pneumotaxic area is inactive. When the pneumotaxic area is active, it overrides the apneustic area.

REGULATION OF RESPIRATORY CENTER ACTIVITY

Although the basic rhythm of respiration is set and coordinated by the respiratory center, the rhythm can be modified in response to the demands of the body by neural input to the center.

Cortical Influences

The respiratory center has connections with the cerebral cortex, which means we can voluntarily alter our pattern of breathing. We can even refuse to breathe at all for a short time. Voluntary control is protective because it enables us to prevent water or irritating gases from entering the lungs. The ability to stop breathing is limited by the buildup of CO_2 in the blood, however. When the pCO_2 increases to a certain level, the inspiratory area is stimulated, impulses are sent to inspiratory muscles, and breathing resumes whether or not the person wishes. It is impossible for people to kill themselves by holding their breath.

Inflation Reflex

Located in the walls of bronchi and bronchioles throughout the lungs are **stretch receptors.** When the receptors become overstretched, impulses are sent along the vagus nerves to the inspiratory area and apneustic area. In response, the inspiratory area is inhibited and the apneustic area is inhibited from activating the inspiratory area. The result is that expiration follows. As air leaves the lungs during expiration, the lungs deflate and the stretch receptors are no longer stimulated. Thus, the inspiratory and apneustic areas are no longer inhibited and a new inspiration begins. This reflex is referred to as the **inflation (Her-**

ing-Breuer) reflex. Some evidence seems to suggest that the reflex is mainly a protective mechanism for preventing overinflation of the lungs rather than a key component in the regulation of respiration.

Chemical Stimuli

Certain chemical stimuli determine how fast we breathe. The ultimate goal of the respiratory system is to maintain proper levels of carbon dioxide and oxygen and the system is highly responsive to changes in the blood levels of either. Although it is convenient to speak of carbon dioxide as the most important chemical stimulus for regulating the rate of respiration, it is actually hydrogen ions that assume this role. For example, carbon dioxide (CO_2) in the blood combines with water (H_2O) to form carbonic acid (H_2CO_3). But the carbonic acid quickly breaks down into H^+ ions and bicarbonate (HCO_3^-) ions. Any increase in CO_2 will cause an increase in H^+ ions, and any decrease in CO_2 will cause a decrease in H^+ ions. In effect, it is the H^+ ions that alter the rate of respiration rather than the CO_2 molecules. Although the following discussion refers to the levels of CO_2 and their effect on respiration, keep in mind that it is really the H^+ ions that cause the effects.

Within the medulla is an area that is highly sensitive to blood concentrations of CO_2. It is called the **chemosensitive area.** Outside the central nervous system are **chemoreceptors,** the carotid and aortic bodies, located in the carotid and aortic sinuses, respectively. They are sensitive to changes in CO_2 and O_2 levels in the blood.

Under normal circumstances, arterial blood pCO_2 is 40 mm Hg. If there is even a slight increase in pCO_2—a condition called **hypercapnia**—the chemosensitive area in the medulla and chemoreceptors in the carotid and aortic sinuses are stimulated. Stimulation of the chemosensitive area and chemoreceptors causes the inspiratory area to become highly active, and the rate of respiration increases. This increased rate, **hyperventilation,** allows the body to expel more CO_2 until the pCO_2 is lowered to normal. If arterial pCO_2 is lower than 40 mm Hg, the chemosensitive area and chemoreceptors are not stimulated and stimulatory impulses are not sent to the inspiratory area. Consequently, the area sets its own moderate pace until CO_2 accumulates and the pCO_2 rises to 40 mm Hg. A slow rate of respiration is called **hypoventilation.**

The oxygen chemoreceptors are sensitive only to large decreases in the pO_2. If arterial pO_2 falls from a normal of 105 mm Hg to 70 mm Hg, the oxygen chemoreceptors become stimulated and send impulses to the inspiratory area. But if the pO_2 falls much below 70 mm Hg, the cells of the inspiratory area suffer oxygen starvation and do not respond well to any chemical receptors. They send fewer impulses to the inspiratory muscles, and the respiration rate decreases or breathing ceases altogether.

Other Influences

The carotid and aortic sinuses also contain pressoreceptors (baroreceptors) that are stimulated by a rise in blood pressure. Although these pressoreceptors are concerned mainly with the control of circulation (see Figure 20-11), they help control respiration. For example, a sudden rise in blood pressure decreases the rate of respiration, and a drop in blood pressure increases the respiratory rate.

Other factors that control respiration are:

1. An increase in body temperature, as during a fever or vigorous muscular exercise, increases the rate of respiration. A decrease in body temperature decreases respiratory rate. A sudden cold stimulus such as plunging into cold water causes a temporary cessation of breathing called **apnea** (ap-NĒ-a; *a* = without, *pnoia* = breath).

2. A sudden, severe pain brings about apnea, but a prolonged pain triggers the general adaptation syndrome and increases respiration rate.

3. Stretching the anal sphincter muscle increases the respiratory rate. This technique is sometimes employed to stimulate respiration during emergencies.

4. Irritation of the pharynx or larynx by touch or chemicals brings about an immediate cessation of breathing followed by coughing.

INTERVENTION IN RESPIRATORY CRISES

CARDIOPULMONARY RESUSCITATION (CPR)

A serious decrease in respiration or heart rate presents an urgent crisis because the body's cells cannot survive long if they are starved of oxygenated blood. If oxygen is withheld from the cells of the brain for 5 to 6 minutes, there is usually severe and permanent brain injury or death. **Cardiopulmonary resuscitation (CPR)** is the artificial reestablishment of normal or near normal respiration and circulation.

The **A, B, C**'s of cardiopulmonary resuscitation are **Airway, Breathing,** and **Circulation.** The rescuer must establish an airway, provide artificial ventilation if breathing has stopped, and reestablish circulation if there is inadequate cardiac action. Cardiopulmonary resuscitation can be administered at the site of emergency and can be used for any heart or respiratory failure, whether the cause be drowning, strangulation, carbon monoxide or insecticide poisoning, overdose of a drug or anesthesia, electrocution, or myocardial infarction. The success of cardiopulmonary resuscitation is directly related to speed and efficiency. Delay may be fatal. The sequential steps must be continued uniformly and without interruption until the patient recovers or is pronounced dead.

Airway

The first step in cardiopulmonary resuscitation is the immediate opening of the airway. One method of doing this is the **head-tilt maneuver.** With the victim on his or her back, the rescuer places one hand under the victim's neck and the other hand on the forehead. The rescuer simultaneously lifts the neck with one hand and tilts the head backward by pressure on the forehead with the other (Figure 23-23a). The tilted position opens the upper air passageways to their maximum size.

Breathing

If the patient does not resume spontaneous breathing after the head has been tilted backward, artificial ventilation must be given mouth-to-mouth or mouth-to-nose. This is called **external air ventilation** or exhaled air ventilation. In the more usual mouth-to-mouth method, the rescuer maintains the head tilt on the victim and the victim's nostrils are pinched together with the thumb and index finger (Figure 23-23b). The rescuer then opens his or her mouth widely, takes a deep breath, makes a tight seal with his or her mouth around the patient's mouth, and blows in about twice the amount the patient normally breathes. The rescuer then removes his or her mouth and allows the patient to exhale passively. This cycle is repeated approximately 12 times per minute for adults.

Even though the air supplied to the victim is air exhaled by the rescuer, it can still provide sufficient oxygen. Atmospheric air contains about 21 percent O_2 and 0.04 percent of CO_2. Exhaled air still contains about 16 percent O_2 and 4.5 percent CO_2. This amount is more than adequate to maintain a victim's blood O_2 and CO_2 at normal levels if air is given at the prescribed rate and amount.

If the rescuer observes the following three signs, ventilation is adequate: (1) The chest rises and falls with every breath. (2) The lungs can be felt to resist as they expand. (3) Air can be heard escaping during exhalation.

Circulation

One method of reestablishing circulation is called **external cardiac compression,** or **closed-chest cardiac compression** (CCCC). It consists of the application of rhythmic pressure over the sternum (Figure 23-23c). The rescuer places the heels of the hands on the lower half of the sternum and presses down firmly and smoothly at least 60 times a minute. This action compresses the heart and produces an artificial circulation because the heart lies almost in the middle of the chest between the lower portion of the sternum and the spine. When properly done, external cardiac compression can produce systolic blood pressure peaks of over 100 mm Hg. It can also bring carotid arterial blood flow up to 35 percent of normal.

Complications that can occur from the use of cardiac compression include fracture of the ribs and sternum, laceration of the liver, and the formation of fat emboli. They can be minimized by the following precautions.

1. Never compress over the xiphoid process at the tip of the sternum. It extends down over the abdomen, and pressure on it may cause laceration of the liver, which can be fatal.
2. Never let your fingers touch the patient's ribs when you compress. Keep your fingers off the patient, and place the heel of your hand in the middle of the patient's chest over the lower half of the sternum.
3. Never compress the abdomen and chest simultaneously since this action traps the liver and may rupture it.
4. Never use sudden or jerking movements to compress the chest. Compression should be smooth, regular, and uninterrupted, with 50 percent of the cycle compression and 50 percent relaxation.

FIGURE 23-23 Cardiopulmonary resuscitation (CPR). (a) Head-tilt maneuver to open airway. (b) Exhaled air ventilation. Shown is the procedure for mouth-to-mouth resuscitation. (© 1982 by Gerard J. Tortora. Courtesy of Lynne Tortora and James Borghesi.) (c) External cardiac compression technique. On the left is shown the position of the hands. (© 1982 by Gerard J. Tortora. Courtesy of Geraldine C. Tortora and James Borghesi.) On the right is shown the effect of external cardiac compression on the heart.

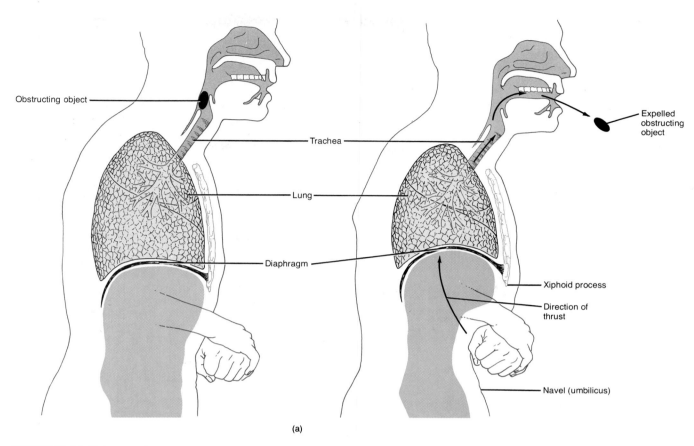

Obstructing object

Trachea

Lung

Diaphragm

Expelled obstructing object

Xiphoid process

Direction of thrust

Navel (umbilicus)

(a)

FIGURE 23-24 Abdominal thrust (Heimlich) maneuver. (a) The application of a quick upward thrust causes sudden elevation of the diaphragm and forceful, rapid expulsion of air in the lungs. Air forced through the trachea and larynx expels the obstructing object.

Compression of the sternum produces some artificial ventilation but not enough for adequate oxygenation of the blood. Therefore exhaled air ventilation must always be used with it. When there are two rescuers, the best technique is to have one rescuer apply at least 60 cardiac compressions per minute. The other rescuer should exhale into the patient's mouth between every fifth and sixth compression.

ABDOMINAL THRUST (HEIMLICH) MANEUVER

Food choking (café coronary), the sixth leading cause of accidental death, is often mistaken for myocardial infarction. However, it can be recognized easily. The victim cannot speak or breathe and may become panic-stricken and run from the room. He or she becomes pale, then deeply cyanotic, and collapses. Without intervention, severe brain damage or death can occur in 5 to 6 minutes.

The food or other obstructing object causing asphyxiation may lodge in the back of the throat or enter the trachea to occlude the airway. Tracheostomy, even when a physician performs it, can be hazardous in a nonclinical setting. An instrument for removing food from the back

of the throat has been developed but is seldom at hand in an emergency. However, there is a first aid procedure that does not require special instruments and can be performed by any informed layman. This procedure is called the **abdominal thrust (Heimlich) maneuver.**

Food choking probably occurs most frequenty during inspiration, which causes the bolus of food to be sucked against the opening into the larynx. At the time of the accident, the lungs are therefore expanded. Even during normal expiration, however, some tidal air (500 ml) and the entire expiratory reserve volume (1,200 ml) are in the lungs. The principle of the abdominal thrust maneuver is to expel the air forcibly in order to dislodge the obstruction.

Essentially, the abdominal thrust maneuver consists of pressing one's fist upward into the epigastric region of the abdomen to elevate the diaphragm suddenly. This thrust compresses the lungs in the rib cage and increases the air pressure in the bronchopulmonary tree. This forces air out through the trachea and will eject the food (or object) blocking the airway. The action can be simulated by inserting a cork in a compressible plastic bottle and then squeezing it suddenly—the cork flies out because of the increased pressure. Figure 23-24 shows how to administer the abdominal thrust maneuver.

(b)

(c)

FIGURE 23-24 (Continued) Abdominal thrust (Heimlich) maneuver. (b) If the victim is upright, stand behind the victim and place both arms around the waist just above the belt line. Then make a fist with one hand and place the thumb side of the fist against the victim's abdomen, slightly above the navel and below the rib cage. Next, grasp your fist with your other hand and press into the victim's abdomen with a quick upward thrust. This should be done by a sharp flexion at the elbows rather than a "bear hug" that compresses the rib cage. The thrust is repeated several times if necessary. (© 1982 by Gerard J. Tortora. Courtesy of Lynne Tortora and James Borghesi.) (c) If the victim is recumbent, kneel astride the victim's hips. Place the hands one on top of the other with the heel of the bottom hand on the victim's abdomen slightly above the navel and below the rib cage. Then apply the quick upward thrust. (© 1982 by Gerard J. Tortora. Courtesy of Lynne Tortora and James Borghesi.)

DISORDERS: HOMEOSTATIC IMBALANCES

Bronchogenic Carcinoma (Lung Cancer)

As part of ordinary breathing, many irritating substances are inhaled. Almost all pollutants, including inhaled smoke, have an irritating effect on the bronchial tubes and lungs and may be regarded as stresses or irritating stimuli. The effects of irritation on the bronchial epithelium are important clinically because a common lung cancer, **bronchogenic carcinoma,** starts in the walls of the bronchi.

Microscopic examination of the pseudostratified epithelium of a bronchial tube reveals three kinds of cells, all of which are in contact with the basement membrane. The columnar cells contain cilia and extend upward from the basement membrane and reach the lumenal surface. At intervals between the ciliated columnar cells are the mucus-secreting goblet cells. The basal cells do not reach the lumenal surface. The basal cells divide continuously, replacing the columnar cells as they wear down and are sloughed off.

The constant irritation by inhaled smoke and pollutants causes an enlargement of the goblet cells of the bronchial epithelium. They respond by secreting excessive mucus. The basal cells also respond to the stress by undergoing cell division so fast that they push into the area occupied by the goblet and columnar cells. As many as 20 rows of basal cells may be produced. Many researchers believe that if the stress is removed at this point, the epithelium can return to normal.

If the stress persists, more and more mucus is secreted and the cilia become less effective. As a result, mucus is not carried toward the throat but remains trapped in the bronchial tubes. The individual then develops a "smoker's cough." Moreover, the constant irritation from the pollutant slowly destroys the alveoli, which are replaced with thick, inelastic connective tissue. Mucus that has accumulated becomes trapped in the air sacs. Millions of sacs rupture, reducing the diffusion surface for the exchange of oxygen and carbon dioxide. The individual has now developed emphysema. If the stress is removed at this point, there is little chance for improvement. Alveolar tissue that has been destroyed cannot be repaired. But removal of the stress can stop further destruction of lung tissue.

If the stress continues, the emphysema gets progressively worse, and the basal cells of the bronchial tubes continue to divide and break through the basement membrane. At this point the stage is set for bronchogenic carcinoma. Columnar and goblet cells disappear and may be replaced with squamous cancer cells. If this happens, the malignant growth spreads throughout the lung and may block a bronchial tube. If the obstruction occurs in a large bronchial tube, very little oxygen enters the lung, and disease-producing bacteria thrive on the mucoid secretions. In the end, the patient may develop emphysema, carcinoma, and a host of infectious diseases. Treatment involves surgical removal of the diseased lung. However, metastasis (spreading) of the growth through the lymphatic or blood system may result in new growths in other parts of the body such as the brain and liver.

Other factors may be associated with lung cancer. For instance, breast, stomach, and prostate malignancies can metastasize to the lungs. People who apparently have not been exposed to pollutants do occasionally develop bronchogenic carcinoma. However, the occurrence of bronchogenic carcinoma is probably over 20 times higher in heavy cigarette smokers than it is in nonsmokers.

Nasal Polyps

Nasal polyps are protruding growths of the mucous membrane that usually hang down from the posterior wall of the nasal septum. The polyps appear as bluish-white tumors and may fill the nasopharynx as they become larger. Nasal polyps usually undergo atrophy if untreated, but they are easily removed by a physician with a nasal snare and cautery.

Bronchial Asthma

Bronchial asthma is a reaction, usually allergic, characterized by attacks of wheezing and difficult breathing. Attacks are brought on by spasms of the smooth muscles that lie in the walls of the smaller bronchi and bronchioles, causing the passageways to close partially. The patient has trouble exhaling, and the alveoli may remain inflated during expiration. Usually the mucous membranes that line the respiratory passageways become irritated and secrete excessive amounts of mucus that may clog the bronchi and bronchioles and worsen the attack. About three out of four asthma victims are allergic to edible or air-borne substances as common as wheat or dust. Others are sensitive to the proteins of harmless bacteria that inhabit the paranasal sinuses, nose, and throat. Asthma might also have a psychosomatic origin. (The anaphylactic mechanism involved in bronchial asthma may be reviewed in Chapter 22.)

Bronchitis

Bronchitis is inflammation of the bronchi. The typical symptom is a productive cough in which a thick greenish-yellow sputum is raised. This signifies an underlying infection that is causing excessive secretion of mucus. Cigarette smoking remains the most important cause of chronic bronchitis, that is, bronchitis that lasts for at least three months of the year for two successive years.

Emphysema

In **emphysema** (em'-fi-SĒ-ma), the alveolar walls lose their elasticity and remain filled with air during expiration. The name means "blown up" or "full of air." Reduced forced expiratory volume is the first symptom. Later, alveoli in other areas of the lungs are damaged. Many alveoli may merge to form larger air sacs with a reduced overall volume. The lungs become permanently inflated because they have lost elasticity. To adjust to the increased lung size, the size of the chest cage increases. The patient has to work voluntarily to exhale. Oxygen diffusion does not occur as easily across the damaged alveolar-capillary membrane, blood O_2 is somewhat lowered, and any mild exercise that raises the oxygen requirements of the cells leaves the patient breathless. As the disease progresses, the alveoli are replaced with thick fibrous connective tissue. Even carbon dioxide does not diffuse easily through this fibrous tissue. If the blood cannot buffer all the hydrogen ions that accumulate, the blood pH drops or unusually high amounts of carbon dioxide may dissolve in the plasma. High carbon dioxide levels produce acid conditions that are toxic to brain cells. Consequently, the inspiratory area becomes less active and the respiration rate slows down, further aggravating the problem. The compressed and damaged capillaries around the deteriorating alveoli may no longer be able to receive blood. As a result, resistance to blood flow increases in the pulmonary trunk and the right ventricle overworks as it attempts to force blood through the remaining capillaries.

Emphysema is generally caused by a long-term irritation. Air pollution, occupational exposure to industrial dust, and cigarette smoke are the most common irritants. Chronic bronchial asthma also may produce alveolar damage. Cases of emphysema are becoming more and more frequent in the United States. The irony is that the disease can be prevented and the progressive deterioration can be stopped by eliminating the harmful stimuli.

Diseases, such as bronchial asthma, bronchitis, and emphysema have in common some degree of obstruction of the air passageways. The term **chronic obstructive pulmonary disease (COPD)** is used to refer to these disorders.

Pneumonia

The term **pneumonia** means an acute infection or inflammation of the alveoli. In this disease, the alveolar sacs fill up with fluid and dead white blood cells, reducing the amount of air space in the lungs. (Remember that one of the cardinal signs of inflammation is edema.) Oxygen has difficulty diffusing through the inflamed alveoli, and the blood O_2 may be drastically reduced. Blood CO_2 usually remains normal because carbon dioxide diffuses through the alveoli more easily than oxygen does. If all the alveoli of a lobe are inflamed, the pneumonia is called *lobar pneumonia.* If only parts of the lobe are involved, it is called *lobular* (*segmental*) *pneumonia.* If both the alveoli and the bronchial tubes are included, it is called *bronchopneumonia.*

The most common cause of pneumonia is the pneumococcus bacterium, but other bacteria or a fungus may be the source of trouble. Viral pneumonia is caused by several viruses, including the influenza virus.

Tuberculosis

The bacterium called *Mycobacterium tuberculosis* produces an inflammation called **tuberculosis.** This disease still ranks as the number-one killer in the communicable disease category. Tuberculosis most often affects the lungs and the pleurae. The bacteria destroy parts of the lung tissue, and the tissue is replaced by fibrous connective tissue. Because the connective tissue is inelastic and thick, the affected areas of the lungs do not snap back during expiration and larger amounts of air are retained. Gases no longer diffuse easily through the fibrous tissue.

Tuberculosis bacteria are spread by inhalation. Although they can withstand exposure to many disinfectants, they die quickly in sunlight. Thus, tuberculosis is sometimes associated with crowded, poorly lit housing. Many drugs are successful in treating tuberculosis. Rest, sunlight, and good diet are vital parts of treatment.

Infant Respiratory Distress Syndrome (RDS)

Sometimes called glassy-lung disease or hyaline membrane disease (HMD), **infant respiratory distress syndrome (RDS)** is responsible for approximately 20,000 newborn infant deaths per year. Before birth, the respiratory passages are filled with fluid. Part of this fluid is amniotic fluid inhaled during respiratory movements in utero. The remainder is produced by the submucosal glands and the goblet cells of the respiratory epithelium.

At birth, this fluid-filled passageway must become an air-filled passageway, and the collapsed primitive alveoli (terminal sacs) must expand and function in gas exchange. The success of this transition depends largely on surfactant, the phospholipid produced by alveolar cells that lowers surface tension in the fluid layer lining the primitive alveoli once air enters the lungs. Surfactant is present in the fetus's lungs as early as 23 weeks and by 28 to 32 weeks the amount present is sufficient to prevent alveolar collapse during breathing. Surfactant is produced continuously by septal alveolar cells. The presence of surfactant can be detected by amniocentesis.

Although in a normal, full-term infant the second and subsequent breaths require less respiratory effort than the first, breathing is not completely normal until about 40 minutes after birth. The entire lung is not inflated fully with the first one or two breaths. In fact, for the first 7 to 10 days, small areas of the lungs may remain uninflated.

In the newborn whose lungs are deficient in surfactant, the effort required for the first breath is essentially the same as that required in normal newborns. However, the surface tension of the alveolar fluid is 7 to 14 times higher than the surface tension of alveolar fluid with a monomolecular layer of surfactant. Consequently, during expiration after the first inspiration, the surface tension of the alveoli increases as the alveoli deflate. The alveoli collapse almost to their original uninflated state.

Idiopathic RDS usually appears within a few hours after

birth. (An idiopathic condition is one that occurs spontaneously in an individual from an unknown or obscure cause.) Affected infants show difficult and labored breathing with withdrawal of the intercostal and subcostal spaces. Death may occur soon after onset of respiratory difficulty or may be delayed for a few days, although many infants survive. At autopsy, the lungs are underinflated and areas of atelectasis (collapse of lung) are prominent. If the infant survives for at least a few hours after developing respiratory distress, the alveoli are often filled with a fluid of high-protein content that resembles a hyaline (or glassy) membrane. RDS occurs frequently in premature infants and also in infants of diabetic mothers, particularly if the diabetes is untreated or poorly controlled.

A treatment called PEEP (positive end expiratory pressure) is now being used to treat RDS. It consists of passing a tube through the air passages to the top of the lungs to provide needed oxygen-rich air at continuous pressures of up to 14 mm Hg. Continuous pressure keeps the baby's alveoli open and available for gas exchange.

Sudden Infant Death Syndrome (SIDS)

Sudden infant death syndrome (SIDS), also called crib death or cot death, kills approximately 10,000 American babies every year. It kills more infants between the ages of 1 week and 12 months than any other disease. SIDS occurs without warning. Although about half of its victims had an upper respiratory infection within two weeks of death, the babies tended to be remarkably healthy. Two other findings are important: the apparently *silent* nature of death and the *disarray* frequently found at the scene.

Given this death scene, it is not surprising that for centuries "crib death" was mistakenly thought to be due to suffocation. Most medical people, however, believe that a healthy child *cannot* smother in its bedclothes. The diaper of a SIDS victim is usually full of stool and urine, and the urinary bladder is almost always empty. External examination reveals only a blood-tinged froth that often exudes from the nostrils. These signs all point to a sudden lethal episode associated with tonic motor activity, not to a gradually deepening coma leading to death.

The similarity of findings in SIDS cases strengthens the theory that there is one underlying cause. The seasonal distribution of SIDS incidences, being lowest during the summer months and highest in the late fall, strongly suggests an infectious agent, viruses being the most likely possibility. Most cases occur at a time in life when gamma globulin levels are low and the infant is at a critical period of susceptibility.

At least thirteen different hypotheses have been proposed as explanations for SIDS. Although the precise pathogenic mechanisms have yet to be clarified, several valuable observations have emerged as a result of investigations during the past decade. Most, if not all, such cases appear to share a common mechanism ("final common pathway"). Thus, SIDS is probably a single disease entity. According to one study, a likely cause of death is laryngospasm, possibly triggered by a previous, mild, nonspecific virus infection of the upper respiratory tract. Another possible factor includes periods of prolonged apnea as a result of malfunction of the respiratory center. Hypoxia may also be involved. Other suspected causes are an allergic reaction or bacterial poisoning.

Coryza (Common Cold) and Influenza (Flu)

Several viruses are responsible for **coryza (common cold).** Typical symptoms include sneezing, excessive nasal secretion, and congestion. The uncomplicated common cold is not usually accompanied by a fever. Complications include sinusitis, ear infections, and laryngitis. Since common colds are caused by viruses, antibiotics are of no use in treatment. Although over-the-counter drugs may lessen the severity of certain symptoms of the common cold, they do not lessen recovery time.

Influenza (flu) is also caused by a virus. Its symptoms include chills, fever (usually higher than 101°F), headache, and muscular aches. Coldlike symptoms appear as the fever subsides. Diarrhea is not a symptom of influenza.

MEDICAL TERMINOLOGY

Asphyxia (*sphyxis* = pulse)   Oxygen starvation due to low atmospheric oxygen or interference with ventilation, external respiration, or internal respiration.

Aspiration (*spirare* = breathe)   The act of breathing, especially breathing in; drawing of a substance in or out by suction.

Bronchiectasis (*ektasis* = dilation)   A chronic dilation of the bronchi or bronchioles.

Cheyne-Stokes respiration   A repeated cycle of irregular breathing beginning with shallow breaths that increase in depth and rapidity, then decrease and cease altogether for 15 to 20 seconds. Cheyne-Stokes is normal in infants. It is also often seen just before death from pulmonary, cerebral, cardiac, and kidney disease and is referred to as the "death rattle."

Diphtheria (*diphthera* = membrane)   An acute bacterial infection that causes the mucous membranes of the oropharynx, nasopharynx, and larynx to enlarge and become leathery. Enlarged membranes may obstruct airways and cause death from asphyxiation.

Dyspnea (*dys* = painful, difficult; *pnoia* = breath)   Labored breathing (short-windedness).

Hypoxia (*hypo* = below, under)   Reduction in oxygen supply to cells.

Orthopnea (*ortho* = straight)   Inability to breathe in a horizontal position.

Pneumonectomy (*pneumo* = lung; *tome* = cutting)   Surgical removal of a lung.

Pulmonary (*pulmo* = lung) **edema**   Excess amounts of interstitial fluid in the lungs producing cough and dyspnea. Common in failure of the left side of the heart.

Rales   Sounds sometimes heard in the lungs that resemble bubbling or rattling. May be caused by air or an abnormal secretion in the lungs.

Respirator   A metal chamber that entombs the chest; also called an "iron lung." Used to produce inspiration and expiration in patient with paralyzed respiratory muscles. Pressure inside the chamber is rhythmically alternated to suck out and push in chest walls.

Rhinitis (*rhino* = nose)   Chronic or acute inflammation of the mucous membrane of the nose.

DRUGS ASSOCIATED WITH THE RESPIRATORY SYSTEM

Many respiratory disorders are treated by means of **nebulization** (neb'-yoo-li-ZĀ-shun). This procedure is the administering of medication in the form of droplets that are suspended in air to selected areas of the respiratory tract. The patient inhales the medication as a fine mist. The number of droplets suspended in the mist and the area of the respiratory tract that the medica-

- Describe the mechanism that regulates food intake.
- Define digestion and distinguish the chemical and mechanical phases.
- Identify the organs of the gastrointestinal (GI) tract and the accessory organs of digestion.
- Discuss the structure of the wall of the GI tract.
- Define the mesentery, mesocolon, falciform ligament, lesser omentum, and greater omentum as extensions of the peritoneum.
- Explain the structure of the mouth and its role in mechanical digestion.
- Describe the location and histology of the salivary glands.
- Explain the composition and functions of saliva in digestion.
- Identify the mechanisms that regulate the secretion of saliva.
- Identify the parts of a typical tooth and compare deciduous and permanent dentitions.
- Discuss the stages of swallowing.
- Describe the role of the esophagus in digestion.
- Describe the anatomy and histology of the stomach and explain the relationship between its structural features and digestion.
- Discuss the role of the stomach in mechanical and chemical digestion and absorption.
- Discuss the factors that control the secretion of gastric juice and gastric emptying.
- Describe the anatomy and histology of the pancreas and its role in digestion.
- Explain how pancreatic secretion is regulated.
- Describe the anatomy and histology of the liver and its role in digestion.
- Explain how bile secretion is regulated.
- Discuss the role of the gallbladder in digestion.
- Discuss the structural features of the small intestine that adapt it for digestion and absorption.
- Identify the digestive activities of the small intestine that reduce carbohydrates, proteins, and fats to their final products.
- Explain the mechanical movements of the small intestine.
- Explain how small intestinal secretions are controlled.
- Define absorption and explain how the end products of digestion are absorbed.
- Describe the anatomy and histology of the large intestine.
- Explain the mechanical movements of the large intestine.
- Discuss the structural features of the large intestine that adapt it for absorption, formation of feces, and defecation.
- Define the processes involved in the formation of feces.
- Discuss the mechanisms involved in defecation.
- Describe the clinical symptoms of the following disorders: dental caries, periodontal disease, peritonitis, peptic ulcers, appendicitis, GI tumors, diverticulitis, cirrhosis, hepatitis, gallstones, anorexia nervosa, and bulimia.
- Define medical terminology associated with the digestive system.
- Explain the actions of selected drugs on the digestive system.

24 The Digestive System

We all know that food is vital to life. It is required for the chemical reactions that occur in every cell—both those that synthesize new enzymes, cell structures, bone, and all the other components of the body and those that release the energy needed for the building processes. However, most of the foods we eat are simply too large to pass through the plasma membranes of the cells. The breaking down of food molecules for use by body cells is called **digestion** and the organs that collectively perform this function comprise the **digestive system.**

REGULATION OF FOOD INTAKE

Basically, the intake of food is regulated by two sensations—hunger and appetite. Whereas **hunger** implies a craving for food in general, **appetite** usually means desire for a specific food. Moreover, hunger is a stronger sensation that is accompanied by a greater degree of discomfort.

The control center for food intake is the hypothalamus. A cluster of nerve cells in the lateral hypothalamus called the **appetite center** discharges impulses that result in increased food intake. A cluster of nerve cells in the medial hypothalamus constitutes the **satiety center.** When this center is stimulated, food intake is inhibited. According to one theory, a low blood glucose level stimulates the appetite center, and a high blood glucose level inhibits it. Decreased amounts of amino acids and fatty acids seem to have the same effect.

Once eating has begun, food intake must continue to be regulated to prevent overfilling the digestive tract. As food fills the upper digestive tract, distension stimulates receptors in its wall. These receptors convey visceral sensory impulses to the satiety center, thus inhibiting food intake until the contents of the tract have been digested.

DIGESTIVE PROCESSES

The digestive system prepares food for consumption by the cells through five basic activities.

1. **Ingestion.** Taking food into the body (eating).
2. **Peristalsis.** The movement of food along the digestive tract.
3. **Digestion.** The breakdown of food by both chemical and mechanical processes.
4. **Absorption.** The passage of digested food from the diges-

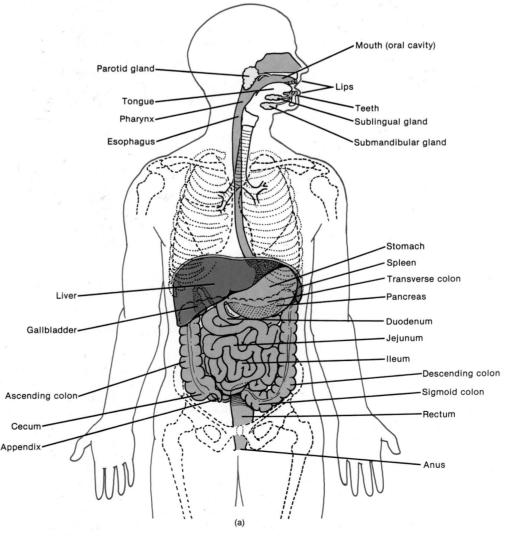

FIGURE 24-1 Organs of the digestive system and related structures. (a) Diagram.

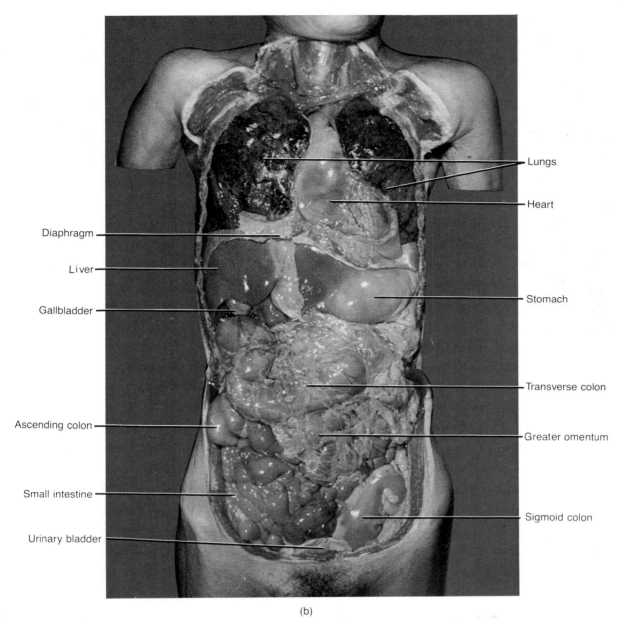

Diaphragm

Liver

Gallbladder

Ascending colon

Small intestine

Urinary bladder

Lungs

Heart

Stomach

Transverse colon

Greater omentum

Sigmoid colon

(b)

FIGURE 24-1 (Continued) Organs of the digestive system and related structures. (b) Photograph. (Courtesy of C. Yokochi and J. W. Rohen, *Photographic Anatomy of the Human Body,* 2nd ed., 1979, IGAKU-SHOIN, Tokyo, New York.)

tive tract into the cardiovascular and lymphatic systems for distribution to cells.

5. Defecation. The elimination of indigestible substances from the body.

Chemical digestion is a series of catabolic reactions that break down the large carbohydrate, lipid, and protein molecules that we eat into molecules usable by body cells. These products of digestion are small enough to pass through the walls of the digestive organs, into the blood and lymph capillaries, and eventually into the body's cells. **Mechanical digestion** consists of various movements that aid chemical digestion. Food is prepared by the teeth before it can be swallowed. Then the smooth muscles of the stomach and small intestine churn the food so it is thoroughly mixed with the enzymes that catalyze the reactions.

ORGANIZATION

The organs of digestion are traditionally divided into two main groups. First is the **gastrointestinal (GI) tract** or **alimentary canal,** a continuous tube running through the ventral body cavity and extending from the mouth to the anus (Figure 24-1). The relationship of the digestive organs to the nine regions of the abdominopelvic cavity may be reviewed in Figure 1-8b. The length of a tract taken from a cadaver is about 9 m (30 ft). In a living person it is somewhat shorter because the muscles in its wall are in a state of tone. Organs composing the gastrointestinal tract include the mouth, pharynx, esophagus, stomach, small intestine, and large intestine. The GI tract contains the food from the time it is eaten until it is digested and prepared for elimination. Muscular contrac-

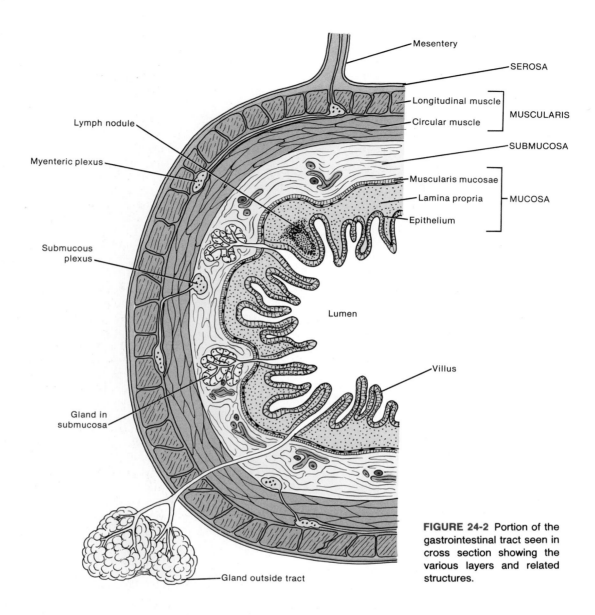

FIGURE 24-2 Portion of the gastrointestinal tract seen in cross section showing the various layers and related structures.

tions in the walls of the GI tract break down the food physically by churning it. Secretions produced by cells along the tract break down the food chemically.

The second group of organs composing the digestive system consists of the **accessory structures**—the teeth, tongue, salivary glands, liver, gallbladder, and pancreas. Teeth protrude into the GI tract and aid in the physical breakdown of food. The other accessory structures, except for the tongue, lie totally outside the tract and produce or store secretions that aid in the chemical breakdown of food. These secretions are released into the tract through ducts.

GENERAL HISTOLOGY

The wall of the GI tract, especially from the esophagus to the anal canal, has the same basic arrangement of tissues. The four coats or tunics of the tract from the inside out are the mucosa, submucosa, muscularis, and serosa or adventitia (Figure 24-2).

Mucosa

The **mucosa,** or inner lining of the tract, is a mucous membrane attached to a thin layer of visceral muscle. Two layers compose the membrane: a *lining epithelium,* which is in direct contact with the contents of the GI tract, and an underlying layer of loose connective tissue called the *lamina propria.* Under the lamina propria is visceral muscle called the *muscularis mucosae.*

The epithelial layer is composed of nonkeratinized cells that are stratified in the mouth and esophagus, but are simple throughout the rest of the tract. The functions of the stratified epithelium are protection and secretion. The functions of the simple epithelium are secretion and absorption. However, the lack of keratin allows some absorption to occur in all parts of the tract.

The lamina propria is made of loose connective tissue containing many blood and lymph vessels and scattered lymph nodules, masses of lymphatic tissue that are not encapsulated. This layer supports the epithelium, binds it to the muscularis mucosae, and provides it with a blood and lymph supply. The blood and lymph vessels are the avenues by which nutrients in the tract reach the other tissues of the body. The lymph tissue also protects against disease. Remember that the GI tract is in contact with the outside environment and contains food which often carries harmful bacteria. Unlike the skin, the mucous membrane of the tract is not protected from bacterial entry by keratin. The lamina propria also contains glandular epithelium that secretes products necessary for chemical digestion.

The muscularis mucosae contains smooth muscle fibers that throw the mucous membrane of the intestine into small folds which increase the digestive and absorptive area. With one exception, which will be described later, the other three coats of the intestine contain no glandular epithelium.

Submucosa

The **submucosa** consists of loose connective tissue that binds the mucosa to the third tunic, the muscularis. It is highly vascular and contains a portion of the *submucous plexus,* which is part of the autonomic nerve supply to the muscularis mucosae. This plexus is important in controlling secretion of the GI tract.

Muscularis

The **muscularis** of the mouth, pharynx, and esophagus consists in part of skeletal muscle that produces voluntary swallowing. Throughout the rest of the tract, the muscularis consists of smooth muscle that is generally found in two sheets: an inner ring of circular fibers and an outer sheet of longitudinal fibers. Contractions of the smooth muscles help to break down food physically, mix it with digestive secretions, and propel it through the tract. The muscularis also contains the major nerve supply to the alimentary tract—the *myenteric plexus,* which consists of fibers from both autonomic divisions. This plexus controls mostly GI motility.

Serosa

The **serosa** is the outermost layer of most portions of the canal. It is a serous membrane composed of connective tissue and epithelium. This tunic is also called the **visceral peritoneum** and forms a portion of the peritoneum, which we will describe in detail.

PERITONEUM

The **peritoneum** (per'-i-tō-NĒ-um) is the largest serous membrane of the body. Serous membranes are also associated with the heart (pericardium) and lungs (pleurae). Serous membranes consist of a layer of simple squamous epithelium (called mesothelium) and an underlying supporting layer of connective tissue. The **parietal peritoneum** lines the wall of the abdominal cavity. The **visceral peritoneum** covers some of the organs and constitutes their serosa. The potential space between the parietal and visceral portions of the peritoneum is called the **peritoneal**

cavity and contains serous fluid. In certain diseases, the peritoneal cavity may become distended by several liters of fluid so that it forms an actual space. Such an accumulation of serous fluid is called **ascites** (a-SĪ-tēz).

Unlike the pericardium and pleurae, the peritoneum contains large folds that weave in between the viscera. The folds bind the organs to each other and to the walls of the cavity and contain the blood and lymph vessels and the nerves that supply the abdominal organs. One extension of the peritoneum is called the **mesentery** (MEZ-en-ter'-ē). It is an outward fold of the serous coat of the small intestine (Figure 24-3). The tip of the fold is attached to the posterior abdominal wall. The mesentery binds the small intestine to the wall. A similar fold of parietal peritoneum, called the **mesocolon** (mez'-ō-KŌ-lon), binds the large intestine to the posterior body wall. It also carries blood vessels and lymphatics to the intestines.

Other important peritoneal folds are the falciform ligament, the lesser omentum, and the greater omentum. The **falciform** (FAL-si-form) **ligament** attaches the liver to the anterior abdominal wall and diaphragm. The **lesser omentum** (ō-MENT-um) arises as two folds in the serosa of the stomach and duodenum suspending the stomach and duodenum from the liver. The **greater omentum** is a four-layered fold in the serosa of the stomach that hangs down like an apron over the front of the intestines. It then passes up to part of the large intestine (the transverse colon), wraps itself around it, and finally attaches to the parietal peritoneum of the posterior wall of the abdominal cavity. Because the greater omentum contains large quantities of adipose tissue, it commonly is called the "fatty apron." The greater omentum contains numerous lymph nodes. If an infection occurs in the intestine, plasma cells formed in the lymph nodes combat the infection and help prevent it from spreading to the peritoneum.

MOUTH (ORAL CAVITY)

The **mouth,** also referred to as the **oral** or **buccal** (BUK-al) **cavity,** is formed by the cheeks, hard and soft palates, and tongue (Figure 24-4). Forming the lateral walls of the oral cavity are the **cheeks**—muscular structures covered on the outside by skin and lined by nonkeratinized stratified squamous epithelium. The anterior portions of the cheeks terminate in the superior and inferior lips.

The **lips (labia)** are fleshy folds surrounding the orifice of the mouth. They are covered on the outside by skin and on the inside by a mucous membrane. The transition zone where the two kinds of covering tissue meet is called the **vermilion** (ver-MIL-yon). This portion of the lips is nonkeratinized and the color of the blood in the underlying blood vessels is visible through the transparent surface layer of the vermilion. The inner surface of each lip is attached to its corresponding gum by a midline fold of mucous membrane called the **labial frenulum** (LĀ-bē-al FREN-yoo-lum).

The orbicularis oris muscle and connective tissue lie between the external integumentary covering and the internal mucosal lining. During chewing, the cheeks and lips help to keep food between the upper and lower teeth. They also assist in speech.

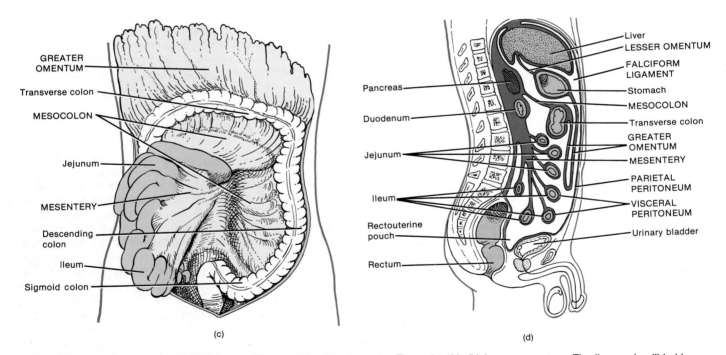

FIGURE 24-3 Extensions of the peritoneum. (a) Greater omentum (see also Figure 24-1b). (b) Lesser omentum. The liver and gallbladder have been lifted. (c) Mesentery. The greater omentum has been lifted. (d) Sagittal section through the abdomen and pelvis indicating the relationship of the peritoneal extensions to each other.

The **vestibule** of the oral cavity is bounded externally by the cheeks and lips and internally by the gums and teeth. The **oral cavity proper** extends from the vestibule to the **fauces** (FAW-sēs), the opening between the oral cavity and the pharynx or throat.

The **hard palate,** the anterior portion of the roof of

the mouth, is formed by the maxillae and palatine bones, covered by mucous membrane, and forms a bony partition between the oral and nasal cavities. The **soft palate** forms the posterior portion of the roof of the mouth. It is an arch-shaped muscular partition between the oropharynx and nasopharynx and is lined by mucous membrane.

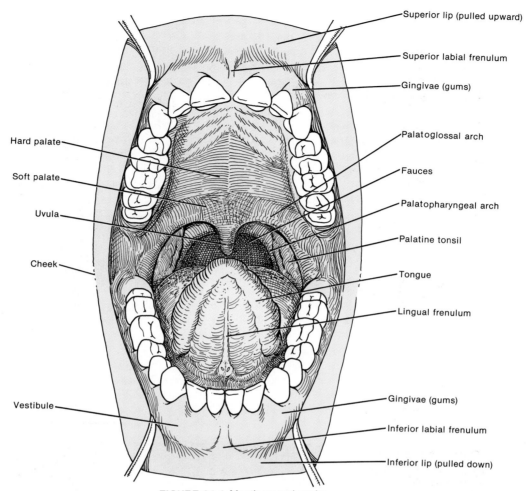

FIGURE 24-4 Mouth or oral cavity.

Hanging from the free border of the soft palate is a conical muscular process called the **uvula** (OO-vyoo-la). On either side of the base of the uvula are two muscular folds that run down the lateral side of the soft palate. Anteriorly, the **palatoglossal arch (anterior pillar)** extends inferiorly, laterally, and anteriorly to the side of the base of the tongue. Posteriorly, the **palatopharyngeal** (PAL-a-tō-fa-rin'-jē-al) **arch (posterior pillar)** projects inferiorly, laterally, and posteriorly to the side of the pharynx. The palatine tonsils are situated between the arches, and the lingual tonsil is situated at the base of the tongue. At the posterior border of the soft palate, the mouth opens into the oropharynx through the fauces.

TONGUE

The **tongue,** together with its associated muscles, forms the floor of the oral cavity. It is an accessory structure of the digestive system composed of skeletal muscle covered with mucous membrane (Figure 24-5). The tongue is divided into symmetrical lateral halves by a median septum that extends throughout its entire length and is attached inferiorly to the hyoid bone. Each half of the tongue consists of an identical complement of extrinsic and intrinsic muscles.

The **extrinsic muscles** of the tongue originate outside the tongue and insert into it. They include the hyoglossus, chondroglossus, genioglossus, styloglossus, and palatoglossus (see Figure 11-7). The extrinsic muscles move the tongue from side to side and in and out. These movements maneuver food for chewing, shape the food into a rounded mass, called a **bolus,** and force the food to the back of the mouth for swallowing. They also form the floor of the mouth and hold the tongue in position. The **intrinsic muscles** originate and insert within the tongue and alter the shape and size of the tongue for speech and swallowing. The intrinsic muscles include the longitudinalis superior, longitudinalis inferior, transversus linguae, and verticalis linguae (Figure 24-5d). The **lingual frenulum,** a fold of mucous membrane in the midline of the undersurface of the tongue, aids in limiting the movement of the tongue posteriorly.

CLINICAL APPLICATION

If the lingual frenulum is too short, tongue movements are restricted, speech is faulty, and the person is said to be "tongue-tied." This congenital problem is referred to as **ankyloglossia** (ang-ki-lō-GLOSS-ē-a). It can be corrected by cutting the lingual frenulum.

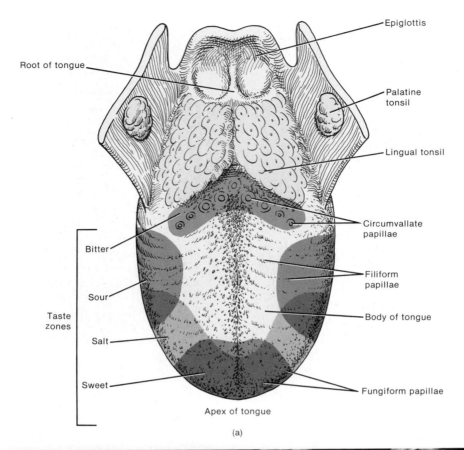

Epiglottis

Root of tongue

Palatine tonsil

Lingual tonsil

Circumvallate papillae

Bitter

Filiform papillae

Sour

Taste zones

Body of tongue

Salt

Sweet

Fungiform papillae

Apex of tongue

(a)

Fungiform papilla

Filiform papillae

(b)

FIGURE 24-5 Tongue. (a) Locations of the papillae and the four taste zones. (b) Scanning electron micrograph of a fungiform papilla surrounded by filiform papillae at a magnification of 245×. (Courtesy of Richard K. Kessel and Randy H. Kardon, *Tissues and Organs: A Text-Atlas of Scanning Electron Microscopy.* Copyright © 1979 by Scientific American, Inc.)

Gustatory
(taste)
pore

(c)

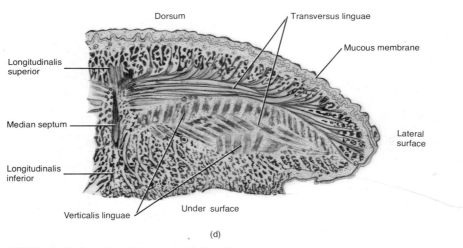

Dorsum

Transversus linguae

Mucous membrane

Longitudinalis
superior

Median septum

Longitudinalis
inferior

Lateral
surface

Verticalis linguae

Under surface

(d)

FIGURE 24-5 (*Continued*) Tongue. (c) Scanning electron micrograph of a circumvallate papilla at
a magnification of 1020×. (d) Frontal section through the left side of the tongue showing its intrinsic
muscles. (Courtesy of Richard K. Kessel and Randy H. Kardon, *Tissues and Organs: A Text-Atlas
of Scanning Microscopy.* Copyright © 1979 by Scientific American, Inc.)

The upper surface and sides of the tongue are covered
with **papillae** (pa-PIL-ē), projections of the lamina propria
covered with epithelium (Figure 24-5a–c). **Filiform papillae** are conical projections distributed in parallel rows
over the anterior two-thirds of the tongue. They contain
no taste buds. **Fungiform papillae** are mushroomlike elevations distributed among the filiform papillae and more
numerous near the tip of the tongue. They appear as
red dots on the surface of the tongue, and most of them
contain taste buds. **Circumvallate papillae** are arranged
in the form of an inverted V on the posterior surface of
the tongue, and all of them contain taste buds. Note the
taste zones of the tongue in Figure 24-5a.

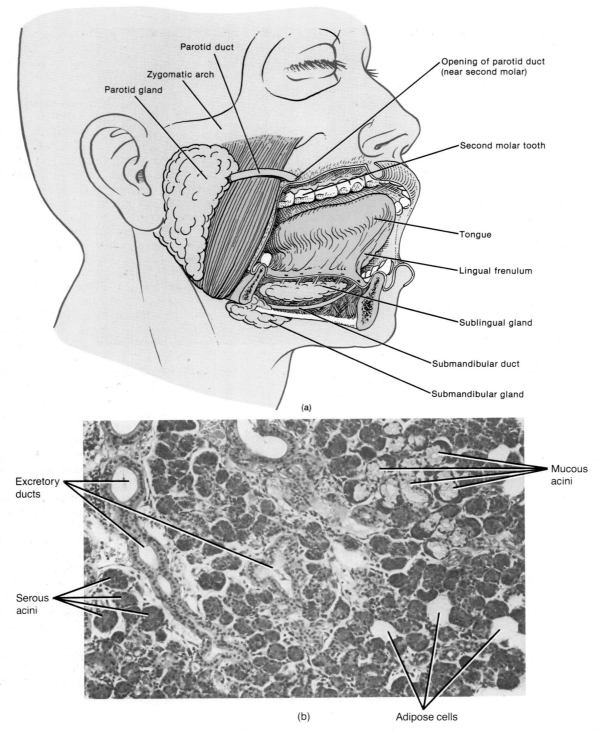

FIGURE 24-6 Salivary glands. (a) Location of the salivary glands. (b) Photomicrograph of the submandibular gland showing mostly serous acini and a few mucous acini at a magnification of 120×. The parotids consist of all serous acini and the sublinguals consist of mostly mucous acini and a few serous acini. (© by Michael H. Ross. Used by permission.)

SALIVARY GLANDS

Saliva is a fluid that is continuously secreted by glands in or near the mouth. Ordinarily, just enough saliva is secreted to keep the mucous membranes of the mouth moist, but when food enters the mouth, secretion increases so the saliva can lubricate, dissolve, and begin the chemical breakdown of the food. The mucous membrane lining the mouth contains many small glands, the **buccal glands,** that secrete small amounts of saliva. However, the major portion of saliva is secreted by the **salivary glands,** accessory structures that lie outside the mouth and pour their contents into ducts that empty into the oral cavity. There are three pairs of salivary glands: parotid, submandibular (submaxillary), and sublingual glands (Figure 24-6a).

The **parotid glands** are located under and in front of the ears between the skin and the masseter muscle. They are compound tubuloacinar glands. Each secretes into the oral cavity vestibule via a duct, called the parotid duct, that pierces the buccinator muscle to open into the vestibule opposite the upper second molar tooth. The **submandibular glands,** which are compound acinar glands, are found beneath the base of the tongue in the posterior part of the floor of the mouth (Figure 24-6b). Their ducts, the submandibular ducts, run superficially under the mucosa on either side of the midline of the floor of the mouth and enter the oral cavity proper just behind the central incisors. The **sublingual glands,** also compound acinar glands, are anterior to the submandibular glands, and their ducts, the lesser sublingual ducts, open into the floor of the mouth in the oral cavity proper.

CLINICAL APPLICATION

Although any of the salivary glands may become infected as a result of a nasopharyngeal infection, the parotids are typically the target of the mumps viruses. **Mumps** is an inflammation and enlargement of the parotid glands accompanied by fever and extreme pain during swallowing. In about 20 to 35 percent of males past puberty, the testes may also become inflamed, and although it rarely occurs, sterility is a possible consequence. In some people, the pancreas is also involved.

Composition of Saliva

The fluids secreted by the buccal glands and the three pairs of salivary glands constitute **saliva.** Amounts of saliva secreted daily vary considerably but range from 1,000 to 1,500 ml. Chemically, saliva is 99.5 percent water and 0.5 percent solutes. Among the solutes are salts—chlorides, bicarbonates, and phosphates of sodium and potassium. Some dissolved gases and various organic substances including urea and uric acid, serum albumin and globulin, mucin, the bacteriolytic enzyme lysozyme, and the digestive enzyme salivary amylase are also present.

Each saliva-producing gland supplies different ingredients to saliva. The parotids contain cells that secrete a watery serous liquid containing the enzyme salivary amylase. The submandibular glands contain cells similar to those found in the parotids plus some mucous cells. Therefore, they secrete a fluid that is thickened with mucus, but still contains quite a bit of enzyme. The sublingual glands contain mostly mucous cells, so they secrete a much thicker fluid that contributes only a small amount of enzyme to the saliva.

The water in saliva provides a medium for dissolving foods so they can be tasted and digestive reactions can take place. The chlorides in the saliva activate the salivary amylase. The bicarbonates and phosphates buffer chemicals that enter the mouth and keep the saliva at a slightly acidic pH of 6.35 to 6.85. Urea and uric acid are found in saliva because the saliva-producing glands (like the sweat glands of the skin) help the body to get rid of wastes. Mucin is a protein that forms mucus when dissolved in water. Mucus lubricates the food so it can be easily turned in the mouth, formed into a ball, and swallowed. The enzyme lysozyme destroys bacteria, thereby protecting the mucous membrane from infection and thus the teeth from decay.

Secretion of Saliva

Salivation is entirely under nervous control. Normally, moderate amounts of saliva are continuously secreted in response to parasympathetic stimulation to keep the mucous membranes moist and to lubricate the movements of the tongue and lips during speech. The saliva is then swallowed and reabsorbed to prevent fluid loss. Dehydration causes the salivary and buccal glands to cease secreting saliva to conserve water. The subsequent feeling of dryness in the mouth promotes sensations of thirst. This phenomenon is also noted during fear or anxiety, when sympathetic stimulation dominates.

Food stimulates the glands to secrete heavily. When food is taken into the mouth, chemicals in the food stimulate receptors in taste buds on the tongue. Rolling a dry, indigestible object over the tongue produces friction, which also may stimulate the receptors. Impulses are conveyed from the receptors to two salivary nuclei in the brain stem called the **superior** and **inferior salivatory nuclei.** The nuclei are located at about the junction of the medulla and pons. Returning parasympathetic autonomic impulses from the nuclei activate the secretion of saliva (see Figure 16-2).

The smell, sight, touch, or sound of food preparation also stimulates increased saliva secretion. These stimuli constitute psychological activation and involve learned behavior. Memories stored in the cerebral cortex that associate the stimuli with food are stimulated. The cortex sends impulses to the nuclei in the brain stem via extrapyramidal pathways, and the salivary glands are activated. Psychological activation of the glands has some benefit to the body because it allows the mouth to start chemical digestion as soon as the food is ingested.

Salivation also occurs in response to swallowing irritating foods or during nausea. Reflexes originating in the stomach and upper small intestine stimulate salivation. This mechanism presumably helps to dilute or neutralize the irritating substance.

Saliva continues to be secreted heavily some time after food is swallowed. This flow of saliva washes out the mouth and dilutes and buffers the chemical remnants of irritating substances.

TEETH

The **teeth,** or **dentes,** are accessory structures of the digestive system located in sockets of the alveolar processes of the mandible and maxillae. The alveolar processes are covered by the **gingivae** (jin-JI-vē) or gums, which extend slightly into each socket forming the gingival sulcus (Figure 24-7). The sockets are lined by the **periodontal ligament,** which consists of dense fibrous connective tissue and is attached to the socket walls and the cemental surface of the roots. Thus it anchors the teeth in position and also acts as a shock absorber to dissipate the forces of chewing.

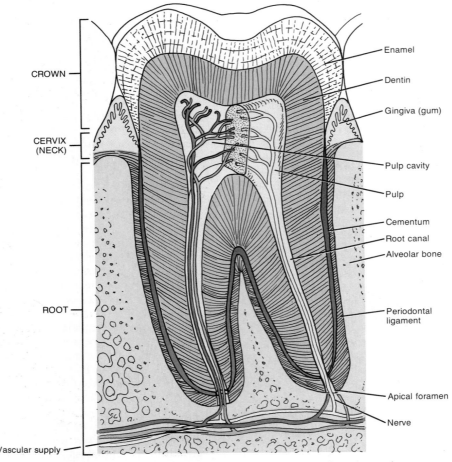

CROWN

CERVIX (NECK)

ROOT

Enamel

Dentin

Gingiva (gum)

Pulp cavity

Pulp

Cementum

Root canal

Alveolar bone

Periodontal ligament

Apical foramen

Nerve

Vascular supply

FIGURE 24-7 Parts of a typical tooth as seen in a section through a molar.

A typical tooth consists of three principal portions. The **crown** is the portion above the level of the gums. The **root** consists of one to three projections embedded in the socket. The **cervix** is the constricted junction line of the crown and the root.

Teeth are composed primarily of **dentin,** a bonelike substance that gives the tooth its basic shape and rigidity. The dentin encloses a cavity. The enlarged part of the cavity, the **pulp cavity,** lies in the crown and is filled with **pulp,** a connective tissue containing blood vessels, nerves, and lymphatics. Narrow extensions of the pulp cavity run through the root of the tooth and are called **root canals.** Each root canal has an opening at its base, the **apical foramen.** Through the foramen enter blood vessels bearing nourishment, lymphatics affording protection, and nerves providing sensation. The dentin of the crown is covered by **enamel** that consists primarily of calcium phosphate and calcium carbonate. Enamel is the hardest substance in the body and protects the tooth from the wear of chewing. It is also a barrier against acids that easily dissolve the dentin. The dentin of the root is covered by **cementum,** another bonelike substance, which attaches the root to the periodontal ligament.

Dental Terminology

Because of the curvature of the dental arches, it is necessary to use terms other than anterior, posterior, medial, and lateral in describing the surfaces of the teeth. Accordingly, the following directional terms are used. **Labial** refers to the surface of a tooth in contact with or directed toward the lips. **Buccal** refers to the surface in contact with or directed toward the cheeks. **Lingual** is restricted to the teeth of the lower jaw and refers to the surface directed toward the tongue. **Palatal,** on the other hand, is restricted to the teeth of the upper jaw and refers to the surface directed toward the palate. The term **mesial** designates the anterior or medial side of the tooth relative to its position in the dental arch. **Distal** refers to the posterior or lateral side of the tooth relative to its position in the dental arch. Essentially, mesial and distal refer to the sides of adjacent teeth that are in contact with each other. Finally, **occlusal** refers to the biting surface of a tooth.

Dentitions

Everyone has two **dentitions,** or sets of teeth. The first of these—the **deciduous teeth, milk teeth,** or **baby teeth**— begin to erupt at about 6 months of age, and one pair appears at about each month thereafter until all 20 are present. Figure 24-8a illustrates the deciduous teeth. The **incisors,** which are closest to the midline, are chisel-shaped and adapted for cutting into food. They are referred to as **central** or **lateral incisors** on the basis of their position. Next to the incisors, moving posteriorly,

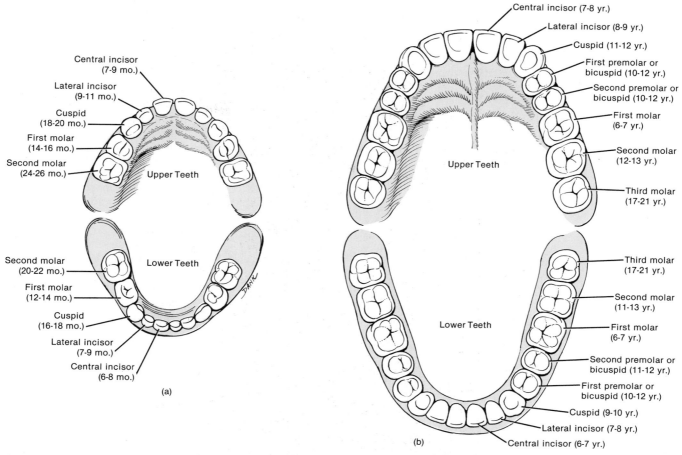

FIGURE 24-8 Dentitions and times of eruptions. The times of eruptions are indicated in parentheses. (a) Deciduous dentition. (b) Permanent dentition.

are the **cuspids (canines),** which have a pointed surface called a cusp. Cuspids are used to tear and shred food. The incisors and cuspids have only one root apiece. Behind them lie the **first** and **second molars,** which have four cusps. Upper molars have three roots; lower molars have two roots. The molars crush and grind food.

All the deciduous teeth are lost—generally between 6 and 12 years of age—and are replaced by the **permanent dentition** (Figure 24-8b). The permanent dentition contains 32 teeth that appear between the age of 6 and adulthood. It resembles the deciduous dentition with the following exceptions. The deciduous molars are replaced with the **first** and **second premolars (bicuspids)** that have two cusps and one root (upper first bicuspids have two roots) and are used for crushing and grinding. The permanent molars erupt into the mouth behind the bicuspids. They do not replace any deciduous teeth and erupt as the jaw grows to accommodate them—the **first molars** at age 6, the **second molars** at age 12, the **third molars (wisdom teeth)** after age 18. The human jaw has become smaller through time and often does not afford enough room behind the second molars for the eruption of the third molars. In this case, the third molars remain embedded in the alveolar bone and are said to be "impacted." Most often they cause pressure and pain and must be surgically removed. In some individuals, third molars may be dwarfed in size or may not develop at all.

DIGESTION IN THE MOUTH

Mechanical

Through chewing, or **mastication,** the tongue manipulates food, the teeth grind it, and the food is mixed with saliva. As a result, the food is reduced to a soft, flexible bolus that is easily swallowed.

Chemical

The enzyme **salivary amylase** (AM-i-lās) initiates the breakdown of starch. This is the only chemical digestion that occurs in the mouth. Carbohydrates are either monosaccharide and disaccharide sugars or polysaccharide starches (see Chapter 2). Most of the carbohydrates we eat are polysaccharides. Since only monosaccharides can be absorbed into the bloodstream, ingested disaccharides and polysaccharides must be broken down. The function of salivary amylase is to break the chemical bonds between some of the monosaccharides in the starches to reduce the long-chain polysaccharides to the disaccharide maltose. Food usually is swallowed too quickly for all of the starches to be reduced to disaccharides in the mouth. However, salivary amylase in the swallowed food continues to act on starches for another 15 to 30 minutes in

EXHIBIT 24-1
DIGESTION IN THE MOUTH

STRUCTURE	ACTIVITY	RESULT
Cheeks	Keep food between teeth during mastication.	Foods uniformly chewed.
Lips	Keep food between teeth during mastication.	Foods uniformly chewed.
Tongue Extrinsic muscles	Move tongue from side to side and in and out.	Food maneuvered for mastication, shaped into bolus, and maneuvered for deglutition.
Intrinsic muscles	Alter shape of tongue.	Deglutition and speech.
Taste buds	Serve as receptors for food stimulus.	Secretion of saliva stimulated by nerve impulses from taste buds to salivatory nuclei in brain stem to salivary glands.
Buccal glands	Secrete saliva.	Lining of mouth and pharynx moistened and lubricated.
Salivary glands	Secrete saliva.	Lining of mouth and pharynx moistened and lubricated. Saliva softens, moistens, and dissolves food, coats food with mucin, cleanses mouth and teeth. Salivary amylase reduces polysaccharides to the disaccharide maltose.
Teeth	Cut, tear, and pulverize food.	Solid foods reduced to smaller particles for swallowing.

the stomach before the stomach acids eventually inactivate it.

Exhibit 24-1 summarizes digestion in the mouth.

DEGLUTITION

Swallowing, or **deglutition** (dē-gloo-TISH-un), is a mechanism that moves food from the mouth to the stomach. It is facilitated by saliva and mucus and involves the mouth, pharynx, and esophagus. Swallowing is conveniently divided into three stages: (1) the voluntary stage, in which the bolus is moved into the oropharynx, (2) the pharyngeal stage, the involuntary passage of the bolus through the pharynx into the esophagus, and (3) the esophageal stage, the involuntary passage of the bolus through the esophagus into the stomach.

Swallowing starts when the bolus is forced to the back of the mouth cavity and into the oropharynx by the movement of the tongue upward and backward against the palate. This represents the **voluntary stage** of swallowing (Figure 24-9).

With the passage of the bolus into the oropharynx, the involuntary **pharyngeal stage** of swallowing begins. The respiratory passageways close and breathing is temporarily interrupted. The bolus stimulates receptors in the oropharynx, which send impulses to the **deglutition center** in the medulla and lower pons of the brain stem. The returning impulses cause the soft palate and uvula to move upward to close off the nasopharynx, and the larynx is pulled forward and upward under the tongue. As the larynx rises, it meets the epiglottis, which seals off the glottis. The movement of the larynx also pulls the vocal cords together, further sealing off the respiratory tract, and widens the opening between the laryngopharynx and esophagus. The bolus passes through the laryngopharynx and enters the esophagus in 1 to 2 seconds. The respiratory passageways then reopen and breathing resumes.

ESOPHAGUS

The **esophagus** (e-SOF-a-gus), the third organ involved in deglutition, is a muscular, collapsible tube that lies behind the trachea. It is about 23 to 25 cm (10 inches) long and begins at the end of the laryngopharynx, passes through the mediastinum anterior to the vertebral column, pierces the diaphragm through an opening called the *esophageal hiatus,* and terminates in the superior portion of the stomach.

The *mucosa* of the esophagus consists of nonkeratinized stratified squamous epithelium, lamina propria, and a muscularis mucosae (Figure 24-10a, b). The *submucosa* contains connective tissue and blood vessels. The *muscularis* of the upper third is striated, the middle third is striated and smooth, and the lower third is smooth. The outer layer is known as the *adventitia* (ad-ven-TISH-a) and not the serosa because the loose connective tissue of the layer is not covered by epithelium (mesothelium) and because the connective tissue merges with the connective tissue of surrounding structures.

The esophagus does not produce digestive enzymes and does not carry on absorption. It secretes mucus and transports food to the stomach. The passage of food from the laryngopharynx into the esophagus is regulated by a sphincter (thick circle of muscle around an opening) at the entrance to the esophagus called the **upper esophageal** (e-sof'-a-JĒ-al) **sphincter.** It consists of the cricopharyngeus muscle attached to the cricoid cartilage. The elevation of the larynx during the pharyngeal stage of swallowing causes the sphincter to relax and the bolus enters the esophagus. The sphincter also relaxes during expiration.

During the **esophageal phase** of swallowing, food is pushed through the esophagus by involuntary muscular movements called **peristalsis** (per'-i-STAL-sis) (Figure 24-10c, d). Peristalsis is a function of the muscularis and is controlled by the medulla. In the section of the esophagus lying just above and around the top of the bolus,

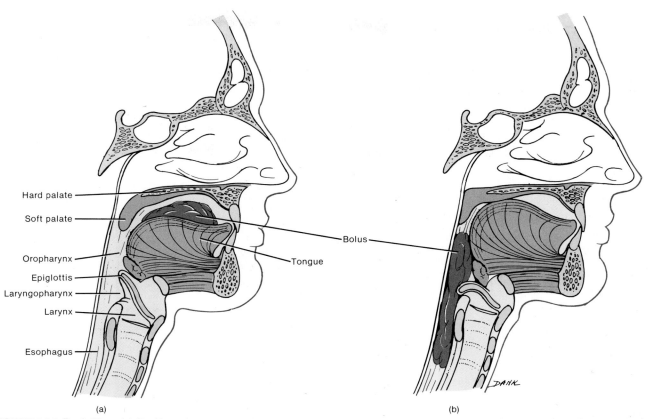

(a)   (b)

FIGURE 24-9 Deglutition. (a) Position of structures prior to deglutition. (b) During deglutition, the tongue rises against the palate, the nose is closed off, the larynx rises, the epiglottis seals off the larynx, and the bolus is passed into the esophagus.

the circular muscle fibers contract. The contraction constricts the esophageal wall and squeezes the bolus downward. Meanwhile, longitudinal fibers lying around the bottom of and just below the bolus also contract. Contraction of the longitudinal fibers shortens this lower section, pushing its walls outward so it can receive the bolus. The contractions are repeated in a wave that moves down the esophagus, pushing the food toward the stomach. Passage of the bolus is further facilitated by glands secreting mucus. The passage of solid or semisolid food from the mouth to the stomach takes 4 to 8 seconds. Very soft foods and liquids pass through in about 1 second.

Just above the level of the diaphragm, the esophagus is slightly narrowed. This narrowing has been attributed to a physiological sphincter in the inferior part of the esophagus known as the **lower esophageal (gastroesophageal) sphincter.** The lower esophageal sphincter relaxes during swallowing and thus aids the passage of the bolus from the esophagus into the stomach.

CLINICAL APPLICATION

If the lower esophageal sphincter fails to relax normally as food approaches, the condition is called **achalasia** (ak'-a-LĀ-zē-a; *a* = without; *chalasis* = relaxation). As a result, food passage from the esophagus into the stomach is greatly impeded. A whole meal may become lodged in the esophagus, entering the stomach very slowly. Distension of the esophagus results in chest pain that is often confused with pain originating from the heart. The condition is caused by malfunction of the myenteric plexus.

If, on the other hand, the lower esophageal sphincter fails to close adequately after food has entered the stomach, the stomach contents can enter the lower esophagus. Hydrochloric acid from the stomach contents can irritate the esophageal wall, resulting in a burning sensation. The sensation is known as **heartburn** because it is experienced in the region over the heart, but it is not, in fact, related to any cardiac problem. Heartburn can be treated by taking antacids (Tums, Gelusil, Rolaids, Maalox) that neutralize the hydrochloric acid and lessen the severity of the burning sensation and discomfort.

Exhibit 24-2 summarizes the digestion-related activities of the pharynx and esophagus.

STOMACH

The **stomach** is a J-shaped enlargement of the GI tract directly under the diaphragm in the epigastric, umbilical, and left hypochondriac regions of the abdomen (see Figure 1-8b). The superior portion of the stomach is a continuation of the esophagus. The inferior portion empties into the duodenum, the first part of the small intestine. Within each individual, the position and size of the stomach vary continually. For instance, the diaphragm pushes the stomach downward with each inspiration and pulls it upward with each expiration. Empty, it is about the size of a large sausage, but it can stretch to accommodate large amounts of food.

FIGURE 24-10 Esophagus. (a) Photomicrograph of a portion of the wall of the esophagus at a magnification of 60×. (b) Photomicrograph of an enlarged aspect of the mucosa of the esophagus at a magnification of 180×. (© 1983 by Michael H. Ross. Used by permission.) (c) Diagram of peristalsis. (d) Anteroposterior projection of peristalsis made during fluoroscopic examination while a patient was swallowing barium. (Courtesy of Lester W. Paul and John H. Juhl, *The Essentials of Roentgen Interpretation,* 3rd ed., Harper & Row, Publishers, Inc., New York, 1972.)

EXHIBIT 24-2

DIGESTIVE ACTIVITIES OF THE PHARYNX AND ESOPHAGUS

STRUCTURE	ACTIVITY	RESULT
Pharynx	Deglutition.	Moves bolus from oropharynx to laryngopharynx and into esophagus; closes air passageways.
	Relaxation of upper esophageal sphincter.	Moves bolus from laryngopharynx into esophagus.
Esophagus	Peristalsis.	Forces bolus down esophagus.
	Relaxation of lower esophageal sphincter.	Moves bolus into stomach.
	Secretion of mucus.	Lubricates esophagus for smooth passage of bolus.

ANATOMY

The stomach is divided into four areas: cardia, fundus, body, and pylorus (Figure 24-11). The **cardia** surrounds the lower esophageal sphincter. The rounded portion above and to the left of the cardia is the **fundus.** Below the fundus is the large central portion of the stomach, called the **body.** The narrow, inferior region is the **pylorus (antrum).** The concave medial border of the stomach is called the **lesser curvature,** and the convex lateral border is the **greater curvature.** The pylorus communicates with the duodenum of the small intestine via a sphincter called the **pyloric sphincter (valve).**

CLINICAL APPLICATION

Two abnormalities of the pyloric sphincter can occur in infants. **Pylorospasm** is characterized by failure of the muscle fibers encircling the opening to relax normally. It can be caused by hypertrophy or continuous spasm of the sphincter and usually occurs between the second and twelfth weeks of life. Ingested food does not pass easily from the stomach to the small intestine, the stomach becomes overly full, and the infant vomits frequently to relieve the pressure. Pylorospasm is treated by adrenergic drugs that relax the muscle fibers of the sphincter. **Pyloric stenosis** is a narrowing of the pyloric sphincter caused by a tumorlike mass that apparently is formed by enlargement of the circular muscle fibers. It must be surgically corrected.

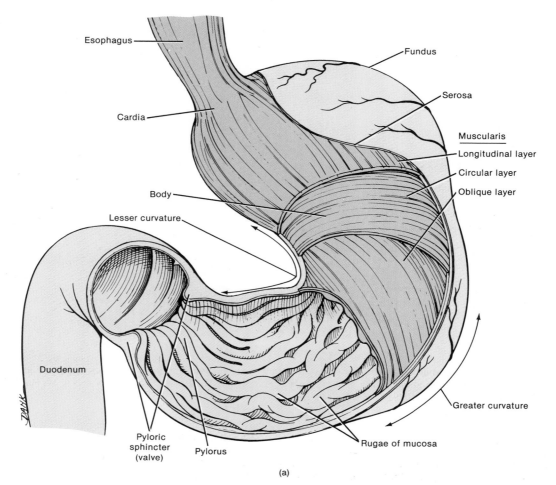

(a)

FIGURE 24-11 External and internal anatomy of the stomach. (a) Diagram in anterior view.

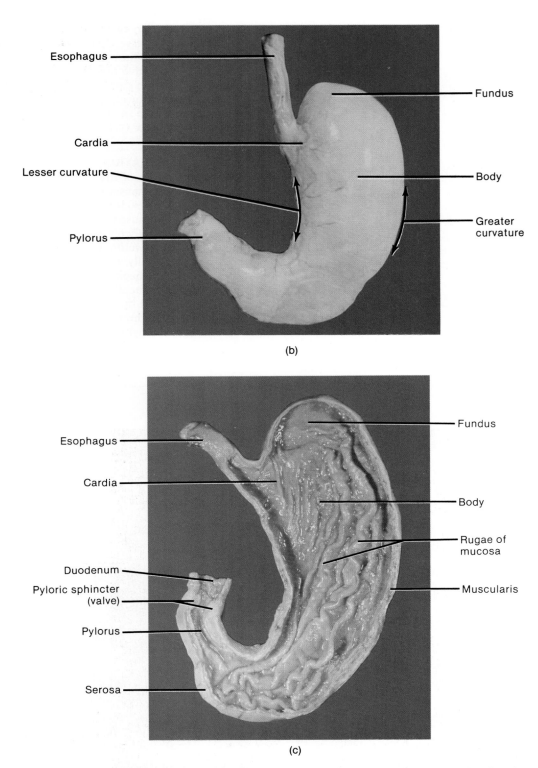

(b)

(c)

FIGURE 24-11 (*Continued*) External and internal anatomy of the stomach. (b) Photograph of the external surface in anterior view. (c) Photograph of the internal surface in anterior view showing rugae. (Photographs courtesy of C. Yokochi and J. W. Rohen, *Photographic Anatomy of the Human Body,* 2nd ed., 1979, IGAKU-SHOIN, Ltd., Tokyo, New York.)

HISTOLOGY

The stomach wall is composed of the same four basic layers as the rest of the alimentary canal, with certain modifications. When the stomach is empty, the *mucosa* lies in large folds, called **rugae** (ROO-jē), that can be seen with the naked eye. Microscopic inspection of the mucosa reveals a layer of simple columnar epithelium containing many narrow openings that extend down into the lamina propria (Figure 24-12a–d). These pits—**gastric**

Ruga

Mucosa

Submucosa

Muscularis

(a)

FIGURE 24-12 Histology of the stomach. (a) Photomicrograph of a portion of the fundic wall of the stomach at a magnification of 18×. (© 1983 by Michael H. Ross. Used by permission.) (b) Scanning electron micrograph of the stomach wall at a magnification of 55×. (Courtesy of Richard K. Kessel and Randy H. Kardon, *Tissues and Organs: A Text-Atlas of Scanning Electron Microscopy.* Copyright © 1979 by Scientific American, Inc.)

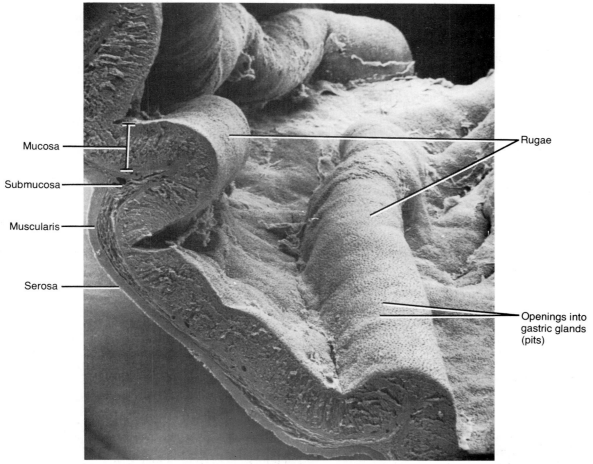

Mucosa

Submucosa

Muscularis

Serosa

Rugae

Openings into gastric glands (pits)

(b)

Surface
epithelial cells

Gastric gland (pit)

Mucus secreting cells
of gastric gland (pit)

Mucous cells

Parietal cells

Parietal cells

Zymogenic
(chief) cells

Zymogenic
(chief cells)

(c)

(d)

FIGURE 24-12 (*Continued*) Histology of the stomach. (c) Diagram of gastric glands from the fundic wall of the stomach. (d) Photomicrograph of an enlarged aspect of the mucosa of the fundic wall of the stomach at a magnification of 180×. (© 1983 by Michael H. Ross. Used by permission.)

glands—are lined with several kinds of secreting cells: zymogenic, parietal, mucous, and enteroendocrine. The **zymogenic (chief) cells** secrete the principal gastric enzyme precursor, pepsinogen. Hydrochloric acid, involved in the conversion of pepsinogen to the active enzyme pepsin, and intrinsic factor, involved in the absorption of vitamin B_{12} for red blood cell production, are produced by the **parietal cells.** You may recall from Chapter 19 that inability to produce intrinsic factor can result in pernicious anemia. The **mucous cells** secrete mucus. The **enteroendocrine (argentaffin) cells** secrete stomach gastrin, a hormone that stimulates secretion of hydrochloric acid and pepsinogen, contracts the lower esophageal sphincter, mildly increases motility of the GI tract, and relaxes the pyloric sphincter. Secretions of the gastric glands are together called **gastric juice.**

CLINICAL APPLICATION

Viewing of the mucosa of the stomach through an illuminated tube fitted with a lens system is called **gastroscopy.** During the procedure, the patient is anesthetized and the gastroscope is passed into the stomach, which is then inflated with air. A flexible fiberoptic instrument makes possible direct visualization of different parts of the gastric mucosa. It is also possible to perform a gastric mucosal biopsy through a gastroscope.

The *submucosa* of the stomach is composed of loose areolar connective tissue, which connects the mucosa to the muscularis.

The *muscularis,* unlike that in other areas of the alimentary canal, has three layers of smooth muscle: an outer longitudinal layer, a middle circular layer, and an inner oblique layer. This arrangement of fibers allows the stomach to contract in a variety of ways to churn food, break it into small particles, mix it with gastric juice, and pass it to the duodenum.

The *serosa* covering the stomach is part of the visceral peritoneum. At the lesser curvature the two layers of the visceral peritoneum come together and extend upward to the liver as the lesser omentum. At the greater curvature, the visceral peritoneum continues downward as the greater omentum hanging over the intestines.

DIGESTION IN THE STOMACH

Mechanical

Several minutes after food enters the stomach, gentle, rippling, peristaltic movements called **mixing waves** pass over the stomach every 15 to 25 seconds. These waves macerate food, mix it with the secretions of the gastric glands, and reduce it to a thin liquid called **chyme** (kīm). Few mixing waves are observed in the fundus, which is primarily a storage area. Foods may remain in the fundus for an hour or more without becoming mixed with gastric juice. During this time, salivary digestion continues.

As digestion proceeds in the stomach, more vigorous mixing waves begin at the body of the stomach and intensify as they reach the pylorus. The pyloric sphincter normally remains almost, but not completely, closed. As food reaches the pylorus, each mixing wave forces a small amount of the gastric contents into the duodenum through the pyloric sphincter. Most of the food is forced back into the body of the stomach where it is subjected to further mixing. The next wave pushes it forward again and forces a little more into the duodenum. The forward and backward movement of the gastric contents are responsible for almost all of the mixing in the stomach.

Chemical

The principal chemical activity of the stomach is to begin the digestion of proteins. In the adult, digestion is achieved primarily through the enzyme **pepsin.** Pepsin breaks certain peptide bonds between the amino acids making up proteins. Thus a protein chain of many amino acids is broken down into smaller fragments called *peptides.* Pepsin is most effective in the very acidic environment of the stomach (pH 2). It becomes inactive in an alkaline environment.

What keeps pepsin from digesting the protein in stomach cells along with the food? First, pepsin is secreted in an inactive form called **pepsinogen,** so it cannot digest the proteins in the zymogenic cells that produce it. It is not converted into active pepsin unil it comes in contact with the hydrochloric acid secreted by the parietal cells. Second, the stomach cells are protected by mucus, especially after pepsin has been activated. The mucus coats the mucosa to form a barrier between it and the gastric juices.

Another enzyme of the stomach is **gastric lipase.** Gastric lipase splits the butterfat molecules found in milk. This enzyme operates best at a pH of 5 to 6 and has a limited role in the adult stomach. Adults rely almost ex-

clusively on an enzyme found in the small intestine to digest fats.

The infant stomach also secretes **rennin,** which is important in the digestion of milk. Rennin and calcium act on the casein of milk to produce a curd. The coagulation prevents too rapid a passage of milk from the stomach. Rennin is absent in the gastric secretions of adults.

The principal activities of gastric digestion are summarized in Exhibit 24-3.

REGULATION OF GASTRIC SECRETION

Stimulation

The secretion of gastric juice is regulated by both nervous and hormonal mechanisms (Figure 24-13). Parasympathetic impulses from nuclei in the medulla are transmitted via the vagus (X) nerves and stimulate the gastric glands to secrete pepsinogen, hydrochloric acid, mucus, and stomach gastrin. Stomach gastrin is also secreted by gastric glands in response to certain foods that enter the stomach.

● *Cephalic (Reflex) Phase*   The **cephalic phase** of gastric secretions occurs before food enters the stomach and prepares the stomach for digestion. The sight, smell, taste, or thought of food initiates this reflex. Nerve impulses from the cerebral cortex or feeding center in the hypothalamus send impulses to the medulla. The medulla relays impulses over the parasympathetic fibers in the vagus (X) nerve to stimulate the gastric glands to secrete.

● *Gastric Phase*   Once the food reaches the stomach, both nervous and hormonal mechanisms ensure that gastric secretion continues. This is the **gastric phase** of secretion. Food of any kind causes distension and stimulates receptors in the walls of the stomach. These receptors send impulses to the medulla and back to the gastric glands, and they may send messages directly to the glands as well. The impulses stimulate the flow of gastric juice.

EXHIBIT 24-3
SUMMARY OF GASTRIC DIGESTION

STRUCTURE	ACTIVITY	RESULT
Mucosa		
Zymogenic (chief) cells	Secrete pepsinogen.	Precursor of pepsin is produced.
Parietal cells	Secrete hydrochloric acid.	Converts pepsinogen into pepsin, which digests proteins into peptides.
	Secrete intrinsic factor.	Required for absorption of vitamin B_{12} and erythrocyte formation.
Mucous cells	Secrete mucus.	Prevents digestion of stomach wall.
Enteroendocrine cells	Secrete stomach gastrin.	Stimulates gastric secretion, contracts lower esophageal sphincter, increases motility of the stomach, and relaxes pyloric sphincter.
Muscularis	Mixing waves.	Macerate food, mix it with gastric juice, reduce food to chyme, and force chyme through pyloric sphincter.
Pyloric sphincter	Opens to permit passage of chyme into duodenum.	Prevents backflow of food from duodenum to stomach.

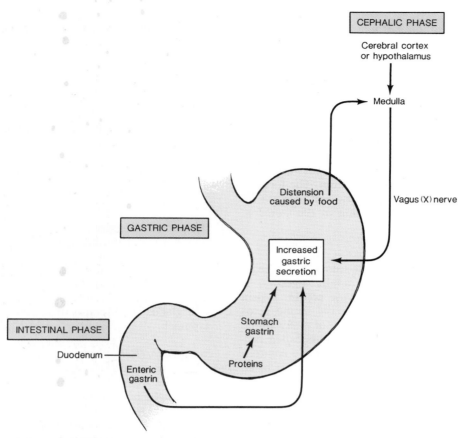

FIGURE 24-13 Summary of factors that stimulate gastric secretion.

Emotions such as anger, fear, and anxiety may slow down digestion in the stomach because they stimulate the sympathetic nervous system, which inhibits the impulses of the parasympathetic fibers.

Protein foods and alcohol stimulate the pyloric mucosa to secrete the hormone **stomach gastrin.** It is absorbed into the bloodstream, circulated through the body and finally reaches its target cells, the gastric glands, where it stimulates secretion of large amounts of gastric juice. It also contracts the lower esophageal sphincter, increases motility of the GI tract, and relaxes the pyloric sphincter and ileocecal sphincter.

● *Intestinal Phase*   Some investigators believe that when partially digested proteins leave the stomach and enter the duodenum, they stimulate the duodenal mucosa to release **enteric gastrin,** a hormone that stimulates the gastric glands to continue their secretion. This constitutes the **intestinal phase** of secretion. However, this mechanism produces relatively small amounts of gastric juice.

Inhibition

Even though chyme stimulates gastric secretion during the intestinal phase, it can inhibit secretion during the gastric phase. For example, the presence of food in the small intestine during the gastric phase initiates an **enterogastric reflex** in which nerve impulses carried to the medulla from the duodenum return to the stomach and inhibit gastric secretion. These impulses ultimately inhibit

parasympathetic stimulation and stimulate sympathetic activity. Stimuli that initiate this reflex are distension of the duodenum, the presence of acid or partially digested proteins in food in the duodenum, or irritation of the duodenal mucosa.

Several intestinal hormones also inhibit gastric secretion. In the presence of acid, partially digested proteins, fats, hypertonic or hypotonic fluids, or irritating substances in chyme, the intestinal mucosa releases **secretin, cholecystokinin (CCK),** and **gastric inhibiting peptide (GIP).** All three hormones inhibit gastric secretion and decrease motility of the GI tract (Exhibit 24-4). Secretin and cholecystokinin are also important in the control of pancreatic and intestinal secretion, and cholecystokinin also helps regulate secretion of bile from the gallbladder (see Exhibit 24-5).

GASTRIC EMPTYING

Gastric emptying is stimulated by two principal factors: nerve impulses in response to distension and stomach gastrin released in the presence of certain types of foods. As noted earlier, during the gastric phase of secretion, distension and the presence of partially digested proteins and alcohol stimulate secretion of gastric juice and stomach gastrin. In the presence of stomach gastrin, the lower esophageal sphincter contracts, the motility of the stomach increases, and the pyloric sphincter relaxes. The net effect of these actions is stomach emptying (Figure 24-14a).

EXHIBIT 24-4

HORMONAL CONTROL OF GASTRIC SECRETION, PANCREATIC SECRETION, AND SECRETION AND RELEASE OF BILE

HORMONE	WHERE PRODUCED	STIMULANT	ACTION
Stomach gastrin	Pyloric mucosa.	Partially digested proteins in stomach.	Stimulates secretion of gastric juice, constricts lower esophageal sphincter, increases motility of GI tract, and relaxes pyloric sphincter and ileocecal sphincter.
Enteric gastrin	Intestinal mucosa.	Partially digested proteins in chyme in small intestine.	Same as above.
Secretin	Intestinal mucosa.	Acid, partially digested proteins, fats, hypertonic or hypotonic fluids, or irritants in chyme in small intestine.	Inhibits secretion of gastric juice, decreases motility of the GI tract, stimulates secretion of pancreatic juice rich in sodium bicarbonate ions, stimulates secretion of bile by hepatic cells of liver, and stimulates secretion of intestinal juice.
Cholecystokinin (CCK)	Intestinal mucosa.	Same as for secretin.	Inhibits secretion of gastric juice, decreases motility of the GI tract, stimulates the secretion of pancreatic juice rich in digestive enzymes, causes ejection of bile from the gallbladder and opening of the sphincter of the hepatopancreatic ampulla, and stimulates secretion of intestinal juice.
Gastric inhibiting peptide (GIP)	Intestinal mucosa.	Same as for secretin.	Inhibits secretion of gastric juice and decreases motility of the GI tract.

The stomach empties all its contents into the duodenum 2 to 6 hours after ingestion. Food rich in carbohydrate leaves the stomach in a few hours. Protein foods are somewhat slower, and emptying is slowest after a meal containing large amounts of fat.

Stomach emptying is inhibited by the enterogastric reflex and hormones released in response to certain constituents in chyme. The enterogastric reflex not only inhibits gastric secretion, it also inhibits gastric motility. The hormones secretin, cholecystokinin (CCK) and gastric inhibiting peptide (GIP) also inhibit gastric secretion and inhibit gastric motility (Figure 24-14b). The rate of stomach emptying is limited to the amount of chyme that the small intestine can process.

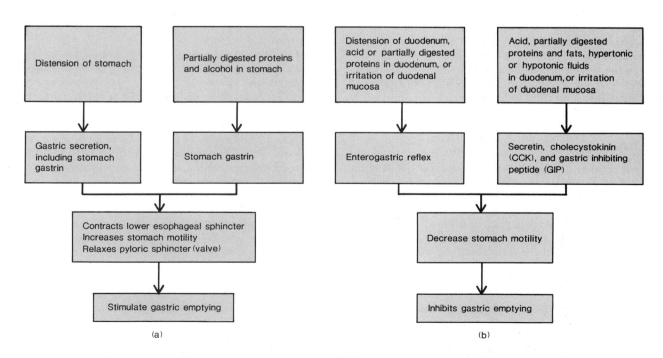

FIGURE 24-14 Factors that (a) stimulate gastric emptying and (b) inhibit gastric emptying.

CLINICAL APPLICATION

Excessive gastric emptying in the wrong direction sometimes occurs. **Vomiting** is the forcible expulsion of the contents of the upper GI tract (stomach and sometimes duodenum) through the mouth. The strongest stimuli for vomiting are irritation and distension of the stomach. Other stimuli include unpleasant sights and dizziness. Impulses are transmitted to the vomiting center in the medulla and returning impulses to the upper GI tract organs, diaphragm, and abdominal muscles bring about the vomiting act. The steps typically involved in the act are: (1) a deep breath inspiration, (2) opening of the upper esophageal sphincter, (3) closure of the glottis, (4) elevation of the soft palate to close off the internal nares, (5) contraction of the diaphragm and abdominal muscles, and (6) opening of the lower esophageal sphincter. Basically, vomiting involves squeezing the stomach between the diaphragm and abdominal muscles and expulsion of the contents through open esophageal sphincters.

Prolonged vomiting, especially in infants and elderly people, can be serious because the loss of gastric juice and fluids can lead to disturbances in fluid and acid–base balance.

ABSORPTION

The stomach wall is impermeable to the passage of most materials into the blood, so most substances are not absorbed until they reach the small intestine. However, the stomach does participate in the absorption of some water, electrolytes, certain drugs, and alcohol.

PANCREAS

The next organ of the GI tract involved in the breakdown of food is the small intestine. Chemical digestion in the small intestine depends not only on its own secretions, but on activities of three accessory structures of digestion outside the alimentary canal: the pancreas, liver, and gallbladder. We will first consider the activities of the accessory structures and then examine their contributions to digestion in the small intestine.

ANATOMY

The **pancreas** is a soft, oblong tubuloacinar gland about 12.5 cm (6 inches) long and 2.5 cm (1 inch) thick. It lies posterior to the greater curvature of the stomach and is connected by a duct (sometimes two) to the duodenum (Figure 24-15a). The pancreas is divided into a head, body, and tail. The **head** is the expanded portion near the C-shaped curve of the duodenum. Moving superiorly and to the left of the head are the centrally located **body** and the terminal tapering **tail.**

The pancreas is linked to the small intestine by a series of ducts. The products of its secreting cells are dumped into small ducts attached to the cells and eventually leave the pancreas through a large main tube called the **pancreatic duct.** In most people the pancreatic duct unites with the common bile duct from the liver and gallbladder and enters the duodenum in a common duct, called the **hepatopancreatic ampulla.** The ampulla opens on an elevation of the duodenal mucosa known as the **duodenal papilla,** about 10 cm (4 inches) below the pylorus of the stomach. An **accessory duct** may also lead from the pancreas and empty into the duodenum about 2.5 cm (1 inch) above the ampulla.

HISTOLOGY

The pancreas is made up of small clusters of glandular epithelial cells. About 1 percent of the cells, the **islets of Langerhans,** form the endocrine portion of the pancreas and consist of alpha, beta, and delta cells that secrete hormones (glucagon, insulin, and somatostatin, respectively). The functions of these hormones may be reviewed in Chapter 18. The remaining 99 percent of the cells, called **acini** (AS-i-nē), are the exocrine portions of the organ (see Figure 18-18). Secreting cells of the acini release a mixture of digestive enzymes called pancreatic juice.

PANCREATIC JUICE

Each day the pancreas produces 1,200 to 1,500 ml (about 1.2 to 1.5 qt) of **pancreatic juice,** a clear, colorless liquid. It consists mostly of water, some salts, sodium bicarbonate, and enzymes. The sodium bicarbonate gives pancreatic juice a slightly alkaline pH (7.1 to 8.2) that stops the action of pepsin from the stomach and creates the proper environment for the enzymes in the small intestine. The enzymes in pancreatic juice include a carbohydrate-digesting enzyme called *pancreatic amylase;* several protein-digesting enzymes called *trypsin* (TRIP-sin), *chymotrypsin* (kī'-mō-TRIP-sin) and *carboxypolypeptidase* (kar-bok'-sē-polē'-PEP-ti-dās); the principal fat-digesting enzyme in the adult body called *pancreatic lipase;* and nucleic acid–digesting enzymes called *ribonuclease* and *deoxyribonuclease.*

Just as pepsin is produced in the stomach in an inactive form (pepsinogen), so too are the protein-digesting enzymes of the pancreas. This prevents the enzymes from digesting cells of the pancreas. The active enzyme trypsin is secreted in an inactive form called *trypsinogen* (trip-SIN-ō-jen). Its activation to trypsin is accomplished in the small intestine by an enzyme secreted by the intestinal mucosa when chyme comes in contract with the mucosa. The activating enzyme is called *enterokinase* (en'-ter-ō-KĪ-nās). Chymotrypsin is activated in the small intestine by trypsin from its inactive form, *chymotrypsinogen.* Carboxypolypeptidase is also activated in the small intestine by trypsin. Its inactive form is called procarboxypolypeptidase.

REGULATION OF PANCREATIC SECRETIONS

Pancreatic secretion, like gastric secretion, is regulated by both nervous and hormonal mechanisms (Figure 24-16). When the cephalic and gastric phases of gastric secretion occur, parasympathetic impulses are simultaneously transmitted along the vagus (X) nerves to the pancreas that result in the secretion of pancreatic enzymes.

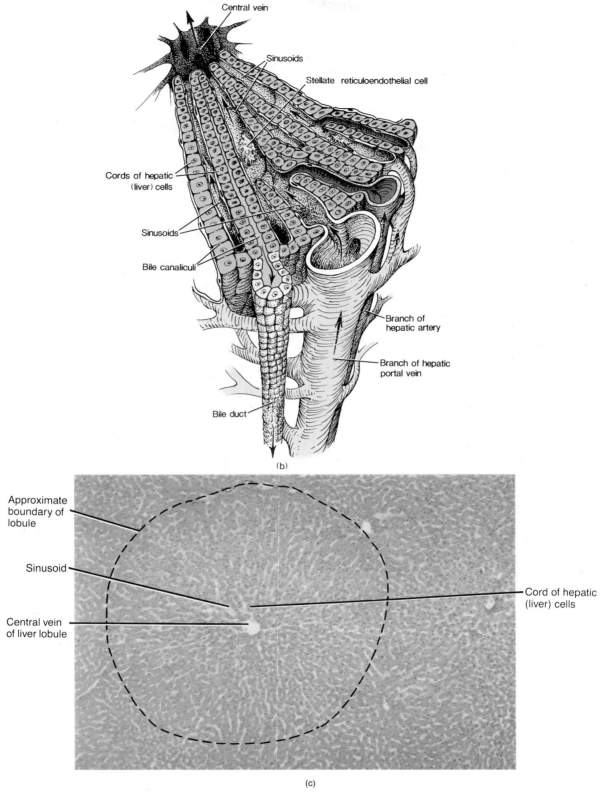

Central vein

Sinusoids

Stellate reticuloendothelial cell

Cords of hepatic (liver) cells

Sinusoids

Bile canaliculi

Branch of hepatic artery

Branch of hepatic portal vein

Bile duct

(b)

Approximate boundary of lobule

Sinusoid

Central vein of liver lobule

Cord of hepatic (liver) cells

(c)

FIGURE 24-18 (*Continued*) Histology of the liver. (b) Enlarged aspect showing details. (c) Photomicrograph of a liver lobule at a magnification of 65×. (© 1983 by Michael H. Ross. Used by permission.)

gallstone formation and the diagnosis and treatment of gallstones are discussed in detail at the end of the chapter.

If the liver is unable to remove bilirubin from the blood because of increased destruction of red blood cells or obstruction of bile ducts, large amounts of bilirubin circulate through the bloodstream and collect in other tissues, giving the skin and eyes a yellow color. This condition is called **jaundice. If** the jaundice is due to damaged red blood cells, it is called **hemolytic jaundice;** if it is due to obstruction in the biliary system, it is known as **obstructive jaundice.**

REGULATION OF BILE SECRETION

The rate at which bile is secreted is determined by several factors. Vagal stimulation can increase the production of bile to more than twice the normal rate. Secretin, the hormone that stimulates the synthesis of pancreatic juice rich in sodium bicarbonate, also stimulates the secretion of bile (see Exhibit 24-4). Within limits, as blood flow through the liver increases, so does the secretion of bile. Finally, the presence of large amounts of bile salts in the blood also increases the rate of bile production.

FUNCTIONS OF THE LIVER

The liver performs many vital functions. Among these are the following.

1. The liver manufactures bile salts which are used in the small intestine for the emulsification and absorption of fats.

2. The liver, together with mast cells, manufactures the anticoagulant heparin and most of the other plasma proteins, such as prothrombin, fibrinogen, and albumin.

3. The stellate reticuloendothelial cells of the liver phagocytose worn-out red and white blood cells and some bacteria.

4. Liver cells contain enzymes that either break down poisons or transform them into less harmful compounds. When amino acids are burned for energy, for example, they leave behind toxic nitrogenous wastes (such as ammonia), that are converted to urea by the liver cells. Moderate amounts of urea are harmless to the body and are easily excreted by the kidneys and sweat glands.

5. Newly absorbed nutrients are collected in the liver. Depending on the body's needs, it can change any excess monosaccharides into glycogen or fat, both of which can be stored, or it can transform glycogen, fat, and protein into glucose.

6. The liver stores glycogen, copper, iron, and vitamins A, D, E, and K. It also stores some poisons that cannot be broken down and excreted. (High levels of DDT are found in the livers of animals, including humans, who eat sprayed fruits and vegetables.)

7. The liver and kidneys participate in the activation of vitamin D.

GALLBLADDER

The **gallbladder** is a pear-shaped sac about 7 to 10 cm (3 to 4 inches) long. It is located in a fossa of the visceral surface of the liver (see Figures 24-15a and 24-17).

HISTOLOGY

The inner wall of the gallbladder consists of a mucous membrane arranged in rugae resembling those of the stomach. The middle, muscular coat of the wall consists of smooth muscle fibers. Contraction of these fibers by hormonal stimulation ejects the contents of the gallbladder into the cystic duct. The outer coat is the visceral peritoneum.

FUNCTION

The function of the gallbladder is to store and concentrate bile (up to 10-fold) until it is needed in the small intestine. In the concentration process, water and many ions are absorbed by the gallbladder mucosa. Bile from the liver enters the small intestine through the common bile duct. When the small intestine is empty, a valve in the duct, called the **sphincter of the hepatopancreatic ampulla,** closes, and the backed-up bile overflows into the cystic duct to the gallbladder.

EMPTYING OF THE GALLBLADDER

In order for the gallbladder to eject bile into the small intestine to participate in the digestive process, the muscularis must contract to force bile into the common bile duct and the sphincter of the hepatopancreatic ampulla must relax. Chyme entering the duodenum that contains particularly high concentrations of fats or partially digested proteins stimulates the intestinal mucosa to secrete cholecystokinin (CCK). This hormone brings about contraction of the muscularis coupled with relaxation of the sphincter of the hepatopancreatic ampulla, resulting in emptying of the gallbladder (see Exhibit 24-4).

SMALL INTESTINE

The major portions of digestion and absorption occur in a long tube called the **small intestine.** The small intestine begins at the pyloric sphincter of the stomach, coils through the central and lower part of the abdominal cavity, and eventually opens into the large intestine. It averages 2.5 cm (1 inch) in diameter and about 6.35 m (21 ft) in length.

ANATOMY

The small intestine is divided into three segments (see Figure 24-1). The **duodenum** (doo'-ō-DĒ-num), the shortest part, originates at the pyloric sphincter of the stomach and extends about 25 cm (10 inches) until it merges with the jejunum. The **jejunum** (jē-JOO-num) is about 2.5 m (8 ft) long and extends to the ileum. The final portion of the small intestine, the **ileum** (IL-ē-um), measures about 3.6 m (12 ft) and joins the large intestine at the **ileocecal** (il'-ē-ō-SĒ-kal) **valve.**

HISTOLOGY

The wall of the small intestine is composed of the same four tunics that make up most of the GI tract. However, both the mucosa and the submucosa are modified to allow the small intestine to complete the processes of digestion and absorption (Figure 24-19).

The mucosa contains many pits lined with glandular epithelium. These pits—the **intestinal glands**—secrete the intestinal digestive enzymes, collectively called intestinal juice. The submucosa of the duodenum contains **Brunner's glands,** which secrete an alkaline mucus to protect the wall of the small intestine from the action of the enzymes and to aid in neutralizing acid in the chyme. Some of the epithelial cells in the mucosa and submucosa have been transformed to goblet cells, which secrete additional mucus.

Since almost all the absorption of nutrients occurs in the small intestine, its structure is specially adapted for this function. Its length alone provides a large surface

(a)

FIGURE 24-19 Small intestine. Shown are various structures that adapt the small intestine for digestion and absorption. (a) Photograph of a section of the jejunum cut open to expose the plicae circulares. (Courtesy of C. Yokochi and J. W. Rohen, *Photographic Anatomy of the Human Body,* 2nd ed., 1979, IGAKU-SHOIN, Ltd., Tokyo, New York.) (b) Villi in relation to the layers. (c) Enlarged aspect of several villi. (d) Scanning electron micrograph of a cross section of the small intestine at a magnification of 30×. (Courtesy of Richard K. Kessel and Randy H. Kardon, *Tissues and Organs: A Text-Atlas of Scanning Electron Microscopy.* Copyright © 1979 by Scientific American, Inc.)

FIGURE 24-20 Histology of the small intestine. (a) Photomicrograph of a portion of the wall of the duodenum at a magnification of 40×. (b) Photomicrograph of an enlarged aspect of the duodenal mucosa at a magnification of 165×. (© 1983 by Michael H. Ross. Used by permission.)

area for absorption and that area is further increased by modifications in the structure of its wall. The epithelium covering and lining the mucosa consists of simple columnar epithelium. These epithelial cells, except those transformed into goblet cells, contain **microvilli,** fingerlike projections of the plasma membrane. Larger amounts of digested nutrients diffuse into the intestinal wall because the microvilli increase the surface area of the plasma membrane. They also increase the surface area for digestion.

The mucosa lies in a series of **villi,** projections 0.5 to 1 mm high, giving the intestinal mucosa its velvety appearance. The enormous number of villi (4 to 5 million)

vastly increases the surface area of the epithelium available for absorption and digestion. Each villus has a core of lamina propria, the connective tissue layer of the mucosa. Embedded in this connective tissue are an arteriole, a venule, a capillary network, and a **lacteal** (LAK-tē-al) or lymphatic vessel. Nutrients that diffuse through the epithelial cells which cover the villus are able to pass through the capillary walls and the lacteal and enter the cardiovascular and lymphatic systems.

In addition to the microvilli and villi, a third set of projections called **plicae circulares** (PLĪ-kē SER-kyoo-lar-es) further increases the surface area for absorption and digestion. The plicae are permanent deep folds in

(c)

FIGURE 24-20 (*Continued*) Histology of the small intestine. (c) Photomicrograph of an enlarged aspect of two villi from the ileum at a magnification of 120×. (© 1983 by Michael H. Ross. Used by permission.) (d) Scanning electron micrograph of several villi at a magnification of 120×. (Courtesy of Biophoto Associates.)

Simple columnar epithelium

Lamina propria

Goblet cells

Intestinal glands

Muscularis mucosae

(d)

the mucosa and submucosa. Some of the folds extend all the way around the intestine, and others extend only part way around. The plicae circulares enhance absorption by causing the chyme to spiral, rather than moving in a straight line, as it passes through the small intestine.

The muscularis of the small intestine consists of two layers of smooth muscle. The outer, thinner layer contains longitudinally arranged fibers. The inner, thicker layer contains circularly arranged fibers. Except for a major portion of the duodenum, the serosa (or visceral peritoneum) completely covers the small intestine. Additional histological aspects of the small intestine are shown in Figure 24-20.

There is an abundance of lymphatic tissue in the form of lymph nodules, masses of lymphatic tissue not covered by a capsule wall. **Solitary lymph nodules** are most numerous in the lower part of the ileum. Aggregated lymph nodules, referred to as **Peyer's (PĪ-erz) patches,** are numerous in the ileum.

INTESTINAL JUICE

Intestinal juice is a clear yellow fluid secreted in amounts of about 2 to 3 liters (about 2 to 3 qt) a day. It has a pH of 7.6, which is slightly alkaline, and contains water, mucus, and several enzymes. The intestinal enzymes include three carbohydrate-digesting enzymes called *maltase, sucrase,* and *lactase;* several protein-digesting en-

zymes called *peptidases;* and two nucleic acid–digesting enzymes, *ribonuclease* and *deoxyribonuclease.* Much of the digestion of foods by enzymes produced by the small intestine actually occurs in epithelial cells lining the small intestine rather than in a fluid outside the cells in the lumen of the tube.

DIGESTION IN THE SMALL INTESTINE

Mechanical

The movements of the small intestine are arbitrarily divided into two types: segmentation and peristalsis. **Segmentation** is the major movement of the small intestine. It is strictly a localized contraction in areas containing food. It mixes chyme with the digestive juices and brings the particles of food into contact with the mucosa for absorption. It does not push the intestinal contents along the tract. Segmentation starts with the contractions of circular muscle fibers in a portion of the small intestine, an action that constricts the intestine into segments. Next, muscle fibers that encircle the middle of each segment also contract, dividing each segment again. Finally, the fibers that contracted first relax, and each small segment unites with an adjoining small segment so that large segments are reformed. This sequence of events is repeated 12 to 16 times a minute, sloshing the chyme back and forth (Figure 24-21). Segmentation depends mainly on

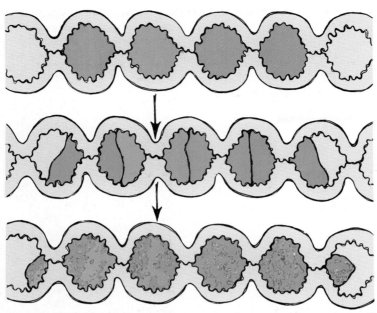

FIGURE 24-21 Segmentation. Localized contractions thoroughly mix the small intestinal contents.

intestinal distension which initiates impulses to the central nervous system. Returning parasympathetic impulses increase motility. Sympathetic impulses decrease intestinal motility.

Peristalsis propels the chyme onward through the intestinal tract. Peristaltic contractions in the small intestine are normally very weak compared to those in the esophagus or stomach. Chyme moves through the small intestine at a rate of about 1 cm/min. Thus, chyme remains in the small intestine for 3 to 5 hours. Peristalsis, like segmentation, is initiated by distension and controlled by the autonomic nervous system.

Chemical

In the mouth, salivary amylase converts starch (polysaccharide) to maltose (disaccharide). In the stomach, pepsin converts proteins to peptides (small proteins). Thus, chyme entering the small intestine contains partially digested carbohydrates, partially digested proteins, and essentially undigested lipids. The completion of the digestion of carbohydrates, proteins, and lipids is a collective effort of pancreatic juice, bile, and intestinal juice in the small intestine.

● *Carbohydrates* Even though the action of **salivary amylase** may continue in the stomach for some time, its activity is blocked by the acid pH of the stomach. Thus, few starches are reduced to maltose by the time chyme leaves the stomach. Any starches not already broken down into the disaccharide are converted by **pancreatic amylase,** an enzyme in pancreatic juice that acts in the small intestine.

Sucrose and lactose, two disaccharides, are ingested as such and are not acted upon until they reach the small intestine. Three enzymes in the intestinal juice digest the disaccharides into monosaccharides. **Maltase** splits maltose into two molecules of glucose. **Sucrase** breaks sucrose into a molecule of glucose and a molecule of fructose. **Lactase** digests lactose into a molecule of glucose and a molecule of galactose. This completes the digestion of carbohydrates.

CLINICAL APPLICATION
In some individuals the mucosal cells of the small intestine fail to produce lactase, which is essential for the digestion of lactose. This condition is called **lactose intolerance.** The deficiency can lead to an electrolyte imbalance. Its symptoms include diarrhea, gas and bloating, and abdominal cramps. The deficiency occurs in 30 to 70 percent of blacks and 15 percent of Caucasians. It sometimes occurs following gastric surgery.

● *Proteins* Protein digestion starts in the stomach, where proteins are fragmented by the action of **pepsin** into peptides. Enzymes found in pancreatic juice continue the digestion. **Trypsin** and **chymotrypsin** continue to break down proteins into peptides. Although pepsin, trypsin, and chymotrypsin all convert whole proteins into peptides, their actions differ somewhat since each splits peptide bonds between different amino acids. **Carboxypolypeptidase** acts on peptides and breaks the peptide bond that attaches the terminal amino acid to the carboxyl (acid) end of the peptide. Protein digestion is completed by the **peptidases. Aminopeptidase** acts on peptides and breaks the peptide bonds that attach amino acids to the amino end of the peptide. **Dipeptidase** splits dipeptides (two amino acids joined by a peptide bond) into amino acids that can be absorbed.

EXHIBIT 24-5

SUMMARY OF DIGESTIVE ENZYMES

ENZYME	SOURCE	SUBSTRATE	PRODUCT
Salivary amylase	Salivary glands.	Starches (polysaccharides).	Maltose (disaccharide).
Pepsin (activated from pepsinogen by hydrochloric acid)	Stomach (zymogenic cells).	Proteins.	Peptides.
Pancreatic amylase	Pancreas.	Starches (polysaccharides).	Maltose (disaccharide).
Trypsin (activated from trypsinogen by enterokinase)	Pancreas.	Proteins.	Peptides.
Chymotrypsin (activated from chymotrypsinogen by trypsin)	Pancreas.	Proteins.	Peptides.
Carboxypolypeptidase (activated from procarboxypolypeptidase by trypsin)	Pancreas.	Terminal amino acid at carboxyl (acid) end of peptides.	Peptides and amino acids.
Pancreatic lipase	Pancreas.	Neutral fats (triglycerides) that have been emulsified by bile salts.	Fatty acids and monoglycerides.
Maltase	Small intestine.	Maltose.	Glucose.
Sucrase	Small intestine.	Sucrose.	Glucose and fructose.
Lactase	Small intestine.	Lactose.	Glucose and galactose.
Peptidases Aminopeptidase	Small intestine.	Terminal amino acids at amino end of peptides.	Amino acids.
Dipeptidase	Small intestine.	Dipeptides.	Amino acids.
Ribonuclease	Pancreas and small intestine.	Ribonucleic acid nucleotides.	Pentoses and nitrogen bases.
Deoxyribonuclease	Pancreas and small intestine.	Deoxyribonucleic acid nucleotides.	Pentoses and nitrogen bases.

● *Lipids*   In an adult, almost all lipid digestion occurs in the small intestine. The first step in the process involves the preparation of neutral fats (triglycerides) by bile salts. Neutral fats, or just simply fats, are the most abundant lipids in the diet. They are called triglycerides because they consist of a molecule of glycerol and three molecules of fatty acid (see Figure 2-11). Bile salts break the globules of fat into droplets about 1 μm in diameter. This process is called **emulsification.** It is necessary so that the fat-splitting enzyme can get at the lipid molecules. In the second step, **pancreatic lipase,** an enzyme found in pancreatic juice, hydrolyzes each fat molecule into fatty acids and monoglycerides, end products of fat digestion. Lipase removes two of the three fatty acids from glycerol; the third remains attached to the glycerol, thus forming monoglycerides.

● *Nucleic Acids*   Both intestinal juice and pancreatic juice contain **nucleases** that digest nucleotides into their constituent pentoses and nitrogen bases. **Ribonuclease** acts on ribonucleic acid nucleotides and **deoxyribonuclease** acts on deoxyribonucleic acid nucleotides.

A summary of digestive enzymes in terms of source, substrate acted on, and product is presented in Exhibit 24-5.

REGULATION OF INTESTINAL SECRETION

The most important means for regulating small intestinal secretion is local reflexes in response to the presence of chyme. Also, secretin and cholecystokinin (CCK) stimulate the production of intestinal juice.

ABSORPTION

All the chemical and mechanical phases of digestion from the mouth down through the small intestine are directed toward changing food into forms that can pass through the epithelial cells lining the mucosa into the underlying blood and lymph vessels. These forms are monosaccharides (glucose, fructose, and galactose), amino acids, fatty acids, glycerol, and glycerides. Passage of these digested nutrients from the alimentary canal into the blood or lymph is called **absorption.**

About 90 percent of all absorption of nutrients takes place throughout the length of the small intestine. The other 10 percent occurs in the stomach and large intestine. Any undigested or unabsorbed material left in the small intestine is passed on to the large intestine. Absorption of materials in the small intestine occurs specifically through the villi and depends on diffusion, facilitated diffusion, osmosis, and active transport.

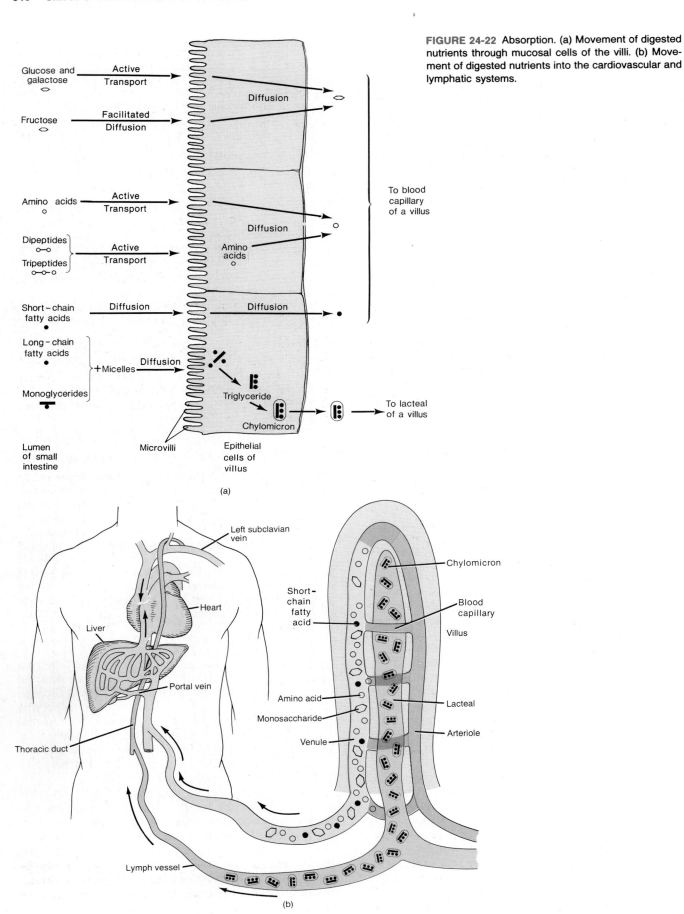

FIGURE 24-22 Absorption. (a) Movement of digested nutrients through mucosal cells of the villi. (b) Movement of digested nutrients into the cardiovascular and lymphatic systems.

Carbohydrates

Essentially all carbohydrates are absorbed as monosaccharides. Glucose and galactose are transported into epithelial cells of the villi by an active process that is somehow coupled with the active transport of sodium. Fructose is transported by facilitated diffusion. Transported monosaccharides then move out of the epithelial cells by diffusion and enter the capillaries of the villi. From here they are transported in the bloodstream to the liver via the hepatic portal system. After their passage through the liver, they move through the heart and then enter general circulation (Figure 24-22).

Proteins

Most proteins are absorbed as amino acids and the process occurs mostly in the duodenum and jejunum. Amino acid transport into epithelial cells of the villi is an active transport process coupled with active sodium transport. Amino acids move out of the epithelial cells by diffusion to enter the bloodstream. They follow the same route as that taken by monosaccharides. At times, dipeptides and tripeptides can be taken in by epithelial cells by active transport. Most of them are hydrolyzed to amino acids in the cells and then passed into the capillaries of the villi (Figure 24-22).

Lipids

As a result of emulsification and fat digestion, neutral fats (triglycerides) are broken down into monoglycerides and fatty acids. Lipase effectively removes two of the three fatty acids from glycerol during fat digestion; the other fatty acid remains attached to glycerol, thus forming monoglycerides. Short-chain fatty acids (those with less than 10 to 12 carbon atoms) pass into the epithelial cells by diffusion and follow the same route taken by monosaccharides and amino acids (Figure 24-22).

Most fatty acids are long-chain fatty acids. They and the monoglycerides are transported differently. Bile salts form spherical aggregates called **micelles** (mī-SELZ). They are about 25 Å in diameter and consist of 20 to 50 molecules of bile salt. Despite their large size, micelles have the ability to dissolve in water in the intestinal fluid. During fat digestion, fatty acids and monoglycerides dissolve in the center of the micelles, and it is in this form that they reach the epithelial cells of the villi. On coming into contact with the surfaces of the epithelial cells, fatty acids and monoglycerides diffuse into the cells, leaving the micelles behind in chyme. The micelles continually repeat this ferrying function. The majority of bile salts in the small intestine are ultimately reabsorbed in the ileum to be returned by the blood to the liver for resecretion. This cycle is called **enterohepatic circulation.**

Within the epithelial cells, many monoglycerides are further digested by lipase in the cells to glycerol and fatty acids. Then, the fatty acids and glycerol are recombined to form triglycerides in the smooth endoplasmic reticulum of the epithelial cell. The triglycerides, still in the endoplasmic reticulum, aggregate into globules along with phospholipids and cholesterol and become coated with proteins. These masses are called **chylomicrons.** The protein coat keeps the chylomicrons suspended and from sticking to each other. The chylomicrons leave the epithelial cell and enter the lacteal of a villus. From here, they are transported by way of lymphatic vessels to the thoracic duct and enter the cardiovascular system at the left subclavian vein. Finally, they arrive at the liver through the hepatic artery (Figure 24-22).

Water

The total volume of fluid that enters the small intestine each day is about 9 liters (about 9 qt). This fluid is derived from ingestion of liquids (about 1.5 liters) and from various gastrointestinal secretions (about 7.5 liters). Roughly 8 to 8.5 liters of the fluid in the small intestine is absorbed; the remainder, about 0.5 to 1.0 liter, passes into the large intestine. There most of it is also absorbed.

The absorption of water by the small intestine occurs by osmosis from the lumen of the small intestine through epithelial cells and into the blood capillaries in the villi. The normal rate of absorption is about 200 to 400 ml/h. Water can move across the intestinal mucosa in both directions. The absorption of water from the small intestine is associated with the absorption of electrolytes and digested foods in order to maintain an osmotic balance with the blood.

Electrolytes

The electrolytes absorbed by the small intestine are mostly constituents of gastrointestinal secretions. Some are also components of ingested foods and liquids. Sodium is able to move in and out of epithelial cells by diffusion. It can also move into mucosal cells by active transport for removal from the small intestine. Chloride, iodide, and nitrate ions can passively follow sodium ions or be actively transported. Calcium ions are also actively transported and their movement depends on parathyroid hormone (PTH) and vitamin D. Other electrolytes such as iron, potassium, magnesium, and phosphate also move by active transport.

Vitamins

Fat-soluble vitamins, such as A, D, E, and K, are absorbed along with ingested dietary fats in micelles. In fact, they cannot be absorbed unless they are ingested with some fat. Most water-soluble vitamins, such as the B vitamins and C, are absorbed by diffusion. Vitamin B_{12}, you may recall, requires combination with intrinsic factor produced by the stomach for its absorption.

A summary of the digestive and absorptive activities of the small intestine is presented in Exhibit 24-6.

LARGE INTESTINE

The overall functions of the large intestine are the completion of absorption, the manufacture of certain vitamins, the formation of feces, and the expulsion of feces from the body.

EXHIBIT 24-6
SUMMARY OF DIGESTION AND ABSORPTION IN THE SMALL INTESTINE

STRUCTURE	ACTIVITY
Pancreas	Delivers pancreatic juice into the duodenum via the pancreatic duct (see Exhibit 24-5 for pancreatic enzymes and their functions).
Liver	Produces bile which is necessary for emulsification of fats.
Gallbladder	Stores, concentrates, and delivers bile into the duodenum via the common bile duct.
Small Intestine Mucosa and submucosa	
Intestinal glands	Secrete intestinal juice (see Exhibit 24-5 for intestinal enzymes and their functions).
Brunner's glands	Secrete mucus for protection and lubrication.
Microvilli	Fingerlike projections of epithelial cells that increase surface area for absorption and digestion.
Villi	Projections of mucosa that are the site of absorption of digested food and also increase the surface area for absorption and digestion.
Plicae circulares	Circular folds of mucosa and submucosa that increase surface area for absorption and digestion.
Muscularis Segmentation	Consists of alternating contractions of circular fibers that produce segmentation and resegmentation of portions of the small intestine; mixes chyme with digestive juices and brings food into contact with the mucosa for absorption.
Peristalsis	Consists of mild waves of contraction and relaxation of circular and longitudinal muscle passing the length of the small intestine; moves chyme toward ileocecal valve.

ANATOMY

The **large intestine** is about 1.5 m (5 ft) in length and averages 6.5 cm (2.5 inches) in diameter. It extends from the ileum to the anus and is attached to the posterior abdominal wall by its **mesocolon** of visceral peritoneum. Structurally, the large intestine is divided into four principal regions: cecum, colon, rectum, and anal canal (Figure 24-23).

The opening from the ileum into the large intestine is guarded by a fold of mucous membrane called the **ileocecal valve.** This structure allows materials from the small intestine to pass into the large intestine. Hanging below the ileocecal valve is the **cecum,** a blind pouch about 6 cm (2.5 inches) long. Attached to the cecum is a twisted, coiled tube, measuring about 8 cm (3 inches) in length, called the **vermiform appendix** (*vermis* = worm). The visceral peritoneum of the appendix, called the **mesoappendix,** attaches the appendix to the inferior part of the ileum and adjacent part of the posterior abdominal wall.

The open end of the cecum merges with a long tube called the **colon.** The colon is divided into ascending, transverse, descending, and sigmoid portions. The **ascending colon** ascends on the right side of the abdomen, reaches the undersurface of the liver, and turns abruptly to the left. Here it forms the **right colic (hepatic) flexure.** The colon continues across the abdomen to the left side as the **transverse colon.** It curves beneath the lower end of the spleen on the left side as the **left colic (splenic) flexure** and passes downward to the level of the iliac crest as the **descending colon.** The **sigmoid colon** begins at the left iliac crest, projects inward to the midline, and terminates as the rectum at about the level of the third sacral vertebra.

The **rectum,** the last 20 cm (8 inches) of GI tract, lies anterior to the sacrum and coccyx. The terminal 2 to 3 cm (1 inch) of the rectum is called the **anal canal** (Figure 24-24). The mucous membrane of the anal canal is arranged in longitudinal folds called **anal columns** that contain a network of arteries and veins. The opening of the anal canal to the exterior is called the **anus.** It is guarded by an internal sphincter of smooth muscle and an external sphincter of skeletal muscle. Normally the anus is closed except during the elimination of the wastes of digestion.

CLINICAL APPLICATION

Inflammation and enlargement of the rectal veins (varicose veins) due to weakening of the valves is known as **hemorrhoids (piles).** Initially contained within the anus (first degree), they gradually enlarge until they prolapse or extend outward on defecation (second degree) and finally remain prolapsed through the anal orifice (third degree).

HISTOLOGY

The wall of the large intestine differs from that of the small intestine in several respects (Figure 24-25). No villi or permanent circular folds are found in the mucosa, which does, however, contain simple columnar epithelium with numerous goblet cells. These cells secrete mucus that lubricates the colonic contents as they pass through the colon. Solitary lymph nodules also are found in the mucosa. The submucosa of the large intestine is similar to that found in the rest of the alimentary canal. The muscularis consists of an external layer of longitudinal muscles

- Define a nutrient and list the functions of the six principal classes of nutrients.
- Define metabolism and contrast the physiological effects of catabolism and anabolism.
- Explain the fate of absorbed carbohydrates.
- Describe the complete oxidation of glucose via glycolysis, the Krebs cycle, and the electron transport chain.
- Define glycogenesis as an example of glucose anabolism into glycogen.
- Define glycogenolysis as an example of glycogen catabolism into glucose.
- Define gluconeogenesis as a conversion of lipids and proteins into glucose.
- Explain the fate of absorbed lipids.
- Discuss fat storage in adipose tissue.
- Explain how glycerol may be converted to glucose.
- Explain the catabolism of fatty acids via beta oxidation.
- Define ketosis and list its effects on the body.
- Define lipogenesis as the synthesis of lipids from glucose or amino acids.
- Explain the fate of absorbed proteins.
- Discuss the catabolism of amino acids.
- Explain the mechanism involved in protein synthesis.
- Distinguish between the absorptive and postabsorptive (fasting) states.
- Explain how various hormones regulate metabolism.
- Compare the sources, functions, and importance of minerals in metabolism.
- Define a vitamin and differentiate between fat-soluble and water-soluble vitamins.
- Compare the sources, functions, and deficiency symptoms and disorders of the principal vitamins.
- Explain how body heat is measured.
- Explain how the caloric value of foods is determined.
- Describe the various mechanisms that produce body heat.
- Define basal metabolic rate (BMR).
- Describe the various ways that body heat is lost.
- Explain how normal body temperature is maintained.
- Describe fever, heat cramp, sunstroke, and heat exhaustion as abnormalities of temperature regulation.
- Define obesity, phenylketonuria (PKU), cystic fibrosis, celiac disease, and kwashiorkor.

25 Metabolism

Nutrients are chemical substances in food that provide energy, form new body components, or assist in the functioning of various body processes. There are six principal classes of nutrients: carbohydrates, lipids, proteins, minerals, vitamins, and water. Carbohydrates, proteins, and lipids are digested by enzymes in the gastrointestinal tract. The end products of digestion that ultimately reach body cells are monosaccharides, amino acids, and fatty acids and monoglycerides. Some are used to synthesize new structural molecules in cells or to synthesize new regulatory molecules, such as hormones and enzymes. Most are used to produce energy to sustain life processes. This energy is used for processes such as active transport, DNA replication, synthesis of proteins and other molecules, muscle contraction, and nerve impulse conduction. Until the energy is needed, it is stored in ATP.

Some minerals and many vitamins are part of enzyme systems that catalyze the reactions undergone by carbohydrates, proteins, and lipids.

Water has five major functions. It is an excellent solvent and suspending medium, it participates in hydrolysis reactions, it acts as a coolant, it lubricates, and it helps to maintain a constant body temperature due to its ability to release and absorb heat slowly.

METABOLISM

Metabolism (me-TAB-ō-lizm) refers to all the chemical reactions of the body. The body's metabolism may be thought of as an energy-balancing act between catabolic reactions, which provide energy, and anabolic reactions, which require energy.

CATABOLISM

Catabolism (ka-TAB-ō-lizm) is the term for decomposition chemical reactions that provide energy (Figure 25-1). Digestion is a catabolic process in which the breaking of bonds of food molecules releases energy. Another example is a process called oxidation (cellular respiration). It involves the breakdown of absorbed nutrients resulting in a release of energy. **Oxidation** is the removal of electrons and hydrogen ions (hydrogen atoms) from a molecule or, less commonly, the addition of oxygen to a molecule. Most oxidations are actually dehydrogenations, that is, they involve the loss of hydrogen atoms. As you will see later, each time one substance is oxidized, another is almost simultaneously reduced. **Reduction** is the addition of electrons and hydrogen ions (hydrogen atoms) to a molecule or, less commonly, the removal of oxygen from a molecule. It is the opposite of oxidation. Glucose is the body's favorite nutrient for oxidation, but fats and proteins are also oxidized. Oxidation also involves the synthesis of ATP from ADP. As substances are oxidized, energy is produced. The energy is removed in a stepwise fashion and ultimately trapped in ATP for storage. When body cells require energy, ATP is broken down and the energy is released.

ANABOLISM

Anabolism (a-NAB-ō-lizm) is just the opposite of catabolism—a series of synthetic reactions whereby small molecules are built up into larger ones that form the body's structural and functional components. Anabolic reactions require energy and the energy is supplied by catabolic reactions of the body (Figure 25-1). One example of an anabolic process is the formation of peptide bonds between amino acids, thereby building up the amino acids into the protein portions of cytoplasm, enzymes, hormones, and antibodies. Fats also participate in the body's anabolism. For instance, fats can be built into the phospholipids that form the plasma membrane. They are also part of the steroid hormones.

We will now consider the metabolism of carbohydrates, lipids, and proteins in body cells.

CARBOHYDRATE METABOLISM

During the process of digestion, polysaccharides and disaccharides are hydrolyzed to become the monosaccharides—glucose, fructose, and galactose—which are absorbed into the capillaries of the villi of the small intestine. They are then carried through the hepatic portal vein to the liver, where fructose and galactose are converted to glucose. The liver is the only organ that has the necessary enzymes to make this conversion. Thus the story of carbohydrate metabolism is really the story of glucose metabolism.

FATE OF CARBOHYDRATES

Since glucose is the body's preferred source of energy, the fate of absorbed glucose depends on the body cells' energy needs. If the cells require immediate energy, the glucose is oxidized by the cells. Each gram of carbohydrate produces about 4.0 Calories (Cal). (The caloric content of a food is a measure of the heat it releases upon

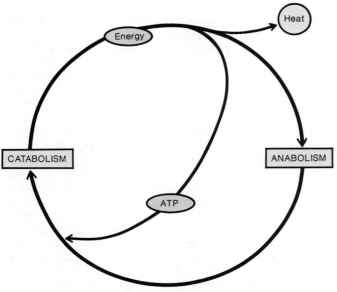

FIGURE 25-1 Relationship of ATP to catabolism and anabolism.

oxidation. Determination of caloric value is described later in the chapter.) The glucose not needed for immediate use is handled in several ways. First, the liver can convert excess glucose to glycogen and then store it. Skeletal muscle cells can also store glycogen. Second, if the glycogen storage areas are filled up, the liver cells can transform the glucose to fat that can be stored in adipose tissue. Later, when the cells need more energy, the glycogen and fat can be converted back to glucose, which is released into the bloodstream so it can be transported to cells for oxidation. Third, excess glucose can be excreted in the urine. Normally, this happens only when a meal containing mostly carbohydrates and no fats is eaten. Without the inhibiting effect of fats, the stomach empties its contents quickly and the carbohydrates are all digested at the same time. As a result, large numbers of monosaccharides suddenly flood into the bloodstream. Since the liver is unable to process all of them simultaneously, the blood glucose level rises and the condition of hyperglycemia may result in glucose in the urine (glycosuria).

GLUCOSE MOVEMENT INTO CELLS

Before glucose can be used by body cells, it must first pass through the plasma membrane and then enter the cytoplasm. The process by which this occurs is facilitated diffusion (Chapter 3) and the rate of glucose transport is greatly increased by insulin (Chapter 18). Immediately upon entry into cells, glucose combines with a phosphate group, produced by the breakdown of ATP. This addition of a phosphate group to a molecule is called *phosphorylation*. The product of this enzymatically catalyzed reaction is glucose-6-phosphate (see Figure 25-7). In most cells of the body, phosphorylation serves to capture glucose in the cell so that it cannot move back out. Liver cells, kidney tubule cells, and intestinal epithelial cells have the necessary enzyme (phosphatase) to remove the phosphate group and transport glucose out of the cell (see Figure 25-7).

GLUCOSE CATABOLISM

The **oxidation** of glucose is also known as **cellular respiration.** It occurs in every cell in the body and provides the cell's chief source of energy. The complete oxidation of glucose to carbon dioxide and water produces large amounts of energy. It occurs in three successive stages: glycolysis, the Krebs cycle, and the electron transport chain.

Glycolysis

The term **glycolysis** (glī-KOL-i-sis; *glyco* = sugar, *lysis* = breakdown) refers to a series of chemical reactions in the cytoplasm of a cell that convert a six-carbon molecule of glucose into two three-carbon molecules of pyruvic acid. The simplified overall reaction for glycolysis is:

$$C-C-C-C-C-C \longrightarrow 2\ C-C-C + 2\ ATP$$

Glucose   Pyruvic
                                   acid

The details of glycolysis are shown in Figure 25-2a. The essential features of the process are:

1. Each of the 10 reactions is catalyzed by a specific enzyme. (The names of the enzymes are indicated in the rust color.)

2. All of the reactions are reversible.

3. The first three reactions involve the addition of a phosphate group (phosphorylation) to glucose, its conversion to fructose, and the addition of another phosphate group to fructose to prime the pathway. This involves an input of energy in which two molecules of ATP are converted to ADP.

4. The double phosphorylated molecule of fructose splits into two three-carbon compounds, glyceraldehyde-3-phosphate and dihydroxyacetone phosphate. These compounds are interconvertible.

5. Each of the two three-carbon compounds is degraded to form a molecule of pyruvic acid. In the process, each three-carbon compound generates two molecules of ATP.

6. Since two molecules of ATP are needed to get glycolysis started and four molecules of ATP are generated to complete the process, there is a net gain of two molecules of ATP for each molecule of glucose that is oxidized (Figure 25-2b).

7. Most of the energy produced during glycolysis is used to generate ATP. The remainder is expended as heat energy, some of which helps to maintain body temperature.

The fate of pyruvic acid depends on the availability of oxygen. If anaerobic conditions exist in cells (as during strenuous exercise), pyruvic acid is reduced by the addition of two hydrogen atoms to form lactic acid (Figure 25-2a). The lactic acid may be transported to the liver, where it is converted back to pyruvic acid. Or, it may remain in the cells until aerobic conditions are restored (recall the mechanisms for the repayment of the oxygen debt described in Chapter 10) and it is then converted to pyruvic acid in the cells.

Under aerobic conditions, the process of the complete oxidation of glucose continues, and pyruvic acid is oxidized to form carbon dioxide and water in two sets of reactions: the Krebs cycle and the electron transport chain. However, before pyruvic acid can enter the Krebs cycle, it must be prepared. This process occurs in mitochondria.

Formation of Acetyl Coenzyme A

Although all enzymes are proteins, many enzymes are attached to nonprotein compounds. If a nonprotein group detaches from the enzyme and acts as a carrier molecule, it is called a **coenzyme.** Enzymes work together with their conezymes. The enzyme catalyzes the reaction, and the coenzyme attaches to the end product of the reaction and carries it to the next reaction. Each step in the oxidation of glucose requires a different enzyme and often a coenzyme as well. We are interested in only one coenzyme at this point: a substance called **coenzyme A (CoA).**

During the transitional step between glycolysis and the Krebs cycle, pyruvic acid is prepared for entrance into the cycle. Essentially, pyruvic acid is converted to a two-carbon compound by the loss of carbon dioxide. The loss of a molecule of CO_2 by a substance is called **decarboxylation** (dē-kar-bok'-si-LĀ-shun). (It is indicated in the diagram by the arrow labeled CO_2.) This two-carbon fragment, called an **acetyl group,** attaches itself to coenzyme A, and the whole complex is called **acetyl coenzyme A** (Figure 25-3). It is in this form that pyruvic acid enters the Krebs cycle.

(b)

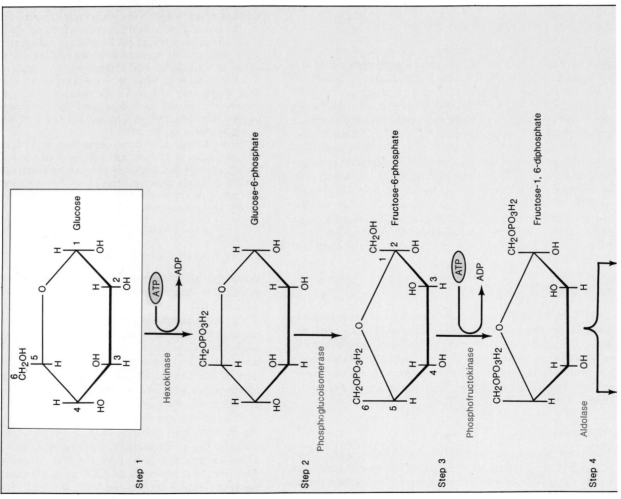

Step 1

Glucose

Hexokinase

Step 2

Glucose-6-phosphate

Phosphoglucoisomerase

Step 3

Fructose-6-phosphate

Phosphofructokinase

Step 4

Fructose-1, 6-diphosphate

Aldolase

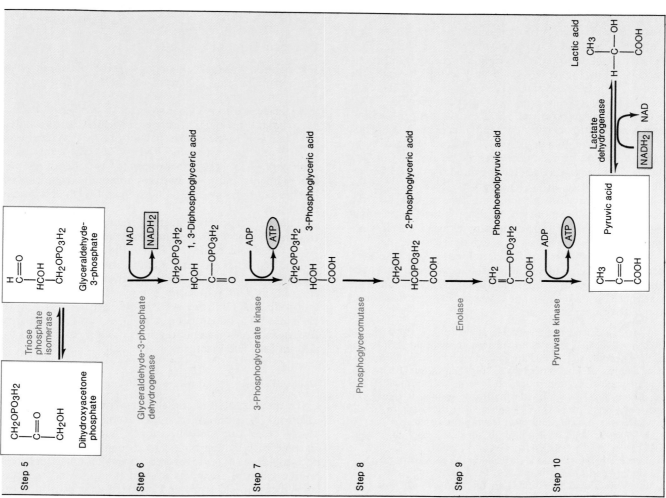

FIGURE 25-2 Glycolysis. As a result of glycolysis, there is a net gain of two molecules of ATP. (a) Detailed version. (b) Simplified version.

(a)

FIGURE 25-3 Formation of acetyl coenzyme A from pyruvic acid for entrance into the Krebs cycle.

Krebs Cycle

The **Krebs cycle** is also called the **citric acid cycle** or the **tricarboxylic acid (TCA) cycle**. It is a series of reactions that occur in the matrix of mitochondria of cells (Figure 25-4). Coenzyme A carries an acetyl group to a mitochondrion, detaches itself, and goes back into the cytoplasm to pick up another fragment. The acetyl group combines with oxaloacetic acid to form citric acid. From this point the Krebs cycle consists mainly of a series of decarboxylation and oxidation-reduction reactions, each controlled by a different enzyme.

Let us look at the decarboxylation reactions first. As we have seen (Figure 25-3), in preparation for entering the cycle, pyruvic acid is decarboxylated to form an acetyl group. In the cycle isocitric acid loses a molecule of CO_2 to form α-ketoglutaric acid. In the next step, α-ketoglutaric acid is decarboxylated (and picks up a molecule of CoA) to form succinyl CoA. Thus each time pyruvic acid enters the Krebs cycle three molecules of CO_2 are liberated by decarboxylation. The molecules of CO_2 leave the mitochondria, diffuse through the cytoplasm of the cell to the plasma membrane, and then diffuse into the blood by internal respiration. Eventually the CO_2 is transported by the blood to the lungs by external respiration. Once it reaches the lungs, it is exhaled.

Now let us look at the oxidation-reduction reactions. Remember, when a molecule is oxidized, it loses hydrogen atoms (electrons and hydrogen ions). When a molecule is reduced, it gains hydrogen atoms. Every oxidation is coupled with a reduction.

1. In the conversion of pyruvic acid to acetyl CoA (Figure 25-3), pyruvic acid loses two hydrogen atoms, that is, it is oxidized. The hydrogen atoms are picked up by a coenzyme called **nicotinamide adenine dinucleotide (NAD)**. NAD contains two adenine nucleotides and is derived from a B vitamin called niacin. Since NAD picks up the hydrogen atoms, it is reduced and represented as $NADH_2$ (indicated in the diagram by the curved arrow entering and then leaving the reaction).

2. Isocitric acid is oxidized to form α-ketoglutaric acid. NAD is reduced to $NADH_2$.

3. α-Ketoglutaric acid is oxidized in the conversion to succinyl CoA. Again, NAD is reduced to $NADH_2$.

4. Succinic acid is oxidized to fumaric acid. In this case, the coenzyme that picks up the two hydrogen atoms is **flavin adenine dinucleotide (FAD)**. FAD, like NAD, contains two adenine nucleotides, but it is derived from vitamin B_2 (riboflavin). It belongs to a class of compounds called **flavoproteins**. FAD is reduced to $FADH_2$.

5. Malic acid is oxidized to oxalocetic acid. NAD is reduced to $NADH_2$.

Thus, each time a molecule of pyruvic acid enters the Krebs cycle, four molecules of $NADH_2$ and one molecule of $FADH_2$ are produced by the various oxidation-reduction reactions. These reduced coenzymes are very important because they now contain the stored energy originally in glucose and then in pyruvic acid. In the next stage in the complete oxidation of glucose, the electron transport chain, the energy in the coenzymes is transferred to ATP for storage.

Electron Transport Chain

The **electron transport chain** is a series of oxidation-reduction reactions that occur on mitochondrial cristae in which the energy in $NADH_2$ and $FADH_2$ is liberated and transferred to ATP for storage. In the electron transport chain, three types of carrier molecules are alternately oxidized and reduced and participate in the generation of ATP: (1) the coenzyme **FAD**, (2) a coenzyme called **coenzyme Q**, and (3) **cytochromes** (SĪ-tō-krōms), red protein pigments that have an iron-containing group capable of alternating between a reduced form (Fe^{2+}) and an oxidized form (Fe^{3+}).

The first step in the electron transport chain involves the transfer of hydrogen atoms from $NADH_2$ (mostly from the Krebs cycle) to FAD (Figure 25-5). $NADH_2$ is oxidized to NAD and FAD is reduced to $FADH_2$. The importance of the hydrogen transfer from $NADH_2$ to FAD is that it releases energy. This energy is used to produce ATP from ADP. The hydrogen atoms are then used to reduce coenzyme Q. After that step, the hydrogen atoms do not stay intact. They ionize into hydrogen ions and electrons.

$$H \quad \longrightarrow \quad H^+ \quad + \quad e^-$$

Hydrogen atom   Hydrogen ion   Electron

In the next step of the electron transport system, the electrons (e^-) from the hydrogen atoms are passed successively from one cytochrome to another—from cytochrome b to cytochrome c to cytochrome a and finally to cytochrome a_3 (cytochrome oxidase). Each cytochrome in the electron transport system is alternately reduced as it picks up electrons and oxidized as it gives up electrons. Because of the involvement of cytochromes in the electron transport chain, it is also known as the **cytochrome system**.

In the process of electron transfer between cytochromes, more energy is liberated and then stored in ATP. The formation of ATP from the liberated energy occurs

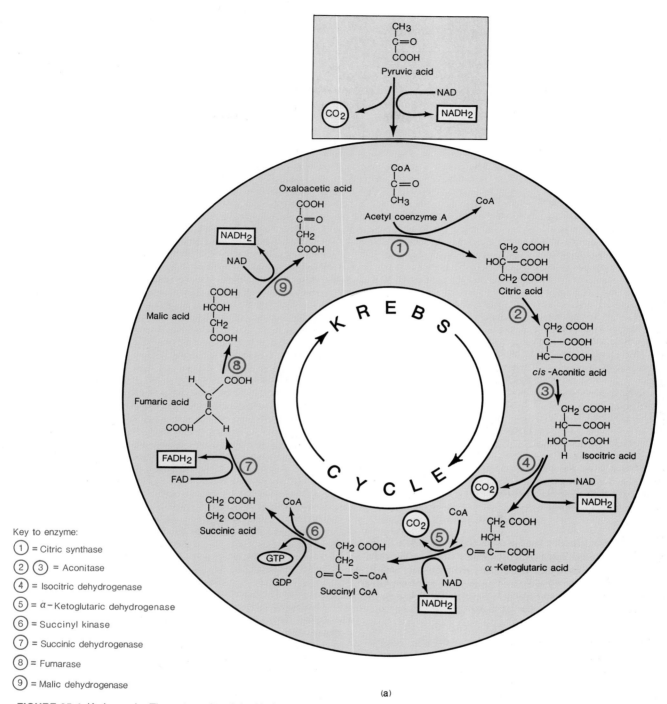

FIGURE 25-4 Krebs cycle. The net results of the Krebs cycle are the production of reduced coenzymes ($NADH_2$ and $FADH_2$), which contain stored energy; the generation of GTP, a high-energy compound that is used to produce ATP; and the formation of CO_2, which is transported to the lungs for expiration. (a) Detailed version.

between cytochrome b and cytochrome c and between cytochrome a and cytochrome a_3. Thus, ATP molecules are formed at three different places in the electron transport chain. At the end of the electron transport chain, the electrons are passed to oxygen, which becomes negatively charged. This oxygen, required for the complete oxidation of glucose, is supplied to your body cells through inspiration, external respiration, and internal res-

piration. The oxygen combines with the H^+ ions released from the breakdown of hydrogen atoms and forms water.

ATP Production

Let's add up the number of ATP molecules formed during the complete oxidation of glucose. During glycolysis there is a gain of two molecules of ATP (see Figure 25-2).

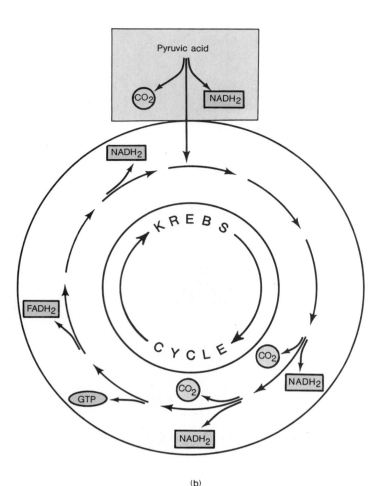

(b)

FIGURE 25-4 (*Continued*) Krebs cycle. (b) Simplified version.

Each molecule of glucose that is oxidized yields two molecules of pyruvic acid, both of which enter the Krebs cycle. Although the Krebs cycle does not produce any ATP, it does generate a substance called **guanosine triphosphate (GTP).** GTP is produced from guanosine diphosphate (GDP), using the energy of succinyl coenzyme A (see Figure 25-4). GTP is a high-energy compound that may be used to convert ADP to ATP. Accordingly, we will assume that there is a production of two molecules of ATP in the Krebs cycle since the energy in GTP is incorporated into ATP. The bulk of the ATP is generated in the electron transport chain, where a total of 34 are produced. This ATP accounting can be summarized as follows:

Process	Net Gain of ATP
Glycolysis	2
Krebs cycle	2
Electron transport chain	34
Total	38

The complete oxidation of a molecule of glucose can be summarized as follows (Figure 25-6):

$$C_6H_{12}O_6 + 6O_2 \rightarrow 38ATP + 6CO_2 + 6H_2O$$

Glucose   Oxygen   Adenosine   Carbon   Water
triphosphate   dioxide

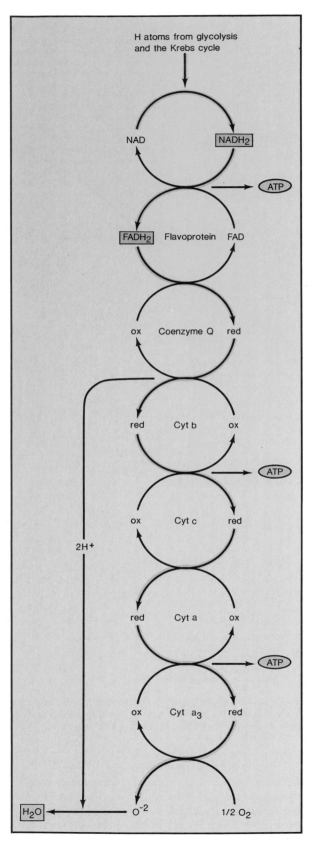

FIGURE 25-5 Electron transport chain. The successive transfer of hydrogen atoms and electrons results in the formation of ATP at three different places.

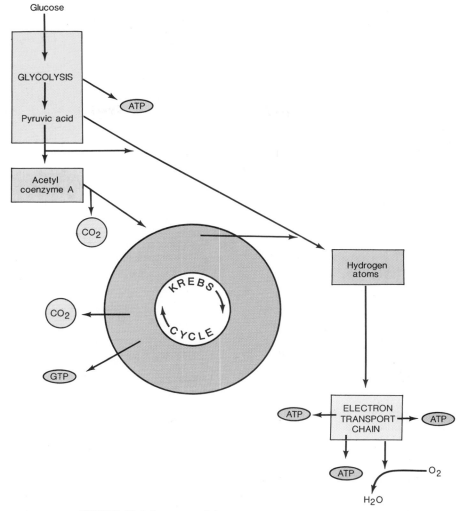

FIGURE 25-6 Summary of the complete oxidation of glucose.

About 43 percent of the energy originally in glucose is captured by ATP, the remainder is given off as heat.

Glycolysis, the Krebs cycle, and especially the electron transport chain provide all the ATP for cellular activities. And because the Krebs cycle and electron transport chain are aerobic processes, the cells cannot carry on their activities for long without sufficient oxygen.

GLUCOSE ANABOLISM

Most of the glucose in the body is catabolized to supply energy. However, some glucose participates in a number of anabolic reactions. One is the synthesis of glycogen from many glucose molecules. Another is the manufacture of glucose from the breakdown products of proteins and lipids.

Glucose Storage: Glycogenesis

If glucose is not needed immediately for energy, it is combined with many other molecules of glucose to form a long-chain molecule called glycogen. This process is called **glycogenesis** (*glyco* = sugar; *genesis* = origin). The

body can store about 500 g (about 1.1 lb) of glycogen in the liver and skeletal muscle cells. Roughly 80 percent of the glycogen is stored in the skeletal muscles.

In the process of glycogenesis (Figure 25-7), glucose that enters cells is first phosphorylated to glucose-6-phosphate. This is then converted to glucose-1-phosphate, then to uridine diphosphate glucose, and finally to glycogen. Glycogenesis is stimulated by insulin from the pancreas.

Glucose Release: Glycogenolysis

When the body needs energy, the glycogen stored in the liver is broken down into glucose and released into the bloodstream to be transported to cells, where it will be catabolized. The process of converting glycogen back to glucose is called **glycogenolysis** (*lysis* = breakdown). Glycogenolysis usually occurs between meals.

Glycogenolysis does not occur exactly by the reverse processes involved in glycogenesis (Figure 25-7). The first step in glycogenolysis involves the splitting off of glucose molecules from the branched glycogen molecule by phosphorylation to form glucose-1-phosphate. Phosphorylase, the enzyme that catalyzes this reaction, is activated by

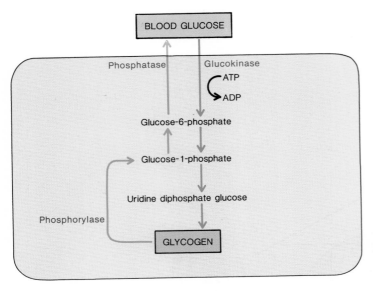

FIGURE 25-7 Glycogenesis and glycogenolysis. The glycogenesis pathway, the conversion of glucose into glycogen, is indicated by red arrows. The glycogenolysis pathway, the conversion of glycogen into glucose, is indicated by blue arrows. Some of the enzymes involved are indicated by parentheses.

the hormones glucagon from the pancreas and epinephrine from the adrenal medulla. Glucose-1-phosphate is then converted to glucose-6-phosphate and finally to glucose. Phosphatase, the enzyme that converts glucose-6-phosphate into glucose and releases glucose into the blood, is present in liver cells, but absent in skeletal muscle cells.

Formation of Glucose from Proteins and Fats: Gluconeogenesis

When your liver runs low on glycogen, it is time to eat. If you do not eat, your body starts catabolizing fats and proteins. Actually, the body normally catabolizes some of its fats and a few of its proteins. But large-scale fat and protein catabolism does not happen unless you are starving, eating meals that contain very few carbohydrates, or suffering from an endocrine disorder.

Both fat molecules and protein molecules may be converted in the liver to glucose. The process by which glucose is formed from noncarbohydrate sources is called **gluconeogenesis** (gloo'-kō-nē'-o-JEN-e-sis).

In the process of gluconeogenesis, moderate quantities of glucose can be formed from amino acids and the glycerol portion of fat molecules (Figure 25-8). About 60 percent of the amino acids in the body can undergo this conversion. Amino acids such as alanine, cysteine, glycine, serine, and threonine are converted to pyruvic acid. The pyruvic acid may be resynthesized into glucose or enter the Krebs cycle. Glycerol may be converted into glyceraldehyde-3-phosphate, which may also be resynthesized into glucose or form pyruvic acid. Figure 25-8 also shows how gluconeogenesis is related to other metabolic reactions.

Gluconeogenesis is stimulated by cortisol, one of the glucocorticoid hormones of the adrenal cortex, and thyroxine from the thyroid gland. Cortisol mobilizes proteins from body cells, making them available in the form of amino acids, thus supplying a pool of amino acids for gluconeogenesis. Thyroxine also mobilizes proteins and may mobilize fats from fat depots by making glycerol available for gluconeogenesis. Gluconeogenesis is also

stimulated by epinephrine, glucagon, and growth hormone (GH).

LIPID METABOLISM

Lipids are second to carbohydrates as a source of energy. More frequently they are used as building blocks to form essential structures. When neutral fats (triglycerides) are eaten, they are ultimately digested into fatty acids and monoglycerides. Short-chain fatty acids enter the blood capillaries in villi, while long-chain fatty acids and monoglycerides are carried in micelles to epithelial cells of the villi for entrance. Once inside, they are further digested to glycerol and fatty acids, recombined to form triglycerides, and transported as chylomicrons through the lacteals of villi into the thoracic duct, subclavian vein, and to the liver.

FATE OF LIPIDS

Lipids, such as fatty acids and glycerol, like carbohydrates, may be oxidized to produce ATP. Each gram of fat produces about 9.0 Cal. If the body has no immediate need to utilize fats this way, they are stored in adipose tissue (fat depots) throughout the body and in the liver. Other lipids are used as structural molecules or to synthesize other essential substances. For example, phospholipids are constituents of plasma membranes and substrates for the synthesis of prostaglandins, lipoproteins are used to transport cholesterol throughout the body, thromboplastin is needed for blood clotting, and myelin sheaths speed up nerve impulse conduction. Cholesterol, another lipid, is used in the synthesis of bile salts and steroid hormones (adrenocortical hormones and sex hormones). The various functions of lipids in the body may be reviewed in Exhibit 2-5.

FAT STORAGE

The major function of adipose tissue is to store fats until they are needed for energy in other parts of the body.

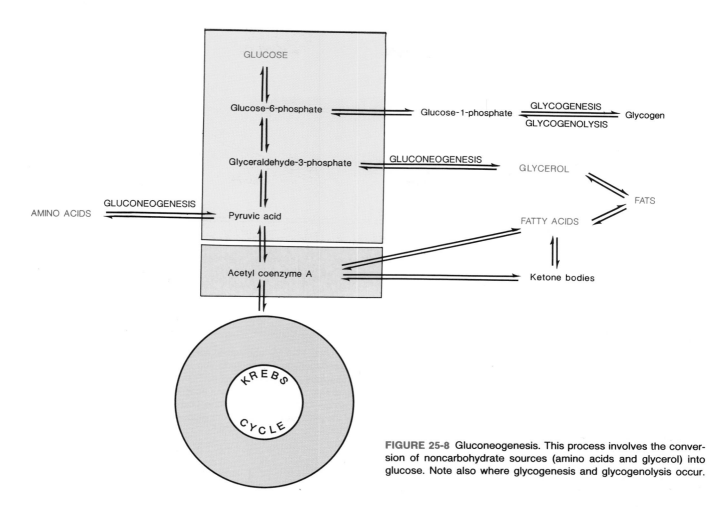

FIGURE 25-8 Gluconeogenesis. This process involves the conversion of noncarbohydrate sources (amino acids and glycerol) into glucose. Note also where glycogenesis and glycogenolysis occur.

It also insulates and protects. About 50 percent of stored fat is deposited in subcutaneous tissue—approximately 12 percent around the kidneys, 10 to 15 percent in the omenta, 20 percent in genital areas, and 5 to 8 percent between muscles. Fat is also stored behind the eyes, in the sulci of the heart, and in the folds of the large intestine.

Adipose cells contain lipases that catalyze the deposition of fats from chylomicrons and hydrolyze fats into fatty acids and glycerol. Because of the rapid exchange of fatty acids, fats are renewed about once every two to three weeks. Thus, the fat stored in your adipose tissue today is not the same fat that was present last month. Moreover, fat is continually released from storage, transported through the blood, and redeposited in other adipose tissue cells.

LIPID CATABOLISM

Fats stored in adipose tissue constitute the largest reserve of energy. The body can store much more fat than it can glycogen. Moreover, the energy yield of fats is more than twice that of carbohydrates. Nevertheless, fats are only the body's second-favorite source of energy because they are more difficult to catabolize than carbohydrates.

Glycerol

Before fat molecules can be metabolized as an energy source, they must first be released from fat depots and split into glycerol and fatty acids. Both these processes are stimulated by growth hormone (GH). The glycerol and fatty acids are then catabolized separately (Figure 25-9).

Glycerol is converted easily by many cells of the body to glyceraldehyde-3-phosphate, one of the compounds also formed during the catabolism of glucose. The cells then transform glyceraldehyde-3-phosphate into glucose

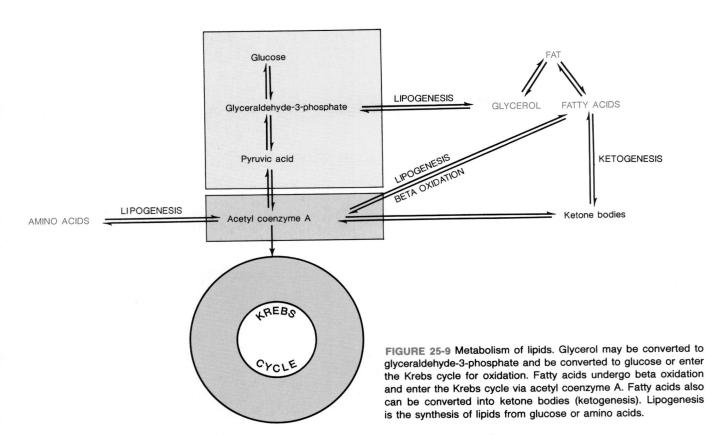

FIGURE 25-9 Metabolism of lipids. Glycerol may be converted to glyceraldehyde-3-phosphate and be converted to glucose or enter the Krebs cycle for oxidation. Fatty acids undergo beta oxidation and enter the Krebs cycle via acetyl coenzyme A. Fatty acids also can be converted into ketone bodies (ketogenesis). Lipogenesis is the synthesis of lipids from glucose or amino acids.

for catabolism. This is one example of gluconeogenesis. Since glyceraldehyde-3-phosphate is an intermediate product in the conversion of glycerol to glucose, glycerol thus enters the glycolytic pathway to be used for energy.

Fatty Acids

Fatty acids are catabolized differently and the process occurs in the matrix of mitochondria. The first step in fatty-acid catabolism involves a series of reactions called **beta oxidation.** Through a series of complex reactions involving dehydrogenation, hydration, and cleavage, enzymes remove pairs of carbon atoms at a time from the long chain of carbon atoms comprising a fatty acid. The result of beta oxidation is a series of two-carbon fragments, **acetyl coenzyme A (CoA).** Since the majority of fatty acids have even numbers of carbon atoms, the number of acetyl CoA molecules produced is easily calculated by dividing the number of carbon atoms in the fatty acid by two. Thus, palmitic acid, a fatty acid with 18 carbon atoms, will produce 9 acetyl CoA molecules upon beta oxidation.

In the second step of fatty acid catabolism, the acetyl CoA formed as a result of beta oxidation enters the Krebs cycle (Figure 25-9). In terms of energy yield, an 18-carbon fatty acid, such as palmitic acid, can yield a net of 129 ATPs upon its complete oxidation via the Krebs cycle and electron transport chain.

As part of normal fatty acid catabolism, the liver can take acetyl CoA molecules, two at a time, and condense them to form a substance called **acetoacetic acid** which is converted mostly into β-hydroxybutyric acid and partially into **acetone.** These substances are collectively known as **ketone bodies,** and their formation is called **ketogenesis** (Figure 25-9). They then leave the liver, enter the bloodstream, and diffuse into other body cells where they are broken down into acetyl CoA, which enters the Krebs cycle for oxidation. During periods of excessive beta oxidation, large amounts of acetyl CoA are produced. This might occur following a meal rich in fats or during fasting or starvation because essentially no carbohydrates are available for catabolism. It may also occur in diabetes mellitus because insulin is not available to stimulate glycogenesis and glucose transport into cells. When a diabetic becomes seriously insulin deficient, one of the telltale signs is a sweet smell of acetone on the breath.

> **CLINICAL APPLICATION**
> Since the body prefers glucose as a source of energy, ketone bodies are generally produced in very small quantities (1.5 to 2.0 mg/100 ml blood). When the number of ketone bodies in the blood rises above normal—a condition called **ketosis**—the ketone bodies, most of which are acids, must be buffered by the body. If too many accumulate, they use up the body's buffers and the blood pH falls. Thus extreme or prolonged ketosis can lead to **acidosis,** or abnormally low blood pH.

EXHIBIT 25-2

SUMMARY OF HORMONAL REGULATION OF THE METABOLISM OF CARBOHYDRATES, LIPIDS, AND PROTEINS

HORMONE	ACTIONS
Insulin	Increases glucose uptake by cells, especially muscle and adipose cells. Stimulates conversion of glucose into glycogen (glycogenesis). Stimulates fat synthesis and inhibits fat breakdown. Stimulates active transport of amino acids into cells, especially muscle cells. Stimulates protein synthesis.
Glucagon	Stimulates conversion of glycogen into glucose (glycogenolysis). Stimulates conversion of noncarbohydrates into glucose (gluconeogenesis). Stimulates breakdown of fats (fat mobilization).
Epinephrine	Stimulates conversion of glycogen into glucose (glycogenolysis). Stimulates conversion of noncarbohydrates into glucose (gluconeogenesis). Stimulates breakdown of fats (fat mobilization).
Growth hormone (GH)	Stimulates active transport of amino acids into body cells, especially muscle cells. Stimulates protein synthesis. Stimulates conversion of noncarbohydrates into glucose (gluconeogenesis). Stimulates breakdown of fats (fat mobilization).
Thyroxine	Stimulates protein synthesis. Stimulates conversion of noncarbohydrates into glucose (gluconeogenesis). Stimulates breakdown of fats (fat mobilization).
Cortisol	Stimulates conversion of noncarbohydrates into glucose (gluconeogenesis).
Testosterone	Increases protein deposition in body cells, especially muscle cells.

MINERALS

Minerals are inorganic substances. They may appear in combination with each other or in combination with organic compounds. Minerals constitute about 4 percent of the total body weight, and they are concentrated most heavily in the skeleton. Minerals known to perform functions essential to life include calcium, phosphorus, sodium, chlorine, potassium, magnesium, iron, sulfur, iodine, manganese, cobalt, copper, zinc, selenium, and chromium. Other minerals—aluminum, silicon, arsenic, nickel—are present in the body, but their functions have not yet been determined.

Calcium and phosphorus form part of the structure of bone. But since minerals do not form long-chain compounds, they are otherwise poor building materials. Their chief role is to help regulate body processes. Calcium, iron, magnesium, and manganese are constituents of some coenzymes. Magnesium also serves as a catalyst for the conversion of ADP to ATP. Without these minerals, metabolism would stop and the body would die. Minerals such as sodium and phorphorus work in buffer systems. Sodium helps regulate the osmosis of water and, along with other ions, is involved in the generation of nerve impulses. Exhibit 25-3 describes the functions of some minerals vital to the body. Note that the body generally uses the ions of the minerals rather than the nonionized form. Some minerals, such as chlorine, are toxic or even fatal to the body if ingested in the nonionized form.

VITAMINS

Organic nutrients required in minute amounts to maintain growth and normal metabolism are called **vitamins.** Un-like carbohydrates, fats, or proteins, vitamins do not provide energy or serve as building materials. The essential function of vitamins is the regulation of physiological processes. Of the vitamins whose functions are known, most serve as coenzymes.

Most vitamins cannot be synthesized by the body. They may be ingested in foods or pills. Other vitamins, such as vitamin K, are produced by bacteria in the gastrointestinal tract. The body can assemble some vitamins if the raw materials called *provitamins* are provided. Vitamin A is produced by the body from the provitamin carotene, a chemical present in spinach, carrots, liver, and milk. No single food contains all the required vitamins—one of the best reasons for eating a balanced diet.

CLINICAL APPLICATION

The term **avitaminosis** refers to a deficiency of any vitamin in the diet. **Hypervitaminosis** refers to an excess of one or more vitamins. An excess of vitamin A, called hypervitaminosis A, results in peeling and itching of the skin, hair loss, headache, and dizziness. An excess of vitamin D, called hypervitaminosis D, results in weakness, fatigue, weight loss, and nausea.

On the basis of solubility, vitamins are divided into two principal groups: fat-soluble and water-soluble. **Fat-soluble** vitamins are absorbed along with ingested dietary fats by the small intestine as micelles. In fact, they cannot be absorbed unless they are ingested with some fat. Fat-soluble vitamins are generally stored in cells, particularly liver cells, so reserves can be built up. Examples of fat-soluble vitamins are vitamins A, D, E, and K. **Water-**

soluble vitamins, by contrast, are absorbed along with water in the gastrointestinal tract and dissolve in the body fluids. Excess quantities of these vitamins are excreted in the urine. Thus the body does not store water-soluble vitamins well. Examples of water-soluble vitamins are the B vitamins and vitamin C.

Exhibit 25-4 lists the principal vitamins, their sources, functions, and related disorders.

METABOLISM AND BODY HEAT

We will now consider the relationship of foods to body heat, mechanisms of heat gain and loss, and the regulation of body temperature.

MEASURING HEAT

Heat is a form of energy that can be measured as **temperature** and expressed in units called calories. A **calorie (cal)** is the amount of heat energy required to raise the temperature of 1 g of water 1°C, from 14° to 15°C. Since the calorie is a small unit and large amounts of energy are stored in foods, the **Calorie** or **kilocalorie** is used instead. A Calorie is equal to 1,000 cal and is defined as the amount of heat required to raise the temperature of 1,000 g of water 1°C. The Calorie is the unit we use to express the heating value of foods and to measure the body's metabolic rate.

The apparatus used to determine the caloric value of foods is called a **calorimeter.** A weighed sample of a dehydrated food is burned completely in an insulated metal container. The energy released by the burning food is absorbed by the container and then transferred to a known volume of water that surrounds the container. The change in the water's temperature is directly related to the number of Calories released by the food. Knowing the caloric value of foods is important—if we know the amount of energy the body uses for various activities, we can adjust our food intake. In this way we can control body weight by taking in only enough Calories to sustain our activities.

PRODUCTION OF BODY HEAT

Most of the heat produced by the body comes from oxidation of the food we eat. The rate at which this heat is produced—the **metabolic rate**—is also measured in Calories. Among the factors that affect metabolic rate are the following:

EXHIBIT 25-3
MINERALS VITAL TO THE BODY

MINERAL	COMMENTS	IMPORTANCE
Calcium	Most abundant cation in body. Appears in combination with phosphorus in ratio of 2:1.5. About 99 percent is stored in bone and teeth. Remainder stored in muscle, other soft tissues, and blood plasma. Blood calcium level controlled by calcitonin (CT) and parathyroid hormone (PTH). Absorption occurs only in the presence of vitamin D. Most is excreted in feces and small amount in urine. Sources are milk, egg yolk, shellfish, green leafy vegetables.	Formation of bones and teeth, blood clotting, normal muscles and nerve activity, endocytosis and exocytosis, cellular motility, chromosome movement prior to cell division, glycogen metabolism, and synthesis and release of transmitter substances.
Phosphorus	About 80 percent found in bones and teeth. Remainder distributed in muscle, brain cells, blood. More functions than any other mineral. Blood phosphorus level controlled by calcitonin (CT) and parathyroid hormone (PTH). Most excreted in urine, small amount eliminated in feces. Sources are dairy products, meat, fish, poultry, nuts.	Formation of bones and teeth. Constitutes a major buffer system of blood. Plays important role in muscle contraction and nerve activity. Component of many enzymes. Involved in transfer and storage of energy (ATP). Component of DNA and RNA.
Iron	About 66 percent found in hemoglobin of blood. Remainder distributed in skeletal muscles, liver, spleen, enzymes. Normal losses of iron occur by shedding of hair, epithelial cells, and mucosal cells and in sweat, urine, feces, and bile. Sources are meat, liver, shellfish, egg yolk, beans, legumes, dried fruits, nuts, cereals.	As component of hemoglobin, carries O_2 to body cells. Component of cytochromes involved in formation of ATP from catabolism.
Iodine	Essential component of thyroid hormones. Excreted in urine. Sources are seafood, cod-liver oil, and vegetables grown in iodine-rich soils or iodized salt.	Required by thyroid gland to synthesize thyroid hormones, hormones that regulate metabolic rate.
Copper	Some stored in liver and spleen. Most excreted in feces. Sources include eggs, whole-wheat flour, beans, beets, liver, fish, spinach, asparagus.	Required with iron for synthesis of hemoglobin. Component of enzyme necessary for melanin pigment formation.
Sodium	Most found in extracellular fluids, some in bones. Excreted in urine and perspiration. Normal intake of NaCl (table salt) supplies required amounts.	As most abundant cation in extracellular fluid, strongly affects distribution of water through osmosis. Part of bicarbonate buffer system. Functions in nerve impulse conduction.

1. **Exercise.** During strenuous exercise, the metabolic rate may increase to as much as 50 times the normal rate.

2. **Nervous system.** In a stress situation, the sympathetic nervous system is stimulated and the nerves release norepinephrine (NE), which increases the metabolic rate of body cells. Strong sympathetic stimulation may increase the metabolic rate by 160 times, but only for a few minutes.

3. **Hormones.** In addition to norepinephrine (NE), two other hormones affect metabolic rate: epinephrine produced by the adrenal glands and thyroid hormones produced by the thyroid gland. Epinephrine is secreted in stress situations. Increased secretions of thyroid hormones increase the metabolic rate.

4. **Body temperature.** The higher the body temperature, the higher the metabolic rate. Each 1°C rise in temperature increases the rate of biochemical reactions by about 10 percent. The metabolic rate may be substantially increased during fever.

BASAL METABOLIC RATE

Since many factors affect metabolic rate, it is measured under standard conditions designed to reduce or eliminate those factors as much as possible. These conditions of the body are called the **basal state** and the measurement obtained is the **basal metabolic rate (BMR).** The person should not exercise for 30 to 60 minutes before the measurement is taken. The individual must be completely at rest, but awake. Air temperature should be comfortable. The person should fast for at least 12 hours. Body temperature should be normal. Basal metabolic rate is a measure of the rate at which the body breaks down foods (and therefore releases heat). It is expressed in Calories per square meter of body surface area per hour. BMR is also a measure of how much thyroxine the thyroid gland is producing, since thyroxine regulates the rate of food breakdown and is not a controllable factor under basal conditions.

The basal metabolic rate is most often measured indirectly by measuring oxygen consumption. If a given amount of food releases a given amount of heat energy when it is oxidized, it must combine with a given amount of oxygen. Thus, by measuring the amount of oxygen needed for the metabolism of foods, we can determine how many Calories are produced. The amount of heat energy released when 1 liter of oxygen combines with carbohydrates is 5.05 Calories; with fats, the heat released is 4.70 Calories; with proteins, the heat released is 4.60 Calories. The average of the three values is 4.825 Calories. Therefore, every time a liter of oxygen is consumed, 4.825 Calories are produced. For the normal adult, the average BMR is about 70 Calories per hour.

MINERAL	COMMENTS	IMPORTANCE
Potassium	Principal cation in intracellular fluid. Most is excreted in urine. Normal food intake supplies required amounts.	Functions in transmission of nerve impulses and muscle contraction.
Chlorine	Found in extracellular and intracellular fluids. Principal anion of extracellular fluid. Most excreted in urine. Normal intake of NaCl supplies required amounts.	Assumes role in acid–base balance of blood, water balance, and formation of HCl in stomach.
Magnesium	Component of soft tissues and bone. Excreted in urine and feces. Widespread in various foods.	Required for normal functioning of muscle and nervous tissue. Participates in bone formation. Constituent of many coenzymes.
Sulfur	Constituent of many proteins (such as insulin) and some vitamins (thiamine and biotin). Excreted in urine. Sources include beef, liver, lamb, fish, poultry, eggs, cheese, beans.	As component of hormones and vitamins, regulates various body activities.
Zinc	Important component of certain enzymes. Widespread in many foods, especially meats.	As a component of carbonic anhydrase, important in carbon dioxide metabolism. Necessary for normal growth, proper functioning of prostate gland, normal taste sensations and appetite, and normal sperm counts in males. As a component of peptidases, it is involved in protein digestion.
Fluorine	Component of bones, teeth, other tissues.	Appears to improve tooth structure and inhibit cariogenesis.
Manganese	Some stored in liver and spleen. Most excreted in feces.	Activates several enzymes. Needed for hemoglobin synthesis, urea formation, growth, reproduction, and lactation.
Cobalt	Constituent of vitamin B_{12}.	As part of B_{12}, required for maturation of erythropoeisis.
Chromium	Found in high concentrations in brewer's yeast. Also found in wine and some brands of beer.	Necessary for the proper utilization of dietary sugars and other carbohydrates by optimizing the production and effects of insulin. Helps increase blood levels of HDL, while decreasing levels of LDL.
Selenium		An antioxidant. Prevents chromosome breakage and may assume a role in preventing certain birth defects.

EXHIBIT 25-4
THE PRINCIPAL VITAMINS

VITAMIN	COMMENT AND SOURCE	FUNCTION	DEFICIENCY SYMPTOMS AND DISORDERS
FAT-SOLUBLE			
A	Formed from provitamin carotene (and other provitamins) in intestinal tract. Requires bile salts and fat for absorption. Stored in liver. Sources of carotene and other provitamins include yellow and green vegetables; sources of vitamin A include fish-liver oils, milk, butter.	Maintains general health and vigor of epithelial cells.	Deficiency results in atrophy and keratinization of epithelium, leading to dry skin and hair, increased incidence of ear, sinus, respiratory, urinary, and digestive infections, inability to gain weight, drying of cornea with ulceration (*xerophthalmia*), nervous disorders, and skin sores.
		Essential for formation of rhodopsin, light-sensitive chemical in rods of retina.	*Night blindness* or decreased ability for dark adaptation.
		Growth of bones and teeth by apparently helping to regulate activity of osteoblasts and osteoclasts.	Slow and faulty development of bones and teeth.
D	In presence of sunlight, provitamin D_3 (derivative of cholesterol) converted to vitamin D in the skin, liver, and kidneys. Dietary vitamin D requires moderate amounts of bile salts and fat for absorption. Stored in tissues to slight extent. Most excreted via bile. Sources include fish-liver oils, egg yolk, fortified milk.	Essential for absorption and utilization of calcium and phosphorus from gastrointestinal tract. May work with parathyroid hormone (PTH) that controls calcium metabolism.	Defective utilization of calcium by bones leads to *rickets* in children and *osteomalacia* in adults. Possible loss of muscle tone.
E (tocopherols)	Stored in liver, adipose tissue, and muscles. Requires bile salts and fat for absorption. Sources include fresh nuts and wheat germ, seed oils, green leafy vegetables.	Believed to inhibit catabolism of certain fatty acids that help form cell structures, especially membranes. Involved in formation of DNA, RNA, and red blood cells. May promote wound healing, contribute to the normal structure and functioning of the nervous system, prevent scarring and reduce the severity of visual loss associated with retrolental fibroplasia (an eye disease in premature infants caused by too much oxygen in incubators) by functioning as an antioxidant. Believed to help protect liver from toxic chemicals like carbon tetrachloride.	May cause the oxidation of unsaturated fats resulting in abnormal structure and function of mitochondria, lysosomes, and plasma membranes, a possible consequence being hemolytic anemia. Deficiency also causes muscular dystrophy in monkeys and sterility in rats.
K	Produced in considerable quantities by intestinal bacteria. Requires bile salts and fat for absorption. Stored in liver and spleen. Other sources include spinach, cauliflower, cabbage, liver.	Coenzyme believed essential for synthesis of prothrombin by liver and several clotting factors. Also known as antihemorrhagic vitamin.	Delayed clotting time results in excessive bleeding.

LOSS OF BODY HEAT

Body heat is produced by the oxidation of foods we eat. This heat must be removed continuously or body temperature would rise steadily. The principal routes of heat loss include radiation, conduction, convection, and evaporation.

Radiation

Radiation is the transfer of heat as infrared heat rays from one object to another without physical contact. Your body loses heat by the radiation of heat waves to cooler objects nearby such as ceilings, floors, and walls. If these objects are at a higher temperature, you absorb heat by

EXHIBIT 25-4 Continued

VITAMIN	COMMENT AND SOURCE	FUNCTION	DEFICIENCY SYMPTOMS AND DISORDERS
WATER-SOLUBLE			
B₁ (thiamine)	Rapidly destroyed by heat. Not stored in body. Excessive intake eliminated in urine. Sources include whole-grain products, eggs, pork, nuts, liver, yeast.	Acts as coenzyme for many different enzymes involved in carbohydrate metabolism of pyruvic acid to CO_2 and H_2O. Essential for synthesis of acetylcholine.	Improper carbohydrate metabolism leads to buildup of pyruvic and lactic acids and insufficient energy for muscle and nerve cells. Deficiency leads to two syndromes: (1) *beriberi*—partial paralysis of smooth muscle of GI tract causing digestive disturbances, skeletal muscle paralysis, atrophy of limbs; (2) *polyneuritis*—due to degeneration of myelin sheaths reflexes related to kinesthesia are impaired, impairment of sense of touch, decreased intestinal motility, stunted growth in children, and poor appetite.
B₂ (riboflavin)	Not stored in large amounts in tissues. Most is excreted in urine. Small amounts supplied by bacteria of GI tract. Other sources include yeast, liver, beef, veal, lamb, eggs, whole-grain products, asparagus, peas, beets, peanuts.	Component of certain coenzymes (e.g., FAD) concerned with carbohydrate and protein metabolism, especially in cells of eye, integument, mucosa of intestine, blood.	Deficiency may lead to improper utilization of oxygen resulting in blurred vision, cataracts, and corneal ulcerations. Also dermatitis and cracking of skin, lesions of intestinal mucosa, and development of one type of anemia.
Niacin (nicotinamide)	Derived from amino acid tryptophan. Sources include yeast, meats, liver, fish, whole-grain products, peas, beans, nuts.	Essential component of coenzyme (NAD) concerned with energy-releasing reactions. In lipid metabolism, inhibits production of cholesterol and assists in fat breakdown.	Principal deficiency is *pellagra*, characterized by dermatitis, diarrhea, and psychological disturbances.
B₆ (pyridoxine)	Formed by bacteria of GI tract. Stored in liver, muscle, brain. Other sources include salmon, yeast, tomatoes, yellow corn, spinach, whole-grain products, liver, yogurt.	May function as coenzyme in fat metabolism. Essential coenzyme for normal amino acid metabolism. Assists production of circulating antibodies.	Most common deficiency symptom is dermatitis of eyes, nose, and mouth. Other symptoms are retarded growth and nausea.
B₁₂ (cyanocobalamin)	Only B vitamin not found in vegetables; only vitamin containing cobalt. Absorption from GI tract dependent on HCl and intrinsic factor secreted by gastric mucosa. Sources include liver, kidney, milk, eggs, cheese, meat.	Coenzyme necessary for red blood cell formation, formation of amino acid methionine, entrance of some amino acids into Krebs cycle, and manufacture of choline (chemical similar in function to acetylcholine).	Pernicious anemia and malfunction of nervous system due to degeneration of axons of spinal cord.
Pantothenic acid	Stored primarily in liver and kidneys. Some produced by bacteria of GI tract. Other sources include kidney, liver, yeast, green vegetables, cereal.	Constituent of coenzyme A essential for transfer of pyruvic acid into Krebs cycle, conversion of lipids and amino acids into glucose, and synthesis of cholesterol and steroid hormones.	Experimental deficiency tests indicate fatigue, muscle spasms, neuromuscular degeneration, insufficient production of adrenal steroid hormones.
Folic acid	Synthesized by bacteria of GI tract. Other sources include green leafy vegetables and liver.	Component of enzyme systems synthesizing purines and pyrimidines built into DNA and RNA. Essential for normal production of red and white blood cells.	Production of abnormally large red blood cells (macrocytic anemia).
Biotin	Synthesized by bacteria of GI tract. Other sources include yeast, liver, egg yolk, kidneys.	Essential coenzyme for conversion of pyruvic acid to oxaloacetic acid and synthesis of fatty acids and purines.	Mental depression, muscular pain, dermatitis, fatigue, nausea.
C (ascorbic acid)	Rapidly destroyed by heat. Some stored in glandular tissue and plasma. Sources include citrus fruits, tomatoes, green vegetables.	Exact role not understood. Promotes many metabolic reactions, particularly protein metabolism, including laying down of collagen in formation of connective tissue. As coenzyme may combine with poisons, rendering them harmless until excreted. Works with antibodies. Promotes wound healing.	Scurvy; anemia; many symptoms related to poor connective tissue growth and repair including tender swollen gums, loosening of teeth (alveolar processes also deteriorate), poor wound healing, bleeding (vessel walls fragile because of connective tissue degeneration), and retardation of growth.

radiation. Incidentally, the air temperature has no relationship to the radiation of heat to and from objects. Skiers can remove their shirts in bright sunshine, even though the air temperature is very low, because the radiant heat from the sun is adequate to warm them. In a room at 21°C (70°F), about 60 percent of heat loss is by radiation.

Conduction

Another method of heat transfer is **conduction.** In this process, body heat is transferred to a substance or object in contact with the body, such as chairs, clothing, or jewelry. About 3 percent of body heat is lost via conduction.

Convection

Convection is the transfer of heat by the movement of a fluid between areas of different temperature. When cool air makes contact with the body, it becomes warmed and is carried away by convection currents. Then more cool air makes contact with the body and is carried away. The faster the air moves, the faster the rate of conduction. About 15 percent of body heat is lost to the air by convection.

Evaporation

Evaporation is the conversion of a liquid to a vapor. Water has a high heat of evaporation. The *latent heat of evaporation* is the amount of heat necessary to evaporate 1 g of water at 30°C (86°F). Because of water's high latent heat of evaporation, every gram of water evaporating from

EXHIBIT 25-5
SUMMARY OF HEAT-PRODUCING AND HEAT-LOSING MECHANISMS

MECHANISM	COMMMENT
Heat Production	Most body heat is produced by the oxidation of foods, and the rate at which it is produced is called the metabolic rate. The rate is affected by exercise, strong sympathetic stimulation, hormones, and body temperature.
Heat Loss Radiation	Transfer of heat from the body to an object without physical contact. Example is losing heat to a cool object such as a floor.
Conduction	Transfer of heat from the body to any object in physical contact with the body such as clothing.
Convection	When cool air makes contact with the body, it is warmed and carried away by convection currents.
Evaporation	The conversion of a liquid to a vapor in which the evaporating substance (e.g., perspiration) removes heat from the body.

the skin takes with it a great deal of heat—about 0.58 Calorie per gram of water. Under normal conditions, about 22 percent of heat loss occurs through evaporation. The evaporation of only 150 ml of water per hour is enough to remove all the heat produced by the body under basal conditions. Under extreme conditions, about 4 liters (1 gal) of perspiration is produced each hour and this volume can remove 2,000 Calories of heat from the body. This is approximately 32 times the basal level of heat production. The rate of evaporation is inversely related to relative humidity, the ratio of the actual amount of moisture in the air to the greatest amount it can hold at a given temperature. The higher the relative humidity, the lower the rate of evaporation.

A summary of heat-producing and heat-losing mechanisms is presented in Exhibit 25-5.

BODY TEMPERATURE REGULATION

If the amount of heat production equals the amount of heat loss, you maintain a constant body temperature near 37°C (98.6°F). If your heat-producing mechanisms generate more heat than is lost by your heat-losing mechanisms, your body temperature rises. If your heat-losing mechanisms give off more heat than is generated by heat-producing mechanisms, your temperature falls.

Hypothalamic Thermostat

Body temperature is regulated by mechanisms that attempt to keep heat production and heat loss in balance. A center of control of these mechanisms is found in the hypothalamus in a group of neurons in the anterior portion referred to as the *preoptic area.* If blood temperature increases, the neurons of the preoptic area fire impulses more rapidly. If something causes the blood's temperature to decrease, these neurons fire impulses more slowly. The preoptic area is adjusted to maintain normal body temperature and thus serves as your thermostat.

Impulses from the preoptic area are sent to other portions of the hypothalamus known as the heat-losing center and the heat-promoting center. The *heat-losing center,* when stimulated by the preoptic area, sets into operation a series of responses that lower body temperature. The *heat-promoting center,* when stimulated by the preoptic area, sets into operation a series of responses that raise body temperature. The heat-losing center is mainly parasympathetic in function; the heat-promoting center is primarily sympathetic.

Mechanisms of Heat Production

Suppose the environmental temperature is low or blood temperature falls below normal. Both stresses stimulate the preoptic area. The preoptic area, in turn, activates the heat-promoting center. In response, the heat-promoting center discharges impulses that automatically set into operation a number of responses designed to increase body heat and bring body temperature back up to normal.

● *Vasoconstriction* Impulses from the heat-promoting center stimulate sympathetic nerves that cause blood ves-

STUDENT OBJECTIVES

- Identify the external and internal gross anatomical features of the kidneys.
- Define the structural adaptations of a nephron for urine formation.
- Describe the blood and nerve supply to the kidneys.
- Describe the structure and function of the juxtaglomerular apparatus.
- Discuss the process of urine formation.
- Define glomerular filtration, tubular reabsorption, and tubular secretion.
- Compare the chemical composition of plasma, glomerular filtrate, and urine.
- Define the forces that support and oppose the filtration of blood in the kidneys.
- Explain the mechanism and importance of tubular reabsorption.
- Compare the obligatory and facultative reabsorption of water.
- Explain the mechanism and importance of tubular secretion.
- Explain the role of the countercurrent multiplier mechanism in producing dilute and concentrated urine.

- Compare the effects of blood pressure, blood concentration, temperature, diuretics, and emotions on urine volume.
- List and define the physical characteristics of urine.
- List the normal chemical constituents of urine.
- Define albuminuria, glycosuria, hematuria, pyuria, ketosis, bilirubinuria, urobilinogenuria, casts, and renal calculi.
- Discuss the structure and physiology of the ureters.
- Describe the structure and physiology of the urinary bladder.
- Explain the physiology of the micturition reflex.
- Compare the causes of incontinence, retention, and suppression.
- Explain the structure and physiology of the urethra.
- Discuss the causes of gout, glomerulonephritis, pyelitis, pyelonephritis, cystitis, nephrosis, and polycystic disease.
- Discuss the operational principle of hemodialysis.
- Define medical terminology associated with the urinary system.
- Explain the actions of selected drugs on the urinary system.

26 The Urinary System

The metabolism of nutrients results in the production of wastes by body cells, including carbon dioxide and excess water and heat. Protein catabolism produces toxic nitrogenous wastes such as ammonia and urea. In addition, many of the essential ions such as sodium, chloride, sulfate, phosphate, and hydrogen tend to accumulate in excess of the body's needs. All the toxic materials and the excess essential materials must be eliminated.

The primary function of the **urinary system** is to help keep the body in homeostasis by controlling the composition and volume of blood. It does so by removing and restoring selected amounts of water and solutes. Two kidneys, two ureters, one urinary bladder, and a single urethra make up the system (Figure 26-1). The kidneys regulate the composition and volume of the blood and remove wastes from the blood in the form of urine. They excrete selected amounts of various wastes, assume a role in erythropoiesis by forming renal erythropoietic factor, help to control blood pH, help to regulate blood pressure by secreting renin that activates the renin-angiotensin pathway, and participate in the activation of vitamin D. Urine is excreted from each kidney through its ureter and is stored in the urinary bladder until it is expelled from the body through the urethra. Other systems that aid in waste elimination are the respiratory, integumentary, and digestive systems (see Exhibit 26-3).

KIDNEYS

The paired **kidneys** are reddish organs that resemble kidney beans in shape. They are found just above the waist between the parietal peritoneum and the posterior wall of the abdomen. Since they are external to the peritoneal lining of the abdominal cavity, their placement is described as *retroperitoneal* (re'-trō-per-i-tō-NĒ-al). Other retroperitoneal structures include the ureters and suprarenal glands. Relative to the vertebral column, the kidneys are located between the levels of the last thoracic and third lumbar vertebrae and are partially protected by the eleventh and twelfth pairs of ribs. The right kidney is slightly lower than the left because of the large area occupied by the liver.

EXTERNAL ANATOMY

The average adult kidney measures about 10 to 12 cm (4 to 5 inches) long, 5.0 to 7.5 cm (2 to 3 inches) wide, and 2.5 cm (1 inch) thick. Its concave medial border faces the vertebral column. Near the center of the concave border is a notch called the **hilus,** through which the ureter leaves the kidney. Blood and lymph vessels and nerves also enter and exit the kidney through the hilus (Figure 26-2). The hilus is the entrance to a cavity in the kidney called the **renal sinus.**

Three layers of tissue surround each kidney. The innermost layer, the **renal capsule,** is a smooth, transparent, fibrous membrane that can easily be stripped off the kidney and is continuous with the outer coat of the ureter at the hilus. It serves as a barrier against trauma and the spread of infection to the kidney. The second layer, the **adipose capsule,** is a mass of fatty tissue surrounding the renal capsule. It also protects the kidney from trauma and holds it firmly in place in the abdominal cavity. The outermost layer, the **renal fascia,** is a thin layer of fibrous connective tissue that anchors the kidney to its surrounding structures and to the abdominal wall.

CLINICAL APPLICATION
Floating kidney or **ptosis** (TŌ-sis) occurs when the kidney is no longer held in place securely by the adjacent organs or its covering of fat and slips from its normal position. Individuals, especially thin people, in whom either the adipose capsule or renal fascia is deficient may develop ptosis. It is dangerous because it may cause kinking of the ureter with reflux of urine and retrograde pressure. Pain occurs if the ureter is twisted. Also, if the kidneys drop below the rib cage, they become susceptible to blows and penetrating injuries.

INTERNAL ANATOMY

A coronal (frontal) section through a kidney reveals an outer, reddish area called the **cortex** and an inner, reddish-brown region called the **medulla** (Figure 26-2b, c). Within the medulla are 8 to 18 striated, triangular structures termed **renal (medullary) pyramids.** The striated appearance is due to the presence of straight tubules and blood vessels. The bases of the pyramids face the cortical area, and their apices, called **renal papillae,** are directed toward the center of the kidney. The cortex is the smooth-textured area extending from the renal capsule to the bases of the pyramids and into the spaces between them. It is thus divided into an outer cortical zone and an inner

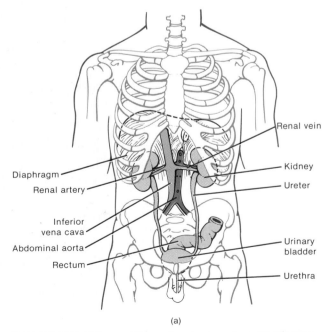

(a)

FIGURE 26-1 Organs of the male urinary system. (a) Diagram.

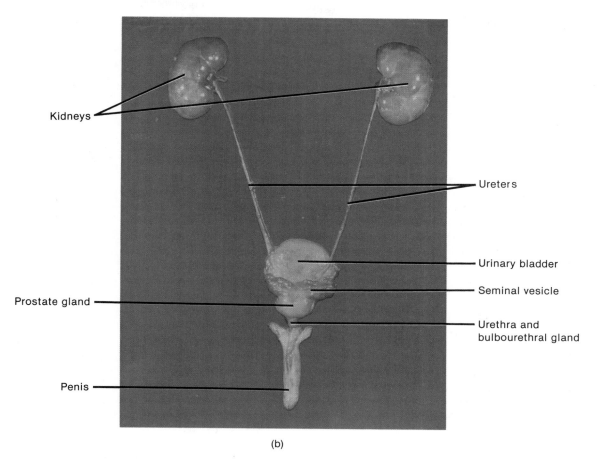

(b)

FIGURE 26-1 (*Continued*) Organs of the male urinary system. (b) Photograph. (Courtesy of C. Yokochi and J. W. Rohen, *Photographic Anatomy of the Human Body,* 2nd ed., 1979, IGAKU-SHOIN, Ltd., Tokyo, New York.)

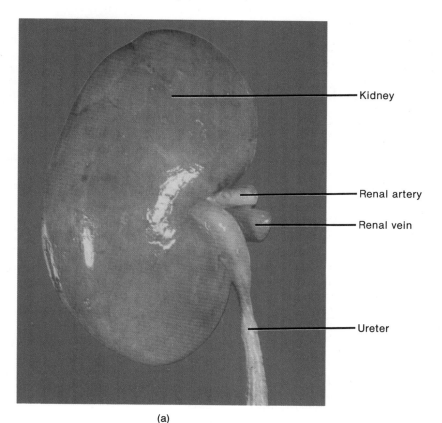

(a)

FIGURE 26-2 Kidney. (a) Photograph of the external view of the right kidney. (Courtesy of C. Yokochi and J. W. Rohen, *Photographic Anatomy of the Human Body,* 2nd ed., 1979, IGAKU-SHOIN, Ltd., Tokyo, New York.)

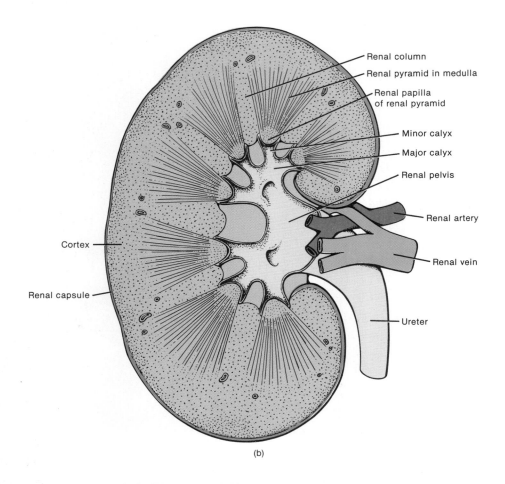

Renal column
Renal pyramid in medulla
Renal papilla
of renal pyramid
Minor calyx
Major calyx
Renal pelvis
Renal artery
Renal vein
Ureter
Cortex
Renal capsule

(b)

Renal column
Cortex
Minor calyx
Renal capsule

Renal pyramid in medulla
Renal papilla of renal pyramid
Renal artery
Renal vein
Renal pelvis
Ureter

(c)

FIGURE 26-2 (*Continued*) Kidney. (b) Diagram of a coronal section of the right kidney illustrating the internal anatomy. (c) Photograph of a coronal and a cross section of the right kidney illustrating the internal anatomy. (Courtesy of C. Yokochi and J. W. Rohen, *Photographic Anatomy of the Human Body,* 2nd ed., 1979, IGAKU-SHOIN, Ltd., Tokyo, New York.)

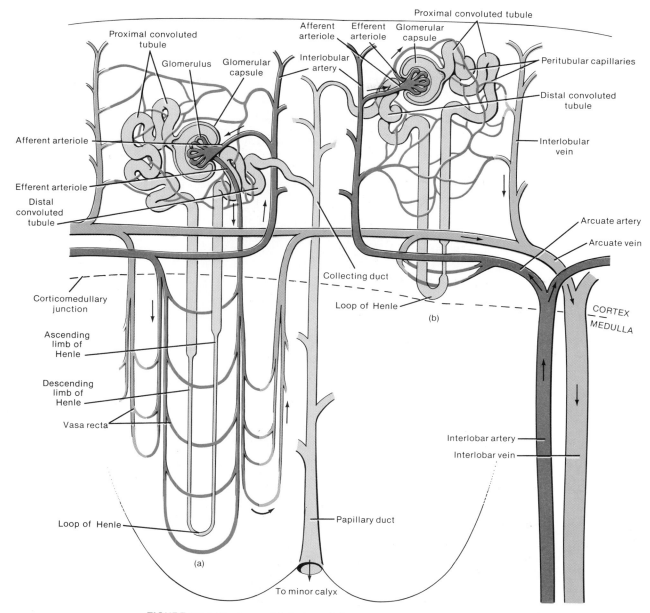

FIGURE 26-3 Nephrons. (a) Juxtamedullary nephron. (b) Cortical nephron.

juxtamedullary zone. The cortical substance between the renal pyramids forms the **renal columns.**

Together the cortex and renal pyramids constitute the parenchyma of the kidney. Structurally, the parenchyma of each kidney consists of approximately 1 million microscopic units called nephrons, collecting ducts, and their associated vascular supply. Nephrons are the functional units of the kidney. They help regulate blood composition and form urine.

In the renal sinus of the kidney is a large cavity called the **renal pelvis.** The edge of the pelvis contains cuplike extensions called **major** and **minor calyces** (KĀ-li-sēz).

There are 2 or 3 major calyces and 8 to 18 minor calyces. Each minor calyx collects urine from collecting ducts of the pyramids. From the major calyces, the urine drains into the pelvis and out through the ureter.

NEPHRON

The functional unit of the kidney is the **nephron** (NEF-ron) (Figure 26-3). Essentially, a nephron is a **renal tubule** and its vascular component. It begins as a double-walled cup, called the **glomerular (Bowman's) capsule,** lying in the cortex of the kidney. The outer wall, or *parietal layer,*

FIGURE 26-4 Endothelial-capsular membrane. (a) Parts of a renal corpuscle. (b) Enlarged aspect of a portion of the endothelial-capsular membrane.

Capsular space   Parietal layer of glomerular capsule

Podocyte of visceral layer of glomerular capsule

Podocytes of visceral layer of glomerular capsule   (c)

Pedicels   (d)   Filtration slits

FIGURE 26-4 (*Continued*) Endothelial-capsular membrane. (c) Scanning electron micrograph of a renal corpuscle at a magnification of 1,230×. (Courtesy of Richard K. Kessel and Randy H. Kardon, *Tissues and Organs: A Text-Atlas of Scanning Electron Microscopy.* Copyright © 1979 by Scientific American, Inc.) (d) Scanning electron micrograph of a podocyte at a magnification of 7,800×. (Courtesy of Richard K. Kessel and Randy H. Kardon, *Tissues and Organs: A Text-Atlas of Scanning Electron Microscopy.* Copyright © 1979 by Scientific American, Inc.)

is composed of simple squamous epithelium (Figure 26-4). It is separated from the inner wall, known as the *visceral layer,* by the *capsular space.* The visceral layer consists of epithelial cells called podocytes. It surrounds a capillary network called the **glomerulus** (glō-MER-yoo-lus). Collectively, the glomerular capsule and its enclosed glomerulus constitute a **renal corpuscle** (KŌR-pus-sul).

The visceral layer of the glomerular capsule and the endothelium of the glomerulus form an **endothelial-capsular membrane.** This membrane consists of the following parts in the order in which substances filtered by the kidney must pass through.

1. **Endothelium of the glomerulus.** This single layer of endothelial cells has completely open pores (fenestrated) averaging 500 to 1,000 Å in diameter.
2. **Basement membrane of the glomerulus.** This extracellular membrane lies beneath the endothelium and contains no pores. It consists of fibrils in a glycoprotein matrix. It serves as the dializing membrane.
3. **Epithelium of the visceral layer of the glomerular capsule.** These epithelial cells, because of their peculiar shape, are called **podocytes.** The podocytes contain footlike structures called **pedicels** (PED-i-sels). The pedicels are arranged parallel to the circumference of the glomerulus and cover the basement membrane except for spaces between them called **filtration slits (slit pores).**

The endothelial-capsular membrane filters water and solutes in the blood. Large molecules, such as proteins, and the formed elements in blood do not normally pass

through it. The water and solutes that are filtered out of the blood pass into the capsular space between the visceral and parietal layers of the glomerular capsule and then into the renal tubule.

The glomerular capsule opens into the first section of the renal tubule, called the **proximal convoluted tubule,** which also lies in the cortex. Convoluted means the tubule is coiled rather than straight; proximal signifies that the glomerular capsule is the origin of the tubule. The wall of the proximal convoluted tubule consists of cuboidal epithelium with microvilli. These cytoplasmic extensions, like those of the small intestine, increase the surface area for reabsorption and secretion.

Nephrons are frequently classified into two kinds. A **cortical nephron** usually has its glomerulus in the outer cortical zone, and the remainder of the nephron rarely penetrates the medulla. A **juxtamedullary nephron** usually has its glomerulus close to the corticomedullary junction, and other parts of the nephron penetrate deeply into the medulla (Figure 26-3).

In the juxtamedullary nephron, the renal tubule straightens, becomes thinner, and dips into the medulla, where it is called the **descending limb of Henle** (HEN-lē). This section consists of squamous epithelium. The tubule then increases in diameter as it bends into a U-shaped structure called the **loop of Henle.** It then ascends toward the cortex as the **ascending limb of Henle,** which consists of cuboidal and low columnar epithelium.

Renal corpuscle

Capsular space

Macula densa

Glomerulus

Distal convoluted tubule

Glomerular capsule

(a)

FIGURE 26-5 Histology of a nephron. (a) Photomicrograph of the cortex of the kidney showing a renal corpuscle and surrounding renal tubules at a magnification of 400×. (© 1983 by Michael H. Ross. Used by permission.) (b) Photomicrograph of the cortex of the kidney showing renal tubules at a magnification of 400×. (© 1983 by Michael H. Ross. Used by permission.) (c) Scanning electron micrograph of several renal tubules at a magnification of 500×. (Courtesy of Fisher Scientific Company and S.T.E.M. Laboratories, Inc., Copyright, 1975.)

Distal convoluted tubules

Proximal convoluted tubules

(b)

(c)

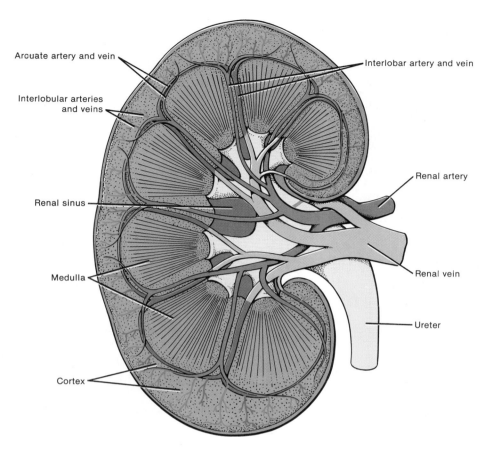

FIGURE 26-6 Blood supply of the right kidney seen in coronal section.

In the cortex, the tubule again becomes convoluted. Because of its distance from the point of origin at the glomerular capsule, this section is referred to as the **distal convoluted tubule.** The cells of the distal tubule, like those of the proximal tubule, are cuboidal. Unlike the cells of the proximal tubule, however, the cells of the distal tubule have few microvilli. In a cortical nephron, the proximal section runs into the distal tubule without the intervening limb of Henle. In both types of nephron, the distal tubule terminates by merging with a straight **collecting duct.**

In the medulla, the collecting ducts receive the distal tubules of several nephrons, pass through the renal pyramids, and open at the renal papillae into the minor calyces through a number of large **papillary ducts.** On the average, there are 30 papillary ducts per renal papilla. Cells of the collecting ducts are cuboidal; those of the papillary ducts are columnar.

The histology of a nephron and glomerulus is shown in Figure 26-5.

BLOOD AND NERVE SUPPLY

The nephrons are largely responsible for removing wastes from the blood and regulating its fluid and electrolyte content. Thus they are abundantly supplied with blood vessels. The right and left **renal arteries** transport about one-fourth the total cardiac output to the kidneys (Figure 26-6). Approximately 1,200 ml passes through the kidneys every minute.

Before or immediately after entering through the hilus, the renal artery divides into several branches that enter the parenchyma and pass as the **interlobar arteries** between the renal pyramids in the renal columns. At the base of the pyramids, the interlobar arteries arch between the medulla and cortex; here they are known as the **arcuate arteries.** Divisions of the arcuate arteries produce a series of **interlobular arteries,** which enter the cortex and divide into **afferent arterioles** (see Figure 26-3).

One afferent arteriole is distributed to each glomerular capsule, where the arteriole divides into the tangled capillary network termed the **glomerulus.** The glomerular capillaries then reunite to form an **efferent arteriole,** which leads away from the capsule and is smaller in diameter than the afferent arteriole. This variation in diameter helps raise the glomerular pressure. The afferent-efferent arteriole situation is unique because blood usually flows out of capillaries into venules and not into other arterioles.

Each efferent arteriole of a cortical nephron divides to form a network of capillaries, called the **peritubular capillaries,** around the convoluted tubules. The efferent arteriole of a juxtamedullary nephron also forms peritubular capillaries. In addition, it forms long loops of thin-walled vessels called **vasa recta** that dip down alongside the loop of Henle into the medullary region of the papilla.

The peritubular capillaries eventually reunite to form

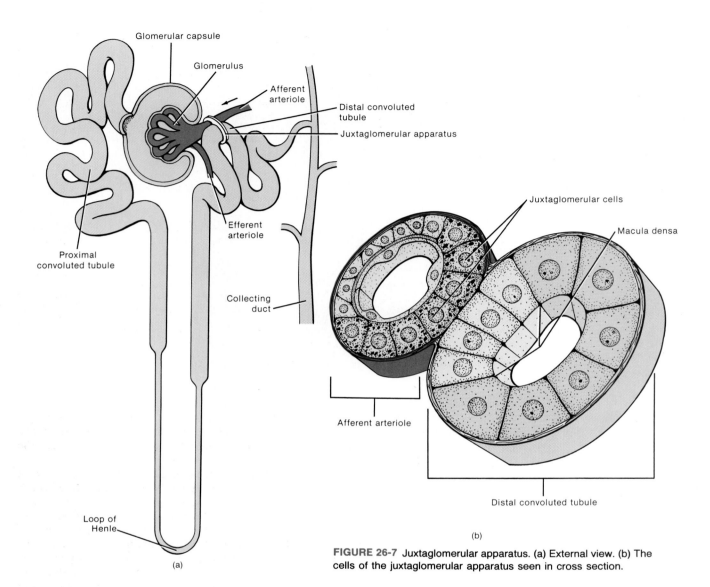

FIGURE 26-7 Juxtaglomerular apparatus. (a) External view. (b) The cells of the juxtaglomerular apparatus seen in cross section.

interlobular veins. The blood then drains through the **arcuate veins** to the **interlobar veins** running between the pyramids and leaves the kidney through a single **renal vein** that exits at the hilus. The vasa recta pass blood into the interlobular veins. From here, it goes to the arcuate veins, the interlobar veins, and then into the renal vein.

The nerve supply to the kidneys is derived from the **renal plexus** of the autonomic system. Nerves from the plexus accompany the renal arteries and their branches and are distributed to the vessels. Because the nerves are vasomotor, they regulate the circulation of blood in the kidney by regulating the diameters of the arterioles.

JUXTAGLOMERULAR APPARATUS

As the afferent arteriole approaches the renal corpuscle, the smooth muscle cells of the tunica media become modi-

fied. Their nuclei become rounded (instead of elongated), and their cytoplasm contains granules (instead of myofibrils). Such modified cells are called **juxtaglomerular cells.** The cells of the distal convoluted tubule adjacent to the afferent arteriole become considerably narrower. Collectively, these cells are known as the **macula densa.** Together with the modified cells of the afferent arteriole they constitute the **juxtaglomerular apparatus** (Figure 26-7). The juxtaglomerular apparatus helps to regulate renal blood pressure.

PHYSIOLOGY

The major work of the urinary system is done by the nephrons. The other parts of the system are primarily passageways and storage areas. Nephrons carry out three important functions. They control blood concentration

EXHIBIT 26-1
CHEMICALS IN PLASMA, FILTRATE, AND URINE DURING 24-HOUR PERIOD*

CHEMICAL	PLASMA	FILTRATE IMMEDIATELY AFTER GLOMERULAR CAPSULE	REABSORBED FROM FILTRATE	URINE
Water	180,000 ml	180,000 ml	178,000–179,000 ml	1,000–2,000 ml
Proteins	7,000–9,000	10–20	10–20	0[†]
Chloride (Cl⁻)	630	630	625	5
Sodium (Na⁺)	540	540	537	3
Bicarbonate (HCO₃⁻)	300	300	299.7	0.3
Glucose	180	180	180	0
Urea	53	53	28	25
Potassium (K⁺)	28	28	24	4
Uric acid	8.5	8.5	7.7	0.8
Creatinine	1.5	1.5	0	1.5

* All values, except for water, are expressed in grams. The chemicals are arranged in sequence from highest to lowest concentration in plasma.
† Although trace amounts of protein (170 to 250 mg) normally appear in urine, we will assume for purposes of discussion that all of it is reabsorbed from filtrate.

and volume by removing selected amounts of water and solutes. They help regulate blood pH. And they remove toxic wastes from the blood. As the nephrons go about these activities, they remove many materials from the blood, return the ones that the body requires, and eliminate the remainder. The eliminated materials are collectively called **urine.** The entire volume of blood in the body is filtered by the kidneys approximately 60 times a day. Urine formation requires three principal processes: glomerular filtration, tubular reabsorption, and tubular secretion.

Glomerular Filtration

The first step in the production of urine is called **glomerular filtration.** Filtration—the forcing of fluids and dissolved substances through a membrane by pressure—occurs in the renal corpuscle of the kidneys across the endothelial-capsular membrane. When blood enters the glomerulus, the blood pressure forces water and dissolved blood components (plasma) through the endothelial pores of the capillaries, basement membrane, and on through the filtration slits of the adjoining visceral wall of the glomerular capsule (see Figure 26-4). The resulting fluid is the **filtrate.** In a healthy person, the filtrate consists of all the materials present in the blood except for the formed elements and most proteins, which are too large to pass through the endothelial-capsular barrier. Exhibit 26-1 compares the constituents of plasma, glomerular filtrate, and urine during a 24-hour period. Although the values shown are typical, they vary considerably according to diet. The chemicals listed in plasma are those present in glomerular blood plasma before filtration. The chemicals listed in the filtrate immediately after the glo-

merular capsule are those that pass from the glomerular blood plasma through the endothelial-capsular membrane before reabsorption. The chemicals in the filtrate are the ones that have been filtered.

Renal corpuscles are especially structured for filtering blood. First, each capsule contains a tremendous length of highly coiled glomerular capillaries presenting a vast surface area for filtration. Second, the endothelial-capsular membrane is structurally adapted for filtration. Although the endothelial pores generally do not restrict the passage of substances, the basement membrane permits the passage of smaller molecules. Thus water, glucose, vitamins, amino acids, small proteins, nitrogenous wastes, and ions pass into the glomerular capsule. Large proteins and the formed elements in blood do not normally pass through the basement membrane. The filtration slits permit only the occasional passage of very small plasma proteins such as albumins. Third, the efferent arteriole is smaller in diameter than the afferent arteriole, so there is resistance to the outflow of blood from the glomerulus. Consequently, blood pressure is higher in the glomerular capillaries than in other capillaries. Glomerular blood pressure averages about 60 mm Hg, whereas the blood pressure of other capillaries averages only 30 mm Hg. Fourth, the endothelial-capsular membrane separating the blood from the space in the glomerular capsule is very thin (0.1 μm).

The filtering of the blood depends on a number of opposing pressures (Figure 26-8). The chief one is the **glomerular blood hydrostatic pressure.** *Hydrostatic (hydro = water) pressure* is the force that a fluid under pressure exerts against the walls of its container. Glomerular blood hydrostatic pressure means the blood pressure in the glomerulus. This pressure tends to move fluid out

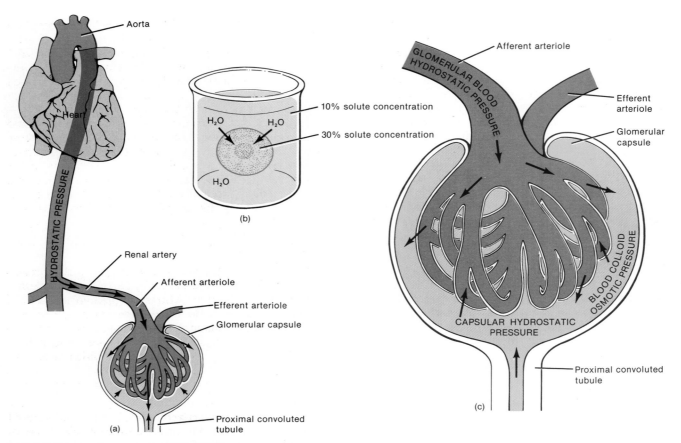

FIGURE 26-8 Glomerular filtration. (a) Heart action and resistance of walls of blood vessels provide the hydrostatic pressure for filtration. The hydrostatic pressure is counteracted by the wall of the glomerular capsule and the already present filtrate. (b) The development of osmotic pressure. When water moves into the cell, it swells as its osmotic pressure increases. (c) Application of hydrostatic pressure and osmotic pressure to glomerular filtration.

of the glomeruli at a force averaging about 60 mm Hg.

However, glomerular blood hydrostatic pressure is opposed by two other forces. The first of these, **capsular hydrostatic pressure,** develops in the following way. When the filtrate is forced into the space between the walls of the glomerular capsule, it meets with two forms of resistance: the walls of the capsule and the fluid that has already filled the renal tubule. As a result, some filtrate is pushed back into the capillary. The amount of "push" is the capsular hydrostatic pressure. It usually measures about 20 mm Hg (Figure 26-8c).

The second force opposing filtration into the glomerular capsule is the **blood colloid osmotic pressure.** *Osmotic pressure* is the pressure that develops because of water movement into a contained solution. Suppose we place a cell in a hypotonic solution. As water moves from the solution into the cell, the volume inside the cell increases, forcing the cell membrane outward (Figure 26-8b). Osmotic pressure always develops in the solution with the higher concentration of solutes. Hydrostatic pressure develops because of a force outside a solution. Osmotic pressure develops because of the concentration of the solution itself. Since the blood contains a much higher concentration of proteins than the filtrate does, water moves out

of the filtrate and back into the blood vessel. This blood osmotic pressure is normally about 30 mm Hg.

To determine how much filtration finally occurs, we have to subtract the forces that oppose filtration from the glomerular blood hydrostatic pressure. The net result is called the **effective filtration pressure,** which is abbreviated P_{eff}.

$$P_{eff} = \begin{pmatrix} \text{glomerular} \\ \text{blood} \\ \text{hydrostatic} \\ \text{pressure} \end{pmatrix} - \begin{pmatrix} \text{capsular} \\ \text{hydrostatic} + \\ \text{pressure} \end{pmatrix} \begin{matrix} \text{blood} \\ \text{colloid} \\ \text{osmotic} \\ \text{pressure} \end{matrix}$$

By substituting the values just discussed, a normal P_{eff} may be calculated as follows.

$$P_{eff} = (60 \text{ mm Hg}) - (20 \text{ mm Hg} + 30 \text{ mm Hg})$$
$$= (60 \text{ mm Hg}) - (50 \text{ mm Hg})$$
$$= 10 \text{ mm Hg}$$

This means that a pressure of about 10 mm Hg causes a normal amount of plasma to filter from the glomerulus into the glomerular capsule. This is about 125 ml of filtrate per minute.

Certain conditions may alter these pressures and thus the P_{eff}. In some forms of kidney disease, such as glomerulonephritis, glomerular capillaries become so permeable that the plasma proteins are able to pass from the blood into the filtrate. As a result, the capsular filtrate exerts an osmotic pressure that draws water out of the blood. Thus, if a capsular osmotic pressure develops, the P_{eff} will increase. At the same time, blood colloid osmotic pressure decreases, further increasing the P_{eff}.

The P_{eff} also is affected by changes in the general arterial blood pressure. Severe hemorrhaging produces a drop in general blood pressure, which also decreases the glomerular blood hydrostatic pressure. If the blood pressure falls to the point where the hydrostatic pressure in the glomeruli reaches 50 mm Hg, no filtration occurs, because the glomerular blood hydrostatic pressure equals the opposing forces. Such a condition is called renal suppression.

CLINICAL APPLICATION
Renal suppression or **anuria** (a-NOO-rē-a) is the failure of the kidneys to secrete urine. It may be caused by insufficient pressure to accomplish filtration or by inflammation of the glomeruli so that plasma is prevented from reaching the glomerulus.

A final factor that may affect the P_{eff} is the regulation of the size of the afferent and efferent arterioles. In this case, glomerular blood hydrostatic pressure is regulated separately from the general blood pressure. Sympathetic impulses and small doses of epinephrine cause constriction of both afferent and efferent arterioles. However, intense sympathetic impulses and large doses of epinephrine cause greater constriction of afferent than efferent arterioles. This intense stimulation results in a decrease in glomerular hydrostatic pressure even though blood pressure in other parts of the body may be normal or even higher than normal. Intense sympathetic stimulation is most likely to occur during the alarm reaction of the general adaptation syndrome. Blood may also be shunted away from the kidneys during hemorrhage.

Tubular Reabsorption

The amount of filtrate that flows out of all the renal corpuscles of both kidneys every minute is called the **glomerular filtration rate (GFR)**. In the normal adult, this rate is about 125 ml/min—about 180 liters (48 gal) a day. But as the filtrate passes through the renal tubules, it is reabsorbed into the blood at a rate of about 123 to 124 ml/min—about 178 to 179 liters/day. Thus only about 1 percent of the filtrate actually leaves the body—about 1 to 1.5 ml/min, or 1 to 2 liters/day. The movement of the filtrate back into the blood of the peritubular capillaries or vasa recta is called **tubular reabsorption.**

Tubular reabsorption is carried out by epithelial cells throughout the renal tubule. It is a very discriminating process. Only specific amounts of certain substances are reabsorbed, depending on the body's needs at the time. The maximum amount of a substance that can be reabsorbed under any condition is called the substance's **tubular maximum (Tm).** Materials that are reabsorbed include water, glucose, amino acids, and ions such as Na$^+$, K$^+$,

Ca^{2+}, Cl$^-$, HCO$_3$$^-$, and HPO$_4$$^{2-}$. Tubular reabsorption allows the body to retain most of its nutrients. Wastes such as urea are only partially reabsorbed. Exhibit 26-1 compares the values for the chemicals in the filtrate immediately after the glomerular capsule with those reabsorbed from the filtrate. It will give you an idea of how much of the various substances the kidneys reabsorb.

Reabsorption is carried out through both passive and active transport mechanisms. It is believed that glucose is reabsorbed by an active process involving a carrier system. The carrier, probably an enzyme, exists in the membranes of the tubular epithelial cells in a fixed and limited amount. Normally, all the glucose filtered by the glomeruli (125 mg/100 ml of filtrate/min) is reabsorbed by the tubules.

CLINICAL APPLICATION
The capacity of the glucose carrier system is limited. If the plasma concentration of glucose is significantly above normal, the glucose transport mechanism cannot reabsorb it all and the excess remains in the urine. If there is a malfunction in the tubular carrier mechanism, glucose appears in the urine even though the blood sugar level is normal. This condition is called **glycosuria.**

Sodium ions are actively transported (reabsorbed) mostly from the proximal and also from the distal convoluted tubules and collecting ducts. In the distal convoluted tubules and collecting ducts, Na$^+$ reabsorption varies with its concentration in extracellular fluid. When the Na$^+$ concentration of the blood is low, there is a drop in blood pressure, and the **renin-angiotensin pathway** goes into operation. The juxtaglomerular cells of the kidneys secrete an enzyme called renin, which converts angiotensin (synthesized by the liver) into angiotensin I. As it passes through the lungs, angiotensin I is converted into angiotensin II. Angiotensin II stimulates the adrenal zona glomerulosa of the cortex to produce aldosterone. Aldosterone brings about increased Na$^+$ and water reabsorption (Figure 26-9). Extracellular fluid volume increases and blood pressure is restored to normal.

When Na$^+$ ions move out of the tubules into the peritubular blood, the blood momentarily becomes more electropositive than the filtrate. Chloride, a negatively charged ion, follows the positive sodium ion out of the tubule by electrostatic attraction. Thus the movement of Na$^+$ ions influences the movement of Cl$^-$ ions and other anions into the blood.

Cl$^-$ ions, and not Na$^+$ ions, are actively transported from the ascending limb of Henle. In this portion of the nephron, then, Cl$^-$ ions move actively and Na$^+$ ions follow passively. Whether Na$^+$ ions move actively and Cl$^-$ ions follow passively or Cl$^-$ ions move actively and Na$^+$ ions follow passively, the end result is the same.

Water reabsorption is controlled both by sodium transport and by water carrier molecules. As Na$^+$ ions are transported from the proximal convoluted tubules into the blood, the osmotic pressure of the blood becomes higher than that of the filtrate. Water follows the Na$^+$

FIGURE 26-9 Role of the juxtaglomerular apparatus in maintaining normal blood volume.

ions into the blood in order to reestablish the osmotic equilibrium. About 80 percent of the water is reabsorbed by this method from the proximal convoluted tubule. This portion of reabsorbed water occurs as a function of the law of osmosis, so it is referred to as **obligatory reabsorption.** The proximal convoluted tubule has no control over osmosis because it is always permeable to water (Figure 26-10).

Passage of most of the remaining water in the filtrate can be regulated. The permeability of the cells of the distal and collecting tubules is controlled by the antidiuretic hormone (ADH), produced by the hypothalamus. When the blood-water concentration is low, the posterior pituitary releases ADH, a hormone that increases the permeability of the plasma membranes of the distal tubule and collecting duct cells. As the membranes become more permeable, more water molecules pass into the cells. This type of absorption is called *facultative* (FAK-ul-tā'-tiv), meaning that it occurs under some conditions but not others. **Facultative reabsorption** is responsible for about 20 percent of the water in the filtrate. It is a major mechanism for controlling water content of the blood (Figure 26-10). Aldosterone has an effect similar to ADH.

Tubular Secretion

The third process involved in urine formation is **tubular secretion.** Whereas tubular reabsorption removes substances from the filtrate into the blood, tubular secretion adds materials to the filtrate from the blood. These secreted substances include potassium and hydrogen ions, ammonia, creatinine, and the drugs penicillin and para-aminohippuric acid. Tubular secretion has two principal effects. It rids the body of certain materials, and it controls the blood pH.

The body has to maintain normal blood pH (7.35 to 7.45) despite the fact that a normal diet provides more acid-producing foods than alkali-producing foods. To raise blood pH, the renal tubules secrete hydrogen and ammonium ions into the filtrate. Both these substances make the urine acidic.

Secretion of the hydrogen ions occurs through the formation of carbonic acid (Figure 26-11a). A certain amount of carbon dioxide normally diffuses from the peritubular blood into the cells of the distal tubules and collecting ducts. Once inside the epithelial cells, the CO_2 combines with water to form carbonic acid (H_2CO_3), which then dissociates into H^+ ions and bicarbonate ions (HCO_3^-). A low blood pH stimulates the cells to secrete the H^+ ion into the urine. As the H^+ enters the urine, it displaces another positive ion, usually Na^+, and forms a weak acid or a salt of the acid that is eliminated in urine. The displaced Na^+ or other positive ion diffuses from the urine into the tubule cell, where it combines with the bicarbonate ion to form sodium bicarbonate

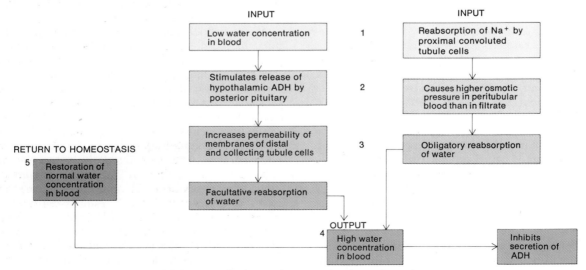

FIGURE 26-10 Factors that control water reabsorption.

FIGURE 26-11 Role of the kidneys in maintaining blood pH. (a) Acidification of urine and conservation of sodium bicarbonate ($NaHCO_3$) by the elimination of hydrogen ions (H^+). (b) Acidification of urine and conservation of sodium bicarbonate ($NaHCO_3$) by the elimination of ammonium ions (NH_4^+).

($NaHCO_3$), which then is absorbed into the blood. Not only is H^+ eliminated by the body and Na^+ conserved, but Na^+ is conserved in the form of $NaHCO_3$, which can buffer other H^+ ions in the blood.

A second mechanism for raising blood pH is secretion of the ammonium ion (Figure 26-11b). Ammonia, in certain concentrations, is a poisonous waste product derived from the deamination of amino acids. The liver converts much of the ammonia to a less toxic compound called urea. Urea and ammonia both become part of the glomerular filtrate and are subsequently expelled from the body. Any ammonia produced by the deamination of amino acids in the tubule is secreted into the urine. When ammonia (NH_3) forms in the distal and collecting tubule cells, it combines with H^+ to form the ammonium ion (NH_4^+). (The H^+ may come from the dissociation of H_2CO_3.) The cells secrete NH_4^+ into the filtrate, where it takes the place of a positive ion, usually Na^+, in a salt and is eliminated. The displaced Na^+ diffuses into the renal cells and combines with HCO_3^- to form $NaHCO_3$.

As a result of hydrogen and ammonium ion secretion, urine normally has an acidic pH of 6. The relationship of renal tubule ion excretion to blood pH level is summarized in Figure 26-12. Exhibit 26-2 summarizes filtration, reabsorption, and secretion in the nephrons.

Countercurrent Multiplier Mechanism

The various metabolic wastes in urine must be eliminated from the body regardless of its state of hydration. During dehydration, little urine is excreted, but it is concentrated. Such urine is said to be *hypertonic (hyperosmotic)* to the blood plasma. When there are large amounts of water in the body, more urine is excreted, but it is diluted. Such urine is said to be *hypotonic (hyposmotic)* to blood plasma. The ability of the kidneys to produce either hyperosmotic or hyposmotic urine depends in part on the **countercurrent multiplier mechanism.**

The countercurrent multipler mechanism is based on the anatomical arrangements of the juxtamedullary nephrons and vasa recta. Both begin in the cortex, dip deeply into the medulla, and return to the cortex. The normal concentration of the glomerular filtrate as it enters the proximal convoluted tubule is about 300 milliosmols (mOsm) per liter. A milliosmol represents the concentration of substances in the filtrate, mainly NaCl. It is a measure of the osmotic properties of a solution. This concentration of glomerular filtrate is isotonic to plasma. Figure 26-13 shows that the concentration of the interstitial fluid increases from 300 mOsm in the cortex to 1,200 mOsm in the inner medulla. As the filtrate passes through the ascending limb of Henle, Cl^- is actively transported from the filtrate into the interstitial fluid. Na^+ follows passively and the concentration of NaCl becomes about four times greater at the bottom of the interstitial fluid than at the top.

RETURN TO HOMEOSTASIS

FIGURE 26-12 Summary of kidney mechanisms that maintain the homeostasis of blood pH.

EXHIBIT 26-2
FILTRATION, REABSORPTION, AND SECRETION

REGION OF NEPHRON	ACTIVITY
Renal corpuscle (endothelial-capsular membrane)	Filtration of glomerular blood under hydrostatic pressure results in formation of filtrate free of plasma proteins and cellular elements of blood.
Proximal convoluted tubule and descending and ascending limbs of Henle	Reabsorption of physiologically important solutes such as Na^+, K^+, Cl^-, HCO_3^-, and glucose. Obligatory reabsorption of water by osmosis (proximal convoluted tubule).
Distal convoluted tubule	Reabsorption of Na^+. Facultative reabsorption of water under control of ADH. Secretion of H^+, NH_3, K^+, creatinine, and certain drugs. Conservation of $NaHCO_3$.
Collecting duct	Facultative reabsorption of water under control of ADH.

The descending limb of Henle is relatively permeable to Na^+ and Cl^-, while the ascending limb has a mechanism that actively transports Cl^- from the filtrate into the interstitial fluid of the medulla. When the Cl^- is actively transported, Na^+ follows passively. Almost immedi-ately they diffuse into the descending limb of Henle. This diffusion increases the concentration of NaCl in the filtrate that flows *downward* toward the loop of Henle. This continual active transport of Cl^- and passive movement of Na^+ out of the ascending limb of Henle into the interstitial

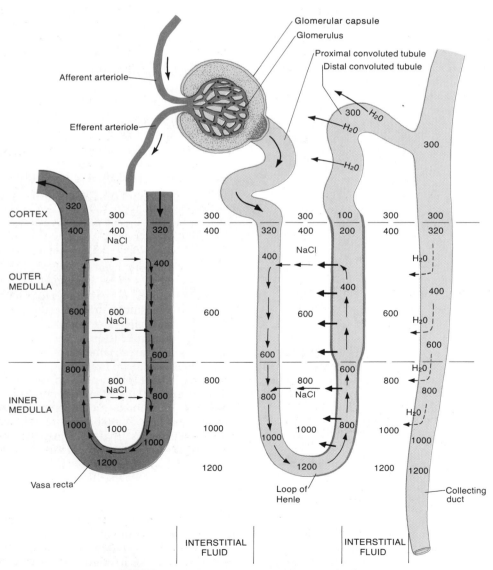

FIGURE 26-13 Countercurrent multiplier mechanism.

fluid of the medulla and then into the descending limb of Henle causes the high concentration of NaCl in the interstitial fluid of the medulla. Thus two parallel streams of liquid flowing opposite each other (*countercurrent*) result in an increased NaCl concentration in the medullary interstitial fluid (*multiplier*).

The high concentration of NaCl is also the result of the countercurrent flow of blood in the vasa recta. These vessels run parallel to the loop tubules and are freely permeable. As blood circulates through the descending limb of the vasa recta, Na⁺ and Cl⁻ diffuse into it from the interstitial fluid of the medulla. The concentration of ᵗhe blood reaches 1,200 mOsm at the tip of the vessel. As the blood flows through the ascending limb of the vasa recta, almost all of the NaCl diffuses back into the interstitial fluid of the medulla. By the time the blood in the vasa recta leaves the medulla, its concentration is only slightly higher than it was when it entered. Thus the countercurrent flow through the vasa recta permits most of the NaCl to remain in the interstitial fluid of the medulla.

The ascending limb of Henle is very impermeable to water. Yet large amounts of Cl⁻ are actively transported out of the ascending limb, and Na⁺ follows passively into the interstitial fluid of the medulla. This means that the fluid in the ascending limb becomes more dilute as it approaches the cortex. In fact, the concentration of fluid decreases to about 100 mOsm. Thus the fluid in the ascending limb is hypotonic (hyposmotic) to the blood plasma. In the absence of ADH, no water is reabsorbed from the distal convoluted tubule and collecting duct as the fluid passes through. The fluid empties into the papillary ducts, calyces, and renal pelvis. The fluid thus entering the pelvis and passing through the ureter into the urinary bladder is very dilute. In the absence of ADH, the urine may be only one-fourth as concentrated as the blood plasma and glomerular filtrate.

In the presence of ADH, the epithelial pores of the distal convoluted tubule and collecting duct enlarge and they become permeable to water. As the dilute fluid passes through the distal convoluted tubule and collecting duct, most of the water is reabsorbed by osmosis. The loss of this water raises the concentration of the dilute fluid from

100 to 300 mOsm—the same concentration as the interstitial fluid of the cortex. As the fluid passes into the medulla through the collecting duct, it is exposed to the hyperosmotic interstitial fluid of the medulla. As a result, large amounts of water are absorbed by osmosis into the interstitial fluid of the medulla from the collecting duct. The highly concentrated interstitial fluid of the medulla causes the rapid osmosis of water when the epithelial pores are opened by ADH. Thus the concentration of fluid in the collecting duct approaches 1,200 mOsm—the concentration of the interstitial fluid of the medulla. The fluid that enters the papillary ducts, calyces, renal pelvis, and ureters and flows into the urinary bladder is a very concentrated urine. In the presence of ADH, it may be four times more concentrated than the blood plasma and glomerular filtrate.

HOMEOSTASIS

Excretion is one of the primary ways in which the volume, pH, and chemistry of the body fluids are kept in homeostasis. The kidneys assume a good deal of the burden for excretion, but they share this responsibility with several other organ systems.

OTHER EXCRETORY ORGANS

The lungs, integument, and alimentary canal all perform special excretory functions (Exhibit 26-3). A major responsibility for regulating body temperature through the excretion of water is assumed by sudoriferous (sweat) glands of the skin. The lungs maintain blood-gas homeostasis through the elimination of carbon dioxide. One way in which the kidneys maintain homeostasis is by coordinating their activities with other excretory organs. When the integument increases its excretion of water, the renal tubules increase their reabsorption of water, and blood volume is maintained. When the lungs fail to eliminate enough carbon dioxide, the kidneys attempt to compensate. They change some of the carbon dioxide into sodium bicarbonate, which becomes part of the blood buffer systems.

URINE

The by-product of the kidneys' activities is **urine.** In a healthy person its volume, pH, and solute concentration vary with the needs of the internal environment. During certain pathological conditions, the characteristics of urine may change drastically. An analysis of the volume and physical and chemical properties of urine tells us much about the state of the body.

Volume

The volume of urine eliminated per day in the normal adult varies between 1,000 and 2,000 ml (1 to 2 qt). Urine volume is influenced by a number of factors: blood pressure, blood concentration, diet, temperature, diuretics, mental state, and general health.

EXHIBIT 26-3
EXCRETORY ORGANS AND PRODUCTS ELIMINATED

EXCRETORY ORGANS	PRODUCTS ELIMINATED	
	PRIMARY	SECONDARY
Kidneys	Water, nitrogenous wastes from protein catabolism, and inorganic salts.	Heat and carbon dioxide.
Lungs	Carbon dioxide.	Heat and water.
Skin (sudoriferous glands)	Heat.	Carbon dioxide, water, salts, and urea.
Alimentary canal	Solid wastes and secretions.	Carbon dioxide, water, salts, and heat.

● *Blood Pressure*   The cells of the juxtaglomerular apparatus are particularly sensitive to changes in blood pressure. When renal blood pressure falls below normal, the juxtaglomerular apparatus secretes renin and the renin-angiotensin pathway is activated (see Figure 26-9). As a result, obligatory reabsorption of water increases, blood volume increases, and urine volume decreases. By raising the blood pressure, the juxtaglomerular apparatus ensures that the kidney cells receive enough oxygen and that the glomerular hydrostatic pressure is high enough to maintain a normal P_{eff}. The juxtaglomerular apparatus also regulates blood pressure throughout the body.

● *Blood Concentration*   The concentrations of water and solutes in the blood also affect urine volume. If you have gone without water all day and the water concentration of your blood becomes low, osmotic receptors in the hypothalamus stimulate the posterior pituitary to release ADH. The hormone stimulates the cells of the distal convoluted tubule and collecting duct to transport water out of the filtrate and into the blood by facultative reabsorption. Thus urine volume decreases and water is conserved.

If you have just drunk an excessive amount of liquid, urine volume may be increased through two mechanisms. First, the blood-water concentration increases above normal. This means that the osmoreceptors in the hypothalamus are no longer stimulated to secrete ADH and facultative water reabsorption stops. Second, the excess water causes the blood pressure to rise. In response, the renal vessels dilate, more blood is brought to the glomeruli, and the filtration rate increases.

The concentration of sodium ions in the blood also influences urine volume. Sodium concentration affects aldosterone secretion, which in turn affects both sodium reabsorption and the obligatory reabsorption of water.

● *Temperature*   When the internal or external temperature rises above normal, the cutaneous vessels dilate and fluid diffuses from the capillaries to the surface of the skin. As water volume decreases, ADH is secreted and facultative reabsorption increases. In addition, the increase in temperature stimulates the abdominal vessels to constrict, so the blood flow in the glomeruli and filtration decrease. Both mechanisms reduce the volume of urine.

If the body is exposed to low temperatures, the cutaneous vessels constrict and the abdominal vessels dilate. More blood is shunted to the glomeruli, glomerular blood hydrostatic pressure increases, and urine volume increases.

● *Diuretics*   Certain chemicals increase urine volume by inhibiting the facultative reabsorption of water. Such chemicals are called **diuretics,** and the abnormal increase in urine flow is called *diuresis*. Some diuretics act directly on the tubular epithelium as they are carried through the kidneys. Others act indirectly by inhibiting the secretion of ADH as they circulate through the brain. Coffee, tea, and alcoholic beverages are diuretics.

● *Emotions*   Some emotional states can affect urine volume. Extreme nervousness, for example, can cause an enormous discharge of urine because impulses from the brain cause an increase in blood pressure, resulting in an increased glomerular filtration rate.

Physical Characteristics

Normal urine is usually a yellow or amber-colored, transparent liquid with a characteristic odor. The color is caused by urochrome, pigments derived from the metabolism of bile. It varies considerably with the ratio of solutes to water in the urine. The less water there is, the darker the urine. Fever decreases urine volume in the same way that high environmental temperatures do, sometimes making the urine quite concentrated. It is not uncommon for a feverish person to have dark yellow or brown urine. The color of urine may also be affected by diet, such as a reddish color from beets, and by the presence of abnormal constituents, such as certain drugs. A red or brown to black color may indicate the presence of red blood cells or hemoglobin from bleeding in the urinary system.

Fresh urine is usually transparent. Turbid (cloudy) urine does not necessarily indicate a pathological condition, since turbidity may result from mucin secreted by the lining of the urinary tract. The presence of mucin above a critical level usually denotes an abnormality, however.

The odor of urine may vary. For example, the digestion of asparagus adds a substance called methyl mercaptan that gives urine a characteristic odor. In cases of diabetes, urine has a "sweetish" odor because of the presence of acetone. Stale urine develops the odor of ammonia due to ammonium carbonate formation as a result of urea decomposition.

Normal urine is slightly acid. It ranges between pH 5.0 and 7.8 and rarely becomes more acid than 4.5 or more alkaline than 8. Variations in urine pH are closely related to diet. These variations are due to differences in the end products of metabolism. Whereas a high-protein diet increases acidity, a diet composed largely of vegetables increases alkalinity. High altitude, fasting, and exercise also cause variations in urinary pH. Ammonium carbonate forms in standing urine. Since it can dissociate into ammonium ion and form a strong base, the presence

EXHIBIT 26-4
PHYSICAL CHARACTERISTICS OF NORMAL URINE

CHARACTERISTIC	DESCRIPTION
Volume	1–2 liters in 24 hours, but varies considerably.
Color	Yellow or amber, but varies with concentration and diet.
Turbidity	Transparent when freshly voided, but becomes turbid upon standing.
Odor	Aromatic, but becomes ammonialike upon standing.
pH	5.0–7.8, average 6.0; varies considerably with diet.
Specific gravity	1.008–1.030.

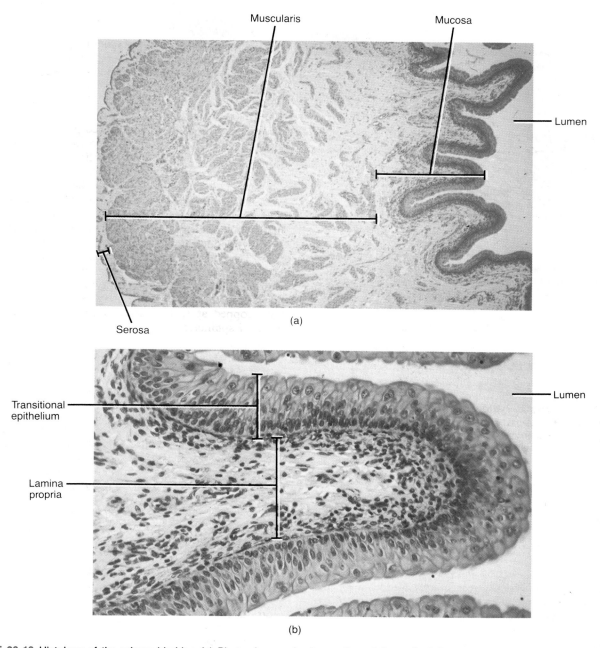

Muscularis

Mucosa

Lumen

Serosa

(a)

Transitional
epithelium

Lamina
propria

Lumen

(b)

FIGURE 26-16 Histology of the urinary bladder. (a) Photomicrograph of a portion of the wall of the urinary bladder at a magnification of 50×. (b) Photomicrograph of an enlarged aspect of the mucosa of the urinary bladder at a magnification of 200×. (© 1983 by Michael H. Ross. Used by permission.)

female urethra is directed obliquely inferiorly and anteriorly. The opening of the urethra to the exterior, the **urethral orifice** is located between the clitoris and vaginal opening.

In males, the urethra is about 20 cm (8 inches) long. Immediately below the urinary bladder it passes vertically through the prostate gland, then pierces the urogenital diaphragm, and finally pierces the penis and takes a curved course through its body (see Figures 28-1 and 28-8a).

STRUCTURE

The wall of the female urethra consists of three coats: an inner mucous coat that is continuous externally with that of the vulva, an intermediate thin layer of spongy tissue containing a plexus of veins, and an outer muscular coat that is continuous with that of the urinary bladder and consists of circularly arranged fibers of smooth muscle.

The male urethra is composed of two coats. An inner mucous membrane is continuous with the mucous membrane of the urinary bladder. An outer submucous tissue connects the urethra with the structures through which it passes.

PHYSIOLOGY

The urethra is the terminal portion of the urinary system. It serves as the passageway for discharging urine from the body. The male urethra also serves as the duct through which reproductive fluid (semen) is discharged from the body.

DISORDERS: HOMEOSTATIC IMBALANCES

Gout
Gout is a hereditary condition associated with an excessively high level of uric acid in the blood. When nucleic acids are catabolized, a certain amount of uric acid is produced as a waste. Some people seem to produce excessive amounts of uric acid, and others seem to have trouble excreting normal amounts. In either case, uric acid accumulates in the body and tends to solidify into crystals that are deposited in the joints and kidney tissue. When the crystals are deposited in the joints, the condition is called gouty arthritis. Gout is aggravated by excessive use of diuretics, dehydration, and starvation.

Glomerulonephritis (Bright's Disease)
Glomerulonephritis (Bright's disease) is an inflammation of the kidney that involves the glomeruli. One of the most common causes of glomerulonephritis is an allergic reaction to the toxins given off by streptococci bacteria that have recently infected another part of the body, especially the throat. The glomeruli become so inflamed, swollen, and engorged with blood that the glomerular membranes become highly permeable and allow blood cells and proteins to enter the filtrate. Thus the urine contains many erythrocytes and much protein. The glomeruli may be permanently changed, leading to chronic renal disease and renal failure.

Pyelitis and Pyelonephritis
Pyelitis is an inflammation of the kidney pelvis and its calyces. **Pyelonephritis** is the interstitial inflammation of one or both kidneys. It usually involves both the parenchyma and the renal pelvis and is due to bacterial invasion from the middle and lower urinary tracts or the bloodstream.

Cystitis is an inflammation of the urinary bladder involving principally the mucosa and submucosa. It may be caused by bacterial infection, chemicals, or mechanical injury.

Nephrosis
Nephrosis is a condition in which the glomerular membrane leaks, allowing large amounts of protein to escape from the blood into the urine. Water and sodium then accumulate in the body creating edema especially around the ankles and feet, abdomen, and eyes. Nephrosis is more common in children than adults, but occurs in all ages. Although it cannot always be cured, certain synthetic steroid hormones such as cortisone and prednisone, which are similar to the natural hormones secreted by the adrenal glands, can suppress some forms of it.

Polycystic Disease
Polycystic disease may be caused by a defect in the renal tubular system that deforms nephrons and results in cystlike dilations along their course. It is the most common inherited disorder of the kidneys. The kidney tissue is riddled with cysts, small holes, and fluid-filled bubbles ranging in size from a pinhead to the diameter of an egg. These cysts gradually increase until they squeeze out the normal tissue, interfering with kidney function and causing uremia. The chief symptom is weight gain.

The kidneys themselves may enlarge from their normal ½ lb to as much as 30 lb. Many people lead normal lives without ever knowing they have the disease, however, and the condition is sometimes discovered only after death. Although the disease is progressive, its advance can be slowed by diet, drugs, and fluid intake. Kidney failure as a result of polycystic disease seldom occurs before the mid-40s and frequently can be postponed until the 60s.

Hemodialysis Therapy
If the kidneys are so impaired by disease or injury that they are unable to excrete nitrogenous wastes and regulate pH and electrolyte concentration of the plasma, the blood must be filtered by an artificial device. Such filtering of the blood is called **hemodialysis.** *Dialysis* means using a semipermeable membrane to separate large nondiffusible particles from smaller diffusible ones. One of the best-known devices for accomplishing dialysis is the kidney machine (Figure 26-17). A tube connects it with the patient's radial artery. The blood is pumped from the artery through the tubes to one side of a semipermeable dialyzing membrane made of cellophane sheets. The other side of the membrane is continually washed with an artificial solution called the dialyzing solution. The blood that passes through the artificial kidney is treated with an anticoagulant. Only about 500 ml of the patient's blood is in the machine at a time. This volume is easily compensated for by vasoconstriction and increased cardiac output.

All substances (including wastes) in the blood except protein molecules and blood cells can diffuse back and forth across the semipermeable membrane. The electrolyte level of the plasma is controlled by keeping the dialyzing solution electrolytes at the same concentration found in normal plasma. Any excess plasma electrolytes move down the concentration gradient and into the dialyzing solution. If the plasma electrolyte

FIGURE 26-17 Operation of an artificial kidney. The blood route is indicated in red and blue. The route of the dialyzing solution is indicated in gold.

UNIT V

Continuity

This unit is designed to show you how the human organism is adapted for reproduction. It also traces the developmental sequence involved in pregnancy. Principles of inheritance and birth control conclude the unit.

- Define reproduction and classify the organs of reproduction by function.

- Descibe the structure and function of the scrotum.

- Explain the structure, histology, and functions of the testes.

- Describe the parts of a spermatozoon and explain their functions.

- Describe the physiological effects of testosterone.

- Describe the seminiferous tubules, straight tubules, and rete testis as components of the duct system of the testes.

- Describe the location, structure, histology, and functions of the ductus epididymis.

- Explain the structure and functions of the ductus deferens.

- Describe the structure and function of the ejaculatory duct.

- Describe the three anatomical subdivisions of the male urethra.

- Explain the location and functions of the seminal vesicles, prostate gland, and bulbourethral glands, the accessory reproductive glands.

- Discuss the chemical composition of semen.

- Trace the course of sperm cells through the system of ducts that lead from the testes to the exterior.

- Explain the structure and functions of the penis.

- Describe the location, histology, and functions of the ovaries.

- Explain the location, structure, histology, and functions of the uterine tubes.

- Describe the location and ligamentous attachments of the uterus.

- Explain the histology and blood supply of the uterus.

- Compare the principal events of the menstrual and ovarian cycles.

- Discuss the physiological effects of estrogens and progesterone.

- Correlate the activities of the menstrual and ovarian cycles.

- Discuss the hormonal interactions that control the menstrual and ovarian cycles and menopause.

- Describe the location, structure, and functions of the vagina.

- Describe the components of the vulva and explain their functions.

- Describe the anatomical boundaries of the perineum.

- Explain the structure, development, and histology of the mammary glands.

- Explain the symptoms and causes of sexually transmitted diseases (STDs) such as gonorrhea, syphilis, genital herpes, trichomoniasis, and nongonococcal urethritis.

- Describe the symptoms and causes of male disorders (prostate dysfunctions, impotence, and infertility) and female disorders [amenorrhea, dysmenorrhea, premenstrual syndrome (PMS), toxic shock syndrome (TSS), ovarian cysts, endometriosis, infertility, breast tumors, cervical cancer, and pelvic inflammatory disease or PID].

- Define medical terminology associated with the reproductive systems.

28 The Reproductive Systems

Reproduction is the mechanism by which the thread of life is sustained. It is the process by which a single cell duplicates its genetic material, allowing an organism to grow and repair itself. In this sense, reproduction maintains the life of the individual. But reproduction is also the process by which genetic material is passed from generation to generation. In this regard, reproduction maintains the continuation of the species.

The organs of the male and female reproductive systems may be grouped by function. The testes and ovaries, also called **gonads,** function in the production of gametes—sperm cells and ova, respectively. The gonads also secrete hormones. The production of gametes and their discharge into ducts classifies the gonads as exocrine glands, whereas their production of hormones classifies them as endocrine glands. The **ducts** transport, receive, and store gametes. Still other reproductive organs, called **accessory glands,** produce materials that support gametes.

MALE REPRODUCTIVE SYSTEM

The organs of the male reproductive system are the testes, or male gonads, which produce sperm; a number of ducts that either store or transport sperm to the exterior; accessory glands that add secretions constituting the semen;

and several supporting structures, including the penis (Figure 28-1).

SCROTUM

The **scrotum** is a cutaneous outpouching of the abdomen consisting of loose skin and superficial fascia. It is the supporting structure for the testes. Externally, it looks like a single pouch of skin separated into lateral portions by a median ridge called the **raphe** (RĀ-fē). Internally, it is divided by a septum into two sacs, each containing a single testis. The septum consists of superficial fascia and contractile tissue called the **dartos** (DAR-tōs), which consists of bundles of smooth muscle fibers. The dartos is also found in the subcutaneous tissue of the scrotum and is directly continuous with the subcutaneous tissue of the abdominal wall. The dartos causes wrinkling of the skin of the scrotum.

The location of the scrotum and the contraction of its muscle fibers regulate the temperature of the testes, organs that produce sperm and the hormones testosterone and inhibin. The production and survival of sperm require a temperature that is lower than body temperature. Because the scrotum is outside the body cavities, it supplies an environment about 3°F below body temperature. The

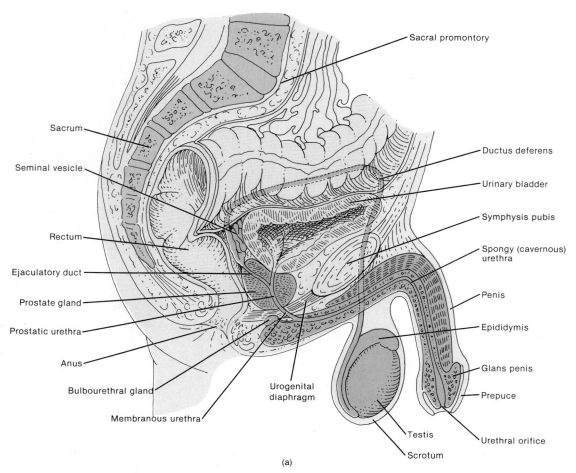

(a)

FIGURE 28-1 Male organs of reproduction and surrounding structures seen in sagittal section. (a) Diagram.

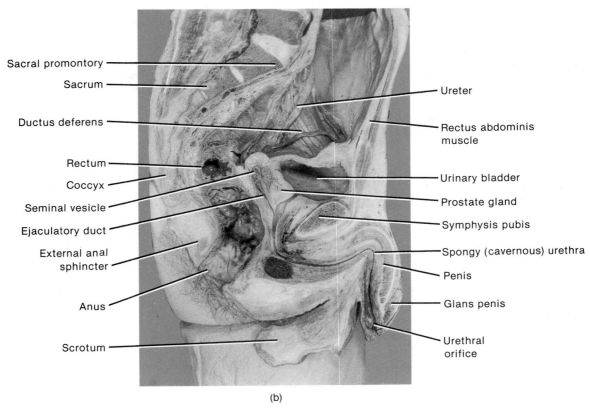

Sacral promontory

Sacrum

Ductus deferens

Rectum

Coccyx

Seminal vesicle

Ejaculatory duct

External anal
sphincter

Anus

Scrotum

Ureter

Rectus abdominis
muscle

Urinary bladder

Prostate gland

Symphysis pubis

Spongy (cavernous) urethra

Penis

Glans penis

Urethral
orifice

(b)

FIGURE 28-1 (*Continued*) Male organs of reproduction and surrounding structures seen in sagittal section. (b) Photograph. (Courtesy of C. Yokochi and J. W. Rohen, *Photographic Anatomy of the Human Body*, 2nd ed., 1979, IGAKU-SHOIN, Ltd., Tokyo, New York.)

cremaster (krē-MAS-ter) **muscle,** a small, circular band of skeletal muscle, elevates the testes upon exposure to cold, moving them closer to the pelvic cavity where they can absorb body heat. Exposure to warmth reverses the process.

TESTES

The **testes** are paired oval glands measuring about 5 cm (2 inches) in length and 2.5 cm (1 inch) in diameter (Figure 28-2). They weigh between 10 and 15 g. The testes develop high on the embryo's posterior abdominal wall, and usually enter the scrotum by 32 weeks. Full descent is not complete until just prior to birth.

CLINICAL APPLICATION

When the testes do not descend, the condition is referred to as **cryptorchidism** (krip-TOR-ki-dizm). The condition occurs in about 3 percent of full-term infants and about 30 percent of premature infants. Cryptorchidism results in sterility because the cells involved in the initial development of sperm cells are destroyed by the higher body temperature of the pelvic cavity. Undescended testes can be placed in the scrotum by administering hormones or by surgical means prior to puberty without ill effects.

The testes are covered by a dense layer of white fibrous tissue, the **tunica albuginea** (al'-byoo-JIN-ē-a), that extends inward and divides each testis into a series of inter-

nal compartments called **lobules.** Each of the 200 to 300 lobules contains one to three tightly coiled tubules, the convoluted **seminiferous tubules,** that produce sperm by a process called **spermatogenesis.** This process is considered in detail in Chapter 29.

A cross section through a seminiferous tubule reveals that it is packed with sperm cells in various stages of development (Figure 28-3). The most immature cells, the **spermatogonia,** are located against the basement membrane. These cells collectively constitute the germinal epithelium. Toward the lumen of the tube, one can see layers of progressively more mature cells. In order of advancing maturity, these are primary spermatocytes, secondary spermatocytes, and spermatids. By the time a **sperm cell,** or **spermatozoon** (sper'-ma-tō-ZŌ-on), has reached full maturity, it is in the lumen of the tubule and begins to be moved through a series of ducts. Embedded between the developing sperm cells in the tubules are **sustentacular** (sus'-ten-TAK-yoo-lar) **cells.** These cells produce secretions that supply nutrients to the spermatozoa and secrete the hormone inhibin. Between the seminiferous tubules are clusters of **interstitial endocrinocytes.** These cells secrete the male hormone testosterone, the most important androgen.

Spermatozoa

Spermatozoa are produced or matured at the rate of about 300 million per day and, once ejaculated, have a life expectancy of about 48 hours within the female reproductive

(a)

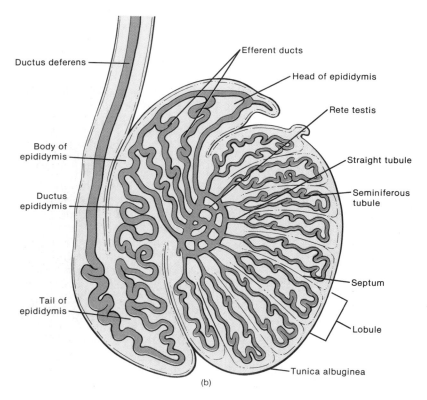

(b)

FIGURE 28-2 External and internal anatomy of testes. (a) Photograph of external view. (Courtesy of C. Yokochi and J. W. Rohen, *Photographic Anatomy of the Human Body,* 2nd ed., 1979, IGAKU-SHOIN, Tokyo, New York.) (b) Diagram of a sagittal section illustrating the internal anatomy.

(a)

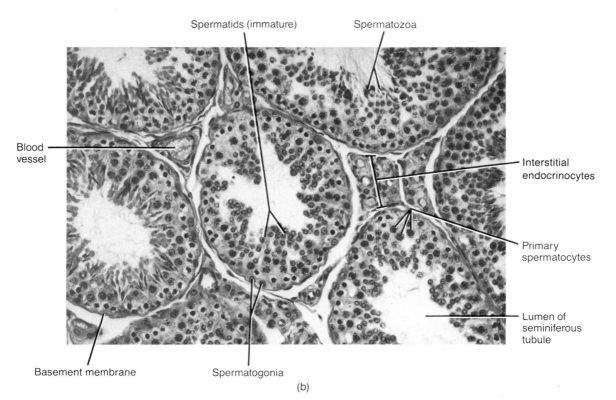

(b)

FIGURE 28-3 Histology of the testes. (a) Diagram of a cross section of a portion of a seminiferous tubule showing the stages of spermatogenesis. (b) Photomicrograph of an enlarged aspect of several seminiferous tubules at a magnification of 350×. (© 1983 by Michael H. Ross. Used by permission.)

FIGURE 28-3 (*Continued*) Histology of the testes. (c) Scanning electron micrograph of a seminiferous tubule at a magnification of 663×. (Courtesy of Richard K. Kessel and Randy H. Kardon, *Tissues and Organs: A Text-Atlas of Scanning Electron Microscopy.* Copyright © 1979 by Scientific American, Inc.)

tract. A spermatozoon is highly adapted for reaching and penetrating a female ovum. It is composed of a head, a midpiece, and a tail (Figure 28-4). Within the **head** are the nuclear material and the *acrosome,* which contains enzymes (hyaluronidase and proteinases) that effect penetration of the sperm cell into the ovum. Numerous mitochondria in the **midpiece** carry on the metabolism that provides energy for locomotion. The **tail,** a typical flagellum, propels the sperm along its way.

Testosterone and Inhibin

Secretions of the anterior pituitary gland assume a major role in the developmental changes associated with puberty. At the onset of puberty the anterior pituitary starts to secrete gonadotropic hormones called follicle-stimulating hormone (FSH) and luteinizing hormone (LH). Their release is controlled from the hypothalamus by gonadotropin releasing factor (GnRF). Once secreted, the gonadotropic hormones have profound effects on male reproductive organs. FSH acts on the seminiferous tubules to initiate spermatogenesis and stimulate sustentacular cells. LH also assists the seminiferous tubules to develop mature sperm, but its chief function is to stimulate the interstitial endocrinocytes to secrete the hormone testosterone (tes-TOS-te-rōn).

Testosterone is synthesized from cholesterol or acetyl coenzyme A in the testes. It is the principal male hormone (androgen) and has a number of effects on the body. It controls the development, growth, and maintenance of the male sex organs. It also stimulates bone growth, protein anabolism, sexual behavior, final maturation of sperm, and the development of male secondary sex characteristics. These characteristics, which appear at puberty, include muscular and skeletal development resulting in wide shoulders and narrow hips; body hair patterns that include pubic hair, axillary and chest hair (within hereditary limits), facial hair, and temporal hairline recession; and enlargement of the thyroid cartilage of the larynx producing deepening of the voice. Testosterone also stimulates descent of the testes just prior to birth.

The interaction of LH with testosterone illustrates the operation of another negative feedback system (Figure 28-5). LH stimulates the production of testosterone. But once the testosterone concentration in the blood reaches a certain level, it inhibits the release of GnRF by the hypothalamus. This inhibition, in turn, inhibits the release of LH by the anterior pituitary. Thus testosterone production is decreased. However, once the testosterone concentration in the blood decreases to a certain level, GnRF is released by the hypothalamus. This release of LH stimulates the release of LH by the anterior pituitary, and

(b)

FIGURE 28-4 Spermatozoa. (a) Diagram of the parts of a spermato-zoon. (b) Scanning electron micrograph of several spermatozoa at a magnification of 2,000×. (Courtesy of Fisher Scientific Company and S.T.E.M. Laboratories, Inc., Copyright 1975.)

stimulates testosterone production. Thus the testosterone-LH cycle is complete.

Inhibin is a protein hormone that has a direct effect on the anterior pituitary by inhibiting the secretion of FSH. FSH brings about spermatogenesis and stimulates sustentacular cells. Once the degree of spermatogenesis required for male reproductive functions has been achieved, sustentacular cells secrete inhibin. The hormone feeds back negatively to the anterior pituitary to inhibit FSH and thus to decrease spermatogenesis (Figure

FIGURE 28-5 Secretion, physiological effects, and control of testosterone and inhibin.

28-5). If spermatogenesis is proceeding too slowly, lack of inhibin production permits FSH secretion and an increased rate of spermatogenesis.

DUCTS

Ducts of the Testis

When the sperm mature, they are moved through the convoluted seminiferous tubules to the **straight tubules** (see Figure 28-2b). The straight tubules lead to a network of ducts in the testis called the **rete** (RĒ-tē) **testis.** Some of the cells lining the rete testis possess cilia that probably help move the sperm along. The sperm are next transported out of the testis.

Epididymis

The sperm are transported out of the testis through a series of coiled **efferent ducts** in the epididymis that empty into a single tube called the ductus epididymis. At this point, the sperm are morphologically mature.

The two **epididymides** (ep'-i-DID-i-mi-dēz') are comma-shaped organs. Each lies along the posterior border of the testis (see Figures 28-1 and 28-2) and consists mostly of a tightly coiled tube, the **ductus epididymis** (ep'-i-DID-i-mis). The larger, superior portion of the epididymis is known as the **head.** It consists of the efferent ducts that empty into the ductus epididymis. The **body** of the epididymis contains the ductus epididymis. The **tail** is the smaller, inferior portion. Within the tail, the ductus epididymis continues as the ductus deferens.

The ductus epididymis measures about 6 m (20 ft) in length and 1 mm in diameter. It is tightly packed within the epididymis, which measures only about 3.8 cm (1.5 inches). The ductus epididymis is lined with pseudostratified columnar epithelium, and its wall contains smooth muscle. The free surfaces of the columnar cells contain long, branching microvilli called *stereocilia* (Figure 28-6).

Functionally, the ductus epididymis is the site of sperm maturation. They require between 18 hours and 10 days to complete their maturation, that is, to become capable of fertilizing an ovum. The ductus epididymis also stores spermatozoa and propels them toward the urethra during ejaculation by peristaltic contraction of its smooth muscle. Spermatozoa may remain in storage in the ductus epididymis for up to four weeks. After that, they are reabsorbed.

Ductus (Vas) Deferens

Within the tail of the epididymis, the ductus epididymis becomes less convoluted, its diameter increases, and at this point it is referred to as the **ductus (vas) deferens** or **seminal duct** (see Figure 28-2). The ductus deferens, about 45 cm long (18 inches), ascends along the posterior border of the testis, penetrates the inguinal canal, and enters the pelvic cavity, where it loops over the side and down the posterior surface of the urinary bladder (see Figure 28-1). The dilated terminal portion of the ductus deferens is known as the **ampulla** (am-POOL-la). The ductus deferens is lined with pseudostratified epithelium and contains a heavy coat of three layers of muscle. Functionally, the ductus deferens stores sperm for up to several months and propels them toward the urethra during ejaculation by peristaltic contractions of the muscular coat.

CLINICAL APPLICATION

One method of sterilization of males is called **vasectomy,** in which a portion of each ductus deferens is removed. In the procedure, an incision is made in the scrotum, the ducts are located, each is tied in two places, and the portion between the ties is excised (see Figure 29-17a). Although sperm production continues in the testes, the sperm cannot reach the exterior because the ducts are cut and the sperm degenerate.

Traveling with the ductus deferens as it ascends in the scrotum are the testicular artery, autonomic nerves, veins that drain the testes, lymphatics, and the cremaster muscle. These structures constitute the **spermatic cord,** a supporting structure of the male reproductive system.

FIGURE 28-6 Histology of the ductus epididymis. (a) Photomicrograph of the ductus epididymis seen in cross section at a magnification of 160×. (b) Photomicrograph of an enlarged aspect of the mucosa of the ductus epididymis at a magnification of 350×. (© 1983 by Michael H. Ross. Used by permission.)

Pseudostratified epithelium with stereocilia

Lumen of ductus epididymis

Blood vessels

(a)

Stereocilia of pseudostratified epithelium

Nuclei of columnar cells

Smooth muscle

Basal cell

Connective tissue

(b)

The cremaster muscle, which also surrounds the testes, elevates the testes during sexual stimulation and exposure to cold. The spermatic cord passes through the **inguinal** (IN-gwin-al) **canal,** a slitlike passageway in the anterior abdominal wall just superior to the medial half of the inguinal ligament.

CLINICAL APPLICATION

The area of the inguinal canal is a weak spot in the abdominal wall. It is frequently the site of an **inguinal hernia**—a rupture or separation of a portion of the abdominal wall resulting in the protrusion of a part of an organ.

Ejaculatory Duct

Posterior to the urinary bladder are the **ejaculatory** (ē-JAK-yoo-la-tō'-rē) **ducts** (Figure 28-7). Each duct is about 2 cm (1 inch) long and is formed by the union of the duct from the seminal vesicle and ductus deferens. The ejaculatory ducts eject spermatozoa into the prostatic urethra.

Urethra

The **urethra** is the terminal duct of the system, serving as a passageway for spermatozoa or urine. In the male, the urethra passes through the prostate (PROS-tāt) gland, the urogenital diaphragm, and the penis. It measures about 20 cm (8 inches) in length and is subdivided into three parts (see Figures 28-1 and 28-8). The **prostatic urethra** is 2 to 3 cm (1 inch) long and passes through the prostate gland. It continues inferiorly and as it passes through the urogenital diaphragm, a muscular partition between the two ischiopubic rami, it is known as the **membranous** (MEM-bra-nus) **urethra.** The membranous portion is about 1 cm (½ inch) in length. As it passes through the corpus spongiosum of the penis, it is known as the **spongy (cavernous) urethra.** This portion is about 15 cm (6 inches) long. The spongy urethra enters the bulb of the penis and terminates at the external **urethral orifice.**

ACCESSORY GLANDS

Whereas the ducts of the male reproductive system store and transport sperm cells, the **accessory glands** secrete

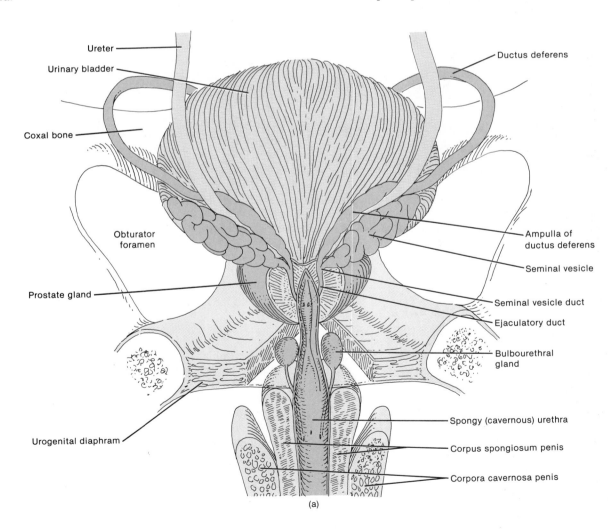

(a)

FIGURE 28-7 Male reproductive organs in relation to surrounding structures seen in posterior view. (a) Diagram.

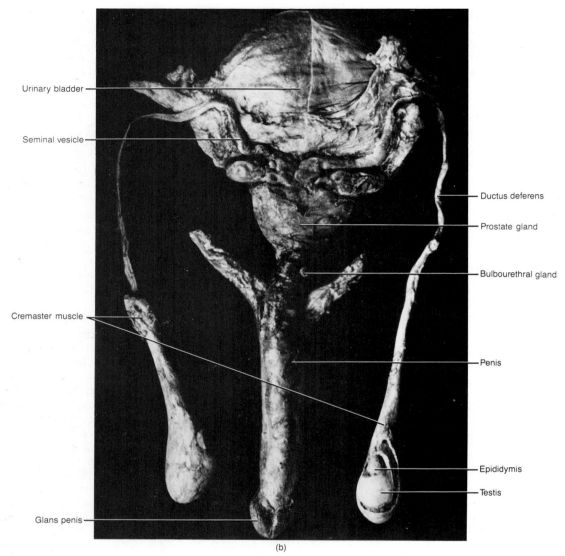

Urinary bladder

Seminal vesicle

Cremaster muscle

Glans penis

Ductus deferens

Prostate gland

Bulbourethral gland

Penis

Epididymis

Testis

(b)

FIGURE 28-7 (*Continued*) Male reproductive organs in relation to surrounding structures seen in posterior view. (b) Photograph. (Courtesy of C. Yokochi and J. W. Rohen, *Photographic Anatomy of the Human Body,* 2nd ed., 1979, IGAKU-SHOIN, Ltd., Tokyo, New York.)

the liquid portion of semen. The paired **seminal vesicles** (VES-i-kuls) are convoluted pouchlike structures, about 5 cm (2 inches) in length, lying posterior to and at the base of the urinary bladder in front of the rectum (see Figure 28-7). They secrete an alkaline, viscous fluid, rich in the sugar fructose, and pass it into the ejaculatory duct. This secretion contributes to sperm viability. It constitutes about 60 percent of the volume of semen.

The **prostate gland** is a single, doughnut-shaped gland about the size of a chestnut (see Figure 28-7). It is inferior to the urinary bladder and surrounds the superior portion of the urethra. The prostate secretes an alkaline fluid into the prostatic urethra. The prostatic secretion constitutes 13 to 33 percent of the volume of semen and contributes to sperm motility.

The paired **bulbourethral** (bul'-bō-yoo-RĒ-thral) **glands** are about the size of peas. They are located beneath the prostate on either side of the membranous urethra (see Figure 28-7). The bulbourethral glands secrete mucus

for lubrication and a substance that neutralizes urine. Their ducts open into the spongy urethra.

SEMEN

Semen (seminal fluid) is a mixture of sperm and the secretions of the seminal vesicles, the prostate gland, and the bulbourethral glands. The average volume of semen for each ejaculation is 2.5 to 6 ml, and the average range of spermatozoa ejaculated is 50 to 100 million/ml. When the number of spermatozoa falls below 20 million/ml, the male is likely to be sterile. The very large number is required because only a small percentage eventually reach the ovum. And, although only a single spermatozoon fertilizes an ovum, fertilization seems to require the combined action at the ovum of a larger number of them. The intercellular material of the cells covering the ovum presents a barrier to the sperm. This barrier is digested by the hyaluronidase and proteinases secreted by the acrosomes of sperm. However, it appears that a single sperm does not produce enough of these enzymes to dissolve the barrier. A passageway through which one may enter can be created only by the action of many.

Semen has a slightly alkaline pH of 7.35 to 7.50. The prostatic secretion gives semen a milky appearance, and fluids from the seminal vesicles and bulbourethral glands give it a mucoid consistency. Semen provides spermatozoa with a transportation medium and nutrients. It neutralizes the acid environment of the male urethra and the female vagina. It also contains enzymes which activate sperm after ejaculation.

Semen has been shown to contain an antibiotic, called *seminalplasmin,* which has the ability to destroy a number of bacteria. Its antimicrobial activity has been described as similar to that currently exerted by penicillin, streptomycin, and tetracyclines. Since both semen and the lower female reproductive tract contain bacteria, seminalplasmin may keep these bacteria under control to help ensure fertilization.

Once ejaculated into the vagina, liquid semen coagu-lates rapidly because of a clotting enzyme produced by the prostate that acts on a substance produced by the seminal vesicle. This clot liquefies in a few minutes because of another enzyme produced by the prostate gland. It is not clear why semen coagulates and liquefies in this manner.

PENIS

The **penis** is used to introduce spermatozoa into the vagina (Figure 28-8). The distal end of the penis is a slightly enlarged region called the **glans,** which means shaped like an acorn. Covering the glans is the loosly fitting **prepuce** (PRĒ-pyoos), or **foreskin.**

CLINICAL APPLICATION
Circumcision (*circumcido* = to cut around) is a surgical procedure in which part or all of the prepuce is removed. It is usually performed by the third or fourth day after birth (or on the eighth day as part of a Jewish religious rite).

Internally, the penis is composed of three cylindrical masses of tissue bound together by fibrous tissue. The two dorsolateral masses are called the **corpora cavernosa penis.** The smaller midventral mass, the **corpus spongiosum penis,** contains the spongy urethra. All three masses of tissue are erectile and contain blood sinuses. Under the influence of sexual stimulation, the arteries supplying the penis dilate, and large quantities of blood enter the blood sinuses. Expansion of these spaces compresses the veins draining the penis so most entering blood is retained. These vascular changes result in an **erection,** a parasympathetic reflex. The penis returns to its flaccid state when the arteries constrict and pressure on the veins is relieved (Chapter 29). During ejaculation, a sympathetic reflex, the smooth muscle sphincter at the base of the urinary bladder is closed because of the higher pressure in the urethra caused by expansion of the corpus spongiosum penis. Thus urine is not expelled during ejaculation and semen does not enter the urinary bladder.

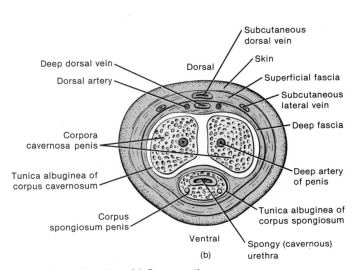

FIGURE 28-8 Internal structure of the penis. (a) Coronal section. (b) Cross section.

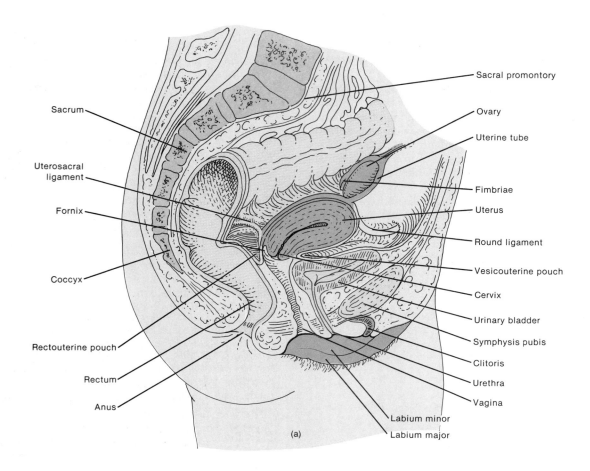

Sacral promontory
Sacrum
Ovary
Uterine tube
Uterosacral ligament
Fimbriae
Fornix
Uterus
Round ligament
Coccyx
Vesicouterine pouch
Cervix
Urinary bladder
Rectouterine pouch
Symphysis pubis
Rectum
Clitoris
Anus
Urethra
Vagina
Labium minor
Labium major

(a)

Sacrum
Sigmoid colon
Coccyx
Rectouterine pouch
Rectum
Vagina
Anus
Uterus
Urinary bladder
Symphysis pubis
Clitoris
Urethra
Hymen
Labium major
Labium minor

(b)

FIGURE 28-9 Female organs of reproduction and surrounding structures seen in sagittal section. (a) Diagram. (b) Photograph. (Courtesy of C. Yokochi and J. W. Rohen, *Photographic Anatomy of the Human Body,* 2nd ed., 1979, IGAKU-SHOIN, Ltd., Tokyo, New York.)

FEMALE REPRODUCTIVE SYSTEM

The female organs of reproduction include the ovaries, which produce ova (eggs); the uterine (fallopian) tubes, which transport the ova to the uterus (womb); the vagina; and external organs that constitute the vulva, or pudendum (Figure 28-9). The mammary glands also are considered part of the female reproductive system.

OVARIES

The **ovaries,** or female gonads, are paired glands resembling unshelled almonds in size and shape. They are positioned in the upper pelvic cavity, one on each side of the uterus. The ovaries are maintained in position by a series of ligaments (Figures 28-10 and 28-11). They are attached to the broad ligament of the uterus, which is itself part of the parietal peritoneum, by a double-layered fold of peritoneum called the **mesovarium** which surrounds the ovary and ovarian ligament. The ovaries are anchored to the uterus by the **ovarian ligament** and are attached to the pelvic wall by the **suspensory ligament.** Each ovary also contains a hilus, the point of entrance for blood vessels and nerves.

The microscope reveals that each ovary consists of the following parts (Figure 28-12).

1. Germinal epithelium. A layer of simple cuboidal epithelium that covers the free surface of the ovary and serves as a source of ovarian follicles.

2. Tunica albuginea. A capsule of collagenous connective tissue immediately deep to the germinal epithelium.

3. Stroma. A region of connective tissue deep to the tunica albuginea and composed of an outer, dense layer called the *cortex* and an inner, loose layer known as the *medulla.* The cortex contains ovarian follicles.

4. Ovarian follicles. Ova and their surrounding tissues in various stages of development.

5. Graafian (GRAF-ē-an) follicle. An endocrine gland made up of a mature ovum and its surrounding tissues. The Graafian follicle secretes hormones called estrogens.

6. Corpus luteum. Glandular body that develops from a Graafian follicle after extrusion of an ovum (ovulation). The corpus luteum produces the hormones progesterone, estrogens, and relaxin.

The ovaries produce ova, discharge ova (ovulation), and secrete the female sex hormones, progesterone, estrogens, and relaxin. They are analogous to the testes of the male reproductive system.

UTERINE TUBES

The female body contains two **uterine (fallopian) tubes** that extend laterally from the uterus and transport the ova from the ovaries to the uterus (see Figure 28-10). Measuring about 10 cm (4 inches) long, the tubes are positioned between the folds of the broad ligaments of the uterus. The funnel-shaped open distal end of each tube, called the **infundibulum,** lies close to the ovary but is not attached to it and is surrounded by a fringe of fingerlike projections called **fimbriae (FIM-brē-ē).** From the infundibulum the uterine tube extends medially and inferiorly and attaches to the superior lateral angle of

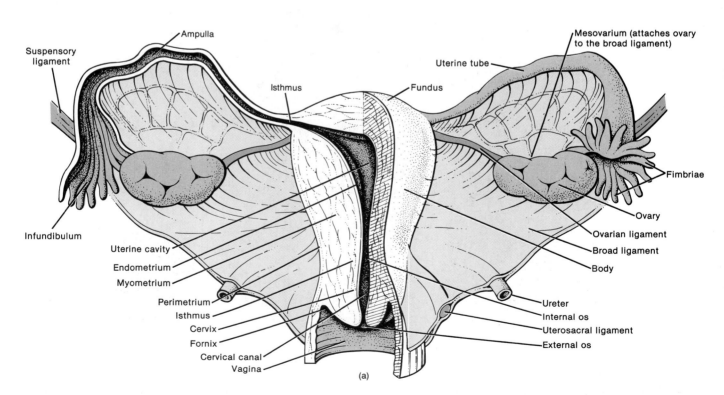

FIGURE 28-10 Uterus and associated structures seen in anterior view. (a) Diagram. The left side of the figure has been sectioned to show internal structures.

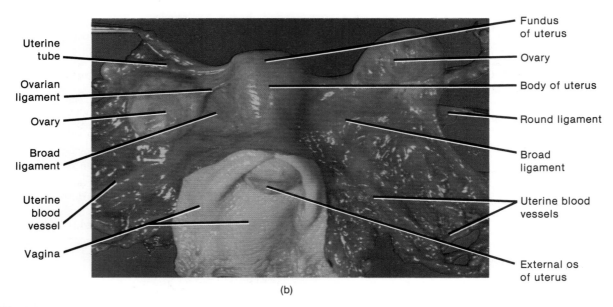

Uterine tube

Ovarian ligament

Ovary

Broad ligament

Uterine blood vessel

Vagina

Fundus of uterus

Ovary

Body of uterus

Round ligament

Broad ligament

Uterine blood vessels

External os of uterus

(b)

FIGURE 28-10 (*Continued*) Uterus and associated structures seen in anterior view. (b) Photograph. (Courtesy of C. Yokochi and J. W. Rohen, *Photographic Anatomy of the Human Body,* 2nd ed., 1979, IGAKU-SHOIN, Ltd., Tokyo, New York.)

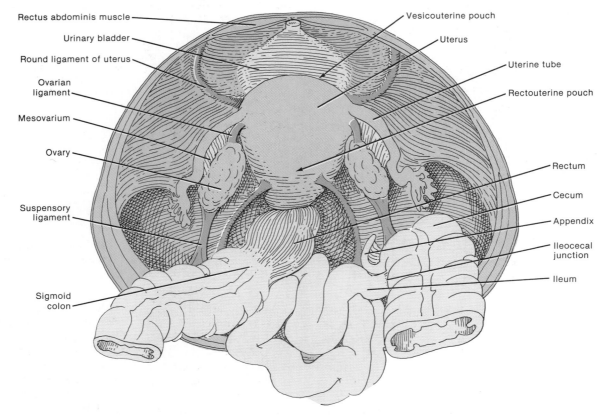

Rectus abdominis muscle

Urinary bladder

Round ligament of uterus

Ovarian ligament

Mesovarium

Ovary

Suspensory ligament

Sigmoid colon

Vesicouterine pouch

Uterus

Uterine tube

Rectouterine pouch

Rectum

Cecum

Appendix

Ileocecal junction

Ileum

FIGURE 28-11 Female pelvis viewed from above.

(a)

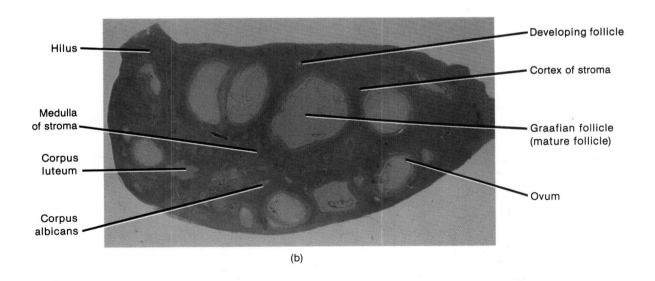

(b)

FIGURE 28-12 Histology of the ovary. (a) Diagram of the parts of an ovary seen in sectional view. The arrows indicate the sequence of developmental stages that occur as part of the ovarian cycle. (b) Photograph of an ovary in sectional view. (Courtesy of C. Yokochi and J. W. Rohen, *Photographic Anatomy of the Human Body,* 2nd ed., 1979, IGAKU-SHOIN, Ltd., Tokyo, New York.)

the uterus. The **ampulla** of the uterine tube is the widest, longest portion, making up about two-thirds of its length. The **isthmus** of the uterine tube is the short, narrow, thick-walled portion that joins the uterus.

Histologically, the uterine tubes are composed of three layers. The internal **mucosa** contains ciliated columnar cells and secretory cells, which are believed to aid the movement and nutrition of the ovum. The middle layer, the **muscularis,** is composed of a thick, circular region

of smooth muscle and an outer, thin, longitudinal region of smooth muscle. Peristaltic contractions of the muscularis and the ciliated action of the mucosa help move the ovum down into the uterus. The outer layer of the uterine tubes is a serous membrane, the **serosa.**

About once a month an immature ovum ruptures from the surface of the ovary near the infundibulum of the uterine tube, a process called **ovulation.** The ovum is swept into the tube by the ciliary action of the epithelium of

Germinal epithelium

Tunica albuginea

Primary follicles

Primordial follicles

Cortex of stroma

(c)

Cytoplasm of oocyte

Follicle cells (2 layers)

Nucleus of oocyte

(d)

FIGURE 28-12 (*Continued*) Histology of the ovary. (c) Photomicrograph of the cortex of the stroma of an ovary at a magnification of 60×. (d) Photomicrograph of an enlarged aspect of a primary follicle at a magnification of 300×. (e) Photomicrograph of an enlarged aspect of a secondary follicle at a magnification of 160×. The theca interna and theca externa are connective tissue coverings around the secondary follicle. (Photomicrographs © 1983 by Michael H. Ross. Used by permission.)

Follicular cavity

Follicle cells

Theca interna

Theca externa

Oocyte

Zona pellucida

(e)

the infundibulum. It is then moved along the tube by ciliary action supplemented by the peristaltic contractions of the muscularis. If the ovum is fertilized by a sperm cell, it usually occurs in the ampulla of the uterine tube. Fertilization may occur at any time up to about 24 hours following ovulation. The fertilized ovum, now referred to as a blastocyst, descends into the uterus within 7 days. An unfertilized ovum disintegrates.

CLINICAL APPLICATION

Fertilization may occur while the ovum is free in the pelvic cavity, and implantation may take place on one of the pelvic viscera. *Pelvic implantations* usually fail because the developing fertilized ovum does not make vascular connection with the maternal blood supply. On occasion, a fertilized ovum fails to descend to the uterus and implants in the uterine tube. In the case of a *tubular implantation,* the pregnancy must be terminated surgically before the tube ruptures. Both pelvic and tubular implantations are referred to as **ectopic (ek-TOP-ik) pregnancies.**

UTERUS

The site of menstruation, implantation of a fertilized ovum, development of the fetus during pregnancy, and labor is the **uterus.** Situated between the urinary bladder and the rectum, the uterus is shaped like an inverted pear (see Figures 28-9, 28-10, and 28-11). Before the first pregnancy, the adult uterus measures approximately 7.5 cm (3 inches) long, 5 cm (2 inches) wide, and 2.5 cm (1 inch) thick.

Anatomical subdivisions of the uterus include the dome-shaped portion above the uterine tubes called the **fundus,** the major tapering central portion called the **body,** and the inferior narrow portion opening into the vagina called the **cervix.** Between the body and the cervix is the **isthmus** (IS-mus), a constricted region about 1 cm (½ inch) long. The interior of the body of the uterus is called the **uterine cavity,** and the interior of the narrow cervix is called the **cervical canal.** The junction of the isthmus with the cervical canal is the **internal os.** The **external os** is the place where the cervix opens into the vagina.

Normally the uterus is flexed between the uterine body and the cervix. This is called **anteflexion.** In this position, the body of the uterus projects anteriorly and slightly superiorly over the urinary bladder, and the cervix projects inferiorly and posteriorly and enters the anterior wall of the vagina at nearly a right angle. Several structures that are either extensions of the parietal peritoneum or fibromuscular cords, referred to as ligaments, maintain the position of the uterus. The paired **broad ligaments** are double folds of parietal peritoneum attaching the uterus to either side of the pelvic cavity. Uterine blood vessels and nerves pass through the broad ligaments. The paired **uterosacral ligaments,** also peritoneal extensions, lie on either side of the rectum and connect the uterus to the sacrum. The **cardinal (lateral cervical) ligaments** extend below the bases of the broad ligaments between the pelvic wall and the cervix and vagina. These ligaments contain smooth muscle, uterine blood vessels, and nerves

and are the chief ligaments that maintain the position of the uterus and help to keep it from dropping down into the vagina. The **round ligaments** are bands of fibrous connective tissue between the layers of the broad ligament. They extend from a point on the uterus just below the uterine tubes to a portion of the external genitalia.

CLINICAL APPLICATION

Although the ligaments normally maintain the anteflexed position of the uterus, they also afford the uterine body some movement. As a result, the uterus may become malpositioned. A posterior tilting of the uterus is called **retroflexion.**

Histologically, the uterus consists of three layers of tissue. The outer layer, the **perimetrium** or **serosa,** is part of the visceral peritoneum. Laterally, it becomes the broad ligament. Anteriorly, it is reflected over the urinary bladder and forms a shallow pouch, the **vesicouterine** (ves'-i-kō-YOO-ter-in) **pouch** (see Figure 28-9). Posteriorly, it is reflected onto the rectum and forms a deep pouch, the **rectouterine** (rek-tō-YOO-ter-in) **pouch**—the lowest point in the pelvic cavity.

The middle layer of the uterus, the **myometrium,** forms the bulk of the uterine wall (Figure 28-13). This layer consists of three layers of smooth muscle fibers and is thickest in the fundus and thinnest in the cervix. During childbirth, coordinated contractions of the muscles help to expel the fetus from the body of the uterus.

The inner layer of the uterus, the **endometrium,** is a mucous membrane composed of two principal layers. The *stratum functionalis,* the layer closer to the uterine cavity, is shed during menstruation. The second layer, the *stratum basalis* (bā-SAL-is), is permanent and produces a new functionalis following menstruation. The endometrium contains numerous glands.

CLINICAL APPLICATION

More and more physicians are beginning to use **colposcopy** (kol-POS-ko-pē) in order to evaluate the status of the mucosa of the vagina and cervix. Colposcopy is the direct examination of the vaginal and cervical mucosa with a magnifying device (a colposcope) similar to a low-power binocular microscope. Various instruments that magnify the mucous membrane from about 6 to 40 times its actual size are commercially available. The application of a 3 percent solution of acetic acid removes mucus and enhances the appearance of mucosal columnar epithelium.

Blood is supplied to the uterus by branches of the internal iliac artery called *uterine arteries.* Branches called *arcuate arteries* are arranged in a circular fashion in the myometrium and give off *radial arteries* that penetrate deeply into the myometrium (Figure 28-14). Just before the branches enter the endometrium, they divide into two kinds of arterioles. The *straight arteriole* terminates in the basalis and supplies it with the materials necessary to regenerate the functionalis. The *spiral arteriole* penetrates the functionalis and changes markedly during the menstrual cycle. The uterus is drained by the *uterine veins.*

Simple columnar epithelium

Lumen of uterus

Endometrial glands

Stratum functionalis

Stratum basalis

Myometrium

(a)

Simple columnar epithelium

Lumen of uterus

Stratum functionalis

Stratum basalis

Endometrial glands

(b)

FIGURE 28-13 Histology of the uterus. (a) Photomicrograph of a portion of the uterine wall in which the stratum functionalis is in an early stage of proliferation at a magnification of 25×. (b) Photomicrograph of a portion of the uterine wall in which the stratum functionalis is in an advanced stage of proliferation at a magnification of 25×. Note the increased thickness of the stratum functionalis and the long, straight nature of the glands. (© 1983 by Michael H. Ross. Used by permission.)

ENDOCRINE RELATIONS: MENSTRUAL AND OVARIAN CYCLES

The principal events of the menstrual cycle can be correlated with those of the ovarian cycle and changes in the endometrium. All are hormonally controlled events.

The **menstrual cycle** is a series of changes in the endometrium of a nonpregnant female. Each month the endometrium is prepared to receive a fertilized ovum. An implanted fertilized ovum eventually develops into an embryo and then a fetus, which normally remains in the uterus until delivery. If no fertilization occurs, the stratum

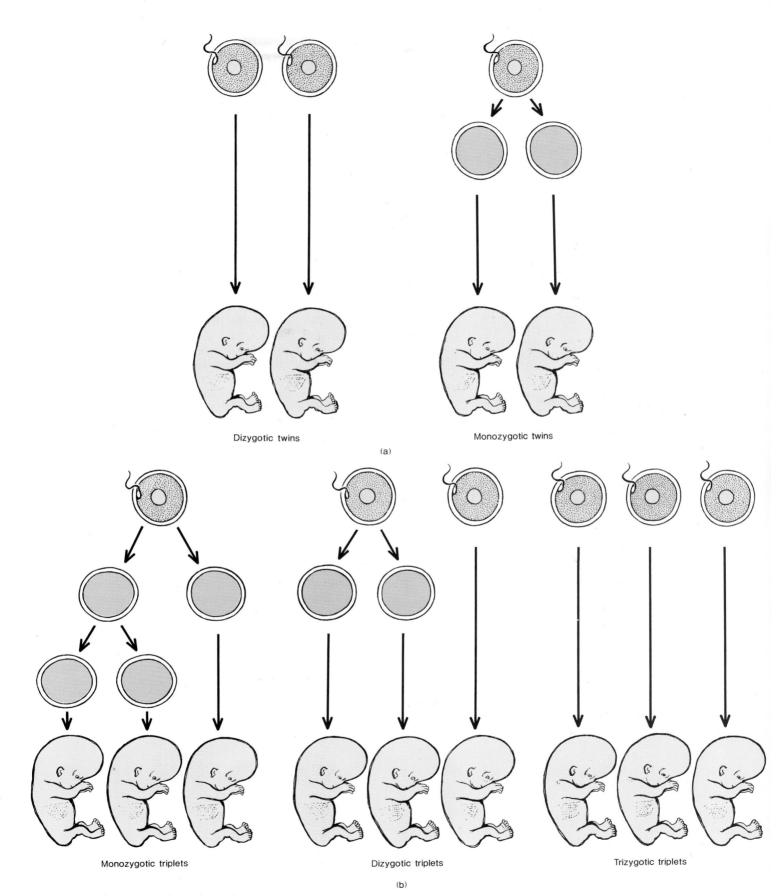

FIGURE 29-5 Possible embryonic origins of (a) twins and (b) triplets.

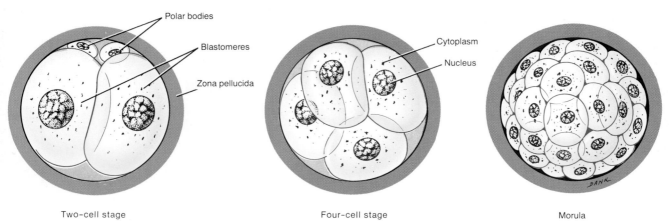

Two-cell stage

Four-cell stage

Morula

FIGURE 29-6 Formation of the morula.

creases the number of cells, it does not result in an increase in the size of the developing organism.

The first cleavage is completed after about 36 hours, and each succeeding division takes slightly less time (Figure 29-6). By the second day after conception, the second cleavage is completed. By the end of the third day there are 16 cells. The progressively smaller cells produced are called **blastomeres** (BLAS-tō-mērz). A few days after fertilization the successive cleavages have produced a solid mass of cells, the **morula** (MOR-yoo-la) or mulberry, which is about the same size as the original zygote.

Development of the Blastocyst

As the number of cells in the morula increases, it moves from the original site of fertilization down through the ciliated uterine tube toward the uterus and enters the uterine cavity. By this time, the dense cluster of cells is altered to form a hollow ball of cells. The mass is now referred to as a **blastocyst** (Figure 29-7).

The blastocyst is differentiated into an outer covering of cells called the **trophectoderm** (trō-FEK-tō-derm), an **inner cell mass,** and an internal fluid-filled cavity called the **blastocoel** (BLAS-tō-sēl). The trophectoderm ultimately forms part of the membranes composing the fetal portion of the placenta; the inner cell mass develops into the embryo.

Implantation

The attachment of the blastocyst to the endometrium occurs 7 to 8 days after fertilization and is called **implantation** (Figure 29-8). At this time, the endometrium is in its postovulatory phase. During implantation, the cells of the trophectoderm secrete an enzyme that enables the blastocyst to penetrate the uterine lining and become buried in the endometrium, usually on the posterior wall of the fundus or body of the uterus. The blastocyst, now only about $\frac{1}{100}$ of an inch in diameter, becomes oriented so that the inner cell mass is toward the endometrium. Implantation enables the blastocyst to absorb nutrients from the glands and blood vessels of the endometrium for its subsequent growth and development.

A summary of the principal events associated with fertilization and implantation is shown in Figure 29-9.

CLINICAL APPLICATION

In the early months of pregnancy, some females develop **hyperemesis gravidarum (morning sickness),** characterized by nausea and vomiting. Although the exact cause is unknown, it is possible that the degenerative products of digested portions of the endometrium during implantation may be responsible. Another possible cause is high levels of estrogens secreted by the placenta.

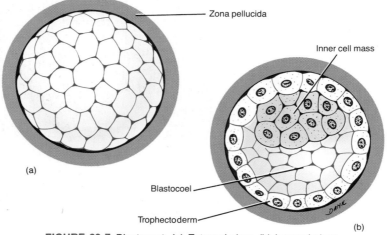

FIGURE 29-7 Blastocyst. (a) External view. (b) Internal view.

FIGURE 29-8 Implantation. (a) External view of the blastocyst in relation to the endometrium of the uterus about 5 days after fertilization. (b) Internal view of the blastocyst in relation to the endometrium about 6 days after fertilization. (c) Internal view of the blastocyst at implantation about 7 days after fertilization. (d) Photomicrograph of implantation. (Courtesy of Roberts Rugh and Landrum B. Shettles, M.D., with Richard N. Einhorn, *From Conception to Birth: The Drama of Life's Beginnings,* Harper & Row, Publishers, Inc., New York, 1971.)

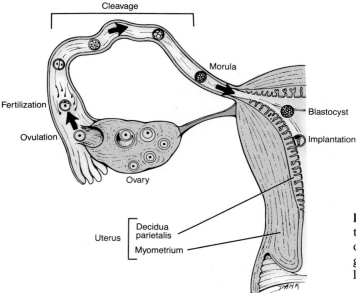

FIGURE 29-9 Summary of events associated with fertilization and implantation.

External Human Fertilization

On July 12, 1978, Louise Joy Brown was born near Manchester, England. Her birth was the first recorded case of **external human fertilization,** fertilization in a glass dish. The procedure developed for external human fertilization is believed to have been carried out as follows. The female is given FSH soon after menstruation, so that several ova, rather than the typical single one, will be produced (superovulation). Administration of LH may also ensure the maturation of the ova. Next, a small incision is made near the umbilicus, and the ova are aspirated from the follicles and placed in a medium that simulates the fluids in the female reproductive tract. The ova are then transferred to a solution of the male's sperm. Once fertilization has taken place, the fertilized ovum is put in another medium and is observed for cleavage. When the fertilized ovum reaches the 8-cell or 16-cell stage, it is introduced into the uterus for implantation and subsequent growth. The growth and development sequences that occur are similar to those in internal fertilization.

EMBRYONIC DEVELOPMENT

The first two months of development are generally considered the **embryonic period.** During this period the developing human is called an **embryo.** The months of development after the second month are considered the **fetal period,** and during this time the developing human is called a **fetus.** By the end of the embryonic period the rudiments of all the principal adult organs are present, the embryonic membranes are developed, and the placenta is functioning.

Beginnings of Organ Systems

Following implantation, the inner cell mass of the blastocyst begins to differentiate into the three **primary germ layers:** ectoderm, endoderm, and mesoderm. They are the embryonic tissues from which all tissues and organs of the body will develop. The various movements of cell groups leading to the establishment of the primary germ layers are referred to as **gastrulation.**

In the human, the germ layers form so quickly that it is difficult to determine the exact sequence of events. Before implantation, a layer of **ectoderm** (the trophectoderm) already has formed around the blastocoel (see Figure 29-8b). The trophectoderm will become part of the chorion, one of the fetal membranes. Within 8 days after fertilization, the inner cell mass moves downward so a space called the **amniotic cavity** lies between the inner cell mass and the trophectoderm. The bottom layer of the inner cell mass develops into an **endodermal** germ layer.

About the twelfth day after fertilization, striking changes appear (Figure 29-10a). A layer of cells from the inner cell mass has grown around the top of the amniotic cavity. These cells will become the amnion, another fetal membrane. The cells below the cavity are called the **embryonic disc.** They will form the embryo. At this stage, the embryonic disc contains ectodermal and endodermal cells; the mesodermal cells are scattered external to the disc. The cells of the endodermal layer have been dividing rapidly, so that groups of them now extend downward in a circle, forming the yolk sac, another fetal membrane. The **mesodermal** cells also have been dividing, and many have left the area of the embryonic disc and can be seen around the structures that are becoming fetal membranes.

About the fourteenth day, the cells of the embryonic disc differentiate into three distinct layers: the upper ectoderm, the middle mesoderm, and the lower endoderm (Figure 29-10b). At this time the two ends of the embryonic disc draw together, squeezing off the yolk sac. The resulting cavity inside the disc is the endoderm-lined **primitive gut.** The mesoderm in the disc soon splits into two layers, and the space between the layers becomes the **extraembryonic coelom.**

As the embryo develops (Figure 29-10c), the endoderm becomes the epithelial lining the digestive tract and respiratory tract and a number of other organs. The mesoderm forms the peritoneum, muscle, bone, and other connective tissue. The ectoderm develops into the skin and nervous system. Exhibit 29-1 provides more details about the fates of these primary germ layers.

EXHIBIT 29-1

STRUCTURES PRODUCED BY THE THREE PRIMARY GERM LAYERS

ENDODERM	MESODERM	ECTODERM
Epithelium of digestive tract (except the oral cavity and anal canal) and the epithelium of its glands.	All skeletal, most smooth, and all cardiac muscle.	All nervous tissue.
Epithelium of urinary bladder, gallbladder, and liver.	Cartilage, bone, and other connective tissues.	Epidermis of skin.
Epithelium of pharynx, auditory tube, tonsils, larynx, trachea, bronchi, and lungs.	Blood, bone marrow, and lymphoid tissue.	Hair follicles, arrector pili muscles, nails, and epithelium of skin glands (sebaceous and sudoriferous).
Epithelium of thyroid, parathyroid, pancreas, and thymus glands.	Endothelium of blood vessels and lymphatics.	Lens, cornea, and optic nerve of eye and internal eye muscles.
Epithelium of prostate and bulbourethral glands, vagina, vestibule, urethra, and associated glands such as the greater vestibular and lesser vestibular glands.	Dermis of skin.	Internal and external ear.
	Fibrous tunic and vascular tunic of eye.	Neuroepithelium of sense organs.
	Middle ear.	Epithelium of oral cavity, nasal cavity, paranasal sinuses, salivary glands, and anal canal.
	Mesothelium of coelomic and joint cavities.	Epithelium of pineal gland, hypophysis, and adrenal medulla.
	Epithelium of kidneys and ureters.	
	Epithelium of adrenal cortex.	
	Epithelium of gonads and genital ducts.	

Embryonic Membranes

During the embryonic period, the **embryonic membranes** form (Figure 29-11). These membranes lie outside the embryo and protect and nourish the embryo and later the fetus. The membranes are the yolk sac, amnion, chorion (KŌ-rē-on), and allantois (a-LAN-tō-is).

The **yolk sac** is an endoderm-lined membrane that, in many species, provides the primary or exclusive nutrient for the embryo. However, the human embryo re-

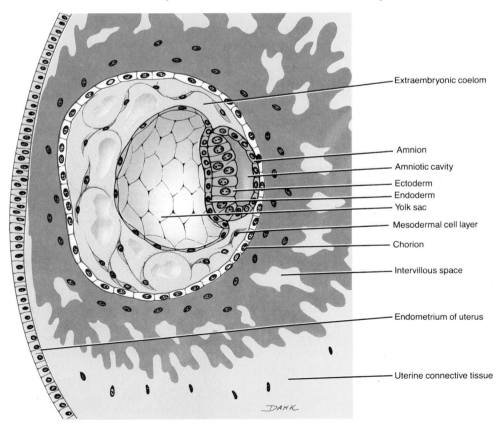

(a)

FIGURE 29-10 Formation of the primary germ layers and associated structures. (a) Internal view of the developing embryo about 12 days after fertilization.

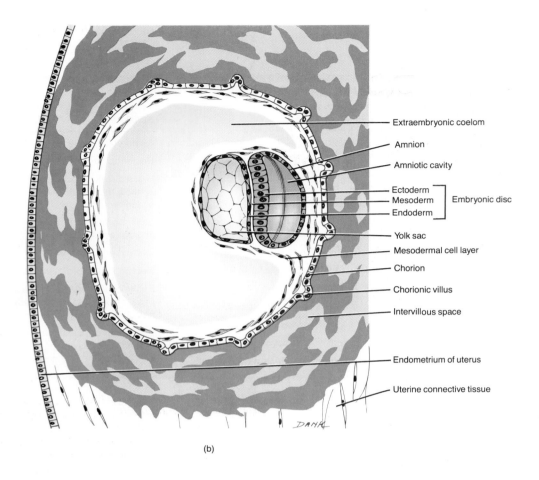

Extraembryonic coelom

Amnion

Amniotic cavity

Ectoderm
Mesoderm } Embryonic disc
Endoderm

Yolk sac

Mesodermal cell layer

Chorion

Chorionic villus

Intervillous space

Endometrium of uterus

Uterine connective tissue

(b)

Embryo
Head   Heart   Tail

Extraembryonic coelom

Amniotic cavity
Chorion

Body stalk (future umbilical cord)

Yolk sac

Chorionic villi

Intervillous space

(c)

FIGURE 29-10 (Continued) Formation of the primary germ layers and associated structures. (b) Internal view of the developing embryo about 14 days after fertilization. (c) External view of the developing embryo about 25 days after fertilization.

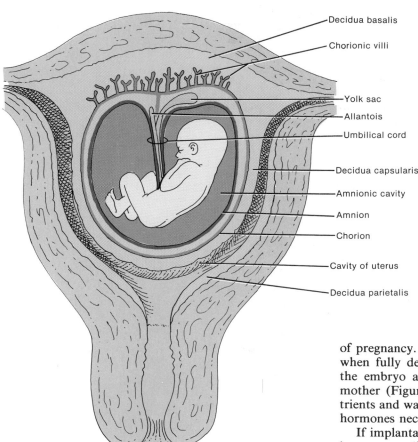

Decidua basalis

Chorionic villi

Yolk sac

Allantois

Umbilical cord

Decidua capsularis

Amnionic cavity

Amnion

Chorion

Cavity of uterus

Decidua parietalis

FIGURE 29-11 Embryonic membranes.

ceives its nourishment from the endometrium and the yolk sac remains small (see Figure 29-14b–d). During an early stage of development it becomes a nonfunctional part of the umbilical cord.

The **amnion** is a thin, protective membrane that initially overlies the embryonic disc and is formed by the eighth day following fertilization. As the embryo grows, the amnion entirely surrounds the embryo and becomes filled with **amniotic fluid** (see Figure 29-14c–g). Amniotic fluid serves as a shock absorber for the fetus. The amnion usually ruptures just before birth and with its fluid constitutes the "bag of waters."

The **chorion** is derived from the trophectoderm of the blastocyst and its associated mesoderm. It surrounds the embryo and, later, the fetus. Eventually the chorion becomes the principal part of the placenta, the structure through which materials are exchanged between mother and fetus. The amnion also surrounds the fetus and eventually fuses to the inner layer of the chorion.

The **allantois** is a small vascularized membrane. Later its blood vessels serve as connections in the placenta between mother and fetus. This connection is the umbilical cord.

Placenta and Umbilical Cord

Development of the placenta, the third major event of the embryonic period, is accomplished by the third month of pregnancy. The **placenta** has the shape of a flat cake when fully developed and is formed by the chorion of the embryo and a portion of the endometrium of the mother (Figure 29-12). It provides an exchange of nutrients and wastes between fetus and mother and secretes hormones necessary to maintain pregnancy.

If implantation occurs, a portion of the endometrium becomes modified and is known as the **decidua** (dē-SID-yoo-a). The decidua includes all but the deepest layer of the endometrium and is shed when the fetus is delivered. Different regions of the decidua are named on the basis of their positions relative to the site of the implanted ovum (see Figures 29-11 and 29-13). The *decidua parietalis* (pa-rī-e-TAL-is) is the portion of the modified endometrium that lines the entire pregnant uterus, except for the area where the placenta is forming. The *decidua capsularis* is the portion of the endometrium between the embryo and the uterine cavity. The *decidua basalis* is the portion of the endometrium between the chorion and the muscularis of the uterus. The decidua basalis becomes the maternal part of the placenta.

During embryonic life, fingerlike projections of the chorion, called **chorionic villi** (kō'-rē-ON-ik VIL-ī), grow into the decidua basalis of the endometrium (see Figure 29-12a). These will contain fetal blood vessels of the allantois. They continue growing until they are bathed in maternal blood sinuses called **intervillous** (in-ter-VIL-us) **spaces.** Thus, maternal and fetal blood vessels are brought into proximity. It should be noted, however, that maternal and fetal blood do not normally mix. Oxygen and nutrients from the mother's blood diffuse into the capillaries of the villi. From the capillaries the nutrients circulate into the umbilical vein. Wastes leave the fetus through the umbilical arteries, pass into the capillaries of the villi, and diffuse into the maternal blood. The **umbilical cord** consists of an outer layer of amnion containing the umbilical arteries and umbilical vein, supported internally by mucous connective tissue from the allantois called Wharton's jelly (see Figure 29-14b–h).

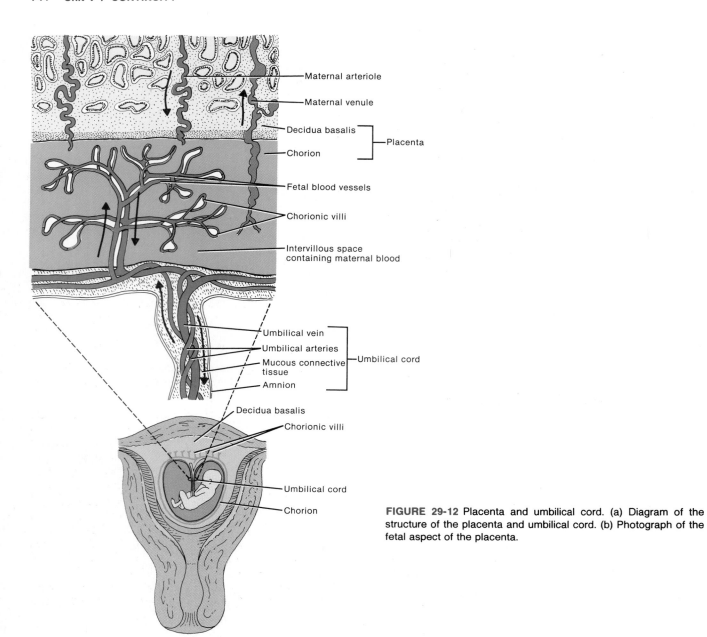

(a)

FIGURE 29-12 Placenta and umbilical cord. (a) Diagram of the structure of the placenta and umbilical cord. (b) Photograph of the fetal aspect of the placenta.

(b)

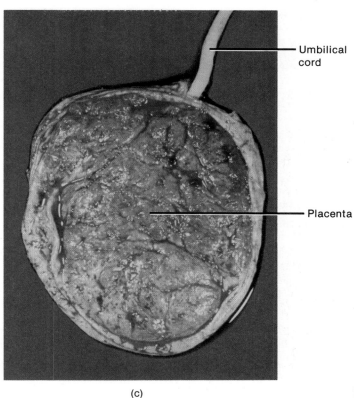

Umbilical cord

Placenta

(c)

FIGURE 29-12 (*Continued*) Placenta and umbilical cord. (c) Photograph of the maternal aspect of the placenta. (Courtesy of C. Yokochi and J. W. Rohen, *Photographic Anatomy of the Human Body,* 2nd ed., 1979, IGAKU-SHOIN, Ltd., Tokyo, New York.)

FETAL GROWTH

During the **fetal period,** organs established by the primary germ layers grow rapidly. The organism takes on a human appearance. A summary of changes associated with the fetal period is presented in Exhibit 29-2. Representative photos of embryos and fetuses are shown in various stages of development in Figure 29-14.

CLINICAL APPLICATION

At delivery, the placenta detaches from the uterus and is referred to as the **afterbirth.** At this time, the umbilical cord is severed, leaving the baby on its own. The scar that marks the site of the entry of the fetal umbilical cord into the abdomen is the **umbilicus (navel).**

Pharmaceutical houses use human placenta as a source of hormones, drugs, and blood. Portions of placentas are also used for burn coverage. The placental and umbilical cord veins are used in blood vessel grafts.

CLINICAL APPLICATION

Physicians can see into the uteri of pregnant women without exposing them to the known dangers of x-rays and without pain or intrusion. One such technique, called **ultrasound** (also **sonography** or **ultrasonography**), uses high frequency, inaudible sound waves, which are directed into the abdomen of the mother-to-be and then reflected back to a receiver. The reflected waves give a visual "echo" of what's inside the uterus. This echo is transformed electronically into an image on a screen. Another technique is the **Doppler Detector,** a fetal monitor that registers the baby's heart rate. It is used routinely in many hospitals during labor.

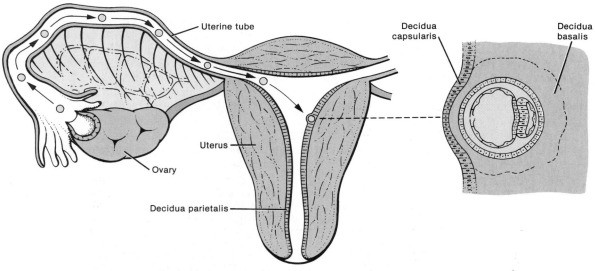

Uterine tube

Decidua capsularis

Decidua basalis

Uterus

Ovary

Decidua parietalis

FIGURE 29-13 Regions of the decidua.

EXHIBIT 29-2
CHANGES ASSOCIATED WITH FETAL GROWTH

END OF MONTH	APPROXIMATE SIZE AND WEIGHT	REPRESENTATIVE CHANGES
1	0.6 cm (³⁄₁₆ inch)	Eyes, nose, and ears not yet visible. Backbone and vertebral canal form. Small buds that will develop into arms and legs form. Heart forms and starts beating. Body systems begin to form.
2	3 cm (1¼ inches) 1 g (¹⁄₃₀ oz)	Eyes far apart, eyelids fused, nose flat. Ossification begins. Limbs become distinct as arms and legs. Digits are well formed. Major blood vessels form. Many internal organs continue to develop.
3	7.5 cm (3 inches) 28 g (1 oz)	Eyes almost fully developed but eyelids still fused, nose develops bridge, and external ears are present. Ossification continues. Appendages are fully formed and nails develop. Heartbeat can be detected. Body systems continue to develop.
4	18 cm (6½–7 inches) 113 g (4 oz)	Head large in proportion to rest of body. Face takes on human features and hair appears on head. Skin bright pink. Many bones ossified, and joints begin to form. Continued development of body systems.
5	25–30 cm (10–12 inches) 227–454 g (½–1 lb)	Head less disproportionate to rest of body. Fine hair (lanugo) covers body. Skin still bright pink. Rapid development of body systems.
6	27–35 cm (11–14 inches) 567–681 g (1¼—1½ lb)	Head becomes even less disproportionate to rest of body. Eyelids separate and eyelashes form. Skin wrinkled and pink.
7	32–42 cm (13–17 inches) 1,135–1,362 g (2½–3 lb)	Head and body become more proportionate. Skin wrinkled and pink. Seven-month fetus (premature baby) is capable of survival.
8	41–45 cm (16½–18 inches) 2,043–2,270 g (4½–5 lb)	Subcutaneous fat deposited. Skin less wrinkled. Testes descend into scrotum. Bones of head are soft. Chances of survival much greater at end of eighth month.
9	50 cm (20 inches) 3,178–3,405 g (7–7½ lb)	Additional subcutaneous fat accumulates. Lanugo shed. Nails extend to tips of fingers and maybe even beyond.

In the United States today, most obstetricians prescribe ultrasound examinations only when there is some clinical question about the normal progress of the pregnancy. By far the most common use of diagnostic ultrasound is to determine true fetal age when the date of conception is unknown or mistaken by the mother.

Ultrasound has gained a wide application beyond obstetrics as well, in detecting the presence of tumors, gallstones, and other abnormal internal masses.

HORMONES OF PREGNANCY

The corpus luteum is maintained throughout pregnancy. For about 8 to 10 weeks after fertilization it continues to secrete **estrogens** and **progesterone.** Both these hormones maintain the lining of the uterus during pregnancy and prepare the mammary glands to secrete milk. The amount secreted by the corpus luteum, however, is only slightly more than that produced after ovulation in a normal menstrual cycle. The high levels of estrogens and progesterone needed to maintain pregnancy and develop the mammary glands for lactation are provided by the placenta.

The chorion of the placenta secretes **human chorionic gonadotropin (HCG).** The primary role of HCG seems to be to maintain the activity of the corpus luteum, especially with regard to continuous progesterone secretion—an activity necessary for the continued attachment of the fetus to the lining of the uterus (Figure 29-15).

CLINICAL APPLICATION
Human chorionic gonadotropin is excreted in the urine of pregnant women from about the eighth day of pregnancy reaching its peak of excretion about the eighth week of pregnancy. The HCG level decreases sharply during the fourth and fifth months and then levels off until childbirth. Excretion of HCG in the urine serves as the basis for most **pregnancy tests.**

The placenta begins to secrete estrogens and progesterone no later than the sixtieth day of pregnancy. They are secreted in increasing quantities until the time of birth. By the fourth month, when the placenta is established, the secretion of HCG is greatly reduced because the secretions of the corpus luteum are no longer essential. The placenta supplies the levels of estrogens and progesterone needed to maintain the pregnancy. The fetal hormones thus take over management of the mother's body in preparation for parturition (birth) and lactation. Following delivery, estrogens and progesterone in the blood decrease to normal levels.

Recent evidence indicates that the placenta produces a regulating factor called **placental luteotropic releasing factor (pLRF).** It is chemically similar to the luteotropic releasing factor (LRF) produced by the hypothalamus, which causes the synthesis and release of luteinizing hormone (LH) by the anterior pituitary. The proposed function of pLRF is to stimulate the secretion of HCG by the placenta.

(a)

(b)   (c)

FIGURE 29-14 Human embryos and fetuses in various stages of development. (a) A 40-day embryo seen in anterior view (left) and posterior view (right). Note the brain, spinal cord, and upper and lower extremities. (b) Seven-week embryo with umbilical cord and yolk sac. (c) Eight-week embryo showing yolk sac, amnion, and placenta. (Courtesy of Roberts Rugh and Landrum B. Shettles, M.D., with Richard N. Einhorn, *From Conception to Birth: The Drama of Life's Beginnings,* Harper & Row, Publishers, Inc., New York, 1971.)

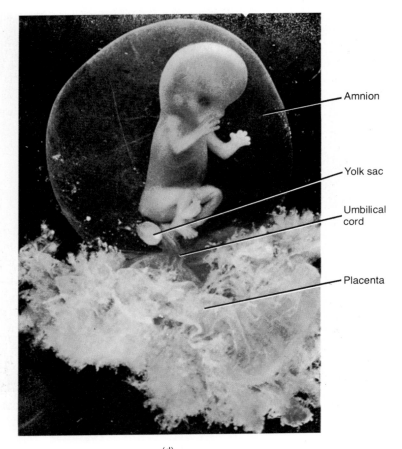

Amnion

Yolk sac

Umbilical
cord

Placenta

(d)

Placenta   Umbilical cord   Amnion

(e)

FIGURE 29-14 (*Continued*) Human embryos and fetuses in various stages of development. (d) Ten-week fetus. (e) Twelve-week fetus. Note the progressive development of the extremities. (Courtesy of Roberts Rugh and Landrum B. Shettles, M.D., with Richard N. Einhorn, *From Conception to Birth: The Drama of Life's Beginnings,* Harper & Row, Publishers, Inc., New York, 1971.)

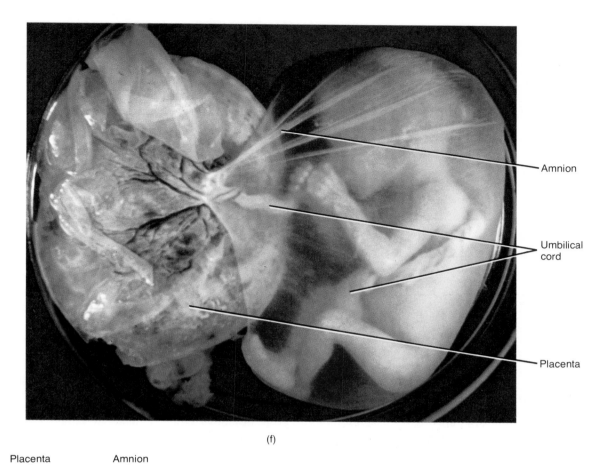

Amnion

Umbilical cord

Placenta

(f)

Placenta   Amnion

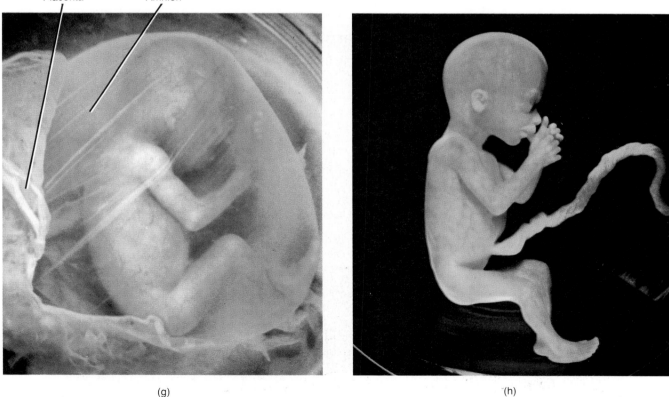

(g)   (h)

FIGURE 29-14 (*Continued*) Human embryos and fetuses in various stages of development. (f) Fourteen-week fetus. (g) Sixteen-week fetus. (h) Seventeen-week fetus sucking its thumb. (Courtesy of Roberts Rugh and Landrum B. Shettles, M.D., with Richard N. Einhorn, *From Conception to Birth: The Drama of Life's Beginnings,* Harper & Row, Publishers, Inc., New York, 1971.)

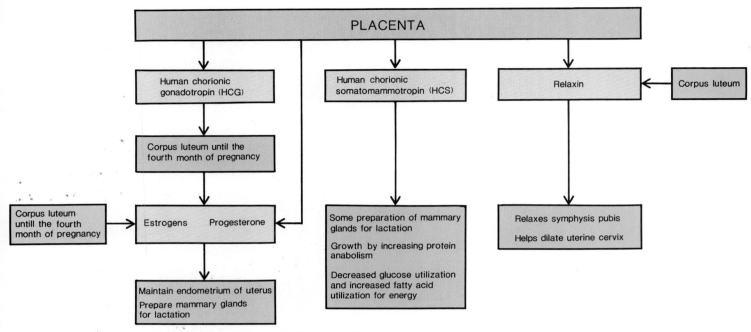

FIGURE 29-15 Hormones of pregnancy.

Another hormone produced by the chorion of the placenta is **human chorionic somatomammotropin (HCS).** Its secretion begins at about the same time as that of HCG, but its pattern of secretion is quite different. The rate of secretion of HCS increases in proportion to placental mass, reaching maximum levels after 32 weeks and remaining relatively constant after that. HCS is believed to stimulate some development of breast tissue for lactation, enhance growth by causing protein deposition in tissues, and regulate certain aspects of metabolism. For example, HCS causes decreased utilization of glucose by the mother, thus making more available for fetal metabolism. Also, HCS promotes the release of fatty acids from fat depots, providing an alternative source of energy for the mother's metabolism.

Relaxin is a hormone produced by the placenta and ovaries. Its physiological role is to relax the symphysis pubis and ligaments of the sacroiliac and sacrococcygeal joints and help dilate the uterine cervix toward the end of pregnancy. Both of these actions assist in delivery.

PARTURITION AND LABOR

The time the embryo or fetus is carried in the uterus is called **gestation** (jes-TĀ-shun). The total human gestation period is about 280 days from the beginning of the last menstrual period. The term **parturition** (par'-too-RISH-un) refers to birth. Parturition is accompanied by a sequence of events commonly called **labor.**

The **onset of labor** is apparently related to a complex interaction of many factors. Just prior to birth, the muscles of the uterus contract rhythmically and forcefully. Both placental and ovarian hormones seem to play a role in these contractions. Since progesterone inhibits uterine contractions, labor cannot take place until its effects are diminished. At the end of gestation, the level of estrogens

in the mother's blood is sufficient to overcome the inhibiting effects of progesterone and labor commences. It has been suggested that some factor released by the placenta, fetus, or mother rather suddenly overcomes the inhibiting effects of progesterone so that estrogens can exert their effect. Prostaglandins may also play a role in labor. Oxytocin (OT) from the posterior pituitary gland also stimulates uterine contractions. And, relaxin assists by relaxing the symphysis pubis and helping to dilate the uterine cervix.

Uterine contractions occur in waves, quite similar to peristaltic waves, that start at the top of the uterus and move downward. These waves expel the fetus. **True labor** begins when pains occur at regular intervals. The pains correspond to uterine contractions. As the interval between contractions shortens, the contractions intensify. Another sign of true labor in some females is localization of pain in the back, which is intensified by walking. A reliable indication of true labor is the "show" and dilation of the cervix. The "show" is a discharge of a blood-containing mucus that accumulates in the cervical canal during pregnancy. In **false labor,** pain is felt in the abdomen at irregular intervals. The pain does not intensify and is not altered significantly by walking. There is no "show" and no cervical dilation.

Labor can be divided into three stages (Figure 29-16).

1. The **stage of dilation** is the time from the onset of labor to the complete dilation of the cervix (Fig. 29-16). During this stage there are regular contractions of the uterus, usually a rupturing of the amniotic sac, and complete dilation (10 cm) of the cervix. If the amniotic sac does not rupture spontaneously, it is done artificially.

2. The **stage of expulsion** is the time from complete cervical dilation to delivery.

3. The **placental stage** is the time after delivery until the placenta or "afterbirth" is expelled by powerful uterine contrac-

FIGURE 29-16 Parturition. (a) Fetal position prior to birth. (b) Dilation. Protrusion of amnionic sac through partly dilated cervix (left). Amnionic sac ruptured and complete dilation of cervix (right). (c) Stage of expulsion. (d) Placental stage.

tions. These contractions also constrict blood vessels that were torn during delivery. In this way, the possibility of hemorrhage is reduced.

CLINICAL APPLICATION
Pudendal (pyoo-DEN-dal) **nerve block** is used for procedures such as episiotomy. The primary innervation to the skin and muscles of the perineum is the pudendal nerve. In the transvaginal approach, the needle is passed through the lateral vaginal wall to a point just medial to the ischial spine. The anesthesia results in loss of the anal reflex, relaxation of the muscles of the floor of the pelvis, and loss of sensation to the vulva and lower one-third of the vagina.

(e)

FIGURE 29-16 (*Continued*) Parturition. (e) Sequential photographs of parturition. Top: Emergence of the infant's head. (Courtesy of Kinne, Photo Researchers.) Bottom left: Delivery of the infant. (Courtesy of Thomas, Photo Researchers.) Bottom right: Delivery of the placenta. (Courtesy of McCartney, Photo Researchers.)

Various deformities of the female pelvis may be responsible for **dystocia,** (dis-TŌ-sē-a), that is, difficult labor. Pelvic deformities may be congenital or acquired from disease, fractures, or poor posture. Among other conditions associated with difficult labor are malposition of the fetus, malpresentation of the fetus, and premature rupture of the fetal membranes.

ADJUSTMENTS OF THE INFANT AT BIRTH

During pregnancy, the embryo and later the fetus is totally dependent on the mother for its existence. The mother supplies the fetus with oxygen and nutrients, eliminates its carbon dioxide and other wastes, and protects it against shocks, temperature changes, and certain harmful microbes. At birth the baby becomes self-supporting, and the newborn's body systems must make various adjustments. Following are some changes that occur in the respiratory and cardiovascular systems.

RESPIRATORY SYSTEM

The respiratory system is fairly well developed at least 2 months before birth as evidenced by the fact that prema-

ture babies delivered at 7 months are able to breathe and cry. The fetus depends entirely on the mother for obtaining oxygen and eliminating carbon dioxide. The fetal lungs are either collapsed or partially filled with amniotic fluid, which is absorbed at birth. After delivery the baby's supply of oxygen from the mother is stopped. Circulation in the baby continues and as the blood level of carbon dioxide increases, the respiratory center in the medulla is stimulated. This causes the respiratory muscles to contract and the baby draws its first breath. Since the first inspiration is unusually deep because the lungs contain no air, the baby exhales vigorously and naturally cries. A full-term baby may breathe 45 times a minute for the first two weeks after exposure to air. The rate is gradually reduced until it approaches a normal rate.

CARDIOVASCULAR SYSTEM

Following the first inspiration by the baby, the cardiovascular system must make several adjustments. The foramen ovale of the fetal heart between atria closes at the moment of birth. This diverts deoxygenated blood to the lungs for the first time. The foramen ovale is closed by two flaps of heart tissue that fold together and permanently fuse. The remnant of the foramen ovale is the fossa ovalis. Once the lungs begin to function, the ductus arteriosus is shut off by contractions of the muscles in its wall. The ductus arteriosus generally does not completely and irreversibly close for about 3 months following birth. Incomplete closing, as you already know, results in patent ductus arteriosus.

The ductus venosus of fetal circulation connects the umbilical vein directly with the inferior vena cava. It forces any remaining umbilical blood directly into the fetal liver and from there to the heart. When the umbilical cord is severed, all visceral blood of the fetus goes directly to the fetal heart via the inferior vena cava. This shunting of blood usually occurs within minutes after birth, but may take a week or two to complete. The ligamentum venosum, the remnant of the ductus venosus, is well established by the eighth postnatal week.

At birth, the infant's pulse may be from 120 to 160 per minute and may go as high as 180 following excitation. Several days after birth, there is a greater independent need for oxygen which stimulates an increase in the rate of erythrocyte and hemoglobin production. This increase usually lasts for only a few days. Moreover, the white blood cell count at birth is very high, sometimes as much as 45,000 cells per cubic millimeter, but this decreases rapidly by the seventh day.

Finally, the infant's liver may not be adjusted at birth to control the production of bile pigment. As a result of this and other complicating factors, a temporary jaundice may result in as many as 50 percent of normal newborns by the third or fourth day after birth.

POTENTIAL HAZARDS TO THE DEVELOPING EMBRYO AND FETUS

The developing fetus is susceptible to a number of potential hazards that can be transmitted from the mother. Such hazards include infectious microbes, chemicals and drugs, alcohol, and cigarette smoking. In addition, certain environmental conditions and pollutants can damage the fetus or even cause fetal death.

CHEMICALS AND DRUGS

Since the placenta is known to be an ineffective barrier between the maternal and fetal circulations, actually any drug or chemical dangerous to the infant may be considered potentially dangerous to the fetus when given to the mother. Many chemicals and drugs have been proven to be toxic and teratogenic to the developing embryo and fetus. A **teratogen** (*terato* = monster) is any agent or influence that causes physical defects in the developing embryo. Examples are pesticides, defoliants, industrial chemicals, some hormones, antibiotics, oral anticoagulants, anticonvulsants, antitumor agents, thyroid drugs, thalidomide, diethylstilbestrol (DES), LSD, and marijuana. In addition, accumulating evidence suggests that many agents assumed to be nonteratogenic can produce long-term adverse effects on reproductive ability and neurobehavioral development.

IRRADIATION

Ionizing radiations are potent teratogens. Treatment of pregnant mothers with large doses of x-rays and radium during the embryo's susceptible period of development may cause microcephaly (small size of head in relation to the rest of the body), mental retardation, and skeletal malformations. Caution is advised for diagnostic x-rays during the first trimester of pregnancy.

ALCOHOL

Alcohol has been a suspected teratogen for centuries, but only recently has a relationship been recognized between maternal alcohol intake and the characteristic pattern of malformations in the fetus. The term applied to the effects of intrauterine exposure to alcohol is **fetal alcohol syndrome (FAS).** Studies to date have indicated that in the general population the incidence of FAS may exceed 1 per 1,000 live births, and may be by far the number one fetal teratogen. The symptoms shown by children may include slow growth before and after birth, small head, facial irregularities such as narrow eye slits and sunken nasal bridge, defective heart and other organs, malformed arms and legs, genital abnormalities, and mental retardation. There are also behavioral problems, such as hyperactivity, extreme nervousness, and a poor attention span.

CIGARETTE SMOKING

The latest evidence not only indicates a causal relationship between cigarette smoking during pregnancy and low infant birth weight, but also points to a strong probable association between smoking and a higher fetal and infant mortality. Infants nursing from smoking mothers have also been found to have an increased incidence of gastrointestinal disturbances. Other pathologies of infants of smoking mothers include an increased incidence of respiratory problems during the first year of life, including bronchitis and pneumonia. Cigarette smoking may be teratogenic and cause cardiac abnormalities and anencephaly (a developmental anomaly with absence of neural tissue in the cranium). Maternal smoking also appears to be a significant etiologic factor in the development of cleft lip and palate and has been tentatively linked with sudden infant death syndrome (SIDS).

LACTATION

The term **lactation** refers to the secretion and ejection of milk by the mammary glands. The major hormone in promoting lactation is **prolactin (PRL)** from the anterior pituitary gland. It is released in response to the hypothalamic prolactin releasing factor (PRF). Even though PRL levels increase as the pregnancy progresses, there is no milk secretion because estrogens and progesterone cause the hypothalamus to release prolactin-inhibiting factor (PIF). Following delivery, the levels of estrogens and progesterone in the mother's blood decrease and the inhibition is removed.

The principal stimulus in maintaining prolactin secretion during lactation is the sucking action of the infant. Sucking initiates impulses from receptors in the nipples to the hypothalamus. The impulses inhibit PIF production, and PRL is released by the anterior pituitary. The sucking action also initiates impulses to the posterior pituitary via the hypothalamus. These impulses stimulate the release of the hormone **oxytocin (OT)** by the posterior pituitary gland. This hormone induces certain cells surrounding the outer walls of the alveoli to contract, thereby compressing the alveoli. The compression moves milk from the alveoli of the mammary gland into the ducts, where it can be sucked. This process is referred to as **milk letdown.** Lactation often prevents the occurrence of female ovarian cycles for the first few months following delivery by inhibiting FSH and LH release by the anterior pituitary.

During late pregnancy and the first few days after birth, the mammary glands secrete a cloudy fluid called **colostrum.** Although it is not as nutritious as true milk, since it contains less lactose and virtually no fat, it serves adequately until the appearance of true milk on about the fourth day. Colostrum is thought to contain antibodies that protect the infant during the first few months of life.

Following birth of the infant, PRL level starts to return to the nonpregnant level, but each time the mother nurses the infant, nerve impulses from the nipples to the hypothalamus cause the release of PRF and the secretion of a tenfold increase in PRL by the anterior pituitary that lasts about an hour. PRL acts on the mammary glands to provide milk for the next nursing period. If this surge of PRL is blocked by injury or disease or if nursing is discontinued, the mammary glands lose their ability to secrete milk in a few days. However, milk secretion can continue for several years if the child continues to suckle. Milk secretion normally decreases considerably within seven to nine months.

CLINICAL APPLICATION

In the 1940s about 65 percent of the infants in the United States were breast fed, but by 1972 only 15 percent were nursed as long as two months. Now, more and more mothers are **breast feeding.** Advocates of breast feeding feel that it offers the following advantages to the infant.

1. It establishes early and prolonged contact between mother and infant.

2. The infant is more in control of intake.

3. Fats and iron in human milk are better absorbed than those in cow's milk and the amino acids in human milk are more readily metabolized. Also, the lower sodium content of human milk is more suited to the infant's needs.

4. Breast feeding provides important antibodies that prevent gastroenteritis. Immunity to respiratory infections and meningitis is also greater.

5. Premature infants benefit from breast feeding because the milk produced by mothers of premature infants seems to be specially adapted to the infant's needs by having a higher protein content than the milk of mothers of full-term infants.

6. There is less likelihood of an allergic reaction in the baby to the milk of the mother.

7. The act of sucking on the breast promotes development of the jaw, facial muscles, and teeth.

BIRTH CONTROL

Methods of **birth control** include removal of the gonads and uterus, sterilization, and contraception.

REMOVAL OF GONADS AND UTERUS

Castration (removal of the testes), **hysterectomy** (removal of the uterus), and **oophorectomy** (ō'-of-ō-REK-tō-mē;

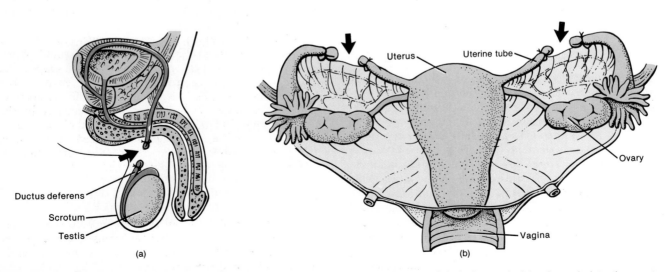

FIGURE 29-17 Sterilization. (a) Vasectomy. The ductus deferens of each testis is cut and tied after an incision is made into the scrotum. (b) Tubal ligation. Each uterine tube is cut and tied after an incision is made into the abdomen.

removal of the ovaries) are all absolute preventive methods. Once performed, these operations cannot be reversed and it is impossible to produce offspring. However, removal of the testes or ovaries has adverse effects because of the importance of these organs in the endocrine system. Generally these operations are performed only if the organs are diseased. Castration before puberty prevents the development of secondary sex characteristics.

STERILIZATION

One means of **sterilization** of males is **vasectomy**—a simple operation in which a portion of each ductus (vas) deferens is removed (Figure 29-17a). An incision is made in the scrotum, the tubes are located, and each is tied in two places. Then the portion between the ties is cut out. Sperm production can continue in the testes, but the sperm cannot reach the exterior.

Sterilization in females generally is achieved by performing a **tubal ligation** (lī-GĀ-shun). An incision is made into the abdominal cavity, the uterine tubes are squeezed, and a small loop called a knuckle is made (Figure 29-17b). A suture is tied tightly at the base of the knuckle and the knuckle is then cut. After four or five days the suture is digested by body fluids and the two severed ends of the tubes separate. The ovum thus is prevented from passing to the uterus, and the sperm cannot reach the ovum. Sterilization normally does not affect sexual performance or enjoyment.

Another method of sterilizing women is the **laparoscopic** (lap'-a-rō-SKŌ-pik) **technique.** After a woman receives local or general anesthesia, a harmless gas is introduced into her abdomen to create a gas bubble. The bubble expands the abdominal cavity and pushes the intestines away from the pelvic organs, permitting safe, easy access to the uterine tubes. The doctor makes a small incision at the lower rim of the umbilicus and inserts a laparoscope to view the inside of the abdominal cavity and the uterine tubes. The tubes can be closed with this instrument or a second incision can be made at the pubic hairline to insert a cautery forceps. Once the uterine tubes are sealed, the instrument is removed, the gas is released, and the incision is covered with a bandage. After a few hours the patient can usually go home.

CLINICAL APPLICATION
Microsurgery is any operation done with the aid of microscopes. Surgeons work with spectacle-mounted magnifiers or with microscopes. Laparoscopy is one such technique and several others have been mentioned throughout the text.

In the United States alone, more than 10 million people have been sterilized by artificial closures. For men who have had vasectomies, reversal through microsurgery has meant about a 90 percent chance of regaining a normal sperm count. Limited success has been attained in reversing women who have undergone sterilization surgery through microsurgery.

Microsurgery's impact on curing fertility problems has been nothing short of revolutionary. Using this technique, surgeons can remove a prolactin-secreting adenoma, a hormone-producing tumor of the pituitary gland that causes infertility in women, without damaging the pituitary gland itself. Microsurgery is also used to open blocked uterine tubes and spermatic ducts. Natural obstructions are a major cause of sterility in both sexes.

Neurosurgeons and ear and eye surgeons all rely on microsurgery, and thousands of people today already owe their health, children, limbs, and lives to microsurgery.

CONTRACEPTION

Contraception is the prevention of fertilization without destroying fertility by natural, mechanical, or chemical means.

Natural

The **natural methods** include complete or periodic abstinence. An example of periodic abstinence is the rhythm method, which takes advantage of the fact that a fertilizable ovum is available only during a period of three to five days in each menstrual cycle. During this time the couple refrains from intercourse. Its effectiveness is limited by the fact that few women have absolutely regular cycles. Moreover, some women occasionally ovulate during the "safe" times of the month, such as during menstruation.

Mechanical

Mechanical methods of contraception include the use of the condom by the male and the diaphragm by the female. The **condom** is a nonporous, elastic (rubber or similar material) covering placed over the penis that prevents deposition of sperm in the female reproductive tract. The **diaphragm** (Figure 29-18) is a dome-shaped structure that fits over the cervix and is generally used in conjunction with a sperm-killing chemical. The diaphragm stops the sperm from passing into the cervix. The chemical kills the sperm cells.

Another mechanical method of contraception is an **intrauterine device (IUD).** The device is a small object made of plastic, copper, or stainless steel and shaped like a loop, coil, T, or 7. It is inserted into the cavity of the uterus (Figure 29-19). It is not clear how IUDs operate. Some investigators believe they cause changes in the uterus lining that, in turn, produce a substance which destroys either the sperm or the fertilized ovum. Some IUDs release contraceptive agents in minute amounts; one new device secretes progesterone.

Chemical

Chemical methods of contraception include spermicidal and hormonal methods. Various foams, creams, jellies, suppositories, and douches make the vagina and cervix unfavorable for sperm survival.

The hormonal method, **oral contraception** (the pill), has found rapid and widespread use. Although several

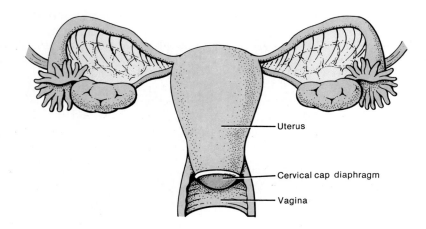

FIGURE 29-18 Cervical cap diaphragm. This particular type of diaphragm fits directly over the cervix of the uterus. The thin spring around the margin of the diaphragm opens outward, presses against the wall of the vagina, and stretches across the cervix.

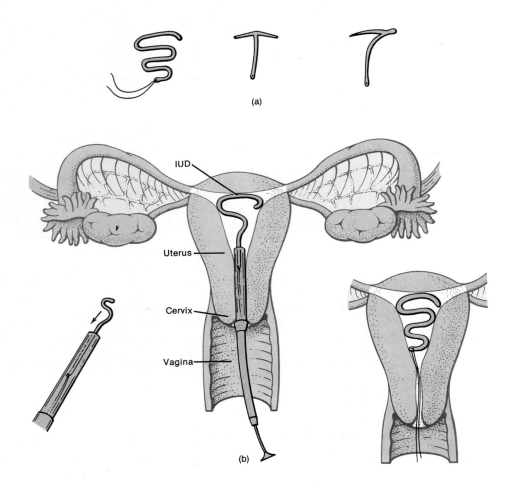

FIGURE 29-19 Intrauterine devices (IUDs). (a) Representative designs of three IUDs. (b) Procedure for insertion. The device is compressed to fit into a long, narrow-bore tube that is passed through the slightly dilated cervix. Once in position in the uterus, it spreads out to its former shape. IUDs have a thread or chain projecting into the vagina that is used as an indication that the device is still in place and for removal by the physician.

EXHIBIT 29-3

BIRTH CONTROL METHODS

METHOD	COMMENTS
Removal of gonads and uterus	Irreversible sterility. Generally performed if organs are diseased rather than as contraceptive method because of importance of hormones produced by gonads.
Sterilization	Procedure involving severing ductus (vas) deferens in males (vasectomy) and uterine tubes in females (tubal ligation and laparascopic technique).
Natural contraception	Abstinence from intercourse during time of month woman is fertile. Under ideal circumstances, effectiveness in women with regular menstrual cycles may approach that of mechanical and chemical contraceptives. Extremely difficult to determine fertile period. Effectiveness can be increased by recording body temperature each morning before getting up; a small rise in temperature indicates that ovulation has occurred one or two days earlier and ovum can no longer be fertilized.
Mechanical contraception Condom	Thin, strong sheath of rubber or similar material worn by male to prevent sperm from entering vagina. Failures caused by sheath tearing or slipping off after climax or not putting the sheath on soon enough. If used correctly and consistently, effectiveness similar to that of diaphragm.
Diaphragm	Flexible rubber dome inserted into vagina to cover cervix, providing barrier to sperm. Usually used with spermicidal cream or jelly. Must be left in place at least 6 hours after intercourse and may be left in place as long as 24 hours. Must be fitted by physician or other trained personnel and refitted every two years and after each pregnancy. Offers high level of protection if used with spermicide; rate of 2 to 3 pregnancies per 100 women per year is estimated for consistent users. If not used consistently, much higher pregnancy rates must be expected. Occasional failures caused by improper insertion or displacement during sexual intercourse.
Intrauterine device (IUD)	Small object (loop, coil, T, or 7) made of plastic, copper, or stainless steel and inserted into uterus by physician. May be left in place for long periods of time (some must be changed every two to three years). Does not require continued attention by user. Some women cannot use them because of expulsion, bleeding, or discomfort. Not recommended for women who have not had child because uterus is too small and cervical canal too narrow. Infrequently, insertion may lead to inflammation of pelvic organs.
Chemical contraception Foams, creams, jellies, suppositories, vaginal douches	Sperm-killing chemicals inserted into vagina to coat vaginal surfaces and cervical opening. Provide protection for about 1 hour. Effective when used alone, but significantly more effective when used with diaphragm or condom.
Oral contraceptive (OC)	Except for total abstinence or surgical sterilization, most effective contraceptive known. Side effects include nausea, occasional light bleeding between periods, breast tenderness or enlargement, fluid retention, and weight gain. Should not be used by women who have cardiovascular conditions (thromboembolic disorders, cerebrovascular disease, heart disease, hypertension), liver malfunction, cancer or neoplasia of breast or reproductive organs, or by women who smoke.

pills are available, the one most commonly used contains a high concentration of progesterone and a low concentration of estrogens. These two hormones act on the anterior pituitary to decrease the secretion of FSH and LH by inhibiting GnRF by the hypothalamus. The low levels of FSH and LH are not adequate to initiate follicle maturation or ovulation. In the absence of a mature ovum, pregnancy cannot occur.

Women for whom all oral contraceptives are contraindicated include those with a history of thromboembolic disorders (predisposition to blood clotting), cerebral blood vessel damage, hypertension, liver malfunction, heart disease, or cancer of the breast or reproductive system. About 40 percent of all pill users experience side effects— generally minor problems such as nausea, weight gain, headache, irregular menses, spotting between periods, and amenorrhea. The statistics on the life-threatening conditions associated with the pill such as blood clots, heart attacks, liver tumors, and gallbladder disease are somewhat more reassuring. For all the problems combined,

fewer than 2 deaths occurred per 100,000 users under age 30, 4 among those 30 to 35, 10 among those 35 to 39, and 18 among women over 40.

The major exception is that women who take the pill and smoke face far higher odds of developing heart attack and stroke than do nonsmoking pill users.

A summary of birth control methods is presented in Exhibit 29-3.

CLINICAL APPLICATION

The quest for an efficient male oral contraceptive has been disappointing until recently. An oral contraceptive, **gossypol,** which is derived from the cotton plant, has achieved an efficacy (power to produce effects) rate of 99.9 percent in clinical tests in more than 4,000 healthy males in China. Its action is to inhibit an enzyme required for spermatogenesis. Levels of blood luteinizing hormone (LH) and testosterone remained

FIGURE 29-20 Inheritance of phenylketonuria (PKU).

unchanged and potency was not impaired. The number and morphology of spermatozoa gradually recover and fertility is restored to normal within three months after termination of therapy.

INHERITANCE

Inheritance is the passage of hereditary traits from one generation to another. It is the process by which you acquired your characteristics from your parents and will transmit your characteristics to your children. The branch of biology that deals with inheritance is called **genetics.**

GENOTYPE AND PHENOTYPE

The nuclei of all human cells except gametes contain 23 pairs of chromosomes—the diploid number. One chromosome from each pair comes from the mother, and the other comes from the father. Homologues, the two chromosomes in a pair, contain genes that control the same traits. If a chromosome contains a gene for height, its homologue will contain a gene for height.

The relationship of genes to heredity is illustrated admirably by the disorder called phenylketonuria or **PKU** (see Figure 29-20). People with PKU are unable to manufacture the enzyme phenylalanine hydroxylase. It is believed the PKU is brought about by an abnormal gene, which can be symbolized as p. The normal gene will be symbolized as P. The chromosome concerned with directions for phenylalanine hydroxylase production will have either p or P on it. Its homologue will also have p or P. Thus every individual will have one of the following genetic makeups, or **genotypes** (JĒ-nō-tīps): PP, Pp, or pp. Although people with genotypes of Pp have the abnor-

mal gene, only those with genotype pp suffer from the disorder because the normal gene, when present, dominates over and inhibits the abnormal one. A gene that dominates is called the **dominant gene,** and the trait expressed is said to be a dominant trait. The gene that is inhibited is called the **recessive gene,** and the trait it controls is called the recessive trait.

By tradition, we symbolize the dominant gene with a capital letter and the recessive one with a lowercase letter. An individual with the same genes on homologous chromosomes (for example, PP or pp) is said to be **homozygous** for the trait. An individual with different genes on homologous chromosomes (for example, Pp), is said to be **heterozygous** for the trait. **Phenotype** (FĒ-nō-tīp) refers to how the genetic makeup is expressed in the body. A person with Pp has a different genotype from one with PP, but both have the same phenotype—which in this case is normal production of phenylalanine hydroxylase.

To determine how gametes containing haploid chromosomes unite to form diploid fertilized eggs, special charts called **Punnett squares** are used. Usually, the male gametes (sperm cells) are placed at the side of the chart and the female gametes (ova) at the top (as in Figure 29-20). The four spaces on the chart represent the possible combinations of male and female gametes that could form fertilized eggs.

Exhibit 29-4 lists some of the variety of simple inherited structural and functional traits in humans. The classification of traits in this manner is rather arbitrary and somewhat artificial, since structure and function are intimately related. Nevertheless, some organizational scheme is

EXHIBIT 29-4
HEREDITARY TRAITS IN HUMANS

DOMINANT	RECESSIVE
Curly hair	Straight hair
Dark brown hair	All other colors
Coarse body hair	Fine body hair
Pattern baldness (dominant in males)	Baldness (recessive in females)
Normal skin pigmentation	Albinism
Brown eyes	Blue or gray eyes
Near or farsightedness	Normal vision
Normal hearing	Deafness
Normal color vision	Color blindness
Broad lips	Thin lips
Large eyes	Small eyes
Polydactylism (extra digits)	Normal digits
Brachydactylism (short digits)	Normal digits
Syndactylism (webbed digits)	Normal digits
Hypertension	Normal blood pressure
Diabetes insipidus	Normal excretion
Huntington's chorea	Normal nervous system
Normal mentality	Schizophrenia
Migraine headaches	Normal
Normal resistance to disease	Susceptibility to disease
Enlarged spleen	Normal spleen
Enlarged colon	Normal colon
A or B blood factor	O blood factor
Rh blood factor	No Rh blood factor

(a)

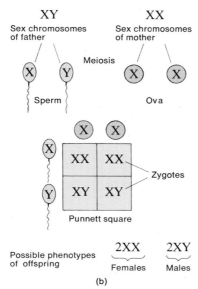

(b)

FIGURE 29-21 Inheritance of sex. (a) Normal human male chromosomes. Sex chromosomes are the 23rd pair indicated in the blue colored box. (b) Sex determination.

needed that gives an indication of the wide spectrum of inherited human characteristics. Many traits were at one time thought to be inherited in a simple Mendelian manner and were classified as "dominant" or "recessive." Then further research revealed that some traits were actually complex and that the genetic basis was correspondingly complicated. No wonder students of human genetics prefer to work with enzymes and antigens, in which the path from a gene to its product is much shorter, more direct, and less influenced by other genes and by environment. (If you would like to find out about the inheritance of some particular traits, the first place to look is McKusick's *Mendelian Inheritance in Man* [1971]).

Normal traits do not always dominate over abnormal ones, but genes for severe disorders are more frequently recessive than dominant. People who have severe disorders often do not live long enough to pass the abnormal gene on to the next generation. In this way, expression of the gene tends to be weeded out of the population.

CLINICAL APPLICATION

An exception is **Huntington's chorea**—a major disorder caused by a dominant gene and characterized by degeneration of nervous tissue, usually leading to mental disturbance and death. The first signs of Huntington's chorea do not occur until adulthood, very often after the person has already produced offspring.

INHERITANCE OF SEX

Microscopic examination of the chromosomes in cells reveals that one pair differs in males and in females (Figure 29-21a). In females, the pair consists of two rod-shaped chromosomes designated as X chromosomes. One X chromosome is present in males, but its mate is a hook-shaped structure called a Y chromosome. The XX pair in the female and the XY pair in the male are called the **sex chromosomes.** All other chromosomes are called **autosomes.**

The sex chromosomes are responsible for the sex of the individual (Figure 29-21b). When a spermatocyte undergoes meiosis to reduce its chromosome number, one daughter cell will contain the X chromosome and the other contain the Y chromosome. Oocytes have no Y chromosomes and produce only X-containing ova. If the ovum is subsequently fertilized by an X-bearing sperm, the offspring normally will be female (XX). Fertilization by a Y sperm normally produces a male (XY). Thus, sex is determined at fertilization. Although X and Y sperm are produced in equal amounts, more males are born than females. It is speculated that Y sperm swim faster than X sperm because X sperm contain more genetic material.

COLOR BLINDNESS AND X-LINKED INHERITANCE

The sex chromosomes also are responsible for the transmission of a number of nonsexual traits. Genes for these traits appear on X chromosomes, but many of these genes are absent from Y chromosomes. This feature produces a pattern of heredity that is different from the pattern described earlier. Let us consider color blindness. The gene for **color blindness** is a recessive one designated c. Normal color vision, designated C, dominates. The C/c genes are located on the X chromosome. The Y chromosome does not contain the segment of DNA that programs this aspect of vision. Thus the ability to see colors depends entirely on the X chromosomes. The genetic possibilities are:

$X^C X^C$	Normal female
$X^C X^c$	Normal female carrying the recessive gene
$X^c X^c$	Color-blind female
$X^C Y$	Normal male
$X^c Y$	Color-blind male

Only females who have two X^c chromosomes are color blind. In $X^C X^c$ females the trait is inhibited by the normal, dominant gene. Males, on the other hand, do not have a second X chromosome that would inhibit the trait. Therefore all males with an X^c chromosome will be color blind. The inheritance of color blindness is illustrated in Figure 29-22.

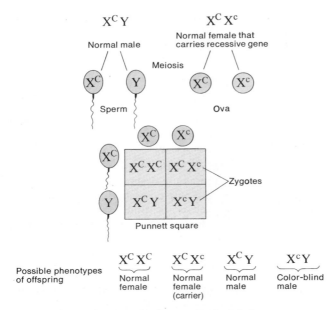

FIGURE 29-22 Inheritance of color blindness.

Traits inherited in the manner just described are called **X-linked traits.** Another X-linked trait is **hemophilia**—a condition in which the blood fails to clot or clots very slowly after an injury (Chapter 19). Like the trait for color blindness, hemophilia is caused by a recessive gene. If H represents normal clotting and h represents abnormal clotting, then X^hX^h females will have the disorder. Males with X^HY will be normal; males with X^hY will be hemophiliac. Actually, clotting time varies somewhat among hemophiliacs, so the condition may be affected by other genes as well.

A few other X-linked traits in humans are nonfunc-tional sweat glands, certain forms of diabetes, some types of deafness, uncontrollable rolling of the eyeballs, absence of central incisors, night blindness, one form of cataract, white forelocks, juvenile glaucoma, and juvenile muscular dystrophy.

AMNIOCENTESIS

Amniocentesis is a technique of withdrawing some of the amniotic fluid that bathes the developing fetus to diagnose genetic disorders or to determine fetal maturity or well-being. The fluid is removed by hypodermic needle puncture of the uterus, usually 16 to 20 weeks after conception. Cells in the fluid are examined for biochemical defects and abnormalities in chromosome number or structure. Over 50 inheritable biochemical defects and close to 300 chromosomal disorders can be detected through amniocentesis, including hemophilia, certain muscular dystrophies, Tay-Sachs disease, myelocytic leukemia, Klinefelter's and Turner's syndromes, sickle cell anemia, thalassemia, and cystic fibrosis. When both parents are known or suspected to be genetic carriers of any one of these disorders, amniocentesis is advised.

CLINICAL APPLICATION

One chromosome disorder that may be diagnosed through amniocentesis is **Down's syndrome.** This disorder is characterized by mental retardation, retarded physical development (short stature and stubby fingers), distinctive facial structures (large tongue, broad skull, slanting eyes, and round head), and malformation of the heart, ears, hands, and feet. Sexual maturity is rarely attained. Individuals with the disorder usually have 47 chromosomes instead of the normal 46 (an extra chromosome in the twenty-first pair).

MEDICAL TERMINOLOGY

Abortion   Premature expulsion from the uterus of the products of conception—embryo or nonviable fetus.

Cautery   Application of a caustic (burning) substance or instrument for the destruction of tissue.

Cesarean section (*caedere* = to cut)   Removal of the baby and placenta through an abdominal incision in the uterine wall.

Colpotomy (*colp* = vagina; *tome* = cutting)   Incision of the vagina.

Culdoscopy (*skopein* = to examine)   A procedure in which a culdoscope (endoscope) is used to view the pelvic cavity. The approach is through the vagina.

Hermaphroditism   Presence of both male and female sex organs in one individual.

Karyotype (*karyon* = nucleus)   The chromosomal elements typical of a cell, drawn in their true proportions, based on the average of measurements determined in a number of cells. Useful in judging whether or not chromosomes are normal in number and structure.

Lethal gene   A gene that, when expressed, results in death either in the embryonic state or shortly after birth.

Lochia   The discharge from the birth canal consisting initially of blood and later of serous fluid occurring after childbirth. The discharge is derived from the former placental site and may last up to about a week and a half.

Mutation   A permanent heritable change in a gene that causes it to have a different effect than it had previously.

Puerperal fever   Infectious disease of childbirth, also called puerperal sepsis and childbed fever. The disease results from an infection originating in the birth canal and affects the endometrium.

STUDY OUTLINE

Gamete Formation (p. 732)
Diploid and Haploid Cells
1. Ova and sperm are collectively called gametes or sex cells and are produced in gonads.
2. Uninucleated somatic cells divide by mitosis, the process in which each daughter cell receives the full complement of 23 chromosome pairs (46 chromosomes). Somatic cells are said to be diploid ($2n$).
3. Immature gametes divide by meiosis in which the pairs of chromosomes are split so that the mature gamete has only 23 chromosomes. It is said to be haploid (n).

Spermatogenesis

1. Spermatogenesis occurs in the testes. It results in the formation of four haploid spermatozoa.
2. The spermatogenesis sequence consists of reduction division, equatorial division, and spermiogenesis.

Oogenesis

1. Oogenesis occurs in the ovaries. It results in the formation of a single haploid ovum.
2. The oogenesis sequence consists of reduction division, equatorial division, and maturation.

Sexual Intercourse (p. 734)

1. The role of the male in the sex act involves erection, lubrication, and orgasm.
2. The female role also involves erection, lubrication, and orgasm (climax).

Pregnancy (p. 735)

1. Pregnancy is a sequence of events that includes fertilization, implantation, embryonic growth, fetal growth, and birth.
2. Its various events are hormonally controlled.

Fertilization and Implantation

1. Fertilization refers to the penetration of the ovum by a sperm cell and the subsequent union of the sperm and ovum nuclei to form a zygote.
2. Penetration is facilitated by hyaluronidase and proteinases produced by sperm.
3. Normally only one sperm fertilizes an ovum.
4. Early rapid cell division of a zygote is called cleavage, and the cells produced by cleavage are called blastomeres.
5. The solid mass of cells produced by cleavage is a morula.
6. The morula develops into a blastocyst, a hollow ball of cells differentiated into a trophectoderm (future embryonic membranes) and inner cell mass (future embryo).
7. The attachment of a blastocyst to the endometrium is called implantation.
8. It occurs by enzymatic degradation of the endometrium.

Embryonic Development

1. During embryonic growth, the primary germ layers and embryonic membranes are formed and the placenta is functioning.
2. The primary germ layers—ectoderm, mesoderm, and endoderm—form all tissues of the developing organism.
3. Embryonic membranes include the yolk sac, amnion, chorion, and allantois.
4. Fetal and maternal materials are exchanged through the placenta.
5. During the fetal period, organs established by the primary germ layers grow rapidly.

Hormones of Pregnancy

1. Pregnancy is maintained by human chorionic gonadotropin (HCG), estrogens, and progesterone.
2. Placental luteotropic releasing factor (pLRF) stimulates secretion of HCG.
3. Human chorionic somatomammotropin (HCS) assumes a role in breast development, protein anabolism, and glucose and fatty acid catabolism.
4. Relaxin relaxes the symphysis pubis and helps dilate the uterine cervix toward the end of pregnancy.

Parturition and Labor (p. 750)

1. The time an embryo or fetus is carried in the uterus is called gestation.
2. Parturition refers to birth and is accompanied by a sequence of events called labor.
3. The birth of a baby involves dilation of the cervix, expulsion of the fetus, and delivery of the placenta.

Adjustments of the Infant at Birth (p. 752)

1. The fetus depends on the mother for oxygen and nutrients, removal of wastes, and protection.
2. Following birth the respiratory and cardiovascular systems undergo changes in adjusting to self-supporting postnatal life.

Potential Hazards to the Developing Embryo and Fetus (p. 753)

1. The developing embryo and fetus is susceptible to many potential hazards that can be transmitted from the mother.
2. Examples are infections, microbes, chemicals and drugs, alcohol, and smoking.

Lactation (p. 753)

1. Lactation refers to the secretion and ejection of milk by the mammary glands.
2. Secretion is influenced by prolaction (PRL), estrogens, and progesterone.
3. Ejection is influenced by oxytocin (OT).

Birth Control (p. 754)

1. Methods include removal of gonads and uterus, sterilization (vasectomy, tubal ligation, laparascopic technique), and contraception (natural, mechanical, and chemical).
2. Contraceptive pills of the combination type contain estrogens and progesterone in concentrations that decrease the secretion of FSH and LH and thereby inhibit ovulation.

Inheritance (p. 758)

1. Inheritance is the passage of hereditary traits from one generation to another.
2. The genetic makeup of an organism is called its genotype. The traits expressed are called its phenotype.
3. Dominant genes control a particular trait; expression of recessive genes is inhibited by dominant genes.
4. Amniocentesis is the withdrawal of amniotic fluid. It can be used to diagnose inherited biochemical defects and chromosomal disorders, such as hemophilia, Tay-Sachs disease, sickle cell anemia, and Down's syndrome.
5. Down's syndrome is a chromosomal abnormality characterized by mental retardation and retarded physical development.
6. Sex is determined by the Y chromosome of the male at fertilization.
7. Color blindness and hemophilia primarily affect males because there are no counterbalancing dominant genes on the Y chromosomes.

REVIEW QUESTIONS

1. Define developmental anatomy.
2. Explain the terms haploid and diploid. What is the importance of meiosis?
3. Compare the events associated with spermatogenesis and oogenesis. How are the processes similar? How do they differ?
4. Explain the role of the male's erection, lubrication, and orgasm in the sex act. How do the female's erection, lubrica-

tion, and orgasm (climax) contribute to the sex act?
5. Define fertilization. Where does it normally occur? How is a morula formed?
6. Explain how twins, triplets, and quadruplets are produced.
7. Describe the components of a blastocyst.
8. What is implantation? How does the fertilized ovum implant itself? What causes morning sickness?
9. Describe the procedure for external human fertilization.
10. Define the embryonic period and the fetal period.
11. List several body structures formed by the endoderm, mesoderm, and ectoderm.
12. What is an embryonic membrane? Describe the functions of the four embryonic membranes.
13. Explain the importance of the placenta and umbilical cord to fetal growth.
14. Outline some of the major developmental changes during fetal growth.
15. List the hormones involved in pregnancy and describe the functions of each.
16. Define gestation and parturition.
17. Distinguish between false and true labor. Describe what happens during the stage of dilation, the stage of expulsion, and the placental stage of delivery.
18. Discuss the principal respiratory and cardiovascular adjustments made by an infant at birth.
19. Explain in detail some of the potential hazards for the developing embryo and fetus.
20. What is lactation? Name the hormones involved and their functions.
21. Briefly describe the following methods of birth control: removal of the gonads, sterilization, rhythm.
22. Distinguish between a condom, a diaphragm, and an IUD as methods of mechanical contraception.
23. List several examples and functions of chemical contraceptives.
24. Explain the operation of the combination contraceptive pill.
25. Define inheritance. What is genetics?
26. Define the following terms: genotype, phenotype, dominant, recessive, homozygous, heterozygous.
27. What is a Punnett square?
28. List several dominant and recessive traits inherited in humans.
29. Set up Punnett squares to show the inheritance of the following traits: sex, color blindness, hemophilia.
30. What is amniocentesis? What is its value?
31. What is X-linked inheritance?
32. Refer to the glossary of medical terminology associated with development and inheritance. Be sure that you can define each term.

Appendix A: Measurements

UNITS OF MEASUREMENT

When you measure something, you are comparing it with some standard scale to determine its *magnitude*. How long is it? How much does it weigh? How fast is it going? Some measurements are made directly by comparing the unknown quantity with the known unit of the same kind, for example, weighing a patient on a scale and taking the reading directly in pounds. Other measurements are indirect and are done by calculation, for example, counting a person's blood cells in a certain number of squares on a microscope slide and then calculating the total blood count.

Regardless of how a measurement is taken, it always requires two things: a *number* and a *unit*. When recording the weight of a patient, you would not just say 145. You have to give both the number (145) and the unit (pounds). When you count blood cells, you report the measurement as 10,000 (number) white blood cells per cubic millimeter of blood (unit).

All the units in use can be expressed in terms of one of three special units called *fundamental units*. These fundamental units have been established arbitrarily as length, mass, and time. Mass is perhaps an unfamiliar term to you. *Mass* is the amount of matter an object contains. The mass of this textbook is the same whether it is measured in a laboratory, under the sea, on top of a mountain, or even on the moon. No matter where you take it, it still has the same quantity of matter. *Weight,* on the other hand, is determined by the pull of gravity on an object. This textbook will not have the same weight on earth as on the moon because of the differences in gravitation. However, as long as we are dealing only with earthbound objects, weight and mass may be considered synonymous terms because the force of gravity on the surface of the earth is nearly constant. Thus weight remains nearly the same regardless of where the measurements are taken.

All units other than the fundamental ones are *derived units*—they can always be written as some combination of the three fundamental units. For example, units of volume are derived from units of length (the volume of a cube = length × width × height). Units of speed are combinations of distance and time (miles per hour).

Units are grouped into systems of measurement. The two principal systems of measurement commonly used in this country are the U.S. and the metric systems. The apothecary system is used by physicians and pharmacists.

U.S. SYSTEM

The **U.S. system** of measurement is used in everyday household work, industry, and some fields of engineering. The fundamental units in the U.S. system are the foot (length), the pound (mass), and the second (time).

The basic problem with the U.S. system is that there is no *uniform* progression from one unit to another. If you want to convert a measurement of 2½ yd to feet,

EXHIBIT A-1

U.S. UNITS OF MEASUREMENT

FUNDA-MENTAL OR DERIVED UNIT	UNITS AND U.S. EQUIVALENTS
Length	1 inch = 0.083 foot
	1 foot (ft) = 12 inches
	= 0.333 yard
	1 yard (yd) = 3 ft = 36 inches
	1 mile (mi) = 1,760 yd = 5,280 ft
Mass	1 grain (gr) = 0.002285 ounce
	1 dram (dr) = 27.34 gr
	= 0.063 ounce
	1 ounce (oz) = 16 dr = 437.5 gr
	1 pound (lb) = 16 oz = 7,000 gr
	1 ton = 2,000 lb
Time	1 second (sec) = 1/86,400 of a day
	1 minute (min) = 60 sec
	1 hour (hr) = 60 min = 3,600 sec
	1 day = 24 hr = 1,440 min
	= 86,400 sec
Volume	1 fluidram
	(fl dr) = 0.125 fluidounce
	1 fluidounce
	(fl oz) = 8 fl dr
	= 0.0625 quart = 0.008 gallon
	1 pint (pt) = 16 fl oz = 128 fl dr
	1 quart (qt) = 2 pt = 32 fl oz = 256 fl dr
	= 0.25 gallon
	1 gallon (gal) = 4 qt = 8 pt = 128 fl oz
	= 1,024 fl dr

you have to multiply it by 3 because there are 3 ft in a yard. If you want to convert the same length to inches, you have to multiply by 3 and then by 12 (or by 36) because there are 12 inches in a foot. In other words, to convert one unit of length to another, it is necessary to use *different* numbers each time. Conversions in the metric system are much easier since they are based on progressions of the number 10.

Exhibit A-1 lists U.S. units of measurement.

METRIC SYSTEM

The **metric system** introduced in France in 1790, is now used by all major countries except the United States. Scientific observations are almost universally expressed in metric units.

LENGTH

The standard of length in the metric system is the *meter* (m). It was originally defined in 1790 as one ten-millionth of the distance from the North Pole to the Equator. In 1889 it was redefined as the distance measured at 0°C between two lines on a bar of platinum-iridium kept at the International Bureau of Weights and Measures in France. Finally, in 1960 the meter was redefined as the length of 1,650,763.73 light waves emitted by atoms of the gas krypton under strictly specified conditions. The meter is equal to 39.37 inches.

A major advantage of the metric system is that units are related to one another by factors of 10. Thus 1 m is 10 decimeters (dm) or 100 centimeters (cm) or 1,000 millimeters (mm). Conversion from one unit to another is simple. Figure A-1 illustrates the differences between metric and U.S. conversions by comparing the meter stick and the yardstick. Exhibit A-2 lists the metric units of length with U.S. equivalents.

Since numbers with many zeros (very large numbers or very small fractions) are cumbersome to work with, they are expressed in *exponential form,* that is, as powers of 10. The form of exponential notation is

$$M \times 10^n$$

You can determine M and n in two steps. For example, how is 0.0000000001 written in exponential form? First, determine M by moving the decimal point so that only one nonzero digit is to the left of it:

0.0000000001.

The digit to the left of the decimal is 1; therefore $M = 1$. Second, determine n by counting the number of places you moved the decimal point. If you moved the point to the left, make the number positive; if you moved it to the right, it is negative. Since you moved the decimal point 10 places to the right, $n = -10$. Thus

$$0.0000000001 = 1 \times 10^{-10}$$

Now do a problem on your own. The wavelength of yellow light is about 0.000059 cm. Convert the centimeters to exponential form. If your answer is 5.9×10^{-5}, you are ready to continue. If you got the wrong answer, reread the discussion.

When we are working with a very large number, the same rules apply, but our exponential value will be positive rather than negative. Refer to Exhibit A-2. Note that 1 km equals 1,000 m. Even though 1,000 is not a cumbersome number, we can still convert it into exponential form. First, move the decimal point so there is only one nonzero digit to the left of it to determine M:

1.000.

Now, because the decimal has been moved three places to the left, n equals +3 or simply 3. Thus

$$1 \text{ km} = 1 \times 10^3 \text{ m}$$

EXHIBIT A-2
METRIC UNITS OF LENGTH AND SOME U.S. EQUIVALENTS

METRIC UNIT	MEANING OF PREFIX	METRIC EQUIVALENT	U.S. EQUIVALENT
1 kilometer (km)	kilo = 1,000	1,000 m	3,280.84 ft = 0.62 mi 1 mi = 1.61 km
1 hectometer (hm)	hecto = 100	100 m	328 ft
1 dekameter (dam)	deka = 10	10 m	32.8 ft
1 meter (m)	Standard unit of length		39.37 inches = 3.28 ft = 1.09 yd
1 decimeter (dm)	deci = $\frac{1}{10}$	0.1 m	3.94 inches
1 centimeter (cm)	centi = $\frac{1}{100}$	0.01 m	0.394 inch 1 inch = 2.54 cm
1 millimeter (mm)	milli = $\frac{1}{1,000}$	0.001 m = $\frac{1}{10}$ cm	0.0394 inches
1 micrometer (μm) [formerly micron (μ)]	micro = $\frac{1}{1,000,000}$	0.000,001 m = $\frac{1}{10,000}$ cm	3.94×10^{-5} inches
1 nanometer (nm) [formerly millimicron (mμ)]	nano = $\frac{1}{1,000,000,000}$	0.000,000,001 m = $\frac{1}{10,000,000}$ cm	3.94×10^{-8} inches
1 angstrom (Å)		0.000,000,000,1 m = $\frac{1}{100,000,000}$ cm	3.94×10^{-9} inches

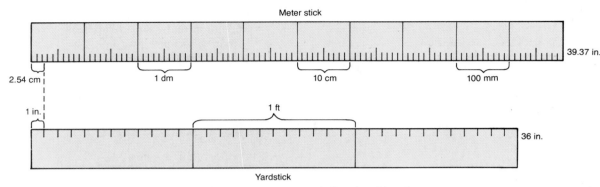

FIGURE A-1 Metric and U.S. units of length.

Do another problem on your own. The speed of light is about 30,000,000,000 cm/sec. Convert the centimeters to exponential form. Your answer should be 3×10^{10} cm.

Review Exhibit A-2 and note some common metric and U.S. equivalents. Note also the exponential forms.

MASS

Now let us look at the second fundamental unit of the metric system: mass. The standard unit of mass is the *kilogram* (kg). A kilogram is defined as the mass of a platinum-iridium cylinder kept at the International Bureau of Weights and Measures in France. The standard pound is defined in terms of the standard kilogram: 1 lb equals 0.4536 kg.

Exhibit A-3 lists metric units of mass and some U.S. equivalents.

EXHIBIT A-3
METRIC UNITS OF MASS AND SOME U.S. EQUIVALENTS

METRIC UNIT	METRIC EQUIVALENT	U.S. EQUIVALENT
1 kilogram (kg)	1,000 g	2.205 lb 1 ton = 907 kg
1 hectogram (hg)	100 g	
1 dekagram (dag)	10 g	0.0353 oz
1 gram (g)	1 g	1 lb = 453.6 g 1 oz = 28.35 g
1 decigram (dg)	0.1 g	
1 centigram (cg)	0.01 g	
1 milligram (mg)	0.001 g	0.015 gr
1 microgram (μg)	0.000,001 g	

TIME

The third fundamental unit of both the metric and the U.S. systems is time. The standard of time is the *second*. Formerly, the second was defined as 1/86,400 of a mean solar day. (A mean solar day is the average of the lengths of all days throughout the year.) Currently, the second is defined as the time required for 9,192,631,770 vibrations of cesium atoms when they are vibrating in a specific manner. Units of time are used in measuring pulse and heart rate, metabolic rate, x-ray exposure, and intervals between medications.

Exhibit A-1 lists the units of time.

VOLUME

Units of volume, or capacity, are derived units based on length. *Volume* in the U.S. system may be expressed as cubic feet (ft³), cubic inches (inch³), and cubic yards (yd³) or as a unit of volume such as the quart. Volume in the metric system may be expressed in cubic units of length such as cubic centimeters (cm³) or in terms of the basic unit of volume, the *liter*. A liter is defined as the volume occupied by 100 g of pure water at 4°C. Since 1 cm³ of water at this temperature weighs 1 g, then 1,000 g of water occupies a volume of 1,000 cm³. This means that a liter is equal to 1,000 cm³ and 1 milliliter (ml) is equal to 1 cm³. Because of this relationship, many volume-measuring devices, such as hypodermic needles, may be graduated in either milliliters or cubic centimeters.

Exhibit A-4 lists metric units of volume and some U.S. equivalents.

EXHIBIT A-4
METRIC UNITS OF VOLUME AND SOME EQUIVALENTS

METRIC UNIT	METRIC EQUIVALENT	U.S. EQUIVALENT
1 liter	1,000 ml	33.81 fl oz = 1.057 qt 946 ml = 1 qt
1 milliliter (ml)	0.001 liter	0.0338 fl oz 30 ml = 1 fl oz 5 ml = 1 teaspoon
1 cubic centimeter (cm³)	0.999972 ml	0.0338 fl oz

APOTHECARY SYSTEM

In addition to the U.S. and metric systems, there is the **apothecary system.** This system is commonly used by physicians prescribing medications and by pharmacists preparing them. Exhibit A-5 lists the important units and equivalents of the apothecary system. Note that the units of mass have the same names as in the U.S. system, but they are not equivalent (1 oz = 28.35 g; 1 oz ap = 30 g) and they do not have the same relationship to one another (1 oz = 16 dr; 1 oz ap = 8 dr ap). The units of volume are the same in both systems.

EXHIBIT A-5

APOTHECARY SYSTEM OF MASS AND VOLUME WITH METRIC EQUIVALENTS

FUNDAMENTAL UNIT	APOTHECARY UNIT AND CONVERSION	METRIC EQUIVALENT
Mass	1 grain (gr) = 0.002083 ounce	1 g = 15 gr
	1 dram (dr ap) = 60 gr	4 g = 1 dr
	1 ounce (oz ap) = 8 dr ap	30 g = 1 oz
	1 pound (lb ap) = 12 oz ap	1 kg = 32 oz
Volume	1 fluidram (fl dr) = 60 minims (min)	1 ml (or cm³) = 15 min
	1 fluidounce (fl oz) = 8 fl dr	4 ml (or cm³) = 1 fl dr
	1 pint (pt) = 16 fl oz	30 ml (or cm³) = 1 fl oz
		500 ml (or cm³) = 1 pt
		1,000 ml (or cm³) = 1 qt

FREQUENTLY USED CONVERSIONS BASED ON MILLIGRAM

1,000 mg (1 g)	= 15 grains
600 mg (0.6 g)	= 10 grains
300 mg (0.3 g)	= 5 grains
60 mg (0.06 g)	= 1 grain
30 mg (0.03 g)	= 0.50 (1/2) grain
20 mg (0.02 g)	= 0.33 (1/3) grain
10 mg (0.01 g)	= 0.166 (1/6) grain
5 mg (0.005 g)	= 0.083 (1/12) grain
4 mg (0.004 g)	= 0.66 (1/15) grain
1 mg (0.001 g)	= 0.016 (1/60) grain
0.5 mg (0.0005 g) =	0.0083 (1/120) grain
0.1 mg (0.0001 g) =	0.0016 (1/600) grain

Appendix B: Abbreviations

Many terms, especially medical terms, are commonly expressed in abbreviated form. In order to familiarize you with some of these abbreviations, an alphabetical list has been prepared. Most of the terms listed have been used in the book; some terms not referred to in the book have been included because of their frequent use.

AID	automatic implantable defibrillation; artificial insemination by donor
AIDS	acquired immune deficiency syndrome
AMI	acute myocardial infarction
ANS	autonomic nervous system
ARF	acute renal failure
AV	atrioventricular
BBB	blood–brain barrier
BEAM	brain electrical activity mapping
BMR	basal metabolic rate
BP	blood pressure
BS	blood sugar
BUN	blood urea nitrogen
CABG	coronary artery bypass grafting
CAC	cardioacceleratory center
CAD	coronary artery disease
CAPD	continuous ambulatory peritoneal dialysis
CBC	complete blood count
CCCC	closed-chest cardiac compression
CF	cystic fibrosis
CHD	coronary heart disease
CHF	congestive heart failure
CIC	cardioinhibitory center
CNS	central nervous system
CO	cardiac output; carbon monoxide
COAD	chronic obstructive airways disease
COPD	chronic obstructive pulmonary disease
CPR	cardiopulmonary resuscitation
C-section	cesarean section
CSF	cerebrospinal fluid
CT (CAT)	computed tomography (computed axial tomography)
CVA	cerebrovascular accident
CVP	central venous pressure
D & C	dilation and curettage
DBP	diastolic blood pressure
DES	diethylstilbestrol
DH	delayed hypersensitivity
DMD	Duchenne muscular dystrophy
DMSO	dimethyl sulfoxide
DSR	dynamic spatial reconstructor
DVT	deep vein thrombosis
EB virus	Epstein-Barr virus
ECF	extracellular fluid
ECG (EKG)	electrocardiogram
EDV	end-diastolic volume
EEG	electroencephalogram
EM	electron micrograph
EMG	electromyogram
EOG	electrooculogram
EPSP	excitatory postsynaptic potential
ER	endoplasmic reticulum
ESR	erythrocyte sedimentation rate
ESRD	end-stage renal disease
ESV	end-systolic volume
FAS	fetal alcohol syndrome
GFR	glomerular filtration rate
GI	gastrointestinal
Hb	hemoglobin
HBO	hyperbaric oxygenation
HBV	hepatitis B virus
HDCV	human diploid cell vaccine
HDL	high-density lipoprotein
HF	heart failure
HR	heart rate
HSV	herpes simplex virus
ICF	intracellular fluid
ID	intradermal
IM	intramuscular
INF	interferon
IPSP	inhibitory postsynaptic potential
IUCD	intrauterine contraceptive device
IUD	intrauterine device
IV	intravenous
LDL	low-density lipoprotein
LLQ	left lower quadrant
LUQ	left upper quadrant
LV	left ventricular
meq/liter	milliequivalents per liter
MG	myasthenia gravis
MI	myocardial infarction
mm³	cubic millimeter
mm Hg	millimeters of mercury
MS	multiple sclerosis
MVP	mitral valve prolapse
NGU	nongonococcal urethritis
NLMC	nocturnal leg muscle cramping
NMR	nuclear magnetic resonance
NPN	nonprotein nitrogen

NREM	nonrapid eye movement	**RUQ**	right upper quadrant
OC	oral contraceptive	**SA**	sinoatrial (sinuatrial)
OTC	over-the-counter	**SCA**	sickle cell anemia
PCT	positron computed tomography	**SIDS**	sudden infant death syndrome
PE	pulmonary embolism	**SLE**	systemic lupus erythematosus
P_{eff}	effective filtration pressure	**SNS**	somatic nervous system
PEMFs	pulsating electromagnetic fields	**SPF**	sun protection factor
PG	prostaglandin	**STD**	sexually transmitted disease
PID	pelvic inflammatory disease	**SV**	stroke volume
PKU	phenylketonuria	**TIA**	transient ischemic attack
PMNs	polymorphonuclear leucocytes	**Tm**	tubular maximum
PMS	premenstrual syndrome	**TM**	transcendental meditation
PNS	peripheral nervous system	**TM joint**	temporomandibular joint
PPNG	penicillinase-producing *Neisseria gonorrhea*	**TPN**	total parenteral nutrition
PTA	percutaneous transluminal angioplasty	**TSS**	toxic shock syndrome
RA	rheumatoid arthritis	**UA**	urinalysis
RAS	reticular activating system	**URI**	upper respiratory infection
RBC	red blood cell	**UTI**	urinary tract infection
RDS	respiratory distress syndrome	**UV**	ultraviolet
REM	rapid eye movement	**VD**	venereal disease
RES	reticuloendothelial system	**VF**	ventricular fibrillation
Rh	*Rhesus*	**VSD**	ventricular septal defect
RIG	rabies immune globulin	**VT**	ventricular tachycardia
RLQ	right lower quadrant	**WBC**	white blood cell

Glossary of Prefixes, Suffixes, and Combining Forms

Many medical terms are "compound" words, that is, they are made up of one or more word roots or combining forms of word roots with prefixes or suffixes. For example, *leucocyte* (white blood cell) is a combination of *leuco,* the combining form for the word root meaning "white," and *cyt,* the word root meaning "cell." Learning the medical meanings of the fundamental word parts will enable you to analyze many long, complicated terms.

The following list includes some of the most commonly used word roots, combining forms, prefixes, and suffixes used in making medical terms and an example for each.

Word Roots and Combining Forms

Acou-, Acu- hearing Acoustics (a-KOO-stiks), the science of sounds or hearing.

Acr-, Acro- extremity Acromegaly (ak'-rō-MEG-a-lē), hyperplasia of the nose, jaws, fingers, and toes.

Aden-, Adeno- gland Adenoma (ad-en-Ō-ma), a tumor with a glandlike structure.

Alg-, Algia- pain Neuralgia (nyoo-RAL-ja), pain along the course of a nerve.

Angi- vessel Angiocardiography (an'-jē-ō-kard-ē-OG-ra-fē), roentgenography of the great blood vessels and heart after intravenous injection of radiopaque fluid.

Arthr-, Arthro- joint Arthropathy (ar-THROP-a-thē), disease of a joint.

Aut-, Auto- self Autolysis (aw-TOL-i-sis), destruction of cells of the body by their own enzymes, even after death.

Bio- life, living Biopsy (BĪ-op-sē), examination of tissue removed from a living body.

Blast- germ, bud Blastocyte (BLAS-tō-sīt), an embryonic or undifferentiated cell.

Blephar- eyelid Blepharitis (blef-a-RĪt-is), inflammation of the eyelids.

Brachi- arm Brachialis (brā-kē-AL-is), muscle that flexes the forearm.

Bronch- trachea, windpipe Bronchoscopy (bron-KOS-kō-pē), direct visual examination of the bronchi.

Bucc-, cheek Buccocervical (bū-kō-SER-vi-kal), pertaining to the cheek and neck.

Capit- head Decapitate (dē-KAP-i-tāt), to remove the head.

Carcin- cancer Carcinogenic (kar-sin-ō-JEN-ik), causing cancer.

Cardi-, Cardia-, Cardio- heart Cardiogram (KARD-ē-o-gram), a recording of the force and form of the heart's movements.

Cephal- head Hydrocephalus (hī-drō-SEF-a-lus), enlargement of the head due to an abnormal accumulation of fluid.

Cerebro- brain Cerebrospinal (se-rē'-brō-SPĪN-al) fluid, fluid contained within the cranium and spinal canal.

Cheil- lip Cheilosis (kī-LŌ-sis), dry scaling of the lips.

Chole- bile, gall Cholecystogram (kō-lē-SIS-tō-gram), roentgenogram of the gallbladder.

Chondr-, Chondri-, Chondrio- cartilage Chondrocyte (KON-drō-sīt), a cartilage cell.

Chrom-, Chromat-, Chromato- color Hyperchromic (hī-per-KRŌ-mik), highly colored.

Crani- skull Craniotomy (krā-nē-OT-o-mē), surgical opening of the skull.

Cry-, Cryo- cold Cryosurgery (krī-ō-SERJ-e-rē), surgical procedure using a very cold liquid nitrogen probe.

Cut- skin Subcutaneous (sub-kyoo-TĀ-nē-us), under the skin.

Cysti-, Cysto- sac, bladder Cystoscope (SIS-tō-skōp), instrument for interior examination of the bladder.

Cyt-, Cyto-, Cyte- cell Cytology (sī-TOL-ō-jē), the study of cells.

Dactyl-, Dactylo- digits (usually fingers, but sometimes toes) Polydactylism (pol-ē-DAK-til-ism), above normal number of fingers or toes.

Derma-, Dermato- skin Dermatosis (der-ma-TŌ-sis), any skin disease.

Entero- intestine Enteritis (ent-e-RĪT-is), inflammation of the intestine.

Erythro- red Erythrocyte (e-RITH-rō-sīt), red blood cell.

Galacto- milk Galactose (ga-LAK-tose), a milk sugar.

Gastr- stomach Gastrointestinal (gas'-trō-in-TES-tin-al), pertaining to the stomach and intestine.

Gloss-, Glosso- tongue Hypoglossal (hī'-pō-GLOS-al), located under the tongue.

Glyco- sugar Glycosuria (glī'-kō-SHUR-ē-a), sugar in the urine.

Gyn-, Gyne-, Gynec- female, women Gynecology (gīn'-e-KOL-ō-jē), the medical specialty dealing with disorders of the female.

Hem-, Hemat- blood Hematoma (hē'-ma-TŌ-ma), a tumor or swelling filled with blood.

Hepar-, Hepato- liver Hepatitis (hep-a-TĪT-is), inflammation of the liver.

Hist-, Histio- tissue Histology (his-TOL-ō-jē), the study of tissues.

Hydr- water Hydrocele (HĪ-drō-sēl), accumulation of fluid in a saclike cavity.

Hyster- uterus Hysterectomy (his'-te-REK-tō-mē), surgical removal of the uterus.

Ileo- ileum Ileocecal valve (il'-ē-ō-SĒ-kal), folds at the opening between ileum and cecum.

Ilio- ilium Iliosacral (il'-ē-ō-SA-kral), pertaining to ilium and sacrum.

Lachry-, Lacri- tears Nasolacrimal (nā-zō-LAK-rim-al), pertaining to the nose and lacrimal apparatus.

Laparo- loin, flank, abdomen Laparoscopy (lap'-a-ROS-kō-pē), examination of the interior of the abdomen by means of a laparoscope.

Leuco-, Leuko- white Leucocyte (LYOO-kō-sīt), white blood cell.

Lip-, Lipo- fat Lipoma (lī-PŌ-ma), a fatty tumor.

Mamm- breast Mammography (ma-MOG-ra-fē), roentgenography of the mammary gland.

Mast- breast Mastitis (ma-STĪT-is), inflammation of the mammary gland.

Meningo- membrane Meningitis (men-in-JĪT-is), inflammation of the membranes of spinal cord and brain.

Metro- uterus Endometrium (en'-dō-MĒ-trē-um), lining of the uterus.

Morpho- form, shape Morphology (mor-FOL-o-jē), the study of form and structure of things.

Myelo- marrow, spinal cord Poliomyelitis (pō'-lē-ō-mī'-a-LĪT-is), inflammation of the gray matter of the spinal cord.

Myo- muscle Myocardium (mī'-ō-KARD-ē-um), heart muscle.

Necro- corpse, dead Necrosis (ne-KRŌ-sis), death of areas of tissue surrounded by healthy tissue.

Nephro- kidney Nephrosis (ne-FRŌ-sis), degeneration of kidney tissue.

Neuro- nerve Neuroblastoma (nyoor'-ō-blas-TŌ-ma), malignant tumor of the nervous system composed of embryonic nerve cells.

Oculo- eye Binocular (bī-NOK-yoo-lar), pertaining to the two eyes.

Odont- tooth Orthodontic (or-thō-DONT-ik), pertaining to the proper positioning and relationship of the teeth.

Oo- egg Oocyte (Ō-ō-sīt), original egg cell.

Ophthalm- eye Ophthalmology (of'-thal-MOL-o-jē), the study of the eye and its diseases.

Oss-, Osseo-, Osteo- bone Osteoma (os-tē-Ō-ma), bone tumor.

Oto- ear Otosclerosis (ō'-tō-skle-RŌ-sis), formation of bone in the labyrinth of the ear.

Patho- disease Pathogenic (path'-ō-JEN-ik), causing disease.

Ped- children Pediatrician (pēd'-ē-a-TRISH-an), medical specialist in the treatment of children.

Phag-, Phago- to eat Phagocytosis (fag'-ō-sī-TŌ-sis), the process by which cells ingest particulate matter.

Philic-, Philo- to like, have an affinity for Hydrophilic (hī'-drō-FIL-ik), having an affinity for water.

Phleb- vein Phlebitis (fli-BĪT-is), inflammation of the veins.

Pneumo- lung, air Pneumothorax (nyoo-mō-THŌR-aks), air in the thoracic cavity.

Pod- foot Podiatry (po-DĪ-a-trē), the diagnosis and treatment of foot disorders.

Procto- anus, rectum Proctoscopy (prok-TOS-kō-pē), instrumental examination of the rectum.

Psycho- soul, mind Psychiatry (sī-KĪ-a-trē), treatment of mental disorders.

Pyo- pus Pyuria (pī-YOOR-ē-a), pus in the urine.

Rhin- nose Rhinitis (rī-NĪT-is), inflammation of nasal mucosa.

Scler-, Sclero- hard Atherosclerosis (ath'-er-ō-skle-RŌ-sis), hardening of the arteries.

Sep-, Septic- toxic condition due to microorganisms Septicemia (sep'-ti-SĒ-mē-a), presence of bacterial toxins in the blood (blood poisoning).

Soma-, Somato- body Somatotropic (sō-mat-ō-TRŌ-pik), having a stimulating effect on body growth.

Sten- narrow Stenosis (ste-NŌ-sis), narrowing of a duct or canal.

Stasis-, Stat- stand still Homeostasis (hō'-mē-ō-STĀ-sis), achievement of a steady state.

Therm- heat Thermometer (ther-MOM-et-er), instrument used to measure and record heat.

Tox-, Toxic- poison Toxemia (tok-SĒ-mē-a), poisonous substances in the blood.

Trich- hair Trichosis (trik-Ō-sis), disease of the hair.

Viscer- organ Visceral (VIS-e-ral), pertaining to the abdominal organs.

Zoo- animal Zoology (zō-OL-o-jē), the study of animals.

Prefixes

A-, An- without, lack of, deficient Anesthesia (an'-es-THĒ-zha), without sensation.

Ab- away from, from Abnormal (ab-NOR-mal), away from normal.

Ad- to, near, toward Adduction (a-DUK-shun), movement of an extremity toward the axis of the body.

Ambi- both sides Ambidextrous (am'-bi-DEK-strus), able to use either hand.

Ante- before Antepartum (ant-ē-PAR-tum), before delivery of a baby.

Anti- against Anticoagulant (an-tī-kō-AG-yoo-lant), a substance that prevents coagulation of blood.

Bi- two, double, both Biceps (BĪ-seps), a muscle with two heads of origin.

Brachy- short Brachyesophagus (brā-kē-e-SOF-a-gus), short esophagus.

Brady- slow Bradycardia (brād'-ē-KARD-ē-a), abnormal slowness of the heartbeat.

Cata- down, lower, under, against Catabolism (ka-TAB-a-lizm), metabolic breakdown into simpler substances.

Circum- around Circumrenal (ser-kum-RĒN-al), around the kidney.

Col-, Con-, Com- with, together Congenital (kon-JEN-i-tal), existing at birth.

Contra- against, opposite Contraception (kon-tra-SEP-shun), the prevention of conception.

Crypt- hidden, concealed Cryptorchidism (krip-TOR-ka-dizm'), undescended or hidden testes.

Di-, Diplo- two Diploid (DIP-loyd), having double the haploid number of chromosomes.

Dis- separation, apart, away from Disarticulate (dis'-ar-TIK-yoo-lāt'), to separate at a joint.

Dys- painful, difficult Dyspnea (disp-NĒ-a), difficult breathing.

E-, Ec-, Ex- out from, out of Eccentric (ek-SEN-trik), not located at the center.

Ecto-, Exo- outside Ectopic pregnancy (ek-TOP-ik), gestation outside the uterine cavity.

Em-, En- in, on Empyema (em'-pī-Ē-ma), pus in a body cavity.

End-, Endo- inside Endocardium (en'-dō-KARD-ē-um), membrane lining the inner surface of the heart.

Epi- upon, on, above Epidermis (ep'-i-DER-mis), outermost layer of skin.

Eu- well Eupnea (YOOP-nē-a), normal breathing.

Ex-, Exo- out, away from Exocrine (EK-sō-krin), excreting outwardly or away from.

Extra- outside, beyond, in addition to Extracellular (ek'-stra-SEL-yoo-lar), outside of the cell.

Hemi- half Hemiplegia (hem-ē-PLĒ-jē-a), paralysis of only half of the body.

Heter-, Hetero- other, different Heterogeneous (het'-e-rō-JEN-ē-us), composed of different substances.

Homeo-, Homo- unchanging, the same, steady Homeostasis (hō'-mē-ō-STĀ-sis), achievement of a steady state.

Hyper- beyond, excessive Hyperglycemia (hī-per-glī-SĒ-mē-a), excessive amount of sugar in the blood.

ing bony surfaces are separated by an elastic substance to which both are attached, so that the mobility is slight, but may be exerted in all directions.

Anabolism (a-NAB-ō-lizm) Synthetic reactions whereby small molecules are built up into larger ones.

Anaerobic (an-AIR-ō-bik) Not requiring molecular oxygen.

Anal (Ā-nal) **canal** The terminal 2 or 3 cm of the rectum; opens to the exterior at the anus.

Anal column A longitudinal fold in the mucous membrane of the anal canal that contains a network of arteries and veins.

Analgesia (an-al-JĒ-zē-a) Absence of normal sense of pain.

Anal triangle The subdivision of the male or female perineum that contains the anus.

Anamnestic (an'-am-NES-tik) **response** Accelerated, more intense production of antibodies upon a subsequent exposure to an antigen after the initial exposure.

Anaphase (AN-a-fāz) The third stage of mitosis in which the chromatids that have separated at the centromeres move to opposite poles.

Anaphylaxis (an'-a-fi-LAK-sis) A hypersensitivity reaction involving IgE antibodies, mast cells, and basophils; examples are hay fever, bronchial asthma, hives, and anaphylactic shock.

Anastomosis (a-nas-tō-MŌ-sis) An end-to-end union or joining together of blood vessels, lymphatics, or nerves.

Anatomical (an'-a-TOM-i-kal) **position** A position of the body universally used in anatomical descriptions in which the body is erect, facing the observer, the upper extremities are at the sides, and palms of the hands are facing forward.

Anatomy (a-NAT-ō-mē) The structure or study of structure of the body and the relationship of its parts to each other.

Androgen (AN-drō-jen) Substance producing or stimulating male characteristics, such as the male hormone testosterone.

Anemia (a-NĒ-mē-a) Condition of the blood in which the number of functional red blood cells or their hemoglobin content is below normal.

Anesthesia (an'-es-THĒ-zē-a) A total or partial loss of feeling or sensation, usually defined with respect to loss of pain sensation.

Aneurysm (AN-yoo-rizm) A saclike enlargement of a blood vessel caused by a weakening of the wall.

Angina pectoris (an-JĪ-na PEK-tō-ris) A pain in the chest related to reduced coronary circulation that may or may not involve heart or artery disease.

Angiography (an-jē-OG-ra-fē) X-ray examination of blood vessels after injection of a radiopaque substance into the common carotid or vertebral artery; used to demonstrate cerebral blood vessels and may detect brain tumors with specific vascular patterns.

Angiotensin (an-jē-ō-TEN-sin) Either of two forms of a protein associated with regulation of blood pressure. Angiotensin I is produced by the action of renin on angiotensinogen and is converted by the action of a plasma enzyme into angiotensin II, which controls aldosterone secretion by the adrenal cortex.

Ankyloglossia (ang'-ki-lō-GLOSS-ē-a) "Tongue-tied"; restriction of tongue movements by a short lingual frenulum.

Anomaly (a-NOM-a-lē) An abnormality that may be a developmental (congenital) defect; a variant from the usual standard.

Anopsia (an-OP-sē-a) A defect of vision.

Anorexia nervosa (an-ō-REK-sē-a ner-VŌ-sa) A disorder characterized by loss of appetite and bizarre patterns of eating.

Anosmia (an-OZ-mē-a) Loss of the sense of smell.

Anoxia (an-OK-sē-a) Deficiency of oxygen.

Antagonist (an-TAG-ō-nist) A muscle that has an action oppo-

site that of the agonist and yields to the movement of the agonist.

Antepartum (an-tē-PAR-tum) Before delivery of the child; occurring (to the mother) before childbirth.

Anterior (an-TĒR-ē-or) Nearer to or at the front of the body. Also called **ventral.**

Anterior root The structure composed of axons of motor or efferent fibers that emerges from the anterior aspect of the spinal cord and extends laterally to join a posterior root, forming a spinal nerve. Also called a **ventral root.**

Antibiotic (an'-ti-bī-OT-ik) Literally, "antilife"; a chemical produced by a microorganism that is able to inhibit the growth of or kill other microorganisms.

Antibody (AN-ti-bod'-ē) A substance produced by certain cells in the presence of a specific antigen that combines with that antigen to neutralize, inhibit, or destroy it.

Anticoagulant (an-tī-cō-AG-yoo-lant) A substance that is able to delay, suppress, or prevent the clotting of blood.

Antidiuretic hormone (ADH) Hormone produced by neurosecretory cells in the paraventricular nucleus of the hypothalamus that stimulates water reabsorption from kidney cells into the blood and vasoconstriction of arterioles.

Antigen (AN-ti-jen) Any substance that when introduced into the tissues or blood induces the formation of antibodies or reacts with them.

Anulus fibrosus (AN-yoo-lus fi-BRŌ-sus) A ring of fibrous tissue and fibrocartilage that encircles the pulpy substance (nucleus pulposus) of an intervertebral disc.

Anus (Ā-nus) The distal end and outlet of the rectum.

Aorta (ā-OR-ta) The main systemic trunk of the arterial system of the body; emerges from the left ventricle.

Aortic (ā-OR-tik) **reflex** A reflex concerned with maintaining normal general systemic blood pressure.

Aphasia (a-FĀ-zē-a) Loss of ability to express oneself properly through speech or verbal comprehension.

Apnea (ap-NĒ-a) Temporary cessation of breathing.

Apneustic (ap-NOO-stik) **area** Portion of the respiratory center in the pons that helps coordinate the transition between inspiration and expiration.

Apocrine (AP-ō-krin) **gland** A type of gland in which the secretory products gather at the free end of the secreting cell and are pinched off, along with some of the cytoplasm, to become the secretion, as in mammary glands.

Aponeurosis (ap'-ō-nyoo-RŌ-sis) A sheetlike layer of dense, regularly arranged connective tissue joining one muscle with another or with bone.

Appendicitis (a-pen-di-SĪ-tis) Inflammation of the vermiform appendix.

Appetite A desire for food stimulated by a psychological need; may not be related to hunger, which is a response to a nutritional or energy requirement.

Appositional growth Growth due to surface deposition of material, as in the growth in diameter of cartilage and bone. Also called **exogenous growth.**

Aqueous humor (ĀK-wē-us HYOO-mor) The watery fluid that fills the anterior cavity of the eye.

Arachnoid (a-RAK-noyd) The middle of the three coverings (meninges) of the brain.

Arachnoid villus (VIL-us) Berrylike tuft of arachnoid that protrudes into the superior sagittal sinus and through which the cerebrospinal fluid enters the bloodstream.

Arbor vitae (AR-bor VĒ-tē) The treelike appearance of the white matter tracts of the cerebellum when seen in midsagittal section. A series of branching ridges within the cervix of the uterus.

Arch of the aorta (ā-OR-ta) The most superior portion of the aorta, lying between the ascending and descending segments of the aorta.

Areflexia (a'-rē-FLEK-sē-a)   Absence of reflexes.

Areola (a-RĒ-ō-la)   Any tiny space in a tissue. The pigmented ring around the nipple of the breast.

Arm   The portion of the upper extremity from the shoulder to the elbow.

Arrector pili (a-REK-tor PI-lē)   Smooth muscles attached to hairs; contraction pulls the hairs into a more vertical position, resulting in "goose bumps."

Arrhythmia (a-RITH-mē-a)   Irregular heart rhythm.

Arteriogram (ar-TĒR-ē-ō-gram)   A roentgenogram of an artery after injection of a radiopaque substance into the blood.

Arteriole (ar-TĒ-rē-ōl)   A small arterial branch that delivers blood to a capillary.

Artery (AR-ter-ē)   A blood vessel that carries blood away from the heart.

Arthritis (ar-THRĪ-tis)   Inflammation of a joint.

Arthrology (ar-THROL-ō-jē)   The study or description of joints.

Arthroscopy (ar-THROS-co-pē)   Surgical technique in which an arthroscope is inserted into a small incision, usually in the knee, to repair torn cartilage.

Articular (ar-TIK-yoo-lar) **capsule**   Sleevelike structure around a synovial joint composed of a fibrous capsule and a synovial membrane.

Articular cartilage (KAR-ti-lij)   Hyaline cartilage attached to articular bone surfaces.

Articular disc   Fibrocartilage pad between articular surfaces of bones of some synovial joints. Also called a **meniscus** (men-IS-cus).

Articulation (ar-tik'-yoo-LA-shun)   A joint.

Ascending colon (KŌ-lon)   The portion of the large intestine that passes upward from the cecum to the lower edge of the liver where it bends at the right colic (hepatic) flexure to become the transverse colon.

Aseptic (ā-SEP-tik)   Free from any infectious or septic material.

Asphyxia (as-FIX-ē-a)   Unconsciousness due to interference with the oxygen supply of the blood.

Aspirate (AS-pir-āt)   To remove by suction.

Association area   A portion of the cerebral cortex connected by many motor and sensory fibers to other parts of the cortex. The association areas are concerned with motor patterns, memory, concepts of word-hearing and word-seeing, reasoning, will, judgment, and personality traits.

Association neuron (NOO-ron)   A nerve cell lying completely within the central nervous system that carries impulses from sensory neurons to motor neurons. Also called an **internuncial** or **connecting neuron.**

Astereognosis (as-ter'-ē-ōg-NŌ-sis)   Inability to recognize objects or forms by touch.

Astigmatism (a-STIG-ma-tizm)   An irregularity of the lens or cornea of the eye causing the image to be out of focus and producing faulty vision.

Astrocyte (AS-trō-sīt)   A neuroglial cell having a star shape.

Ataxia (a-TAK-sē-a)   A lack of muscular coordination, lack of precision.

Atelectasis (at'-ē-LEK-ta-sis)   A collapsed or airless state of all or part of the lung, which may be acute or chronic.

Atherosclerosis (ath'-er-ō-skle-RŌ-sis)   A process in which fatty substances (cholesterol and triglycerides) are deposited in the walls of medium and large arteries in response to certain stimuli (hypertension, carbon monoxide, dietary cholesterol). Following endothelial damage, platelets release chemicals that stimulate uptake of cholesterol and proliferation of smooth muscle cells in the tunica media, resulting in formation of an atherosclerotic plaque that decreases the size of the arterial lumen.

Atom   Unit of matter that comprises a chemical element; consists of a nucleus and electrons.

Atresia (a-TRĒ-zē-a)   Abnormal closure of a passage, or absence of a normal body opening.

Atrial fibrillation (Ā-trē-al fib-ri-LĀ-shun)   Asynchronous contraction of the atria that results in the cessation of atrial pumping.

Atrioventricular (AV) (ā'-trē-ō-ven-TRIK-yoo-lar) **bundle**   The portion of the conduction system of the heart that begins at the atrioventricular node and passes through the cardiac skeleton separating the atria and the ventricles, then runs a short distance down the interventricular septum before splitting into right and left bundle branches.

Atrioventricular (AV) node   The portion of the conduction system of the heart made up of a compact mass of conducting cells located near the orifice of the coronary sinus in the right atrial wall.

Atrioventricular (AV) valve   A structure made up of membranous flaps or cusps that allows blood to flow in one direction only, from an atrium into a ventricle.

Atrium (Ā-trē-um)   A superior chamber of the heart.

Atrophy (AT-rō-fē)   Wasting away or decrease in size of a part, due to a failure, abnormality of nutrition, or lack of use.

Auditory ossicle (AW-di-tō-rē OS-si-kul)   One of three small bones of the middle ear called the malleus, incus, and stapes.

Auditory tube   The tube that connects the middle ear with the nose and nasopharynx of the throat. Also called the **Eustachian tube.**

Auricle (OR-i-kul)   The flap or pinna of the ear. An appendage of an atrium of the heart.

Auscultation (aws-kul-TĀ-shun)   Examination by listening to sounds in the body.

Autoimmunity   An immunologic response against a person's own tissue antigens.

Autonomic ganglion (aw'-tō-NOM-ik GANG-lē-on)   A cluster of sympathetic or parasympathetic cell bodies located outside the central nervous system.

Autonomic nervous system (ANS)   Visceral efferent neurons, both sympathetic and parasympathetic, that transmit impulses from the central nervous system to smooth muscle, cardiac muscle, and glands; so named because this portion of the nervous system was thought to be self-governing or spontaneous.

Autonomic plexus (PLEK-sus)   An extensive network of sympathetic and parasympathetic fibers; the cardiac, celiac, and pelvic plexuses are located in the thorax, abdomen, and pelvis, respectively.

Autopsy (AW-top-sē)   The examination of the body after death; a postmortem study of the corpse.

Autoregulation (aw-tō-reg-yoo-LĀ-shun)   A local, automatic adjustment of blood flow in a given region of the body in response to tissue needs.

Autosome   Any chromosome other than the pair of sex chromosomes.

Avitaminosis (ā-vī-ta-min-Ō-sis)   A deficiency of a vitamin in the diet.

Axilla (ak-SIL-a)   The small hollow beneath the arm where it joins the body at the shoulders. Also called the **armpit.**

Axon (AK-son)   The process of a nerve cell that carries an impulse away from the cell body.

Babinski (ba-BIN-skē) **sign**   Extension of the great toe, with or without fanning of the other toes, in response to stimulation of the outer margin of the sole of the foot; normal up to 1½ years of age.

Back   The posterior part of the body; the dorsum.

Bainbridge (BĀN-bridge) **reflex**   The increased heart rate that follows increased pressure or distension of the right atrium.

Ball-and-socket joint   A synovial joint in which the rounded surface of one bone moves within a cup-shaped depression or fossa of another bone, as in the shoulder or hip joint. Also called a **spheroid joint.**

Basal metabolic (BĀ-sal met'-a-BOL-ik) rate (BMR)   The rate of metabolism measured under standard or basal conditions.

Basal ganglia (GANG-glē-a)   Paired clusters of cell bodies that make up the central gray matter in each cerebral hemisphere, including the caudate nucleus, lentiform nucleus, claustrum, and amygdaloid body. Also called **cerebral nuclei.**

Base   The broadest part of a pyramidal structure. A nonacid, or a proton acceptor, characterized by excess of hydroxide ions and a pH greater than 7. A ring-shaped, nitrogen-containing organic molecule which is one of the components of a nucleotide, for example, adenine, guanine, cytosine, thymine, and uracil.

Basement membrane   Thin, extracellular layer consisting of basal lamina secreted by epithelial cells and reticular lamina secreted by connective tissue cells.

Basilar (BAS-i-lar) membrane   A membrane in the cochlea of the inner ear that separates the cochlear duct from the scala tympani and on which the spiral organ rests.

Basophil (BĀ-sō-fil)   A type of white blood cell characterized by a pale nucleus and large granules which stain readily with basic dyes.

B cell   A lymphocyte that develops into a plasma cell that produces antibodies or a memory cell.

Belly   The abdomen. The gaster or prominent, fleshy part of a skeletal muscle.

Benign (bē-NĪN)   Not malignant.

Beta (BĀ-ta) cell   A cell in the islets of Langerhans in the pancreas that secretes insulin.

Bicuspid (bī-KUS-pid) valve   Atrioventricular valve on the left side of the heart. Also called the **mitral valve.**

Bilateral (bī-LAT-er-al)   Pertaining to two sides of the body.

Bile (bīl)   A secretion of the liver.

Biliary calculi (BIL-ē-er-ē CAL-kyoo-lē)   Gallstones formed by the crystallization of cholesterol in bile.

Bilirubin (bil-ē-ROO-bin)   A red pigment that is one of the end products of hemoglobin breakdown by the liver cells and is excreted as a waste material in the bile.

Bilirubinuria (bil-ē-roo-bi-NOO-rē-a)   The presence of above normal levels of bilirubin in urine.

Biliverdin (bil-ē-VER-din)   A green pigment that is one of the first products of hemoglobin breakdown in the liver cells and is converted to bilirubin or excreted as a waste material in bile.

Biofeedback   Process by which an individual gets constant signals (feedback) about various visceral body functions.

Biopsy (BĪ-op-sē)   Removal of tissue or other material from the living body for examination, usually microscopic.

Blastocoel (BLAS-tō-sēl)   The fluid-filled cavity within the blastocyst.

Blastocyst (BLAS-tō-sist)   In the development of an embryo, a hollow ball of cells that consists of a blastocoel (the internal cavity), trophectoderm (outer cells), and inner cell mass.

Blastomere (BLAS-tō-mēr)   One of the cells resulting from the cleavage of a fertilized ovum.

Bleeding time   The time required for the cessation of bleeding from a small skin puncture as a result of platelet disintegration and blood vessel constriction; ranges from 1 to 4 minutes.

Blind spot   Area in the retina at the end of the optic nerve in which there are no light receptor cells.

Blood   The fluid that circulates through the heart, arteries, capillaries, and veins and that constitutes the chief means of transport within the body.

Blood–brain barrier (BBB)   A special mechanism that prevents the passage of materials from the blood to the cerebrospinal fluid and brain.

Blood pressure (BP)   Force exerted by blood as it presses against and attempts to stretch blood vessels, especially arteries; clinically, a measure of the pressure in arteries during ventricular systole and ventricular diastole.

Body cavity   A space within the body that contains various internal organs.

Bolus (BŌ-lus)   A soft, rounded mass, usually food, that is swallowed.

Bony labyrinth (LAB-ī-rinth)   A series of cavities within the petrous portion of the temporal bone, forming the vestibule, cochlea, and semicircular canals of the inner ear.

Bradycardia (brad-ē-KAR-dē-a)   A slow heartbeat or pulse rate.

Brain   A mass of nerve tissue located in the cranial cavity.

Brain electrical activity mapping (BEAM)   Noninvasive procedure that measures and displays the electrical activity of the brain; used primarily to diagnose epilepsy.

Brain sand   Calcium deposits in the pineal gland that are laid down starting at puberty.

Brain stem   The portion of the brain immediately superior to the spinal cord, made up of the medulla oblongata, pons varolii, and midbrain.

Broca's (BRŌ-kaz) area   Motor area of the brain in the frontal lobe that translates thoughts into speech. Also called the **motor speech area.**

Bronchi (BRONG-kē)   Branches of the respiratory passageway including primary bronchi (the two divisions of the trachea), secondary or lobar bronchi (divisions of the primary that are distributed to the lobes of the lung), and tertiary or segmental bronchi (divisions of the secondary that are distributed to bronchopulmonary segments of the lung).

Bronchial asthma (BRONG-kē-al AZ-ma)   Usually allergic reaction characterized by smooth muscle spasms in bronchi resulting in wheezing and difficult breathing.

Bronchial tree   The trachea, bronchi, and their branching structures.

Bronchiole (BRONG-kē-ōl)   Branch of a tertiary bronchus further dividing into terminal bronchioles (distributed to lobules of the lung), which divide into respiratory bronchioles (distributed to alveolar sacs).

Bronchitis (BRONG-kī-tis)   Inflammation of the bronchi characterized by a productive cough.

Bronchogenic carcinoma (brong'-kō-JEN-ik kar'-si-NŌ-ma)   Cancer originating in the bronchi.

Bronchogram (BRONG-kō-gram)   A roentgenogram of the lungs and bronchi

Bronchopulmonary (brong'-kō-PUL-mō-ner'-ē) segment   One of the smaller divisions of a lobe of a lung supplied by its own branches of a bronchus.

Bronchoscope (BRONG-kō-skōp)   An instrument used to examine the interior of the bronchi of the lungs.

Buccal (BUK-al)   Pertaining to the cheek or mouth.

Buffer (BUF-er) system   A pair of chemicals, one a weak acid and one a weak base, that resists changes in pH.

Bulbourethral (bul'-bō-yoo-RĒ-thral) gland   One of a pair of glands located inferior to the prostate on either side of the urethra that secretes an alkaline fluid into the cavernous urethra. Also called a **Cowpers' gland.**

Bulimia (boo-LIM-ē-a)   A disorder characterized by uncontrollable overeating followed by forced vomiting or overdoses of laxatives.

Bundle branch   One of the two branches of the atrioventricular bundle, made up of specialized muscle fibers that transmit electrical impulses to the ventricles.

Bunion (BUN-yun) Inflammation and thickening of the bursa of a toe joint due to displacement of the great toe.

Burn An injury caused by heat (fire, steam), chemicals, electricity, or the ultraviolet rays of the sun.

Bursa (BUR-sa) A sac or pouch of synovial fluid located at friction points, especially about joints.

Bursitis (bur-SĪ-tis) Inflammation of a bursa.

Buttocks (BUT-oks) The two fleshy masses on the posterior aspect of the lower trunk, formed by the gluteal muscles.

Calcitonin (kal-si-TŌ-nin) **(CT)** A hormone produced by the thyroid gland that lowers the calcium and phosphate levels of the blood by inhibiting bone breakdown and accelerating calcium absorption by bones.

Calculus (KAL-kyoo-lus) A stone, or insoluble mass of crystallized salts or other material, formed within the body, as in the gallbladder, kidney, or urinary bladder.

Callus (KAL-lus) A growth of new bone tissue in and around a fractured area, ultimately replaced by mature bone. An acquired, localized thickening.

Calorie (KAL-ō-rē) A unit of heat. The small calorie (cal) is the standard unit and is the amount of heat necessary to raise 1 g of water 1°C from 14° to 15°C. The large Calorie (kilocalorie), used in metabolic and nutrition studies, is the amount of heat necessary to raise 1 kg of water 1°C, or 1,000 cal.

Calyx (KĀL-iks) Any cuplike division of the kidney pelvis. *Plural,* **calyces** (KĀ-li-sēz).

Canaliculus (kan'-a-LIK-yoo-lus) A small channel or canal, as in bones, where they connect the lacunae. *Plural,* **canaliculi** (kan'-a-LIK-yoo-lī).

Canal of Schlemm (shlem) A circular venous sinus located at the junction of the sclera and the cornea through which aqueous humor drains from the anterior chamber of the eyeball into the blood.

Cancer (KAN-ser) A malignant tumor of epithelial origin tending to infiltrate and give rise to new growths or metastases. Also called **carcinoma** (kar'-si-NŌ-ma).

Canthus (KAN-thus) The angular junction of the eyelids at either corner of the eyes.

Capillary (KAP-i-lar'-ē) A microscopic blood vessel located between an arteriole and venule through which materials are exchanged between blood and body cells.

Carbohydrate (kar'-bō-HĪ-drāt) An organic compound containing carbon, hydrogen, and oxygen in a particular amount and arrangement and comprised of sugar subunits; usually has the formula $(CH_2O)n$.

Carbon monoxide poisoning Hypoxia due to increased levels of carbon monoxide as a result of its preferential and tenacious combination with hemoglobin compared to oxygen.

Carcinogen (kar-SIN-ō-jen) Any substance that causes cancer.

Cardiac (KAR-dē-ak) **cycle** A complete heartbeat consisting of systole and diastole of both atria plus systole and diastole of both ventricles.

Cardiac muscle An organ specialized for contraction, composed of striated muscle cells, forming the walls of the heart, and stimulated by an intrinsic conduction system and visceral efferent neurons.

Cardiac notch An angular notch in the anterior border of the left lung.

Cardiac output (CO) The volume of blood pumped from one ventricle of the heart (usually measured from the left ventricle) in 1 min; about 5.2 liters/min under normal resting conditions.

Cardiac reserve The maximum percentage that cardiac output can increase above normal.

Cardioacceleratory (kar-dē-ō-ak-SEL-er-a-tō'-rē) **center** A group of neurons in the medulla from which cardiac nerves (sympathetic) arise; impulses along the nerves release epi-

nephrine that increases the rate and force of heartbeat.

Cardioinhibitory (kar-dē-ō-in-HIB-i-tō'-rē) **center** A group of neurons in the medulla from which parasympathetic fibers that reach the heart via the vagus (X) nerve arise; impulses along the nerves release acetylcholine that decreases the rate and force of heartbeat.

Cardiopulmonary resuscitation (rē-sus-i-TĀ-shun) **(CPR)** A technique employed to restore life or consciousness to a person apparently dead or dying; includes external respiration (exhaled air respiration) and external cardiac massage.

Caries (KĀ-rēz) Decay of a tooth or bone.

Carotid (ka-ROT-id) **body** Receptor in the carotid sinus that responds to alterations in blood levels of oxygen, carbon dioxide, and hydrogen ions.

Carotid sinus A dilated region of the internal carotid artery immediately above the bifurcation of the common carotid artery that contains receptors that monitor blood pressure.

Carotid sinus reflex A reflex concerned with maintaining normal blood pressure in the brain.

Carpus (KAR-pus) A collective term for the eight bones of the wrist.

Cartilage (KAR-ti-lij) A type of connective tissue consisting of chondrocytes in lacunae embedded in a dense network of collagenous and elastic fibers and a matrix of chondroitin sulfate.

Cartilaginous (kar'-ti-LAJ-i-nus) **joint** A joint without a joint cavity where the articulating bones are held tightly together by cartilage, allowing little or no movement.

Cast A small mass of hardened material formed within a cavity in the body and then discharged from the body; can originate in different areas and be composed of various materials.

Castration (kas-TRĀ-shun) The removal of the testes.

Catabolism (ka-TAB-ō-lizm) Decomposition processes that provide energy.

Cataract (KAT-a-rakt) Loss of transparency of the crystalline lens of the eye or its capsule or both.

Catheter (KATH-i-ter) A tube that can be inserted into a body cavity through a canal or into a blood vessel; used to remove fluids, such as urine or blood, and to introduce diagnostic materials or medication.

Cauda equina (KAW-da ē-KWĪ-na) A taillike collection of roots of spinal nerves at the inferior end of the spinal canal.

Cecum (SĒ-kum) A blind pouch at the proximal end of the large intestine to which the ileum is attached.

Cell The basic structural and functional unit of all organisms; the smallest structure capable of performing all the activities vital to life.

Cell inclusion A lifeless, often temporary, constituent in the cytoplasm of a cell as opposed to an organelle.

Cellular immunity That component of immunity in which specially sensitized lymphocytes (T cells) attach to antigens to destroy them.

Cementum (se-MEN-tum) Calcified tissue covering the root of a tooth.

Center of ossification (os'-i-fi-KĀ-shun) An area in the cartilage model of a future bone where the cartilage cells hypertrophy, secrete enzymes that result in the calcification of their matrix resulting in the death of the cartilage cells, followed by the invasion of the area by osteoblasts that then lay down bone.

Central canal A microscopic tube running the length of the spinal cord in the gray commissure.

Central fovea (FŌ-vē-a) A cuplike depression in the center of the macula lutea of the retina, containing cones only; the area of clearest vision.

Central nervous system (CNS) Brain and spinal cord.

Centrioles (SEN-trē-ōlz) Paired, cylindrical organelles within a centrosome, each consisting of a ring of microtubules and arranged at right angles to each other; function in cell division.

Centromere (SEN-trō-mēr) The clear, constricted portion of a chromosome where the two chromatids are joined; serves as the point of attachment for the chromosomal microtubules.

Centrosome (SEN-trō-sōm) A rather dense area of cytoplasm, near the nucleus of a cell, containing a pair of centrioles.

Cerebellar peduncle (ser-e-BEL-ar pe-DUNG-kul) A bundle of nerve fibers connecting the cerebellum with the brain stem.

Cerebellum (ser-e-BEL-um) The portion of the hindbrain lying posterior to the medulla and pons, concerned with coordination of movements.

Cerebral aqueduct (SER-e-bral AK-we-dukt) A channel through the midbrain connecting the third and fourth ventricles and containing cerebrospinal fluid.

Cerebral cortex The surface of the cerebral hemispheres, 2 to 4 mm thick, consisting of six layers of nerve cell bodies (gray matter) in most areas.

Cerebral palsy (PAL-zē) A group of nonprogressive motor disorders caused by damage to motor areas of the brain (cerebral cortex, basal ganglia, and cerebellum) during fetal life, birth, or infancy.

Cerebral peduncle (pe-DUNG-kul) One of a pair of nerve fiber bundles located on the ventral surface of the midbrain, conducting impulses between the pons and the cerebral hemispheres.

Cerebrospinal (se-rē'-brō-SPĪ-nal) **fluid (CSF)** A fluid produced in the choroid plexuses of the ventricles of the brain that circulates in the ventricles and the subarachnoid space around the brain and spinal cord.

Cerebrovascular (se-rē'-brō-VAS-kyoo-lar) **accident (CVA)** Destruction of brain tissue (infarction) resulting from disorders of blood vessels that supply the brain. Also called a **stroke.**

Cerebrum (SER-ē-brum) The two hemispheres of the forebrain, making up the largest part of the brain.

Ceruminous (se-ROO-mi-nus) **gland** A modified sudoriferous (sweat) gland in the external auditory meatus that secretes cerumen (ear wax).

Cervix (SER-viks) Neck; any constricted portion of an organ, especially the lower cylindrical part of the uterus.

Chemoreceptor (kē'-mō-rē-SEP-tor) Receptor that detects the presence of chemicals.

Chemotaxis (kē'-mō-TAK-sis) Attraction of phagocytes to a chemical stimulus.

Chemotherapy (kē'-mō-THER-a-pē) The treatment of illness or disease by chemicals.

Chiropractic (kī'-rō-PRAK-tik) A system of treating disease by using one's hands to manipulate body parts, mostly the vertebral column.

Chloride shift Diffusion of bicarbonate ions from red blood cells into plasma and of chloride ions from plasma into red blood cells that maintains ionic balance between red blood cells and plasma.

Cholecystectomy (kō'-lē-sis-TEK-tō-mē) Surgical removal of the gallbladder.

Cholesterol (kō-LES-te-rol) Classified as a lipid, the most abundant steroid in animal tissues; located in cell membranes and used for the synthesis of steroid hormones and bile salts.

Cholinergic (kō'-lin-ER-jik) **fiber** A nerve ending that liberates acetylcholine at a synapse.

Chondrocyte (KON-drō-sīt) Cell of mature cartilage.

Chondroitin (kon-DROY-tin) **sulfate** An amorphous matrix material found outside the cell.

Chordae tendineae (KOR-dē TEN-di-nē) Cords that connect the heart valves with the papillary muscles.

Chorion (KŌ-rē-on) The outermost fetal membrane; serves a protective and nutritive function.

Choroid (KŌ-royd) One of the vascular coats of the eyeball.

Choroid plexus (PLEK-sus) A vascular structure located in the roof of each of the four ventricles of the brain; produces cerebrospinal fluid.

Chromaffin (krō-MAF-in) **cell** Cell that has an affinity for chrome salts, due in part to the presence of the precursors of the chemical transmitter epinephrine; found, among other places, in the adrenal medulla.

Chromatid (KRŌ-ma-tid) One of a pair of identical connected nucleoprotein strands that are joined at the centromere and separate during cell division, each becoming a chromosome of one of the two daughter cells.

Chromatin (KRŌ-ma-tin) The threadlike mass of the genetic material consisting principally of DNA, which is present in the nucleus of a nondividing or interphase cell.

Chromosome (KRŌ-mō-sōm) One of the 46 small, dark-staining bodies that appear in the nucleus of a human diploid cell during cell division.

Chronic (KRON-ik) Long-term or frequently recurring; applied to a disease that is not acute.

Chylomicron (kī'-lō-MĪ-kron) Aggregate of triglycerides, phospholipids, and cholesterol coated with protein that enters a lacteal of a villus during absorption.

Chyme (kīm) The semifluid mixture of partly digested food and digestive secretions found in the stomach and small intestine during digestion of a meal.

Ciliary (SIL-ē-ar'-ē) **body** One of the three portions of the vascular tunic of the eyeball, the others being the choroid and the iris; includes the ciliary muscle and the ciliary processes.

Cilium (SIL-ē-um) A hair or hairlike process projecting from a cell that may be used to move the entire cell or to move substances along the surface of the cell.

Circadian (ser-KĀ-dē-an) **rhythm** A cycle of active and nonactive periods in organisms determined by internal mechanisms and repeating about every 24 hours.

Circle of Willis A ring of arteries forming an anastomosis at the base of the brain between the internal carotid and vertebral arteries and arteries supplying the brain.

Circulation time Time required for blood to pass from the right atrium, through pulmonary circulation, back to the left ventricle, through systemic circulation to the foot, and back again to the right atrium; normally about 1 min.

Circulatory (SER-kyoo-la-to'-rē) **shock** A condition in which body tissues, including critical organs, do not receive an adequate blood supply because of reduced cardiac output or reduced blood volume; signs include lowered blood pressure, feeble and rapid pulse, shallow and rapid breathing, paleness or cyanosis, and mental confusion or unconsciousness.

Circumcision (ser'-kum-SIZH-un) Removal of the foreskin, the fold over the glans penis.

Circumduction (ser'-kum-DUK-shun) A movement at a synovial joint in which the distal end of a bone moves in a circle while the proximal end remains relatively stable.

Circumvallate papilla (ser'-kum-VAL-āt pa-PIL-a) One of the circular projections which are arranged in an inverted V-shaped row at the posterior portion of the tongue; the largest of the elevations on the upper surface of the tongue and contains taste buds.

Cirrhosis (si-RŌ-sis) A liver disorder in which the parenchymal cells are destroyed and replaced by connective tissue.

Cisterna chyli (sis-TER-na KĪ-lē) The origin of the thoracic duct.

Cleavage The rapid mitotic divisions following the fertilization of an ovum, resulting in an increased number of progressively smaller cells, called blastomeres, so that the overall size of the zygote remains the same.

Cleft palate Condition in which the palative processes of the maxilla do not unite before birth; cleft lip, a split in the upper lip, is often associated with cleft palate.

Climacteric (klī-mak-TER-ik) Cessation of the reproductive function in the female or diminution of testicular activity in the male.

Climax The peak period or moments of greatest intensity during sexual excitement.

Clitoris (KLI-to-ris) An erectile organ of the female that is homologous to the male penis.

Clot The end result of a series of biochemical reactions that changes liquid plasma into a gelatinous mass; specifically, the conversion of fibrinogen into a tangle of polymerized fibrin molecules.

Clot retraction (rē-TRAK-shun) The consolidation of a fibrin clot to pull damaged tissue together.

Clotting time The time required for blood to coagulate as a result of platelet disintegration; ranges from 5 to 15 minutes.

Coagulation (cō-ag-yoo-LĀ-shun) Process by which a blood clot is formed.

Coccyx (KOK-six) The fused bones at the end of the vertebral column.

Cochlea (KŌK-lē-a) A winding, cone-shaped tube forming a portion of the inner ear and containing the spiral organ.

Cochlear duct The membranous cochlea consisting of a spirally arranged tube enclosed in the bony cochlea and lying along its outer wall. Also called the **scala media.**

Coenzyme A nonprotein substance that is associated with and activates an enzyme.

Coitus (KŌ-i-tus) Sexual intercourse. Also called **copulation.**

Collagen (KOL-a-jen) A protein that is the main organic constituent of connective tissue.

Collateral circulation The alternate route taken by blood through an anastomosis.

Colon The division of the large intestine consisting of ascending, transverse, descending, and sigmoid portions.

Color blindness Any deviation in the normal perception of colors, resulting from the lack of one or more of the photopigments of the cones.

Colostomy (kō-LOS-tō-mē) The surgical creation of a new opening from the colon to the body surface.

Colostrum (kō-LOS-trum) A thin, cloudy fluid secreted by the mammary glands a few days prior to or after delivery before true milk is secreted.

Colposcopy (kol-POS-kō-pē) Direct examination of the vaginal and cervical mucosa using a magnifying device.

Coma (KŌ-ma) Profound unconsciousness from which one cannot be roused.

Common bile duct A tube formed by the union of the common hepatic duct and the cystic duct that empties bile into the duodenum at the hepatopancreatic ampulla.

Complement A group of interacting plasma proteins that is activated by certain antigen-antibody reactions and follows a specific sequence of reactions on the surface of a cell on which the antibody is fixed so that the membrane is lysed.

Compliance The ease with which the lungs and thoracic wall can be expanded.

Computed tomography (tō-MOG-ra-fē) **(CT)** X-ray technique that provides a cross-sectional picture of any area of the body.

Concha (KONG-ka) A scroll-like bone found in the skull. *Plural,* **conchae** (KONG-kē).

Concussion (kon-KUSH-un) Traumatic injury to the brain that produces no visible bruising, but may result in abrupt, temporary loss of consciousness.

Cone The light-sensitive receptor in the retina concerned with color vision.

Congenital (kon-JEN-i-tal) Present at the time of birth.

Congestive heart failure (CHF) Chronic or acute state that results when the heart is not capable of supplying the oxygen demands of the body.

Conjunctiva (kon'-junk-TĪ-va) The delicate membrane covering the eyeball and lining the eyelids.

Conjunctivitis (kon-junk'-ti-VĪ-tis) Inflammation of the delicate membrane covering the eyeball and lining the eyelids.

Connective tissue The most abundant of the four tissue types in the body, performing the functions of binding and supporting; consists of relatively few cells in a great deal of intercellular substance.

Constipation (con-sti-PĀ-shun) Infrequent or difficult defecation caused by decreased motility of the intestines.

Contact inhibition Phenomenon by which migration of a growing cell is stopped when it makes contact with another cell of its own kind.

Contraception (kon'-tra-SEP-shun) The prevention of conception or impregnation without destroying fertility.

Contractility (kon'-trak-TIL-i-tē) The ability of muscle tissue to shorten.

Contralateral (kon'-tra-LAT-er-al) On the opposite side; affecting the opposite side of the body.

Conus medullaris (KŌ-nus med-yoo-LAR-is) The tapered portion of the spinal cord below the lumbar enlargement.

Convergence (con-VER-jens) An anatomical arrangement in which the synaptic knobs of several presynaptic neurons terminate on one postsynaptic neuron. The medial movement of the two eyeballs so that both are directed toward a near object being viewed in order to produce a single image.

Convulsion (con-VUL-shun) Violent, involuntary, tetanic contractions of an entire group of muscles.

Cornea (KŌR-nē-a) The transparent fibrous coat that covers the iris of the eye.

Corona radiata Innermost layer of follicle cells surrounding an ovum.

Coronary (KOR-ō-na-rē) **artery disease (CAD)** A condition in which the heart muscle receives inadequate blood due to an interruption of its blood supply.

Coronary artery spasm A condition in which the smooth muscle of a coronary artery undergoes a sudden contraction, resulting in vasoconstriction.

Coronary circulation The pathway followed by the blood from the ascending aorta through the blood vessels supplying the heart and returning to the right atrium. Also called **cardiac circulation.**

Coronary sinus (SĪ-nus) A wide venous channel on the posterior surface of the heart that collects the blood from the coronary circulation and returns it to the right atrium.

Corpora quadrigemina (KOR-por-a kwad-ri-JEM-in-a) Four small elevations on the dorsal region of the midbrain concerned with visual and auditory functions.

Corpus albicans (KOR-pus AL-bi-kanz) A white fibrous patch in the ovary that forms after the corpus luteum regresses.

Corpus callosum (ka-LŌ-sum) The great commissure of the brain between the cerebral hemispheres.

Corpus luteum (LOO-tē-um) A yellow endocrine gland in the ovary formed when a follicle has discharged its ovum; secretes estrogens, progesterone, and relaxin.

Cortex (KOR-teks) An outer layer of an organ. The convoluted layer of gray matter covering each cerebral hemisphere.

Costal cartilage (KOS-tal KAR-ti-lij) Hyaline cartilage that attaches a rib to the sternum.

Countercurrent multiplier mechanism The ability of the kidneys to produce either hyperosmotic or hyposmotic urine.

Cramp A spasmodic, especially a tonic, contraction of one or many muscles, usually painful.

Cranial (KRĀ-nē-al) **cavity** A subdivision of the dorsal body cavity formed by the cranial bones and containing the brain.

Cranial nerve One of 12 pairs of nerves that leave the brain, pass through foramina in the skull, and supply the head, neck, and part of the trunk; each is designated by a Roman numeral and a name.

Craniosacral (krā-nē-ō-SĀ-kral) **outflow** The fibers of parasympathetic preganglionic neurons, which have their cell bodies located in nuclei in the brain stem and in the lateral gray matter of the sacral portion of the spinal cord.

Craniotomy (krā'-nē-OT-ō-mē) Any operation on the skull, as for surgery on the brain or decompression of the fetal head in difficult labor.

Cranium (KRĀ-nē-um) The skeleton of the skull that protects the brain and the organs of sight, hearing, and balance; includes the frontal, parietal, temporal, occipital, sphenoid, and ethmoid bones.

Crenation (krē-NĀ-shun) The shrinkage of red blood cells into knobbed, starry forms when placed in a hypertonic solution.

Cretinism (KRĒ-tin-izm) Severe congenital thyroid deficiency during childhood leading to physical and mental retardation.

Crista (KRIS-ta) A crest or ridged structure. A small elevation in the ampulla of each semicircular duct that serves as a receptor for dynamic equilibrium.

Cross-matching A test performed prior to blood transfusion to check for compatibility of blood types. Red blood cells from donor are mixed with a sample of recipient's serum and then rbc's from recipient are mixed with a sample of donor's serum. If no agglutination occurs, the bloods are deemed compatible.

Cryosurgery (KRĪ-ō-ser-jer-ē) The destruction of tissue by application of extreme cold.

Cryptorchidism (krip-TOR-ki-dizm) The condition of undescended testes.

Cupula (KUP-yoo-la) A mass of gelatinous material covering the hair cells of a crista, a receptor in the ampulla of a semicircular canal stimulated when the head moves.

Curvature (KUR-va-tūr) A nonangular deviation of a straight line, as in the greater and lesser curvatures of the stomach. Abnormal curvatures of the vertebral column include kyphosis, lordosis, and scoliosis.

Cushing's syndrome Condition caused by a hypersecretion of glucocorticoids characterized by spindly legs, "moon face," "buffalo hump," pendulous abdomen, flushed facial skin, and poor wound healing.

Cutaneous (kyoo-TĀ-nē-us) Pertaining to the skin.

Cyanosis (sī'-a-NŌ-sis) Slightly bluish or dark purple discoloration of the skin and the mucous membrane due to an oxygen deficiency.

Cyclic adenosine-3',5'-monophosphate (cyclic AMP) Molecule formed from ATP by the action of the enzyme adenyl cyclase; serves as an intracellular messenger for some hormones.

Cystic (SIS-tik) **duct** The duct that transports bile from the gallbladder to the common bile duct.

Cystitis (sis-TĪ-tis) Inflammation of the urinary bladder.

Cystoscope (SIS-ti-skōp) An instrument used to examine the inside of the urinary bladder.

Cytochrome (SĪ-tō-krōm) A protein containing an iron portion capable of alternating between a reduced form and an oxidized form.

Cytokinesis (sī'-tō-ki-NĒ-sis) Division of the cytoplasm.

Cytology (sī-TOL-o-jē) The study of cells.

Cytoplasm (SĪ-tō-plazm) Substance within a cell's plasma membrane and external to its nucleus. Also called **protoplasm.**

Cytoskeleton Complex internal structure of cytoplasm consisting of microfilaments and microtubules.

Dead air volume The volume of air that is inhaled but remains in spaces in the upper respiratory system and does not reach the alveoli for participation in gaseous exchange; about 150 ml.

Deafness Lack of the sense of hearing or a significant hearing loss.

Debility (dē-BIL-i-tē) Weakness of tonicity in functions or organs of the body.

Decibel (DES-i-bel) **(db)** A unit that measures sound intensity (loudness).

Decidua (dē-SID-yoo-a) That portion of the endometrium of the uterus (all but the deepest layer) that is modified for pregnancy and shed after childbirth.

Deciduous (dē-SID-yoo-us) Falling off or being shed seasonally or at a particular stage of development. In the body, referring to the first set of teeth.

Decompression sickness A condition characterized by joint pains and neurologic symptoms; follows from a too-rapid reduction of environmental pressure or decompression so that nitrogen that dissolved in body fluid under pressure comes out of solution as bubbles that form air emboli and occlude crucial blood vessels. Also called **caisson disease** or **bends.**

Decubitus (dē-KYOO-bi-tus) **ulcer** Tissue destruction due to a constant deficiency of blood to tissues overlying a bony projection that has been subjected to prolonged pressure against an object such as a bed, cast, or splint. Also called **bedsore, pressure sore,** or **trophic ulcer.**

Decussation (dē'-ku-SĀ-shun) A crossing over; usually refers to the crossing of most of the fibers in the large motor tracts to opposite sides in the medullary pyramids.

Deep Away from the surface of the body.

Deep fascia (FASH-ē-a) A sheet of connective tissue wrapped around a muscle to hold it in place.

Defecation (def-e-KĀ-shun) The discharge of feces from the rectum.

Defibrillation (dē-fib-ri-LĀ-shun) Delivery of a very strong electrical current to the heart in an attempt to stop ventricular fibrillation.

Deglutition (dē-gloo-TISH-un) The act of swallowing.

Dehydration (dē-hī-DRĀ-shun) Excessive loss of water from the body or its parts.

Delta cell A cell in the islets of Langerhans in the pancreas that secretes somatostatin.

Demineralization (de-min'-er-al-i-ZĀ-shun) Loss of calcium and phosphorus from bones.

Dendrite (DEN-drīt) A nerve cell process carrying an impulse toward the cell body.

Dentin (DEN-tin) The osseous tissues of a tooth enclosing the pulp cavity.

Dentition (den-TI-shun) The eruption of teeth. The number, shape, and arrangement of teeth.

Deoxyribonucleic (dē-ok'-sē-ri'-bō-nyoo-KLĒ-ik) **acid (DNA)** A nucleic acid in the shape of a double helix constructed of nucleotides consisting of one of four nitrogen bases (adenine, cytosine, guanine, or thymine), deoxyribose, and a phosphate group; encoded in the nucleotides is genetic information.

Depolarization (dē-pō-lar-i-ZĀ-shun) Used in neurophysiology to describe the reduction of voltage across a cell membrane; expressed as a movement toward less negative (more positive) voltages on the interior side of the cell membrane.

Depression (dē-PRESS-shun) Movement in which a part of the body moves downward.

Dermal papilla (pa-PILL-a) Fingerlike projection of the papillary region of the dermis that may contain blood capillaries or Meissner's corpuscles, nerve endings sensitive to touch.

Dermatology (der-ma-TOL-ō-jē) The medical specialty dealing with diseases of the skin.

Dermatome (DER-ma-tōm) An instrument for incising the

skin or cutting thin transplants of skin. The cutaneous area developed from one embryonic spinal cord segment and receiving most of its innervation from one spinal nerve.

Dermis (DER-mis) A layer of dense connective tissue lying deep to the epidermis; the true skin or corium.

Descending colon (KŌ-lon) The part of the large intestine descending from the left colic (splenic) flexure to the level of the left iliac crest.

Developmental anatomy The study of development from the fertilized egg to the adult form. The branch of anatomy called embryology is generally restricted to the study of development from the fertilized egg through the eighth week in utero.

Diabetes insipidus (dī-a-BĒ-tēz in-SIP-i-dus) Condition caused by hyposecretion of antidiuretic hormone (ADH) and characterized by excretion of large amounts of urine and thirst.

Diabetes mellitus (MEL-i-tus) Condition caused by hyposecretion of insulin and characterized by hyperglycemia, increased urine production, excessive thirst, and excessive eating.

Diagnosis (dī-ag-NŌ-sis) Recognition of disease states from signs and symptoms by inspection, palpation, laboratory tests, and other means.

Dialysis (dī-AL-i-sis) The process of separating crystalloids (smaller particles) from colloids (larger particles) by the difference in their rates of diffusion through selectively permeable membrane.

Diapedesis (dī'-a-pe-DĒ-sis) The passage of white blood cells through intact blood vessel walls.

Diaphragm (DĪ-a-fram) Any partition that separates one area from another, especially the dome-shaped skeletal muscle between the thoracic and abdominal cavities.

Diaphysis (dī-AF-i-sis) The shaft of a long bone.

Diarrhea (dī-a-RĒ-a) Frequent defecation of liquid feces caused by increased motility of the intestines.

Diarthrosis (dī'-ar-THRŌ-sis) Articulation in which opposing bones move freely, as in a hinge joint.

Diastole (dī-AS-tō-lē) In the cardiac cycle, the phase of relaxation or dilation of the heart muscle, especially of the ventricles.

Diastolic (di-as-TOL-ik) **blood pressure** The force exerted by blood on arterial walls during ventricular relaxation; the lowest blood pressure measured in the large arteries, about 80 mm Hg under normal conditions for a young, adult male.

Diencephalon (dī-'en-SEF-a-lon) A part of the brain consisting primarily of the thalamus and the hypothalamus.

Differential (dif-fer-EN-shal) **count** Determination of the number of each kind of white blood cell in a sample of 100 cells for diagnostic purposes.

Diffusion (dif-YOO-zhun) A passive process in which there is a net or greater movement of molecules or ions from a region of high concentration to a region of low concentration until equilibrium is reached.

Digestion (di-JES-chun) The mechanical and chemical breakdown of food to simple molecules that can be absorbed by the body.

Dilate (DĪ-lāte) To expand or swell.

Diploid (DIP-loyd) Having the number of chromosomes characteristically found in the somatic cells of an organism.

Diplopia (di-PLŌ-pē-a) Double vision.

Dislocation (dis-lō-KĀ-shun) Displacement of a bone from a joint with tearing of ligaments, tendons, and articular capsules. Also called **luxation** (luks-Ā-shun).

Dissect (DĪ-sekt) To separate tissues and parts of a cadaver (corpse) or an organ for anatomical study.

Dissociation (dis'-sō-sē-Ā-shun) Separation of inorganic acids, bases, and salts into ions when dissolved in water. Also called **ionization.**

Distal (DIS-tal) Farther from the attachment of an extremity to the trunk or a structure; farther from the point of origin.

Diuretic (dī-yoo-RET-ik) A chemical that increases urine volume by inhibiting facultative reabsorption of water.

Divergence (di-VER-jens) An anatomical arrangement in which the synaptic knobs of one presynaptic neuron terminate on several postsynaptic neurons.

Diverticulitis (dī-ver-tik-yoo-LĪ-tis) Inflammation of diverticula, saclike outpouchings of the colonic wall when the muscularis becomes weak.

Dominant gene A gene that is able to override the influence of the complementary gene on the homologous chromosome; the gene that is expressed.

Dorsal body cavity Cavity near the dorsal surface of the body that consists of a cranial cavity and vertebral canal.

Dorsiflexion (dor'-si-FLEK-shun) Flexion of the foot at the ankle.

Down's syndrome An inherited defect due to an extra copy of chromosome 21. Symptoms include mental retardation; a small skull, flattened front to back; a short, flat nose; short fingers; and a widened space between first two digits of the hand and foot. Also called **trisomy 21.**

Ductus arteriosus (DUK-tus ar-tē-rē-Ō-sus) A small vessel connecting the pulmonary trunk with the aorta; found only in the fetus.

Ductus deferens (DEF-er-ens) The duct that conducts spermatozoa from the epididymis to the ejaculatory duct. Also called the **vas deferens** or **seminal duct.**

Ductus venosus (ve-NŌ-sus) A small vessel in the fetus that helps the circulation bypass the liver.

Duodenal papilla (doo-ō-DĒ-nal pa-PILL-a) An elevation on the duodenal mucosa that receives the hepatopancreatic ampulla.

Duodenum (doo'-ō-DĒ-num) The first portion of the small intestine.

Dura mater (DYOO-ra MĀ-ter) The outer membrane (meninx) covering the brain and spinal cord.

Dynamic equilibrium (ē-kwi-LIB-rē-um) The maintenance of body position, mainly the head, in response to sudden movements such as rotation, acceleration, and deceleration.

Dynamic spatial reconstructor (DSR) An x-ray machine that has the ability to construct moving, three-dimensional, life size images of all or part of an internal organ from any view desired.

Dysfunction (dis-FUNK-shun) Absence of complete normal function.

Dyslexia (dis-LEK-sē-a) Impairment of ability to comprehend written language.

Dysmenorrhea (dis'-men-ō-RĒ-a) Painful menstruation.

Ectoderm The outermost of the three primary germ layers which gives rise to the nervous system and the epidermis of skin and its derivatives.

Edema (e-DĒ-ma) An abnormal accumulation of fluid in body tissues.

Effector (e-FEK-tor) The organ of the body, either a muscle or a gland, that responds to a motor neuron impulse.

Efferent (EF-er-ent) **ducts** A series of coiled tubes that transport spermatozoa from the rete testis to the epididymis.

Efferent neuron (NOO-ron) A neuron that conveys impulses from the brain and spinal cord to effectors which may be either muscles or glands. Also called a **motor neuron.**

Ejaculation (ē-jak-yoo-LĀ-shun) The reflex ejection or expulsion of semen from the penis.

Ejaculatory (e-JAK-yoo-la-tor'-ē) **duct** A tube that transports spermatozoa from the ductus deferens to the prostatic urethra.

Elasticity (e-las-TIS-i-tē) The ability of tissue to return to its original shape after contraction or extension.

Electrocardiogram (e-lek'-trō-KAR-dē-ō-gram) **(ECG or EKG)** A recording of the electrical changes that accompany the cardiac cycle.

Electroencephalogram (e-lek'-trō-en-SEF-a-lō-gram) **(EEG)** A recording of the electrical impulses of the brain.

Electrolyte (ē-LEK-trō-līt) Any compound that separates into ions when dissolved in water and is able to conduct electricity.

Electron transport chain A series of oxidation-reduction reactions in the catabolism of glucose in which energy is released and transferred for storage to ATP.

Elevation (el-e-VĀ-shun) Movement in which a part of the body moves upward.

Ellipsoidal (e-lip-SOY-dal) **joint** A synovial joint structured so that an oval-shaped condyle of one bone fits into an elliptical cavity of another bone, permitting side-to-side and back-and-forth movements, as at the joint at the wrist between the radius and carpals. Also called a **condyloid joint.**

Embolism (EM-bō-lizm) Obstruction or closure of a vessel by an embolus.

Embolus (EM-bō-lus) A blood clot, bubble of air, fat from broken bones, mass of bacteria, or other debris or foreign material transported by the blood.

Embryo (EM-brē-ō) The young of any organism in an early stage of development; in humans, the developing organism from fertilization to the end of the eighth week in utero.

Embryology (em'-brē-OL-ō-jē) The study of development from the fertilized egg to the end of the eighth week in utero.

Emesis (EM-e-sis) Vomiting.

Emmetropia (em'-e-TRŌ-pē-a) The ideal optical condition of the eyes.

Emphysema (em'-fi-SĒ-ma) A swelling or inflation of air passages due to loss of elasticity in the alveoli.

Emulsification (ē-mul'-si-fi-KĀ-shun) The dispersion of large fat globules to smaller uniformly distributed particles.

Enamel (e-NAM-el) The hard, white substance covering the crown of a tooth.

End-diastolic (dī-a-STOL-ik) **volume (EDV)** The volume of blood that enters a ventricle during diastole.

Endocardium (en-dō-KAR-dē-um) The layer of the heart wall, composed of endothelium and smooth muscle, that lines the inside of the heart and covers the valves and tendons that hold the valves open.

Endochondral ossification (en'-dō-KON-dral os'-i-fi-KĀ-shun) The replacement of cartilage by bone. Also called **intracartilaginous ossification.**

Endocrine (EN-dō-krin) **gland** A gland that secretes hormones into the blood; a ductless gland.

Endocytosis (en'-dō-sī-TŌ-sis) The uptake into a cell of substances that are unable to penetrate the cell membrane; includes both phagocytosis and pinocytosis, processes in which the membrane first invaginates and then pinches off to enclose some of the surrounding medium.

Endoderm The innermost of the three primary germ layers of the developing embryo which gives rise to the digestive tract, urinary bladder and urethra, and respiratory tract.

Endogenous (en-DOJ-e-nus) Growing from or beginning within the organism.

Endolymph (EN-dō-lymf') The fluid within the membranous labyrinth of the inner ear.

Endometriosis (en'-dō-MĒ-trē-ō'-sis) The growth of endometrial tissue outside the uterus.

Endometrium (en'-dō-MĒ-trē-um) The mucous membrane lining the uterus.

Endomysium (en'-dō-MIZ-ē-um) Invaginations of the perimysium separating each individual muscle cell.

Endoneurium (en'-dō-NYOO-rē-um) Connective tissue wrapping around individual nerve fibers.

Endoplasmic reticulum (en'-dō-PLAZ-mik re-TIK-yoo-lum) **(ER)** A network of channels running through the cytoplasm of a cell that serves in intracellular transportation, support, storage, synthesis, and packaging of molecules. Portions of ER where ribosomes are attached to the outer surface are called granular or rough reticulum; portions that have no ribosomes are called agranular or smooth reticulum.

Endorphin (en-DOR-fin) A peptide in the central nervous system that acts as a painkiller.

Endoscope (EN-dō-skōp') An instrument used to look inside hollow organs such as the stomach (gastroscope) or urinary bladder (cystoscope).

Endosteum (en-DOS-tē-um) The membrane that lines the medullary cavity of bones.

Endothelial-capsular (en-dō-THĒ-lē-al) **membrane** A filtration membrane in a nephron of a kidney consisting of the endothelium and basement membrane of the glomerulus and the epithelium of the visceral layer of the glomerular capsule.

Endothelium (en'-dō-THĒ-lē-um) The layer of simple squamous epithelium that lines the cavities of the heart and blood and lymphatic vessels.

End-systolic (si-STOL-ik) **volume (ESV)** The volume of blood remaining in a ventricle following its systole.

Enkephalin (en-KEF-a-lin) A peptide found in the central nervous system that acts as a painkiller.

Enteroendocrine (en-ter-ō-ENDŌ-krin) **cell** A stomach cell that secretes the hormone stomach gastrin. Also called an **argentaffin cell.**

Enterogastric (en-te-rō-GAS-trik) **reflex** A reflex that inhibits gastric secretion; initiated by food in the small intestine.

Enzyme (EN-zīm) A substance that affects the speed of chemical changes; an organic catalyst, usually a protein.

Eosinophil (ē'-ō-SIN-ō-fil) A type of white blood cell characterized by granular cytoplasm readily stained by eosin.

Ependyma (e-PEN-de-ma) Neuroglial cells that line ventricles of the brain.

Epicardium (ep'-i-KAR-dē-um) The thin outer layer of the heart wall, composed of serous tissue and mesothelium. Also called the **visceral pericardium.**

Epidemiology (ep'-i-DĒ-mē-ol'-ō-jē) Medical science concerned with the occurrence and distribution of disease in human populations.

Epidermis (ep'-i-DERM-is) The outermost layer of skin, composed of stratified squamous epithelium.

Epididymis (ep'-i-DID-i-mis) An elongated tube along the posterior bladder of the testis in which sperm are stored. *Plural,* **epididymides** (ep'-i-DID-i-mi-dēz).

Epiglottis (ep'-i-GLOT-is) A large, leaf-shaped piece of cartilage lying on top of the larynx, with its "stem" attached to the thyroid cartilage and its "leaf" portion unattached and free to move up and down to cover the glottis.

Epilepsy (EP-i-lep'-sē) Neurological disorder characterized by short, periodic attacks of motor, sensory, or psychological malfunction.

Epimysium (ep'-i-MIZ-ē-um) Fibrous connective tissue around muscles.

Epinephrine (ep-ē-NEF-rin) Hormone secreted by the adrenal medulla that produces actions similar to those that result from sympathetic stimulation. Also called **adrenaline.**

Epineurium (ep'-i-NYOO-rē-um) The outermost covering around the entire nerve.

Epiphyseal (ep'-i-FIZ-ē-al) **plate** The plate between the epiphysis and diaphysis.

Epiphysis (ē-PIF-i-sis) The end of a long bone, usually larger in diameter than the shaft (the diaphysis).

Episiotomy (ē-piz'-ē-OT-ō-mē) Incision of the clinical perineum at the end of second stage of labor to avoid tearing.

Epistaxis (ep'-i-STAK-sis) Hemorrhage from the nose; nosebleed.

Epithelial (ep'-i-THĒ-lē-al) **tissue** The tissue that forms glands or the outer part of the skin and lines blood vessels, hollow organs, and passages that lead externally from the body.

Erection (ē-REK-shun) The enlarged and stiff state of the penis (or clitoris) resulting from the engorgement of the spongy erectile tissue with blood.

Eructation (e-ruk'-TĀ-shun) The forceful expulsion of gas from the stomach. Also called **belching.**

Erythema (er'-e-THĒ-ma) Skin redness usually caused by engorgement of the capillaries in the lower layers of the skin.

Erythrocyte (e-RITH-rō-sīt) Red blood cell.

Erythropoiesis (e-rith'-rō-poy-Ē-sis) The process by which erythrocytes are formed.

Erythropoietin (ē-rith'-rō-POY-ē-tin) A hormone formed from a plasma protein that stimulates red blood cell production.

Esophagus (e-SOF-a-gus) A hollow muscular tube connecting the pharynx and the stomach.

Essential amino acids Those 10 amino acids that cannot be synthesized by the human body at an adequate rate to meet its needs and therefore must be obtained from the diet.

Estrogens (ES-tro-jens) Female sex hormones produced by the ovaries concerned with the development and maintenance of female reproductive structures and secondary sex characteristics, fluid and electrolyte balance, and protein anabolism. Examples are β-estradiol, estrone, and estriol.

Etiology (ē'-tē-OL-ō-jē) The study of the causes of disease, including theories of origin and the organisms, if any, involved.

Euphoria (yoo-FŌR-ē-a) A subjectively pleasant feeling of well-being marked by confidence and assurance.

Eupnea (yoop-NĒ-a) Normal quiet breathing.

Euthanasia (yoo'-tha-NĀ-zē-ā) The practice of ending a life in case of incurable disease.

Eversion (ē-VER-zhun) The movement of the sole outward at the ankle joint.

Exacerbation (eg-zas'-er-BĀ-shun) An increase in the severity of symptoms or of disease.

Excitability (ek-sīt'-a-BIL-i-tē) The ability of muscle tissue to receive and respond to stimuli; the ability of nerve cells to respond to stimuli and convert them into impulses.

Excitatory postsynaptic potential (EPSP) The slight decrease in negative voltage seen on the postsynaptic membrane when it is stimulated by a presynaptic terminal. The EPSP is a localized event that decreases in strength from the point of excitation.

Excrement (EKS-kre-ment) Material cast out from the body as waste, especially fecal matter.

Excretion (eks-KRĒ-shun) The process of eliminating waste products from a cell, tissue, or the entire body; or, the products excreted.

Exocrine (EK-sō-krin) **gland** A gland that secretes substances into ducts that empty at covering or lining epithelium or directly onto a free surface.

Exocytosis (ex'-ō-sī-TŌ-sis) A process of discharging cellular products too big to go through the membrane. Particles for export are enclosed by Golgi membranes when they are synthesized. Vesicles pinch off from the Golgi complex and carry the enclosed particles to the interior surface of the cell membrane where the vesicle membrane and cell membrane fuse and the contents of the vesicle are discharged.

Exogenous (ex-SOJ-e-nus) Originating outside an organ or part.

Exophthalmic goiter (ek'-sof-THAL-mik GOY-ter) Condition caused by hypersecretion of thyroid hormones characterized by protrusion of the eyeballs and an enlarged thyroid.

Expiration (ek-spi-RĀ-shun) Breathing out; expelling air from the lungs into the atmosphere. Also called **exhalation.**

Expiratory (eks-PĪ-ra-tō-rē) **reserve volume** The volume of air in excess of tidal volume that can be forcibly exhaled; about 1,200 ml.

Extensibility (ek-sten'-si-BIL-i-tē) The ability of tissue to be stretched when pulled.

Extension (ek-STEN-shun) An increase in the anterior angle between two bones, except in extension of the knee and toes, in which the posterior angle is involved; restoring a body part to its anatomical position after flexion.

External auditory (AW-di-tōr-ē) **canal** or **meatus** (mē-Ā-tus) A canal in the temporal bone that leads to the middle ear.

External ear The outer ear, consisting of the pinna, the external auditory canal, and the tympanic membrane or eardrum.

External nares (NA-rēz) The external nostrils, or the openings into the nasal cavity on the exterior of the body.

Exteroceptor (eks'-ter-ō-SEP-tor) A receptor adapted for the reception of stimuli from outside the body.

Extracellular fluid (ECF) Fluid outside of body cells, such as interstitial fluid and plasma.

Extrinsic (ek-STRIN-sik) Of external origin.

Extrinsic clotting pathway Sequence of reactions leading to blood clotting that is initiated by the release of tissue thromboplastin by damaged tissue.

Exudate (EKS-yoo-dāt) Escaping fluid or semifluid material that oozes from a space that may contain serum, pus, and cellular debris.

Face The anterior aspect of the head.

Facilitated diffusion (fa-SIL-i-tā-ted dif-YOO-zhun) Diffusion in which a substance not soluble by itself in lipids is transported across a semipermeable membrane by combining with a carrier substance.

Facilitation (fa-sil-i-TĀ-shun) The process in which a nerve cell membrane is partially depolarized by a subliminal stimulus so that a subsequent subliminal stimulus can further depolarize the membrane to reach the threshold of impulse initiation.

Facultative (FAK-ul-tā-tive) **water reabsorption** The absorption of water from distal convoluted tubules and collecting ducts of nephrons in response to antidiuretic hormone (ADH).

Falx cerebelli (falks ser'-e-BEL-lē) A small triangular process of the dura mater attached to the occipital bone in the posterior cranial fossa and projecting inward between the two cerebellar hemispheres.

Falx cerebri (SER-e-brē) A fold of the dura mater extending down into the longitudinal fissure between the two cerebral hemispheres.

Fascia (FASH-ē-a) A fibrous membrane covering, supporting, and separating muscles.

Fascicle (FAS-i-kul) A small bundle or cluster, especially of nerve or muscle fibers. Also called **fasciculus** (fa-SIK-yoo-lus); *plural,* **fasciculi** (fa-SIK-yoo-lī).

Fat A lipid compound formed from one molecule of glycerol and three molecules of fatty acids; the body's most highly concentrated source of energy. Adipose tissue, composed of loose connective tissue and cells specialized for fat storage and present in the form of soft pads between various organs for support, protection, and insulation.

Fauces (FAW-sēs)   The opening from the mouth into the pharynx.

Febrile (FĒ-bril)   Feverish; pertaining to a fever.

Feces (FĒ-sēz)   Material discharged from the rectum and made up of bacteria, excretions, and food residue. Also called **stool.**

Fenestra cochlea (fe-NES-tra KŌK-lē-a)   A small opening between the middle and inner ear, directly below the oval window, covered by the second tympanic membrane. Also called the **round window.**

Fenestra vestibuli (ves-TIB-yoo-lī)   A small opening between the middle and inner ear into which the footplate of the stapes fits. Also called the **oval window.**

Fertilization (fer'-ti-li-ZĀ-shun)   Penetration of an ovum by a spermatozoon and subsequent union of the nuclei of the cells.

Fetal (FĒ-tal) **alcohol syndrome (FAS)**   Term applied to the effects of intrauterine exposure to alcohol, such as slow growth, defective organs, and mental retardation.

Fetal circulation   The circulatory system of the fetus, including the placenta and special blood vessels involved in the exchange of materials between fetus and mother.

Fetus (FĒ-tus)   The latter stages of the developing young of an animal; in humans, the developing organism in utero from the beginning of the third month to birth.

Fever   An elevation in body temperature above its normal temperature of 37°C (98.6°F).

Fibrillation (fi-bre-LĀ-shun)   Irregular twitching of individual muscle cells (fibers) or small groups of muscle fibers preventing effective action by an organ or muscle.

Fibrin (FĪ-brin)   An insoluble protein that is essential to blood clotting; formed from fibrinogen by action of thrombin.

Fibrinogen (fī-BRIN-ō-jen)   A high-molecular-weight protein in the blood plasma that by the action of thrombin is converted to fibrin.

Fibrinolysis (fī-bri-NOL-i-sis)   Dissolution of a blood clot by the action of a proteolytic enzyme that converts insoluble fibrin into a soluble substance.

Fibroblast (FĪ-brō-blast)   A large, flat cell that forms collagenous and elastic fibers and the viscous ground substance of loose connective tissue.

Fibrocyte (FĪ-brō-sīt)   A mature fibroblast that no longer produces fibers or matrix in connective tissue.

Fibromyositis (fi'-brō-mī-ō'-SĪ-tis)   A group of symptoms including pain, tenderness, and stiffness of joints, muscles, or adjacent structures. Called **"charleyhorse"** when it involves the thigh.

Fibrosis (fī-BRŌ-sis)   Abnormal formation of fibrous tissue.

Fibrous (FĪ-brus) **joint**   A joint that allows little or no movement, such as a suture and syndesmosis.

Fibrous tunic (TOO-nik)   The outer coat of the eyeball, made up of the posterior sclera and the anterior cornea.

Fight-or-flight response   The effect of the stimulation of the sympathetic division of the autonomic nervous system.

Filiform papilla (FIL-i-form pa-PIL-a)   One of the conical projections which are distributed in parallel rows over the anterior two-thirds of the tongue and contain no taste buds.

Filtration (fil-TRĀ-shun)   The passage of a liquid through a filter or membrane that acts like a filter.

Filum terminale (FĪ-lum ter-mi-NAL-ē)   Nonnervous fibrous tissue of the spinal cord that extends inferiorly from the conus medullaris to the coccyx.

Fimbriae (FIM-brē-ē)   Fringelike structures, especially the lateral ends of the uterine tubes.

Fissure (FISH-ur)   A groove, fold, or slit that may be normal or abnormal.

Fistula (FIS-choo-la)   An abnormal passage between two organs or between an organ cavity and the outside.

Fixed macrophage (MAK-rō-fāj)   Stationary phagocytic cell of the reticuloendothelial system (RES) found in the liver, lungs, brain, spleen, lymph nodes, subcutaneous tissue, and bone marrow. Also called a **histiocyte.**

Flaccid (FLAK-sid)   Relaxed, flabby, or soft; lacking muscle tone.

Flagellum (fla-JEL-um)   A hairlike, motile process on the extremity of a bacterium or protozoon. *Plural,* **flagella** (fla-JEL-a).

Flatfoot   A condition in which the ligaments and tendons of the arches of the foot are weakened and the height of the longitudinal arch decreases.

Flatus (FLĀ-tus)   Gas or air in the digestive tract; commonly used to denote passage of gas rectally.

Flexion (FLEK-shun)   A folding movement in which there is a decrease in the angle between two bones anteriorly, except in flexion of the knee and toes, in which the bones are approximated posteriorly.

Flexor reflex   A polysynaptic reflex arc that withdraws a part of the body from a harmful stimulus.

Fluoroscope (FLOOR-ō-skōp)   An instrument for visual observation of the body by means of x-ray.

Follicle-stimulating (FOL-i-kul) **hormone (FSH)**   Hormone secreted by the adenohypophysis (anterior lobe) of the pituitary gland that initiates development of ova and stimulates the ovaries to secrete estrogens in females and initiates sperm production in males.

Fontanel (fon'-ta-NEL)   A membrane-covered spot where bone formation is not yet complete, especially between the cranial bones of an infant's skull.

Foot   The terminal part of the lower extremity.

Foramen (fo-RĀ-men)   A passage or opening; a communication between two cavities of an organ or a hole in a bone for passage of vessels or nerves.

Foramen ovale (ō-VAL-ē)   An opening in the fetal heart in the septum between the right and left atria. A hole in the greater wing of the sphenoid bone that transmits the mandibular branch of the trigeminal (V) nerve.

Forearm (FOR-arm)   The part of the upper extremity between the elbow and the wrist.

Fossa (FOS-a)   A furrow or shallow depression.

Fourth ventricle (VEN-tri-kul)   A cavity within the brain lying between the cerebellum and the medulla and pons.

Fracture (FRAK-chur)   Any break in a bone.

Frontal plane   A plane at a right angle to a midsagittal plane that divides the body or organs into anterior and posterior portions. Also called **coronal plane.**

Fulminate (FUL-mi-nāt')   To occur suddenly with great intensity.

Functional residual (re-ZID-yoo-al) **volume**   The sum of residual volume plus expiratory reserve volume; about 2,400 ml.

Fungiform papilla (FUN-ji-form pa-PILL-a)   A mushroomlike elevation on the upper surface of the tongue appearing as a red dot; most contain taste buds.

Gallbladder   A small pouch that stores bile, is located under the liver, and is filled and emptied of bile via the cystic duct.

Gamete (GAM-ēt)   A male or female reproductive cell; the spermatozoon or ovum.

Ganglion (GANG-glē-on)   A group of nerve cell bodies that lie outside the central nervous system. *Plural,* **ganglia** (GANG-glē-a).

Gangrene (GANG-rēn)   Death and rotting of a considerable mass of tissue that usually is caused by interruption of blood supply followed by bacterial invasion (*Clostridium*).

Gastrointestinal (gas-trō-in-TES-ti-nal) **(GI) tract**   A continuous tube running through the ventral body cavity extending from the mouth to the anus. Also called the **alimentary canal.**

Gavage (ga-VAZH)   Feeding through a tube passed through the esophagus and into the stomach.

Gene (jēn)   Biological unit of heredity; an ultramicroscopic, self-reproducing DNA particle located in a definite position on a particular chromosome.

General adaptation syndrome   Wide ranging set of bodily changes triggered by a stressor that gears the body to meet an emergency.

Generator potential   The graded depolarization that occurs in the region of an afferent neuron adjacent to a receptor cell.

Genetics   The study of heredity.

Genital herpes (JEN-i-tal HER-pēz)   A sexually transmitted disease caused by type II herpes simplex virus.

Genotype (JĒ-nō-tīp)   The total hereditary information carried by an individual; the genetic makeup of an organism.

Germinal (JER-mi-nal) **epithelium**   A layer of epithelial cells that covers the ovaries and produces ova and lines the seminiferous tubules of the testes and produces spermatozoa.

Gerontology (jer'-on-TOL-o-jē)   The study of old age.

Gestation (jes-TĀ-shun)   The period of intrauterine fetal development.

Giantism (GĪ-an-tizm)   Condition caused by hypersecretion of growth hormone (GH) during childhood characterized by excessive bone growth and body size.

Gingivae (jin-JI-vē)   Gums. They cover the alveolar processes of the mandible and maxilla and extend slightly into each socket.

Gingivitis (jin'-je-VĪ-tis)   Inflammation of the gums.

Gland   Single or group of specialized epithelial cells that secrete substances.

Glans penis (glanz PĒ-nis)   The slightly enlarged region at the distal end of the penis.

Glaucoma (glaw-KŌ-ma)   An eye disorder in which there is increased pressure due to an excess of fluid within the eye.

Gliding joint   A synovial joint having articulating surfaces that are usually flat, permitting only side-to-side and back-and-forth movements, as between carpal bones, tarsal bones, and the scapula and clavicle. Also called an **arthrodial joint.**

Glomerular (glō-MER-yoo-lar) **capsule**   A double-walled globe at the proximal end of a nephron that encloses the glomerulus. Also called **Bowman's capsule.**

Glomerular filtration   The first step in urine formation in which substances in blood are filtered at the endothelial-capsular membrane and the filtrate enters the proximal convoluted tubule of a nephron.

Glomerular filtration rate (GFR)   The total volume of fluid that enters all the glomerular capsules of the kidneys in 1 min; about 125 ml/min.

Glomerulonephritis (glō-mer-yoo-lō-nef-RĪ-tis)   Inflammation of the glomeruli of the kidney. Also called **Bright's disease.**

Glomerulus (glō-MER-yoo-lus)   A rounded mass of nerves or blood vessels, especially the microscopic tuft of capillaries that is surrounded by the glomerular capsule of each kidney tubule.

Glottis (GLOT-is)   The air passageway between the vocal folds in the larynx.

Glucagon (GLOO-ka-gon)   A hormone produced by the pancreas that increases the blood glucose level.

Glucocorticoids (gloo-kō-KOR-ti-koyds)   A group of hormones of the adrenal cortex.

Gluconeogenesis (gloo'-kō-nē'-ō-JEN-e-sis)   The conversion of a substance other than carbohydrate into glucose.

Glucose (GLOO-kōs)   A six-carbon sugar, $C_6H_{12}O_6$; the major energy source for every cell type in the body. Its metabolism is possible by every known living cell for the production of ATP.

Glycogen (GLĪ-ko-jen)   A highly branched polymer of glucose containing thousands of subunits; functions as a compact store of glucose molecules in liver and muscle cells.

Glycogenesis (glī'-kō-JEN-e-sis)   The process by which many molecules of glucose combine to form a molecule called glycogen.

Glycogenolysis (glī'-kō-je-NOL-i-sis)   The process of converting glycogen to glucose.

Glycolysis (glī-KOL-i-sis)   A series of chemical reactions that break down glucose into pyruvic acid, with a net gain of two molecules of ATP.

Glycosuria (glī'-kō-SOO-rē-a)   The presence of glucose in the urine.

Gnostic (NOS-tik)   Pertaining to the faculties of perceiving and recognizing.

Gnostic area   Sensory area of the cerebral cortex that receives and integrates sensory input from various parts of the brain so that a common thought can be formed.

Goblet cell   A goblet-shaped unicellular gland that secretes mucus.

Goiter (GOY-ter)   An enlargement of the thyroid gland.

Golgi (GOL-jē) **complex**   An organelle in the cytoplasm of cells consisting of four to eight flattened channels, stacked upon one another, with expanded areas at their ends; functions in packaging secreted proteins, lipid secretion, and carbohydrate synthesis.

Gomphosis (gom-FŌ-sis)   A fibrous joint in which a cone-shaped peg fits into a socket.

Gonad (GŌ-nad)   A gland that produces gametes and hormones; the ovary in the female and the testis in the male.

Gonadocorticoids (gō-na-dō-KOR-ti-koyd)   Sex hormones secreted by the adrenal cortex.

Gonadotropic (go'-nad-ō-TRŌ-pik) **hormone**   A hormone that regulates the functions of the gonads.

Gonorrhea (gon'-ō-RĒ-a)   Infectious, sexually transmitted disease caused by the bacterium *Neisseria gonorrhoeae* and characterized by inflammation of the urogenital mucosa, discharge of pus, and painful urination.

Gout (gowt)   Hereditary condition associated with excessive uric acid in the blood; the acid crystallizes and deposits in joints and the kidneys.

Graafian (GRAF-ē-an) **follicle**   An endocrine gland consisting of a mature ovum and its surrounding tissue that secretes estrogens.

Gray matter   Area in the central nervous system consisting of nonmyelinated nerve tissue.

Greater omentum (ō-MEN-tum)   A large fold in the serosa of the stomach that hangs down like an apron over the front of the intestines.

Greater vestibular (ves-TIB-yoo-lar) **glands**   A pair of glands on either side of the vaginal orifice that open by a duct into the space between the hymen and the labia minora. Also called **Bartholin's glands.**

Groin (groyn)   The depression between the thigh and the trunk; the inguinal region.

Gross anatomy   The branch of anatomy that deals with structures that can be studied without using a microscope. Also called **macroscopic anatomy.**

Growth hormone (GH)   Hormone secreted by the adenohypophysis (anterior lobe) of the pituitary that brings about growth of body tissues, especially skeletal and muscular. Also known as **somatotropin** and **somatotropic hormone (STH).**

Gustatory (GUS-ta-tō'-rē)   Pertaining to taste.

Gynecology (gī'-ne-KOL-ō-jē)   The branch of medicine dealing with the study and treatment of disorders of the female reproductive system.

Gyrus (JĪ-rus)   One of the convolutions of the cerebral cortex of the brain. *Plural,* **gyri** (JĪ-rī).

Hair A threadlike structure produced by hair follicles that develops in the dermis. Also called **pilus.**

Hair follicle (FOL-li-kul) Structure composed of epithelium surrounding the root of a hair from which hair develops.

Hallucination (ha-loo'-sī-NĀ-shun) A sensory perception of something that does not really exist in the world, that is, a sensory experience created from within the brain.

Hand The terminal portion of an upper extremity, including the carpus, metacarpus, and phalanges.

Haploid (HAP-loyd) Having half the number of chromosomes characteristically found in the somatic cells of an organism; characteristic of mature gametes.

Hard palate (PAL-at) The anterior portion of the roof of the mouth, formed by the maxillae and palatine bones and lined by mucous membrane.

Haustra (HAWS-tra) The sacculated elevations of the colon.

Haversian (ha-VER-shun) **system** The basic unit of structure in adult compact bone, consisting of a Haversian canal with its concentrically arranged lamellae, lacunae, osteocytes, and canaliculi. Also called an **osteon.**

Head The superior part of a human, cephalic to the neck. The superior or proximal part of a structure.

Heart A hollow muscular organ lying slightly to the left of the midline of the chest that pumps the blood through the cardiovascular system.

Heart block An arrhythmia of the heart in which the atria and ventricles contract independently because of a blocking of electrical impulses through the heart at a critical point in the conduction system.

Heat exhaustion Condition characterized by cool, clammy skin, profuse perspiration, and salt loss that results in muscle cramps, dizziness, vomiting, and fainting. Also called **heat prostration.**

Hematocrit (hē'-MAT-ō-krit) An expression of the percentage of the volume occupied by blood cells. Usually calculated by centrifuging a blood sample in a graduated tube and then reading off the volumes of cells and total blood.

Hematology (hē'-ma-TOL-ō-jē) The study of the blood.

Hematoma (hē'-ma-TŌ-ma) A tumor or swelling filled with blood.

Hematopoiesis (hem'-a-tō-poy-Ē-sis) Blood cell production occurring in the red marrow of bones. Also called **hemopoiesis** (hē-mō-poy-Ē-sis).

Hematuria (hē'-ma-TOOR-ē-a) Blood in the urine.

Hemiballismus (hem'-i-ba-LIZ-mus) Violent muscular restlessness of half of the body, especially of the upper extremity.

Hemocytoblast (hē'-mō-SĪ-tō-blast) Immature stem cell in bone marrow that develops along different lines into all the different mature blood cells.

Hemodialysis (hē'-mō-dī-AL-i-sis) Filtering of the blood by means of an artificial device so that certain substances are removed from the blood as a result of the difference in rates of their diffusion through a semipermeable membrane while the blood is being circulated outside the body.

Hemodynamics (hē-mō-dī-NA-miks) The study of factors and forces that govern the flow of blood through blood vessels.

Hemoglobin (hē'-mō-GLŌ-bin) **(Hb)** A substance in erythrocytes consisting of the protein globin and the iron-containing, red pigment heme and constituting about 33 percent of the cell volume; involved in the transport of oxygen and carbon dioxide.

Hemolysis (hē-MOL-i-sis) The escape of hemoglobin from the interior of the red blood cell into the surrounding medium; results from disruption of the integrity of the cell membrane by toxins or drugs, freezing or thawing, or hypotonic solutions.

Hemolytic disease of the newborn A hemolytic anemia of a newborn child that results from the destruction of the infant's rbc's by antibodies produced by the mother; usually the antibodies are due to an Rh blood type incompatibility. Also called **erythroblastosis fetalis.**

Hemophilia (hē'-mō-FĒL-ē-a) A hereditary blood disorder where there is a deficient production of certain factors involved in blood clotting, resulting in excessive bleeding into joints, deep tissues, and elsewhere.

Hemorrhage (HEM-or-rij) Bleeding; the escape of blood from blood vessels, especially when it is profuse.

Hemorrhoids (HEM-ō-royds) Dilated or varicosed blood vessels (usually veins) in the anal region. Also called **piles.**

Hemostasis (hē-MŌS-ta-sis) The stoppage of bleeding.

Hepatic (he-PAT-ik) Refers to the liver.

Hepatic duct A duct that receives bile from the bile capillaries. Small hepatic ducts merge to form the larger right and left hepatic ducts that unite to leave the liver as the common hepatic duct.

Hepatic portal circulation The flow of blood from the digestive organs to the liver before returning to the heart.

Hepatitis (hep-a-TĪ-tis) Inflammation of the liver due to a virus, drugs, and chemicals.

Hepatopancreatic (hep'-a-tō-pan'-krē-A-tik) **ampulla** A small, raised area in the duodenum where the combined common bile duct and main pancreatic duct empty into the duodenum. Also called the **ampulla of Vater.**

Hernia (HER-nē-a) The protrusion or projection of an organ or part of an organ through the wall of the cavity containing it.

Herniated (her'-nē-Ā-ted) **disc** A rupture of an intervertebral disc so that the nucleus pulposus protrudes into the vertebral cavity. Also called a **slipped disc.**

Heterozygous (he-ter-ō-ZĪ-gus) Possessing a pair of different genes on homologous chromosomes for a particular hereditary characteristic.

Hilus (HĪ-lus) An area, depression, or pit where blood vessels and nerves enter or leave the organ. Also called a **hilum.**

Hinge joint A synovial joint in which a convex surface of one bone fits into a concave surface of another bone, such as the elbow, knee, ankle, and interphalangeal joints. Also called a **ginglymus joint.**

Histamine (HIS-ta-mēn) Substance found in many cells, especially mast cells, basophils, and platelets, released when the cells are injured; results in vasodilation, increased permeability of blood vessels, and bronchiole constriction.

Histology (his-TOL-ō-jē) Microscopic study of the structure of tissues.

Holocrine (HŌL-ō-krin) **gland** A type of gland in which the entire secreting cell, along with its accumulated secretions, makes up the secretory product of the gland, as in the sebaceous (oil) glands.

Holter monitor Electrocardiograph worn by a person while going about everyday routines.

Homeostasis (hō'-mē-ō-STĀ-sis) The condition in which the body's internal environment remains relatively constant, within limits.

Homologous (hō-MOL-ō-gus) Similar in structure and origin, but not necessarily in function.

Homozygous (hō'-mō-ZĪ-gus) Possessing a pair of identical genes on homologous chromosomes for a particular hereditary characteristic.

Horizontal plane A plane that runs parallel to the ground and divides the body or organs into superior and inferior portions. Also called **transverse plane.**

Hormone (HOR-mōn) A secretion of an endocrine gland that alters the physiological activity of target cells of the body.

Human chorionic gonadotropin (kō-rē-ON-ik gō-nad-ō-TRŌ-pin) **(HCG)** A hormone produced by the developing placenta that maintains the corpus luteum.

Human chorionic somatomammotropin (sō-mat-ō-mam-ō-TRŌ-pin) **(HCS)** A hormone produced by the chorion of the placenta that may stimulate breast tissue for lactation, enhance body growth, and regulate metabolism.

Humoral (YOO-mor-al) **immunity** That component of immunity in which lymphocytes (B cells) develop into plasma cells that produce antibodies that destroy antigens.

Hunger A desire for food stimulated by a nutritional or energy requirement, as opposed to appetite, which is a response to a psychological need.

Hyaluronic (hī-a-loo-RON-ik) **acid** An amorphous matrix material found outside the cell.

Hyaluronidase (hī-a-loo-RON-i-dās) An enzyme that breaks down hyaluronic acid, increasing the permeability of connective tissues by dissolving the substances that hold body cells together.

Hydrocele (HĪ-drō-sēl) A fluid-containing sac or tumor. Specifically, a collection of fluid formed in the space along the spermatic cord and in the scrotum.

Hydrocephalus (hī-drō-SEF-a-lus) Abnormal accumulation of cerebrospinal fluid on the brain.

Hymen (HĪ-men) A thin fold of vascularized mucous membrane at the vaginal orifice.

Hyperbaric oxygenation (hī'-per-BA-rik ok'-sē-je-NĀ-shun) **(HBO)** Using pressure supplied by a hyperbaric chamber to cause more oxygen to dissolve in blood to treat patients infected with anaerobic bacteria (tetanus and gangrene bacteria). Also used to treat carbon monoxide poisoning, asphyxia, smoke inhalation, and certain heart disorders.

Hypercalcemia (hī'-per-kal-SĒ-mē-a) An excess of calcium in the blood.

Hypercapnia (hī'-per-KAP-nē-a) An abnormal increase in the amount of carbon dioxide in the blood.

Hyperemia (hī'-per-Ē-mē-a) An excess of blood in an area or part of the body.

Hyperextension (hī'-per-ek-STEN-shun) Continuation of extension beyond the anatomical position, as in bending the head backward.

Hyperglycemia (hī'-per-glī-SĒ-mē-a) An elevated blood sugar level.

Hypermetropia (hī'-per-mē-TRŌ-pē-a) A condition in which visual images are focused behind the retina with resulting defective vision of near objects; farsightedness.

Hyperplasia (hī'-per-PLĀ-zē-a) An abnormal increase in the number of normal cells in a tissue or organ, increasing its size.

Hyperpolarization (hī'-per-PŌL-a-ri-zā'-shun) Increase in the internal negativity across a cell membrane, thus increasing the voltage and moving it farther away from the threshold value.

Hypersecretion (hī'-per-se-KRĒ-shun) Overactivity of glands resulting in excessive secretion.

Hypersensitivity (hī'-per-sen-si-TI-vi-tē) Reaction to an antigen that results in pathological changes. Also called **allergy.**

Hypertension (hī'-per-TEN-shun) High blood pressure.

Hyperthermia (hī'-per-THERM-ē-a) An elevated body temperature.

Hypertonic (hī'-per-TON-ik) Having an osmotic pressure greater than that of a solution with which it is compared.

Hypertrophy (hī-PER-trō-fē) An excessive enlargement or overgrowth of tissue without cell division.

Hyperventilation (hī'-per-ven-ti-LĀ-shun) A rate of respiration higher than that required to maintain a normal level of plasma pCO_2.

Hypervitaminosis (hī'-per-vī'-ta-min-Ō-sis) An excess of one or more vitamins.

Hypocalcemia (hī'-pō-kal-SĒ-mē-a) A below normal level of calcium in the blood.

Hypochloremia (hī'-pō-klō-RĒ-mē-a) Deficiency of chloride in the blood.

Hypoglycemia (hī'-pō-glī-SĒ-mē-a) An abnormally low concentration of glucose in the blood; can result from excess insulin (injected or secreted).

Hypokalemia (hī'-pō-kā-LĒ-mē-a) Deficiency of potassium in the blood.

Hypomagnesemia Deficiency of magnesium in the blood.

Hyponatremia (hī'-pō-na-TRĒ-mē-a) Deficiency of sodium in the blood.

Hyposecretion (hī'-pō-se-KRĒ-shun) Underactivity of glands resulting in diminished secretion.

Hypothalamic-hypophyseal (hī'-pō-thal-AM-ik hī'-po-FIZ-ē-al) **tract** A bundle of nerve processes made up of fibers that have their cell bodies in the hypothalamus but release their neurosecretions in the posterior pituitary gland or neurohypophysis.

Hypothalamus (hī'-pō-THAL-a-mus) A portion of the diencephalon, lying beneath the thalamus and forming the floor and part of the wall of the third ventricle.

Hypothermia (hī'-pō-THER-mē-a) A low body temperature.

Hypotonic (hī'-pō-TON-ik) Having an osmotic pressure lower than that of a solution with which it is compared.

Hypoventilation (hī-pō-ven-ti-LĀ-shun) A rate of respiration lower than that required to maintain a normal level of plasma pCO_2.

Hypoxia (hī-POKS-ē-a) Lack of adequate oxygen. Also called **anoxia.**

Hysterectomy (his-te-REK-to-mē) The surgical removal of the uterus.

Ileocecal (il'-ē-ō-SĒ-kal) **valve** A fold of mucous membrane that guards the opening from the ileum into the large intestine.

Ileum (IL-ē-um) The terminal portion of the small intestine.

Immunity (i-MYOON-i-tē) The state of being resistant to injury, particularly by poisons, foreign proteins, and invading parasites, due to the presence of antibodies.

Immunogenicity (im-yoo-nō-jen-IS-it-ē) Ability of an antigen to stimulate antibody production.

Immunoglobulin (im-yoo-nō-GLOB-yoo-lin) **(Ig)** An antibody synthesized by special lymphocytes (plasma cells) in response to the introduction of antigen. Immunoglobulins are divided into five kinds (IgG, IgM, IgA, IgD, IgE) based primarily on the larger protein component present in the immunoglobulin.

Immunologic (im'-yoo-nō-LOJ-ik) **tolerance** A state of specific nonresponsiveness to an antigen, such as a self antigen, following exposure to the antigen.

Immunosuppression (im'-yoo-nō-su-PRESH-un) Inhibition of the immune response.

Impetigo (im'-pe-TĪ-go) A contagious skin disorder characterized by pustular eruptions.

Implantation (im-plan-TĀ-shun) The insertion of a tissue or a part into the body. The attachment of the blastocyst to the lining of the uterus seven to eight days after fertilization.

Impotence (IM-pō-tens) Weakness; inability to copulate; failure to maintain an erection.

Incontinence (in-KON-ti-nens) Inability to retain urine, semen, or feces, through loss of spincter control.

Infant respiratory distress syndrome (RDS) A disease of newborn infants, especially premature ones, in which insufficient amounts of surfactant are produced and breathing is labored. Also called **hyaline membrane disease (HMD).**

Infarction (in-FARK-shun) The presence of a localized area of necrotic tissue, produced by inadequate oxygenation of the tissue.

Infectious mononucleosis (mon-ō-nook'-lē-Ō-sis) Contagious disease caused by a virus and characterized by an elevated mononucleocyte and lymphocyte count, fever, sore throat, stiff neck, cough, and malaise.

Inferior (in-FĒR-ē-or) Away from the head or toward the lower part of a structure. Also called **caudad.**

Inferior vena cava (VĒ-na CĀ-va) Large vein that collects blood from parts of the body inferior to the heart and returns it to the right atrium.

Infertility Inability to conceive or to cause conception. Also called **sterility.**

Inflammation (in'-fla-MĀ-shun) Localized, protective response to tissue injury designed to destroy, dilute, or wall off the infecting agent or injured tissue; characterized by redness, pain, heat, swelling, and sometimes loss of function.

Inflation reflex Reflex that prevents overinflation of the lungs. Also called the **Hering-Breuer reflex.**

Infundibulum (in'-fun-DIB-yoo-lum) The stalklike structure that attaches the pituitary gland (hypophysis) to the hypothalamus of the brain. The funnel-shaped, open, distal end of the uterine tube.

Ingestion (in-JES-chun) The taking in of food, liquids, or drugs, by mouth.

Inheritance The acquisition of body characteristics and qualities by transmission of genetic information from parents to offspring.

Inhibin A male sex hormone secreted by sustentacular cells that inhibits FSH release by the adenohypophysis (anterior pituitary) and thus spermatogenesis.

Inhibitory postsynaptic potential (IPSP) The increase in the internal negativity of the membrane potential so that the voltage moves further from the threshold value.

Insertion (in-SER-shun) The manner or place of attachment of a muscle to the bone that it moves.

Inspiration (in-spi-RĀ-shun) The act of drawing air into the lungs.

Inspiratory (in-SPĪ-ra-tō-rē) **capacity** The total inspiratory ability of the lungs; the sum of tidal volume and inspiratory reserve; about 3,600 ml.

Inspiratory reserve volume The volume of air in excess of tidal volume that can be inhaled by forced inspiration; about 3,100 ml.

Insula (IN-su-la) A triangular area of cerebral cortex that lies deep within the lateral cerebral fissure, under the parietal, frontal, and temporal lobes, and cannot be seen in an external view of the brain. Also called the **island** or **isle of Reil.**

Insulin (IN-su-lin) A hormone produced by the pancreas that decreases the blood glucose level.

Integumentary (in-teg'-yoo-MEN-tar-ē) Relating to the skin.

Intercalated (in-TER-ka-lāt-ed) **disc** An irregular transverse thickening of sarcolemma that separates cardiac muscle cells from each other.

Intercellular fluid That portion of extracellular fluid that bathes the cells of the body; the internal environment of the body. Also called **interstitial fluid.**

Interferon (in'-ter-FĒR-on) A protein naturally produced by virus-infected host cells that inhibits intracellular viral replication in uninfected host cells; artificially synthesized through recombinant DNA techniques.

Intermediate Between two structures, one of which is medial and one of which is lateral.

Internal capsule A thick sheet of white matter made up of myelinated fibers connecting various parts of the cerebral cortex and lying between the thalamus and the caudate and lentiform nuclei of the basal ganglia.

Internal ear The inner ear or labyrinth, lying inside the temporal bone, containing the organs of hearing and balance.

Internal nares (NA-rēz) The two openings posterior to the nasal cavities opening into the nasopharynx. Also called the **choanae.**

Interphase (IN-ter-fāz) The period during its life cycle when a cell is carrying on every life process except division; the stage between two mitotic divisions.

Interstitial (in'-ter-STISH-al) **endocrinocyte** A cell located in the connective tissue between seminiferous tubules in a mature testis that secretes testosterone. Also called an **interstitial cell of Leydig.**

Interstitial growth Growth from within, as in the growth of cartilage. Also called **endogenous growth.**

Intervertebral (in'-ter-VER-te-bral) **disc** A pad of fibrocartilage located between the bodies of two vertebrae.

Intracellular (in'-tra-SEL-yoo-lar) **fluid (ICF)** Fluid inside of body cells.

Intramembranous ossification (in'-tra-MEM-bra-nus os'-i-fi-KĀ-shun) The method of bone formation in which the bone is formed directly in membranous tissue.

Intraocular (in-tra-OC-yoo-lar) **pressure** Pressure in the eyeball, produced mainly by aqueous humor.

Intrapleural pressure Air pressure between the two pleural layers of the lungs, usually subatmospheric. Also called **intrathoracic pressure.**

Intrapulmonic pressure Air pressure within the lungs. Also called **intraalveolar pressure.**

Intrauterine device (IUD) A small metal or plastic object inserted into the uterus for the purpose of preventing pregnancy.

Intrinsic clotting pathway Sequence of reactions leading to blood clotting that is initiated by the release of chemicals by disintegrated platelets.

Intrinsic factor A glycoprotein synthesized and secreted by the gastric mucosa that facilitates vitamin B_{12} absorption.

Intubation (in'-too-BĀ-shun) Insertion of a tube through the nose or mouth into the larynx and trachea for entrance of air or to dilate a stricture.

In utero (YOO-ter-ō) Within the uterus.

Inversion (in-VER-zhun) The movement of the sole inward at the ankle joint.

In vitro (VĒ-trō) Literally, in glass; outside the living body and in an artificial environment such as a laboratory test tube.

In vivo (VĒ-vō) In the living body.

Ion (Ī-on) Any charged atom or group of atoms; usually formed when a substance, such as a salt, dissolves and dissociates.

Ipsilateral (ip'-si-LAT-er-al) On the same side, affecting the same side of the body.

Ischemia (is-KĒ-mē-a) A lack of sufficient blood to a part due to obstruction of circulation.

Islet of Langerhans (Ī-let of LAHNG-er-hanz) A cluster of endocrine gland cells in the pancreas that secretes insulin, glucagon, and somatostatin.

Isometric contraction A muscle contraction in which tension on the muscle increases, but there is minimal muscle shortening so that no movement is produced.

Isotonic (ī'-sō-TON-ik) Having equal tension or tone. Having equal osmotic pressure between two different solutions or between two elements in a solution.

Isotonic contraction A muscle contraction in which tension remains constant, but the muscle shortens and pulls on another structure to produce movement.

Isotope (Ī-sō-tōpe') A chemical element that has the same atomic number as another but a different atomic weight. Radioactive isotopes change into other elements with the emission of certain radiations.

Jaundice (JAWN-dis) A condition characterized by yellowness of skin, white of eyes, mucous membranes, and body fluids.

Jejunum (jē-JOO-num)   The middle portion of the small intestine.

Joint kinesthetic (kin'-es-THET-ik) **receptor**   A proprioceptive receptor located in a joint, stimulated by joint movement.

Juxtaglomerular (juks-ta-glō-MER-yoo-lar) **apparatus**   Consists of the macula densa (cells of the distal convoluted tubule adjacent to the afferent arteriole) and juxtaglomerular cells (modified cells of the afferent arteriole); secretes renin when blood pressure starts to fall.

Karyotype (KAR-ē-ō-tīp)   Chromosome characteristics of an individual or a group of cells.

Keratin (KER-a-tin)   An insoluble protein found in the hair, nails, and other keratinized tissues of the epidermis.

Ketone (KĒ-ton) **bodies**   Substances produced primarily during excessive fat metabolism, such as acetone, acetoacetic acid, and β-hydroxybutyric acid.

Ketosis (kē-TŌ-sis)   Abnormal condition marked by excessive production of ketone bodies.

Kidney (KID-nē)   One of the paired reddish organs located in the lumbar region that regulates the composition and volume of blood and produces urine.

Kinesthesia (kin-is-THĒ-szē-a)   Ability to perceive extent, direction, or weight of movement; muscle sense.

Krebs cycle   A series of energy-yielding chemical reactions in which carbon dioxide is formed and energy is transferred to carrier molecules for subsequent liberation. Also called the **citric acid cycle** and **tricarboxylic acid (TCA) cycle.**

Kyphosis (kī-FŌ-sis)   An exaggeration of the thoracic curve of the vertebral column, resulting in a "round-shouldered" or hunchback appearance.

Labial frenulum (LĀ-bē-al FREN-yoo-lum)   A medial fold of mucous membrane between the inner surface of the lip and the gums.

Labia majora (LĀ-bē-a ma-JOR-a)   Two longitudinal folds of skin extending downward and backward from the mons pubis of the female.

Labia minora (min-OR-a)   Two small folds of mucous membrane lying medial to the labia majora of the female.

Labium (LĀ-bē-um)   A lip. A liplike structure. *Plural,* **labia** (LĀ-bē-a).

Labyrinthine (lab-i-RIN-thēn) **disease**   Malfunction of the internal ear characterized by deafness, tinnitus, vertigo, nausea, and vomiting.

Laceration (las'-er-Ā-shun)   A torn, ragged, or mangled wound due to trauma.

Lacrimal (LAK-ri-mal) **canal**   A duct, one on each eyelid, commencing at the punctum at the medial margin of an eyelid and conveying the tears medially into the nasolacrimal sac.

Lacrimal gland   Secretory cells located at the superior lateral portion of each orbit that secrete tears into excretory ducts that open onto the surface of the conjunctiva.

Lacrimal sac   The superior expanded portion of the nasolacrimal duct that receives the tears from a lacrimal canal.

Lactation (lak-TĀ-shun)   The secretion and ejection of milk by the mammary glands.

Lacteal (LAK-tē-al)   One of many intestinal lymph vessels in villi that absorb fat from digested food.

Lacuna (la-KOO-na)   A small, hollow space, such as that found in bones in which the osteoblasts lie. *Plural,* **lacunae** (la-KOO-nē).

Lamellae (la-MEL-ē)   Concentric rings found in compact bone.

Lamina (LAM-i-na)   A thin, flat layer or membrane, as the flattened part of either side of the arch of a vertebra. *Plural,* **laminae** (LAM-i-nē).

Lamina propria (PRO-prē-a)   The connective tissue layer of a mucous membrane.

Lanugo (lan-YOO-gō)   Fine downy hairs that cover the fetus.

Large intestine   The portion of the digestive tract extending from the ileum of the small intestine to the anus, divided structurally into the cecum, colon, rectum, and anal canal.

Laryngitis (la-rin-JĪ-tis)   Inflammation of the mucous membrane lining the larynx.

Laryngopharynx (la-rin'-gō-FAR-inks)   The inferior portion of the pharynx, extending downward from the level of the hyoid bone to divide posteriorly into the esophagus and anteriorly into the larynx.

Larynx (LAR-inks)   The voice box, a short passageway that connects the pharynx with the trachea.

Lateral (LAT-er-al)   Farther from the midline of the body or a structure.

Lateral ventricle (VEN-tri-kul)   A cavity within a cerebral hemisphere that communicates with the lateral ventricle in the other cerebral hemisphere and with the third ventricle by way of the interventricular foramen.

Leg   The part of the lower extremity between the knee and the ankle.

Lens   A transparent organ lying posterior to the pupil and iris of the eyeball and anterior to the vitreous humor.

Lesion (LĒ-zhun)   Any localized, abnormal change in tissue formation.

Lesser omentum (ō-MEN-tum)   A fold of the peritoneum that extends from the liver to the lesser curvature of the stomach and the commencement of the duodenum.

Lesser vestibular (ves-TIB-yoo-lar) **gland**   One of the paired mucus-secreting glands that have ducts that open on either side of the urethral orifice in the vestibule of the female.

Leucocyte (LOO-kō-sīt)   A white blood cell.

Leucocytosis (loo-kō-sī-TŌ-sis)   An increase in the number of white blood cells, characteristic of many infections and other disorders.

Leucopenia (loo-kō-PĒ-nē-a)   A decrease of the number of white blood cells below 5,000/mm³.

Leukemia (loo-KĒ-mē-a)   A cancerlike disease of the blood-forming organs characterized by a rapid and abnormal increase in the number of white blood cells plus many immature cells in the circulating blood.

Leukoplakia (loo-kō-PLĀ-kē-a)   A disorder in which there are white patches in the mucous membranes of the tongue, gums, and cheeks.

Libido (li-BĒ-dō)   The sexual drive, conscious or unconscious.

Ligament (LIG-a-ment)   Dense, regularly arranged connective tissue that attaches bone to bone.

Limbic system   A portion of the forebrain, sometimes termed the visceral brain, concerned with various aspects of emotion and behavior, that includes the limbic lobe (hippocampus and associated areas of gray matter plus the cingulate gyrus), certain parts of the temporal and frontal cortex, some thalamic and hypothalamic nuclei, and parts of the basal ganglia.

Lingual frenulum (LIN-gwal FREN-yoo-lum)   A fold of mucous membrane that connects the tongue to the floor of the mouth.

Lipase (LĪ-pās)   A fat-splitting enzyme.

Lipid   An organic compound composed of carbon, hydrogen, and oxygen that is usually insoluble in water, but soluble in alcohol, ether, and chloroform; examples include fats, phospholipids, steroids, and prostaglandins.

Lipogenesis (li-pō-GEN-e-sis)   The synthesis of lipids from glucose or amino acids by liver cells.

Liver   Large gland under the diaphragm that occupies most of the right hypochondriac region and part of the epigastric region; functionally, it produces bile salts, heparin, and plasma proteins; converts one nutrient into another; detoxifies substances; stores glycogen, minerals, and vitamins; carries on phagocytosis of blood cells and bacteria; and helps to activate vitamin D.

Local   Pertaining to or restricted to one spot or part.

Lordosis (lor-DŌ-sis) An exaggeration of the lumbar curve of the vertebral column.

Lower extremity The appendage attached at the pelvic girdle, consisting of the thigh, knee, leg, ankle, foot, and toes.

Lumbar (LUM-bar) Region of the back and side between the ribs and pelvis; loins.

Lumen (LOO-men) The space within an artery, vein, intestine, or tube.

Lung One of the two main organs of respiration, lying on either side of the heart in the thoracic cavity.

Lunula (LOO-nyoo-la) The moon-shaped white area at the base of a nail.

Luteinizing (LOO-tē-in'-īz-ing) **hormone (LH)** A hormone secreted by the adenohypophysis (anterior lobe) of the pituitary gland that stimulates ovulation, progesterone secretion by the corpus luteum, and readies the mammary glands for milk secretion in females and stimulates testosterone secretion by the testes in males.

Lymph (limf) Fluid confined in lymphatic vessels and flowing through the lymphatic system to be returned to the blood.

Lymphangiography (lim-fan'-jē-OG-ra-fē) A procedure by which lymphatic vessels and lymph organs are filled with a radiopaque substance in order to be x-rayed.

Lymphatic (lim-FAT-ik) Pertaining to lymph. A large vessel that collects lymph from lymph capillaries and converges with other lymphatics to form the thoracic and right lymphatic ducts.

Lymph capillary Blind-ended microscopic lymph vessel that begins in spaces between cells and converges with other lymph capillaries to form lymphatics.

Lymph node An oval or bean-shaped structure located along the lymphatic vessels.

Lymphocyte (LIM-fō-sīt) A type of white blood cell, found in lymph nodes, associated with the immune system.

Lysosome (LĪ-sō-sōm) An organelle in the cytoplasm of a cell, enclosed by a single membrane and containing powerful digestive enzymes.

Lysozyme (LĪ-sō-zīm) A bactericidal enzyme found in tears, saliva, and perspiration.

Macula (MAK-yoo-la) A discolored spot or a colored area. A small, flat region on the wall of the utricle and saccule that serves as a receptor for static equilibrium.

Macula lutea (LOO-tē-a) The yellow spot in the center of the retina.

Malaise (ma-LĀYZ) Discomfort, uneasiness, and indisposition, often indicative of infection.

Malignant (ma-LIG-nant) Referring to diseases that tend to become worse and cause death; especially the invasion and spreading of cancer.

Mammary (MAM-ar-ē) **gland** Modified sweat gland of the female that secretes milk for the nourishment of the young.

Margination (mar'-ji-NĀ-shun) Accumulation and adhesion of neutrophils to the endothelium of blood vessels at the site of injury during the early stages of inflammation.

Marrow (MAR-ō) Soft, spongelike material in the cavities of bone. Red marrow produces blood cells; yellow marrow, formed mainly of fatty tissue, has no blood-producing function.

Mast cell A cell found in loose connective tissue along blood vessels that produces heparin, an anticoagulant. The name given to a basophil after it has left the bloodstream and entered the tissues.

Mastectomy (mas-TEK-tō-mē) Surgical removal of breast tissue.

Mastication (mas'-ti-KĀ-shun) Chewing.

Meatus (mē-Ā-tus) A passage or opening, especially the external portion of a canal.

Mechanoreceptor (me-KAN-ō-rē'-sep-tor) Receptor that detects mechanical deformation of the receptor itself or adjacent cells; stimuli so detected include touch, pressure, proprioception, hearing, and equilibrium.

Medial (MĒ-dē-al) Nearer the midline of the body or a structure.

Medial lemniscus (lem-NIS-kus) A flat band of myelinated nerve fibers extending through the medulla, pons, and midbrain and terminating in the thalamus on the opposite side. Second-order sensory neurons in this tract transmit impulses for discriminating touch, pressure, and vibration sensations.

Mediastinum (mē'-dē-as-TĪ-num) A mass of tissue found between the pleurae of the lungs that extends from the sternum to the vertebral column.

Medulla (me-DUL-la) An inner layer of an organ, such as the medulla of the kidneys.

Medulla oblongata (ob'-long-GA-ta) The most inferior part of the brain stem.

Medullary (MED-yoo-lar'-ē) **cavity** The space within the diaphysis of a bone that contains yellow marrow.

Medullary rhythmicity (rith-MIS-i-tē) **area** Portion of the respiratory center in the medulla that controls the basic rhythm of respiration.

Meiosis (mē-Ō-sis) A type of cell division restricted to sex-cell production involving two successive nuclear divisions which result in daughter cells with the haploid number of chromosomes.

Meissner's (MĪS-nerz) **corpuscle** The sensory receptor for the sensation of touch; found in the dermal papillae, especially in palms and soles.

Melanin (MEL-a-nin) A dark black, brown, or yellow pigment found in some parts of the body such as the skin.

Melanocyte (MEL-a-nō-sīt') A pigmented cell located between or beneath cells of the deepest layer of the epidermis that synthesizes melanin.

Melanocyte-stimulating hormone (MSH) A hormone secreted by the adenohypophysis (anterior lobe) of the pituitary gland that stimulates the dispersion of melanin granules in melanocytes.

Melanoma (mel'-a-NŌ-ma) A usually dark, malignant tumor of the skin containing melanin.

Melatonin (mel-a-TŌN-in) A hormone secreted by the pineal gland that may inhibit reproductive activities.

Membrane A thin, flexible sheet of tissue composed of an epithelial layer and an underlying connective tissue layer, as in an epithelial membrane, or of loose connective tissue only, as in a synovial membrane.

Membrane potential The voltage present, at any instant, across the cell membrane; measured with microelectrodes inside and outside the cell, it usually registers "resting" values of about −70 mV. Also called a **resting potential.**

Membranous labyrinth (mem-BRA-nus LAB-i-rinth) The portion of the labyrinth of the inner ear that is located inside the bony labyrinth and separated from it by the perilymph; made up of the membranous semicircular canals, the saccule and utricle, and the cochlear duct.

Memory The ability to recall thoughts; commonly classified as activated and long-term.

Menarche (me-NAR-kē) Beginning of the menstrual function.

Ménière's (men-YAIRZ) **syndrome** A type of labyrinthine disease characterized by fluctuating loss of hearing, vertigo, and tinnitus.

Meninges (me-NIN-jēz) Three membranes covering the brain and spinal cord, called the dura mater, arachnoid, and pia mater. *Singular,* **meninx** (MEN-inks).

Meningitis (men-in-JĪ-tis) Inflammation of the meninges, most commonly the pia mater and arachnoid.

Menopause (MEN-ō-pawz) The termination of the menstrual cycles.

Menstrual (MEN-stroo-al) **cycle** A series of changes in the endometrium of a nonpregnant female that prepares the lining of the uterus to receive a fertilized ovum.

Menstruation (men'-stroo-Ā-shun) Periodic discharge of blood, tissue fluid, mucus, and epithelial cells that usually lasts for 5 days; caused by a sudden reduction in estrogens and progesterone. Also called the **menstrual phase** or **menses.**

Merkel's (MER-kuls) **disc** An encapsulated, cutaneous receptor for touch located in deeper layers of epidermal cells.

Merocrine (MER-ō-krin) **gland** A secretory cell that remains intact throughout the process of formation and discharge of the secretory product, as in the salivary and pancreatic glands.

Mesenchyme (MEZ-en-kīm) An embryonic connective tissue from which all other connective tissues arise.

Mesentery (MEZ-en-ter'-ē) A fold of peritoneum attaching the small intestine to the posterior abdominal wall.

Mesocolon (mez'-ō-KŌ-lon) A fold of peritoneum attaching the colon to the posterior abdominal wall.

Mesoderm The middle of the three primary germ layers which gives rise to connective tissues, blood and blood vessels, and muscles.

Mesothelium (mez'-ō-THĒ-lē-um) The layer of simple squamous epithelium that lines serous cavities.

Metabolism (me-TAB-ō-lizm) The sum of all the biochemical reactions that occur within an organism, including the synthetic (anabolic) reactions and decomposition (catabolic) reactions.

Metacarpus (met'-a-KAR-pus) A collective term for the five bones that make up the palm of the hand.

Metaphase (MET-a-phāz) The second stage of mitosis in which chromatid pairs line up on the equatorial plane of the cell.

Metaphysis (me-TAF-i-sis) Growing portion of a bone.

Metastasis (me-TAS-ta-sis) The transfer of disease from one organ or part of the body to another.

Metatarsus (met'-a-TAR-sus) A collective term for the five bones located in the foot between the tarsals and the phalanges.

Micelle (mī-SEL) A spherical aggregate of bile salts that dissolves fatty acids and monosaccharides so that they can be transported into small intestinal epithelial cells.

Microcephalus (mi-kro-SEF-a-lus) An abnormally small head; premature closing of the anterior fontanel so that the brain has insufficient room for growth, resulting in mental retardation.

Microfilament (mī-krō-FIL-a-ment) Rodlike cytoplasmic structure, ranging in diameter from 30 Å to 120 Å; comprises contractile units in muscle cells and provides support, shape, and movement in nonmuscle cells.

Microglia (mi-krō-GLĒ-a) Neuroglial cells that carry on phagocytosis. Also called **brain macrophages.**

Microphage (MĪK-rō-fāj) Granular leucocyte that carries on phagocytosis, especially neutrophils and eosinophils.

Microtubule (mī-krō-TOOB-yool') Cylindrical cytoplasmic structure, ranging in diameter from 180 Å to 300 Å, consisting of the protein tubulin; provides support, structure, and transportation.

Microvilli (mī'-krō-VIL-ē) Microscopic, fingerlike projections of the cell membranes of small intestinal cells that increase surface area for absorption.

Micturition (mik'-too-RISH-un) The act of expelling urine from the bladder. Also called **urination.**

Midbrain The part of the brain between the pons and the diencephalon. Also called the **mesencephalon.**

Middle ear A small, epithelial-lined cavity hollowed out of the temporal bone, separated from the external ear by the eardrum and from the internal ear by a thin bony partition containing the oval and round windows; extending across the middle ear are the three auditory ossicles. Also called the **tympanic cavity.**

Midsagittal plane A plane through the midline of the body running vertical to the ground and dividing the body or organs into equal right and left sides. Also called a **median plane.**

Milking Contraction of skeletal muscle around veins causing venous valves to open and thus helping to drive blood back to the heart.

Milk letdown Contraction of alveolar cells to force milk into ducts of mammary glands, stimulated by oxytocin which is released from the posterior pituitary in response to suckling action.

Mineral Inorganic, homogeneous solid substance that may perform a function vital to life; examples include calcium, sodium, potassium, iron, phosphorus, and chlorine.

Mineralocorticoids (min'-er-al-ō-KOR-ti-koyds) A group of hormones of the adrenal cortex.

Minimal volume The volume of air in the lungs even after the thoracic cavity has been opened forcing out some of the residual volume.

Minute volume of respiration (MVR) Total volume of air taken into the lungs per minute; about 6,000/ml.

Mitochondrion (mi'-tō-KON-drē-on) A double-membraned organelle that plays a central role in the production of ATP; known as the powerhouse of the cell.

Mitosis (mī-TŌ-sis) The orderly division of the nucleus of a cell that ensures that each new daughter nucleus has the same number and kind of chromosomes as the original parent nucleus. The process includes the replication of chromosomes and the distribution of the two sets of chromosomes into two separate and equal nuclei.

Mitotic apparatus Collective term for continuous and chromosomal microtubules and centrioles; involved in cell division.

Modality (mō-DAL-i-tē) Any of the specific sensory entities, such as vision, smell, or taste.

Modiolus (mō-DĪ-ō'-lus) The central pillar or column of the cochlea.

Molecule (MOL-e-kyool) The chemical combination of two or more atoms.

Monoclonal antibody Antibody produced by in vitro clones of B cells hydridized with cancerous cells.

Monocyte (MON-ō-sīt') A type of white blood cell characterized by agranular cytoplasm; the largest of the leucocytes.

Mons pubis (monz PYOO-bis) The rounded, fatty prominence over the symphysis pubis, covered by coarse pubic hair.

Morbid (MOR-bid) Diseased; pertaining to disease.

Morula (MOR-yoo-la) A solid mass of cells produced by successive cleavages of a fertilized ovum a few days after fertilization.

Motor area The region of the cerebral cortex that governs muscular movement, particularly the precentral gyrus of the frontal lobe.

Motor end plate The portion of the muscle cell membrane directly under the termination of a motor neuron where the neuron releases its transmitter substance.

Motor unit A motor neuron together with the muscle cells it stimulates.

Mucous (MYOO-kus) **cell** A unicellular gland that secretes mucus. Also called a **goblet cell.**

Mucous membrane A membrane that lines a body cavity that opens to the exterior. Also called the **mucosa.**

Mucus The thick fluid secretion of the mucous glands and mucous membranes.

Multiple sclerosis (skler-Ō-sis) Progressive destruction of myelin sheaths of neurons in the central nervous system, short-circuiting conduction pathways.

Mumps Inflammation and enlargement of the parotid glands accompanied by fever and extreme pain during swallowing.

Murmur An unusual heart sound; may indicate a disorder such as a malfunctioning mitral valve or may have no clinical significance.

Muscle An organ composed of one of three types of muscle tissue (skeletal, cardiac, or visceral), specialized for contraction to produce voluntary or involuntary movement of parts of the body.

Muscle spindle An encapsulated receptor in a skeletal muscle, consisting of specialized muscle cells and nerve endings, stimulated by changes in length or tension of muscle cells; a proprioceptor.

Muscle tissue A tissue specialized to produce motion in response to nerve impulses by its qualities of contractility, extensibility, elasticity, and excitability.

Muscle tone A sustained, partial contraction of portions of a skeletal muscle in response to activation of stretch receptors.

Muscular dystrophy (DIS-trō-fē') Inherited myopathy characterized by degeneration of muscle cells that leads to progressive atrophy.

Muscularis (MUS-kyoo-la'-ris) A muscular layer or tunic of an organ.

Muscularis mucosae (myoo-KŌ-sē) A thin layer of smooth muscle cells located in the outermost layer of the mucosa of the alimentary canal, underlying the lamina propria of the mucosa.

Mutation (myoo-TĀ-shun) Any change in the sequence of bases in the DNA molecule resulting in a permanent alteration in some inheritable characteristic.

Myasthenia (mī-as-THĒ-nē-a) **gravis** Weakness of skeletal muscles caused by antibodies directed against acetylcholine receptors that inhibit muscle contraction.

Myelin (MĪ-e-lin) **sheath** A white, phospholipid, segmented covering, formed by Schwann cells, around the axons and dendrites of many peripheral neurons.

Myocardial infarction (mī'-ō-KAR-dē-al in-FARK-shun) Gross necrosis of myocardial tissue due to interrupted blood supply. Also called a **heart attack.**

Myocardium (mī'-ō-KAR-dē-um) The middle layer of the heart wall, made up of cardiac muscle, comprising the bulk of the heart, and lying between the epicardium and the endocardium.

Myofibril (mī'-ō-FĪ-bril) A threadlike structure, running longitudinally through a muscle cell, consisting mainly of thick myofilaments (myosin) and thin myofilaments (actin).

Myoglobin (mī-ō-GLŌ-bin) The oxygen-binding, iron-containing conjugated protein complex present in the sarcoplasm of muscle cells; contributes the red color to muscle.

Myogram (MĪ-ō-gram) The record or tracing produced by the myograph, the apparatus that measures and records the effects of muscular contractions.

Myology (mī-OL-ō-jē) The study of the muscles and their parts.

Myometrium (mī'-ō-MĒ-trē-um) The smooth muscle coat or tunic of the uterus.

Myopia (mī-Ō-pē-a) Defect in vision so that objects can be seen distinctly only when very close to the eyes; nearsightedness.

Myosin (MĪ-ō-sin) The contractile protein that makes up the thick myofilaments of muscle cells.

Myotonia (mī-ō-TO-nē-a) A continuous spasm of muscle; increased muscular irritability and tendency to contract and less ability to relax.

Myxedema (mix-e-DĒ-ma) Condition caused by hypothyroidism during the adult years characterized by swelling of facial tissues.

Nail A hard plate, composed largely of keratin, that develops from the epidermis of the skin to form a protective covering on the dorsal surface of the distal phalanges of the fingers and toes.

Nail matrix (MĀ-triks) The part of the nail beneath the body and root from which the nail is produced.

Nasal (NĀ-zal) **cavity** A mucosa-lined cavity on either side of the nasal septum that opens onto the face at an external naris and into the nasopharynx at an internal naris.

Nasal septum (SEP-tum) A vertical partition composed of bone and cartilage, covered with a mucous membrane, separating the nasal cavity into left and right sides.

Nasolacrimal (nā'-zō-LAK-ri-mal) **duct** A canal that transports the lacrimal secretion from the nasolacrimal sac into the nose.

Nasopharynx (nā'-zō-FAR-inks) The uppermost portion of the pharynx, lying posterior to the nose and extending down to the soft palate.

Neck The part of the body connecting the head and the trunk. A constricted portion of an organ such as the neck of a femur or the neck of the uterus.

Necrosis (ne-KRŌ-sis) Death of a cell or group of cells as a result of disease or injury.

Negative feedback The principle governing most control systems; a mechanism of response in which a stimulus initiates actions which reverse or reduce the stimulus.

Neoplasm (NĒ-ō-plazm) A mass of new, abnormal tissue; a tumor.

Nephritis (ne-FRĪT-is) Inflammation of the kidney.

Nephron (NEF-ron) The functional unit of the kidney.

Nephrosis (nef-RŌ-sis) A degenerative disease of the kidney in which glomerular damage results in leakage of protein into urine.

Nerve A cordlike bundle of nerve fibers and their associated connective tissue coursing together outside the central nervous system.

Nerve impulse A wave of negativity that sweeps along the outside of the membrane of a neuron. Also called an **action potential.**

Nervous tissue Tissue that initiates and transmits nerve impulses to coordinate homeostasis.

Neuritis (noo-RĪ-tis) Inflammation of a nerve.

Neuroeffector (noo-rō-e-FEK-tor) **junction** Cumulative term for neuromuscular and neuroglandular junctions.

Neurofibril (noo-rō-FĪ-bril) One of the delicate threads that forms a complicated network in the cytoplasm of the cell body and processes of a neuron.

Neuroglandular (noo-rō-GLAND-yoo-lar) **junction** Area of contact between a motor neuron and a gland.

Neuroglia (noo-ROG-lē-a) Cells of the nervous system that are specialized to perform the functions of connective tissue. The neuroglia of the central nervous system are the astrocytes, oligodendrocytes, microglia, and ependyma; neuroglia of the peripheral nervous system include the Schwann cells and the ganglion satellite cells. Also called **glial cells.**

Neurohypophysis (noo-rō-hī-POF-i-sis) The posterior lobe of the pituitary gland.

Neuromuscular (noo-rō-MUS-kyoo-lar) **junction** The area of contact between a motor neuron and a muscle fiber. Also called a **myoneural junction.**

Neuron (NOO-ron) A nerve cell, consisting of a cell body, dendrites, and an axon.

Neurosecretory (noo-rō-SĒC-re-tō-rē) **cell** A cell in a nucleus in the hypothalamus that produces oxytocin (**OT**) or antidiuretic hormone (ADH), hormones stored in the neurohypophysis of the pituitary gland.

Neutrophil (NOO-trō-fil) A type of white blood cell characterized by granular cytoplasm which stains as readily with acid or basic dyes. Also called a **polymorph.**

Night blindness Poor or no vision in dim light or at night, although good vision present during bright illumination; frequently caused by a deficiency of vitamin A.

Nipple A pigmented, wrinkled projection on the surface of the mammary gland which is the location of the openings of the lactiferous ducts for milk release.

Nissl (NIS-l) **bodies** Rough endoplasmic reticulum in the cell bodies of neurons that functions in protein synthesis.

Nociceptor (nō'-sē-SEP-tor) A receptor that detects pain.

Node of Ranvier (ron-vē-Ā) A space, along a myelinated nerve fiber, between the individual Schwann cells that form the myelin sheath and the neurilemma.

Nonessential amino acid An amino acid that can be synthesized by body cells through transamination, the transfer of an amino group from an amino acid to another substance.

Nongonococcal urethritis (NGU) A sexually transmitted disease characterized by inflammation of the urethra caused by a microorganism other than *Neisseria gonorrhoeae* or by nonmicrobial factors.

Norepinephrine (nor'-ep-ē-NEF-rin) **(NE)** A hormone secreted by the adrenal medulla that produces actions similar to those that result from sympathetic stimulation. Also called **noradrenaline.**

Nuclear magnetic resonance (NMR) A diagnostic procedure that focuses on the nuclei of atoms of a single element in a tissue, usually hydrogen, to determine if they behave normally in the presence of an external magnetic force; used to indicate the biochemical activity of a tissue.

Nucleic (noo-KLĒ-ic) **acid** An organic compound that is a long polymer of nucleotides, with each nucleotide containing a pentose sugar, a phosphate group, and one of four possible nitrogen bases (adenine, cytosine, guanine, and thymine or uracil).

Nucleosome (NOO-klē-ō-sōm) Elementary structural subunit of a chromosome consisting of histones and DNA.

Nucleus (NOO-klē-us) A spherical or oval organelle of a cell that contains the hereditary factors of the cell, called genes. A cluster of nerve cell bodies in the central nervous system. The central portion of an atom made up of protons and neutrons.

Nystagmus (nis-TAG-mus) Constant, involuntary, rhythmic movement of the eyeballs; horizontal, rotary, or vertical.

Obesity (ō-BĒS-i-tē) Body weight 10 to 20 percent over a desirable standard as a result of excessive accumulation of fat.

Obligatory water reabsorption The absorption of water from proximal convoluted tubules of nephrons as a function of osmosis.

Occlusion (ō-KLOO-zhun) The act of closure or state of being closed.

Olfactory (ōl-FAK-tō-rē) Pertaining to smell.

Olfactory bulb A mass of gray matter at the termination of an olfactory nerve, lying beneath the frontal lobe of the cerebrum on either side of the crista galli of the ethmoid bone.

Olfactory tract A bundle of axons that extends from the olfactory bulb posteriorly to the olfactory portion of the cortex.

Oligodendrocyte (o-lig-ō-DEN-drō-sīt) A neuroglial cell that supports neurons and produces a myelin sheath around axons of neurons of the central nervous system.

Oncogene (ONG-kō-jēn) Gene that has the ability to transform a normal cell into a cancerous cell.

Oncology (ong-KOL-ō-jē) The study of tumors.

Oogenesis (ō'-ō-JEN-e-sis) Formation and development of the ovum.

Oophorectomy (ō'-of-ō-REK-tō-mē) The surgical removal of the ovaries.

Opsonization (op-sō-ni-ZĀ-shun) The action of some antibodies that renders bacteria and other foreign cells more susceptible to phagocytosis. Also called **immune adherence.**

Optic (OP-tik) Refers to the eye, vision, or properties of light.

Optic chiasma (kī-AZ-ma) A crossing point of the optic nerves, anterior to the pituitary gland.

Optic disc A small area of the retina containing openings through which the fibers of the ganglion neurons emerge as the optic nerve. Also called the **blind spot.**

Optic tract A bundle of axons that transmits impulses from the retina of the eye between the optic chiasma and the thalamus.

Oral contraceptive (OC) A hormonal compound that is swallowed and that prevents ovulation, and thus pregnancy. Also called **"the pill."**

Orbit (OR-bit) The bony, pyramid-shaped cavity of the skull that holds the eyeball.

Organ A structure of definite form and function composed of two or more different kinds of tissues.

Organelle (or-gan-EL) A component or structure within a cell that is specialized to serve a specific function in cellular activities.

Organism (OR-ga-nizm) A total living form; one individual.

Orgasm (OR-gazm) Sensory and motor events involved in ejaculation for the male and involuntary contraction of the perineal muscles in the female at the climax of sexual intercourse.

Orifice (OR-i-fis) Any aperture or opening.

Origin (OR-i-jin) The place of attachment of a muscle to the more stationary bone, or the end opposite the insertion.

Oropharynx (or'-ō-FAR-inks) The second portion of the pharynx, lying posterior to the mouth and extending from the soft palate down to the hyoid bone.

Osmosis (os-MŌ-sis) The net movement of water molecules through a selectively permeable membrane from an area of high water concentration to an area of lower water concentration until an equilibrium is reached.

Osmotic pressure The force under which a solvent (water) moves through a selectively permeable membrane from higher to lower water concentration.

Ossicle (OS-si-kul) Small bone, as in the middle ear (malleus, incus, stapes).

Ossification (os'-i-fi-KĀ-shun) Formation of bone. Also called **osteogenesis.**

Osteoblast (OS-tē-ō-blast') A bone-forming cell.

Osteoclast (OS-tē-ō-clast') A large, multinuclear cell that destroys or resorbs bone tissue.

Osteocyte (OS-tē-ō-sīt') A mature osteoblast that has lost its ability to produce new bone tissue.

Osteology (os'-tē-OL-ō-jē) The study of bones.

Osteomalacia (os'-tē-ō-ma-LĀ-shē-a) A deficiency of vitamin D causing demineralization and softening of bone.

Osteomyelitis (os'-tē-ō-mī-i-LĪ-tis) Inflammation of bone marrow or of the bone and marrow.

Osteoporosis (os'-tē-ō-pō-RŌ-sis) Increased porosity of bone.

Otic (Ō-tik) Pertaining to the ear.

Otitis media (ō-TĪ-tus MĒ-dē-a) Acute infection of the middle ear cavity characterized by an inflamed tympanic membrane, subject to rupture.

Otolith (Ō-tō-lith) A particle of calcium carbonate embedded in the otolithic membrane that functions in maintaining static equilibrium.

Otolithic (ō-tō-LITH-ik) **membrane** Thick, gelatinous, glycoprotein layer located directly over hair cells of the macula in the saccule and utricle of the inner ear.

Ovarian (ō-VAR-ē-an) **cycle**   A monthly series of events in the ovary designed to produce a mature ovum.

Ovarian follicle (FOL-i-kul)   A general name for an ovum in any stage of development, along with its surrounding group of epithelial cells.

Ovary (Ō-var-ē)   Female gonad that produces ova and the hormones estrogens, progesterone, and relaxin.

Ovulation (ō-vyoo-LĀ-shun)   The rupture of a Graafian follicle with discharge of an ovum into the pelvic cavity.

Ovum (Ō-vum)   The female reproductive or germ cell; an egg cell.

Oxidation (ok-si-DĀ-shun)   The removal of electrons or the addition of oxygen to a molecule. The oxidation of glucose in the body is also called **cellular respiration.**

Oxygen debt   The volume of oxygen required to oxidize the lactic acid produced by muscular exercise.

Oxyhemoglobin (ok'-sē-HĒ-mō-glō-bin) **(HbO₂)**   Hemoglobin combined with oxygen.

Oxyphil cell   A cell found in the parathyroid gland that secretes parathyroid hormone (PTH).

Oxytocin (ok'-sē-TŌ-sin) **(OT)**   A hormone secreted by neurosecretory cells in the paraventricular nucleus of the hypothalamus that stimulates contraction of the smooth muscle cells in the pregnant uterus and contractile cells around the ducts of mammary glands.

Pacinian (pa-SIN-ē-an) **corpuscle**   Oval pressure receptor located in subcutaneous tissue and consisting of concentric layers of connective tissue wrapped around an afferent nerve fiber.

Paget's (PAJ-ets) **disease**   A disorder characterized by an irregular thickening and softening of the bones. The cause is unknown, but the bone-producing osteoblasts and the bone-destroying osteoclasts apparently become uncoordinated.

Palate (PAL-at)   The horizontal structure separating the oral and the nasal cavities; the roof of the mouth.

Palliative (PAL-ē-a-tiv)   Serving to relieve or alleviate without curing.

Palpate (PAL-pāt)   To examine by touch; to feel.

Pancreas (PAN-krē-as)   A soft, oblong organ lying along the greater curvature of the stomach and connected by a duct to the duodenum. It is both exocrine (secreting pancreatic juice) and endocrine (secreting insulin, glucagon, and somatostatin).

Pancreatic (pan'-krē-AT-ik) **duct**   A single, large tube that drains pancreatic juice into the duodenum at the hepatopancreatic ampulla.

Papanicolaou (pap'-a-NIK-ō-la-oo) **test**   A cytological staining test for the detection and diagnosis of premalignant and malignant conditions of the female genital tract. Cells scraped from the genital epithelium are smeared, fixed, stained, and examined microscopically. Also called a **Pap smear.**

Paralysis (pa-RAL-a-sis)   Loss or impairment of motor function due to a lesion of nervous or muscular origin.

Paranasal sinus (par'-a-NĀ-zal SĪ-nus)   A mucus-lined air cavity in a skull bone that communicates with the nasal cavity. Paranasal sinuses are located in the frontal, maxillary, ethmoid, and sphenoid bones.

Paraplegia (par-a-PLĒ-jē-a)   Paralysis of the lower limbs only.

Parasympathetic (par'-a-sim-pa-THET-ik) **division**   One of the two subdivisions of the autonomic nervous system, having cell bodies of preganglionic neurons in nuclei in the brain stem and in the lateral gray matter of the sacral portion of the spinal cord; primarily concerned with activities that restore and conserve body energy. Also called the **craniosacral division.**

Parathyroid (par'-a-THĪ-royd) **gland**   One of four small endocrine glands embedded on the posterior surfaces of the lateral lobes of the thyroid.

Parathyroid hormone (PTH)   A hormone secreted by the parathyroid glands that decreases blood phosphate level and increases blood calcium level.

Parenchyma (par-EN-ki-ma)   The functional parts of any organ, as opposed to tissue that forms its ground substance or framework.

Parietal (pa-RĪ-e-tal)   Pertaining to or forming the outer wall of a body cavity.

Parietal cell   The secreting cell of the gastric glands that produces hydrochloric acid and intrinsic factor.

Parietal pleura (PLOO-ra)   The outer layer of the serous pleural membrane that encloses and protects the lungs; the layer that is attached to the walls of the pleural cavity.

Parkinsonism (PARK-in-so-nizm')   Progressive degeneration of basal ganglia of the cerebrum resulting in decreased production of dopamine that leads to tremor, slowing of voluntary movements, and muscle weakness. Also called **Parkinson's disease.**

Parotid (pa-ROT-id) **gland**   One of the paired salivary glands located inferior and anterior to the ears connected to the oral cavity via a duct that opens into the inside of the cheek opposite the upper second molar tooth.

Paroxysm (PAR-ok-sizm)   A sudden periodic attack or recurrence of symptoms of a disease.

Pars intermedia   A small avascular zone between the adenohypophysis and neurohypophysis of the pituitary gland.

Parturition (par'-too-RISH-un)   Act of giving birth to young; childbirth, delivery.

Patellar (pa-TELL-ar) **reflex**   Extension of the lower leg by contraction of the quadriceps femoris muscle in response to tapping the patellar ligament.

Patent ductus arteriosus   Congenital anatomical heart defect in which the fetal connection between the aorta and pulmonary trunk remains open instead of closing completely after birth.

Pathogenesis (path'-ō-JEN-e-sis)   The development of disease or a morbid or pathological state.

Pathogen (PATH-ō-jen)   A disease-producing organism.

Pathological (path'-ō-LOJ-i-kal) **anatomy**   The study of structural changes caused by disease.

Pediatrician (pē'-dē-a-TRISH-un)   A physician who specializes in the care and treatment of children and their illnesses.

Pedicel (PED-i-sel)   Footlike structure, as on podocytes of a glomerular.

Pelvic (PEL-vik) **cavity**   Inferior portion of the abdominopelvic cavity that contains the urinary bladder, sigmoid colon, rectum, and internal female and male reproductive structures.

Pelvic inflammatory disease (PID)   Collective term for any extensive bacterial infection of the pelvic organs, especially the uterus, uterine tubes, and ovaries.

Pelvimetry (pel-VIM-e-trē)   Measurement of the size of the inlet and outlet of the birth canal.

Pelvis   The basinlike structure formed by the two pelvic bones, the sacrum, and the coccyx. The expanded, proximal portion of the ureter, lying within the kidney and into which the major calyces open.

Penis (PĒ-nis)   The male copulary organ, used to introduce spermatozoa into the female vagina.

Pepsin   Protein-digesting enzyme secreted by zymogenic (chief) cells of the stomach as the inactive form pepsinogen which is converted to active pepsin by hydrochloric acid.

Peptic ulcer   An ulcer that develops in areas of the gastrointestinal tract exposed to hydrochloric acid; classified as a gastric ulcer if in the lesser curvature of the stomach and as a duodenal ulcer if in the first part of the duodenum.

Pericardial (per'-i-KAR-dē-al) **cavity** Small potential space between the visceral and parietal pericardium.

Pericardium (per'-i-KAR-dē-um) A loose-fitting serous membrane that encloses the heart, consisting of an outer fibrous layer and an internal serous layer. Also called **parietal pericardium, pericardial sac.**

Perichondrium (per'-i-KON-drē-um) The membrane that covers cartilage.

Perikaryon (per'-i-KAR-ē-on) The nerve cell body that contains the nucleus and other organelles.

Perilymph (PER-i-lymf') The fluid contained between the bony and membranous labyrinths of the inner ear.

Perimetrium (per-i-MĒ-trē-um) The serosa of the uterus.

Perimysium (per'-i-MIZ-ē-um) Invagination of the epimysium that divides muscles into bundles.

Perineum (per-i-NĒ-um) The pelvic floor; the space between the anus and the scrotum in the male and between the anus and the vulva in the female.

Perineurium (per'-i-NYOO-rē-um) Connective tissue wrapping each muscle bundle.

Periodontal (per-ē-ō-DON-tal) **disease** A collective term for conditions characterized by degeneration of gingivae, alveolar bone, periodontal ligament, and cementum.

Periosteum (per'-ē-OS-tē-um) The membrane that covers bone and is essential for bone growth, repair, and nutrition.

Peripheral (pe-RIF-er-al) Located on the outer part or a surface of the body.

Peripheral nervous system (PNS) The part of the nervous system that lies outside the central nervous system—nerves and ganglia.

Peripheral resistance The hindrance to blood flow due to friction between blood and blood vessels; related to viscosity of blood and diameter and length of blood vessels.

Peristalsis (per'-i-STAL-sis) Successive muscular contractions along the wall of a hollow muscular structure.

Peritoneum (per'-i-tō-NĒ-um) The largest serous membrane of the body which lines the abdominal cavity and covers the viscera.

Peritonitis (per'-i-tō-NĪ-tis) Inflammation of the peritoneum.

Peroxisome (pe-ROKS-ī-sōm) Organelle similar in structure to a lysosome that contains enzymes related to hydrogen peroxide metabolism; abundant in liver cells.

Perspiration Substance produced by sudoriferous (sweat) glands containing water, salts, urea, uric acid, amino acids, ammonia, sugar, lactic acid, and ascorbic acid; helps maintain body temperature and eliminate wastes.

Peyer's (PĪ-erz) **patches** Aggregated lymph nodules that are most numerous in the ileum.

pH A symbol of the measure of the concentration of hydrogen ions in a solution. The pH scale extends from 0 to 14 with a value of 7 expressing neutrality, values lower than 7 expressing increasing acidity, and values higher than 7 expressing increased alkalinity.

Phagocytosis (fag'-ō-sī-TŌ-sis) The process by which cells ingest particulate matter; especially the ingestion and destruction of microbes, cell debris, and other foreign matter.

Phalanx (FĀ-lanks) The bone of a finger or toe. *Plural,* **phalanges** (fa-LAN-jēz).

Phantom pain A sensation of pain as originating in a limb that has been amputated.

Pharynx (FAR-inks) The throat; a tube that starts at the internal nares and runs partway down the neck where it opens into the esophagus posteriorly and into the larynx anteriorly.

Pheochromocytoma (fē-ō-krō'-mō-sī-TŌ-ma) Tumor of the chromaffin cells of the adrenal medulla.

Phenotype (FĒ-nō-tīp) The observable expression of genotype; physical characteristics of an organism determined by genetic makeup and influenced by interaction between genes and internal and external environmental factors.

Phenylketonuria (fen'-il-kē'-tō-NOO-rē-a) **(PKU)** A disorder characterized by an elevation of the amino acid phenylalanine in the blood.

Photoreceptor Receptor that detects light on the retina of the eye. Also called **electromagnetic receptor.**

Physiology (fiz'-ē-OL-ō-jē) Science that deals with the functions of an organism or its parts.

Pia mater (PĒ-a MĀ-ter) The inner membrane (meninx) covering the brain and spinal cord.

Pineal (PIN-ē-al) **gland** The cone-shaped gland located in the roof of the third ventricle. Also called the **epiphysis cerebri.**

Pinealocyte (pin-ē-AL-o-sīt) Secretory cell of the pineal gland that produces hormones.

Pinna (PIN-na) The projecting part of the external ear. Also called the **auricle.**

Pinocytosis (pi'-nō-sī-TŌ-sis) The process by which cells ingest liquid.

Pituicyte (pi-TOO-i-sīt) Supporting cell of the posterior lobe of the pituitary gland.

Pituitary (pi-TOO-i-tar'-ē) **dwarfism** Condition caused by hyposecretion of growth hormone (GH) during the growth years and characterized by childlike physical traits in an adult.

Pituitary gland A small endocrine gland lying in the sella turcica of the sphenoid bone and attached to the hypothalamus by the infundibulum; nicknamed the "master gland." Also called the **hypophysis.**

Pivot joint A synovial joint in which a rounded, pointed, or conical surface of one bone articulates with a shallow depression of another bone, as in the joint between the atlas and axis and between the proximal ends of the radius and ulna. Also called a **trochoid joint.**

Placenta (pla-SEN-ta) The special structure through which the exchange of materials between fetal and maternal circulations occurs. Also called the **afterbirth.**

Plantar flexion (PLAN-tar FLEK-shun) Extension of the foot at the ankle joint.

Plaque (plak) A cholesterol-containing mass in the tunica media of arteries. A mass of bacterial cells, dextran (polysaccharide), and other debris that adheres to teeth.

Plasma (PLAZ-ma) The extracellular fluid found in blood vessels; blood minus the formed elements.

Plasma cell Cell that produces antibodies and develops from a B cell (lymphocyte).

Plasma (cell) membrane Outer, limiting membrane that separates the cell's internal parts from extracellular fluid and the external environment.

Platelet plug Aggregation of thrombocytes at a damaged blood vessel to prevent blood loss.

Pleura (PLOOR-a) The serous membrane that enfolds the lungs and lines the walls of the chest and diaphragm.

Pleural cavity Small potential space between the visceral and parietal pleurae.

Plexus (PLEK-sus) A network of nerves, veins, or lymphatic vessels.

Plicae circulares (PLĪ-kē SER-kyoo-lar-es) Permanent, deep, transverse folds in the mucosa and submucosa of the small intestine that increase the surface area for absorption.

Pneumonia (noo-MŌ-nē-a) Acute infection or inflammation of the alveoli of the lungs.

Pneumotaxic (noo-mō-TAK-sik) **area** Portion of the respiratory center in the pons that helps coordinate the transition between inspiration and expiration.

Podiatry (pō-DĪ-a-trē) The diagnosis and treatment of foot disorders.

Renal corpuscle (KOR-pus'-l)   A glomerular capsule and its enclosed glomerulus.

Renal erythropoietic (ē-rith'-rō-poy-Ē-tik) **factor**   An enzyme released by the kidneys and liver in response to hypoxia that acts on a plasma protein to bring about the production of erythropoietin, which stimulates red blood cell production.

Renal pelvis   A cavity in the center of the kidney formed by the expanded, proximal portion of the ureter, lying within the kidney, and into which the major calyces open.

Renal pyramid   A triangular structure in the renal medulla composed of the straight segments of renal tubules.

Renal suppression   The sudden stoppage of urine production and excretion. Also called **anuria.**

Renin (RĒ-nin)   An enzyme released by the kidney into the plasma where it converts angiotensinogen into angiotensin I.

Renin-angiotensin (an'-jē-ō-TEN-sin) **pathway**   A mechanism for the control of aldosterone secretion by angiotensin II, initiated by the secretion of renin by the kidney in response to low blood pressure.

Residual (re-ZID-yoo-al) **volume**   The volume of air still contained in the lungs after a maximal expiration; about 1,200 ml.

Resistance   Ability to ward off disease. The hindrance encountered by an electrical charge as it moves through a substance from one point to another. The hindrance encountered by blood as it flows through the vascular system or by air through respiratory passageways.

Respiration (res-pi-RĀ-shun)   Overall exchange of gases between the atmosphere, blood, and body cells consisting of pulmonary ventilation, external respiration, and internal respiration.

Respiratory center   Neurons in the reticular formation of the brain stem that regulate the rate of respiration.

Resting potential   The voltage that exists between the inside and outside of a cell membrane when the cell is not responding to a stimulus; about −70 to −90 mV, with the inside of the cell negative.

Resuscitation (rē-sus'-i-TĀ-shun)   Act of bringing a person back to full consciousness.

Retention (rē-TEN-shun)   A failure to void urine due to obstruction, nervous contraction of urethra, or absence of sensation of desire to urinate.

Rete (RĒ-tē) **testis**   The network of ducts in the testes.

Reticular (re-TIK-yoo-lar) **activating system (RAS)**   An extensive network of branched nerve cells running through the core of the brain stem. When these cells are activated, a generalized alert or arousal behavior results.

Reticular formation   A network of small groups of nerve cells scattered among bundles of fibers beginning in the medulla as a continuation of the spinal cord and extending upward through the central part of the brain stem.

Reticulocyte (rē-TIK-yoo-lō-sīt)   An immature red blood cell.

Retina (RET-i-na)   The inner coat of the eyeball, lying only in the posterior portion of the eye and consisting of nervous tissue and a pigmented layer comprised of epithelial cells lying in contact with the choroid. Also called the **nervous tunic.**

Retinene (re'-ti-NĒN)   The pigment portion of the photopigment rhodopsin. Also called **visual yellow.**

Retraction (rē-TRAK-shun)   The movement of a protracted part of the body backward on a plane parallel to the ground, as in pulling the lower jaw back in line with the upper jaw.

Retroflexion (re-trō-FLEK-shun)   A malposition of the uterus in which it is tilted posteriorly.

Retroperitoneal (re'-trō-per-i-tō-NĒ-al)   External to the peritoneal lining of the abdominal cavity.

Rheumatism (ROO-ma-tizm')   Any painful state of the supporting structures of the body—bones, ligaments, joints, tendons, or muscles.

Rh factor   An inherited agglutinogen on the surface of red blood cells.

Rhodopsin (rō-DOP-sin)   A photosensitive, reddish-purple pigment in rods of the retina, consisting of the protein scotopsin plus retinene. Also called **visual purple.**

Ribonucleic (rī'-bō-nyoo-KLĒ-ik) **acid (RNA)**   A single-stranded nucleic acid constructed of nucleotides consisting of one of four possible nitrogen bases (adenine, cytosine, guanine, or uracil), ribose, and a phosphate group; three types are messenger RNA (mRNA), transfer RNA (tRNA), and ribosomal RNA (rRNA), each of which cooperates with DNA for protein synthesis.

Ribosome (RĪ-bō-sōm)   An organelle in the cytoplasm of cells, composed of ribosomal RNA and ribosomal proteins, that synthesizes proteins; nicknamed the "protein factory."

Rickets (RIK-ets)   Condition affecting children characterized by soft and deformed bones resulting from inadequate calcium metabolism due to a vitamin D deficiency.

Right heart (atrial) reflex   A reflex concerned with maintaining normal venous blood pressure.

Right lymphatic (lim-FAT-ik) **duct**   A vessel of the lymphatic system that drains lymph from the upper right side of the body and empties it into the right subclavian vein.

Rigor mortis   State of partial contraction of muscles following death due to lack of ATP that causes cross bridges of thick myofilaments to remain attached to thin myofilaments, thus preventing relaxation.

Rod   A visual receptor in the retina of the eye that is specialized for vision in dim light.

Roentgenogram (RENT-gen-ō-gram)   A photographic image.

Root canal   A narrow extension of the pulp cavity lying within the root of a tooth.

Root hair plexus (PLEK-sus)   A network of dendrites arranged around the root of a hair as free or naked nerve endings that are stimulated when a hair shaft is moved.

Rotation (rō-TĀ-shun)   Moving a bone around its own axis, with no other movement.

Rugae (ROO-gē)   Large folds in the mucosa of an empty hollow organ, such as the stomach and vagina.

Saccule (SAK-yool)   The lower and smaller of the two chambers in the membranous labyrinth inside the vestibule of the inner ear containing a receptor organ for static equilibrium.

Sacral promontory (PROM-on-tor'-ē)   The superior surface of the body of the first sacral vertebra that projects anteriorly into the pelvic cavity; a line from the sacral promontory to the superior border of the symphysis pubis divides the abdominal and pelvic cavities.

Saddle joint   A synovial joint in which the articular surfaces of both of the bones are saddle-shaped or concave in one direction and convex in the other direction, as in the joint between the trapezium and the metacarpal of the thumb. Also called a **sellaris joint.**

Sagittal plane   A plane parallel to a midsagittal plane that divides the body or organs into unequal left and right portions. Also called a **parasagittal plane.**

Saliva (sa-LĪ-va)   A clear, alkaline, somewhat viscous secretion produced by the three pairs of salivary glands; contains various salts, mucin, lysozyme, and salivary amylase.

Salivary amylase (SAL-i-ver-ē AM-i-lās)   An enzyme in saliva that initiates the chemical breakdown of starch, mostly in the mouth.

Salivary gland   One of three pairs of glands that lie outside the mouth and pour their secretory product (called saliva) into ducts that empty into the oral cavity; the parotid, submandibular, and sublingual glands.

Salpingitis (sal'-pin-JĪ-tis)   Inflammation of the uterine or auditory tube.

Saltatory (sal-ta-TŌ-rē) **conduction**   The propagation of an action potential (nerve impulse) along the exposed portions of a myelinated nerve fiber. The action potential appears at successive nodes of Ranvier and therefore seems to jump or leap from node to node.

Sarcolemma (sar'-kō-LEM-ma)   The cell membrane of a muscle cell, especially of a skeletal muscle cell.

Sarcoma (sar-KŌ-ma)   A connective tissue tumor, often highly malignant.

Sarcomere (SAR-kō-mēr)   A contractile unit in a striated muscle cell extending from one Z line to the next Z line.

Sarcoplasm (SAR-kō-plazm)   The cytoplasm of a muscle cell.

Sarcoplasmic reticulum (sar'-kō-PLAZ-mik re-TIK-yoo-lum)   A network of saccules and tubes surrounding myofibrils of a muscle cell, comparable to endoplasmic reticulum; functions to reabsorb calcium ions during relaxation and to release them to cause contraction.

Satiety (sa-TĪ-e-tē)   Fullness or gratification, as of hunger or thirst.

Satiety center   A collection of nerve cells located in the medial portion of the hypothalamus, close to the third ventricle, that inhibit the feeding center.

Saturated fat   A fat whose carbon atoms are bonded to the maximum number of hydrogen atoms; found naturally in animal foods such as meat, milk, milk products, and eggs.

Scala tympani (SKA-la TIM-pan-ē)   The lower spiral-shaped channel of the bony cochlea, filled with perilymph.

Scala vestibuli (ves-TIB-u-lē)   The upper spiral-shaped channel of the bony cochlea, filled with perilymph.

Schwann (shwon) **cell**   A neuroglial cell of the peripheral nervous system that forms the myelin sheath and neurilemma of a nerve fiber by wrapping around a nerve fiber in a jelly-roll fashion.

Sciatica (sī-AT-i-ka)   Inflammation and pain along the sciatic nerve; felt at the back of the thigh running down the inside of the leg.

Sclera (SKLĒ-ra)   The white coat of fibrous tissue that forms the outer protective covering over the eyeball except in the area of the anterior cornea; the posterior portion of the fibrous tunic.

Sclerosis (skle-RŌ-sis)   A hardening with loss of elasticity of the tissues.

Scoliosis (skō'-lē-Ō-sis)   An abnormal lateral curvature from the normal vertical line of the spine.

Scotopsin (skō-TOP-sin)   The protein portion of the visual pigment rhodopsin found in rods of the retina.

Scrotum (SKRŌ-tum)   A skin-covered pouch that contains the testes and their accessory structures.

Sebaceous (se-BĀ-shus) **gland**   An exocrine gland in the dermis of the skin, almost always associated with a hair follicle, that secretes sebum. Also called an **oil gland.**

Sebum (SĒ-bum)   Secretion of sebaceous (oil) glands.

Secondary sex characteristic   A feature characteristic of the male or female body that develops at puberty under the stimulation of sex hormones, but is not directly involved in sexual reproduction, such as distribution of body hair, voice pitch, body shape, and muscle development.

Secretion (se-KRĒ-shun)   Production and release from a gland cell of a fluid, especially a functionally useful product as opposed to a waste product.

Selectively permeable membrane   A membrane that permits the passage of certain substances, but restricts the passage of others. Also called a **semipermeable membrane.**

Semen (SĒ-men)   A fluid discharge at ejaculation by a male that consists of a mixture of spermatozoa and the secretions of the seminal vesicles, the prostate gland, and the bulbourethral glands. Also called **seminal fluid.**

Semicircular canals   Three bony channels projecting superiorly and posteriorly from the vestibule of the inner ear, filled with perilymph, in which lie the membranous semicircular canals filled with endolymph. They contain receptors for equilibrium.

Semicircular ducts   The membranous semicircular canals filled with endolymph and floating in the perilymph of the bony semicircular canals. They contain cristae that are concerned with dynamic equilibrium.

Semilunar (sem'-ē-LOO-nar) **valve**   A valve guarding the entrance into the aorta or the pulmonary trunk from a ventricle of the heart.

Seminal vesicle (SEM-i-nal VES-i-kul)   One of a pair of convoluted, pouchlike structures, lying posterior and inferior to the urinary bladder and anterior to the rectum, that secrete a component of semen into the ejaculatory ducts.

Seminiferous tubule (sem'-i-NI-fer-us TOO-byool)   A tightly coiled duct, located in a lobule of the testis, where spermatozoa are produced.

Senescence (se-NES-ens)   The process of growing old; the period of old age.

Senility (se-NIL-i-tē)   A loss of mental or physical ability due to old age.

Sensation   A state of awareness of external or internal conditions of the body.

Sensory area   A region of the cerebral cortex concerned with the interpretation of sensory impulses.

Septal defect   An opening in the septum between the left and right sides of the heart.

Septum (SEP-tum)   A wall dividing two cavities.

Serosa (ser-Ō-sa)   Any serous membrane. The outermost layer or tunic of an organ formed by a serous membrane. The membrane that lines the pleural, pericardial, and peritoneal cavities.

Serous (SIR-us) **membrane**   A membrane that lines body cavities that does not open to the exterior. Also called the **serosa.**

Serum   Plasma minus its clotting proteins.

Sex chromosomes   The twenty-third pair of chromosomes, designated X and Y, which determine the genetic sex of an individual; in males, the pair is XY; in females, XX.

Sexual intercourse   The insertion of the erect penis of a male into the vagina of a female. Also called **coitus.**

Sexually transmitted disease (STD)   General term for any of a large number of diseases spread by sexual contact.

Shingles   Acute infection of the peripheral nervous system caused by a virus.

Shoulder   A synovial or diarthrotic joint where the humerus joins the scapula.

Sigmoid colon (SIG-moyd KŌ-lon)   The S-shaped portion of the large intestine that begins at the level of the left iliac crest, projects inward to the midline, and terminates at the rectum at about the level of the third sacral vertebra.

Sign   Any objective evidence of disease such as a lesion, swelling, or fever.

Sinoatrial (si-nō-Ā-trē-al) **(SA) node**   A compact mass of cardiac muscle cells specialized for conduction, located in the right atrium beneath the opening of the superior vena cava. Also called the **sinuatrial node** or **pacemaker.**

Sinus (SĪ-nus)   A hollow in a bone (paranasal sinus) or other tissue; a channel for blood; any cavity having a narrow opening.

Sinusitis (sīn-yoo-SĪT-is)   Inflammation of the mucous membrane of a paranasal sinus.

Sinusoid (SĪN-yoo-soyd)   A microscopic space or passage for blood in certain organs such as the liver or spleen.

Skeletal muscle   An organ specialized for contraction, composed of striated muscle cells, supported by connective tissue, attached to a bone by a tendon or an aponeurosis, and stimulated by somatic efferent neurons.

Skull The skeleton of the head consisting of the cranial and facial bones.

Sliding-filament theory The most commonly accepted explanation for muscle contraction in which actin and myosin myofilaments move into interdigitation with each other, decreasing the length of the sarcomeres.

Small intestine A long tube of the gastrointestinal tract that begins at the pyloric sphincter of the stomach, coils through the central and lower part of the abdominal cavity, and ends at the large intestine; divided into three segments: duodenum, jejunum, and ileum.

Sodium-potassium pump An active transport system located in the cell membrane that transports sodium ions out of the cell and potassium ions into the cell at the expense of cellular ATP. It functions to keep the ionic concentrations of these elements at physiological levels.

Soft palate (PAL-at) The posterior portion of the roof of the mouth, extending posteriorly from the palatine bones and ending at the uvula. It is a muscular partition lined with mucous membrane.

Solution A homogeneous molecular or ionic dispersion of one or more substances (solutes) in a (usually liquid) dissolving medium (solvent).

Somatic (sō-MAT-ik) nervous system (SNS) The portion of the peripheral nervous system made up of the somatic efferent fibers that run between the central nervous system and the skeletal muscles and skin.

Somesthetic (sō'-mes-THET-ik) Pertaining to sensations and sensory structures of the body.

Spasm (spazm) An involuntary, convulsive, muscular contraction.

Spermatic (sper-MAT-ik) cord A supporting structure of the male reproductive system, extending from a testis to the deep inguinal ring, that includes the ductus deferens, arteries, veins, lymphatics, nerves, cremaster muscle, and connective tissue.

Spermatogenesis (sper'-ma-tō-JEN-e-sis) The formation and development of the spermatozoa.

Spermatozoon (sper'-ma-tō-ZŌ-on) A mature sperm cell.

Spermicide (SPER-mi-sīd') An agent that kills spermatozoa.

Spermiogenesis (sper'-mē-ō-JEN-e-sis) The maturation of spermatids into spermatozoa.

Sphincter (SFINGK-ter) A circular muscle constricting an orifice.

Sphincter of the hepatopancreatic ampulla A circular muscle at the opening of the common bile and main pancreatic ducts in the duodenum. Also called the **sphincter** of **Oddi.**

Sphygmomanometer (sfig'-mō-ma-NOM-e-ter) An instrument for measuring arterial blood pressure.

Spina bifida (SPĪ-na BIF-i-da) A congenital defect of the vertebral column in which the two halves of the neural arch of a vertebra fail to fuse in midline.

Spinal (SPĪ-nal) cord A mass of nerve tissue located in the vertebral canal from which 31 pairs of spinal nerves originate.

Spinal (lumbar) puncture Withdrawal of some of the cerebrospinal fluid from the subarachnoid space in the lumbar region.

Spinal nerve One of the 31 pairs of nerves that originate on the spinal cord from posterior and anterior roots.

Spinous (SPĪ-nus) process A sharp or thornlike process or projection. Also called a **spine.** A sharp ridge running diagonally across the posterior surface of the scapula.

Spiral organ The organ of hearing, consisting of supporting cells and hair cells that rest on the basilar membrane and extend into the endolymph of the cochlear duct. Also called the **organ of Corti.**

Spirometer (spī-ROM-e-ter) An apparatus used to measure air capacity of the lungs. Also called a **respirometer.**

Spleen (SPLĒN) Large mass of lymphatic tissue between the fundus of the stomach and the diaphragm that functions in phagocytosis, production of lymphocytes, and blood storage.

Sprain Forcible wrenching or twisting of a joint with partial rupture or other injury to its attachments without dislocation.

Sputum (SPYOO-tum) Substance ejected from the mouth containing saliva and mucus.

Starling's law of the capillaries The movement of fluid between plasma and interstitial fluid is in a state of near equilibrium at the arterial and venous ends of a capillary, that is, filtered fluid and absorbed fluid plus that returned to the lymphatic system are nearly equal.

Starling's law of the heart The force of muscular contraction is determined by the length of the cardiac muscle fibers.

Stasis (STĀ-sis) Stagnation or halt of normal flow of fluids, as blood, urine, or of the intestinal mechanism.

Static equilibrium (ē-kwi-LIB-rē-um) The maintenance of posture in response to changes in the orientation of the body, mainly the head, relative to the ground.

Stellate reticuloendothelial (STEL-āte re-tik'-yoo-lō-en'-dō-THĒ-lē-al) cell Phagocytic cell that lines a sinusoid of the liver. Also called a **Kupffer cell.**

Stenosis (sten-Ō-sis) An abnormal narrowing or constriction of a duct or opening.

Sterile (STE-rīl) Free from any living microorganisms. Unable to conceive or produce offspring.

Sterilization (ster'-i-li-ZĀ-shun) Elimination of all living microorganisms. The rendering of an individual incapable of reproduction (e.g., castration, vasectomy, hysterectomy).

Stereocilia (ste'-rē-ō-SIL-ē-a) Groups of extremely long, slender, nonmotile microvilli projecting from epithelial cells lining the epididymis.

Stereognosis (ste'-rē-og-NŌ-sis) The ability to recognize the size, shape, and texture of an object by touch.

Sternal puncture Introduction of a wide-bore needle into the marrow cavity of the sternum for aspiration of a sample of red bone marrow.

Stimulus Any change in the environment, capable of initiating a response by the nervous system.

Stomach The J-shaped enlargement of the gastrointestinal tract directly under the diaphragm in the epigastric, umbilical, and left hypochondriac regions of the abdomen, between the esophagus and small intestine.

Strabismus (stra-BIZ-mus) A condition in which the visual axes of the two eyes differ, so that they do not both fix on the same object.

Straight tubule (TOO-byool) A duct in a testis leading from a convoluted seminiferous tubule to the rete testis.

Stratum basalis (STRĀ-tum ba-SAL-is) The outer layer of the endometrium, next to the myometrium, that is maintained during menstruation and gestation and produces a new functionalis following menstruation or parturition.

Stratum functionalis (funk'-shun-AL-is) The inner layer of the endometrium, the layer next to the uterine cavity, that is shed during menstruation and that forms the maternal portion of the placenta during gestation.

Stressor A stress that is extreme, unusual, or long-lasting and triggers the general adaptation syndrome.

Stretch receptor Receptor in the walls of bronchi, bronchioles, and lungs that sends impulses to the respiratory center that prevents overinflation of the lungs.

Stretch reflex A monosynaptic reflex triggered by a sudden stretch of a muscle and ending with a contraction of that same muscle.

Stricture (STRIK-cher) A local contraction of a tubular structure.

Stroke volume The volume of blood ejected by either ventricle in one systole; about 70 ml.

Stroma (STRŌ-ma)   The tissue that forms the ground substance, foundation, or framework of an organ, as opposed to its functional parts.

Subarachnoid (sub'-a-RAK-noyd) **space**   A space between the arachnoid and the pia mater that surrounds the brain and spinal cord and through which cerebrospinal fluid circulates.

Subcutaneous (sub'-kyoo-TĀ-nē-us)   Beneath the skin. Also called **hypodermic.**

Subcutaneous layer   A continuous sheet of loose connective tissue and adipose tissue between the dermis of the skin and the deep fascia of the muscles. Also called the **superficial fascia.**

Sublingual (sub-LING-gwal) **gland**   One of a pair of salivary glands situated in the floor of the mouth under the mucous membrane and to the side of the lingual frenulum, with a duct that opens into the floor of the mouth.

Submandibular gland   One of a pair of salivary glands found beneath the base of the tongue under the mucous membrane in the posterior part of the floor of the mouth, posterior to the sublingual glands, with a duct situated to the side of the lingual frenulum. Also called the **submaxillary gland.**

Submucosa (sub-myoo-KŌ-sa)   A layer of connective tissue located beneath a mucous membrane, as in the gastrointestinal tract or the urinary bladder where a submucosa connects the mucosa to the muscularis tunic.

Substrate   A substance with which an enzyme reacts.

Subthreshold stimulus   A stimulus of such weak intensity that it cannot initiate muscle contraction. Also called a **subliminal stimulus.**

Sudoriferous (soo'-dor-IF-er-us) **gland**   An apocrine or eccrine exocrine gland in the dermis or subcutaneous layer that produces perspiration. Also called a **sweat gland.**

Sulcus (SUL-kus)   A groove or depression between parts, especially between the convolutions of the brain. *Plural,* **sulci.**

Summation (sum-MĀ-shun)   The algebraic addition of the excitatory and inhibitory effects of many stimuli applied to a nerve cell body. The increased strength of muscle contraction that results when stimuli follow in rapid succession.

Sunstroke   Condition produced when the body cannot easily lose heat and characterized by reduced perspiration and elevated body temperature. Also called **heatstroke.**

Superficial (soo'-per-FISH-al)   Located on or near the surface of the body.

Superior (soo-PĒR-ē-or)   Toward the head or upper part of a structure. Also called **cephalad** or **craniad.**

Superior vena cava (VĒ-na CĀ-va)   Large vein that collects blood from parts of the body superior to the heart and returns it to the right atrium.

Supination (soo-pī-NĀ-shun)   A movement of the forearm in which the palm of the hand is turned anteriorly.

Suppuration (sup'-yoo-RĀ-shun)   Pus formation and discharge.

Surface anatomy   The study of the structures that can be identified from the outside of the body.

Surfactant (sur-FAK-tant)   A phospholipid substance produced by the lungs that decreases surface tension.

Susceptibility (sus-sep'-ti-BIL-i-tē)   Lack of resistance of a body to the deleterious or other effects of an agent such as pathogenic microorganisms.

Sustentacular (sus'-ten-TAK-yoo-lar) **cell**   A supporting cell of seminiferous tubules that produces secretions for supplying nutrients to spermatozoa and the hormone inhibin. Also called a **Sertoli cell.**

Suture (SOO-cher)   A fibrous joint, especially in the skull, where bone surfaces are closely united.

Sympathetic (sim'-pa-THET-ik) **division**   One of the two subdivisions of the autonomic nervous system, having cell bodies of preganglionic neurons in the lateral gray columns of the thoracic segment and first two or three lumbar segments of the spinal cord; primarily concerned with processes involving the expenditure of energy. Also called the **thoracolumbar division.**

Sympathomimetic (sim'-pa-thō-mi-MET-ik)   Producing effects that mimic those brought about by the sympathetic division of the autonomic nervous system.

Symphysis (SIM-fi-sis)   A line of union. A slightly movable cartilaginous joint such as the symphysis pubis between the anterior surfaces of the pubic bones.

Symptom (SIMP-tum)   An observable abnormality that indicates the presence of a disease or disorder of the body.

Synapse (SIN-aps)   The junction between the processes of two adjacent neurons; the place where the activity of one neuron affects the activity of another.

Synaptic (sin-AP-tik) **cleft**   The narrow gap that separates the membrane of an axon terminal of one nerve cell and the membrane of another nerve cell and across which the transmitter substance diffuses to affect the postsynaptic cell.

Synaptic delay   The length of time between the arrival of the action potential at the axon terminal and the membrane potential (IPSP or EPSP) change on the postsynaptic membrane; usually less than 1 msec.

Synaptic knob   Expanded distal end of a telodendrium that contains synaptic vesicles. Also called an **end foot.**

Synaptic vesicle   Membrane-enclosed sac in a synaptic knob that stores transmitter substances.

Synarthrosis (sin'-ar-THRŌ-sis)   An immovable joint.

Synchondrosis (sin'-kon-DRŌ-sis)   A cartilaginous joint in which the connecting material is hyaline cartilage.

Syndesmosis (sin'-dez-MŌ-sis)   A fibrous joint in which articulating bones are united by dense fibrous tissue.

Syndrome (SIN-drōm)   A group of signs and symptoms that occur together in a pattern characteristic of a particular disease or abnormal condition.

Syneresis (si-NER-e-sis)   The process of clot retraction.

Synergist (SIN-er-jist)   A muscle that assists the prime mover or agonist by reducing undesired action or unnecessary movement. Also called a **fixator.**

Synostosis (sin'-os-TŌ-sis)   A joint in which the dense fibrous connective tissue that unites bones at a suture has been replaced by bone, resulting in a complete fusion across the suture line.

Synovial (si-NŌ-vē-al) **cavity**   The space between the articulating bones of a synovial or diarthrotic joint, filled with synovial fluid.

Synovial fluid   Secretion of synovial membranes that lubricates joints and nourishes articular cartilage.

Synovial joint   A fully movable or diarthrotic joint in which a joint or synovial cavity is present between the two articulating bones.

Synovial membrane   The inner of the two layers of the articular capsule of a synovial joint, composed of loose connective tissue covered with epithelium that secretes synovial fluid into the joint cavity.

Syphilis (SIF-i-lis)   A sexually transmitted disease caused by the bacterium *Treponema pallidum.*

System   An association of organs that have a common function.

Systemic (sis-TEM-ik)   Affecting the whole body; generalized.

Systemic anatomy   The study of particular systems of the body, such as the skeletal, muscular, nervous, cardiovascular, or urinary systems.

Systemic circulation   The routes through which oxygenated blood flows from the left ventricle through the aorta to all the organs of the body and deoxygenated blood returns to the right atrium.

Systemic lupus erythematosus (er-i-them-a-TŌ-sus) **(SLE)**   An autoimmune, inflammatory disease that may affect every tissue of the body.

Systole (SIS-tō-lē)   In the cardiac cycle, the phase of contraction of the heart muscle, especially of the ventricles.

Systolic (sis-TO-lik) **blood pressure**   The force exerted by blood on arterial walls during ventricular contraction; the highest pressure measured in the large arteries, about 120 mm Hg under normal conditions for a young, adult male.

Tachycardia (tak'-i-KAR-dē-a)   A rapid heartbeat or pulse rate.

Tactile (TAK-tīl)   Pertaining to the sense of touch.

Taenia coli (TĒ-nē-a KŌ-lī)   One of three flat bands of muscles running the length of the large intestine.

Target cell   A cell whose activity is affected by a particular hormone.

Tarsus (TAR-sus)   A collective term for the seven bones of the ankle.

T cell   A lymphocyte that can differentiate into one of four kinds of cells—killer, helper, suppressor, or memory—all of which function in cellular immunity.

Tectorial (tek-TŌ-rē-al) **membrane**   A gelatinous membrane projecting over and in contact with the hair cells of the spiral organ in the chochlear duct.

Telodendria (tel-ō-DEN-drē-a)   The many small terminal branches of a neuron.

Telophase (TEL-ō-fāz)   The final stage of mitosis in which the daughter nuclei become established.

Tendinitis (ten'-din-Ī-tis)   Inflammation of a tendon and synovial membrane at a joint. Also called **tenosynovitis.**

Tendon (TEN-don)   A white fibrous cord of dense, regularly arranged connective tissue that attaches muscle to bone.

Tendon organ   A proprioceptive receptor found chiefly near the junction of tendons and muscles.

Tentorium cerebelli (ten-TŌ-rē-um ser'-e-BEL-ē)   A transverse shelf of dura mater that forms a partition between the occipital part of the cerebral hemispheres and the cerebellum and that covers the cerebellum.

Teratogen (TER-a-tō-jen)   Any agent or factor that causes physical defects in a developing embryo.

Testis (TES-tis)   Male gonad that produces sperm and the hormones testosterone and inhibin.

Testosterone (tes-TOS-te-rōn)   A male sex hormone (androgen) secreted by interstitial endocrinocytes of a mature testis; controls the growth and development of male sex organs, secondary sex characteristics, spermatozoa, and body growth.

Tetanus (TET-a-nus)   An infectious disease caused by the toxin of *Clostridium tetani,* characterized by tonic muscle spasms and exaggerated reflexes, lockjaw, and arching of the back. A smooth, sustained contraction produced by a series of very rapid stimuli to a muscle.

Tetany (TET-a-nē)   A nervous condition caused by hypoparathyroidism and characterized by intermittent or continuous tonic muscular contractions of the extremities.

Tetralogy of Fallot (tet-RAL-ō-jē of fal-Ō)   A combination of four congenital heart defects: (1) constricted pulmonary semilunar valve, (2) interventricular septal opening, (3) emergence of aorta from both ventricles instead of from the left only, and (4) enlarged right ventricle.

Thalamus (THAL-a-mus)   A large, oval structure located above the midbrain, consisting of two masses of gray matter covered by a thin layer of white matter.

Therapy (THER-a-pē)   The treatment of a disease or disorder.

Thermoreceptor (THER-mō-rē-sep-tor)   Receptor that detects changes in temperature.

Thigh   The portion of the lower extremity between the hip and the knee.

Third ventricle (VEN-tri-kul)   A slitlike cavity between the right and left halves of the thalamus and between the lateral ventricles.

Thoracic cavity   Superior component of the ventral body cavity that contains two pleural cavities, the mediastinum, and the pericardial cavity.

Thoracic duct   A lymphatic vessel that begins as a dilation called the cisterna chyli, receives lymph from the left side of the head, neck, and chest, the left arm, and the entire body below the ribs, and empties into the left subclavian vein. Also called the **left lymphatic duct.**

Thoracolumbar outflow   The fibers of the sympathetic preganglionic neurons, which have their cell bodies in the lateral gray columns of the thoracic segment and first two or three lumbar segments of the spinal cord.

Thorax (THŌ-raks)   The chest.

Threshold potential   The membrane voltage that must be reached in order to trigger an action potential (nerve impulse).

Threshold stimulus   Weakest stimulus from a neuron that is capable of initiating muscle contraction. Also called a **liminal stimulus.**

Thrombin (THROM-bin)   The active enzyme formed from prothrombin which acts to convert fibrinogen to fibrin.

Thrombocyte (THROM-bō-sīt)   A fragment of cytoplasm enclosed in a cell membrane and lacking a nucleus; found in the circulating blood; plays a role in blood clotting. Also called a **platelet.**

Thromboplastin (throm-bō-PLAS-tin)   A factor, or collection of factors, whose appearance initiates the blood clotting process.

Thrombosis (throm-BŌ-sis)   The formation of a clot in an unbroken blood vessel.

Thrombus   A clot formed in an unbroken blood vessel.

Thymus (THĪ-mus) **gland**   A bilobed organ, located in the upper mediastinum posterior to the sternum and between the lungs, that plays a role in the immunity mechanism of the body.

Thyroglobulin (thī-rō-GLŌ-byoo-lin) **(TGB)**   A large glycoprotein molecule secreted by follicle cells of the thyroid gland in which iodine is combined with tyrosine to form thyroid hormones.

Thyroid cartilage (THĪ-royd KAR-ti-lij)   The largest single cartilage of the larynx, consisting of two fused plates that form the anterior wall of the larynx. Also called the **Adam's apple.**

Thyroid colloid (KOL-loyd)   A complex in thyroid follicles consisting of thyroglobulin and stored thyroid hormones.

Thyroid follicle (FOL-i-kul)   Spherical sac that forms the parenchyma of the thyroid gland and consists of follicular cells that produce thyroxine (T_4) and triiodothyronine (T_3) and parafollicular cells that produce calcitonin (CT).

Thyroid gland   An endocrine gland with right and left lateral lobes on either side of the trachea connected by an isthmus located in front of the trachea just below the cricoid cartilage.

Thyroid-stimulating hormone (TSH)   A hormone secreted by the adenohypophysis (anterior lobe) of the pituitary gland that stimulates the synthesis and secretion of hormones produced by the thyroid gland.

Thyroxine (thī-ROK-sin) **(T_4)**   A hormone secreted by the thyroid that regulates metabolism, growth and development, and the activity of the nervous system.

Tic   Spasmodic, involuntary twitching by muscles that are ordinarily under voluntary control.

Tidal volume   The volume of air breathed in and out in any one breath; about 500 ml in quiet, resting conditions.

Tinnitus (ti-NĪ-tus)   A ringing or tingling sound in the ears.

Tissue   A group of similar cells and their intercellular substance joined together to perform a specific function.

Tongue   A large skeletal muscle on the floor of the oral cavity.

Tonsil (TON-sil)   A mass of lymphoid tissue embedded in mucous membrane.

Torn cartilage   A tearing of an articular disc in the knee.

Total lung capacity   The sum of tidal volume, inspiratory reserve volume, expiratory reserve volume, and residual volume; about 6,000 ml.

Toxic (TOK-sik)   Pertaining to poison; poisonous.

Toxic shock syndrome (TSS)   A disease caused by the bacterium *Staphylococcus aureus,* occurring among menstruating females who use tampons and characterized by high fever, sore throat, headache, fatigue, irritability, and abdominal pain.

Trabecula (tra-BEK-yoo-la)   Irregular latticework of thin plate of spongy bone. Fibrous cord of connective tissue serving as supporting fiber by forming a septum extending into an organ from its wall or capsule. *Plural,* **trabeculae** (tra-BEK-yoo-lē).

Trachea (TRĀ-kē-a)   Tubular air passageway extending from the larynx to the fifth thoracic vertebra. Also called the **windpipe.**

Tracheostomy (trā-kē-OS-tō-mē)   Creation of an opening into the trachea through the neck, with insertion of a tube to facilitate passage of air or evacuation of secretions.

Trachoma (tra-KŌ-ma)   A chronic infectious disease of the conjunctiva and cornea of the eye caused by a strain of the bacterium *Chlamydia trachomatis.*

Tract   A bundle of nerve fibers in the central nervous system.

Transcription (trans-KRIP-shun)   The first step in the transfer of genetic information in which a single strand of a DNA molecule serves as a template for the formation of an RNA molecule.

Translation (trans-LĀ-shun)   The construction of a new protein on the ribosome of a cell as dictated by the sequence of codons in mRNA.

Transmitter substance   One of a variety of molecules synthesized within the nerve axon terminals, released into the synaptic cleft in response to an action potential, and affecting the membrane potential of the postsynaptic neuron. Also called a **neurotransmitter.**

Transplantation (trans-plan-TĀ-shun)   The replacement of injured or diseased tissues or organs with natural ones.

Transverse colon (trans-VERS KŌ-lon)   The portion of the large intestine extending across the abdomen from the right colic (hepatic) flexure to the left colic (splenic) flexure.

Trauma (TRAW-ma)   An injury, either a physical wound or psychic disorder, caused by an external agent or force, such as a physical blow or emotional shock; the agent or force that causes the injury.

Treppe (TREP-ē)   The gradual increase in the amount of contraction by a muscle caused by rapid, repeated stimuli of the same strength.

Triad (TRĪ-ad)   A complex of three units in a muscle cell composed of a T tubule and the segments of sarcoplasmic reticulum on both sides of it.

Tricuspid (trī-KUS-pid) **valve**   Atrioventricular valve on the right side of the heart.

Trigeminal neuralgia (trī-JEM-i-nal noo-RAL-jē-a)   Pain in one or more of the branches of the trigeminal (V) nerve. Also called **tic douloureux.**

Trigone (TRĪ-gon)   A triangular area at the base of the urinary bladder.

Triiodothyronine (trī-ī-od-ō-THĪ-rō-nēn) (T₃)   A hormone produced by the thyroid gland that regulates metabolism, growth and development, and the activity of the nervous system.

Trophectoderm (trō-FEK-tō-derm)   The outer covering of cells of the blastocyst.

Tropic (TRŌ-pik) **hormone**   A hormone whose target is another endocrine gland.

Trunk   The part of the body to which the upper and lower extremities are attached.

T tubules (TOOB-yools) **(transverse tubules)**   Minute, cylindrical invaginations of the muscle cell membrane that carry the surface action potentials deep into the muscle cell.

Tubal ligation (li-GĀ-shun)   A sterilization procedure in which the uterine tubes are tied and cut.

Tuberculosis (too-berk-yoo-LŌ-sis)   An inflammation of the lungs and pleurae caused by *Mycobacterium tuberculosis* resulting in destruction of lung tissue and its replacement by fibrous connective tissue.

Tubular maximum (Tm)   The maximum amount of a substance that can be reabsorbed by renal tubules under any condition.

Tubular reabsorption   The movement of filtrate from renal tubules back into blood in response to the body's specific needs.

Tubular secretion   The movement of substances in blood back into filtrate in response to the body's specific needs.

Tumor (TOO-mor)   A growth of excess tissue due to an unusually rapid division of cells.

Tunica albuginea (TOO-ni-ka al'-byoo-JIN-ē-a)   A dense layer of white fibrous tissue covering a testis or deep to the surface of an ovary.

Tunica externa (eks-TER-na)   The outer coat of an artery or vein, composed mostly of elastic and collagenous fibers. Also called the **adventitia.**

Tunica interna (in-TER-na)   The inner coat of an artery or vein, consisting of a lining of endothelium and its supporting layer of connective tissue. Also called the **tunica intima.**

Tunica media (MĒ-dē-a)   The middle coat of an artery or vein, composed of smooth muscle and elastic fibers.

T wave   The deflection wave of an electrocardiogram that records ventricular repolarization.

Twitch   Rapid, jerky contraction of a muscle in response to a single stimulus.

Tympanic (tim-PAN-ik) **membrane**   A thin, semitransparent partition of fibrous connective tissue between the external auditory meatus and the middle ear. Also called the **eardrum.**

Ulcer (UL-ser)   An open lesion of the skin or a mucous membrane of the body with loss of substance and necrosis of the tissue.

Umbilical (um-BIL-i-kal) **cord**   The long, ropelike structure, containing the umbilical arteries and vein, which connects the fetus to the placenta.

Umbilicus (um-BIL-i-kus)   A small scar on the abdomen that marks the former attachment of the umbilical cord to the fetus. Also called the **navel.**

Unsaturated fat   A fat whose carbon atoms can still bond to one more hydrogen atom; found naturally in vegetable oils.

Upper extremity   The appendage attached at the shoulder girdle, consisting of the arm, forearm, wrist, hand, and fingers.

Uremia (yoo-RĒ-mē-a)   Accumulation of toxic levels of urea and other nitrogenous waste products in the blood, usually resulting from severe kidney malfunction.

Ureter (yoo-RĒ-ter)   One of two tubes that connect the kidney with the urinary bladder.

Urethra (yoo-RĒ-thra)   The duct from the urinary bladder to the exterior of the body that conveys urine in females and urine and semen in males.

Urinalysis   The physical, chemical, and microscopic analysis or examination of urine.

Urinary (YOO-ri-ner-ē) **bladder**   A hollow, muscular organ situated in the pelvic cavity posterior to the symphysis pubis.

Urine   The fluid produced by the kidneys that contains wastes or excess materials and excreted from the body through the urethra.

Urobilinogenuria The presence of urobilinogen in urine.

Urogenital (YOO'-rō-JEN-i-tal) **triangle** The region of the pelvic floor below the symphysis pubis, bounded by the symphysis pubis and the ischial tuberosities and containing the external genitalia.

Uterine (YOO-ter-in) **tube** Duct that transports ova from the ovary to the uterus. Also called the **fallopian tube** or **oviduct.**

Uterus (YOO-te-rus) The hollow, muscular organ in females that is the site of menstruation, implantation, development of the fetus, and labor. Also called the **womb.**

Utricle (YOO-tri-kul) The larger of the two divisions of the membranous labyrinth located inside the vestibule of the inner ear, containing a receptor organ for static equilibrium.

Uvea (YOO-vē-a) The three structures that make up the vascular tunic of the eye.

Uvula (YOO-vyoo-la) A soft, fleshy mass, especially the V-shaped pendant part, descending from the soft palate.

Vagina (va-JĪ-na) A muscular, tubular organ that leads from the uterus to the vestibule, situated between the urinary bladder and the rectum of the female.

Valvular stenosis (VAL-vyoo-lar STEN-Ō-sis) A narrowing of a heart valve, usually the mitral valve.

Varicose (VAR-i-kōs) Pertaining to an unnatural swelling, as in the case of a varicose vein.

Vasa vasorum (VĀ-sa va-SŌ-rum) Blood vessels supplying nutrients to the larger arteries and veins.

Vascular (VAS-kyoo-lar) Pertaining to or containing many blood vessels.

Vascular spasm Contraction of the smooth muscle in the wall of a damaged blood vessel to prevent blood loss.

Vascular tunic (TOO-nik) The middle layer of the eyeball, composed of the choroid, the ciliary body, and the iris.

Vasectomy (va-SEK-tō-mē) A means of sterilization of males in which a portion of each ductus deferens is removed.

Vasoconstriction (vāz-ō-kon-STRIK-shun) A decrease in the size of the lumen of a blood vessel caused by contraction of the smooth muscle in the wall of the vessel.

Vasodilation (vās'-ō-DĪ-lā-shun) An increase in the size of the lumen of a blood vessel caused by relaxation of the smooth muscle in the wall of the vessel.

Vasomotion (vāz-ō-MŌ-shun) Intermittent contraction and relaxation of the smooth muscle of metarterioles and precapillary sphincters that result in an intermittent blood flow.

Vasomotor (vā-sō-MŌ-tor) **center** A cluster of neurons in the medulla that controls the diameter of blood vessels, especially arteries.

Vein A blood vessel that conveys blood from tissues back to the heart.

Ventral (VEN-tral) **body cavity** Cavity near the ventral aspect of the body that contains viscera and consists of a superior thoracic cavity and an inferior abdominopelvic cavity.

Ventricle (VEN-tri-kul) A cavity in the brain or an inferior chamber of the heart.

Ventricular fibrillation (ven-TRIK-yoo-lar, fib-ri-LĀ-shun) Asynchronous ventricular contractions that result in circulatory failure.

Venule (VEN-yool) A small vein that collects blood from capillaries and delivers it to a vein.

Vermiform appendix (VER-mi-form a-PEN-diks) A twisted, coiled tube attached to the cecum.

Vermis (VER-mis) The central constricted area of the cerebellum that separates the two cerebellar hemispheres.

Vertebral (VER-te-bral) **canal** A cavity within the vertebral column formed by the vertebral foramina of all the vertebrae and containing the spinal cord. Also called the **spinal canal.**

Vertebral column The 26 vertebrae; encloses and protects the spinal cord and serves as a point of attachment for the ribs and back muscles. Also called the **spine, spinal column,** or **backbone.**

Vertigo (VER-ti-go) Sensation of dizziness.

Vestibular (ves-TIB-yoo-lar) **membrane** The membrane that separates the cochlear duct from the scala vestibuli.

Vestibule (VES-ti-byool) A small space or cavity at the beginning of a canal, especially the inner ear, larynx, mouth, nose, and vagina.

Villus (VIL-lus) A projection of the intestinal mucosal cells containing connective tissue, blood vessels, and a lymphatic vessel; functions in the absorption of food. *Plural,* **villi** (VIL-ē).

Viscera (VIS-er-a) The organs inside the ventral body cavity. *Singular,* **viscus.**

Visceral (VIS-er-al) Pertaining to the organs or to the covering of an organ.

Visceral effector (e-FEK-tor) Cardiac muscle, smooth muscle, and glandular epithelium.

Visceral muscle An organ specialized for contraction, composed of smooth muscle cells, located in the walls of hollow internal structures, and stimulated by visceral efferent neurons.

Visceral pleura (PLOO-ra) The inner layer of the serous membrane that covers the lungs.

Visceroceptor (vis'-er-ō-SEP-tor) Receptor that provides information about the body's internal environment.

Vital capacity The sum of inspiratory reserve volume, tidal volume, and expiratory reserve volume; about 4,800 ml.

Vitamin An organic molecule necessary in trace amounts that acts as a catalyst in normal metabolic processes of the body.

Vitreous humor (VIT-rē-us HYOO-mor) A soft, jellylike substance that fills the posterior cavity of the eyeball, lying between the lens and the retina.

Vocal folds Pair of mucous membrane folds below the ventricular folds that function in voice production. Also called **true vocal cords.**

Volkmann's (FOLK-mans) **canal** A minute passageway by means of which blood vessels and nerves from the periosteum penetrate into compact bone.

Vomiting Forcible expulsion of the contents of the upper gastrointestinal tract through the mouth.

Vulva Collective designation for the external genitalia of the female. Also called **pudendum.**

Wallerian (wal-LE-rē-an) **degeneration** Degeneration of the distal portion of the axon and myelin sheath.

Wandering macrophage (MAK-rō-fāj) Phagocytic cell that develops from a monocyte, leaves the blood, and migrates to infected tissues.

Wart Generally benign tumor of epithelial skin cells caused by a virus.

White matter Aggregations or bundles of myelinated axons located in the brain and spinal cord.

Xiphoid (ZĪ-foyd) Sword-shaped. The lowest portion of the sternum.

Yolk sac An extraembryonic membrane that connects with the midgut during early embryonic development, but is nonfunctional in humans.

Zona fasciculata (ZŌ-na fa-sik'-yoo-LA-ta) The middle zone of the adrenal cortex that consists of cells arranged in long, straight cords and that secretes glucocorticoid hormones.

Zona glomerulosa (glo-mer'-yoo-LŌ-sa) The outer zone of the adrenal cortex, directly under the connective tissue covering, that consists of cells arranged in arched loops or round balls and that secretes mineralocorticoid hormones.

Zona pellucida (pe-LOO-si-da) Gelatinous covering that surrounds an ovum.

Zona reticularis (ret-ik'-yoo-LAR-is) The inner zone of the adrenal cortex, consisting of cords of branching cells that secrete sex hormones, chiefly androgens.

Zygote (ZĪ-gōt) The single cell resulting from the union of a male and female gamete; the fertilized ovum.

Zymogenic (zī'-mō-JEN-ik) **cell** A cell that secretes enzymes, for example, the zymogenic (chief) cells of the gastric glands that secrete pepsinogen.

52–58.

84 85 86 9 8 7 6 5 4 3 2